Study and Solutions Guide for

CALCULUS

EIGHTH EDITION

Larson / Hostetler / Edwards

Volume I
Chapters P–11

Bruce H. Edwards
University of Florida

Houghton Mifflin Company   Boston   New York

Vice President and Publisher: Jack Shira
Associate Sponsoring Editor: Cathy Cantin
Development Manager: Maureen Ross
Editorial Assistant: Elizabeth Kassab
Supervising Editor: Karen Carter
Senior Project Editor: Patty Bergin
Editorial Assistant: Neil Reynolds
Art and Design Manager: Gary Crespo
Executive Marketing Manager: Michael Busnach
Senior Marketing Manager: Danielle Potvin
Marketing Coordinator: Nicole Mollica
Senior Manufacturing Coordinator: Karmen Chong

Printed in the United States of America

ISBN: 0-618-52791-5

56789-EB-09 08 07

PREFACE

This *Study and Solutions Guide* is designed as a supplement to *Calculus,* Eighth Edition, by Ron Larson, Robert P. Hostetler, and Bruce H. Edwards. All references to chapters, theorems, and exercises relate to the main text. Solutions to every odd-numbered exercise in the text are given with all essential algebraic steps included. Although this supplement is not a substitute for good study habits, it can be valuable when incorporated into a well-planned course of study. For suggestions that may assist you in the use of this text, your lecture notes, and this *Guide*, please refer to the student website for your text at *college.hmco.com.*

I have made every effort to see that the solutions are correct. However, I would appreciate hearing about any errors or other suggestions for improvement. Good luck with your study of calculus.

Bruce H. Edwards
University of Florida
Gainesville, Florida 32611
(be@math.ufl.edu)

CONTENTS

SOLUTIONS TO ODD-NUMBERED EXERCISES

Chapter P   Preparation for Calculus **1**

Chapter 1   Limits and Their Properties **25**

Chapter 2   Differentiation **50**

Chapter 3   Applications of Differentiation **103**

Chapter 4   Integration **178**

Chapter 5   Logarithmic, Exponential, and Other Transcendental Functions . . .   **223**

Chapter 6   Differential Equations **279**

Chapter 7   Application of Integration **312**

Chapter 8   Integration Techniques, L'Hôpital's Rule, and Improper Integrals . .   **362**

Chapter 9   Infinite Series **436**

Chapter 10   Conics, Parametric Equations, and Polar Coordinates **502**

Chapter 11   Vectors and Geometry of Space **556**

C H A P T E R P
Preparation for Calculus

Section P.1 Graphs and Models 2

Section P.2 Linear Models and Rates of Change 6

Section P.3 Functions and Their Graphs 12

Section P.4 Fitting Models to Data 17

Review Exercises 19

Problem Solving 22

CHAPTER P
Preparation for Calculus

Section P.1   Graphs and Models

1. $y = -\frac{1}{2}x + 2$

x-intercept: $(4, 0)$

y-intercept: $(0, 2)$

Matches graph (b).

3. $y = 4 - x^2$

x-intercepts: $(2, 0), (-2, 0)$

y-intercept: $(0, 4)$

Matches graph (a).

5. $y = \frac{3}{2}x + 1$

x	-4	-2	0	2	4
y	-5	-2	1	4	7

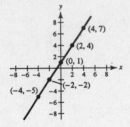

7. $y = 4 - x^2$

x	-3	-2	0	2	3
y	-5	0	4	0	-5

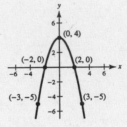

9. $y = |x + 2|$

x	-5	-4	-3	-2	-1	0	1
y	3	2	1	0	1	2	3

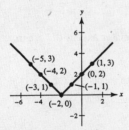

11. $y = \sqrt{x} - 4$

x	0	1	4	9	16
y	-4	-3	-2	-1	0

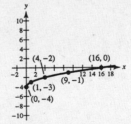

13. $y = \dfrac{2}{x}$

x	-3	-2	-1	0	1	2	3
y	$-\frac{2}{3}$	-1	-2	Undef.	2	1	$\frac{2}{3}$

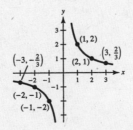

15.
```
Xmin = -3
Xmax = 5
Xscl = 1
Ymin = -3
Ymax = 5
Yscl = 1
```

Note that $y = 4$ when $x = 0$.

17. $y = \sqrt{5 - x}$

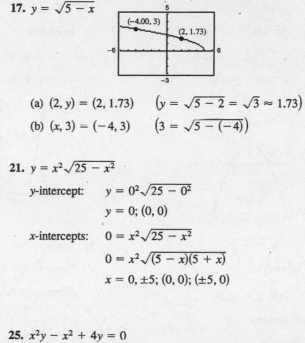

(a) $(2, y) = (2, 1.73)$ $\left(y = \sqrt{5 - 2} = \sqrt{3} \approx 1.73\right)$

(b) $(x, 3) = (-4, 3)$ $\left(3 = \sqrt{5 - (-4)}\right)$

19. $y = x^2 + x - 2$

y-intercept: $y = 0^2 + 0 - 2$

            $y = -2; (0, -2)$

x-intercepts: $0 = x^2 + x - 2$

            $0 = (x + 2)(x - 1)$

            $x = -2, 1; (-2, 0), (1, 0)$

21. $y = x^2\sqrt{25 - x^2}$

y-intercept: $y = 0^2\sqrt{25 - 0^2}$

            $y = 0; (0, 0)$

x-intercepts: $0 = x^2\sqrt{25 - x^2}$

            $0 = x^2\sqrt{(5 - x)(5 + x)}$

            $x = 0, \pm 5; (0, 0); (\pm 5, 0)$

23. $y = \dfrac{3\left(2 - \sqrt{x}\right)}{x}$

y-intercept: None. x cannot equal 0.

x-intercept: $0 = \dfrac{3\left(2 - \sqrt{x}\right)}{x}$

            $0 = 2 - \sqrt{x}$

            $x = 4; (4, 0)$

25. $x^2y - x^2 + 4y = 0$

y-intercept:

     $0^2(y) - 0^2 + 4y = 0$

            $y = 0; (0, 0)$

x intercept:

     $x^2(0) - x^2 + 4(0) = 0$

            $x = 0; (0, 0)$

27. Symmetric with respect to the y-axis since

     $y = (-x)^2 - 2 = x^2 - 2.$

29. Symmetric with respect to the x-axis since

     $(-y)^2 = y^2 = x^3 - 4x.$

31. Symmetric with respect to the origin since

     $(-x)(-y) = xy = 4.$

33. $y = 4 - \sqrt{x + 3}$

No symmetry with respect to either axis or the origin.

35. Symmetric with respect to the origin since

     $-y = \dfrac{-x}{(-x)^2 + 1}$

     $y = \dfrac{x}{x^2 + 1}.$

37. $y = |x^3 + x|$ is symmetric with respect to the y-axis

since $y = |(-x)^3 + (-x)| = |-(x^3 + x)| = |x^3 + x|.$

39. $y = -3x + 2$

Intercepts:

     $\left(\tfrac{2}{3}, 0\right), (0, 2)$

Symmetry: none

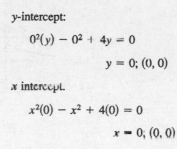

41. $y = \dfrac{1}{2}x - 4$

Intercepts:

     $(8, 0), (0, -4)$

Symmetry: none

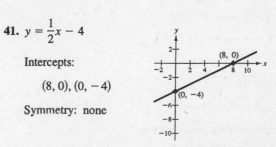

43. $y = 1 - x^2$

Intercepts:

$(1, 0), (-1, 0), (0, 1)$

Symmetry: y-axis

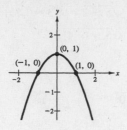

45. $y = (x + 3)^2$

Intercepts:

$(-3, 0), (0, 9)$

Symmetry: none

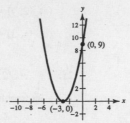

47. $y = x^3 + 2$

Intercepts:

$\left(-\sqrt[3]{2}, 0\right), (0, 2)$

Symmetry: none

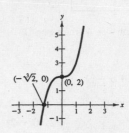

49. $y = x\sqrt{x + 2}$

Intercepts:

$(0, 0), (-2, 0)$

Symmetry: none

Domain: $x \geq -2$

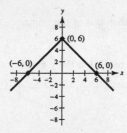

51. $x = y^3$

Intercept: $(0, 0)$

Symmetry: origin

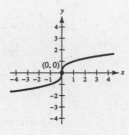

53. $y = \dfrac{1}{x}$

Intercepts: none

Symmetry: origin

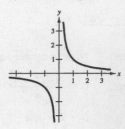

55. $y = 6 - |x|$

Intercepts:

$(0, 6), (-6, 0), (6, 0)$

Symmetry: y-axis

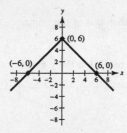

57. $y^2 - x = 9$

$\qquad y^2 = x + 9$

$\qquad y = \pm\sqrt{x + 9}$

Intercepts:

$(0, 3), (0, -3), (-9, 0)$

Symmetry: x-axis

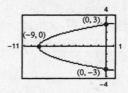

59. $x + 3y^2 = 6$

$\qquad 3y^2 = 6 - x$

$\qquad y = \pm\sqrt{2 - \dfrac{x}{3}}$

Intercepts:

$(6, 0), \left(0, \sqrt{2}\right), \left(0, -\sqrt{2}\right)$

Symmetry: x-axis

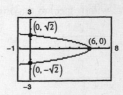

61. $x + y = 2 \Longrightarrow y = 2 - x$

$2x - y = 1 \Longrightarrow y = 2x - 1$

$\qquad 2 - x = 2x - 1$

$\qquad\quad 3 = 3x$

$\qquad\quad 1 = x$

The corresponding y-value is $y = 1$.

Point of intersection: $(1, 1)$

63. $x^2 + y = 6 \Rightarrow y = 6 - x^2$

$x + y = 4 \Rightarrow y = 4 - x$

$6 - x^2 = 4 - x$

$0 = x^2 - x - 2$

$0 = (x - 2)(x + 1)$

$x = 2, -1$

The corresponding y-values are $y = 2$ (for $x = 2$) and $y = 5$ (for $x = -1$).

Points of intersection: $(2, 2), (-1, 5)$

65. $x^2 + y^2 = 5 \Rightarrow y^2 = 5 - x^2$

$x - y = 1 \Rightarrow y = x - 1$

$5 - x^2 = (x - 1)^2$

$5 - x^2 = x^2 - 2x + 1$

$0 = 2x^2 - 2x - 4 = 2(x + 1)(x - 2)$

$x = -1 \text{ or } x = 2$

The corresponding y-values are $y = -2$ and $y = 1$.

Points of intersection: $(-1, -2), (2, 1)$

67.
$$y = x^3$$
$$y = x$$
$$x^3 = x$$
$$x^3 - x = 0$$
$$x(x + 1)(x - 1) = 0$$
$$x = 0, x = -1, \text{ or } x = 1$$

The corresponding y-values are $y = 0$, $y = -1$, and $y = 1$.

Points of intersection: $(0, 0), (-1, -1), (1, 1)$

69.
$$y = x^3 - 2x^2 + x - 1$$
$$y = -x^2 + 3x - 1$$
$$x^3 - 2x^2 + x - 1 = -x^2 + 3x - 1$$
$$x^3 - x^2 - 2x = 0$$
$$x(x - 2)(x + 1) = 0$$
$$x = -1, 0, 2$$
$$(-1, -5), (0, -1), (2, 1)$$

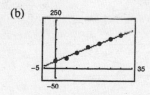

71. $y = \sqrt{x + 6}$

$y = \sqrt{-x^2 - 4x}$

Points of intersection: $(-2, 2), \left(-3, \sqrt{3}\right) \approx (-3, 1.732)$

Analytically, $\sqrt{x + 6} = \sqrt{-x^2 - 4x}$

$$x + 6 = -x^2 - 4x$$
$$x^2 + 5x + 6 = 0$$
$$(x + 3)(x + 2) = 0$$
$$x = -3, y = \sqrt{3} \Rightarrow \left(-3, \sqrt{3}\right)$$
$$x = -2, y = 2 \Rightarrow (-2, 2).$$

73. (a) Using a graphing utility, you obtain

$$y = -0.007t^2 + 4.82t + 35.4.$$

(b)

(c) For 2010, $t = 40$ and $y = 217$.

75.
$$C = R$$
$$5.5\sqrt{x} + 10{,}000 = 3.29x$$
$$\left(5.5\sqrt{x}\right)^2 = (3.29x - 10{,}000)^2$$
$$30.25x = 10.8241x^2 - 65{,}800x + 100{,}000{,}000$$
$$0 = 10.8241x^2 - 65{,}830.25x + 100{,}000{,}000 \quad \text{Use the Quadratic Formula.}$$
$$x \approx 3133 \text{ units}$$

The other root, $x \approx 2949$, does not satisfy the equation $R = C$.

This problem can also be solved by using a graphing utility and finding the intersection of the graphs of C and R.

77. $y = (x + 2)(x - 4)(x - 6)$ (other answers possible)

79. (i) $y = kx + 5$ matches (b).

Use $(1, 7)$:
$$7 = k(1) + 5 \Rightarrow k = 2, \text{ thus, } y = 2x + 5.$$

(ii) $y = x^2 + k$ matches (d).

Use $(1, -9)$:
$$-9 = (1)^2 + k \Rightarrow k = -10, \text{ thus, } y = x^2 - 10.$$

(iii) $y = kx^{3/2}$ matches (a).

Use $(1, 3)$: $3 = k(1)^{3/2} \Rightarrow k = 3$, thus, $y = 3x^{3/2}$.

(iv) $xy = k$ matches (c).

Use $(1, 36)$: $(1)(36) = k \Rightarrow k = 36$, thus, $xy = 36$.

81. False; x-axis symmetry means that if $(1, -2)$ is on the graph, then $(1, 2)$ is also on the graph.

83. True; the x-intercepts are $\left(\dfrac{-b \pm \sqrt{b^2 - 4ac}}{2a}, 0 \right)$.

85.
$$2\sqrt{(x - 0)^2 + (y - 3)^2} = \sqrt{(x - 0)^2 + (y - 0)^2}$$
$$4[x^2 + (y - 3)^2] = x^2 + y^2$$
$$4x^2 + 4y^2 - 24y + 36 = x^2 + y^2$$
$$3x^2 + 3y^2 - 24y + 36 = 0$$
$$x^2 + y^2 - 8y + 12 = 0$$
$$x^2 + (y - 4)^2 = 4$$

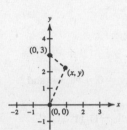

Circle of radius 2 and center $(0, 4)$

Section P.2   Linear Models and Rates of Change

1. $m = 1$

3. $m = 0$

5. $m = -12$

7.

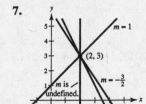

9. $m = \dfrac{2 - (-4)}{5 - 3}$

$= \dfrac{6}{2} = 3$

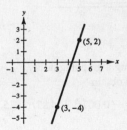

11. $m = \dfrac{5 - 1}{2 - 2}$

$= \dfrac{4}{0}$

undefined

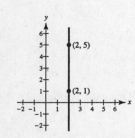

13. $m = \dfrac{2/3 - 1/6}{-1/2 - (-3/4)}$

$= \dfrac{1/2}{1/4} = 2$

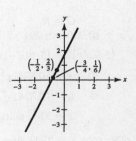

15. Since the slope is 0, the line is horizontal and its equation is $y = 1$.
Therefore, three additional points are $(0, 1)$, $(1, 1)$, and $(3, 1)$.

17. The equation of this line is

$$y - 7 = -3(x - 1)$$

$$y = -3x + 10.$$

Therefore, three additional points are $(0, 10)$, $(2, 4)$, and $(3, 1)$.

19. (a) Slope $= \dfrac{\Delta y}{\Delta x} = \dfrac{1}{3}$

(b)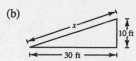

By the Pythagorean Theorem,

$$x^2 = 30^2 + 10^2 = 1000$$

$$x = 10\sqrt{10} \approx 31.623 \text{ feet.}$$

21. (a)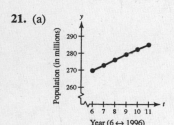

(b) The slopes of the line segments are:

$$\frac{272.9 - 269.7}{7 - 6} = 3.2$$

$$\frac{276.1 - 272.9}{8 - 7} = 3.2$$

$$\frac{279.3 - 276.1}{9 - 8} = 3.2$$

$$\frac{282.3 - 279.3}{10 - 9} = 3.0$$

$$\frac{285.0 - 282.3}{11 - 10} = 2.7$$

The population increased least rapidly between 2000 and 2001.

23. $x + 5y = 20$

$$y = -\tfrac{1}{5}x + 4$$

Therefore, the slope is $m = -\tfrac{1}{5}$ and the y-intercept is $(0, 4)$.

25. $x = 4$

The line is vertical. Therefore, the slope is undefined and there is no y-intercept.

27. $y = \tfrac{3}{4}x + 3$

$$4y = 3x + 12$$

$$0 = 3x - 4y + 12$$

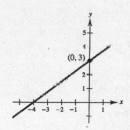

29.
$$y = \tfrac{2}{3}x$$
$$3y = 2x$$
$$2x - 3y = 0$$

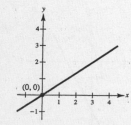

31.
$$y + 2 = 3(x - 3)$$
$$y + 2 = 3x - 9$$
$$y = 3x - 11$$
$$y - 3x + 11 = 0$$

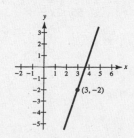

33. $m = \dfrac{6 - 0}{2 - 0} = 3$

$$y - 0 = 3(x - 0)$$
$$y = 3x$$

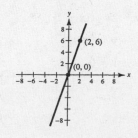

35. $m = \dfrac{1 - (-3)}{2 - 0} = 2$

$y - 1 = 2(x - 2)$

$y - 1 = 2x - 4$

$0 = 2x - y - 3$

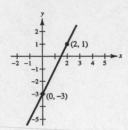

37. $m = \dfrac{8 - 0}{2 - 5} = -\dfrac{8}{3}$

$y - 0 = -\dfrac{8}{3}(x - 5)$

$y = -\dfrac{8}{3}x + \dfrac{40}{3}$

$3y + 8x - 40 = 0$

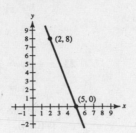

39. $m = \dfrac{8 - 1}{5 - 5}$ Undefined

Vertical line $x = 5$

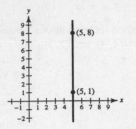

41. $m = \dfrac{7/2 - 3/4}{1/2 - 0} = \dfrac{11/4}{1/2} = \dfrac{11}{2}$

$y - \dfrac{3}{4} = \dfrac{11}{2}(x - 0)$

$y = \dfrac{11}{2}x + \dfrac{3}{4}$

$22x - 4y + 3 = 0$

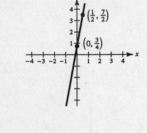

43. $x = 3$

$x - 3 = 0$

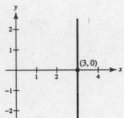

45. $\dfrac{x}{2} + \dfrac{y}{3} = 1$

$3x + 2y - 6 = 0$

47. $\dfrac{x}{a} + \dfrac{y}{a} = 1$

$\dfrac{1}{a} + \dfrac{2}{a} = 1$

$\dfrac{3}{a} = 1$

$a = 3 \Rightarrow x + y = 3$

$x + y - 3 = 0$

49. $y = -3$

$y + 3 = 0$

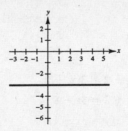

51. $y = -2x + 1$

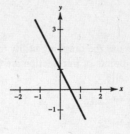

53. $y - 2 = \dfrac{3}{2}(x - 1)$

$y = \dfrac{3}{2}x + \dfrac{1}{2}$

$2y - 3x - 1 = 0$

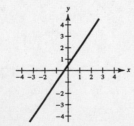

55. $2x - y - 3 = 0$

$y = 2x - 3$

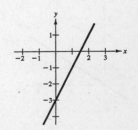

57. (a)

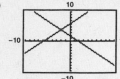

The lines do not appear perpendicular.

(b)

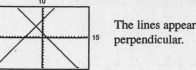

The lines appear perpendicular.

The lines are perpendicular because their slopes 1 and -1 are negative reciprocals of each other.
You must use a square setting in order for perpendicular lines to appear perpendicular. Answers depend on calculator used.

59. $4x - 2y = 3$

$$y = 2x - \frac{3}{2}$$

$$m = 2$$

(a) $\quad y - 1 = 2(x - 2)$

$\quad y - 1 = 2x - 4$

$2x - y - 3 = 0$

(b) $\quad y - 1 = -\frac{1}{2}(x - 2)$

$2y - 2 = -x + 2$

$x + 2y - 4 = 0$

61. $5x - 3y = 0$

$$y = \frac{5}{3}x$$

$$m - \frac{5}{3}$$

(a) $\quad y - \frac{7}{8} = \frac{5}{3}\left(x - \frac{3}{4}\right)$

$24y - 21 = 40x - 30$

$24y - 40x + 9 = 0$

(b) $\quad y - \frac{7}{8} = -\frac{3}{5}\left(x - \frac{3}{4}\right)$

$40y - 35 = -24x + 18$

$40y + 24x - 53 = 0$

63. The given line is vertical.

(a) $x = 2 \Rightarrow x - 2 = 0$

(b) $y = 5 \Rightarrow y - 5 = 0$

65. The slope is 125. $V = 2540$ when $t = 4$.

$$V = 125(t - 4) + 2540 = 125t + 2040$$

67. The slope is -2000. $V = 20,400$ when $t = 4$.

$$V = -2000(t - 4) + 20,400 = -2000t + 28,400$$

69.

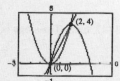

You can use the graphing utility to determine that the points of intersection are $(0, 0)$ and $(2, 4)$. Analytically,

$$x^2 = 4x - x^2$$

$2x^2 - 4x = 0$

$2x(x - 2) = 0$

$\quad x = 0 \Rightarrow y = 0 \Rightarrow (0, 0)$

$\quad x = 2 \Rightarrow y = 4 \Rightarrow (2, 4).$

The slope of the line joining $(0, 0)$ and $(2, 4)$ is $m = (4 - 0)/(2 - 0) = 2$. Hence, an equation of the line is

$$y - 0 = 2(x - 0)$$

$$y = 2x.$$

71. $m_1 = \dfrac{1 - 0}{-2 - (-1)} = -1$

$m_2 = \dfrac{-2 - 0}{2 - (-1)} = -\dfrac{2}{3}$

$m_1 \neq m_2$

The points are not collinear.

73. Equations of perpendicular bisectors:

$$y - \frac{c}{2} = \frac{a - b}{c}\left(x - \frac{a + b}{2}\right)$$

$$y - \frac{c}{2} = \frac{a + b}{-c}\left(x - \frac{b - a}{2}\right)$$

Setting the right-hand sides of the two equations equal and solving for x yields $x = 0$.

Letting $x = 0$ in either equation gives the point of intersection:

$$\left(0, \frac{-a^2 + b^2 + c^2}{2c}\right).$$

This point lies on the third perpendicular bisector, $x = 0$.

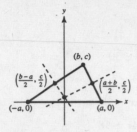

75. Equations of altitudes:

$$y = \frac{a - b}{c}(x + a)$$

$$x = b$$

$$y = -\frac{a + b}{c}(x - a)$$

Solving simultaneously, the point of intersection is

$$\left(b, \frac{a^2 - b^2}{c}\right).$$

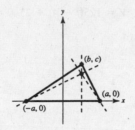

77. Find the equation of the line through the points $(0, 32)$ and $(100, 212)$.

$$m = \frac{180}{100} = \frac{9}{5}$$

$$F - 32 = \frac{9}{5}(C - 0)$$

$$F = \frac{9}{5}C + 32$$

or

$$C = \frac{1}{9}(5F - 160)$$

$$5F - 9C - 160 = 0$$

For $F = 72°$, $C \approx 22.2°$.

79. (a) $W_1 = 0.75x + 12.50$

   $W_2 = 1.30x + 9.20$

(b)

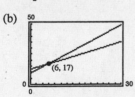

Using a graphing utility, the point of intersection is $(6, 17)$.
Analytically,

$$0.75x + 12.50 = 1.30x + 9.20$$

$$3.3 = 0.55x \Longrightarrow x = 6$$

$$y = 0.75(6) + 12.50 = 17.$$

(c) Both jobs pay $17 per hour if 6 units are produced. For someone who can produce more than 6 units per hour, the second offer would pay more. For a worker who produces less than 6 units per hour, the first offer pays more.

81. (a) Two points are $(50, 580)$ and $(47, 625)$. The slope is

$$m = \frac{625 - 580}{47 - 50} = -15.$$

$$p - 580 = -15(x - 50)$$

$$p = -15x + 750 + 580 = -15x + 1330$$

$$\text{or } x = \frac{1}{15}(1330 - p)$$

(b)

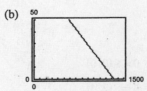

If $p = 655$, $x = \frac{1}{15}(1330 - 655) = 45$ units.

(c) If $p = 595$, $x = \frac{1}{15}(1330 - 595) = 49$ units.

83. The tangent line is perpendicular to the line joining the point (5, 12) and the center (0, 0).

Slope of the line joining (5, 12) and (0, 0) is $\dfrac{12}{5}$.

The equation of the tangent line is

$$y - 12 = \frac{-5}{12}(x - 5)$$

$$y = \frac{-5}{12}x + \frac{169}{12}$$

$$12y + 5x - 169 = 0.$$

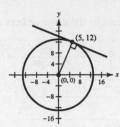

85. $4x + 3y - 10 = 0 \Rightarrow d = \dfrac{|4(0) + 3(0) - 10|}{\sqrt{4^2 + 3^2}} = \dfrac{10}{5} = 2$

87. $x - y - 2 = 0 \Rightarrow d = \dfrac{|1(-2) + (-1)(1) - 2|}{\sqrt{1^2 + 1^2}} = \dfrac{5}{\sqrt{2}} = \dfrac{5\sqrt{2}}{2}$

89. A point on the line $x + y = 1$ is (0, 1). The distance from the point (0, 1) to $x + y - 5 = 0$ is

$$d = \frac{|1(0) + 1(1) - 5|}{\sqrt{1^2 + 1^2}} = \frac{|1 - 5|}{\sqrt{2}} = \frac{4}{\sqrt{2}} = 2\sqrt{2}.$$

91. If $A = 0$, then $By + C = 0$ is the horizontal line $y = -C/B$. The distance to (x_1, y_1) is

$$d = \left| y_1 - \left(\frac{-C}{B} \right) \right| = \frac{|By_1 + C|}{|B|} = \frac{|Ax_1 + By_1 + C|}{\sqrt{A^2 + B^2}}.$$

If $B = 0$, then $Ax + C = 0$ is the vertical line $x = -C/A$. The distance to (x_1, y_1) is

$$d = \left| x_1 - \left(\frac{-C}{A} \right) \right| = \frac{|Ax_1 + C|}{|A|} = \frac{|Ax_1 + By_1 + C|}{\sqrt{A^2 + B^2}}.$$

(Note that A and B cannot both be zero.)

The slope of the line $Ax + By + C = 0$ is $-A/B$. The equation of the line through (x_1, y_1) perpendicular to $Ax + By + C = 0$ is:

$$y - y_1 = \frac{B}{A}(x - x_1)$$

$$Ay - Ay_1 = Bx - Bx_1$$

$$Bx_1 - Ay_1 = Bx - Ay$$

The point of intersection of these two lines is:

$$Ax + By = -C \quad \Longrightarrow \quad A^2x + ABy = -AC \qquad (1)$$

$$Bx - Ay = Bx_1 - Ay_1 \Rightarrow \quad B^2x - ABy = B^2x_1 - ABy_1 \qquad (2)$$

$$(A^2 + B^2)x = -AC + B^2x_1 - ABy_1 \quad \text{(By adding equations (1) and (2))}$$

$$x = \frac{-AC + B^2x_1 - ABy_1}{A^2 + B^2}$$

$$Ax + By = -C \quad \Longrightarrow \quad ABx + B^2y = -BC \qquad (3)$$

$$Bx - Ay = Bx_1 - Ay_1 \Rightarrow \quad -ABx + A^2y = -ABx_1 + A^2y_1 \qquad (4)$$

$$(A^2 + B^2)y = -BC - ABx_1 + A^2y_1 \quad \text{(By adding equations (3) and (4))}$$

$$y = \frac{-BC - ABx_1 + A^2y_1}{A^2 + B^2}$$

$$\left(\frac{-AC + B^2x_1 - ABy_1}{A^2 + B^2}, \frac{-BC - ABx_1 + A^2y_1}{A^2 + B^2} \right) \text{ point of intersection}$$

—CONTINUED—

91. —CONTINUED—

The distance between (x_1, y_1) and this point gives us the distance between (x_1, y_1) and the line $Ax + By + C = 0$.

$$d = \sqrt{\left[\frac{-AC + B^2x_1 - ABy_1}{A^2 + B^2} - x_1\right]^2 + \left[\frac{-BC - ABx_1 + A^2y_1}{A^2 + B^2} - y_1\right]^2}$$

$$= \sqrt{\left[\frac{-AC - ABy_1 - A^2x_1}{A^2 + B^2}\right]^2 + \left[\frac{-BC - ABx_1 - B^2y_1}{A^2 + B^2}\right]^2}$$

$$= \sqrt{\left[\frac{-A(C + By_1 + Ax_1)}{A^2 + B^2}\right]^2 + \left[\frac{-B(C + Ax_1 + By_1)}{A^2 + B^2}\right]^2}$$

$$= \sqrt{\frac{(A^2 + B^2)(C + Ax_1 + By_1)^2}{(A^2 + B^2)^2}}$$

$$= \frac{|Ax_1 + By_1 + C|}{\sqrt{A^2 + B^2}}$$

93. For simplicity, let the vertices of the rhombus be $(0, 0)$, $(a, 0)$, (b, c), and $(a + b, c)$, as shown in the figure. The slopes of the diagonals are then

$$m_1 = \frac{c}{a + b} \text{ and } m_2 = \frac{c}{b - a}.$$

Since the sides of the rhombus are equal, $a^2 = b^2 + c^2$, and we have

$$m_1 m_2 = \frac{c}{a + b} \cdot \frac{c}{b - a} = \frac{c^2}{b^2 - a^2} = \frac{c^2}{-c^2} = -1.$$

Therefore, the diagonals are perpendicular.

95. Consider the figure below in which the four points are collinear. Since the triangles are similar, the result immediately follows.

$$\frac{y_2{}^* - y_1{}^*}{x_2{}^* - x_1{}^*} = \frac{y_2 - y_1}{x_2 - x_1}$$

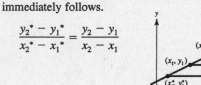
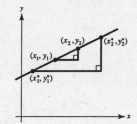

97. True.

$$ax + by = c_1 \Rightarrow y = -\frac{a}{b}x + \frac{c_1}{b} \Rightarrow m_1 = -\frac{a}{b}$$

$$bx - ay = c_2 \Rightarrow y = \frac{b}{a}x - \frac{c_2}{a} \Rightarrow m_2 = \frac{b}{a}$$

$$m_2 = -\frac{1}{m_1}$$

Section P.3   Functions and Their Graphs

1. (a) Domain of f: $-4 \le x \le 4$

　　Range of f: $-3 \le y \le 5$

　　Domain of g: $-3 \le x \le 3$

　　Range of g: $-4 \le y \le 4$

(b) $f(-2) = -1$

　　$g(3) = -4$

(c) $f(x) = g(x)$ for $x = -1$

(d) $f(x) = 2$ for $x = 1$

(e) $g(x) = 0$ for $x = -1, 1$ and 2

3. (a) $f(0) = 2(0) - 3 = -3$

(b) $f(-3) = 2(-3) - 3 = -9$

(c) $f(b) = 2b - 3$

(d) $f(x - 1) = 2(x - 1) - 3 = 2x - 5$

5. (a) $g(0) = 3 - 0^2 = 3$

(b) $g(\sqrt{3}) = 3 - (\sqrt{3})^2 = 3 - 3 = 0$

(c) $g(-2) = 3 - (-2)^2 = 3 - 4 = -1$

(d) $g(t - 1) = 3 - (t - 1)^2 = -t^2 + 2t + 2$

7. (a) $f(0) = \cos(2(0)) = \cos 0 = 1$

(b) $f\left(-\dfrac{\pi}{4}\right) = \cos\left(2\left(-\dfrac{\pi}{4}\right)\right) = \cos\left(-\dfrac{\pi}{2}\right) = 0$

(c) $f\left(\dfrac{\pi}{3}\right) = \cos\left(2\left(\dfrac{\pi}{3}\right)\right) = \cos\dfrac{2\pi}{3} = -\dfrac{1}{2}$

9. $\dfrac{f(x + \Delta x) - f(x)}{\Delta x} = \dfrac{(x + \Delta x)^3 - x^3}{\Delta x} = \dfrac{x^3 + 3x^2\Delta x + 3x(\Delta x)^2 + (\Delta x)^3 - x^3}{\Delta x} = 3x^2 + 3x\Delta x + (\Delta x)^2, \Delta x \neq 0$

11. $\dfrac{f(x) - f(2)}{x - 2} = \dfrac{(1/\sqrt{x - 1} - 1)}{x - 2}$

$= \dfrac{1 - \sqrt{x - 1}}{(x - 2)\sqrt{x - 1}} \cdot \dfrac{1 + \sqrt{x - 1}}{1 + \sqrt{x - 1}} = \dfrac{2 - x}{(x - 2)\sqrt{x - 1}(1 + \sqrt{x - 1})} = \dfrac{-1}{\sqrt{x - 1}(1 + \sqrt{x - 1})}, x \neq 2$

13. $h(x) = -\sqrt{x + 3}$

Domain: $x + 3 \geq 0 \Rightarrow [-3, \infty)$

Range: $(-\infty, 0]$

15. $f(t) = \sec\dfrac{\pi t}{4}$

$\dfrac{\pi t}{4} \neq \dfrac{(2k + 1)\pi}{2} \Rightarrow t \neq 4k + 2$

Domain: all $t \neq 4k + 2, k$ an integer

Range: $(-\infty, -1] \cup [1, \infty)$

17. $f(x) = \dfrac{1}{x}$

Domain: $(-\infty, 0) \cup (0, \infty)$

Range: $(-\infty, 0) \cup (0, \infty)$

19. $f(x) = \sqrt{x} + \sqrt{1 - x}$

$x \geq 0$ and $1 - x \geq 0$

$x \geq 0$ and $x \leq 1$

Domain: $0 \leq x \leq 1$

21. $g(x) = \dfrac{2}{1 - \cos x}$

$1 - \cos x \neq 0$

$\cos x \neq 1$

Domain: all $x \neq 2n\pi, n$ an integer

23. $f(x) = \dfrac{1}{|x + 3|}$

$|x + 3| \neq 0$

$x + 3 \neq 0$

Domain: all $x \neq -3$

25. $f(x) = \begin{cases} 2x + 1, & x < 0 \\ 2x + 2, & x \geq 0 \end{cases}$

(a) $f(-1) = 2(-1) + 1 = -1$

(b) $f(0) = 2(0) + 2 = 2$

(c) $f(2) = 2(2) + 2 = 6$

(d) $f(t^2 + 1) = 2(t^2 + 1) + 2 = 2t^2 + 4$

(Note: $t^2 + 1 \geq 0$ for all t)

Domain: $(-\infty, \infty)$

Range: $(-\infty, 1) \cup [2, \infty)$

27. $f(x) = \begin{cases} |x| + 1, & x < 1 \\ -x + 1, & x \geq 1 \end{cases}$

(a) $f(-3) = |-3| + 1 = 4$

(b) $f(1) = -1 + 1 = 0$

(c) $f(3) = -3 + 1 = -2$

(d) $f(b^2 + 1) = -(b^2 + 1) + 1 = -b^2$

Domain: $(-\infty, \infty)$

Range: $(-\infty, 0] \cup [1, \infty)$

29. $f(x) = 4 - x$

Domain: $(-\infty, \infty)$

Range: $(-\infty, \infty)$

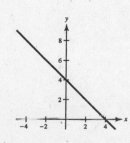

31. $h(x) = \sqrt{x - 1}$

Domain: $[1, \infty)$

Range: $[0, \infty)$

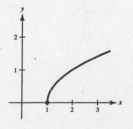

33. $f(x) = \sqrt{9 - x^2}$

Domain: $[-3, 3]$

Range: $[0, 3]$

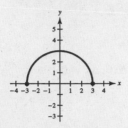

35. $g(t) = 2 \sin \pi t$

Domain: $(-\infty, \infty)$

Range: $[-2, 2]$

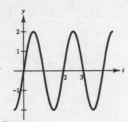

37. The student travels
$$\frac{2 - 0}{4 - 0} = \frac{1}{2} \text{ mi/min during}$$
the first 4 minutes. The student is stationary for the following 2 minutes. Finally, the student travels $\dfrac{6 - 2}{10 - 6} = 1$ mi/min during the final 4 minutes.

39. $x - y^2 = 0 \Longrightarrow y = \pm\sqrt{x}$

y is not a function of x. Some vertical lines intersect the graph twice.

41. y is a function of x. Vertical lines intersect the graph at most once.

43. $x^2 + y^2 = 4 \Longrightarrow y = \pm\sqrt{4 - x^2}$

y is not a function of x since there are two values of y for some x.

45. $y^2 = x^2 - 1 \Longrightarrow y = \pm\sqrt{x^2 - 1}$

y is not a function of x since there are two values of y for some x.

47. $y = f(x + 5)$ is a horizontal shift 5 units to the left. Matches d.

49. $y = -f(-x) - 2$ is a reflection in the y-axis, a reflection in the x-axis, and a vertical shift downward 2 units. Matches c.

51. $y = f(x + 6) + 2$ is a horizontal shift to the left 6 units, and a vertical shift upward 2 units. Matches e.

53. (a) The graph is shifted 3 units to the left.

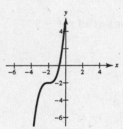

(b) The graph is shifted 1 unit to the right.

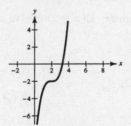

(c) The graph is shifted 2 units upward.

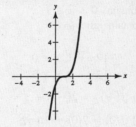

(d) The graph is shifted 4 units downward.

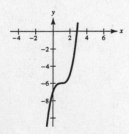

(e) The graph is stretched vertically by a factor of 3.

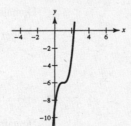

(f) The graph is stretched vertically by a factor of $\frac{1}{4}$.

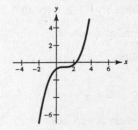

55. (a) $y = \sqrt{x} + 2$ (b) $y = -\sqrt{x}$ (c) $y = \sqrt{x - 2}$

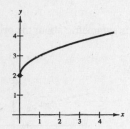   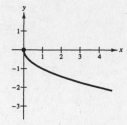   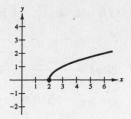

Vertical shift 2 units upward   Reflection about the x-axis   Horizontal shift 2 units to the right

57. (a) $f(g(1)) = f(0) = 0$

   (b) $g(f(1)) = g(1) = 0$

   (c) $g(f(0)) = g(0) = -1$

   (d) $f(g(-4)) = f(15) = \sqrt{15}$

   (e) $f(g(x)) = f(x^2 - 1) = \sqrt{x^2 - 1}$

   (f) $g(f(x)) = g(\sqrt{x}) = (\sqrt{x})^2 - 1 = x - 1, \quad (x \geq 0)$

59. $f(x) = x^2, g(x) = \sqrt{x}$

   $(f \circ g)(x) = f(g(x)) = f(\sqrt{x}) = (\sqrt{x})^2 = x, \quad x \geq 0$

   Domain: $[0, \infty)$

   $(g \circ f)(x) = g(f(x)) = g(x^2) = \sqrt{x^2} = |x|$

   Domain: $(-\infty, \infty)$

   No. Their domains are different.
   $(f \circ g) = (g \circ f)$ for $x \geq 0$.

61. $f(x) = \dfrac{3}{x}, g(x) = x^2 - 1$

   $(f \circ g)(x) = f(g(x)) = f(x^2 - 1) = \dfrac{3}{x^2 - 1}$

   Domain: all $x \neq \pm 1$

   $(g \circ f)(x) = g(f(x)) = g\left(\dfrac{3}{x}\right) = \left(\dfrac{3}{x}\right)^2 - 1 = \dfrac{9}{x^2} - 1 = \dfrac{9 - x^2}{x^2}$

   Domain: all $x \neq 0$

   No, $f \circ g \neq g \circ f$.

63. (a) $(f \circ g)(3) = f(g(3)) = f(-1) = 4$

   (b) $g(f(2)) = g(1) = -2$

   (c) $g(f(5)) = g(-5)$, which is undefined

   (d) $(f \circ g)(-3) = f(g(-3)) = f(-2) = 3$

   (e) $(g \circ f)(-1) = g(f(-1)) = g(4) = 2$

   (f) $f(g(-1)) = f(-4)$, which is undefined

65. $F(x) = \sqrt{2x - 2}$

   Let $h(x) = x - 1, g(x) = 2x$ and $f(x) = \sqrt{x}$.

   Then, $(f \circ g \circ h)(x) = f(g(x - 1)) = f(2(x - 1)) = \sqrt{2(x - 1)} = \sqrt{2x - 2} = F(x)$.

   [Other answers possible]

67. $f(-x) = (-x)^2(4 - (-x)^2) = x^2(4 - x^2) = f(x)$

   Even

69. $f(-x) = (-x)\cos(-x) = -x \cos x = -f(x)$

   Odd

71. (a) If f is even, then $\left(\frac{3}{2}, 4\right)$ is on the graph.

(b) If f is odd, then $\left(\frac{3}{2}, -4\right)$ is on the graph.

73. f is even because the graph is symmetric about the y-axis.

g is neither even nor odd.

h is odd because the graph is symmetric about the origin.

75. $\text{Slope} = \dfrac{3+5}{-4-0} = -2$

$y + 5 = -2(x - 0)$

$y = -2x - 5$

$f(x) = -2x - 5, \quad -4 \le x \le 0$

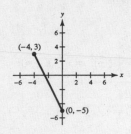

77. $x + y^2 = 0$

$y^2 = -x$

$y = -\sqrt{-x}$

$f(x) = -\sqrt{-x}, \quad x \le 0$

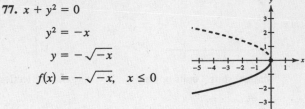

79. Matches (ii). The function is $g(x) = cx^2$. Since $(1, -2)$ satisfies the equation, $c = -2$. Thus, $g(x) = -2x^2$.

81. Matches (iv). The function is $r(x) = c/x$, since it must be undefined at $x = 0$. Since $(1, 32)$ satisfies the equation, $c = 32$. Thus, $r(x) = 32/x$.

83. (a) $T(4) = 16°$, $T(15) \approx 23°$

(b) If $H(t) = T(t - 1)$, then the program would turn on (and off) one hour later.

(c) If $H(t) = T(t) - 1$, then the overall temperature would be reduced 1 degree.

85. (a)

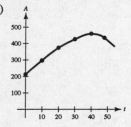

(b) $A(15) \approx 345$ acres/farm

87. $f(x) = |x| + |x - 2|$

If $x < 0$, then $f(x) = -x - (x - 2) = -2x + 2 = 2(1 - x)$.

If $0 \le x < 2$, then $f(x) = x - (x - 2) = 2$.

If $x \ge 2$, then $f(x) = x + (x - 2) = 2x - 2 = 2(x - 1)$.

Thus,

$$f(x) = \begin{cases} 2(1 - x), & x < 0 \\ 2, & 0 \le x < 2. \\ 2(x - 1), & x \ge 2 \end{cases}$$

89. $f(-x) = a_{2n+1}(-x)^{2n+1} + \cdots + a_3(-x)^3 + a_1(-x)$

$= -[a_{2n+1}x^{2n+1} + \cdots + a_3x^3 + a_1x]$

$= -f(x)$

Odd

91. Let $F(x) = f(x)g(x)$ where f and g are even. Then

$$F(-x) = f(-x)g(-x) = f(x)g(x) = F(x).$$

Thus, $F(x)$ is even. Let $F(x) = f(x)g(x)$ where f and g are odd. Then

$$F(-x) = f(-x)g(-x) = [-f(x)][-g(x)] = f(x)g(x) = F(x).$$

Thus, $F(x)$ is even.

93. (a) $V = x(24 - 2x)^2 = 4x(12 - x^2)$

Domain: $0 < x < 12$

(b)

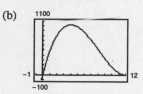

The dimensions for maximum volume are $4 \times 16 \times 16$ cm.

(c)

x	length and width	volume
1	$24 - 2(1)$	$1[24 - 2(1)]^2 = 484$
2	$24 - 2(2)$	$2[24 - 2(2)]^2 = 800$
3	$24 - 2(3)$	$3[24 - 2(3)]^2 = 972$
4	$24 - 2(4)$	$4[24 - 2(4)]^2 = 1024$
5	$24 - 2(5)$	$5[24 - 2(5)]^2 = 980$
6	$24 - 2(6)$	$6[24 - 2(6)]^2 = 864$

The dimensions for maximum volume appear to be $4 \times 16 \times 16$ cm.

95. False; let $f(x) = x^2$.

Then $f(-3) = f(3) = 9$, but $-3 \neq 3$.

97. True, the function is even.

99. First consider the portion of R in the first quadrant: $x \geq 0, 0 \leq y \leq 1$ and $x - y \leq 1$; shown below.

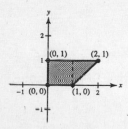

The area of this region is $1 + \frac{1}{2} = \frac{3}{2}$.

By symmetry, you obtain the entire region R:

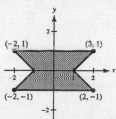

The area of R is $4\left(\frac{3}{2}\right) = 6$.

[49th competition, Problem A1, 1988]

Section P.4   Fitting Models to Data

1. Quadratic function

3. Linear function

5. (a), (b)

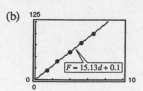

Yes. The cancer mortality increases linearly with increased exposure to the carcinogenic substance.

(c) If $x = 3$, then $y \approx 136$.

7. (a) $d = 0.066F$ or $F = 15.1d + 0.1$

(b)

$F = 15.13d + 0.1$

The model fits well.

(c) If $F = 55$, then $d \approx 0.066(55) = 3.63$ cm.

9. (a) Using a graphing utility, $y = 0.124x + 0.82$.

$r \approx 0.838$ correlation coefficient

(b)

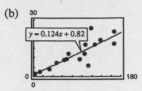

(c) The data indicates that greater per capita electricity consumption tends to correspond to greater per capita gross national product.

The data for Hong Kong, Venezuela and South Korea differ most from the linear model.

(d) Removing the data $(118, 25.59)$, $(113, 5.74)$ and $(167, 17.3)$, you obtain the model $y = 0.134x + 0.28$ with $r \approx 0.968$.

13. (a) Linear: $y_1 = 4.83t + 28.6$

Cubic: $y_2 = -0.1289t^3 + 2.235t^2 - 4.86t + 35.2$

(b)

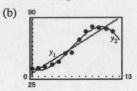

(c) The cubic model is better.

(d) $y = -0.084t^2 + 5.84t + 26.7$

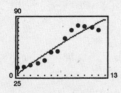

(e) For $t = 14$: Linear model $y_1 \approx 96.2$ million

Cubic model $y_2 \approx 51.5$ million

(f) Answers will vary.

17. (a) Yes, y is a function of t. At each time t, there is one and only one displacement y.

(b) The amplitude is approximately

$$(2.35 - 1.65)/2 = 0.35.$$

The period is approximately

$$2(0.375 - 0.125) = 0.5.$$

(c) One model is $y = 0.35 \sin(4\pi t) + 2$.

(d)

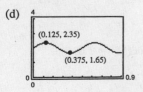

11. (a) $y_1 = 0.0343t^3 - 0.3451t^2 + 0.8837t + 5.6061$

$y_2 = 0.1095t + 2.0667$

$y_3 = 0.0917t + 0.7917$

(b)

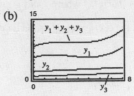

For $t = 12$, $y_1 + y_2 + y_3 \approx 31.06$ cents/mile.

15. (a) $y = -1.806x^3 + 14.58x^2 + 16.4x + 10$

(b)

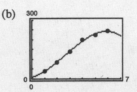

(c) If $x = 4.5$, $y \approx 214$ horsepower.

19. Answers will vary.

Review Exercises for Chapter P

1. $y = 2x - 3$

$x = 0 \Rightarrow y = 2(0) - 3 = -3 \Rightarrow (0, -3)$ y-intercept

$y = 0 \Rightarrow 0 = 2x - 3 \Rightarrow x = \frac{3}{2} \Rightarrow \left(\frac{3}{2}, 0\right)$ x-intercept

3. $y = \dfrac{x - 1}{x - 2}$

$x = 0 \Rightarrow y = \dfrac{0 - 1}{0 - 2} = \dfrac{1}{2} \Rightarrow \left(0, \dfrac{1}{2}\right)$ y-intercept

$y = 0 \Rightarrow 0 = \dfrac{x - 1}{x - 2} \Rightarrow x = 1 \Rightarrow (1, 0)$ x-intercept

5. Symmetric with respect to y-axis since

$$(-x)^2 y - (-x)^2 + 4y = 0$$
$$x^2 y - x^2 + 4y = 0.$$

7. $y = -\frac{1}{2}x + \frac{3}{2}$

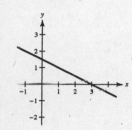

9. $-\frac{1}{3}x + \frac{5}{6}y = 1$

$-\frac{2}{5}x + y = \frac{6}{5}$

$y = \frac{2}{5}x + \frac{6}{5}$

Slope: $\frac{2}{5}$

y-intercept: $\frac{6}{5}$

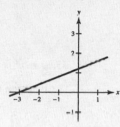

11. $y = 7 - 6x - x^2$

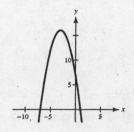

13. $y = \sqrt{5 - x}$

Domain: $(-\infty, 5]$

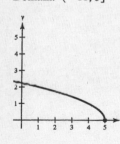

15. $y = 4x^2 - 25$

```
Xmin = -5
Xmax = 5
Xscl = 1
Ymin = -30
Ymax = 10
Yscl = 5
```

17. $3x - 4y = 8$

$\underline{4x + 4y = 20}$

$7x \qquad = 28$

$x = 4$

$y = 1$

Point: $(4, 1)$

19. You need factors $(x + 2)$ and $(x - 2)$. Multiply by x to obtain origin symmetry.

$y = x(x + 2)(x - 2)$

$= x^3 - 4x$

21.

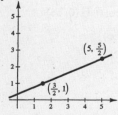

Slope $= \dfrac{(5/2) - 1}{5 - (3/2)} = \dfrac{3/2}{7/2} = \dfrac{3}{7}$

23. $\dfrac{1 - t}{1 - 0} = \dfrac{1 - 5}{1 - (-2)}$

$1 - t = -\dfrac{4}{3}$

$t = \dfrac{7}{3}$

25. $y - (-5) = \frac{3}{2}(x - 0)$

$y = \frac{3}{2}x - 5$

$2y - 3x + 10 = 0$

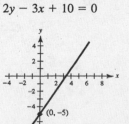

27.
$$y - 0 = -\tfrac{2}{3}(x - (-3))$$
$$y = -\tfrac{2}{3}x - 2$$
$$3y + 2x + 6 = 0$$

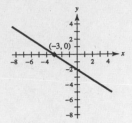

29. (a)
$$y - 4 = \frac{7}{16}(x + 2)$$
$$16y - 64 = 7x + 14$$
$$0 = 7x - 16y + 78$$

(b) Slope of line is $\frac{5}{3}$.

$$y - 4 = \frac{5}{3}(x + 2)$$
$$3y - 12 = 5x + 10$$
$$0 = 5x - 3y + 22$$

(c)
$$m = \frac{4 - 0}{-2 - 0} = -2$$
$$y = -2x$$
$$2x + y = 0$$

(d)
$$x = -2$$
$$x + 2 = 0$$

31. The slope is -850. $V = -850t + 12,500$.
$$V(3) = -850(3) + 12,500 = \$9950$$

33. $x - y^2 = 0$
$$y = \pm\sqrt{x}$$

Not a function of x since there are two values of y for some x.

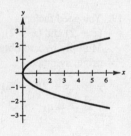

35. $y = x^2 - 2x$

Function of x since there is one value of y for each x.

37. $f(x) = \dfrac{1}{x}$

(a) $f(0)$ does not exist.

(b)
$$\frac{f(1 + \Delta x) - f(1)}{\Delta x} = \frac{\dfrac{1}{1 + \Delta x} - \dfrac{1}{1}}{\Delta x} = \frac{1 - 1 - \Delta x}{(1 + \Delta x)\Delta x}$$
$$= \frac{-1}{1 + \Delta x}, \Delta x \neq -1, 0$$

39. (a) Domain: $36 - x^2 \geq 0 \Rightarrow -6 \leq x \leq 6$ or $[-6, 6]$

Range: $[0, 6]$

(b) Domain: all $x \neq 5$ or $(-\infty, 5) \cup (5, \infty)$

Range: all $y \neq 0$ or $(-\infty, 0) \cup (0, \infty)$

(c) Domain: all x or $(-\infty, \infty)$

Range: all y or $(-\infty, \infty)$

41. (a) $f(x) = x^3 + c, c = -2, 0, 2$

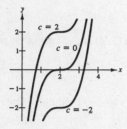

(b) $f(x) = (x - c)^3, c = -2, 0, 2$

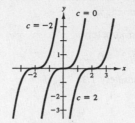

(c) $f(x) = (x - 2)^3 + c, c = -2, 0, 2$

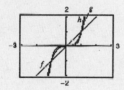

(d) $f(x) = cx^3, c = -2, 0, 2$

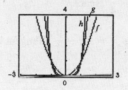

43. (a) Odd powers: $f(x) = x, g(x) = x^3, h(x) = x^5$

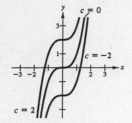

Even powers: $f(x) = x^2, g(x) = x^4, h(x) = x^6$

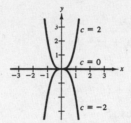

The graphs of f, g, and h all rise to the right and fall to the left. As the degree increases, the graph rises and falls more steeply. All three graphs pass through the points $(0, 0)$, $(1, 1)$, and $(-1, -1)$.

The graphs of f, g, and h all rise to the left and to the right. As the degree increases, the graph rises more steeply. All three graphs pass through the points $(0, 0)$, $(1, 1)$, and $(-1, 1)$.

(b) $y = x^7$ will look like $h(x) = x^5$, but rise and fall even more steeply.

$y = x^8$ will look like $h(x) = x^6$, but rise even more steeply.

45. (a)

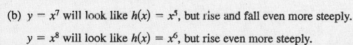

$$2x + 2y = 24$$
$$y = 12 - x$$
$$A = xy = x(12 - x) = 12x - x^2$$

(b) Domain: $0 < x < 12$

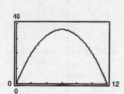

(c) Maximum area is $A = 36$. In general, the maximum area is attained when the rectangle is a square. In this case, $x = 6$.

47. (a) 3 (cubic), negative leading coefficient

(b) 4 (quartic), positive leading coefficient

(c) 2 (quadratic), negative leading coefficient

(d) 5, positive leading coefficient

49. (a) Yes, y is a function of t. At each time t, there is one and only one displacement y.

(b) The amplitude is approximately

$$(0.25 - (-0.25))/2 = 0.25.$$

The period is approximately 1.1.

(c) One model is $y = \dfrac{1}{4}\cos\left(\dfrac{2\pi}{1.1}t\right) \approx \dfrac{1}{4}\cos(5.7t)$

(d)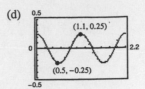

Problem Solving for Chapter P

1. (a)
$$x^2 - 6x + y^2 - 8y = 0$$
$$(x^2 - 6x + 9) + (y^2 - 8y + 16) = 9 + 16$$
$$(x - 3)^2 + (y - 4)^2 = 25$$

Center: $(3, 4)$ Radius: 5

(c) Slope of line from $(6, 0)$ to $(3, 4)$ is $\dfrac{4 - 0}{3 - 6} = -\dfrac{4}{3}$.

Slope of tangent line is $\dfrac{3}{4}$. Hence,

$$y - 0 = \dfrac{3}{4}(x - 6) \implies y = \dfrac{3}{4}x - \dfrac{9}{2} \text{ Tangent line}$$

(b) Slope of line from $(0, 0)$ to $(3, 4)$ is $\dfrac{4}{3}$. Slope of tangent line is $-\dfrac{3}{4}$. Hence,

$$y - 0 = -\dfrac{3}{4}(x - 0) \implies y = -\dfrac{3}{4}x \text{ Tangent line}$$

(d) $-\dfrac{3}{4}x = \dfrac{3}{4}x - \dfrac{9}{2}$

$$\dfrac{3}{2}x = \dfrac{9}{2}$$

$$x = 3$$

Intersection: $\left(3, -\dfrac{9}{4}\right)$

3. $H(x) = \begin{cases} 1, & x \geq 0 \\ 0, & x < 0 \end{cases}$

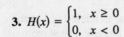

(a) $H(x) - 2$

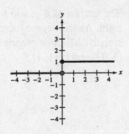

(b) $H(x - 2)$

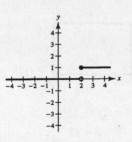

(c) $-H(x)$

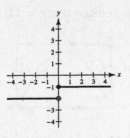

(d) $H(-x)$

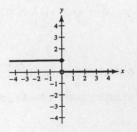

—CONTINUED—

3. —CONTINUED—

(e) $\frac{1}{2}H(x)$

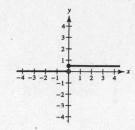

(f) $-H(x - 2) + 2$

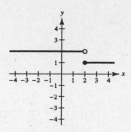

5. (a) $x + 2y = 100 \implies y = \dfrac{100 - x}{2}$

$$A(x) = xy = x\left(\frac{100 - x}{2}\right) = -\frac{x^2}{2} + 50x$$

Domain: $0 < x < 100$

(b)

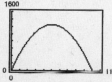

Maximum of 1250 m² at $x = 50$ m, $y = 25$ m.

(c) $A(x) = -\dfrac{1}{2}(x^2 - 100x)$

$\quad\quad\quad = -\dfrac{1}{2}(x^2 - 100x + 2500) + 1250$

$\quad\quad\quad = -\dfrac{1}{2}(x - 50)^2 + 1250$

$A(50) = 1250$ m² is the maximum.

$x = 50$ m, $y = 25$ m

7. The length of the trip in the water is $\sqrt{2^2 + x^2}$, and the length of the trip over land is $\sqrt{1 + (3 - x)^2}$. Hence, the total time is

$$T = \frac{\sqrt{4 + x^2}}{2} + \frac{\sqrt{1 + (3 - x)^2}}{4} \text{ hours.}$$

9. (a) Slope $= \dfrac{9 - 4}{3 - 2} = 5$. Slope of tangent line is less than 5.

(b) Slope $= \dfrac{4 - 1}{2 - 1} = 3$. Slope of tangent line is greater than 3.

(c) Slope $= \dfrac{4.41 - 4}{2.1 - 2} = 4.1$. Slope of tangent line is less than 4.1.

(d) Slope $= \dfrac{f(2 + h) - f(2)}{(2 + h) - 2}$

$\quad\quad\quad = \dfrac{(2 + h)^2 - 4}{h}$

$\quad\quad\quad = \dfrac{4h + h^2}{h}$

$\quad\quad\quad = 4 + h, h \neq 0$

(e) Letting h get closer and closer to 0, the slope approaches 4. Hence, the slope at $(2, 4)$ is 4.

11. (a)
$$\frac{I}{x^2} = \frac{2I}{(x-3)^2}$$

$$x^2 - 6x + 9 = 2x^2$$

$$x^2 + 6x - 9 = 0$$

$$x = \frac{-6 \pm \sqrt{36+36}}{2} = -3 \pm \sqrt{18} \approx 1.2426, -7.2426$$

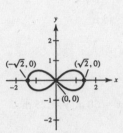

(b)
$$\frac{I}{x^2+y^2} = \frac{2I}{(x-3)^2+y^2}$$

$$(x-3)^2 + y^2 = 2(x^2+y^2)$$

$$x^2 - 6x + 9 + y^2 = 2x^2 + 2y^2$$

$$x^2 + y^2 + 6x - 9 = 0$$

$$(x+3)^2 + y^2 = 18$$

Circle of radius $\sqrt{18}$ and center $(-3, 0)$.

13.
$$d_1 d_2 = 1$$

$$[(x+1)^2 + y^2][(x-1)^2 + y^2] = 1$$

$$(x+1)^2(x-1)^2 + y^2[(x+1)^2 + (x-1)^2] + y^4 = 1$$

$$(x^2-1)^2 + y^2[2x^2 + 2] + y^4 = 1$$

$$x^4 - 2x^2 + 1 + 2x^2y^2 + 2y^2 + y^4 = 1$$

$$(x^4 + 2x^2y^2 + y^4) - 2x^2 + 2y^2 = 0$$

$$(x^2+y^2)^2 = 2(x^2 - y^2)$$

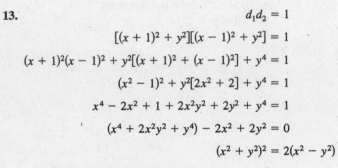

Let $y = 0$. Then $x^4 = 2x^2 \implies x = 0$ or $x^2 = 2$.

Thus, $(0,0)$, $\left(\sqrt{2}, 0\right)$ and $\left(-\sqrt{2}, 0\right)$ are on the curve.

C H A P T E R   1
Limits and Their Properties

Section 1.1   A Preview of Calculus 26

Section 1.2   Finding Limits Graphically and Numerically 26

Section 1.3   Evaluating Limits Analytically 31

Section 1.4   Continuity and One-Sided Limits 36

Section 1.5   Infinite Limits 41

Review Exercises 45

Problem Solving 48

CHAPTER 1
Limits and Their Properties

Section 1.1   A Preview of Calculus

1. Precalculus: $(20 \text{ ft/sec})(15 \text{ seconds}) = 300 \text{ feet}$

3. Calculus required: slope of tangent line at $x = 2$ is rate of change, and equals about 0.16.

5. Precalculus: Area $= \frac{1}{2}bh = \frac{1}{2}(5)(3) = \frac{15}{2}$ sq. units

7. $f(x) = 4x - x^2$

(a)

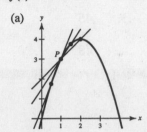

(b) slope $= m = \dfrac{(4x - x^2) - 3}{x - 1}$

$$= \dfrac{(x - 1)(3 - x)}{x - 1} = 3 - x, \quad x \neq 1$$

$x = 2: \ m = 3 - 2 = 1$

$x = 1.5: \ m = 3 - 1.5 = 1.5$

$x = 0.5: \ m = 3 - 0.5 = 2.5$

(c) At $P(1, 3)$ the slope is 2.

You can improve your approximation of the slope at $x = 1$ by considering x-values very close to 1.

9. (a) Area $\approx 5 + \dfrac{5}{2} + \dfrac{5}{3} + \dfrac{5}{4} \approx 10.417$

Area $\approx \dfrac{1}{2}\left(5 + \dfrac{5}{1.5} + \dfrac{5}{2} + \dfrac{5}{2.5} + \dfrac{5}{3} + \dfrac{5}{3.5} + \dfrac{5}{4} + \dfrac{5}{4.5}\right) \approx 9.145$

(b) You could improve the approximation by using more rectangles.

11. (a) $D_1 = \sqrt{(5 - 1)^2 + (1 - 5)^2} = \sqrt{16 + 16} \approx 5.66$

(b) $D_2 = \sqrt{1 + \left(\frac{5}{2}\right)^2} + \sqrt{1 + \left(\frac{5}{2} - \frac{5}{3}\right)^2} + \sqrt{1 + \left(\frac{5}{3} - \frac{5}{4}\right)^2} + \sqrt{1 + \left(\frac{5}{4} - 1\right)^2}$

$\approx 2.693 + 1.302 + 1.083 + 1.031 \approx 6.11$

(c) Increase the number of line segments.

Section 1.2   Finding Limits Graphically and Numerically

1.

x	1.9	1.99	1.999	2.001	2.01	2.1
$f(x)$	0.3448	0.3344	0.3334	0.3332	0.3322	0.3226

$\displaystyle\lim_{x \to 2} \dfrac{x - 2}{x^2 - x - 2} \approx 0.3333$ $\left(\text{Actual limit is } \frac{1}{3}.\right)$

3.

x	-0.1	-0.01	-0.001	0.001	0.01	0.1
$f(x)$	0.2911	0.2889	0.2887	0.2887	0.2884	0.2863

$$\lim_{x \to 0} \frac{\sqrt{x+3} - \sqrt{3}}{x} \approx 0.2887 \quad \left(\text{Actual limit is } 1/(2\sqrt{3}).\right)$$

5.

x	2.9	2.99	2.999	3.001	3.01	3.1
$f(x)$	-0.0641	-0.0627	-0.0625	-0.0625	-0.0623	-0.0610

$$\lim_{x \to 3} \frac{[1/(x+1)] - (1/4)}{x-3} \approx -0.0625 \quad \left(\text{Actual limit is } -\tfrac{1}{16}.\right)$$

7.

x	-0.1	-0.01	-0.001	0.001	0.01	0.1
$f(x)$	0.9983	0.99998	1.0000	1.0000	0.99998	0.9983

$$\lim_{x \to 0} \frac{\sin x}{x} \approx 1.0000 \quad \text{(Actual limit is 1.) (Make sure you use radian mode.)}$$

9. $\lim\limits_{x \to 3} (4 - x) = 1$

11. $\lim\limits_{x \to 2} f(x) = \lim\limits_{x \to 2} (4 - x) = 2$

13. $\lim\limits_{x \to 5} \dfrac{|x - 5|}{x - 5}$ does not exist. For values of x to the left of 5, $|x - 5|/(x - 5)$ equals -1, whereas for values of x to the right of 5, $|x - 5|/(x - 5)$ equals 1.

15. $\lim\limits_{x \to 1} \sin \pi x = 0$

17. $\lim\limits_{x \to 0} \cos(1/x)$ does not exist since the function oscillates between -1 and 1 as x approaches 0.

19. (a) $f(1)$ exists. The black dot at $(1, 2)$ indicates that $f(1) = 2$.

(b) $\lim\limits_{x \to 1} f(x)$ does not exist. As x approaches 1 from the left, $f(x)$ approaches 2.5, whereas as x approaches 1 from the right, $f(x)$ approaches 1.

(c) $f(4)$ does not exist. The hollow circle at $(4, 2)$ indicates that f is not defined at 4.

(d) $\lim\limits_{x \to 4} f(x)$ exists. As x approaches 4, $f(x)$ approaches 2: $\lim\limits_{x \to 4} f(x) = 2$.

21. $\lim\limits_{x \to c} f(x)$ exists for all $c \neq -3$. In particular, $\lim\limits_{x \to 2} f(x) = 2$.

23.

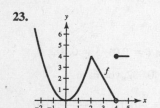

$\lim\limits_{x \to c} f(x)$ exists for all values of $c \neq 4$.

25. One possible answer is

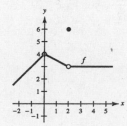

27. $C(t) = 0.75 - 0.50[\![-(t-1)]\!]$

(a)

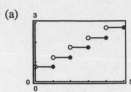

(b)

t	3	3.3	3.4	3.5	3.6	3.7	4
C	1.75	2.25	2.25	2.25	2.25	2.25	2.25

$\lim\limits_{t \to 3.5} C(t) = 2.25$

(c)

t	2	2.5	2.9	3	3.1	3.5	4
C	1.25	1.75	1.75	1.75	2.25	2.25	2.25

$\lim\limits_{t \to 3} C(t)$ does not exist. The values of C jump from 1.75 to 2.25 at $t = 3$.

29. We need $|f(x) - 3| = |(x+1) - 3| = |x - 2| < 0.4$. Hence, take $\delta = 0.4$. If $0 < |x - 2| < 0.4$, then $|x - 2| = |(x+1) - 3| = |f(x) - 3| < 0.4$, as desired.

31. You need to find δ such that $0 < |x - 1| < \delta$ implies

$$|f(x) - 1| = \left|\frac{1}{x} - 1\right| < 0.1. \text{ That is,}$$

$$-0.1 < \frac{1}{x} - 1 < 0.1$$

$$1 - 0.1 < \frac{1}{x} < 1 + 0.1$$

$$\frac{9}{10} < \frac{1}{x} < \frac{11}{10}$$

$$\frac{10}{9} > x > \frac{10}{11}$$

$$\frac{10}{9} - 1 > x - 1 > \frac{10}{11} - 1$$

$$\frac{1}{9} > x - 1 > -\frac{1}{11}.$$

So take $\delta = \dfrac{1}{11}$. Then $0 < |x - 1| < \delta$ implies

$$-\frac{1}{11} < x - 1 < \frac{1}{11}$$

$$-\frac{1}{11} < x - 1 < \frac{1}{9}.$$

Using the first series of equivalent inequalities, you obtain

$$|f(x) - 1| = \left|\frac{1}{x} - 1\right| < 0.1.$$

33. $\lim\limits_{x \to 2} (3x + 2) = 8 = L$

$$|(3x + 2) - 8| < 0.01$$

$$|3x - 6| < 0.01$$

$$3|x - 2| < 0.01$$

$$0 < |x - 2| < \frac{0.01}{3} \approx 0.0033 = \delta$$

Hence, if $0 < |x - 2| < \delta = \dfrac{0.01}{3}$, you have

$$3|x - 2| < 0.01$$

$$|3x - 6| < 0.01$$

$$|(3x + 2) - 8| < 0.01$$

$$|f(x) - L| < 0.01$$

35. $\lim\limits_{x \to 2} (x^2 - 3) = 1 = L$

$$|(x^2 - 3) - 1| < 0.01$$

$$|x^2 - 4| < 0.01$$

$$|(x + 2)(x - 2)| < 0.01$$

$$|x + 2|\,|x - 2| < 0.01$$

$$|x - 2| < \frac{0.01}{|x + 2|}$$

If we assume $1 < x < 3$, then $\delta = 0.01/5 = 0.002$.

Hence, if $0 < |x - 2| < \delta = 0.002$, you have

$$|x - 2| < 0.002 = \frac{1}{5}(0.01) < \frac{1}{|x + 2|}(0.01)$$

$$|x + 2|\,|x - 2| < 0.01$$

$$|x^2 - 4| < 0.01$$

$$|(x^2 - 3) - 1| < 0.01$$

$$|f(x) - L| < 0.01$$

37. $\lim\limits_{x \to 2} (x + 3) = 5$

Given $\varepsilon > 0$:

$$|(x + 3) - 5| < \varepsilon$$
$$|x - 2| < \varepsilon = \delta$$

Hence, let $\delta = \varepsilon$.

Hence, if $0 < |x - 2| < \delta = \varepsilon$, you have

$$|x - 2| < \varepsilon$$
$$|(x + 3) - 5| < \varepsilon$$
$$|f(x) - L| < \varepsilon.$$

39. $\lim\limits_{x \to -4} \left(\tfrac{1}{2}x - 1\right) = \tfrac{1}{2}(-4) - 1 = -3$

Given $\varepsilon > 0$:

$$\left|\left(\tfrac{1}{2}x - 1\right) - (-3)\right| < \varepsilon$$
$$\left|\tfrac{1}{2}x + 2\right| < \varepsilon$$
$$\tfrac{1}{2}|x - (-4)| < \varepsilon$$
$$|x - (-4)| < 2\varepsilon$$

Hence, let $\delta = 2\varepsilon$.

Hence, if $0 < |x - (-4)| < \delta = 2\varepsilon$, you have

$$|x - (-4)| < 2\varepsilon$$
$$\left|\tfrac{1}{2}x + 2\right| < \varepsilon$$
$$\left|\left(\tfrac{1}{2}x - 1\right) + 3\right| < \varepsilon$$
$$|f(x) - L| < \varepsilon.$$

41. $\lim\limits_{x \to 6} 3 = 3$

Given $\varepsilon > 0$:

$$|3 - 3| < \varepsilon$$
$$0 < \epsilon$$

Hence, any $\delta > 0$ will work.

Hence, for any $\delta > 0$, you have

$$|3 - 3| < \varepsilon$$
$$|f(x) - L| < \varepsilon.$$

43. $\lim\limits_{x \to 0} \sqrt[3]{x} = 0$

Given $\varepsilon > 0$: $\left|\sqrt[3]{x} - 0\right| < \varepsilon$

$$\left|\sqrt[3]{x}\right| < \varepsilon$$
$$|x| < \varepsilon^3 = \delta$$

Hence, let $\delta = \varepsilon^3$.

Hence for $0|x - 0|\delta = \varepsilon^3$, you have

$$|x| < \varepsilon^3$$
$$\left|\sqrt[3]{x}\right| < \varepsilon$$
$$\left|\sqrt[3]{x} - 0\right| < \varepsilon$$
$$|f(x) - L| < \varepsilon.$$

45. $\lim\limits_{x \to -2} |x - 2| = |(-2) - 2| = 4$

Given $\varepsilon > 0$:

$$||x - 2| - 4| < \varepsilon$$
$$|-(x - 2) - 4| < \varepsilon \quad (x - 2 < 0)$$
$$|-x - 2| = |x + 2| = |x - (-2)| < \varepsilon$$

Hence, let $\delta = \varepsilon$.

Hence for $0 < |x - (-2)| < \delta = \varepsilon$, you have

$$|x + 2| < \varepsilon$$
$$|-(x + 2)| < \varepsilon$$
$$|-(x - 2) - 4| < \varepsilon$$
$$||x - 2| - 4| < \varepsilon \quad \text{(because } x - 2 < 0)$$
$$|f(x) - L| < \varepsilon.$$

47. $\lim\limits_{x \to 1} (x^2 + 1) = 2$

Given $\varepsilon > 0$:

$$|(x^2 + 1) - 2| < \varepsilon$$
$$|x^2 - 1| < \varepsilon$$
$$|(x + 1)(x - 1)| < \varepsilon$$
$$|x - 1| < \frac{\varepsilon}{|x + 1|}$$

If we assume $0 < x < 2$, then $\delta = \varepsilon/3$.

Hence for $0 < |x - 1| < \delta = \dfrac{\varepsilon}{3}$, you have

$$|x - 1| < \frac{1}{3}\varepsilon < \frac{1}{|x + 1|}\varepsilon$$
$$|x^2 - 1| < \varepsilon$$
$$|(x^2 + 1) - 2| < \varepsilon$$
$$|f(x) - 2| < \varepsilon.$$

49. $f(x) = \dfrac{\sqrt{x+5} - 3}{x - 4}$

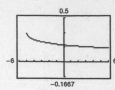

$\lim\limits_{x \to 4} f(x) = \dfrac{1}{6}$

The domain is $[-5, 4) \cup (4, \infty)$. The graphing utility does not show the hole at $\left(4, \frac{1}{6}\right)$.

51. $f(x) = \dfrac{x - 9}{\sqrt{x} - 3}$

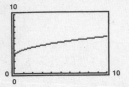

$\lim\limits_{x \to 9} f(x) = 6$

The domain is all $x \geq 0$ except $x = 9$. The graphing utility does not show the hole at $(9, 6)$.

53. $\lim\limits_{x \to 8} f(x) = 25$ means that the values of f approach 25 as x gets closer and closer to 8.

55. No. The fact that $\lim\limits_{x \to 2} f(x) = 4$ has no bearing on the value of f at 2.

57. (a) $C = 2\pi r$

$r = \dfrac{C}{2\pi} = \dfrac{6}{2\pi} = \dfrac{3}{\pi} \approx 0.9549$ cm

(b) If $C = 5.5$, $r = \dfrac{5.5}{2\pi} \approx 0.87535$ cm

If $C = 6.5$, $r = \dfrac{6.5}{2\pi} \approx 1.03451$ cm

Thus $0.87535 < r < 1.03451$

(c) $\lim\limits_{r \to 3/\pi} (2\pi r) = 6$; $\varepsilon = 0.5$; $\delta \approx 0.0796$

59. $f(x) = (1 + x)^{1/x}$

$\lim\limits_{x \to 0} (1 + x)^{1/x} = e \approx 2.71828$

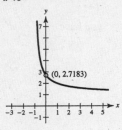

x	$f(x)$	x	$f(x)$
-0.1	2.867972	0.1	2.593742
-0.01	2.731999	0.01	2.704814
-0.001	2.719642	0.001	2.716942
-0.0001	2.718418	0.0001	2.718146
-0.00001	2.718295	0.00001	2.718268
-0.000001	2.718283	0.000001	2.718280

61.

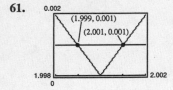

Using the zoom and trace feature, $\delta = 0.001$. That is, for

$$0 < |x - 2| < 0.001, \; \left| \dfrac{x^2 - 4}{x - 2} - 4 \right| < 0.001.$$

63. False; $f(x) = (\sin x)/x$ is undefined when $x = 0$. From Exercise 7, we have

$$\lim\limits_{x \to 0} \dfrac{\sin x}{x} = 1.$$

65. False; let

$$f(x) = \begin{cases} x^2 - 4x, & x \neq 4 \\ 10, & x = 4 \end{cases}.$$

$f(4) = 10$

$\lim\limits_{x \to 4} f(x) = \lim\limits_{x \to 4} (x^2 - 4x) = 0 \neq 10$

67. $f(x) = \sqrt{x}$

(a) $\lim\limits_{x \to 0.25} \sqrt{x} = 0.5$ is true.

As x approaches $0.25 = \frac{1}{4}$, $f(x) = \sqrt{x}$ approaches $\frac{1}{2} = 0.5$.

(b) $\lim\limits_{x \to 0} \sqrt{x} = 0$ is false.

$f(x) = \sqrt{x}$ is not defined on an open interval containing 0 because the domain of f is $x \geq 0$.

69. If $\lim_{x \to c} f(x) = L_1$ and $\lim_{x \to c} f(x) = L_2$, then for every $\varepsilon > 0$, there exists $\delta_1 > 0$ and $\delta_2 > 0$ such that

$|x - c| < \delta_1 \Rightarrow |f(x) - L_1| < \varepsilon$ and $|x - c| < \delta_2 \Rightarrow |f(x) - L_2| < \varepsilon$. Let δ equal the smaller of δ_1 and δ_2.

Then for $|x - c| < \delta$, we have $|L_1 - L_2| = |L_1 - f(x) + f(x) - L_2| \le |L_1 - f(x)| + |f(x) - L_2| < \varepsilon + \varepsilon$.

Therefore, $|L_1 - L_2| < 2\varepsilon$. Since $\varepsilon > 0$ is arbitrary, it follows that $L_1 = L_2$.

71. $\lim_{x \to c} [f(x) - L] = 0$ means that for every $\varepsilon > 0$ there exists $\delta > 0$ such that if

$$0 < |x - c| < \delta,$$

then

$$|(f(x) - L) - 0| < \varepsilon.$$

This means the same as $|f(x) - L| < \varepsilon$ when

$$0 < |x - c| < \delta.$$

Thus, $\lim_{x \to c} f(x) = L$.

73. Answers will vary.

75. The radius OP has a length equal to the altitude z of the triangle plus $\dfrac{h}{2}$. Thus, $z = 1 - \dfrac{h}{2}$.

$$\text{Area triangle} = \frac{1}{2}b\left(1 - \frac{h}{2}\right)$$

$$\text{Area rectangle} = bh$$

Since these are equal, $\dfrac{1}{2}b\left(1 - \dfrac{h}{2}\right) = bh$

$$1 - \frac{h}{2} = 2h$$

$$\frac{5}{2}h = 1$$

$$h = \frac{2}{5}$$

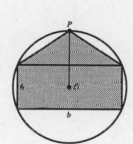

Section 1.3   Evaluating Limits Analytically

1.

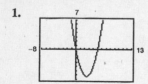

$h(x) = x^2 - 5x$

(a) $\lim_{x \to 5} h(x) = 0$

(b) $\lim_{x \to -1} h(x) = 6$

3.

$f(x) = x \cos x$

(a) $\lim_{x \to 0} f(x) = 0$

(b) $\lim_{x \to \pi/3} f(x) \approx 0.524$

$$\left(= \frac{\pi}{6} \right)$$

5. $\lim_{x \to 2} x^4 = 2^4 = 16$

7. $\lim_{x \to 0} (2x - 1) = 2(0) - 1 = -1$

9. $\lim_{x \to -3} (x^2 + 3x) = (-3)^2 + 3(-3) = 9 - 9 = 0$

11. $\lim\limits_{x\to-3} (2x^2 + 4x + 1) = 2(-3)^2 + 4(-3) + 1 = 18 - 12 + 1 = 7$

13. $\lim\limits_{x\to2} \dfrac{1}{x} = \dfrac{1}{2}$

15. $\lim\limits_{x\to1} \dfrac{x-3}{x^2+4} = \dfrac{1-3}{1^2+4} = \dfrac{-2}{5} = -\dfrac{2}{5}$

17. $\lim\limits_{x\to7} \dfrac{5x}{\sqrt{x+2}} = \dfrac{5(7)}{\sqrt{7+2}} = \dfrac{35}{\sqrt{9}} = \dfrac{35}{3}$

19. $\lim\limits_{x\to3} \sqrt{x+1} = \sqrt{3+1} = 2$

21. $\lim\limits_{x\to-4} (x+3)^2 = (-4+3)^2 = 1$

23. (a) $\lim\limits_{x\to1} f(x) = 5 - 1 = 4$

    (b) $\lim\limits_{x\to4} g(x) = 4^3 = 64$

    (c) $\lim\limits_{x\to1} g(f(x)) = g(f(1)) = g(4) = 64$

25. (a) $\lim\limits_{x\to1} f(x) = 4 - 1 = 3$

    (b) $\lim\limits_{x\to3} g(x) = \sqrt{3+1} = 2$

    (c) $\lim\limits_{x\to1} g(f(x)) = g(3) = 2$

27. $\lim\limits_{x\to\pi/2} \sin x = \sin\dfrac{\pi}{2} = 1$

29. $\lim\limits_{x\to2} \cos\dfrac{\pi x}{3} = \cos\dfrac{2\pi}{3} = -\dfrac{1}{2}$

31. $\lim\limits_{x\to0} \sec 2x = \sec 0 = 1$

33. $\lim\limits_{x\to5\pi/6} \sin x = \sin\dfrac{5\pi}{6} = \dfrac{1}{2}$

35. $\lim\limits_{x\to3} \tan\left(\dfrac{\pi x}{4}\right) = \tan\dfrac{3\pi}{4} = -1$

37. (a) $\lim\limits_{x\to c} [5g(x)] = 5\lim\limits_{x\to c} g(x) = 5(3) = 15$

    (b) $\lim\limits_{x\to c} [f(x) + g(x)] = \lim\limits_{x\to c} f(x) + \lim\limits_{x\to c} g(x) = 2 + 3 = 5$

    (c) $\lim\limits_{x\to c} [f(x)g(x)] = \left[\lim\limits_{x\to c} f(x)\right]\left[\lim\limits_{x\to c} g(x)\right] = (2)(3) = 6$

    (d) $\lim\limits_{x\to c} \dfrac{f(x)}{g(x)} = \dfrac{\lim\limits_{x\to c} f(x)}{\lim\limits_{x\to c} g(x)} = \dfrac{2}{3}$

39. (a) $\lim\limits_{x\to c} [f(x)]^3 = \left[\lim\limits_{x\to c} f(x)\right]^3 = (4)^3 = 64$

    (b) $\lim\limits_{x\to c} \sqrt{f(x)} = \sqrt{\lim\limits_{x\to c} f(x)} = \sqrt{4} = 2$

    (c) $\lim\limits_{x\to c} [3f(x)] = 3\lim\limits_{x\to c} f(x) = 3(4) = 12$

    (d) $\lim\limits_{x\to c} [f(x)]^{3/2} = \left[\lim\limits_{x\to c} f(x)\right]^{3/2} = (4)^{3/2} = 8$

41. $f(x) = -2x + 1$ and $g(x) = \dfrac{-2x^2 + x}{x}$ agree except at $x = 0$.

    (a) $\lim\limits_{x\to0} g(x) = \lim\limits_{x\to0} f(x) = 1$

    (b) $\lim\limits_{x\to-1} g(x) = \lim\limits_{x\to-1} f(x) = 3$

43. $f(x) = x(x + 1)$ and $g(x) = \dfrac{x^3 - x}{x - 1}$ agree except at $x = 1$.

    (a) $\lim\limits_{x\to1} g(x) = \lim\limits_{x\to1} f(x) = 2$

    (b) $\lim\limits_{x\to-1} g(x) = \lim\limits_{x\to-1} f(x) = 0$

45. $f(x) = \dfrac{x^2 - 1}{x + 1}$ and $g(x) = x - 1$ agree except at $x = -1$.

$$\lim\limits_{x\to-1} f(x) = \lim\limits_{x\to-1} g(x) = -2$$

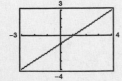

47. $f(x) = \dfrac{x^3 - 8}{x - 2}$ and $g(x) = x^2 + 2x + 4$ agree except at $x = 2$.

$$\lim\limits_{x\to2} f(x) = \lim\limits_{x\to2} g(x) = 12$$

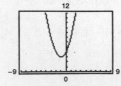

49. $\lim\limits_{x\to 5} \dfrac{x-5}{x^2-25} = \lim\limits_{x\to 5} \dfrac{x-5}{(x+5)(x-5)}$

$\qquad\qquad = \lim\limits_{x\to 5} \dfrac{1}{x+5} = \dfrac{1}{10}$

51. $\lim\limits_{x\to -3} \dfrac{x^2+x-6}{x^2-9} = \lim\limits_{x\to -3} \dfrac{(x+3)(x-2)}{(x+3)(x-3)}$

$\qquad\qquad = \lim\limits_{x\to -3} \dfrac{x-2}{x-3} = \dfrac{-5}{-6} = \dfrac{5}{6}$

53. $\lim\limits_{x\to 0} \dfrac{\sqrt{x+5}-\sqrt{5}}{x} = \lim\limits_{x\to 0} \dfrac{\sqrt{x+5}-\sqrt{5}}{x} \cdot \dfrac{\sqrt{x+5}+\sqrt{5}}{\sqrt{x+5}+\sqrt{5}}$

$\qquad\qquad = \lim\limits_{x\to 0} \dfrac{(x+5)-5}{x\left(\sqrt{x+5}+\sqrt{5}\right)} = \lim\limits_{x\to 0} \dfrac{1}{\sqrt{x+5}+\sqrt{5}} = \dfrac{1}{2\sqrt{5}} = \dfrac{\sqrt{5}}{10}$

55. $\lim\limits_{x\to 4} \dfrac{\sqrt{x+5}-3}{x-4} = \lim\limits_{x\to 4} \dfrac{\sqrt{x+5}-3}{x-4} \cdot \dfrac{\sqrt{x+5}+3}{\sqrt{x+5}+3}$

$\qquad\qquad = \lim\limits_{x\to 4} \dfrac{(x+5)-9}{(x-4)\left(\sqrt{x+5}+3\right)} = \lim\limits_{x\to 4} \dfrac{1}{\sqrt{x+5}+3} = \dfrac{1}{\sqrt{9}+3} = \dfrac{1}{6}$

57. $\lim\limits_{x\to 0} \dfrac{\dfrac{1}{3+x}-\dfrac{1}{3}}{x} = \lim\limits_{x\to 0} \dfrac{3-(3+x)}{(3+x)3(x)} = \lim\limits_{x\to 0} \dfrac{-x}{(3+x)(3)(x)} = \lim\limits_{x\to 0} \dfrac{-1}{(3+x)3} = -\dfrac{1}{9}$

59. $\lim\limits_{\Delta x\to 0} \dfrac{2(x+\Delta x)-2x}{\Delta x} = \lim\limits_{\Delta x\to 0} \dfrac{2x+2\Delta x-2x}{\Delta x}$

$\qquad\qquad = \lim\limits_{\Delta x\to 0} 2 = 2$

61. $\lim\limits_{\Delta x\to 0} \dfrac{(x+\Delta x)^2-2(x+\Delta x)+1-(x^2-2x+1)}{\Delta x} = \lim\limits_{\Delta x\to 0} \dfrac{x^2+2x\Delta x+(\Delta x)^2-2x-2\Delta x+1-x^2+2x-1}{\Delta x}$

$\qquad\qquad = \lim\limits_{\Delta x\to 0} (2x+\Delta x-2) = 2x-2$

63. $\lim\limits_{x\to 0} \dfrac{\sqrt{x+2}-\sqrt{2}}{x} \approx 0.354$

x	-0.1	-0.01	-0.001	0	0.001	0.01	0.1
$f(x)$	0.358	0.354	0.354	?	0.354	0.353	0.349

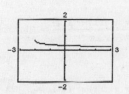

Analytically, $\lim\limits_{x\to 0} \dfrac{\sqrt{x+2}-\sqrt{2}}{x} = \lim\limits_{x\to 0} \dfrac{\sqrt{x+2}-\sqrt{2}}{x} \cdot \dfrac{\sqrt{x+2}+\sqrt{2}}{\sqrt{x+2}+\sqrt{2}}$

$\qquad\qquad = \lim\limits_{x\to 0} \dfrac{x+2-2}{x\left(\sqrt{x+2}+\sqrt{2}\right)} = \lim\limits_{x\to 0} \dfrac{1}{\sqrt{x+2}+\sqrt{2}} = \dfrac{1}{2\sqrt{2}} = \dfrac{\sqrt{2}}{4} \approx 0.354.$

65. $\lim\limits_{x\to 0} \dfrac{\dfrac{1}{2+x}-\dfrac{1}{2}}{x} = -\dfrac{1}{4}$

x	-0.1	-0.01	-0.001	0	0.001	0.01	0.1
$f(x)$	-0.263	-0.251	-0.250	?	-0.250	-0.249	-0.238

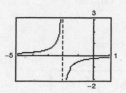

Analytically, $\lim\limits_{x\to 0} \dfrac{\dfrac{1}{2+x}-\dfrac{1}{2}}{x} = \lim\limits_{x\to 0} \dfrac{2-(2+x)}{2(2+x)} \cdot \dfrac{1}{x} = \lim\limits_{x\to 0} \dfrac{-x}{2(2+x)} \cdot \dfrac{1}{x} = \lim\limits_{x\to 0} \dfrac{-1}{2(2+x)} = -\dfrac{1}{4}.$

67. $\lim\limits_{x \to 0} \dfrac{\sin x}{5x} = \lim\limits_{x \to 0} \left[\left(\dfrac{\sin x}{x} \right) \left(\dfrac{1}{5} \right) \right] = (1)\left(\dfrac{1}{5} \right) = \dfrac{1}{5}$

69. $\lim\limits_{x \to 0} \dfrac{\sin x(1 - \cos x)}{2x^2} = \lim\limits_{x \to 0} \left[\dfrac{1}{2} \cdot \dfrac{\sin x}{x} \cdot \dfrac{1 - \cos x}{x} \right]$

$$= \dfrac{1}{2}(1)(0) = 0$$

71. $\lim\limits_{x \to 0} \dfrac{\sin^2 x}{x} = \lim\limits_{x \to 0} \left[\dfrac{\sin x}{x} \sin x \right] = (1)\sin 0 = 0$

73. $\lim\limits_{h \to 0} \dfrac{(1 - \cos h)^2}{h} = \lim\limits_{h \to 0} \left[\dfrac{1 - \cos h}{h}(1 - \cos h) \right]$

$$= (0)(0) = 0$$

75. $\lim\limits_{x \to \pi/2} \dfrac{\cos x}{\cot x} = \lim\limits_{x \to \pi/2} \sin x = 1$

77. $\lim\limits_{t \to 0} \dfrac{\sin 3t}{2t} = \lim\limits_{t \to 0} \left(\dfrac{\sin 3t}{3t} \right)\left(\dfrac{3}{2} \right) = (1)\left(\dfrac{3}{2} \right) = \dfrac{3}{2}$

79. $f(t) = \dfrac{\sin 3t}{t}$

t	-0.1	-0.01	-0.001	0	0.001	0.01	0.1
$f(t)$	2.96	2.9996	3	?	3	2.9996	2.96

Analytically, $\lim\limits_{t \to 0} \dfrac{\sin 3t}{t} = \lim\limits_{t \to 0} 3\left(\dfrac{\sin 3t}{3t} \right) = 3(1) = 3.$

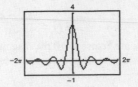

The limit appears to equal 3.

81. $f(x) = \dfrac{\sin x^2}{x}$

x	-0.1	-0.01	-0.001	0	0.001	0.01	0.1
$f(x)$	-0.099998	-0.01	-0.001	?	0.001	0.01	0.099998

Analytically, $\lim\limits_{x \to 0} \dfrac{\sin x^2}{x} = \lim\limits_{x \to 0} x\left(\dfrac{\sin x^2}{x^2} \right) = 0(1) = 0.$

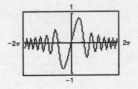

83. $\lim\limits_{\Delta x \to 0} \dfrac{f(x + \Delta x) - f(x)}{\Delta x} = \lim\limits_{\Delta x \to 0} \dfrac{2(x + \Delta x) + 3 - (2x + 3)}{\Delta x} = \lim\limits_{\Delta x \to 0} \dfrac{2x + 2\Delta x + 3 - 2x - 3}{\Delta x} = \lim\limits_{\Delta x \to 0} \dfrac{2\Delta x}{\Delta x} = 2$

85. $\lim\limits_{\Delta x \to 0} \dfrac{f(x + \Delta x) - f(x)}{\Delta x} = \lim\limits_{\Delta x \to 0} \dfrac{\dfrac{4}{x + \Delta x} - \dfrac{4}{x}}{\Delta x} = \lim\limits_{\Delta x \to 0} \dfrac{4x - 4(x + \Delta x)}{(x + \Delta x)x \, \Delta x} = \lim\limits_{\Delta x \to 0} \dfrac{-4}{(x + \Delta x)x} = \dfrac{-4}{x^2}$

87. $\lim\limits_{x \to 0} (4 - x^2) \le \lim\limits_{x \to 0} f(x) \le \lim\limits_{x \to 0} (4 + x^2)$

$$4 \le \lim\limits_{x \to 0} f(x) \le 4$$

Therefore, $\lim\limits_{x \to 0} f(x) = 4.$

89. $f(x) = x \cos x$

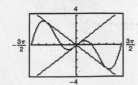

$\lim\limits_{x \to 0} (x \cos x) = 0$

91. $f(x) = |x| \sin x$

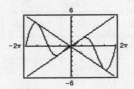

$\lim\limits_{x \to 0} |x| \sin x = 0$

93. $f(x) = x \sin \dfrac{1}{x}$

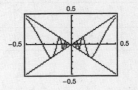

$\lim\limits_{x \to 0} \left(x \sin \dfrac{1}{x} \right) = 0$

95. We say that two functions f and g agree at all but one point (on an open interval) if $f(x) = g(x)$ for all x in the interval except for $x = c$, where c is in the interval.

97. An indeterminant form is obtained when evaluating a limit using direct substitution produces a meaningless fractional expression such as $0/0$. That is,

$$\lim_{x \to c} \frac{f(x)}{g(x)}$$

for which $\displaystyle\lim_{x \to c} f(x) = \lim_{x \to c} g(x) = 0$

99. $f(x) = x$, $g(x) = \sin x$, $h(x) = \dfrac{\sin x}{x}$

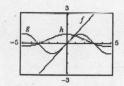

When you are "close to" 0 the magnitude of f is approximately equal to the magnitude of g. Thus, $|g|/|f| \approx 1$ when x is "close to" 0.

101. $s(t) = -16t^2 + 1000$

$$\lim_{t \to 5} \frac{s(5) - s(t)}{5 - t} = \lim_{t \to 5} \frac{600 - (-16t^2 + 1000)}{5 - t} = \lim_{t \to 5} \frac{16(t + 5)(t - 5)}{-(t - 5)} = \lim_{t \to 5} -16(t + 5) = -160 \text{ ft/sec}.$$

Speed = 160 ft/sec

103. $s(t) = -4.9t^2 + 150$

$$\lim_{t \to 3} \frac{s(3) - s(t)}{3 - t} = \lim_{t \to 3} \frac{-4.9(3^2) + 150 - (-4.9t^2 + 150)}{3 - t} = \lim_{t \to 3} \frac{-4.9(9 - t^2)}{3 - t}$$

$$= \lim_{t \to 3} \frac{-4.9(3 - t)(3 + t)}{3 - t} = \lim_{t \to 3} -4.9(3 + t) = -29.4 \text{ m/sec}$$

105. Let $f(x) = 1/x$ and $g(x) = -1/x$. $\displaystyle\lim_{x \to 0} f(x)$ and $\displaystyle\lim_{x \to 0} g(x)$ do not exist.

$$\lim_{x \to 0} [f(x) + g(x)] = \lim_{x \to 0} \left[\frac{1}{x} + \left(-\frac{1}{x} \right) \right] = \lim_{x \to 0} [0] = 0$$

107. Given $f(x) = b$, show that for every $\varepsilon > 0$ there exists a $\delta > 0$ such that $|f(x) - b| < \varepsilon$ whenever $|x - c| < \delta$. Since $|f(x) - b| = |b - b| = 0 < \varepsilon$ for any $\varepsilon > 0$, then any value of $\delta > 0$ will work.

109. If $b = 0$, then the property is true because both sides are equal to 0. If $b \neq 0$, let $\varepsilon > 0$ be given. Since $\displaystyle\lim_{x \to c} f(x) = L$, there exists $\delta > 0$ such that

$$|f(x) - L| < \varepsilon/|b| \text{ whenever } 0 < |x - c| < \delta. \text{ Hence,}$$

wherever $0 < |x - c| < \delta$, we have

$$|b||f(x) - L| < \varepsilon \quad \text{or} \quad |bf(x) - bL| < \varepsilon$$

which implies that $\displaystyle\lim_{x \to c} [bf(x)] = bL$.

111.
$$-M|f(x)| \leq f(x)g(x) \leq M|f(x)|$$

$$\lim_{x \to c} (-M|f(x)|) \leq \lim_{x \to c} f(x)g(x) \leq \lim_{x \to c} (M|f(x)|)$$

$$-M(0) \leq \lim_{x \to c} f(x)g(x) \leq M(0)$$

$$0 \leq \lim_{x \to c} f(x)g(x) \leq 0$$

Therefore, $\displaystyle\lim_{x \to c} f(x)g(x) = 0$.

113. False. As x approaches 0 from the left, $\dfrac{|x|}{x} = -1$.

115. True

117. False. The limit does not exist.

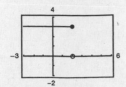

119. Let

$$f(x) = \begin{cases} 4, & \text{if } x \geq 0 \\ -4, & \text{if } x < 0 \end{cases}$$

$$\lim_{x \to 0} |f(x)| = \lim_{x \to 0} 4 = 4.$$

$\lim_{x \to 0} f(x)$ does not exist since for $x < 0$, $f(x) = -4$ and for $x \geq 0$, $f(x) = 4$.

121. $f(x) = \begin{cases} 0, & \text{if } x \text{ is rational} \\ 1, & \text{if } x \text{ is irrational} \end{cases}$

$g(x) = \begin{cases} 0, & \text{if } x \text{ is rational} \\ x, & \text{if } x \text{ is irrational} \end{cases}$

$\lim_{x \to 0} f(x)$ does not exist.

No matter how "close to" 0 x is, there are still an infinite number of rational and irrational numbers so that $\lim_{x \to 0} f(x)$ does not exist.

$$\lim_{x \to 0} g(x) = 0$$

When x is "close to" 0, both parts of the function are "close to" 0.

123. (a) $\displaystyle\lim_{x \to 0} \frac{1 - \cos x}{x^2} = \lim_{x \to 0} \frac{1 - \cos x}{x^2} \cdot \frac{1 + \cos x}{1 + \cos x}$

$\displaystyle = \lim_{x \to 0} \frac{1 - \cos^2 x}{x^2(1 + \cos x)}$

$\displaystyle = \lim_{x \to 0} \frac{\sin^2 x}{x^2} \cdot \frac{1}{1 + \cos x}$

$\displaystyle = (1)\left(\frac{1}{2}\right) = \frac{1}{2}$

(b) Thus, $\dfrac{1 - \cos x}{x^2} \approx \dfrac{1}{2} \implies 1 - \cos x \approx \dfrac{1}{2}x^2$

$\implies \cos x \approx 1 - \dfrac{1}{2}x^2$ for $x \approx 0$.

(c) $\cos(0.1) \approx 1 - \dfrac{1}{2}(0.1)^2 = 0.995$

(d) $\cos(0.1) \approx 0.9950$, which agrees with part (c).

Section 1.4   Continuity and One-Sided Limits

1. (a) $\displaystyle\lim_{x \to 3^+} f(x) = 1$

(b) $\displaystyle\lim_{x \to 3^-} f(x) = 1$

(c) $\displaystyle\lim_{x \to 3} f(x) = 1$

The function is continuous at $x = 3$.

3. (a) $\displaystyle\lim_{x \to 3^+} f(x) = 0$

(b) $\displaystyle\lim_{x \to 3^-} f(x) = 0$

(c) $\displaystyle\lim_{x \to 3} f(x) = 0$

The function is NOT continuous at $x = 3$.

5. (a) $\displaystyle\lim_{x \to 4^+} f(x) = 2$

(b) $\displaystyle\lim_{x \to 4^-} f(x) = -2$

(c) $\displaystyle\lim_{x \to 4} f(x)$ does not exist

The function is NOT continuous at $x = 4$.

7. $\displaystyle\lim_{x \to 5^+} \frac{x - 5}{x^2 - 25} = \lim_{x \to 5^+} \frac{1}{x + 5} = \frac{1}{10}$

9. $\displaystyle\lim_{x \to -3^-} \frac{x}{\sqrt{x^2 - 9}}$ does not exist because $\dfrac{x}{\sqrt{x^2 - 9}}$ decreases without bound as $x \to -3^-$.

11. $\displaystyle\lim_{x \to 0^-} \frac{|x|}{x} = \lim_{x \to 0^-} \frac{-x}{x} = -1$

13. $\displaystyle\lim_{\Delta x \to 0^-} \frac{\frac{1}{x + \Delta x} - \frac{1}{x}}{\Delta x} = \lim_{\Delta x \to 0^-} \frac{x - (x + \Delta x)}{x(x + \Delta x)} \cdot \frac{1}{\Delta x} = \lim_{\Delta x \to 0^-} \frac{-\Delta x}{x(x + \Delta x)} \cdot \frac{1}{\Delta x}$

$$= \lim_{\Delta x \to 0^-} \frac{-1}{x(x + \Delta x)}$$

$$= \frac{-1}{x(x + 0)} = -\frac{1}{x^2}$$

15. $\displaystyle\lim_{x \to 3^-} f(x) = \lim_{x \to 3^-} \frac{x + 2}{2} = \frac{5}{2}$

17. $\displaystyle\lim_{x \to 1^+} f(x) = \lim_{x \to 1^+} (x + 1) = 2$

$$\lim_{x \to 1^-} f(x) = \lim_{x \to 1^-} (x^3 + 1) = 2$$

$$\lim_{x \to 1} f(x) = 2$$

19. $\displaystyle\lim_{x \to \pi} \cot x$ does not exist since

$\displaystyle\lim_{x \to \pi^+} \cot x$ and $\displaystyle\lim_{x \to \pi^-} \cot x$ do not exist.

21. $\displaystyle\lim_{x \to 4^-} (3[\![x]\!] - 5) = 3(3) - 5 = 4$

$([\![x]\!] = 3$ for $3 \le x < 4)$

23. $\displaystyle\lim_{x \to 3} (2 - [\![-x]\!])$ does not exist

because

$$\lim_{x \to 3^-} (2 - [\![-x]\!]) = 2 - (-3) = 5$$

and

$$\lim_{x \to 3^+} (2 - [\![-x]\!]) = 2 - (-4) = 6.$$

25. $f(x) = \dfrac{1}{x^2 - 4}$

has discontinuities at $x = -2$ and $x = 2$ since $f(-2)$ and $f(2)$ are not defined.

27. $f(x) = \dfrac{[\![x]\!]}{2} + x$

has discontinuities at each integer k since $\displaystyle\lim_{x \to k^-} f(x) \ne \lim_{x \to k^+} f(x)$.

29. $g(x) = \sqrt{25 - x^2}$ is continuous on $[-5, 5]$.

31. $\displaystyle\lim_{x \to 0^-} f(x) = 3 = \lim_{x \to 0^+} f(x)$.

f is continuous on $[-1, 4]$.

33. $f(x) = x^2 - 2x + 1$ is continuous for all real x.

35. $f(x) = 3x - \cos x$ is continuous for all real x.

37. $f(x) = \dfrac{x}{x^2 - x}$ is not continuous at $x = 0, 1$. Since $\dfrac{x}{x^2 - x} = \dfrac{1}{x - 1}$ for $x \ne 0$, $x = 0$ is a removable discontinuity, whereas $x = 1$ is a nonremovable discontinuity.

39. $f(x) = \dfrac{x}{x^2 + 1}$ is continuous for all real x.

41. $f(x) = \dfrac{x + 2}{(x + 2)(x - 5)}$

has a nonremovable discontinuity at $x = 5$ since $\displaystyle\lim_{x \to 5} f(x)$ does not exist, and has a removable discontinuity at $x = -2$ since

$$\lim_{x \to -2} f(x) = \lim_{x \to -2} \frac{1}{x - 5} = -\frac{1}{7}.$$

43. $f(x) = \dfrac{|x + 2|}{x + 2}$

has a nonremovable discontinuity at $x = -2$ since $\displaystyle\lim_{x \to -2} f(x)$ does not exist.

45. $f(x) = \begin{cases} x, & x \le 1 \\ x^2, & x > 1 \end{cases}$

has a **possible** discontinuity at $x = 1$.

1. $f(1) = 1$

2. $\left. \begin{array}{l} \lim\limits_{x \to 1^-} f(x) = \lim\limits_{x \to 1^-} x = 1 \\ \lim\limits_{x \to 1^+} f(x) = \lim\limits_{x \to 1^+} x^2 = 1 \end{array} \right\} \lim\limits_{x \to 1} f(x) = 1$

3. $f(1) = \lim\limits_{x \to 1} f(x)$

f is continuous at $x = 1$, therefore, f is continuous for all real x.

47. $f(x) = \begin{cases} \dfrac{x}{2} + 1, & x \le 2 \\ 3 - x, & x > 2 \end{cases}$

has a **possible** discontinuity at $x = 2$.

1. $f(2) = \dfrac{2}{2} + 1 = 2$

2. $\lim\limits_{x \to 2^-} f(x) = \lim\limits_{x \to 2^-} \left(\dfrac{x}{2} + 1 \right) = 2$

$\lim\limits_{x \to 2^+} f(x) = \lim\limits_{x \to 2^+} (3 - x) = 1$

Therefore, f has a nonremovable discontinuity at $x = 2$.

49. $f(x) = \begin{cases} \tan \dfrac{\pi x}{4}, & |x| < 1 \\ x, & |x| \ge 1 \end{cases}$

$= \begin{cases} \tan \dfrac{\pi x}{4}, & -1 < x < 1 \\ x, & x \le -1 \text{ or } x \ge 1 \end{cases}$

has **possible** discontinuities at $x = -1$, $x = 1$.

1. $f(-1) = -1$ $\qquad$ $f(1) = 1$

2. $\lim\limits_{x \to -1} f(x) = -1$ $\qquad$ $\lim\limits_{x \to 1} f(x) = 1$

3. $f(-1) = \lim\limits_{x \to -1} f(x)$ $\qquad$ $f(1) = \lim\limits_{x \to 1} f(x)$

f is continuous at $x = \pm 1$, therefore, f is continuous for all real x.

51. $f(x) = \csc 2x$ has nonremovable discontinuities at integer multiples of $\pi/2$.

53. $f(x) = [\![x - 1]\!]$ has nonremovable discontinuities at each integer k.

55. $\lim\limits_{x \to 0^+} f(x) = 0$

$\lim\limits_{x \to 0^-} f(x) = 0$

f is not continuous at $x = -2$.

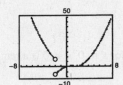

57. $f(2) = 8$

Find a so that $\lim\limits_{x \to 2^+} ax^2 = 8 \implies a = \dfrac{8}{2^2} = 2$.

59. Find a and b such that $\lim\limits_{x \to -1^+} (ax + b) = -a + b = 2$ and $\lim\limits_{x \to 3^-} (ax + b) = 3a + b = -2$.

$\qquad a - b = -2$

$\qquad \underline{(+)\ 3a + b = -2}$

$\qquad\quad 4a \qquad = -4$

$\qquad\qquad a = -1$

$\qquad\qquad b = 2 + (-1) = 1$

$f(x) = \begin{cases} 2, & x \le -1 \\ -x + 1, & -1 < x < 3 \\ -2, & x \ge 3 \end{cases}$

61. $f(g(x)) = (x - 1)^2$

Continuous for all real x.

63. $f(g(x)) = \dfrac{1}{(x^2 + 5) - 6} = \dfrac{1}{x^2 - 1}$

Nonremovable discontinuities at $x = \pm 1$

65. $y = [\![x]\!] - x$

Nonremovable discontinuity at each integer

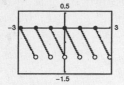

67. $f(x) = \begin{cases} 2x - 4, & x \le 3 \\ x^2 - 2x, & x > 3 \end{cases}$

Nonremovable discontinuity at $x = 3$

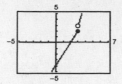

69. $f(x) = \dfrac{x}{x^2 + 1}$

Continuous on $(-\infty, \infty)$

71. $f(x) = \sec \dfrac{\pi x}{4}$

Continuous on:

$\ldots, (-6, -2), (-2, 2), (2, 6), (6, 10), \ldots$

73. $f(x) = \dfrac{\sin x}{x}$

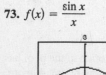

The graph **appears** to be continuous on the interval $[-4, 4]$.

Since $f(0)$ is not defined, we know that f has a discontinuity at $x = 0$. This discontinuity is removable so it does not show up on the graph.

75. $f(x) = \frac{1}{16}x^4 - x^3 + 3$ is continuous on $[1, 2]$.

$f(1) = \frac{33}{16}$ and $f(2) = -4$. By the Intermediate Value Theorem, $f(c) = 0$ for at least one value of c between 1 and 2.

77. $f(x) - x^2 - 2 - \cos x$ is continuous on $[0, \pi]$.

$f(0) = -3$ and $f(\pi) = \pi^2 - 1 > 0$. By the Intermediate Value Theorem, $f(c) = 0$ for the least one value of c between 0 and π.

79. $f(x) = x^3 + x - 1$

$f(x)$ is continuous on $[0, 1]$.

$f(0) = -1$ and $f(1) = 1$

By the Intermediate Value Theorem, $f(x) = 0$ for at least one value of c between 0 and 1. Using a graphing utility, we find that $x \approx 0.6823$.

81. $g(t) = 2 \cos t - 3t$

g is continuous on $[0, 1]$.

$g(0) = 2 > 0$ and $g(1) \approx -1.9 < 0$.

By the Intermediate Value Theorem, $g(t) = 0$ for at least one value c between 0 and 1. Using a graphing utility, we find that $t \approx 0.5636$.

83. $f(x) = x^2 + x - 1$

f is continuous on $[0, 5]$.

$f(0) = -1$ and $f(5) = 29$

$-1 < 11 < 29$

The Intermediate Value Theorem applies.

$$x^2 + x - 1 = 11$$
$$x^2 + x - 12 = 0$$
$$(x + 4)(x - 3) = 0$$
$$x = -4 \text{ or } x = 3$$

$c = 3$ ($x = -4$ is not in the interval.)

Thus, $f(3) = 11$.

85. $f(x) = x^3 - x^2 + x - 2$

f is continuous on $[0, 3]$.

$f(0) = -2$ and $f(3) = 19$

$-2 < 4 < 19$

The Intermediate Value Theorem applies.

$$x^3 - x^2 + x - 2 = 4$$
$$x^3 - x^2 + x - 6 = 0$$
$$(x - 2)(x^2 + x + 3) = 0$$
$$x = 2$$

($x^2 + x + 3$ has no real solution.)

$$c = 2$$

Thus, $f(2) = 4$.

87. (a) The limit does not exist at $x = c$.

(b) The function is not defined at $x = c$.

(c) The limit exists at $x = c$, but it is not equal to the value of the function at $x = c$.

(d) The limit does not exist at $x = c$.

89.

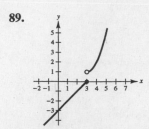

The function is not continuous at $x = 3$ because
$$\lim_{x \to 3^+} f(x) = 1 \neq 0 = \lim_{x \to 3^-} f(x).$$

91. True

1. $f(c) = L$ is defined.

2. $\lim_{x \to c} f(x) = L$ exists.

3. $f(c) = \lim_{x \to c} f(x)$

All of the conditions for continuity are met.

93. False; a rational function can be written as $P(x)/Q(x)$ where P and Q are polynomials of degree m and n, respectively. It can have, at most, n discontinuities.

95. $\lim_{t \to 4^-} f(t) \approx 28$

$\lim_{t \to 4^+} f(t) \approx 56$

At the end of day 3, the amount of chlorine in the pool has decreased to about 28 oz. At the beginning of day 4, more chlorine was added, and the amount was about 56 oz.

97. $C = \begin{cases} 1.04, & 0 < t \leq 2 \\ 1.04 + 0.36[\![t - 1]\!], & t > 2, t \text{ is not an integer} \\ 1.04 + 0.36(t - 2), & t > 2, t \text{ is an integer} \end{cases}$

Nonremovable discontinuity at each integer greater than or equal to 2.

You can also write C as

$$C = \begin{cases} 1.04, & 0 < t \leq 2 \\ 1.04 - 0.36[\![2 - t]\!], & t > 2 \end{cases}.$$

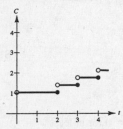

99. Let $s(t)$ be the position function for the run up to the campsite. $s(0) = 0$ ($t = 0$ corresponds to 8:00 A.M., $s(20) = k$ (distance to campsite)). Let $r(t)$ be the position function for the run back down the mountain: $r(0) = k$, $r(10) = 0$. Let
$$f(t) = s(t) - r(t).$$

When $t = 0$ (8:00 A.M.), $f(0) = s(0) - r(0) = 0 - k < 0$.

When $t = 10$ (8:10 A.M.), $f(10) = s(10) - r(10) > 0$.

Since $f(0) < 0$ and $f(10) > 0$, then there must be a value t in the interval $[0, 10]$ such that $f(t) = 0$. If $f(t) = 0$, then $s(t) - r(t) = 0$, which gives us $s(t) = r(t)$. Therefore, at some time t, where $0 \leq t \leq 10$, the position functions for the

101. Suppose there exists x_1 in $[a, b]$ such that $f(x_1) > 0$ and there exists x_2 in $[a, b]$ such that $f(x_2) < 0$. Then by the Intermediate Value Theorem, $f(x)$ must equal zero for some value of x in $[x_1, x_2]$ (or $[x_2, x_1]$ if $x_2 < x_1$). Thus, f would have a zero in $[a, b]$, which is a contradiction. Therefore, $f(x) > 0$ for all x in $[a, b]$ or $f(x) < 0$ for all x in $[a, b]$.

103. If $x = 0$, then $f(0) = 0$ and $\lim_{x \to 0} f(x) = 0$. Hence, f is continuous at $x = 0$.

If $x \neq 0$, then $\lim_{t \to x} f(t) = 0$ for x rational, whereas $\lim_{t \to x} f(t) = \lim_{t \to x} kt = kx \neq 0$ for x irrational. Hence, f is not continuous for all $x \neq 0$.

105. (a)

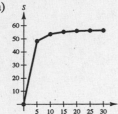

(b) There appears to be a limiting speed and a possible cause is air resistance.

107. $f(x) = \begin{cases} 1 - x^2, & x \le c \\ x, & x > c \end{cases}$

f is continuous for $x < c$ and for $x > c$. At $x = c$, you need $1 - c^2 = c$. Solving $c^2 + c - 1$, you obtain

$$c = \frac{-1 \pm \sqrt{1 + 4}}{2} = \frac{-1 \pm \sqrt{5}}{2}.$$

109. $f(x) = \dfrac{\sqrt{x + c^2} - c}{x}, \; c > 0$

Domain: $x + c^2 \ge 0 \implies x \ge -c^2$ and $x \ne 0$, $[-c^2, 0) \cup (0, \infty)$

$$\lim_{x \to 0} \frac{\sqrt{x + c^2} - c}{x} = \lim_{x \to 0} \frac{\sqrt{x + c^2} - c}{x} \cdot \frac{\sqrt{x + c^2} + c}{\sqrt{x + c^2} + c}$$

$$= \lim_{x \to 0} \frac{(x + c^2) - c^2}{x\left[\sqrt{x + c^2} + c\right]} = \lim_{x \to 0} \frac{1}{\sqrt{x + c^2} + c} = \frac{1}{2c}.$$

Define $f(0) = 1/(2c)$ to make f continuous at $x = 0$.

111. $h(x) = x[\![x]\!]$

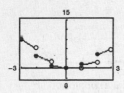

h has nonremovable discontinuities at
$$x = \pm 1, \pm 2, \pm 3, \dots .$$

113. The statement is true.

If $y \ge 0$ and $y \le 1$, then $y(y - 1) \le 0 \le x^2$, as desired. So assume $y > 1$. There are now two cases.

Case 1: If $x \le y - \frac{1}{2}$, then $2x + 1 \le 2y$ and

$$\begin{aligned} y(y - 1) &= y(y + 1) - 2y \\ &\le (x + 1)^2 - 2y \\ &= x^2 + 2x + 1 - 2y \\ &\le x^2 + 2y - 2y \\ &= x^2 \end{aligned}$$

Case 2: If $x \ge y - \frac{1}{2}$

$$\begin{aligned} x^2 &\ge \left(y - \tfrac{1}{2}\right)^2 \\ &= y^2 - y + \tfrac{1}{4} \\ &> y^2 - y \\ &= y(y - 1) \end{aligned}$$

In both cases, $y(y - 1) \le x^2$.

Section 1.5   Infinite Limits

1. $\displaystyle\lim_{x \to -2^+} 2\left|\frac{x}{x^2 - 4}\right| = \infty$

$\displaystyle\lim_{x \to -2^-} 2\left|\frac{x}{x^2 - 4}\right| = \infty$

3. $\displaystyle\lim_{x \to -2^+} \tan \frac{\pi x}{4} = -\infty$

$\displaystyle\lim_{x \to -2^-} \tan \frac{\pi x}{4} = \infty$

5. $f(x) = \dfrac{1}{x^2 - 9}$

x	-3.5	-3.1	-3.01	-3.001	-2.999	-2.99	-2.9	-2.5
$f(x)$	0.308	1.639	16.64	166.6	-166.7	-16.69	-1.695	-0.364

$$\lim_{x \to -3^-} f(x) = \infty$$

$$\lim_{x \to -3^+} f(x) = -\infty$$

7. $f(x) = \dfrac{x^2}{x^2 - 9}$

x	-3.5	-3.1	-3.01	-3.001	-2.999	-2.99	-2.9	-2.5
$f(x)$	3.769	15.75	150.8	1501	-1499	-149.3	-14.25	-2.273

$$\lim_{x \to -3^-} f(x) = \infty$$

$$\lim_{x \to -3^+} f(x) = -\infty$$

9. $\displaystyle\lim_{x \to 0^+} \frac{1}{x^2} = \infty = \lim_{x \to 0^-} \frac{1}{x^2}$

Therefore, $x = 0$ is a vertical asymptote.

11. $\displaystyle\lim_{x \to 2^+} \frac{x^2 - 2}{(x - 2)(x + 1)} = \infty$

$$\lim_{x \to 2^-} \frac{x^2 - 2}{(x - 2)(x + 1)} = -\infty$$

Therefore, $x = 2$ is a vertical asymptote.

$$\lim_{x \to -1^+} \frac{x^2 - 2}{(x - 2)(x + 1)} = \infty$$

$$\lim_{x \to -1^-} \frac{x^2 - 2}{(x - 2)(x + 1)} = -\infty$$

Therefore, $x = -1$ is a vertical asymptote.

13. $\displaystyle\lim_{x \to -2^-} \frac{x^2}{x^2 - 4} = \infty$ and $\displaystyle\lim_{x \to -2^+} \frac{x^2}{x^2 - 4} = -\infty$

Therefore, $x = -2$ is a vertical asymptote.

$$\lim_{x \to 2^-} \frac{x^2}{x^2 - 4} = -\infty \text{ and } \lim_{x \to 2^+} \frac{x^2}{x^2 - 4} = \infty$$

Therefore, $x = 2$ is a vertical asymptote.

15. No vertical asymptote since the denominator is never zero.

17. $f(x) = \tan 2x = \dfrac{\sin 2x}{\cos 2x}$ has vertical asymptotes at

$$x = \frac{(2n + 1)\pi}{4} = \frac{\pi}{4} + \frac{n\pi}{2}, \; n \text{ any integer.}$$

19. $\displaystyle\lim_{t \to 0^+} \left(1 - \frac{4}{t^2} \right) = -\infty = \lim_{t \to 0^-} \left(1 - \frac{4}{t^2} \right)$

Therefore, $t = 0$ is a vertical asymptote.

21. $\lim\limits_{x \to -2^+} \dfrac{x}{(x+2)(x-1)} = \infty$

$\lim\limits_{x \to -2^-} \dfrac{x}{(x+2)(x-1)} = -\infty$

Therefore, $x = -2$ is a vertical asymptote.

$\lim\limits_{x \to 1^+} \dfrac{x}{(x+2)(x-1)} = \infty$

$\lim\limits_{x \to 1^-} \dfrac{x}{(x+2)(x-1)} = -\infty$

Therefore, $x = 1$ is a vertical asymptote.

25. $f(x) = \dfrac{(x-5)(x+3)}{(x-5)(x^2+1)} = \dfrac{x+3}{x^2+1}, x \neq 5$

No vertical asymptotes. The graph has a hole at $x = 5$.

29. $\lim\limits_{x \to -1} \dfrac{x^2-1}{x+1} = \lim\limits_{x \to -1}(x-1) = -2$

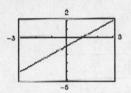

Removable discontinuity at $x = -1$

33. $\lim\limits_{x \to 2^+} \dfrac{x-3}{x-2} = -\infty$

37. $\lim\limits_{x \to -3^-} \dfrac{x^2+2x-3}{x^2+x-6} = \lim\limits_{x \to -3^-} \dfrac{(x-1)(x+3)}{(x-2)(x+3)} = \lim\limits_{x \to -3^-} \dfrac{x-1}{x-2} = \dfrac{4}{5}$

39. $\lim\limits_{x \to 1} \dfrac{x^2-x}{(x^2+1)(x-1)} = \lim\limits_{x \to 1} \dfrac{x}{x^2+1} = \dfrac{1}{2}$

43. $\lim\limits_{x \to 0^+} \dfrac{2}{\sin x} = \infty$

47. $\lim\limits_{x \to (1/2)^-} x \sec(\pi x) = \infty$ and $\lim\limits_{x \to (1/2)^+} x \sec(\pi x) = -\infty.$

Therefore, $\lim\limits_{x \to (1/2)} x \sec(\pi x)$ does not exist.

49. $f(x) = \dfrac{x^2+x+1}{x^3-1} = \dfrac{x^2+x+1}{(x-1)(x^2+x+1)}$

$\lim\limits_{x \to 1^+} f(x) = \lim\limits_{x \to 1^+} \dfrac{1}{x-1} = \infty$

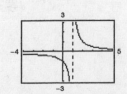

23. $f(x) = \dfrac{x^3+1}{x+1} = \dfrac{(x+1)(x^2-x+1)}{x+1}$

has no vertical asymptote since

$\lim\limits_{x \to -1} f(x) = \lim\limits_{x \to -1}(x^2-x+1) = 3.$

The graph has a hole at $x = -1$.

27. $s(t) = \dfrac{t}{\sin t}$ has vertical asymptotes at $t = n\pi, n$

a nonzero integer. There is no vertical asymptote at $t = 0$ since

$\lim\limits_{t \to 0} \dfrac{t}{\sin t} = 1.$

31. $\lim\limits_{x \to -1^+} \dfrac{x^2+1}{x+1} = \infty$

$\lim\limits_{x \to -1^-} \dfrac{x^2+1}{x+1} = -\infty$

Vertical asymptote at $x = -1$

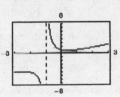

35. $\lim\limits_{x \to 3^+} \dfrac{x^2}{(x-3)(x+3)} = \infty$

41. $\lim\limits_{x \to 0^-}\left(1+\dfrac{1}{x}\right) = -\infty$

45. $\lim\limits_{x \to \pi} \dfrac{\sqrt{x}}{\csc x} = \lim\limits_{x \to \pi}\left(\sqrt{x} \sin x\right) = 0$

51. $f(x) = \dfrac{1}{x^2 - 25}$

$\displaystyle\lim_{x \to 5^-} f(x) = -\infty$

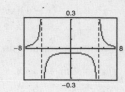

53. A limit in which $f(x)$ increases or decreases without bound as x approaches c is called an infinite limit. ∞ is not a number. Rather, the symbol

$$\lim_{x \to c} f(x) = \infty$$

says how the limit fails to exist.

55. One answer is $f(x) = \dfrac{x - 3}{(x - 6)(x + 2)} = \dfrac{x - 3}{x^2 - 4x - 12}$.

57.

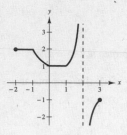

59. (a) $r = 50\pi \sec^2 \dfrac{\pi}{6} = \dfrac{200\pi}{3}$ ft/sec

(b) $r = 50\pi \sec^2 \dfrac{\pi}{3} = 200\pi$ ft/sec

(c) $\displaystyle\lim_{\theta \to (\pi/2)^-} \left[50\pi \sec^2 \theta\right] = \infty$

61. $m = \dfrac{m_0}{\sqrt{1 - (v^2/c^2)}}$

$\displaystyle\lim_{v \to c^-} m = \lim_{v \to c^-} \dfrac{m_0}{\sqrt{1 - (v^2/c^2)}} = \infty$

63. (a) Average speed $= \dfrac{\text{Total distance}}{\text{Total time}}$

$50 = \dfrac{2d}{(d/x) + (d/y)}$

$50 = \dfrac{2xy}{y + x}$

$50y + 50x = 2xy$

$50x = 2xy - 50y$

$50x = 2y(x - 25)$

$\dfrac{25x}{x - 25} = y$

Domain: $x > 25$

(b)

x	30	40	50	60
y	150	66.667	50	42.857

(c) $\displaystyle\lim_{x \to 25^+} \dfrac{25x}{x - 25} = \infty$

As x gets close to 25 mph, y becomes larger and larger.

65. (a) $A = \dfrac{1}{2}bh - \dfrac{1}{2}r^2\theta = \dfrac{1}{2}(10)(10 \tan \theta) - \dfrac{1}{2}(10)^2\theta$

$= 50 \tan \theta - 50 \theta$

Domain: $\left(0, \dfrac{\pi}{2}\right)$

(b)

θ	0.3	0.6	0.9	1.2	1.5
$f(\theta)$	0.47	4.21	18.0	68.6	630.1

(c)

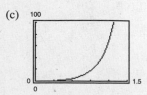

(d) $\displaystyle\lim_{\theta \to \pi/2^-} A = \infty$

67. False; for instance, let

$$f(x) = \frac{x^2 - 1}{x - 1}.$$

The graph of f has a hole at $(1, 2)$, not a vertical asymptote.

69. True

71. Let $f(x) = \frac{1}{x^2}$ and $g(x) = \frac{1}{x^4}$, and $c = 0$.

$$\lim_{x \to 0} \frac{1}{x^2} = \infty \text{ and } \lim_{x \to 0} \frac{1}{x^4} = \infty, \text{ but}$$

$$\lim_{x \to 0}\left(\frac{1}{x^2} - \frac{1}{x^4}\right) = \lim_{x \to 0}\left(\frac{x^2 - 1}{x^4}\right) = -\infty \neq 0.$$

73. Given $\lim_{x \to c} f(x) = \infty$, let $g(x) = 1$. then $\lim_{x \to c} \frac{g(x)}{f(x)} = 0$ by Theorem 1.15.

75. $f(x) = \dfrac{1}{x - 3}$ is defined for all $x > 3$. Let $M > 0$ be given.

We need $\delta > 0$ such that $f(x) = \dfrac{1}{x - 3} > M$ whenever $3 < x < 3 + \delta$

Equivalently, $x - 3 < \dfrac{1}{M}$ whenever $|x - 3| < \delta, x > 3$.

So take $\delta = \dfrac{1}{M}$. Then for $x > 3$ and $|x - 3| < \delta, \dfrac{1}{x - 3} > \dfrac{1}{8} = M$ and hence $f(x) > M$.

Review Exercises for Chapter 1

1. Calculus required. Using a graphing utility, you can estimate the length to be 8.3. Or, the length is slightly longer than the distance between the two points, approximately 8.25.

3. $f(x) = \dfrac{\dfrac{4}{x + 2} - 2}{x}$

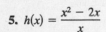

x	-0.1	-0.01	-0.001	0.001	0.01	0.1
$f(x)$	-1.0526	-1.0050	-1.0005	-0.9995	-0.9950	-0.9524

$$\lim_{x \to 0} f(x) \approx -1.0$$

5. $h(x) = \dfrac{x^2 - 2x}{x}$

   (a) $\lim_{x \to 0} h(x) = -2$

   (b) $\lim_{x \to -1} h(x) = -3$

7. $\lim_{x \to 1}(3 - x) = 3 - 1 = 2$

Let $\varepsilon > 0$ be given. Choose $\delta = \varepsilon$. Then for $0 < |x - 1| < \delta = \varepsilon$, you have

$$|x - 1| < \varepsilon$$
$$|1 - x| < \varepsilon$$
$$|(3 - x) - 2| < \varepsilon$$
$$|f(x) - L| < \varepsilon.$$

9. $\lim\limits_{x \to 2} (x^2 - 3) = 1$

Let $\varepsilon > 0$ be given. We need $|x^2 - 3 - 1| < \varepsilon \implies |x^2 - 4| = |(x - 2)(x + 2)| < \varepsilon \implies |x - 2| < \dfrac{1}{|x + 2|}\varepsilon.$

Assuming, $1 < x < 3$, you can choose $\delta = \varepsilon/5$. Hence, for $0 < |x - 2| < \delta = \varepsilon/5$ you have

$$|x - 2| < \frac{\varepsilon}{5} < \frac{1}{|x + 2|}\varepsilon$$

$$|x - 2||x + 2| < \varepsilon$$

$$|x^2 - 4| < \varepsilon$$

$$|(x^2 - 3) - 1| < \varepsilon$$

$$|f(x) - L| < \varepsilon.$$

11. $\lim\limits_{t \to 4} \sqrt{t + 2} = \sqrt{4 + 2} = \sqrt{6} \approx 2.45$

13. $\lim\limits_{t \to -2} \dfrac{t + 2}{t^2 - 4} = \lim\limits_{t \to -2} \dfrac{1}{t - 2} = -\dfrac{1}{4}$

15. $\lim\limits_{x \to 4} \dfrac{\sqrt{x} - 2}{x - 4} = \lim\limits_{x \to 4} \dfrac{\sqrt{x} - 2}{(\sqrt{x} - 2)(\sqrt{x} + 2)}$

$$= \lim\limits_{x \to 4} \frac{1}{\sqrt{x} + 2} = \frac{1}{\sqrt{4} + 2} = \frac{1}{4}$$

17. $\lim\limits_{x \to 0} \dfrac{[1/(x + 1)] - 1}{x} = \lim\limits_{x \to 0} \dfrac{1 - (x + 1)}{x(x + 1)} = \lim\limits_{x \to 0} \dfrac{-1}{x + 1} = -1$

19. $\lim\limits_{x \to -5} \dfrac{x^3 + 125}{x + 5} = \lim\limits_{x \to -5} \dfrac{(x + 5)(x^2 - 5x + 25)}{x + 5}$

$$= \lim\limits_{x \to -5} (x^2 - 5x + 25) = 75$$

21. $\lim\limits_{x \to 0} \dfrac{1 - \cos x}{\sin x} = \lim\limits_{x \to 0} \left(\dfrac{x}{\sin x}\right)\left(\dfrac{1 - \cos x}{x}\right) = (1)(0) = 0$

23. $\lim\limits_{\Delta x \to 0} \dfrac{\sin[(\pi/6) + \Delta x] - (1/2)}{\Delta x} = \lim\limits_{\Delta x \to 0} \dfrac{\sin(\pi/6)\cos \Delta x + \cos(\pi/6)\sin \Delta x - (1/2)}{\Delta x}$

$$= \lim\limits_{\Delta x \to 0} \frac{1}{2} \cdot \frac{(\cos \Delta x - 1)}{\Delta x} + \lim\limits_{\Delta x \to 0} \frac{\sqrt{3}}{2} \cdot \frac{\sin \Delta x}{\Delta x} = 0 + \frac{\sqrt{3}}{2}(1) = \frac{\sqrt{3}}{2}$$

25. $\lim\limits_{x \to c} [f(x) \cdot g(x)] = \left(-\dfrac{3}{4}\right)\left(\dfrac{2}{3}\right) = -\dfrac{1}{2}$

27. $f(x) = \dfrac{\sqrt{2x + 1} - \sqrt{3}}{x - 1}$

(a)

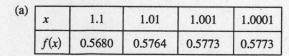

x	1.1	1.01	1.001	1.0001
$f(x)$	0.5680	0.5764	0.5773	0.5773

$$\lim\limits_{x \to 1^+} \frac{\sqrt{2x + 1} - \sqrt{3}}{x - 1} \approx 0.577 \quad \text{(Actual limit is } \sqrt{3}/3.\text{)}$$

(b)

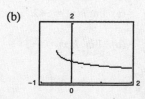

—CONTINUED—

27. —CONTINUED—

(c) $\lim\limits_{x \to 1^+} \dfrac{\sqrt{2x + 1} - \sqrt{3}}{x - 1} = \lim\limits_{x \to 1^+} \dfrac{\sqrt{2x + 1} - \sqrt{3}}{x - 1} \cdot \dfrac{\sqrt{2x + 1} + \sqrt{3}}{\sqrt{2x + 1} + \sqrt{3}}$

$$= \lim\limits_{x \to 1^+} \frac{(2x + 1) - 3}{(x - 1)(\sqrt{2x + 1} + \sqrt{3})}$$

$$= \lim\limits_{x \to 1^+} \frac{2}{\sqrt{2x + 1} + \sqrt{3}}$$

$$= \frac{2}{2\sqrt{3}} = \frac{1}{\sqrt{3}} = \frac{\sqrt{3}}{3}$$

29. $\lim\limits_{t \to a} \dfrac{s(a) - s(t)}{a - t} = \lim\limits_{t \to 4} \dfrac{(-4.9(4)^2 + 200) - (-4.9t^2 + 200)}{4 - t}$

$$= \lim\limits_{t \to 4} \frac{4.9(t - 4)(t + 4)}{4 - t}$$

$$= \lim\limits_{t \to 4} -4.9(t + 4) = -39.2 \text{ m/sec}$$

31. $\lim\limits_{x \to 3^-} \dfrac{|x - 3|}{x - 3} = \lim\limits_{x \to 3^-} \dfrac{-(x - 3)}{x - 3}$

$$= -1$$

33. $\lim\limits_{x \to 2} f(x) = 0$

35. $\lim\limits_{t \to 1} h(t)$ does not exist because

$\lim\limits_{t \to 1^-} h(t) = 1 + 1 = 2$ and

$\lim\limits_{t \to 1^+} h(t) = \frac{1}{2}(1 + 1) = 1.$

37. $f(x) = [\![x + 3]\!]$

$\lim\limits_{x \to k^+} [\![x + 3]\!] = k + 3$ where k is an integer.

$\lim\limits_{x \to k^-} [\![x + 3]\!] = k + 2$ where k is an integer.

Nonremovable discontinuity at each integer k

Continuous on $(k, k + 1)$ for all integers k

39. $f(x) = \dfrac{3x^2 - x - 2}{x - 1} = \dfrac{(3x + 2)(x - 1)}{x - 1}$

$\lim\limits_{x \to 1} f(x) = \lim\limits_{x \to 1} (3x + 2) = 5$

Removable discontinuity at $x = 1$

Continuous on $(-\infty, 1) \cup (1, \infty)$

41. $f(x) = \dfrac{1}{(x - 2)^2}$

$\lim\limits_{x \to 2} \dfrac{1}{(x - 2)^2} = \infty$

Nonremovable discontinuity at $x = 2$

Continuous on $(-\infty, 2) \cup (2, \infty)$

43. $f(x) = \dfrac{3}{x + 1}$

$\lim\limits_{x \to 1^-} f(x) = -\infty$

$\lim\limits_{x \to 1^+} f(x) = \infty$

Nonremovable discontinuity at $x = -1$

Continuous on $(-\infty, -1) \cup (-1, \infty)$

45. $f(x) = \csc \dfrac{\pi x}{2}$

Nonremovable discontinuities at each even integer.

Continuous on

$(2k, 2k + 2)$

for all integers k.

47. $f(2) = 5$

Find c so that $\lim\limits_{x \to 2^+} (cx + 6) = 5.$

$c(2) + 6 = 5$

$2c = -1$

$c = -\dfrac{1}{2}$

49. f is continuous on $[1, 2]$.

$f(1) = -1 < 0$ and

$f(2) = 13 > 0$. Therefore by the Intermediate Value Theorem, there is at least one value c in $(1, 2)$ such that $2c^3 - 3 = 0.$

51. $f(x) = \dfrac{x^2 - 4}{|x - 2|} = (x + 2)\left[\dfrac{x - 2}{|x - 2|}\right]$

(a) $\lim\limits_{x \to 2^-} f(x) = -4$ (b) $\lim\limits_{x \to 2^+} f(x) = 4$ (c) $\lim\limits_{x \to 2} f(x)$ does not exist.

53. $g(x) = 1 + \dfrac{2}{x}$

Vertical asymptote at $x = 0$

55. $f(x) = \dfrac{8}{(x - 10)^2}$

Vertical asymptote at $x = 10$

57. $\displaystyle\lim_{x \to -2^-} \dfrac{2x^2 + x + 1}{x + 2} = -\infty$

59. $\displaystyle\lim_{x \to -1^+} \dfrac{x + 1}{x^3 + 1} = \lim_{x \to -1^+} \dfrac{1}{x^2 - x + 1} = \dfrac{1}{3}$

61. $\displaystyle\lim_{x \to 1^-} \dfrac{x^2 + 2x + 1}{x - 1} = -\infty$

63. $\displaystyle\lim_{x \to 0^+} \left(x - \dfrac{1}{x^3} \right) = -\infty$

65. $\displaystyle\lim_{x \to 0^+} \dfrac{\sin 4x}{5x} = \lim_{x \to 0^+} \left[\dfrac{4}{5} \left(\dfrac{\sin 4x}{4x} \right) \right] = \dfrac{4}{5}$

67. $\displaystyle\lim_{x \to 0^+} \dfrac{\csc 2x}{x} = \lim_{x \to 0^+} \dfrac{1}{x \sin 2x} = \infty$

69. $C = \dfrac{80{,}000p}{100 - p}, \ 0 \le p < 100$

(a) $C(15) \approx \$14{,}117.65$ (b) $C(50) = \$80.000$

(c) $C(90) = \$720{,}000$ (d) $\displaystyle\lim_{p \to 100^-} \dfrac{80{,}000p}{100 - p} = \infty$

Problem Solving for Chapter 1

1. (a) Perimeter $\triangle PAO = \sqrt{x^2 + (y - 1)^2} + \sqrt{x^2 + y^2} + 1$

$\qquad = \sqrt{x^2 + (x^2 - 1)^2} + \sqrt{x^2 + x^4} + 1$

Perimeter $\triangle PBO = \sqrt{(x - 1)^2 + y^2} + \sqrt{x^2 + y^2} + 1$

$\qquad = \sqrt{(x - 1)^2 + x^4} + \sqrt{x^2 + x^4} + 1$

(c) $\displaystyle\lim_{x \to 0^+} r(x) = \dfrac{1 + 0 + 1}{1 + 0 + 1} = \dfrac{2}{2} = 1$

(b) $r(x) = \dfrac{\sqrt{x^2 + (x^2 - 1)^2} + \sqrt{x^2 + x^4} + 1}{\sqrt{(x - 1)^2 + x^4} + \sqrt{x^2 + x^4} + 1}$

x	4	2	1	0.1	0.01
Perimeter $\triangle PAO$	33.02	9.08	3.41	2.10	2.01
Perimeter $\triangle PBO$	33.77	9.60	3.41	2.00	2.00
$r(x)$	0.98	0.95	1	1.05	1.005

3. (a) There are 6 triangles, each with a central angle of $60° = \pi/3$. Hence,

$$\text{Area hexagon} = 6\left[\dfrac{1}{2}bh \right] = 6\left[\dfrac{1}{2}(1) \sin \dfrac{\pi}{3} \right]$$

$$= \dfrac{3\sqrt{3}}{2} \approx 2.598.$$

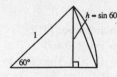
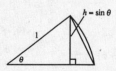

Error: $\pi - \dfrac{3\sqrt{3}}{2} \approx 0.5435$

(b) There are n triangles, each with central angle of $\theta = 2\pi/n$. Hence,

$$A_n = n\left[\dfrac{1}{2}bh \right] = n\left[\dfrac{1}{2}(1) \sin \dfrac{2\pi}{n} \right] = \dfrac{n \sin(2\pi/n)}{2}.$$

(c)

n	6	12	24	48	96
A_n	2.598	3	3.106	3.133	3.139

(d) As n gets larger and larger, $2\pi/n$ approaches 0.

Letting $x = 2\pi/n$,

$$A_n = \dfrac{\sin(2\pi/n)}{2/n} = \dfrac{\sin(2\pi/n)}{(2\pi/n)}\pi = \dfrac{\sin x}{x}\pi$$

which approaches $(1)\pi = \pi$.

5. (a) Slope $= -\dfrac{12}{5}$

(b) Slope of tangent line is $\dfrac{5}{12}$.

$$y + 12 = \frac{5}{12}(x - 5)$$

$$y = \frac{5}{12}x - \frac{169}{12} \text{ Tangent line}$$

(c) $Q = (x, y) = \left(x, -\sqrt{169 - x^2}\right)$

$$m_x = \frac{-\sqrt{169 - x^2} + 12}{x - 5}$$

(d) $\displaystyle\lim_{x \to 5} m_x = \lim_{x \to 5} \frac{12 - \sqrt{169 - x^2}}{x - 5} \cdot \frac{12 + \sqrt{169 - x^2}}{12 + \sqrt{169 - x^2}}$

$$= \lim_{x \to 5} \frac{144 - (169 - x^2)}{(x - 5)(12 + \sqrt{169 - x^2})}$$

$$= \lim_{x \to 5} \frac{x^2 - 25}{(x - 5)(12 + \sqrt{169 - x^2})}$$

$$= \lim_{x \to 5} \frac{(x + 5)}{12 + \sqrt{169 - x^2}} = \frac{10}{12 + 12} = \frac{5}{12}$$

This is the same slope as part (b).

7. (a) $3 + x^{1/3} \geq 0$

$$x^{1/3} \geq -3$$

$$x \geq -27$$

Domain: $x \geq -27, x \neq 1$

(b)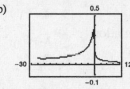

(c) $\displaystyle\lim_{x \to -27^+} f(x) = \frac{\sqrt{3 + (-27)^{1/3}} - 2}{-27 - 1}$

$$= \frac{-2}{-28} = \frac{1}{14} \approx 0.0714$$

(d) $\displaystyle\lim_{x \to 1} f(x) = \lim_{x \to 1} \frac{\sqrt{3 + x^{1/3}} - 2}{x - 1} \cdot \frac{\sqrt{3 + x^{1/3}} + 2}{\sqrt{3 + x^{1/3}} + 2}$

$$= \lim_{x \to 1} \frac{3 + x^{1/3} - 4}{(x - 1)\left(\sqrt{3 + x^{1/3}} + 2\right)}$$

$$= \lim_{x \to 1} \frac{x^{1/3} - 1}{(x^{1/3} - 1)(x^{2/3} + x^{1/3} + 1)\left(\sqrt{3 + x^{1/3}} + 2\right)}$$

$$= \lim_{x \to 1} \frac{1}{(x^{2/3} + x^{1/3} + 1)\left(\sqrt{3 + x^{1/3}} + 2\right)}$$

$$= \frac{1}{(1 + 1 + 1)(2 + 2)} = \frac{1}{12}$$

9. (a) $\displaystyle\lim_{x \to 2} f(x) = 3: g_1, g_4$ (b) f continuous at 2: g_1 (c) $\displaystyle\lim_{x \to 2^-} f(x) = 3: g_1, g_3, g_4$

11.

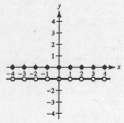

(a) $f(1) = [\![1]\!] + [\![-1]\!] = 1 + (-1) = 0$

$f(0) = 0$

$f\left(\tfrac{1}{2}\right) = 0 + (-1) = -1$

$f(-2.7) = -3 + 2 = -1$

(b) $\displaystyle\lim_{x \to 1^-} f(x) = -1$

$\displaystyle\lim_{x \to 1^+} f(x) = -1$

$\displaystyle\lim_{x \to 1/2} f(x) = -1$

(c) f is continuous for all real numbers except

$x = 0, \pm 1, \pm 2, \pm 3, \ldots$

13. (a)

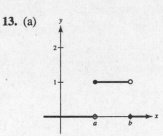

(b) (i) $\displaystyle\lim_{x \to a^+} P_{a,b}(x) = 1$

(ii) $\displaystyle\lim_{x \to a^-} P_{a,b}(x) = 0$

(iii) $\displaystyle\lim_{x \to b^+} P_{a,b}(x) = 0$

(iv) $\displaystyle\lim_{x \to b^-} P_{a,b}(x) = 1$

(c) $P_{a,b}$ is continuous for all positive real numbers except $x = a, b$.

(d) The area under the graph of U, and above the x axis, is 1.

CHAPTER 2
Differentiation

Section 2.1 The Derivative and the Tangent Line Problem 51

Section 2.2 Basic Differentiation Rules and Rates of Change . . . 59

Section 2.3 Product and Quotient Rules and
Higher-Order Derivatives 65

Section 2.4 The Chain Rule 73

Section 2.5 Implicit Differentiation 79

Section 2.6 Related Rates 87

Review Exercises 94

Problem Solving 100

CHAPTER 2
Differentiation

Section 2.1   The Derivative and the Tangent Line Problem

1. (a) At (x_1, y_1), slope $= 0$.

At (x_2, y_2), slope $\approx \frac{5}{2}$.

(b) At (x_1, y_1), slope $\approx -\frac{5}{2}$.

At (x_2, y_2), slope ≈ 2.

3. (a), (b)

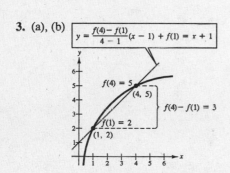

$$y = \frac{f(4) - f(1)}{4 - 1}(x - 1) + f(1) = x + 1$$

(c) $\quad y = \dfrac{f(4) - f(1)}{4 - 1}(x - 1) + f(1)$

$\qquad = \dfrac{3}{3}(x - 1) + 2$

$\qquad = 1(x - 1) + 2$

$\qquad = x + 1$

5. $f(x) = 3 - 2x$ is a line. Slope $= -2$

7. Slope at $(1, -3) = \lim\limits_{\Delta x \to 0} \dfrac{g(1 + \Delta x) - g(1)}{\Delta x}$

$\qquad = \lim\limits_{\Delta x \to 0} \dfrac{(1 + \Delta x)^2 - 4 - (-3)}{\Delta x}$

$\qquad = \lim\limits_{\Delta x \to 0} \dfrac{1 + 2(\Delta x) + (\Delta x)^2 - 1}{\Delta x}$

$\qquad = \lim\limits_{\Delta x \to 0} [2 + (\Delta x)] = 2$

9. Slope at $(0, 0) = \lim\limits_{\Delta t \to 0} \dfrac{f(0 + \Delta t) - f(0)}{\Delta t}$

$\qquad = \lim\limits_{\Delta t \to 0} \dfrac{3(\Delta t) - (\Delta t)^2 - 0}{\Delta t}$

$\qquad = \lim\limits_{\Delta t \to 0} (3 - \Delta t) = 3$

11. $f(x) = 3$

$f'(x) = \lim\limits_{\Delta x \to 0} \dfrac{f(x + \Delta x) - f(x)}{\Delta x}$

$\qquad = \lim\limits_{\Delta x \to 0} \dfrac{3 - 3}{\Delta x}$

$\qquad = \lim\limits_{\Delta x \to 0} 0 = 0$

13. $f(x) = -5x$

$f'(x) = \lim\limits_{\Delta x \to 0} \dfrac{f(x + \Delta x) - f(x)}{\Delta x}$

$\qquad = \lim\limits_{\Delta x \to 0} \dfrac{-5(x + \Delta x) - (-5x)}{\Delta x}$

$\qquad = \lim\limits_{\Delta x \to 0} -5 = -5$

15. $h(s) = 3 + \dfrac{2}{3}s$

$h'(s) = \lim\limits_{\Delta s \to 0} \dfrac{h(s + \Delta s) - h(s)}{\Delta s}$

$\qquad = \lim\limits_{\Delta s \to 0} \dfrac{3 + \frac{2}{3}(s + \Delta s) - \left(3 + \frac{2}{3}s\right)}{\Delta s}$

$\qquad = \lim\limits_{\Delta s \to 0} \dfrac{\frac{2}{3}\Delta s}{\Delta s} = \dfrac{2}{3}$

17. $f(x) = 2x^2 + x - 1$

$$f'(x) = \lim_{\Delta x \to 0} \frac{f(x + \Delta x) - f(x)}{\Delta x}$$

$$= \lim_{\Delta x \to 0} \frac{[2(x + \Delta x)^2 + (x + \Delta x) - 1] - [2x^2 + x - 1]}{\Delta x}$$

$$= \lim_{\Delta x \to 0} \frac{(2x^2 + 4x \Delta x + 2(\Delta x)^2 + x + \Delta x - 1) - (2x^2 + x - 1)}{\Delta x}$$

$$= \lim_{\Delta x \to 0} \frac{4x \Delta x + 2(\Delta x)^2 + \Delta x}{\Delta x} = \lim_{\Delta x \to 0} (4x + 2 \Delta x + 1) = 4x + 1$$

19. $f(x) = x^3 - 12x$

$$f'(x) = \lim_{\Delta x \to 0} \frac{f(x + \Delta x) - f(x)}{\Delta x}$$

$$= \lim_{\Delta x \to 0} \frac{[(x + \Delta x)^3 - 12(x + \Delta x)] - [x^3 - 12x]}{\Delta x}$$

$$= \lim_{\Delta x \to 0} \frac{x^3 + 3x^2 \Delta x + 3x(\Delta x)^2 + (\Delta x)^3 - 12x - 12 \Delta x - x^3 + 12x}{\Delta x}$$

$$= \lim_{\Delta x \to 0} \frac{3x^2 \Delta x + 3x(\Delta x)^2 + (\Delta x)^3 - 12 \Delta x}{\Delta x}$$

$$= \lim_{\Delta x \to 0} (3x^2 + 3x \Delta x + (\Delta x)^2 - 12) = 3x^2 - 12$$

21. $f(x) = \dfrac{1}{x - 1}$

$$f'(x) = \lim_{\Delta x \to 0} \frac{f(x + \Delta x) - f(x)}{\Delta x}$$

$$= \lim_{\Delta x \to 0} \frac{\dfrac{1}{x + \Delta x - 1} - \dfrac{1}{x - 1}}{\Delta x}$$

$$= \lim_{\Delta x \to 0} \frac{(x - 1) - (x + \Delta x - 1)}{\Delta x(x + \Delta x - 1)(x - 1)}$$

$$= \lim_{\Delta x \to 0} \frac{-\Delta x}{\Delta x(x + \Delta x - 1)(x - 1)}$$

$$= \lim_{\Delta x \to 0} \frac{-1}{(x + \Delta x - 1)(x - 1)}$$

$$= -\frac{1}{(x - 1)^2}$$

23. $f(x) = \sqrt{x + 1}$

$$f'(x) = \lim_{\Delta x \to 0} \frac{f(x + \Delta x) - f(x)}{\Delta x}$$

$$= \lim_{\Delta x \to 0} \frac{\sqrt{x + \Delta x + 1} - \sqrt{x + 1}}{\Delta x} \cdot \left(\frac{\sqrt{x + \Delta x + 1} + \sqrt{x + 1}}{\sqrt{x + \Delta x + 1} + \sqrt{x + 1}} \right)$$

$$= \lim_{\Delta x \to 0} \frac{(x + \Delta x + 1) - (x + 1)}{\Delta x \left[\sqrt{x + \Delta x + 1} + \sqrt{x + 1} \right]}$$

$$= \lim_{\Delta x \to 0} \frac{1}{\sqrt{x + \Delta x + 1} + \sqrt{x + 1}} = \frac{1}{\sqrt{x + 1} + \sqrt{x + 1}} = \frac{1}{2\sqrt{x + 1}}$$

25. (a) $f(x) = x^2 + 1$

$$f'(x) = \lim_{\Delta x \to 0} \frac{f(x + \Delta x) - f(x)}{\Delta x}$$

$$= \lim_{\Delta x \to 0} \frac{[(x + \Delta x)^2 + 1] - [x^2 + 1]}{\Delta x}$$

$$= \lim_{\Delta x \to 0} \frac{2x \Delta x + (\Delta x)^2}{\Delta x}$$

$$= \lim_{\Delta x \to 0} (2x + \Delta x) = 2x$$

At $(2, 5)$, the slope of the tangent line is $m = 2(2) = 4$.
The equation of the tangent line is

$$y - 5 = 4(x - 2)$$

$$y - 5 = 4x - 8$$

$$y = 4x - 3.$$

(b)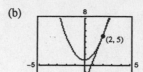

27. (a) $f(x) = x^3$

$$f'(x) = \lim_{\Delta x \to 0} \frac{f(x + \Delta x) - f(x)}{\Delta x}$$

$$= \lim_{\Delta x \to 0} \frac{(x + \Delta x)^3 - x^3}{\Delta x}$$

$$= \lim_{\Delta x \to 0} \frac{3x^2 \Delta x + 3x(\Delta x)^2 + (\Delta x)^3}{\Delta x}$$

$$= \lim_{\Delta x \to 0} (3x^2 + 3x \Delta x + (\Delta x)^2) = 3x^2$$

At $(2, 8)$, the slope of the tangent is $m = 3(2)^2 = 12$.
The equation of the tangent line is

$$y - 8 = 12(x - 2)$$

$$y = 12x - 16.$$

(b)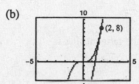

29. (a) $f(x) = \sqrt{x}$

$$f'(x) = \lim_{\Delta x \to 0} \frac{f(x + \Delta x) - f(x)}{\Delta x}$$

$$= \lim_{\Delta x \to 0} \frac{\sqrt{x + \Delta x} - \sqrt{x}}{\Delta x} \cdot \frac{\sqrt{x + \Delta x} + \sqrt{x}}{\sqrt{x + \Delta x} + \sqrt{x}}$$

$$= \lim_{\Delta x \to 0} \frac{(x + \Delta x) - x}{\Delta x(\sqrt{x + \Delta x} + \sqrt{x})}$$

$$= \lim_{\Delta x \to 0} \frac{1}{\sqrt{x + \Delta x} + \sqrt{x}} = \frac{1}{2\sqrt{x}}$$

At $(1, 1)$, the slope of the tangent line is

$$m = \frac{1}{2\sqrt{1}} = \frac{1}{2}.$$

The equation of the tangent line is

$$y - 1 = \frac{1}{2}(x - 1)$$

$$y = \frac{1}{2}x + \frac{1}{2}.$$

(b)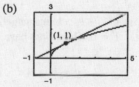

31. (a) $f(x) = x + \dfrac{4}{x}$

$$f'(x) = \lim_{\Delta x \to 0} \frac{f(x + \Delta x) - f(x)}{\Delta x}$$

$$= \lim_{\Delta x \to 0} \frac{(x + \Delta x) + \dfrac{4}{x + \Delta x} - \left(x + \dfrac{4}{x}\right)}{\Delta x}$$

$$= \lim_{\Delta x \to 0} \frac{x(x + \Delta x)(x + \Delta x) + 4x - x^2(x + \Delta x) - 4(x + \Delta x)}{x(\Delta x)(x + \Delta x)}$$

$$= \lim_{\Delta x \to 0} \frac{x^3 + 2x^2(\Delta x) + x(\Delta x)^2 - x^3 - x^2(\Delta x) - 4(\Delta x)}{x(\Delta x)(x + \Delta x)}$$

$$= \lim_{\Delta x \to 0} \frac{x^2(\Delta x) + x(\Delta x)^2 - 4(\Delta x)}{x(\Delta x)(x + \Delta x)}$$

$$= \lim_{\Delta x \to 0} \frac{x^2 + x(\Delta x) - 4}{x(x + \Delta x)}$$

$$= \frac{x^2 - 4}{x^2} = 1 - \frac{4}{x^2}$$

At $(4, 5)$, the slope of the tangent line is

$$m = 1 - \frac{4}{16} = \frac{3}{4}.$$

The equation of the tangent line is

$$y - 5 = \frac{3}{4}(x - 4)$$

$$y = \frac{3}{4}x + 2.$$

(b)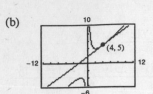

33. From Exercise 27 we know that $f'(x) = 3x^2$. Since the slope of the given line is 3, we have

$$3x^2 = 3$$
$$x = \pm 1.$$

Therefore, at the points $(1, 1)$ and $(-1, -1)$ the tangent lines are parallel to $3x - y + 1 = 0$. These lines have equations

$$y - 1 = 3(x - 1) \qquad \text{and} \qquad y + 1 = 3(x + 1)$$
$$y = 3x - 2 \qquad\qquad\qquad\qquad y = 3x + 2.$$

35. Using the limit definition of derivative,

$$f'(x) = \frac{-1}{2x\sqrt{x}}.$$

Since the slope of the given line is $-\frac{1}{2}$, we have

$$-\frac{1}{2x\sqrt{x}} = -\frac{1}{2}$$
$$x = 1.$$

Therefore, at the point $(1, 1)$ the tangent line is parallel to $x + 2y - 6 = 0$. The equation of this line is

$$y - 1 = -\frac{1}{2}(x - 1)$$

$$y - 1 = -\frac{1}{2}x + \frac{1}{2}$$

$$y = -\frac{1}{2}x + \frac{3}{2}.$$

37. $f(x) = x \implies f'(x) = 1$ Matches (b).

39. $f(x) = \sqrt{x} \implies f'(x)$ Matches (a).

(decreasing slope as $x \to \infty$)

41. $g(5) = 2$ because the tangent line passes through $(5, 2)$.

$$g'(5) = \frac{2 - 0}{5 - 9} = \frac{2}{-4} = -\frac{1}{2}$$

43. The slope of the graph of f is $1 \implies f'(x) = 1$.

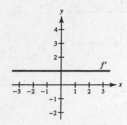

45. The slope of the graph of f is negative for $x < 4$, positive for $x > 4$, and 0 at $x = 4$.

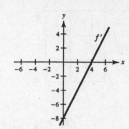

47. Answers will vary. Sample answer: $y = -x$

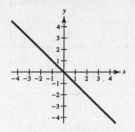

49. $f(x) = 5 - 3x$ and $c = 1$

51. $f(x) = -x^2$ and $c = 6$

53. $f(0) = 2$ and $f'(x) = -3$, $-\infty < x < \infty$

$f(x) = -3x + 2$

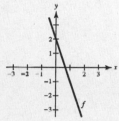

55. $f(0) = 0$; $f'(0) = 0$; $f'(x) > 0$ if $x \neq 0$

$f(x) = x^3$

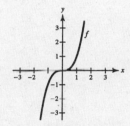

57. Let (x_0, y_0) be a point of tangency on the graph of f. By the limit definition for the derivative, $f'(x) = 4 - 2x$. The slope of the line through $(2, 5)$ and (x_0, y_0) equals the derivative of f at x_0:

$$\frac{5 - y_0}{2 - x_0} = 4 - 2x_0$$

$$5 - y_0 = (2 - x_0)(4 - 2x_0)$$

$$5 - (4x_0 - x_0^2) = 8 - 8x_0 + 2x_0^2$$

$$0 = x_0^2 - 4x_0 + 3$$

$$0 = (x_0 - 1)(x_0 - 3) \implies x_0 = 1, 3$$

Therefore, the points of tangency are $(1, 3)$ and $(3, 3)$, and the corresponding slopes are 2 and -2. The equations of the tangent lines are:

$$y - 5 = 2(x - 2) \qquad y - 5 = -2(x - 2)$$
$$y = 2x + 1 \qquad\qquad y = -2x + 9$$

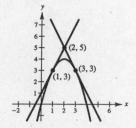

59. (a) $g'(0) = -3$

(b) $g'(3) = 0$

(c) Because $g'(1) = -\frac{8}{3}$, g is decreasing (falling) at $x = 1$.

(d) Because $g'(-4) = \frac{7}{3}$, g is increasing (rising) at $x = -4$.

(e) Because $g'(4)$ and $g'(6)$ are both positive, $g(6)$ is greater than $g(4)$, and $g(6) - g(4) > 0$.

(f) No, it is not possible. All you can say is that g is decreasing (falling) at $x = 2$.

61. $f(x) = \frac{1}{4}x^3$

By the limit definition of the derivative we have $f'(x) = \frac{3}{4}x^2$.

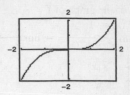

x	-2	-1.5	-1	-0.5	0	0.5	1	1.5	2
$f(x)$	-2	$-\frac{27}{32}$	$-\frac{1}{4}$	$-\frac{1}{32}$	0	$\frac{1}{32}$	$\frac{1}{4}$	$\frac{27}{32}$	2
$f'(x)$	3	$\frac{27}{16}$	$\frac{3}{4}$	$\frac{3}{16}$	0	$\frac{3}{16}$	$\frac{3}{4}$	$\frac{27}{16}$	3

63. $g(x) = \dfrac{f(x + 0.01) - f(x)}{0.01}$

$\quad = [2(x + 0.01) - (x + 0.01)^2 - 2x + x^2]100$

$\quad = 2 - 2x - 0.01$

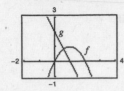

The graph of $g(x)$ is approximately
the graph of $f'(x) = 2 - 2x$.

65. $f(2) = 2(4 - 2) = 4$, $f(2.1) = 2.1(4 - 2.1) = 3.99$

$\quad f'(2) \approx \dfrac{3.99 - 4}{2.1 - 2} = -0.1$ [Exact: $f'(2) = 0$]

67. $f(x) = \dfrac{1}{\sqrt{x}}$ and $f'(x) = \dfrac{-1}{2x^{3/2}}$.

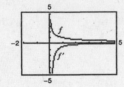

As $x \to \infty$, f is nearly horizontal
and thus $f' \approx 0$.

69. $f(x) = 4 - (x - 3)^2$

$\quad S_{\Delta x}(x) = \dfrac{f(2 + \Delta x) - f(2)}{\Delta x}(x - 2) + f(2)$

$\quad = \dfrac{4 - (2 + \Delta x - 3)^2 - 3}{\Delta x}(x - 2) + 3 = \dfrac{1 - (\Delta x - 1)^2}{\Delta x}(x - 2) + 3 = (-\Delta x + 2)(x - 2) + 3$

(a) $\Delta x = \quad 1$: $S_{\Delta x} = (x - 2) + 3 = x + 1$

$\quad \Delta x = 0.5$: $S_{\Delta x} = \left(\dfrac{3}{2}\right)(x - 2) + 3 = \dfrac{3}{2}x$

$\quad \Delta x = 0.1$: $S_{\Delta x} = \left(\dfrac{19}{10}\right)(x - 2) + 3 = \dfrac{19}{10}x - \dfrac{4}{5}$

(b) As $\Delta x \to 0$, the line approaches the tangent line to f at $(2, 3)$.

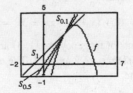

71. $f(x) = x^2 - 1$, $c = 2$

$\quad f'(2) = \lim_{x \to 2} \dfrac{f(x) - f(2)}{x - 2} = \lim_{x \to 2} \dfrac{(x^2 - 1) - 3}{x - 2} = \lim_{x \to 2} \dfrac{(x - 2)(x + 2)}{x - 2} = \lim_{x \to 2}(x + 2) = 4$

73. $f(x) = x^3 + 2x^2 + 1$, $c = -2$

$\quad f'(-2) = \lim_{x \to -2} \dfrac{f(x) - f(-2)}{x + 2} = \lim_{x \to -2} \dfrac{(x^3 + 2x^2 + 1) - 1}{x + 2} = \lim_{x \to -2} \dfrac{x^2(x + 2)}{x + 2} = \lim_{x \to -2} x^2 = 4$

75. $g(x) = \sqrt{|x|}, c = 0$

$$g'(0) = \lim_{x \to 0} \frac{g(x) - g(0)}{x - 0} = \lim_{x \to 0} \frac{\sqrt{|x|}}{x}. \text{ Does not exist.}$$

As $x \to 0^-$, $\dfrac{\sqrt{|x|}}{x} = \dfrac{-1}{\sqrt{|x|}} \to -\infty$.

As $x \to 0^+$, $\dfrac{\sqrt{|x|}}{x} = \dfrac{1}{\sqrt{x}} \to \infty$.

77. $f(x) = (x - 6)^{2/3}, c = 6$

$$f'(6) = \lim_{x \to 6} \frac{f(x) - f(6)}{x - 6} = \lim_{x \to 6} \frac{(x - 6)^{2/3} - 0}{x - 6} = \lim_{x \to 6} \frac{1}{(x - 6)^{1/3}}$$

Does not exist.

79. $h(x) = |x + 5|, c = -5$

$$h'(-5) = \lim_{x \to -5} \frac{h(x) - h(-5)}{x - (-5)} = \lim_{x \to -5} \frac{|x + 5| - 0}{x + 5} = \lim_{x \to -5} \frac{|x + 5|}{x + 5}$$

Does not exist.

81. $f(x)$ is differentiable everywhere except at $x = -1$. (Discontinuity)

83. $f(x)$ is differentiable everywhere except at $x = 3$. (Sharp turn in the graph)

85. $f(x)$ is differentiable on the interval $(1, \infty)$. (At $x = 1$ the tangent line is vertical.)

87. $f(x) = |x + 3|$ is differentiable for all $x \neq -3$. There is a sharp corner at $x = -3$.

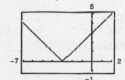

89. $f(x) = x^{2/5}$ is differentiable for all $x \neq 0$. There is a sharp corner at $x = 0$.

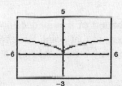

91. $f(x) = |x - 1|$

The derivative from the left is

$$\lim_{x \to 1^-} \frac{f(x) - f(1)}{x - 1} = \lim_{x \to 1^-} \frac{|x - 1| - 0}{x - 1} = -1.$$

The derivative from the right is

$$\lim_{x \to 1^+} \frac{f(x) - f(1)}{x - 1} = \lim_{x \to 1^+} \frac{|x - 1| - 0}{x - 1} = 1.$$

The one-sided limits are not equal. Therefore, f is not differentiable at $x = 1$.

93. $f(x) = \begin{cases} (x - 1)^3, & x \leq 1 \\ (x - 1)^2, & x > 1 \end{cases}$

The derivative from the left is

$$\lim_{x \to 1^-} \frac{f(x) - f(1)}{x - 1} = \lim_{x \to 1^-} \frac{(x - 1)^3 - 0}{x - 1}$$

$$= \lim_{x \to 1^-} (x - 1)^2 = 0.$$

The derivative from the right is

$$\lim_{x \to 1^+} \frac{f(x) - f(1)}{x - 1} = \lim_{x \to 1^+} \frac{(x - 1)^2 - 0}{x - 1}$$

$$= \lim_{x \to 1^+} (x - 1) = 0.$$

These one-sided limits are equal. Therefore, f is differentiable at $x = 1$. ($f'(1) = 0$)

95. Note that f is continuous at $x = 2$. $f(x) = \begin{cases} x^2 + 1, & x \leq 2 \\ 4x - 3, & x > 2 \end{cases}$

The derivative from the left is $\lim\limits_{x \to 2^-} \dfrac{f(x) - f(2)}{x - 2} = \lim\limits_{x \to 2^-} \dfrac{(x^2 + 1) - 5}{x - 2} = \lim\limits_{x \to 2^-} (x + 2) = 4$.

The derivative from the right is $\lim\limits_{x \to 2^+} \dfrac{f(x) - f(2)}{x - 2} = \lim\limits_{x \to 2^+} \dfrac{(4x - 3) - 5}{x - 2} = \lim\limits_{x \to 2^+} 4 = 4$.

The one-sided limits are equal. Therefore, f is differentiable at $x = 2$. $(f'(2) = 4)$

97. (a) The distance from $(3, 1)$ to the line $mx - y + 4 = 0$ is

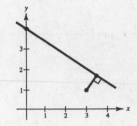

$$d = \frac{|Ax_1 + By_1 + C|}{\sqrt{A^2 + B^2}},$$

$$= \frac{|m(3) - 1(1) + 4|}{\sqrt{m^2 + 1}} = \frac{|3m + 3|}{\sqrt{m^2 + 1}}.$$

(b)

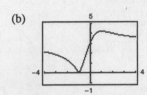

The function d is not differentiable at $m = -1$. This corresponds to the line $y = -x + 4$, which passes through the point $(3, 1)$.

99. False. The slope is $\lim\limits_{\Delta x \to 0} \dfrac{f(2 + \Delta x) - f(2)}{\Delta x}$.

101. False. If the derivative from the left of a point does not equal the derivative from the right of a point, then the derivative does not exist at that point. For example, if $f(x) = |x|$, then the derivative from the left at $x = 0$ is -1 and the derivative from the right at $x = 0$ is 1. At $x = 0$, the derivative does not exist.

103. $f(x) = \begin{cases} x \sin(1/x), & x \neq 0 \\ 0, & x = 0 \end{cases}$

Using the Squeeze Theorem, we have $-|x| \leq x \sin(1/x) \leq |x|$, $x \neq 0$. Thus, $\lim\limits_{x \to 0} x \sin(1/x) = 0 = f(0)$ and f is continuous at $x = 0$. Using the alternative form of the derivative, we have

$$\lim\limits_{x \to 0} \frac{f(x) - f(0)}{x - 0} = \lim\limits_{x \to 0} \frac{x \sin(1/x) - 0}{x - 0} = \lim\limits_{x \to 0} \left(\sin\frac{1}{x} \right).$$

Since this limit does not exist ($\sin(1/x)$ oscillates between -1 and 1), the function is not differentiable at $x = 0$.

$g(x) = \begin{cases} x^2 \sin(1/x), & x \neq 0 \\ 0, & x = 0 \end{cases}$

Using the Squeeze Theorem again, we have $-x^2 \leq x^2 \sin(1/x) \leq x^2$, $x \neq 0$. Thus, $\lim\limits_{x \to 0} x^2 \sin(1/x) = 0 = g(0)$ and g is continuous at $x = 0$. Using the alternative form of the derivative again, we have

$$\lim\limits_{x \to 0} \frac{g(x) - g(0)}{x - 0} = \lim\limits_{x \to 0} \frac{x^2 \sin(1/x) - 0}{x - 0} = \lim\limits_{x \to 0} x \sin\frac{1}{x} = 0.$$

Therefore, g is differentiable at $x = 0$, $g'(0) = 0$.

Section 2.2   Basic Differentiation Rules and Rates of Change

1. (a) $y = x^{1/2}$

   $y' = \frac{1}{2}x^{-1/2}$

   $y'(1) = \frac{1}{2}$

   (b) $y = x^3$

   $y' = 3x^2$

   $y'(1) = 3$

3. $y = 8$

   $y' = 0$

5. $y = x^6$

   $y' = 6x^5$

7. $y = \frac{1}{x^7} = x^{-7}$

   $y' = -7x^{-8} = \frac{-7}{x^8}$

9. $y = \sqrt[5]{x} = x^{1/5}$

   $y' = \frac{1}{5}x^{-4/5} = \frac{1}{5x^{4/5}}$

11. $f(x) = x + 1$

   $f'(x) = 1$

13. $f(t) = -2t^2 + 3t - 6$

   $f'(t) = -4t + 3$

15. $g(x) = x^2 + 4x^3$

   $g'(x) = 2x + 12x^2$

17. $s(t) = t^3 - 2t + 4$

   $s'(t) = 3t^2 - 2$

19. $y = \frac{\pi}{2}\sin\theta - \cos\theta$

   $y' = \frac{\pi}{2}\cos\theta + \sin\theta$

21. $y = x^2 - \frac{1}{2}\cos x$

   $y' = 2x + \frac{1}{2}\sin x$

23. $y = \frac{1}{x} - 3\sin x$

   $y' = -\frac{1}{x^2} - 3\cos x$

Function	Rewrite	Differentiate	Simplify
25. $y = \dfrac{5}{2x^2}$	$y = \dfrac{5}{2}x^{-2}$	$y' = -5x^{-3}$	$y' = \dfrac{-5}{x^3}$
27. $y = \dfrac{3}{(2x)^3}$	$y = \dfrac{3}{8}x^{-3}$	$y' = \dfrac{-9}{8}x^{-4}$	$y' = \dfrac{-9}{8x^4}$
29. $y = \dfrac{\sqrt{x}}{x}$	$y = x^{-1/2}$	$y' = -\dfrac{1}{2}x^{-3/2}$	$y' = -\dfrac{1}{2x^{3/2}}$

31. $f(x) = \frac{3}{x^2} = 3x^{-2}$, $(1, 3)$

   $f'(x) = -6x^{-3} = \frac{-6}{x^3}$

   $f'(1) = -6$

33. $f(x) = -\frac{1}{2} + \frac{7}{5}x^3$, $\left(0, -\frac{1}{2}\right)$

   $f'(x) = \frac{21}{5}x^2$

   $f'(0) = 0$

35. $y = (2x + 1)^2$, $(0, 1)$

   $= 4x^2 + 4x + 1$

   $y' = 8x + 4$

   $y'(0) = 4$

37. $f(\theta) = 4\sin\theta - \theta$, $(0, 0)$

   $f'(\theta) = 4\cos\theta - 1$

   $f'(0) = 4(1) - 1 = 3$

39. $f(x) = x^2 + 5 - 3x^{-2}$

   $f'(x) = 2x + 6x^{-3} = 2x + \frac{6}{x^3}$

41. $g(t) = t^2 - \frac{4}{t^3} = t^2 - 4t^{-3}$

   $g'(t) = 2t + 12t^{-4} = 2t + \frac{12}{t^4}$

43. $f(x) = \frac{x^3 - 3x^2 + 4}{x^2} = x - 3 + 4x^{-2}$

   $f'(x) = 1 - \frac{8}{x^3} = \frac{x^3 - 8}{x^3}$

45. $y = x(x^2 + 1) = x^3 + x$

   $y' = 3x^2 + 1$

47. $f(x) = \sqrt{x} - 6\sqrt[3]{x} = x^{1/2} - 6x^{1/3}$

$f'(x) = \dfrac{1}{2}x^{-1/2} - 2x^{-2/3} = \dfrac{1}{2\sqrt{x}} - \dfrac{2}{x^{2/3}}$

49. $h(s) = s^{4/5} - s^{2/3}$

$h'(s) = \dfrac{4}{5}s^{-1/5} - \dfrac{2}{3}s^{-1/3} = \dfrac{4}{5s^{1/5}} - \dfrac{2}{3s^{1/3}}$

51. $f(x) = 6\sqrt{x} + 5\cos x = 6x^{1/2} + 5\cos x$

$f'(x) = 3x^{-1/2} - 5\sin x = \dfrac{3}{\sqrt{x}} - 5\sin x$

53. (a) $y = x^4 - 3x^2 + 2$

$y' = 4x^3 - 6x$

At $(1, 0)$: $y' = 4(1)^3 - 6(1) = -2$

Tangent line: $y - 0 = -2(x - 1)$

$\qquad\qquad\qquad 2x + y - 2 = 0$

(b)

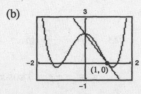

55. (a) $f(x) = \dfrac{2}{\sqrt[4]{x^3}} = 2x^{-3/4}$

$f'(x) = \dfrac{-3}{2}x^{-7/4} = \dfrac{-3}{2x^{7/4}}$

At $(1, 2)$: $f'(1) = \dfrac{-3}{2}$

Tangent line: $y - 2 = -\dfrac{3}{2}(x - 1)$

$\qquad\qquad\qquad y = -\dfrac{3}{2}x + \dfrac{7}{2}$

$\qquad\qquad\qquad 3x + 2y - 7 = 0$

(b)

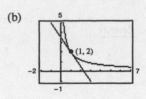

57. $y = x^4 - 8x^2 + 2$

$y' = 4x^3 - 16x$

$\quad = 4x(x^2 - 4)$

$\quad = 4x(x - 2)(x + 2)$

$y' = 0 \implies x = 0, \pm 2$

Horizontal tangents: $(0, 2), (2, -14), (-2, -14)$

59. $y = \dfrac{1}{x^2} = x^{-2}$

$y' = -2x^{-3} = \dfrac{-2}{x^3}$ cannot equal zero.

Therefore, there are no horizontal tangents.

61. $y = x + \sin x, 0 \le x < 2\pi$

$y' = 1 + \cos x = 0$

$\cos x = -1 \implies x = \pi$

At $x = \pi$: $y = \pi$

Horizontal tangent: (π, π)

63. $x^2 - kx = 4x - 9$ Equate functions.

$\quad 2x - k = 4$ Equate derivatives.

Hence, $k = 2x - 4$ and

$x^2 - (2x - 4)x = 4x - 9 \implies -x^2 = -9 \implies x = \pm 3$.

For $x = 3, k = 2$ and for $x = -3, k = -10$.

65. $\dfrac{k}{x} = -\dfrac{3}{4}x + 3$ Equate functions.

$-\dfrac{k}{x^2} = -\dfrac{3}{4}$ Equate derivatives.

Hence, $k = \dfrac{3}{4}x^2$ and

$\dfrac{\frac{3}{4}x^2}{x} = \dfrac{-3}{4}x + 3 \Longrightarrow \dfrac{3}{4}x = -\dfrac{3}{4}x + 3$

$\Longrightarrow \dfrac{3}{2}x = 3 \Longrightarrow x = 2 \Longrightarrow k = 3.$

67. (a) The slope appears to be steepest between A and B.

(b) The average rate of change between A and B is **greater** than the instantaneous rate of change at B.

(c)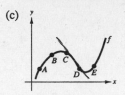

69. $g(x) = f(x) + 6 \Longrightarrow g'(x) = f'(x)$

71.

If f is linear then its derivative is a constant function.

$f(x) = ax + b$

$f'(x) = a$

73. Let (x_1, y_1) and (x_2, y_2) be the points of tangency on $y = x^2$ and $y = -x^2 + 6x - 5$, respectively. The derivatives of these functions are:

$y' = 2x \Longrightarrow m = 2x_1$ and $y' = -2x + 6 \Longrightarrow m = -2x_2 + 6$

$m = 2x_1 = -2x_2 + 6$

$x_1 = -x_2 + 3$

Since $y_1 = x_1^2$ and $y_2 = -x_2^2 + 6x_2 - 5$:

$m = \dfrac{y_2 - y_1}{x_2 - x_1} = \dfrac{(-x_2^2 + 6x_2 - 5) - (x_1^2)}{x_2 - x_1} = -2x_2 + 6$

$\dfrac{(-x_2^2 + 6x_2 - 5) - (-x_2 + 3)^2}{x_2 - (-x_2 + 3)} = -2x_2 + 6$

$(-x_2^2 + 6x_2 - 5) - (x_2^2 - 6x_2 + 9) = (-2x_2 + 6)(2x_2 - 3)$

$-2x_2^2 + 12x_2 - 14 = -4x_2^2 + 18x_2 - 18$

$2x_2^2 - 6x_2 + 4 = 0$

$2(x_2 - 2)(x_2 - 1) = 0$

$x_2 = 1 \text{ or } 2$

$x_2 = 1 \Longrightarrow y_2 = 0, \ x_1 = 2 \text{ and } y_1 = 4$

Thus, the tangent line through $(1, 0)$ and $(2, 4)$ is

$y - 0 = \left(\dfrac{4 - 0}{2 - 1}\right)(x - 1) \Longrightarrow y = 4x - 4.$

$x_2 = 2 \Longrightarrow y_2 = 3, \ x_1 = 1 \text{ and } y_1 = 1$

Thus, the tangent line through $(2, 3)$ and $(1, 1)$ is

$y - 1 = \left(\dfrac{3 - 1}{2 - 1}\right)(x - 1) \Longrightarrow y = 2x - 1.$

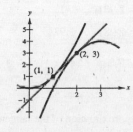

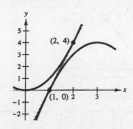

75. $f(x) = 3x + \sin x + 2$

$f'(x) = 3 + \cos x$

Since $|\cos x| \le 1$, $f'(x) \ne 0$ for all x and f does not have a horizontal tangent line.

77. $f(x) = \sqrt{x}$, $(-4, 0)$

$f'(x) = \dfrac{1}{2}x^{-1/2} = \dfrac{1}{2\sqrt{x}}$

$\dfrac{1}{2\sqrt{x}} = \dfrac{0 - y}{-4 - x}$

$4 + x = 2\sqrt{x}\,y$

$4 + x = 2\sqrt{x}\sqrt{x}$

$4 + x = 2x$

$x = 4, y = 2$

The point $(4, 2)$ is on the graph of f.

Tangent line: $y - 2 = \dfrac{0 - 2}{-4 - 4}(x - 4)$

$4y - 8 = x - 4$

$0 = x - 4y + 4$

79. $f'(1) = -1$

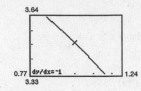

81. (a) One possible secant is between $(3.9, 7.7019)$ and $(4, 8)$:

$y - 8 = \dfrac{8 - 7.7019}{4 - 3.9}(x - 4)$

$y - 8 = 2.981(x - 4)$

$y = S(x) = 2.981x - 3.924$

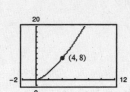

(b) $f'(x) = \dfrac{3}{2}x^{1/2} \implies f'(4) = \dfrac{3}{2}(2) = 3$

$T(x) = 3(x - 4) + 8 = 3x - 4$

$S(x)$ is an approximation of the tangent line $T(x)$.

(c) As you move further away from $(4, 8)$, the accuracy of the approximation T gets worse.

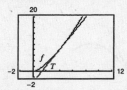

(d)

Δx	-3	-2	-1	-0.5	-0.1	0	0.1	0.5	1	2	3
$f(4 + \Delta x)$	1	2.828	5.196	6.548	7.702	8	8.302	9.546	11.180	14.697	18.520
$T(4 + \Delta x)$	-1	2	5	6.5	7.7	8	8.3	9.5	11	14	17

83. False. Let $f(x) = x^2$ and $g(x) = x^2 + 4$. Then $f'(x) = g'(x) = 2x$, but $f(x) \ne g(x)$.

85. False. If $y = \pi^2$, then $dy/dx = 0$. (π^2 is a constant.)

87. True. If $g(x) = 3f(x)$, then $g'(x) = 3f'(x)$.

89. $f(t) = 2t + 7, [1, 2]$

$f'(t) = 2$

Instantaneous rate of change is the constant 2.
Average rate of change:

$$\frac{f(2) - f(1)}{2 - 1} = \frac{[2(2) + 7] - [2(1) + 7]}{1} = 2$$

(These are the same because f is a line of slope 2.)

91. $f(x) = -\frac{1}{x}, \ [1, 2]$

$f'(x) = \frac{1}{x^2}$

Instantaneous rate of change:

$$(1, -1) \implies f'(1) = 1$$

$$\left(2, -\frac{1}{2}\right) \implies f'(2) = \frac{1}{4}$$

Average rate of change:

$$\frac{f(2) - f(1)}{2 - 1} = \frac{(-1/2) - (-1)}{2 - 1} = \frac{1}{2}$$

93. (a) $s(t) = -16t^2 + 1362$

$v(t) = -32t$

(b) $\dfrac{s(2) - s(1)}{2 - 1} = 1298 - 1346 = -48$ ft/sec

(c) $v(t) = s'(t) = -32t$

When $t = 1$: $v(1) = -32$ ft/sec

When $t = 2$: $v(2) = -64$ ft/sec

(d) $-16t^2 + 1362 = 0$

$$t^2 = \frac{1362}{16} \implies t = \frac{\sqrt{1362}}{4} \approx 9.226 \text{ sec}$$

(e) $v\left(\dfrac{\sqrt{1362}}{4}\right) = -32\left(\dfrac{\sqrt{1362}}{4}\right)$

$$= -8\sqrt{1362} \approx -295.242 \text{ ft/sec}$$

95. $s(t) = -4.9t^2 + v_0 t + s_0$

$= -4.9t^2 + 120t$

$v(t) = -9.8t + 120$

$v(5) = -9.8(5) + 120 = 71$ m/sec

$v(10) = -9.8(10) + 120 = 22$ m/sec

97. From $(0, 0)$ to $(4, 2)$, $s(t) = \frac{1}{2}t \implies v(t) = \frac{1}{2}$ mi/min.

$v(t) = \frac{1}{2}(60) = 30$ mph for $0 < t < 4$

Similarly, $v(t) = 0$ for $4 < t < 6$. Finally, from $(6, 2)$ to $(10, 6)$,

$s(t) = t - 4 \implies v(t) = 1$ mi/in = 60 mph.

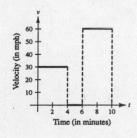

99. $v = 40$ mph $= \frac{2}{3}$ mi/min

$\left(\frac{2}{3} \text{ mi/min}\right)(6 \text{ min}) = 4$ mi

$v = 0$ mph $= 0$ mi/min

$(0 \text{ mi/min})(2 \text{ min}) = 0$ mi

$v = 60$ mph $= 1$ mi/min

$(1 \text{ mi/min})(2 \text{ min}) = 2$ mi

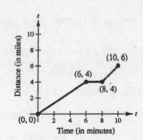

101. (a) Using a graphing utility,

$$R = 0.417v - 0.02.$$

(b) Using a graphing utility,

$$B = 0.00557v^2 + 0.0014v + 0.04.$$

(c) $T = R + B = 0.00557v^2 + 0.418v + 0.02$

(d)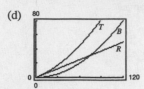

(e) $\dfrac{dT}{dv} = 0.01114v + 0.418$

For $v = 40$, $T'(40) \approx 0.86$.

For $v = 80$, $T'(80) \approx 1.31$.

For $v = 100$, $T'(100) \approx 1.53$.

(f) For increasing speeds, the total stopping distance increases.

103. $V = s^3, \dfrac{dV}{ds} = 3s^2$

When $s = 4$ cm, $\dfrac{dV}{ds} = 48$ cm^2 per cm change in s.

105. $s(t) = -\dfrac{1}{2}at^2 + c$ and $s'(t) = -at$

Average velocity: $\dfrac{s(t_0 + \Delta t) - s(t_0 - \Delta t)}{(t_0 + \Delta t) - (t_0 - \Delta t)} = \dfrac{[-(1/2)a(t_0 + \Delta t)^2 + c] - [-(1/2)a(t_0 - \Delta t)^2 + c]}{2\Delta t}$

$$= \dfrac{-(1/2)a(t_0^2 + 2t_0\Delta t + (\Delta t)^2) + (1/2)a(t_0^2 - 2t_0\Delta t + (\Delta t)^2)}{2\,\Delta t}$$

$$= \dfrac{-2at_0\,\Delta t}{2\,\Delta t}$$

$$= -at_0$$

$$= s'(t_0) \quad \text{instantaneous velocity at } t = t_0$$

107. $N = f(p)$

(a) $f'(1.479)$ is the rate of change of gallons of gasoline sold when the price is \$1.479 per gallon.

(b) $f'(1.479)$ is usually negative. As prices go up, sales go down.

109. $y = ax^2 + bx + c$

Since the parabola passes through $(0, 1)$ and $(1, 0)$, we have:

$(0, 1)$: $1 = a(0)^2 + b(0) + c \Rightarrow c = 1$

$(1, 0)$: $0 = a(1)^2 + b(1) + 1 \Rightarrow b = -a - 1$

Thus, $y = ax^2 + (-a - 1)x + 1$. From the tangent line $y = x - 1$, we know that the derivative is 1 at the point $(1, 0)$.

$$y' = 2ax + (-a - 1)$$

$$1 = 2a(1) + (-a - 1)$$

$$1 = a - 1$$

$$a = 2$$

$$b = -a - 1 = -3$$

Therefore, $y = 2x^2 - 3x + 1$.

111. $y = x^3 - 9x$

$y' = 3x^2 - 9$

Tangent lines through $(1, -9)$:

$$y + 9 = (3x^2 - 9)(x - 1)$$
$$(x^3 - 9x) + 9 = 3x^3 - 3x^2 - 9x + 9$$
$$0 = 2x^3 - 3x^2 = x^2(2x - 3)$$
$$x = 0 \text{ or } x = \tfrac{3}{2}$$

The points of tangency are $(0, 0)$ and $\left(\tfrac{3}{2}, -\tfrac{81}{8}\right)$. At $(0, 0)$, the slope is $y'(0) = -9$. At $\left(\tfrac{3}{2}, -\tfrac{81}{8}\right)$, the slope is $y'\left(\tfrac{3}{2}\right) = -\tfrac{9}{4}$.

Tangent lines:

$$y - 0 = -9(x - 0) \quad \text{and} \qquad y + \tfrac{81}{8} = -\tfrac{9}{4}\left(x - \tfrac{3}{2}\right)$$
$$y = -9x \qquad\qquad\qquad\qquad y = -\tfrac{9}{4}x - \tfrac{27}{4}$$
$$9x + y = 0 \qquad\qquad\quad 9x + 4y + 27 = 0$$

113. $f(x) = \begin{cases} ax^3, & x \le 2 \\ x^2 + b, & x > 2 \end{cases}$

f must be continuous at $x = 2$ to be differentiable at $x = 2$.

$$\left.\begin{array}{l} \lim\limits_{x \to 2^-} f(x) = \lim\limits_{x \to 2^-} ax^3 = 8a \\[2mm] \lim\limits_{x \to 2^+} f(x) = \lim\limits_{x \to 2^+} (x^2 + b) = 4 + b \end{array}\right\} \quad \begin{array}{l} 8a = 4 + b \\[2mm] 8a - 4 = b \end{array}$$

$$f'(x) = \begin{cases} 3ax^2, & x < 2 \\ 2x, & x > 2 \end{cases}$$

For f to be differentiable at $x = 2$, the left derivative must equal the right derivative.

$$3a(2)^2 = 2(2)$$
$$12a = 4$$
$$a = \tfrac{1}{3}$$
$$b = 8a - 4 = -\tfrac{4}{3}$$

115. $f_1(x) = |\sin x|$ is differentiable for all $x \ne n\pi$, n an integer.

$f_2(x) = \sin|x|$ is differentiable for all $x \ne 0$.

You can verify this by graphing f_1 and f_2 and observing the locations of the sharp turns.

Section 2.3   Product and Quotient Rules and Higher-Order Derivatives

1. $g(x) = (x^2 + 1)(x^2 - 2x)$

$g'(x) = (x^2 + 1)(2x - 2) + (x^2 - 2x)(2x)$

$\qquad = 2x^3 - 2x^2 + 2x - 2 + 2x^3 - 4x^2$

$\qquad = 4x^3 - 6x^2 + 2x - 2$

3. $h(t) = \sqrt[3]{t}(t^2 + 4) = t^{1/3}(t^2 + 4)$

$h'(t) = t^{1/3}(2t) + (t^2 + 4)\dfrac{1}{3}t^{-2/3}$

$\qquad = 2t^{4/3} + \dfrac{t^2 + 4}{3t^{2/3}}$

$\qquad = \dfrac{7t^2 + 4}{3t^{2/3}}$

5. $f(x) = x^3 \cos x$

$f'(x) = x^3(-\sin x) + \cos x(3x^2)$

$\quad = 3x^2 \cos x - x^3 \sin x$

7. $f(x) = \dfrac{x}{x^2 + 1}$

$f'(x) = \dfrac{(x^2 + 1)(1) - x(2x)}{(x^2 + 1)^2} = \dfrac{1 - x^2}{(x^2 + 1)^2}$

9. $h(x) = \dfrac{\sqrt[3]{x}}{x^3 + 1} = \dfrac{x^{1/3}}{x^3 + 1}$

$h'(x) = \dfrac{(x^3 + 1)\frac{1}{3}x^{-2/3} - x^{1/3}(3x^2)}{(x^3 + 1)^2}$

$\quad = \dfrac{(x^3 + 1) - x(9x^2)}{3x^{2/3}(x^3 + 1)^2}$

$\quad = \dfrac{1 - 8x^3}{3x^{2/3}(x^3 + 1)^2}$

11. $g(x) = \dfrac{\sin x}{x^2}$

$g'(x) = \dfrac{x^2(\cos x) - \sin x(2x)}{(x^2)^2} = \dfrac{x \cos x - 2 \sin x}{x^3}$

13. $f(x) = (x^3 - 3x)(2x^2 + 3x + 5)$

$f'(x) = (x^3 - 3x)(4x + 3) + (2x^2 + 3x + 5)(3x^2 - 3)$

$\quad = 10x^4 + 12x^3 - 3x^2 - 18x - 15$

$f'(0) = -15$

15. $f(x) = \dfrac{x^2 - 4}{x - 3}$

$f'(x) = \dfrac{(x - 3)(2x) - (x^2 - 4)(1)}{(x - 3)^2}$

$\quad = \dfrac{2x^2 - 6x - x^2 + 4}{(x - 3)^2}$

$\quad = \dfrac{x^2 - 6x + 4}{(x - 3)^2}$

$f'(1) = \dfrac{1 - 6 + 4}{(1 - 3)^2} = -\dfrac{1}{4}$

17. $f(x) = x \cos x$

$f'(x) = (x)(-\sin x) + (\cos x)(1) = \cos x - x \sin x$

$f'\left(\dfrac{\pi}{4}\right) = \dfrac{\sqrt{2}}{2} - \dfrac{\pi}{4}\left(\dfrac{\sqrt{2}}{2}\right) = \dfrac{\sqrt{2}}{8}(4 - \pi)$

Function	*Rewrite*	*Differentiate*	*Simplify*
19. $y = \dfrac{x^2 + 2x}{3}$	$y = \dfrac{1}{3}x^2 + \dfrac{2}{3}x$	$y' = \dfrac{2}{3}x + \dfrac{2}{3}$	$y' = \dfrac{2x + 2}{3}$
21. $y = \dfrac{7}{3x^3}$	$y = \dfrac{7}{3}x^{-3}$	$y' = -7x^{-4}$	$y' = -\dfrac{7}{x^4}$
23. $y = \dfrac{4x^{3/2}}{x}$	$y = 4\sqrt{x}, \; x > 0$	$y' = 2x^{-1/2}$	$y' = \dfrac{2}{\sqrt{x}}$

25. $f(x) = \dfrac{3 - 2x - x^2}{x^2 - 1}$

$f'(x) = \dfrac{(x^2 - 1)(-2 - 2x) - (3 - 2x - x^2)(2x)}{(x^2 - 1)^2}$

$\quad = \dfrac{2x^2 - 4x + 2}{(x^2 - 1)^2} = \dfrac{2(x - 1)^2}{(x^2 - 1)^2}$

$\quad = \dfrac{2}{(x + 1)^2}, \; x \neq 1$

27. $f(x) = x\left(1 - \dfrac{4}{x + 3}\right) = x - \dfrac{4x}{x + 3}$

$f'(x) = 1 - \dfrac{(x + 3)4 - 4x(1)}{(x + 3)^2}$

$\quad = \dfrac{(x^2 + 6x + 9) - 12}{(x + 3)^2}$

$\quad = \dfrac{x^2 + 6x - 3}{(x + 3)^2}$

29. $f(x) = \dfrac{2x+5}{\sqrt{x}} = 2x^{1/2} + 5x^{-1/2}$

$f'(x) = x^{-1/2} - \dfrac{5}{2}x^{-3/2} = x^{-3/2}\left[x - \dfrac{5}{2}\right] = \dfrac{2x-5}{2x\sqrt{x}} = \dfrac{2x-5}{2x^{3/2}}$

31. $h(s) = (s^3 - 2)^2 = s^6 - 4s^3 + 4$

$h'(s) = 6s^5 - 12s^2 = 6s^2(s^3 - 2)$

33. $f(x) = \dfrac{2 - (1/x)}{x-3} = \dfrac{2x-1}{x(x-3)} = \dfrac{2x-1}{x^2-3x}$

$f'(x) = \dfrac{(x^2-3x)2 - (2x-1)(2x-3)}{(x^2-3x)^2} = \dfrac{2x^2 - 6x - 4x^2 + 8x - 3}{(x^2-3x)^2}$

$= \dfrac{-2x^2 + 2x - 3}{(x^2-3x)^2} = -\dfrac{2x^2 - 2x + 3}{x^2(x-3)^2}$

35. $f(x) = (3x^3 + 4x)(x-5)(x+1)$

$f'(x) = (9x^2 + 4)(x-5)(x+1) + (3x^3 + 4x)(1)(x+1) + (3x^3 + 4x)(x-5)(1)$

$= (9x^2 + 4)(x^2 - 4x - 5) + 3x^4 + 3x^3 + 4x^2 + 4x + 3x^4 - 15x^3 + 4x^2 - 20x$

$= 9x^4 - 36x^3 - 41x^2 - 16x - 20 + 6x^4 - 12x^3 + 8x^2 - 16x$

$= 15x^4 - 48x^3 - 33x^2 - 32x - 20$

37. $f(x) = \dfrac{x^2 + c^2}{x^2 - c^2}$

$f'(x) = \dfrac{(x^2 - c^2)(2x) - (x^2 + c^2)(2x)}{(x^2 - c^2)^2}$

$= \dfrac{-4xc^2}{(x^2 - c^2)^2}$

39. $f(t) = t^2 \sin t$

$f'(t) = t^2 \cos t + 2t \sin t$

$= t(t \cos t + 2 \sin t)$

41. $f(t) = \dfrac{\cos t}{t}$

$f'(t) = \dfrac{-t \sin t - \cos t}{t^2} = -\dfrac{t \sin t + \cos t}{t^2}$

43. $f(x) = -x + \tan x$

$f'(x) = -1 + \sec^2 x = \tan^2 x$

45. $g(t) = \sqrt[4]{t} + 8 \sec t = t^{1/4} + 8 \sec t$

$g'(t) = \dfrac{1}{4}t^{-3/4} + 8 \sec t \tan t = \dfrac{1}{4t^{3/4}} + 8 \sec t \tan t$

47. $y = \dfrac{3(1 - \sin x)}{2 \cos x} = \dfrac{3}{2}(\sec x - \tan x)$

$y' = \dfrac{3}{2}(\sec x \tan x - \sec^2 x) = \dfrac{3}{2} \sec x(\tan x - \sec x)$

$= \dfrac{3}{2}(\sec x \tan x - \tan^2 x - 1)$

49. $y = -\csc x - \sin x$

$y' = \csc x \cot x - \cos x$

$= \dfrac{\cos x}{\sin^2 x} - \cos x$

$= \cos x(\csc^2 x - 1)$

$= \cos x \cot^2 x$

51. $f(x) = x^2 \tan x$

$f'(x) = x^2 \sec^2 x + 2x \tan x$

$= x(x \sec^2 x + 2 \tan x)$

53. $y = 2x \sin x + x^2 \cos x$

$y' = 2x \cos x + 2 \sin x + x^2(-\sin x) + 2x \cos x$

$\quad = 4x \cos x + 2 \sin x - x^2 \sin x$

55. $g(x) = \left(\dfrac{x + 1}{x + 2}\right)(2x - 5)$

$g'(x) = \dfrac{2x^2 + 8x - 1}{(x + 2)^2}$ (Form of answer may vary.)

57. $g(\theta) = \dfrac{\theta}{1 - \sin \theta}$

$g'(\theta) = \dfrac{1 - \sin \theta + \theta \cos \theta}{(\sin \theta - 1)^2}$ (Form of answer may vary.)

59. $y = \dfrac{1 + \csc x}{1 - \csc x}$

$y' = \dfrac{(1 - \csc x)(-\csc x \cot x) - (1 + \csc x)(\csc x \cot x)}{(1 - \csc x)^2} = \dfrac{-2 \csc x \cot x}{(1 - \csc x)^2}$

$y'\left(\dfrac{\pi}{6}\right) = \dfrac{-2(2)\left(\sqrt{3}\right)}{(1 - 2)^2} = -4\sqrt{3}$

61. $h(t) = \dfrac{\sec t}{t}$

$h'(t) = \dfrac{t(\sec t \tan t) - (\sec t)(1)}{t^2} = \dfrac{\sec t(t \tan t - 1)}{t^2}$

$h'(\pi) = \dfrac{\sec \pi(\pi \tan \pi - 1)}{\pi^2} = \dfrac{1}{\pi^2}$

63. (a) $f(x) = (x^3 - 3x + 1)(x + 2),\ (1, -3)$

$f'(x) = (x^3 - 3x + 1)(1) + (x + 2)(3x^2 - 3)$

$\quad = 4x^3 + 6x^2 - 6x - 5$

$f'(1) = -1;$ Slope at $(1, -3)$

Tangent line: $y + 3 = -1(x - 1) \Longrightarrow y = -x - 2$

(b)

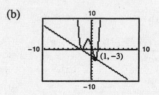

65. (a) $f(x) = \dfrac{x}{x - 1},\ (2, 2)$

$f'(x) = \dfrac{(x - 1)(1) - x(1)}{(x - 1)^2} = \dfrac{-1}{(x - 1)^2}$

$f'(2) = \dfrac{-1}{(2 - 1)^2} = -1;$ Slope at $(2, 2)$

Tangent line: $y - 2 = -1(x - 2) \Longrightarrow y = -x + 4$

(b)

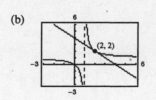

67. (a) $f(x) = \tan x,\ \left(\dfrac{\pi}{4}, 1\right)$

$f'(x) = \sec^2 x$

$f'\left(\dfrac{\pi}{4}\right) = 2;$ Slope at $\left(\dfrac{\pi}{4}, 1\right)$

Tangent line: $\qquad\qquad y - 1 = 2\left(x - \dfrac{\pi}{4}\right)$

$\qquad\qquad\qquad y - 1 = 2x - \dfrac{\pi}{2}$

$\qquad\quad 4x - 2y - \pi + 2 = 0$

(b)

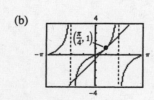

69. $f(x) = \dfrac{8}{x^2 + 4};$ $(2, 1)$

$$f'(x) = \dfrac{(x^2 + 4)(0) - 8(2x)}{(x^2 + 4)^2} = \dfrac{-16x}{(x^2 + 4)^2}$$

$$f'(2) = \dfrac{-16(2)}{(4 + 4)^2} = -\dfrac{1}{2}$$

$$y - 1 = -\dfrac{1}{2}(x - 2)$$

$$y = -\dfrac{1}{2}x + 2$$

$$2y + x - 4 = 0$$

71. $f(x) = \dfrac{16x}{x^2 + 16};$ $\left(-2, -\dfrac{8}{5}\right)$

$$f'(x) = \dfrac{(x^2 + 16)(16) - 16x(2x)}{(x^2 + 16)^2} = \dfrac{256 - 16x^2}{(x^2 + 16)^2}$$

$$f'(-2) = \dfrac{256 - 16(4)}{20^2} = \dfrac{12}{25}$$

$$y + \dfrac{8}{5} = \dfrac{12}{25}(x + 2)$$

$$y = \dfrac{12}{25}x - \dfrac{16}{25}$$

$$25y - 12x + 16 = 0$$

73. $f(x) = \dfrac{x^2}{x - 1}$

$$f'(x) = \dfrac{(x - 1)(2x) - x^2(1)}{(x - 1)^2}$$

$$= \dfrac{x^2 - 2x}{(x - 1)^2} = \dfrac{x(x - 2)}{(x - 1)^2}$$

$f'(x) = 0$ when $x = 0$ or $x = 2$.

Horizontal tangents are at $(0, 0)$ and $(2, 4)$.

75. $f(x) = \dfrac{4x - 2}{x^2}$

$$f'(x) = \dfrac{x^2(4) - (4x - 2)(2x)}{x^4}$$

$$= \dfrac{4x^2 - 8x^2 + 4x}{x^4}$$

$$= \dfrac{4 - 4x}{x^3}$$

$f'(x) = 0$ for $4 - 4x = 0 \Rightarrow x = 1$.

$f(1) = 2$

f has a horizontal tangent at $(1, 2)$.

77. $f(x) = \dfrac{x + 1}{x - 1}$

$$f'(x) = \dfrac{(x - 1) - (x + 1)}{(x - 1)^2} = \dfrac{-2}{(x - 1)^2}$$

$2y + x = 6 \Rightarrow y = -\dfrac{1}{2}x + 3;$ Slope: $-\dfrac{1}{2}$

$$\dfrac{-2}{(x - 1)^2} = -\dfrac{1}{2}$$

$$(x - 1)^2 = 4$$

$$x - 1 = \pm 2$$

$x = -1, 3;$ $f(-1) = 0, f(3) = 2$

$$y - 0 = -\dfrac{1}{2}(x + 1) \Rightarrow y = -\dfrac{1}{2}x - \dfrac{1}{2}$$

$$y - 2 = -\dfrac{1}{2}(x - 3) \Rightarrow y = -\dfrac{1}{2}x + \dfrac{7}{2}$$

79. $f'(x) = \dfrac{(x + 2)3 - 3x(1)}{(x + 2)^2} = \dfrac{6}{(x + 2)^2}$

$$g'(x) = \dfrac{(x + 2)5 - (5x + 4)(1)}{(x + 2)^2} = \dfrac{6}{(x + 2)^2}$$

$$g(x) = \dfrac{5x + 4}{(x + 2)} = \dfrac{3x}{(x + 2)} + \dfrac{2x + 4}{(x + 2)} = f(x) + 2$$

f and g differ by a constant.

81. (a) $p'(x) = f'(x)g(x) + f(x)g'(x)$

$$p'(1) = f'(1)g(1) + f(1)g'(1) = 1(4) + 6\left(-\frac{1}{2}\right) = 1$$

(b) $q'(x) = \dfrac{g(x)f'(x) - f(x)g'(x)}{g(x)^2}$

$$q'(4) = \frac{3(-1) - 7(0)}{3^2} = -\frac{1}{3}$$

83. Area $= A(t) = (2t + 1)\sqrt{t} = 2t^{3/2} + t^{1/2}$

$$A'(t) = 2\left(\frac{3}{2}t^{1/2}\right) + \frac{1}{2}t^{-1/2}$$

$$= 3t^{1/2} + \frac{1}{2}t^{-1/2}$$

$$= \frac{6t + 1}{2\sqrt{t}} \text{ cm}^2/\text{sec}$$

85. $C = 100\left(\dfrac{200}{x^2} + \dfrac{x}{x + 30}\right),\ 1 \le x$

$$\frac{dC}{dx} = 100\left(-\frac{400}{x^3} + \frac{30}{(x + 30)^2}\right)$$

(a) When $x = 10$: $\dfrac{dC}{dx} = -\$38.13$

(b) When $x = 15$: $\dfrac{dC}{dx} = -\$10.37$

(c) When $x = 20$: $\dfrac{dC}{dx} = -\$3.80$

As the order size increases, the cost per item decreases.

87. $P(t) = 500\left[1 + \dfrac{4t}{50 + t^2}\right]$

$$P'(t) = 500\left[\frac{(50 + t^2)(4) - (4t)(2t)}{(50 + t^2)^2}\right]$$

$$= 500\left[\frac{200 - 4t^2}{(50 + t^2)^2}\right]$$

$$= 2000\left[\frac{50 - t^2}{(50 + t^2)^2}\right]$$

$P'(2) \approx 31.55$ bacteria per hour

89. (a) $\sec x = \dfrac{1}{\cos x}$

$$\frac{d}{dx}[\sec x] = \frac{d}{dx}\left[\frac{1}{\cos x}\right] = \frac{(\cos x)(0) - (1)(-\sin x)}{(\cos x)^2} = \frac{\sin x}{\cos x \cos x} = \frac{1}{\cos x} \cdot \frac{\sin x}{\cos x} = \sec x \tan x$$

(b) $\csc x = \dfrac{1}{\sin x}$

$$\frac{d}{dx}[\csc x] = \frac{d}{dx}\left[\frac{1}{\sin x}\right] = \frac{(\sin x)(0) - (1)(\cos x)}{(\sin x)^2} = -\frac{\cos x}{\sin x \sin x} = -\frac{1}{\sin x} \cdot \frac{\cos x}{\sin x} = -\csc x \cot x$$

(c) $\cot x = \dfrac{\cos x}{\sin x}$

$$\frac{d}{dx}[\cot x] = \frac{d}{dx}\left[\frac{\cos x}{\sin x}\right] = \frac{\sin x(-\sin x) - (\cos x)(\cos x)}{(\sin x)^2} = -\frac{\sin^2 x + \cos^2 x}{\sin^2 x} = -\frac{1}{\sin^2 x} = -\csc^2 x$$

91. (a) $n(t) = -3.5806t^3 + 82.577t^2 - 603.60t + 1667.5$

$v(t) = -0.1361t^3 + 3.165t^2 - 23.02t + 59.8$

(c) $A = \dfrac{v(t)}{n(t)} = \dfrac{-0.1361t^3 + 3.165t^2 - 23.02t + 59.8}{-3.5806t^3 + 82.577t^2 - 603.60t + 1667.5}$

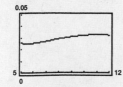

A represents the average retail value (in millions of dollars) per 1000 motor homes.

(b)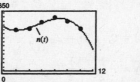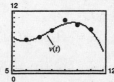

(d) $A'(t)$ represents the rate of change of the average retail value per 1000 motor homes.

93. $f(x) = 4x^{3/2}$

$f'(x) = 6x^{1/2}$

$f''(x) = 3x^{-1/2} = \dfrac{3}{\sqrt{x}}$

95. $f(x) = \dfrac{x}{x - 1}$

$f'(x) = \dfrac{(x - 1)(1) - x(1)}{(x - 1)^2} = \dfrac{-1}{(x - 1)^2}$

$f''(x) = \dfrac{2}{(x - 1)^3}$

97. $f(x) = 3 \sin x$

$f'(x) = 3 \cos x$

$f''(x) = -3 \sin x$

99. $f'(x) = x^2$

$f''(x) = 2x$

101. $f'''(x) = 2\sqrt{x}$

$f^{(4)}(x) = \dfrac{1}{2}(2)x^{-1/2} = \dfrac{1}{\sqrt{x}}$

103.

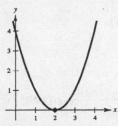

$f(2) = 0$

One such function
is $f(x) = (x - 2)^2$.

105. $f(x) = 2g(x) + h(x)$

$f'(x) = 2g'(x) + h'(x)$

$f'(2) = 2g'(2) + h'(2)$

$\qquad = 2(-2) + 4$

$\qquad = 0$

107. $f(x) = \dfrac{g(x)}{h(x)}$

$f'(x) = \dfrac{h(x)g'(x) - g(x)h'(x)}{[h(x)]^2}$

$f'(2) = \dfrac{h(2)g'(2) - g(2)h'(2)}{[h(2)]^2}$

$\qquad = \dfrac{(-1)(-2) - (3)(4)}{(-1)^2}$

$\qquad = -10$

109.

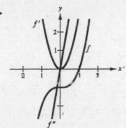

It appears that f is cubic;
so f' would be quadratic
and f'' would be linear.

111.

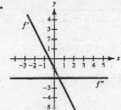

113.

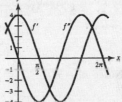

115. $v(t) = 36 - t^2,\ 0 \le t \le 6$

$a(t) = -2t$

$v(3) = 27 \text{ m/sec}$

$a(3) = -6 \text{ m/sec}^2$

The speed of the object is decreasing.

117. $s(t) = -8.25t^2 + 66t$

$v(t) = -16.50t + 66$

$a(t) = -16.50$

t(sec)	0	1	2	3	4
$s(t)$ (ft)	0	57.75	99	123.75	132
$v(t) = s'(t)$ (ft/sec)	66	49.5	33	16.5	0
$a(t) = v'(t)$ (ft/sec^2)	-16.5	-16.5	-16.5	-16.5	-16.5

Average velocity on:

$[0, 1]$ is $\dfrac{57.75 - 0}{1 - 0} = 57.75.$

$[1, 2]$ is $\dfrac{99 - 57.75}{2 - 1} = 41.25.$

$[2, 3]$ is $\dfrac{123.75 - 99}{3 - 2} = 24.75.$

$[3, 4]$ is $\dfrac{132 - 123.75}{4 - 3} = 8.25.$

119. $f(x) = x^n$

$f^{(n)}(x) = n(n-1)(n-2)\cdots(2)(1) = n!$

Note: $n! = n(n-1)\cdots3\cdot2\cdot1$ (read "n factorial")

121. $f(x) = g(x)h(x)$

(a) $f'(x) = g(x)h'(x) + h(x)g'(x)$

$f''(x) = g(x)h''(x) + g'(x)h'(x) + h(x)g''(x) + h'(x)g'(x)$

$\qquad = g(x)h''(x) + 2g'(x)h'(x) + h(x)g''(x)$

$f'''(x) = g(x)h'''(x) + g'(x)h''(x) + 2g'(x)h''(x) + 2g''(x)h'(x) + h(x)g'''(x) + h'(x)g''(x)$

$\qquad = g(x)h'''(x) + 3g'(x)h''(x) + 3g''(x)h'(x) + g'''(x)h(x)$

$f^{(4)}(x) = g(x)h^{(4)}(x) + g'(x)h'''(x) + 3g'(x)h'''(x) + 3g''(x)h''(x) + 3g''(x)h''(x) + 3g'''(x)h'(x)$

$\qquad\quad + g'''(x)h'(x) + g^{(4)}(x)h(x)$

$\qquad = g(x)h^{(4)}(x) + 4g'(x)h'''(x) + 6g''(x)h''(x) + 4g'''(x)h'(x) + g^{(4)}(x)h(x)$

(b) $f^{(n)}(x) = g(x)h^{(n)}(x) + \dfrac{n(n-1)(n-2)\cdots(2)(1)}{1[(n-1)(n-2)\cdots(2)(1)]}g'(x)h^{(n-1)}(x) + \dfrac{n(n-1)(n-2)\cdots(2)(1)}{(2)(1)[(n-2)(n-3)\cdots(2)(1)]}g''(x)h^{(n-2)}(x)$

$\qquad\quad + \dfrac{n(n-1)(n-2)\cdots(2)(1)}{(3)(2)(1)[(n-3)(n-4)\cdots(2)(1)]}g'''(x)h^{(n-3)}(x) + \cdots$

$\qquad\quad + \dfrac{n(n-1)(n-2)\cdots(2)(1)}{[(n-1)(n-2)\cdots(2)(1)](1)}g^{(n-1)}(x)h'(x) + g^{(n)}(x)h(x)$

$\qquad = g(x)h^{(n)}(x) + \dfrac{n!}{1!(n-1)!}g'(x)h^{(n-1)}(x) + \dfrac{n!}{2!(n-2)!}g''(x)h^{(n-2)}(x) + \cdots$

$\qquad\quad + \dfrac{n!}{(n-1)!1!}g^{(n-1)}(x)h'(x) + g^{(n)}(x)h(x)$

Note: $n! = n(n-1)\cdots3\cdot2\cdot1$ (read "n factorial")

123. $f(x) = x^n \sin x$

$f'(x) = x^n \cos x + nx^{n-1}\sin x$

$\qquad = x^{n-1}(x\cos x + n\sin x)$

When $n = 1$: $f'(x) = x\cos x + \sin x$

When $n = 2$: $f'(x) = x(x\cos x + 2\sin x)$

When $n = 3$: $f'(x) = x^2(x\cos x + 3\sin x)$

When $n = 4$: $f'(x) = x^3(x\cos x + 4\sin x)$

For general n, $f'(x) = x^{n-1}(x\cos x + n\sin x)$.

125. $y = \dfrac{1}{x}$, $y' = -\dfrac{1}{x^2}$, $y'' = \dfrac{2}{x^3}$

$x^3y'' + 2x^2y' = x^3\left[\dfrac{2}{x^3}\right] + 2x^2\left[-\dfrac{1}{x^2}\right] = 2 - 2 = 0$

127. $y = 2\sin x + 3$

$\quad y' = 2\cos x$

$\quad y'' = -2\sin x$

$y'' + y = -2\sin x + (2\sin x + 3) = 3$

129. False. If $y = f(x)g(x)$, then

$\dfrac{dy}{dx} = f(x)g'(x) + g(x)f'(x)$.

131. True

$$h'(c) = f(c)g'(c) + g(c)f'(c)$$
$$= f(c)(0) + g(c)(0)$$
$$= 0$$

133. True

135. $f(x) = ax^2 + bx + c$

$f'(x) = 2ax + b$

x-intercept at $(1, 0)$: $0 = a + b + c$

$(2, 7)$ on graph: $7 = 4a + 2b + c$

Slope 10 at $(2, 7)$: $10 = 4a + b$

Subtracting the third equation from the second, $-3 = b + c$. Subtracting this equation from the first, $3 = a$. Then, $10 = 4(3) + b \implies b = -2$. Finally, $-3 = (-2) + c \implies c = -1$.

$f(x) = 3x^2 - 2x - 1$

137. $f(x) = x|x| = \begin{cases} x^2, & \text{if } x \geq 0 \\ -x^2, & \text{if } x < 0 \end{cases}$

$f'(x) = \begin{cases} 2x, & \text{if } x \geq 0 \\ -2x, & \text{if } x < 0 \end{cases} = 2|x|$

$f''(x) = \begin{cases} 2, & \text{if } x > 0 \\ -2, & \text{if } x < 0 \end{cases}$

$f''(0)$ does not exist since the left and right derivatives are not equal.

Section 2.4   The Chain Rule

$y = f(g(x))$	$u = g(x)$	$y = f(u)$
1. $y = (6x - 5)^4$	$u = 6x - 5$	$y = u^4$
3. $y = \sqrt{x^2 - 1}$	$u = x^2 - 1$	$y = \sqrt{u}$
5. $y = \csc^3 x$	$u = \csc x$	$y = u^3$

7. $y = (2x - 7)^3$

$y' = 3(2x - 7)^2(2) = 6(2x - 7)^2$

9. $g(x) = 3(4 - 9x)^4$

$g'(x) = 12(4 - 9x)^3(-9) = -108(4 - 9x)^3$

11. $f(t) = (1 - t)^{1/2}$

$f'(t) = \frac{1}{2}(1 - t)^{-1/2}(-1) = -\frac{1}{2\sqrt{1 - t}}$

13. $y = (9x^2 + 4)^{1/3}$

$y' = \frac{1}{3}(9x^2 + 4)^{-2/3}(18x) = \frac{6x}{(9x^2 + 4)^{2/3}}$

15. $y = 2(4 - x^2)^{1/4}$

$y' = 2\left(\frac{1}{4}\right)(4 - x^2)^{-3/4}(-2x) = \frac{-x}{\sqrt[4]{(4 - x^2)^3}}$

17. $y = (x - 2)^{-1}$

$y' = -1(x - 2)^{-2}(1) = \frac{-1}{(x - 2)^2}$

19. $f(t) = (t - 3)^{-2}$

$f'(t) = -2(t - 3)^{-3} = \frac{-2}{(t - 3)^3}$

21. $y = (x + 2)^{-1/2}$

$\frac{dy}{dx} = -\frac{1}{2}(x + 2)^{-3/2}$

$= -\frac{1}{2(x + 2)^{3/2}}$

23. $f(x) = x^2(x - 2)^4$

$f'(x) = x^2[4(x - 2)^3(1)] + (x - 2)^4(2x)$

$= 2x(x - 2)^3[2x + (x - 2)]$

$= 2x(x - 2)^3(3x - 2)$

25. $y = x\sqrt{1 - x^2} = x(1 - x^2)^{1/2}$

$$y' = x\left[\frac{1}{2}(1 - x^2)^{-1/2}(-2x)\right] + (1 - x^2)^{1/2}(1)$$

$$= -x^2(1 - x^2)^{-1/2} + (1 - x^2)^{1/2}$$

$$= (1 - x^2)^{-1/2}[-x^2 + (1 - x^2)]$$

$$= \frac{1 - 2x^2}{\sqrt{1 - x^2}}$$

27. $y = \dfrac{x}{\sqrt{x^2 + 1}} = x(x^2 + 1)^{-1/2}$

$$y' = x\left[-\frac{1}{2}(x^2 + 1)^{-3/2}(2x)\right] + (x^2 + 1)^{-1/2}(1)$$

$$= -x^2(x^2 + 1)^{-3/2} + (x^2 + 1)^{-1/2}$$

$$= (x^2 + 1)^{-3/2}[-x^2 + (x^2 + 1)]$$

$$= \frac{1}{(x^2 + 1)^{3/2}}$$

29. $g(x) = \left(\dfrac{x + 5}{x^2 + 2}\right)^2$

$$g'(x) = 2\left(\frac{x + 5}{x^2 + 2}\right)\left(\frac{(x^2 + 2) - (x + 5)(2x)}{(x^2 + 2)^2}\right)$$

$$= \frac{2(x + 5)(2 - 10x - x^2)}{(x^2 + 2)^3}$$

31. $f(v) = \left(\dfrac{1 - 2v}{1 + v}\right)^3$

$$f'(v) = 3\left(\frac{1 - 2v}{1 + v}\right)^2\left(\frac{(1 + v)(-2) - (1 - 2v)}{(1 + v)^2}\right)$$

$$= \frac{-9(1 - 2v)^2}{(1 + v)^4}$$

33. $y = \dfrac{\sqrt{x} + 1}{x^2 + 1}$

$$y' = \frac{1 - 3x^2 - 4x^{3/2}}{2\sqrt{x}(x^2 + 1)^2}$$

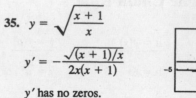

The zero of y' corresponds to the point on the graph of y where the tangent line is horizontal.

35. $y = \sqrt{\dfrac{x + 1}{x}}$

$$y' = -\frac{\sqrt{(x + 1)/x}}{2x(x + 1)}$$

y' has no zeros.

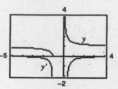

37. $y = \dfrac{\cos \pi x + 1}{x}$

$$\frac{dy}{dx} = \frac{-\pi x \sin \pi x - \cos \pi x - 1}{x^2}$$

$$= -\frac{\pi x \sin \pi x + \cos \pi x + 1}{x^2}$$

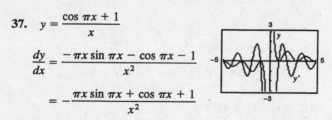

The zeros of y' correspond to the points on the graph of y where the tangent lines are horizontal.

39. (a) $y = \sin x$

$y' = \cos x$

$y'(0) = 1$

1 cycle in $[0, 2\pi]$

(b) $y = \sin 2x$

$y' = 2 \cos 2x$

$y'(0) = 2$

2 cycles in $[0, 2\pi]$

The slope of $\sin ax$ at the origin is a.

41. $y = \cos 3x$

$$\frac{dy}{dx} = -3 \sin 3x$$

43. $g(x) = 3 \tan 4x$

$g'(x) = 12 \sec^2 4x$

45. $y = \sin(\pi x)^2 = \sin(\pi^2 x^2)$

$y' = \cos(\pi^2 x^2)[2\pi^2 x] = 2\pi^2 x \cos(\pi^2 x^2)$

47. $h(x) = \sin 2x \cos 2x$

$$h'(x) = \sin 2x(-2 \sin 2x) + \cos 2x(2 \cos 2x)$$

$$= 2 \cos^2 2x - 2 \sin^2 2x$$

$$= 2 \cos 4x$$

Alternate solution: $h(x) = \frac{1}{2} \sin 4x$

$$h'(x) = \frac{1}{2} \cos 4x(4) = 2 \cos 4x$$

49. $f(x) = \dfrac{\cot x}{\sin x} = \dfrac{\cos x}{\sin^2 x}$

$$f'(x) = \frac{\sin^2 x(-\sin x) - \cos x(2 \sin x \cos x)}{\sin^4 x}$$

$$= \frac{-\sin^2 x - 2 \cos^2 x}{\sin^3 x} = \frac{-1 - \cos^2 x}{\sin^3 x}$$

51. $y = 4 \sec^2 x$

$y' = 8 \sec x \cdot \sec x \tan x$

$\quad = 8 \sec^2 x \tan x$

53. $f(\theta) = \frac{1}{4} \sin^2 2\theta = \frac{1}{4}(\sin 2\theta)^2$

$f'(\theta) = 2(\frac{1}{4})(\sin 2\theta)(\cos 2\theta)(2)$

$\quad = \sin 2\theta \cos 2\theta = \frac{1}{2} \sin 4\theta$

55. $f(t) = 3 \sec^2(\pi t - 1)$

$f'(t) = 6 \sec(\pi t - 1) \sec(\pi t - 1) \tan(\pi t - 1)(\pi)$

$\quad = 6\pi \sec^2(\pi t - 1) \tan(\pi t - 1) = \dfrac{6\pi \sin(\pi t - 1)}{\cos^3(\pi t - 1)}$

57. $y = \sqrt{x} + \frac{1}{4} \sin(2x)^2$

$\quad = \sqrt{x} + \frac{1}{4} \sin(4x^2)$

$\dfrac{dy}{dx} = \frac{1}{2}x^{-1/2} + \frac{1}{4} \cos(4x^2)(8x)$

$\quad = \dfrac{1}{2\sqrt{x}} + 2x \cos(2x)^2$

59. $s(t) = (t^2 + 2t + 8)^{1/2}, \quad (2, 4)$

$s'(t) = \frac{1}{2}(t^2 + 2t + 8)^{-1/2}(2t + 2)$

$\quad = \dfrac{t + 1}{\sqrt{t^2 + 2t + 8}}$

$s'(2) = \dfrac{3}{4}$

61. $f(x) = \dfrac{3}{x^3 - 4} = 3(x^3 - 4)^{-1}, \quad \left(-1, -\dfrac{3}{5}\right)$

$f'(x) = -3(x^3 - 4)^{-2}(3x^2) = -\dfrac{9x^2}{(x^3 - 4)^2}$

$f'(-1) = -\dfrac{9}{25}$

63. $f(t) = \dfrac{3t + 2}{t - 1}, \quad (0, -2)$

$f'(t) = \dfrac{(t - 1)(3) - (3t + 2)(1)}{(t - 1)^2} = \dfrac{-5}{(t - 1)^2}$

$f'(0) = -5$

65. $y = 37 - \sec^3(2x), \quad (0, 36)$

$y' = -3 \sec^2(2x)[2 \sec(2x) \tan(2x)]$

$\quad = -6 \sec^3(2x) \tan(2x)$

$y'(0) = 0$

67. (a) $f(x) = \sqrt{3x^2 - 2}, \quad (3, 5)$

$f'(x) = \frac{1}{2}(3x^2 - 2)^{-1/2}(6x) = \dfrac{3x}{\sqrt{3x^2 - 2}}$

$f'(3) = \dfrac{9}{5}$

Tangent line:

$y - 5 = \dfrac{9}{5}(x - 3) \Longrightarrow 9x - 5y - 2 = 0$

(b)

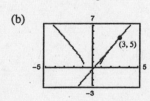

69. (a) $y = (2x^3 + 1)^2, \quad (-1, 1)$

$y' = 2(2x^3 + 1)(6x^2) = 12x^2(2x^3 + 1)$

$y'(-1) = 12(-1) = -12$

Tangent line: $y - 1 = 12(x + 1)$

$\qquad\qquad\qquad y = -12x - 11$

(b)

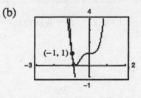

71. (a) $f(x) = \sin 2x, \quad (\pi, 0)$

$f'(x) = 2 \cos 2x$

$f'(\pi) = 2$

Tangent line:

$y = 2(x - \pi) \Longrightarrow 2x - y - 2\pi = 0$

(b)

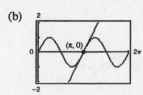

73. (a) $f(x) = \tan^2 x, \quad \left(\dfrac{\pi}{4}, 1\right)$

$f'(x) = 2\tan x \sec^2 x$

$f'\left(\dfrac{\pi}{4}\right) = 2(1)(2) = 4$

Tangent line:

$y - 1 = 4\left(x - \dfrac{\pi}{4}\right) \Rightarrow 4x - y + (1 - \pi) = 0$

(b)
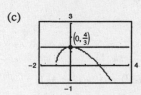

75. (a) $g(t) = \dfrac{3t^2}{\sqrt{t^2 + 2t - 1}}, \quad \left(\dfrac{1}{2}, \dfrac{3}{2}\right)$

$g'(t) = \dfrac{3t(t^2 + 3t - 2)}{(t^2 + 2t - 1)^{3/2}}$

$g'\left(\dfrac{1}{2}\right) = -3$

(b) $y - \dfrac{3}{2} = -3\left(x - \dfrac{1}{2}\right)$

$y = -3x + 3$

(c)
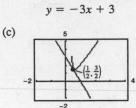

77. (a) $s(t) = \dfrac{(4 - 2t)\sqrt{1 + t}}{3}, \quad \left(0, \dfrac{4}{3}\right)$

$s'(t) = \dfrac{-2\sqrt{1 + t}}{3} + \dfrac{2 - t}{3\sqrt{1 + t}}$

$s'(0) = 0$

(b) $y - \dfrac{4}{3} = 0(x - 0)$

$y = \dfrac{4}{3}$

(c)
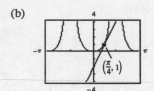

79. $f(x) = \sqrt{25 - x^2}, \quad (3, 4)$

$f'(x) = \dfrac{-x}{\sqrt{25 - x^2}}$

$f'(3) = -\dfrac{3}{4}$

$y - 4 = -\dfrac{3}{4}(x - 3)$

$y = -\dfrac{3}{4}x + \dfrac{25}{4}; \quad \text{Tangent line}$

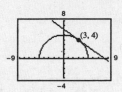

81.

$f(x) = 2\cos x + \sin 2x, \quad 0 < x < 2\pi$

$f'(x) = -2\sin x + 2\cos 2x$

$\quad\quad = -2\sin x + 2 - 4\sin^2 x = 0$

$2\sin^2 x + \sin x - 1 = 0$

$(\sin x + 1)(2\sin x - 1) = 0$

$\sin x = -1 \Rightarrow x = \dfrac{3\pi}{2}$

$\sin x = \dfrac{1}{2} \Rightarrow x = \dfrac{\pi}{6}, \dfrac{5\pi}{6}$

Horizontal tangents at $x = \dfrac{\pi}{6}, \dfrac{3\pi}{2}, \dfrac{5\pi}{6}$

Horizontal tangent at the points $\left(\dfrac{\pi}{6}, \dfrac{3\sqrt{3}}{2}\right), \left(\dfrac{3\pi}{2}, 0\right),$ and

$\left(\dfrac{5\pi}{6}, -\dfrac{3\sqrt{3}}{2}\right)$

83. $f(x) = 2(x^2 - 1)^3$

$f'(x) = 6(x^2 - 1)^2(2x)$

$\quad\quad = 12x(x^4 - 2x^2 + 1)$

$\quad\quad = 12x^5 - 24x^3 + 12x$

$f''(x) = 60x^4 - 72x^2 + 12$

$\quad\quad = 12(5x^2 - 1)(x^2 - 1)$

85. $f(x) = \sin x^2$

$f'(x) = 2x \cos x^2$

$f''(x) = 2x[2x(-\sin x^2)] + 2 \cos x^2$

$\quad = 2[\cos x^2 - 2x^2 \sin x^2]$

87. $h(x) = \frac{1}{9}(3x + 1)^3, \quad \left(1, \frac{64}{9}\right)$

$h'(x) = \frac{1}{9}3(3x + 1)^2(3) = (3x + 1)^2$

$h''(x) = 2(3x + 1)(3) = 6(3x + 1)$

$h''(1) = 24$

89. $f(x) = \cos(x^2), \quad (0, 1)$

$f'(x) = -\sin(x^2)(2x) = -2x \sin(x^2)$

$f''(x) = -2x \cos(x^2)(2x) - 2 \sin(x^2)$

$\quad = -4x^2 \cos(x^2) - 2 \sin(x^2)$

$f''(0) = 0$

91. The zeros of f' correspond to the points where the graph of f has horizontal tangents.

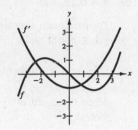

93. The zeros of f' correspond to the points where the graph of f has horizontal tangents.

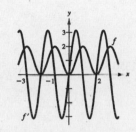

95. $g(x) = f(3x)$

$g'(x) = f'(3x)(3) \implies g'(x) = 3f'(3x)$

97. (a) $f(x) = g(x)h(x)$

$f'(x) = g(x)h'(x) + g'(x)h(x)$

$f'(5) = (-3)(-2) + (6)(3) = 24$

(c) $f(x) = \dfrac{g(x)}{h(x)}$

$f'(x) = \dfrac{h(x)g'(x) - g(x)h'(x)}{[h(x)]^2}$

$f'(5) = \dfrac{(3)(6) \quad (-3)(-2)}{(3)^2} = \dfrac{12}{9} = \dfrac{4}{3}$

(b) $f(x) = g(h(x))$

$f'(x) = g'(h(x))h'(x)$

$f'(5) = g'(3)(-2) = -2g'(3)$

Need $g'(3)$ to find $f'(5)$.

(d) $f(x) = [g(x)]^3$

$f'(x) = 3[g(x)]^2 g'(x)$

$f'(5) = 3(-3)^2(6) = 162$

99. (a) $h(x) = f(g(x)), \ g(1) = 4, \ g'(1) = -\frac{1}{2}, \ f'(4) = -1, \ h'(x) = f'(g(x))g'(x)$

$h'(1) = f'(g(1))g'(1)$

$\quad = f'(4)g'(1)$

$\quad = (-1)\left(-\frac{1}{2}\right) = \frac{1}{2}$

(b) $s(x) = g(f(x)), \ f(5) = 6, \ f'(5) = -1, \ g'(6)$ does not exist.

$s'(x) = g'(f(x))f'(x)$

$s'(5) = g'(f(5))f'(5) = g'(6)(-1)$

Since $g'(6)$ does not exist, $s'(5)$ is not defined.

101. (a) $F = 132,400(331 - v)^{-1}$

$F' = (-1)(132,400)(331 - v)^{-2}(-1)$

$= \dfrac{132,400}{(331 - v)^2}$

When $v = 30, F' \approx 1.461$.

(b) $F = 132,400(331 + v)^{-1}$

$F' = (-1)(132,400)(331 + v)^{-2}(1)$

$= \dfrac{-132,400}{(331 + v)^2}$

When $v = 30, F' \approx -1.016$.

103. $\theta = 0.2 \cos 8t$

The maximum angular displacement is $\theta = 0.2$ (since $-1 \le \cos 8t \le 1$).

$\dfrac{d\theta}{dt} = 0.2[-8 \sin 8t] = -1.6 \sin 8t$

When $t = 3$, $d\theta/dt = -1.6 \sin 24 \approx 1.4489$ radians per second.

105. $S = C(R^2 - r^2)$

$\dfrac{dS}{dt} = C\left(2R \dfrac{dR}{dt} - 2r \dfrac{dr}{dt}\right)$

Since r is constant, we have $dr/dt = 0$ and

$\dfrac{dS}{dt} = (1.76 \times 10^5)(2)(1.2 \times 10^{-2})(10^{-5})$

$= 4.224 \times 10^{-2} = 0.04224$.

107. (a) $x = -1.6372t^3 + 19.3120t^2 - 0.5082t - 0.6162$

(b) $C = 60x + 1350$

$= 60(-1.6372t^3 + 19.3120t^2 - 0.5082t - 0.6162) + 1350$

$\dfrac{dC}{dt} = 60(-4.9116t^2 + 38.624t - 0.5082)$

$= -294.696t^2 + 2317.44t - 30.492$

The function dC/dt is quadratic, not linear. The cost function levels off at the end of the day, perhaps due to fatigue.

109. $f(x + p) = f(x)$ for all x.

(a) Yes, $f'(x + p) = f'(x)$, which shows that f' is periodic as well.

(b) Yes, let $g(x) = f(2x)$, so $g'(x) = 2f'(2x)$. Since f' is periodic, so is g'.

111. (a) $g(x) = \sin^2 x + \cos^2 x = 1 \Longrightarrow g'(x) = 0$

$g'(x) = 2 \sin x \cos x + 2 \cos x(-\sin x) = 0$

(b) $\tan^2 x + 1 = \sec^2 x$

$g(x) + 1 = f(x)$

Taking derivatives of both sides, $g'(x) = f'(x)$. Equivalently, $f'(x) = 2 \sec x \cdot \sec x \cdot \tan x$ and $g'(x) = 2 \tan x \cdot \sec^2 x$, which are the same.

113. $|u| = \sqrt{u^2}$

$\dfrac{d}{dx}[|u|] = \dfrac{d}{dx}[\sqrt{u^2}] = \dfrac{1}{2}(u^2)^{-1/2}(2uu')$

$= \dfrac{uu'}{\sqrt{u^2}} = u' \dfrac{u}{|u|}, \quad u \ne 0$

115. $f(x) = |x^2 - 4|$

$f'(x) = 2x\left(\dfrac{x^2 - 4}{|x^2 - 4|}\right), \quad x \ne \pm 2$

117. $f(x) = |\sin x|$

$f'(x) = \cos x\left(\dfrac{\sin x}{|\sin x|}\right), \quad x \ne k\pi$

119. (a) $f(x) = \sec(2x)$

$f'(x) = 2(\sec 2x)(\tan 2x)$

$f''(x) = 2[2(\sec 2x)(\tan 2x)]\tan 2x + 2(\sec 2x)(\sec^2 2x)(2)$

$\qquad = 4[(\sec 2x)(\tan^2 2x) + \sec^3 2x]$

$f\left(\dfrac{\pi}{6}\right) = \sec\left(\dfrac{\pi}{3}\right) = 2$

$f'\left(\dfrac{\pi}{6}\right) = 2\sec\left(\dfrac{\pi}{3}\right)\tan\left(\dfrac{\pi}{3}\right) = 4\sqrt{3}$

$f''\left(\dfrac{\pi}{6}\right) = 4[2(3) + 2^3] = 56$

$P_1(x) = 4\sqrt{3}\left(x - \dfrac{\pi}{6}\right) + 2$

$P_2(x) = \dfrac{1}{2}(56)\left(x - \dfrac{\pi}{6}\right)^2 + 4\sqrt{3}\left(x - \dfrac{\pi}{6}\right) + 2$

$\qquad = 28\left(x - \dfrac{\pi}{6}\right)^2 + 4\sqrt{3}\left(x - \dfrac{\pi}{6}\right) + 2$

(b)

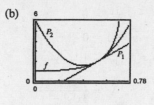

(c) P_2 is a better approximation than P_1.

(d) The accuracy worsens as you move away from $x = \pi/6$.

121. False. If $f(x) = \sin^2 2x$, then

$f'(x) = 2(\sin 2x)(2\cos 2x)$.

123.

$$f(x) = a_1 \sin x + a_2 \sin 2x + \cdots + a_n \sin nx$$

$$f'(x) = a_1 \cos x + 2a_2 \cos 2x + \cdots + na_n \cos nx$$

$$f'(0) = a_1 + 2a_2 + \cdots + na_n$$

$$|a_1 + 2a_2 + \cdots + na_n| = |f'(0)|$$

$$= \lim_{x \to 0}\left|\frac{f(x) - f(0)}{x - 0}\right|$$

$$= \lim_{x \to 0}\left|\frac{f(x)}{\sin x}\right| \cdot \left|\frac{\sin x}{x}\right|$$

$$= \lim_{x \to 0}\left|\frac{f(x)}{\sin x}\right| \le 1$$

Section 2.5   Implicit Differentiation

1. $x^2 + y^2 = 36$

$2x + 2yy' = 0$

$y' = \dfrac{-x}{y}$

3. $x^{1/2} + y^{1/2} = 9$

$\dfrac{1}{2}x^{-1/2} + \dfrac{1}{2}y^{-1/2}y' = 0$

$y' = -\dfrac{x^{-1/2}}{y^{-1/2}}$

$\quad = -\sqrt{\dfrac{y}{x}}$

5. $x^3 - xy + y^2 = 4$

$3x^2 - xy' - y + 2yy' = 0$

$(2y - x)y' = y - 3x^2$

$y' = \dfrac{y - 3x^2}{2y - x}$

7.
$$x^3y^3 - y - x = 0$$
$$3x^3y^2y' + 3x^2y^3 - y' - 1 = 0$$
$$(3x^3y^2 - 1)y' = 1 - 3x^2y^3$$
$$y' = \frac{1 - 3x^2y^3}{3x^3y^2 - 1}$$

9.
$$x^3 - 3x^2y + 2xy^2 = 12$$
$$3x^2 - 3x^2y' - 6xy + 4xyy' + 2y^2 = 0$$
$$(4xy - 3x^2)y' = 6xy - 3x^2 - 2y^2$$
$$y' = \frac{6xy - 3x^2 - 2y^2}{4xy - 3x^2}$$

11.
$$\sin x + 2\cos 2y = 1$$
$$\cos x - 4(\sin 2y)y' = 0$$
$$y' = \frac{\cos x}{4\sin 2y}$$

13.
$$\sin x = x(1 + \tan y)$$
$$\cos x = x(\sec^2 y)y' + (1 + \tan y)(1)$$
$$y' = \frac{\cos x - \tan y - 1}{x\sec^2 y}$$

15.
$$y = \sin(xy)$$
$$y' = [xy' + y]\cos(xy)$$
$$y' - x\cos(xy)y' = y\cos(xy)$$
$$y' = \frac{y\cos(xy)}{1 - x\cos(xy)}$$

17. (a) $x^2 + y^2 = 16$
$$y^2 = 16 - x^2$$
$$y = \pm\sqrt{16 - x^2}$$

(c) Explicitly:
$$\frac{dy}{dx} = \pm\frac{1}{2}(16 - x^2)^{-1/2}(-2x)$$
$$= \frac{\mp x}{\sqrt{16 - x^2}} = \frac{-x}{\pm\sqrt{16 - x^2}} = \frac{-x}{y}$$

(b)

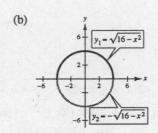

(d) Implicitly: $2x + 2yy' = 0$
$$y' = -\frac{x}{y}$$

19. (a) $16y^2 = 144 - 9x^2$
$$y^2 = \frac{1}{16}(144 - 9x^2) = \frac{9}{16}(16 - x^2)$$
$$y = \pm\frac{3}{4}\sqrt{16 - x^2}$$

(c) Explicitly:
$$\frac{dy}{dx} = \pm\frac{3}{8}(16 - x^2)^{-1/2}(-2x)$$
$$= \mp\frac{3x}{4\sqrt{16 - x^2}} = \frac{-3x}{4(4/3)y} = \frac{-9x}{16y}$$

(b)

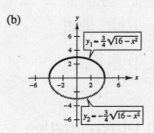

(d) Implicitly: $18x + 32yy' = 0$
$$y' = \frac{-9x}{16y}$$

21.
$$xy = 4$$
$$xy' + y(1) = 0$$
$$xy' = -y$$
$$y' = \frac{-y}{x}$$

At $(-4, -1)$: $y' = -\dfrac{1}{4}$

23.
$$y^2 = \frac{x^2 - 4}{x^2 + 4}$$
$$2yy' = \frac{(x^2 + 4)(2x) - (x^2 - 4)(2x)}{(x^2 + 4)^2}$$
$$2yy' = \frac{16x}{(x^2 + 4)^2}$$
$$y' = \frac{8x}{y(x^2 + 4)^2}$$

At $(2, 0)$: y' is undefined.

25. $x^{2/3} + y^{2/3} = 5$

$\frac{2}{3}x^{-1/3} + \frac{2}{3}y^{-1/3}y' = 0$

$y' = \frac{-x^{-1/3}}{y^{-1/3}} = -\sqrt[3]{\frac{y}{x}}$

At $(8, 1)$: $y' = -\frac{1}{2}$

27. $\tan(x + y) = x$

$(1 + y')\sec^2(x + y) = 1$

$y' = \frac{1 - \sec^2(x + y)}{\sec^2(x + y)}$

$= \frac{-\tan^2(x + y)}{\tan^2(x + y) + 1}$

$= -\sin^2(x + y)$

$= -\frac{x^2}{x^2 + 1}$

At $(0, 0)$: $y' = 0$

29. $(x^2 + 4)y = 8$

$(x^2 + 4)y' + y(2x) = 0$

$y' = \frac{-2xy}{x^2 + 4}$

$= \frac{-2x[8/(x^2 + 4)]}{x^2 + 4}$

$= \frac{-16x}{(x^2 + 4)^2}$

At $(2, 1)$: $y' = \frac{-32}{64} = -\frac{1}{2}$

$\left(\text{Or, you could just solve for } y: y = \frac{8}{x^2 + 4}\right).$

31. $(x^2 + y^2)^2 = 4x^2y$

$2(x^2 + y^2)(2x + 2yy') = 4x^2y' + y(8x)$

$4x^3 + 4x^2yy' + 4xy^2 + 4y^3y' = 4x^2y' + 8xy$

$4x^2yy' + 4y^3y' - 4x^2y' = 8xy - 4x^3 - 4xy^2$

$4y'(x^2y + y^3 - x^2) = 4(2xy - x^3 - xy^2)$

$y' = \frac{2xy - x^3 - xy^2}{x^2y + y^3 - x^2}$

At $(1, 1)$: $y' = 0$

33. $(y - 2)^2 = 4(x - 3)$, $(4, 0)$

$2(y - 2)y' = 4$

$y' = \frac{2}{y - 2}$

At $(4, 0)$: $y' = -1$

Tangent line: $y - 0 = -1(x - 4)$

$y = -x + 4$

35. $xy = 1$, $(1, 1)$

$xy' + y = 0$

$y' = \frac{-y}{x}$

At $(1, 1)$: $y' = -1$

Tangent line: $y - 1 = -1(x - 1)$

$y = -x + 2$

37. $x^2y^2 - 9x^2 - 4y^2 = 0$, $\left(-4, 2\sqrt{3}\right)$

$x^2\,2yy' + 2xy^2 - 18x - 8yy' = 0$

$y' = \frac{18x - 2xy^2}{2x^2y - 8y}$

At $\left(-4, 2\sqrt{3}\right)$: $y' = \frac{18(-4) - 2(-4)(12)}{2(16)(2\sqrt{3}) - 16\sqrt{3}}$

$= \frac{24}{48\sqrt{3}} = \frac{1}{2\sqrt{3}} = \frac{\sqrt{3}}{6}$

Tangent line: $y - 2\sqrt{3} = \frac{\sqrt{3}}{6}(x + 4)$

$y = \frac{\sqrt{3}}{6}x + \frac{8}{3}\sqrt{3}$

39. $3(x^2 + y^2)^2 = 100(x^2 - y^2)$, $(4, 2)$

$6(x^2 + y^2)(2x + 2yy') = 100(2x - 2yy')$

At $(4, 2)$:

$6(16 + 4)(8 + 4y') = 100(8 - 4y')$

$960 + 480y' = 800 - 400y'$

$880y' = -160$

$y' = -\frac{2}{11}$

Tangent line: $y - 2 = -\frac{2}{11}(x - 4)$

$11y + 2x - 30 = 0$

$y = -\frac{2}{11}x + \frac{30}{11}$

41. (a) $\dfrac{x^2}{2} + \dfrac{y^2}{8} = 1$, $(1, 2)$

$$x + \frac{yy'}{4} = 0$$

$$y' = -\frac{4x}{y}$$

At $(1, 2)$: $y' = -2$

Tangent line: $y - 2 = -2(x - 1)$

$$y = -2x + 4$$

Note: From part (a), $\dfrac{1(x)}{2} + \dfrac{2(y)}{8} = 1 \implies \dfrac{1}{4}y = -\dfrac{1}{2}x + 1 \implies y = -2x + 4$, Tangent line.

(b) $\dfrac{x^2}{a^2} + \dfrac{y^2}{b^2} = 1 \implies \dfrac{2x}{a^2} + \dfrac{2yy'}{b^2} = 0 \implies y' = \dfrac{-b^2 x}{a^2 y}$

$$y - y_0 = \frac{-b^2 x_0}{a^2 y_0}(x - x_0), \quad \text{Tangent line at } (x_0, y_0)$$

$$\frac{y_0 y}{b^2} - \frac{y_0{}^2}{b^2} = \frac{-x_0 x}{a^2} + \frac{x_0{}^2}{a^2}$$

Since $\dfrac{x_0{}^2}{a^2} + \dfrac{y_0{}^2}{b^2} = 1$, you have $\dfrac{y_0 y}{b^2} + \dfrac{x_0 x}{a^2} = 1$.

43. $\tan y = x$

$y' \sec^2 y = 1$

$$y' = \frac{1}{\sec^2 y} = \cos^2 y, \quad -\frac{\pi}{2} < y < \frac{\pi}{2}$$

$$\sec^2 y = 1 + \tan^2 y = 1 + x^2$$

$$y' = \frac{1}{1 + x^2}$$

45. $x^2 + y^2 = 36$

$2x + 2yy' = 0$

$$y' = \frac{-x}{y}$$

$$y'' = \frac{y(-1) + xy'}{y^2} = \frac{-y + x(-x/y)}{y^2} = \frac{-y^2 - x^2}{y^3} = \frac{-36}{y^3}$$

47. $x^2 - y^2 = 16$

$2x - 2yy' = 0$

$$y' = \frac{x}{y}$$

$$x - yy' = 0$$

$$1 - yy'' - (y')^2 = 0$$

$$1 - yy'' - \left(\frac{x}{y}\right)^2 = 0$$

$$y^2 - y^3 y'' = x^2$$

$$y'' = \frac{y^2 - x^2}{y^3} = \frac{-16}{y^3}$$

49. $y^2 = x^3$

$2yy' = 3x^2$

$$y' = \frac{3x^2}{2y} = \frac{3x^2}{2y} \cdot \frac{xy}{xy}$$

$$= \frac{3y}{2x} \cdot \frac{x^3}{y^2} = \frac{3y}{2x}$$

$$y'' = \frac{2x(3y') - 3y(2)}{4x^2}$$

$$= \frac{2x[3 \cdot (3y/2x)] - 6y}{4x^2}$$

$$= \frac{3y}{4x^2} = \frac{3x}{4y}$$

51. $\sqrt{x} + \sqrt{y} = 4$

$$\frac{1}{2}x^{-1/2} + \frac{1}{2}y^{-1/2}y' = 0$$

$$y' = \frac{-\sqrt{y}}{\sqrt{x}}$$

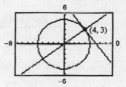

At $(9, 1)$: $y' = -\frac{1}{3}$

Tangent line: $\quad y - 1 = -\frac{1}{3}(x - 9)$

$$y = -\frac{1}{3}x + 4$$

$$x + 3y - 12 = 0$$

53. $x^2 + y^2 = 25$

$$2x + 2yy' = 0$$

$$y' = \frac{-x}{y}$$

At $(4, 3)$:

Tangent line: $y - 3 = \frac{-4}{3}(x - 4) \Longrightarrow 4x + 3y - 25 = 0$

Normal line: $y - 3 = \frac{3}{4}(x - 4) \Longrightarrow 3x - 4y = 0$

At $(-3, 4)$:

Tangent line: $y - 4 = \frac{3}{4}(x + 3) \Longrightarrow 3x - 4y + 25 = 0$

Normal line: $y - 4 = \frac{-4}{3}(x + 3) \Longrightarrow 4x + 3y = 0$

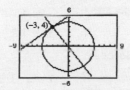

55. $x^2 + y^2 = r^2$

$$2x + 2yy' = 0$$

$$y' = \frac{-x}{y} = \text{slope of tangent line}$$

$$\frac{y}{x} = \text{slope of normal line}$$

Let (x_0, y_0) be a point on the circle. If $x_0 = 0$, then the tangent line is horizontal, the normal line is vertical and, hence, passes through the origin. If $x_0 \neq 0$, then the equation of the normal line is

$$y - y_0 = \frac{y_0}{x_0}(x - x_0)$$

$$y = \frac{y_0}{x_0}x$$

which passes through the origin.

57. $25x^2 + 16y^2 + 200x - 160y + 400 = 0$

$$50x + 32yy' + 200 - 160y' = 0$$

$$y' = \frac{200 + 50x}{160 - 32y}$$

Horizontal tangents occur when $x = -4$:

$$25(16) + 16y^2 + 200(-4) - 160y + 400 = 0$$

$$y(y - 10) = 0 \Rightarrow y = 0, 10$$

Horizontal tangents: $(-4, 0), (-4, 10)$

Vertical tangents occur when $y = 5$:

$$25x^2 + 400 + 200x - 800 + 400 = 0$$

$$25x(x + 8) = 0 \Rightarrow x = 0, -8$$

Vertical tangents: $(0, 5), (-8, 5)$

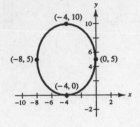

59. Find the points of intersection by letting $y^2 = 4x$ in the equation $2x^2 + y^2 = 6$.

$$2x^2 + 4x = 6 \quad \text{and} \quad (x + 3)(x - 1) = 0$$

The curves intersect at $(1, \pm 2)$.

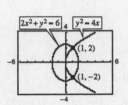

Ellipse:	*Parabola*:
$4x + 2yy' = 0$	$2yy' = 4$
$y' = -\dfrac{2x}{y}$	$y' = \dfrac{2}{y}$

At $(1, 2)$, the slopes are:

$$y' = -1 \qquad\qquad y' = 1$$

At $(1, -2)$, the slopes are:

$$y' = 1 \qquad\qquad y' = -1$$

Tangents are perpendicular.

61. $y = -x$ and $x = \sin y$

Point of intersection: $(0, 0)$

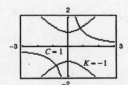

$y = -x$:	$x = \sin y$:
$y' = -1$	$1 = y' \cos y$
	$y' = \sec y$

At $(0, 0)$, the slopes are:

$$y' = -1 \qquad y' = 1$$

Tangents are perpendicular.

63.

$xy = C$	$x^2 - y^2 = K$
$xy' + y = 0$	$2x - 2yy' = 0$
$y' = -\dfrac{y}{x}$	$y' = \dfrac{x}{y}$

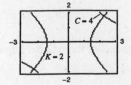

At any point of intersection (x, y) the product of the slopes is $(-y/x)(x/y) = -1$. The curves are orthogonal.

65. $2y^2 - 3x^4 = 0$

   (a) $4yy' - 12x^3 = 0$

$$4yy' = 12x^3$$

$$y' = \frac{12x^3}{4y} = \frac{3x^3}{y}$$

   (b) $4y\dfrac{dy}{dt} - 12x^3\dfrac{dx}{dt} = 0$

$$y\frac{dy}{dt} = 3x^3\frac{dx}{dt}$$

67. $\cos \pi y - 3\sin \pi x = 1$

   (a) $-\pi \sin(\pi y)y' - 3\pi \cos \pi x = 0$

$$y' = \frac{-3\cos \pi x}{\sin \pi y}$$

   (b) $-\pi \sin(\pi y)\dfrac{dy}{dt} - 3\pi \cos(\pi x)\dfrac{dx}{dt} = 0$

$$-\sin(\pi y)\frac{dy}{dt} = 3\cos(\pi x)\frac{dx}{dt}$$

69. A function is in explicit form if y is written as a function of x: $y = f(x)$. For example, $y = x^3$.
An implicit equation is not in the form $y = f(x)$. For example, $x^2 + y^2 = 5$.

71.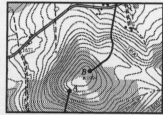

Use starting point B.

73. (a) $x^4 = 4(4x^2 - y^2)$

$$4y^2 = 16x^2 - x^4$$

$$y^2 = 4x^2 - \frac{1}{4}x^4$$

$$y = \pm\sqrt{4x^2 - \frac{1}{4}x^4}$$

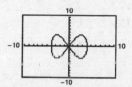

   (b) $y = 3 \Rightarrow 9 = 4x^2 - \dfrac{1}{4}x^4$

$$36 = 16x^2 - x^4$$

$$x^4 - 16x^2 + 36 = 0$$

$$x^2 = \frac{16 \pm \sqrt{256 - 144}}{2} = 8 \pm \sqrt{28}$$

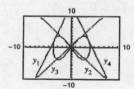

Note that $x^2 = 8 \pm \sqrt{28} = 8 \pm 2\sqrt{7} = \left(1 \pm \sqrt{7}\right)^2$. Hence, there are four values of x:

$$-1 - \sqrt{7}, 1 - \sqrt{7}, -1 + \sqrt{7}, 1 + \sqrt{7}$$

To find the slope, $2yy' = 8x - x^3 \Rightarrow y' = \dfrac{x(8 - x^2)}{2(3)}$.

—CONTINUED—

73. —CONTINUED—

For $x = -1 - \sqrt{7}$, $y' = \frac{1}{3}(\sqrt{7} + 7)$, and the line is

$$y_1 = \frac{1}{3}(\sqrt{7} + 7)(x + 1 + \sqrt{7}) + 3 = \frac{1}{3}[(\sqrt{7} + 7)x + 8\sqrt{7} + 23].$$

For $x = 1 - \sqrt{7}$, $y' = \frac{1}{3}(\sqrt{7} - 7)$, and the line is

$$y_2 = \frac{1}{3}(\sqrt{7} - 7)(x - 1 + \sqrt{7}) + 3 = \frac{1}{3}[(\sqrt{7} - 7)x + 23 - 8\sqrt{7}].$$

For $x = -1 + \sqrt{7}$, $y' = -\frac{1}{3}(\sqrt{7} - 7)$, and the line is

$$y_3 = -\frac{1}{3}(\sqrt{7} - 7)(x + 1 - \sqrt{7}) + 3 = -\frac{1}{3}[(\sqrt{7} - 7)x - (23 - 8\sqrt{7})].$$

For $x = 1 + \sqrt{7}$, $y' = -\frac{1}{3}(\sqrt{7} + 7)$, and the line is

$$y_4 = -\frac{1}{3}(\sqrt{7} + 7)(x - 1 - \sqrt{7}) + 3 = -\frac{1}{3}[(\sqrt{7} + 7)x - (8\sqrt{7} + 23)].$$

(c) Equating y_3 and y_4:

$$-\frac{1}{3}(\sqrt{7} - 7)(x + 1 - \sqrt{7}) + 3 = -\frac{1}{3}(\sqrt{7} + 7)(x - 1 - \sqrt{7}) + 3$$

$$(\sqrt{7} - 7)(x + 1 - \sqrt{7}) = (\sqrt{7} + 7)(x - 1 - \sqrt{7})$$

$$\sqrt{7}x + \sqrt{7} - 7 - 7x - 7 + 7\sqrt{7} = \sqrt{7}x - \sqrt{7} - 7 + 7x - 7 - 7\sqrt{7}$$

$$16\sqrt{7} = 14x$$

$$x = \frac{8\sqrt{7}}{7}$$

If $x = \frac{8\sqrt{7}}{7}$, then $y = 5$ and the lines intersect at $\left(\frac{8\sqrt{7}}{7}, 5\right)$.

75. $y = x^{p/q}$; p, q integers and $q > 0$

$$y^q = x^p$$

$$qy^{q-1}y' = px^{p-1}$$

$$y' = \frac{p}{q} \cdot \frac{x^{p-1}}{y^{q-1}} = \frac{p}{q} \cdot \frac{x^{p-1}y}{y^q}$$

$$= \frac{p}{q} \cdot \frac{x^{p-1}}{x^p}x^{p/q} = \frac{p}{q}x^{p/q-1}$$

Thus, if $y = x^n$, $n = p/q$, then $y' = nx^{n-1}$.

77. $y^4 = y^2 - x^2$

$$4y^3y' = 2yy' - 2x$$

$$2x = (2y - 4y^3)y'$$

$$y' = \frac{2x}{2y - 4y^3} = 0 \implies x = 0$$

Horizontal tangents at $(0, 1)$ and $(0, -1)$

Note: $y^4 - y^2 + x^2 = 0$

$$y^2 = \frac{1 \pm \sqrt{1 - 4x^2}}{2}$$

If you graph these four equations, you will see that these are horizontal tangents at $(0, \pm1)$, but not at $(0, 0)$.

79. $x = y^2$

$1 = 2yy'$

$y' = \dfrac{1}{2y}$, slope of tangent line

Consider the slope of the normal line joining $(x_0, 0)$ and $(x, y) = (y^2, y)$ on the parabola.

$-2y = \dfrac{y - 0}{y^2 - x_0}$

$y^2 - x_0 = -\dfrac{1}{2}$

$y^2 = x_0 - \dfrac{1}{2}$

(a) If $x_0 = \frac{1}{4}$, then $y^2 = \frac{1}{4} - \frac{1}{2} = -\frac{1}{4}$, which is impossible. Thus, the only normal line is the x-axis ($y = 0$).

(b) If $x_0 = \frac{1}{2}$, then $y^2 = 0 \implies y = 0$. Same as part (a).

(c) If $x_0 = 1$, then $y^2 = \frac{1}{2} = x$ and there are three normal lines:

The x-axis, the line joining $(x_0, 0)$ and $\left(\dfrac{1}{2}, \dfrac{1}{\sqrt{2}}\right)$, and the line joining $(x_0, 0)$ and $\left(\dfrac{1}{2}, -\dfrac{1}{\sqrt{2}}\right)$.

If two normals are perpendicular, then their slopes are -1 and 1. Thus,

$-2y = -1 = \dfrac{y - 0}{y^2 - x_0} \implies y = \dfrac{1}{2}$ and $\dfrac{1/2}{(1/4) - x_0} = -1 \implies \dfrac{1}{4} - x_0 = -\dfrac{1}{2} \implies x_0 = \dfrac{3}{4}.$

The perpendicular normal lines are $y = -x + \frac{3}{4}$ and $y = x - \frac{3}{4}$.

Section 2.6   Related Rates

1. $y = \sqrt{x}$

$\dfrac{dy}{dt} = \left(\dfrac{1}{2\sqrt{x}}\right)\dfrac{dx}{dt}$

$\dfrac{dx}{dt} = 2\sqrt{x}\,\dfrac{dy}{dt}$

(a) When $x = 4$ and $dx/dt = 3$,

$\dfrac{dy}{dt} = \dfrac{1}{2\sqrt{4}}(3) = \dfrac{3}{4}.$

(b) When $x = 25$ and $dy/dt = 2$,

$\dfrac{dx}{dt} = 2\sqrt{25}\,(2) = 20.$

3. $xy = 4$

$x\dfrac{dy}{dt} + y\dfrac{dx}{dt} = 0$

$\dfrac{dy}{dt} = \left(-\dfrac{y}{x}\right)\dfrac{dx}{dt}$

$\dfrac{dx}{dt} = \left(-\dfrac{x}{y}\right)\dfrac{dy}{dt}$

(a) When $x = 8$, $y = 1/2$, and $dx/dt = 10$,

$\dfrac{dy}{dt} = -\dfrac{1/2}{8}(10) = -\dfrac{5}{8}.$

(b) When $x = 1$, $y = 4$, and $dy/dt = -6$,

$\dfrac{dx}{dt} = -\dfrac{1}{4}(-6) = \dfrac{3}{2}.$

5. $y = x^2 + 1$

$$\frac{dx}{dt} = 2$$

$$\frac{dy}{dt} = 2x \frac{dx}{dt}$$

(a) When $x = -1$,

$$\frac{dy}{dt} = 2(-1)(2) = -4 \text{ cm/sec.}$$

(b) When $x = 0$,

$$\frac{dy}{dt} = 2(0)(2) = 0 \text{ cm/sec.}$$

(c) When $x = 1$,

$$\frac{dy}{dt} = 2(1)(2) = 4 \text{ cm/sec.}$$

7. $y = \tan x$

$$\frac{dx}{dt} = 2$$

$$\frac{dy}{dt} = \sec^2 x \frac{dx}{dt}$$

(a) When $x = -\pi/3$,

$$\frac{dy}{dt} = (2)^2(2) = 8 \text{ cm/sec.}$$

(b) When $x = -\pi/4$,

$$\frac{dy}{dt} = \left(\sqrt{2}\right)^2(2) = 4 \text{ cm/sec.}$$

(c) When $x = 0$,

$$\frac{dy}{dt} = (1)^2(2) = 2 \text{ cm/sec.}$$

9. (a) $\dfrac{dx}{dt}$ negative $\Longrightarrow \dfrac{dy}{dt}$ positive

(b) $\dfrac{dy}{dt}$ positive $\Longrightarrow \dfrac{dx}{dt}$ negative

11. Yes, y changes at a constant rate.

$$\frac{dy}{dt} = a \cdot \frac{dx}{dt}$$

No, the rate dy/dt is a multiple of dx/dt.

13. $D = \sqrt{x^2 + y^2} = \sqrt{x^2 + (x^2 + 1)^2} = \sqrt{x^4 + 3x^2 + 1}$

$$\frac{dx}{dt} = 2$$

$$\frac{dD}{dt} = \frac{1}{2}(x^4 + 3x^2 + 1)^{-1/2}(4x^3 + 6x)\frac{dx}{dt}$$

$$= \frac{2x^3 + 3x}{\sqrt{x^4 + 3x^2 + 1}}\frac{dx}{dt}$$

$$= \frac{4x^3 + 6x}{\sqrt{x^4 + 3x^2 + 1}}$$

15. $A = \pi r^2$

$$\frac{dr}{dt} = 3$$

$$\frac{dA}{dt} = 2\pi r \frac{dr}{dt}$$

(a) When $r = 6$, $\dfrac{dA}{dt} = 2\pi(6)(3) = 36\pi \text{ cm}^2/\text{min.}$

(b) When $r = 24$, $\dfrac{dA}{dt} = 2\pi(24)(3) = 144\pi \text{ cm}^2/\text{min.}$

17. (a) $\sin\dfrac{\theta}{2} = \dfrac{(1/2)b}{s} \Longrightarrow b = 2s \sin\dfrac{\theta}{2}$

$$\cos\frac{\theta}{2} = \frac{h}{s} \Longrightarrow h = s\cos\frac{\theta}{2}$$

$$A = \frac{1}{2}bh = \frac{1}{2}\left(2s \sin\frac{\theta}{2}\right)\left(s\cos\frac{\theta}{2}\right)$$

$$= \frac{s^2}{2}\left(2\sin\frac{\theta}{2}\cos\frac{\theta}{2}\right) = \frac{s^2}{2}\sin\theta$$

(b) $\dfrac{dA}{dt} = \dfrac{s^2}{2}\cos\theta\dfrac{d\theta}{dt}$ where $\dfrac{d\theta}{dt} = \dfrac{1}{2}$ rad/min.

When $\theta = \dfrac{\pi}{6}$, $\dfrac{dA}{dt} = \dfrac{s^2}{2}\left(\dfrac{\sqrt{3}}{2}\right)\left(\dfrac{1}{2}\right) = \dfrac{\sqrt{3}s^2}{8}$.

When $\theta = \dfrac{\pi}{3}$, $\dfrac{dA}{dt} = \dfrac{s^2}{2}\left(\dfrac{1}{2}\right)\left(\dfrac{1}{2}\right) = \dfrac{s^2}{8}$.

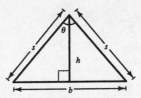

(c) If $\dfrac{d\theta}{dt}$ is constant, $\dfrac{dA}{dt}$ is proportional to $\cos\theta$.

19. $V = \frac{4}{3}\pi r^3$, $\frac{dV}{dt} = 800$

$$\frac{dV}{dt} = 4\pi r^2 \frac{dr}{dt}$$

$$\frac{dr}{dt} = \frac{1}{4\pi r^2}\left(\frac{dV}{dt}\right) = \frac{1}{4\pi r^2}(800)$$

(a) When $r = 30$,

$$\frac{dr}{dt} = \frac{1}{4\pi(30)^2}(800) = \frac{2}{9\pi} \text{ cm/min.}$$

(b) When $r = 60$,

$$\frac{dr}{dt} = \frac{1}{4\pi(60)^2}(800) = \frac{1}{18\pi} \text{ cm/min.}$$

21. $s = 6x^2$

$$\frac{dx}{dt} = 3$$

$$\frac{ds}{dt} = 12x \frac{dx}{dt}$$

(a) When $x = 1$,

$$\frac{ds}{dt} = 12(1)(3) = 36 \text{ cm}^2/\text{sec.}$$

(b) When $x = 10$,

$$\frac{ds}{dt} = 12(10)(3) = 360 \text{ cm}^2/\text{sec.}$$

23. $V = \frac{1}{3}\pi r^2 h = \frac{1}{3}\pi\left(\frac{9}{4}h^2\right)h$ [since $2r = 3h$]

$$= \frac{3\pi}{4}h^3$$

$$\frac{dV}{dt} = 10$$

$$\frac{dV}{dt} = \frac{9\pi}{4}h^2\frac{dh}{dt} \Rightarrow \frac{dh}{dt} = \frac{4\,(dV/dt)}{9\pi h^2}$$

When $h = 15$,

$$\frac{dh}{dt} = \frac{4(10)}{9\pi(15)^2} = \frac{8}{405\pi} \text{ ft/min.}$$

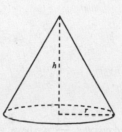

25.

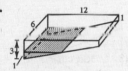

(a) Total volume of pool $= \frac{1}{2}(2)(12)(6) + (1)(6)(12) = 144 \text{ m}^3$

Volume of 1 m of water $= \frac{1}{2}(1)(6)(6) = 18 \text{ m}^3$ (see similar triangle diagram)

% pool filled $= \frac{18}{144}(100\%) = 12.5\%$

(b) Since for $0 \le h \le 2$, $b = 6h$, you have

$$V = \frac{1}{2}bh(6) = 3bh = 3(6h)h = 18h^2$$

$$\frac{dV}{dt} = 36h\frac{dh}{dt} = \frac{1}{4} \Rightarrow \frac{dh}{dt} = \frac{1}{144h} = \frac{1}{144(1)} = \frac{1}{144} \text{ m/min.}$$

27. $x^2 + y^2 = 25^2$

$$2x\frac{dx}{dt} + 2y\frac{dy}{dt} = 0$$

$$\frac{dy}{dt} = \frac{-x}{y} \cdot \frac{dx}{dt} = \frac{-2x}{y} \text{ since } \frac{dx}{dt} = 2.$$

(a) When $x = 7$, $y = \sqrt{576} = 24$, $\frac{dy}{dt} = \frac{-2(7)}{24} = \frac{-7}{12} \text{ ft/sec.}$

—CONTINUED—

27. —CONTINUED—

(a) When $x = 7$, $y = \sqrt{576} = 24$, $\dfrac{dy}{dt} = \dfrac{-2(7)}{24} = \dfrac{-7}{12}$ ft/sec.

When $x = 15$, $y = \sqrt{400} = 20$, $\dfrac{dy}{dt} = \dfrac{-2(15)}{20} = \dfrac{-3}{2}$ ft/sec.

When $x = 24$, $y = 7$, $\dfrac{dy}{dt} = \dfrac{-2(24)}{7} = \dfrac{-48}{7}$ ft/sec.

(b) $A = \dfrac{1}{2}xy$

$\dfrac{dA}{dt} = \dfrac{1}{2}\left(x\,\dfrac{dy}{dt} + y\,\dfrac{dx}{dt}\right)$

From part (a) we have $x = 7$, $y = 24$, $\dfrac{dx}{dt} = 2$, and $\dfrac{dy}{dt} = -\dfrac{7}{12}$. Thus,

$\dfrac{dA}{dt} = \dfrac{1}{2}\left[7\left(-\dfrac{7}{12}\right) + 24(2)\right] = \dfrac{527}{24} \approx 21.96$ ft²/sec.

(c) $\tan\theta = \dfrac{x}{y}$

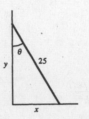

$\sec^2\theta\,\dfrac{d\theta}{dt} = \dfrac{1}{y}\cdot\dfrac{dx}{dt} - \dfrac{x}{y^2}\cdot\dfrac{dy}{dt}$

$\dfrac{d\theta}{dt} = \cos^2\theta\left[\dfrac{1}{y}\cdot\dfrac{dx}{dt} - \dfrac{x}{y^2}\cdot\dfrac{dy}{dt}\right]$

Using $x = 7$, $y = 24$, $\dfrac{dx}{dt} = 2$, $\dfrac{dy}{dt} = -\dfrac{7}{12}$ and $\cos\theta = \dfrac{24}{25}$, we have

$\dfrac{d\theta}{dt} = \left(\dfrac{24}{25}\right)^2\left[\dfrac{1}{24}(2) - \dfrac{7}{(24)^2}\left(-\dfrac{7}{12}\right)\right] = \dfrac{1}{12}$ rad/sec.

29. When $y = 6$, $x = \sqrt{12^2 - 6^2} = 6\sqrt{3}$, and $s = \sqrt{x^2 + (12-y)^2}$

$\qquad\qquad\qquad\qquad\qquad = \sqrt{108 + 36} = 12.$

$$x^2 + (12 - y)^2 = s^2$$

$2x\dfrac{dx}{dt} + 2(12 - y)(-1)\dfrac{dy}{dt} = 2s\dfrac{ds}{dt}$

$x\dfrac{dx}{dt} + (y - 12)\dfrac{dy}{dt} = s\dfrac{ds}{dt}$

Also, $x^2 + y^2 = 12^2$.

$2x\dfrac{dx}{dt} + 2y\dfrac{dy}{dt} = 0 \Longrightarrow \dfrac{dy}{dt} = \dfrac{-x}{y}\dfrac{dx}{dt}$

Thus, $x\dfrac{dx}{dt} + (y - 12)\left(\dfrac{-x}{y}\dfrac{dx}{dt}\right) = s\dfrac{ds}{dt}$.

$\dfrac{dx}{dt}\left[x - x + \dfrac{12x}{y}\right] = s\dfrac{ds}{dt} \Longrightarrow \dfrac{dx}{dt} = \dfrac{sy}{12x}\cdot\dfrac{ds}{dt} = \dfrac{(12)(6)}{(12)\left(6\sqrt{3}\right)}(-0.2) = \dfrac{-1}{5\sqrt{3}} = \dfrac{-\sqrt{3}}{15}$ m/sec (horizontal)

$\dfrac{dy}{dt} = \dfrac{-x}{y}\dfrac{dx}{dt} = \dfrac{-6\sqrt{3}}{6}\cdot\dfrac{\left(-\sqrt{3}\right)}{15} = \dfrac{1}{5}$ m/sec (vertical)

31. (a) $s^2 = x^2 + y^2$

$$\frac{dx}{dt} = -450$$

$$\frac{dy}{dt} = -600$$

$$2s\frac{ds}{dt} = 2x\frac{dx}{dt} + 2y\frac{dy}{dt}$$

$$\frac{ds}{dt} = \frac{x(dx/dt) + y(dy/dt)}{s}$$

When $x = 150$ and $y = 200$, $s = 250$ and $\frac{ds}{dt} = \frac{150(-450) + 200(-600)}{250} = -750$ mph.

(b) $t = \frac{250}{750} = \frac{1}{3}$ hr $= 20$ min

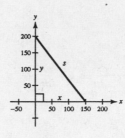

33. $s^2 = 90^2 + x^2$

$x = 30$

$$\frac{dx}{dt} = -28$$

$$2s\frac{ds}{dt} = 2x\frac{dx}{dt} \Rightarrow \frac{ds}{dt} = \frac{x}{s} \cdot \frac{dx}{dt}$$

When $x = 30$, $s = \sqrt{90^2 + 30^2} = 30\sqrt{10}$,

$$\frac{ds}{dt} = \frac{30}{30\sqrt{10}}(-28) = \frac{-28}{\sqrt{10}} \approx -8.85 \text{ ft/sec.}$$

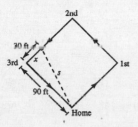

35. (a) $\frac{15}{6} = \frac{y}{y - x} \Rightarrow 15y - 15x - 6y$

$$y = \frac{5}{3}x$$

$$\frac{dx}{dt} = 5$$

$$\frac{dy}{dt} = \frac{5}{3} \cdot \frac{dx}{dt} = \frac{5}{3}(5) = \frac{25}{3} \text{ ft/sec}$$

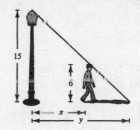

(b) $\frac{d(y - x)}{dt} = \frac{dy}{dt} - \frac{dx}{dt} = \frac{25}{3} - 5 = \frac{10}{3} \text{ ft/sec}$

37. $x(t) = \frac{1}{2}\sin\frac{\pi t}{6}$, $x^2 + y^2 = 1$

(a) Period: $\frac{2\pi}{\pi/6} = 12$ seconds

(b) When $x = \frac{1}{2}$, $y = \sqrt{1^2 - \left(\frac{1}{2}\right)^2} = \frac{\sqrt{3}}{2}$ m.

Lowest point: $\left(0, \frac{\sqrt{3}}{2}\right)$

(c) When $x = \frac{1}{4}$, $y = \sqrt{1 - \left(\frac{1}{4}\right)^2} = \frac{\sqrt{15}}{4}$ and $t = 1$:

$$\frac{dx}{dt} = \frac{1}{2}\left(\frac{\pi}{6}\right)\cos\frac{\pi t}{6} = \frac{\pi}{12}\cos\frac{\pi t}{6}$$

$$x^2 + y^2 = 1$$

$$2x\frac{dx}{dt} + 2y\frac{dy}{dt} = 0 \Rightarrow \frac{dy}{dt} = \frac{-x}{y}\frac{dx}{dt}$$

Thus, $\frac{dy}{dt} = -\frac{1/4}{\sqrt{15}/4} \cdot \frac{\pi}{12}\cos\left(\frac{\pi}{6}\right)$

$$= \frac{-\pi}{\sqrt{15}}\left(\frac{1}{12}\right)\frac{\sqrt{3}}{2} = \frac{-\pi}{24}\frac{1}{\sqrt{5}} = \frac{-\sqrt{5}\pi}{120}.$$

Speed $= \left|\frac{-\sqrt{5}\pi}{120}\right| = \frac{\sqrt{5}\pi}{120}$ m/sec

39. Since the evaporation rate is proportional to the surface area, $dV/dt = k(4\pi r^2)$. However, since $V = (4/3)\pi r^3$, we have

$$\frac{dV}{dt} = 4\pi r^2 \frac{dr}{dt}.$$

Therefore, $k(4\pi r^2) = 4\pi r^2 \dfrac{dr}{dt} \Rightarrow k = \dfrac{dr}{dt}.$

41.
$$pV^{1.3} = k$$

$$1.3pV^{0.3}\frac{dV}{dt} + V^{1.3}\frac{dp}{dt} = 0$$

$$V^{0.3}\left(1.3p\frac{dV}{dt} + V\frac{dp}{dt}\right) = 0$$

$$1.3p\frac{dV}{dt} = -V\frac{dp}{dt}$$

43. $\tan\theta = \dfrac{y}{30}$

$$\frac{dy}{dt} = 3 \text{ m/sec}$$

$$\sec^2\theta \cdot \frac{d\theta}{dt} = \frac{1}{30}\frac{dy}{dt}$$

$$\frac{d\theta}{dt} = \frac{1}{30}\cos^2\theta \cdot \frac{dy}{dt}$$

When $y = 30$, $\theta = \dfrac{\pi}{4}$, and $\cos\theta = \dfrac{\sqrt{2}}{2}$. Thus, $\dfrac{d\theta}{dt} = \dfrac{1}{30}\left(\dfrac{1}{2}\right)(3) = \dfrac{1}{20}$ rad/sec.

45. $\tan\theta = \dfrac{y}{x}, y = 5$

$$\frac{dx}{dt} = -600 \text{ mi/hr}$$

$$(\sec^2\theta)\frac{d\theta}{dt} = -\frac{5}{x^2}\cdot\frac{dx}{dt}$$

$$\frac{d\theta}{dt} = \cos^2\theta\left(-\frac{5}{x^2}\right)\frac{dx}{dt} = \frac{x^2}{L^2}\left(-\frac{5}{x^2}\right)\frac{dx}{dt}$$

$$= \left(-\frac{5^2}{L^2}\right)\left(\frac{1}{5}\right)\frac{dx}{dt} = (-\sin^2\theta)\left(\frac{1}{5}\right)(-600) = 120\sin^2\theta$$

(a) When $\theta = 30°$,

$$\frac{d\theta}{dt} = \frac{120}{4}$$

$$= 30 \text{ rad/hr} = \frac{1}{2} \text{ rad/min}.$$

(b) When $\theta = 60°$,

$$\frac{d\theta}{dt} = 120\left(\frac{3}{4}\right)$$

$$= 90 \text{ rad/hr} = \frac{3}{2} \text{ rad/min}.$$

(c) When $\theta = 75°$,

$$\frac{d\theta}{dt} = 120\sin^2 75°$$

$$\approx 111.96 \text{ rad/hr} \approx 1.87 \text{ rad/min}.$$

47. $\dfrac{d\theta}{dt} = (10 \text{ rev/sec})(2\pi \text{ rad/rev}) = 20\pi \text{ rad/sec}$

(a) $\cos\theta = \dfrac{x}{30}$

$$-\sin\theta\frac{d\theta}{dt} = \frac{1}{30}\frac{dx}{dt}$$

$$\frac{dx}{dt} = -30\sin\theta\frac{d\theta}{dt}$$

$$= -30\sin\theta(20\pi)$$

$$= -600\pi\sin\theta$$

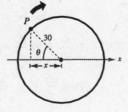

(b)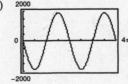

—CONTINUED—

47. —CONTINUED—

(c) $|dx/dt| = |-600\pi \sin \theta|$ is greatest when

$$|\sin \theta| = 1 \implies \theta = \frac{\pi}{2} + n\pi \ (\text{or } 90° + n \cdot 180°).$$

$|dx/dt|$ is least when $\theta = n\pi$ (or $n \cdot 180°$).

(d) For $\theta = 30°$,

$$\frac{dx}{dt} = -600\pi \sin(30°)$$

$$= -600\pi \frac{1}{2} = -300\pi \ \text{cm/sec}.$$

For $\theta = 60°$,

$$\frac{dx}{dt} = -600\pi \sin(60°)$$

$$= -600\pi \frac{\sqrt{3}}{2} = -300\sqrt{3}\pi \ \text{cm/sec}.$$

49. $\tan \theta = \dfrac{x}{50} \implies x = 50 \tan \theta$

$$\frac{dx}{dt} = 50 \sec^2 \theta \ \frac{d\theta}{dt}$$

$$2 = 50 \sec^2 \theta \ \frac{d\theta}{dt}$$

$$\frac{d\theta}{dt} = \frac{1}{25} \cos^2 \theta, \quad -\frac{\pi}{4} \le \theta \le \frac{\pi}{4}$$

51. $x^2 + y^2 = 25$; acceleration of the top of the ladder $-\dfrac{d^2y}{dt^2}$

First derivative: $2x \dfrac{dx}{dt} + 2y \dfrac{dy}{dt} = 0$

$$x \frac{dx}{dt} + y \frac{dy}{dt} = 0$$

Second derivative: $x \dfrac{d^2x}{dt^2} + \dfrac{dx}{dt} \cdot \dfrac{dx}{dt} + y \dfrac{d^2y}{dt^2} + \dfrac{dy}{dt} \cdot \dfrac{dy}{dt} = 0$

$$\frac{d^2y}{dt^2} - \left(\frac{1}{y}\right)\left[-x\frac{d^2x}{dt^2} \ \left(\frac{dx}{dt}\right)^2 \ \left(\frac{dy}{dt}\right)^2\right]$$

When $x = 7, y = 24, \dfrac{dy}{dt} = -\dfrac{7}{12}$, and $\dfrac{dx}{dt} = 2$ (see Exercise 27). Since $\dfrac{dx}{dt}$ is constant, $\dfrac{d^2x}{dt^2} = 0$.

$$\frac{d^2y}{dt^2} = \frac{1}{24}\left[-7(0) - (2)^2 - \left(-\frac{7}{12}\right)^2\right] = \frac{1}{24}\left[-4 - \frac{49}{144}\right] = \frac{1}{24}\left[-\frac{625}{144}\right] \approx -0.1808 \ \text{ft/sec}^2$$

53. (a) $m(s) = 0.3754s^3 - 18.780s^2 + 313.23s - 1707.8$

(b) $\dfrac{dm}{dt} = (1.1262s^2 - 37.560s + 313.23)\dfrac{ds}{dt}$

If $t = 10$ and $\dfrac{ds}{dt} = 0.75$, then $s = 17.8$ and $\dfrac{dm}{dt} \approx 1.1154 \ \text{million/year}$.

Review Exercises for Chapter 2

1. $f(x) = x^2 - 2x + 3$

$$f'(x) = \lim_{\Delta x \to 0} \frac{f(x + \Delta x) - f(x)}{\Delta x}$$

$$= \lim_{\Delta x \to 0} \frac{[(x + \Delta x)^2 - 2(x + \Delta x) + 3] - [x^2 - 2x + 3]}{\Delta x}$$

$$= \lim_{\Delta x \to 0} \frac{(x^2 + 2x(\Delta x) + (\Delta x)^2 - 2x - 2(\Delta x) + 3) - (x^2 - 2x + 3)}{\Delta x}$$

$$= \lim_{\Delta x \to 0} \frac{2x(\Delta x) + (\Delta x)^2 - 2(\Delta x)}{\Delta x} = \lim_{\Delta x \to 0} (2x + \Delta x - 2) = 2x - 2$$

3. $f(x) = \dfrac{x + 1}{x - 1}$

$$f'(x) = \lim_{\Delta x \to 0} \frac{f(x + \Delta x) - f(x)}{\Delta x} = \lim_{\Delta x \to 0} \frac{\dfrac{x + \Delta x + 1}{x + \Delta x - 1} - \dfrac{x + 1}{x - 1}}{\Delta x}$$

$$= \lim_{\Delta x \to 0} \frac{(x + \Delta x + 1)(x - 1) - (x + \Delta x - 1)(x + 1)}{\Delta x(x + \Delta x - 1)(x - 1)}$$

$$= \lim_{\Delta x \to 0} \frac{(x^2 + x\,\Delta x + x - x - \Delta x - 1) - (x^2 + x\,\Delta x - x + x + \Delta x - 1)}{\Delta x(x + \Delta x - 1)(x - 1)}$$

$$= \lim_{\Delta x \to 0} \frac{-2\,\Delta x}{\Delta x(x + \Delta x - 1)(x - 1)} = \lim_{\Delta x \to 0} \frac{-2}{(x + \Delta x - 1)(x - 1)} = \frac{-2}{(x - 1)^2}$$

5. f is differentiable for all $x \neq -1$.

7. $f(x) = 4 - |x - 2|$

   (a) Continuous at $x = 2$

   (b) Not differentiable at $x = 2$ because of the sharp turn in the graph.

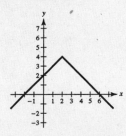

9. Using the limit definition, you obtain $g'(x) = \frac{4}{3}x - \frac{1}{6}$. At $x = -1$,

$$g'(-1) = -\frac{4}{3} - \frac{1}{6} = \frac{-3}{2}.$$

11. (a) Using the limit definition, $f'(x) = 3x^2$. At $x = -1$, $f'(-1) = 3$. The tangent line is

$$y - (-2) = 3(x - (-1))$$

$$y = 3x + 1.$$

   (b)

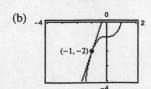

13. $g'(2) = \lim_{x \to 2} \dfrac{g(x) - g(2)}{x - 2}$

$$= \lim_{x \to 2} \frac{x^2(x - 1) - 4}{x - 2}$$

$$= \lim_{x \to 2} \frac{x^3 - x^2 - 4}{x - 2}$$

$$= \lim_{x \to 2} \frac{(x - 2)(x^2 + x + 2)}{x - 2}$$

$$= \lim_{x \to 2} (x^2 + x + 2) = 8$$

15. $y = 25$

$y' = 0$

17. $f(x) = x^8$

$f'(x) = 8x^7$

19. $h(t) = 3t^4$

$h'(t) = 12t^3$

21. $f(x) = x^3 - 3x^2$

$f'(x) = 3x^2 - 6x$

$\qquad = 3x(x - 2)$

23. $h(x) = 6\sqrt{x} + 3\sqrt[3]{x} = 6x^{1/2} + 3x^{1/3}$

$h'(x) = 3x^{-1/2} + x^{-2/3} = \dfrac{3}{\sqrt{x}} + \dfrac{1}{\sqrt[3]{x^2}}$

25. $g(t) = \dfrac{2}{3}t^{-2}$

$g'(t) = \dfrac{-4}{3}t^{-3} = \dfrac{-4}{3t^3}$

27. $f(\theta) = 2\theta - 3\sin\theta$

$f'(\theta) = 2 - 3\cos\theta$

29. $f(\theta) = 3\cos\theta - \dfrac{\sin\theta}{4}$

$f'(\theta) = -3\sin\theta - \dfrac{\cos\theta}{4}$

31.

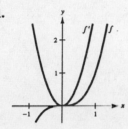

33. $F = 200\sqrt{T}$

$F'(t) = \dfrac{100}{\sqrt{T}}$

(a) When $T = 4$,
$F'(4) = 50$ vibrations/sec/lb.

(b) When $T = 9$,
$F'(9) = 33\frac{1}{3}$ vibrations/sec/lb.

35. $s(t) = -16t^2 + s_0$

$s(9.2) = -16(9.2)^2 + s_0 = 0$

$s_0 = 1354.24$

The building is approximately 1354 feet high (or 415 m).

37. (a)

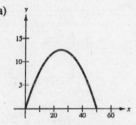

Total horizontal distance: 50

(b) $0 = x - 0.02x^2$

$0 = x\left(1 - \dfrac{x}{50}\right)$ implies $x = 50$.

(c) Ball reaches maximum height when $x = 25$.

(d) $\qquad y = x - 0.02x^2$

$\qquad y' = 1 - 0.04x$

$y'(0) = 1$

$y'(10) = 0.6$

$y'(25) = 0$

$y'(30) = -0.2$

$y'(50) = -1$

(e) $y'(25) = 0$

39. $x(t) = t^2 - 3t + 2 = (t - 2)(t - 1)$

(a) $v(t) = x'(t) = 2t - 3$

(b) $v(t) < 0$ for $t < \frac{3}{2}$

(c) $v(t) = 0$ for $t = \frac{3}{2}$

$x = \left(\frac{3}{2} - 2\right)\left(\frac{3}{2} - 1\right) = \left(-\frac{1}{2}\right)\left(\frac{1}{2}\right) = -\frac{1}{4}$

(d) $x(t) = 0$ for $t = 1, 2$

$|v(1)| = |2(1) - 3| = 1$

$|v(2)| = |2(2) - 3| = 1$

The speed is 1 when the position is 0.

41. $f(x) = (3x^2 + 7)(x^2 - 2x + 3)$

$f'(x) = (3x^2 + 7)(2x - 2) + (x^2 - 2x + 3)(6x)$

$\qquad = 2(6x^3 - 9x^2 + 16x - 7)$

43. $h(x) = \sqrt{x} \sin x = x^{1/2} \sin x$

$h'(x) = \dfrac{1}{2\sqrt{x}} \sin x + \sqrt{x} \cos x$

45. $f(x) = \dfrac{x^2 + x - 1}{x^2 - 1}$

$f'(x) = \dfrac{(x^2 - 1)(2x + 1) - (x^2 + x - 1)(2x)}{(x^2 - 1)^2}$

$\qquad = \dfrac{-(x^2 + 1)}{(x^2 - 1)^2}$

47. $f(x) = (4 - 3x^2)^{-1}$

$f'(x) = -(4 - 3x^2)^{-2}(-6x) = \dfrac{6x}{(4 - 3x^2)^2}$

49. $y = \dfrac{x^2}{\cos x}$

$y' = \dfrac{\cos x(2x) - x^2(-\sin x)}{\cos^2 x} = \dfrac{2x \cos x + x^2 \sin x}{\cos^2 x}$

51. $y = 3x^2 \sec x$

$y' = 3x^2 \sec x \tan x + 6x \sec x$

53. $y = x \cos x - \sin x$

$y' = -x \sin x + \cos x - \cos x = -x \sin x$

55. $f(x) = \dfrac{2x^3 - 1}{x^2} = 2x - x^{-2}, \quad (1, 1)$

$f'(x) = 2 + 2x^{-3}$

$f'(1) = 4$

Tangent line: $y - 1 = 4(x - 1)$

$\qquad\qquad\qquad y = 4x - 3$

57. $f(x) = -x \tan x, \quad (0, 0)$

$f'(x) = -x \sec^2 x - \tan x$

$f'(0) = 0$

Tangent line: $y - 0 = 0(x - 0)$

$\qquad\qquad\qquad y = 0$

59. $v(t) = 36 - t^2, \quad 0 \le t \le 6$

$a(t) = v'(t) = -2t$

$v(4) = 36 - 16 = 20 \text{ m/sec}$

$a(4) = -8 \text{ m/sec}^2$

61. $g(t) = t^3 - 3t + 2$

$g'(t) = 3t^2 - 3$

$g''(t) = 6t$

63. $f(\theta) = 3 \tan \theta$

$f'(\theta) = 3 \sec^2 \theta$

$f''(\theta) = 6 \sec \theta(\sec \theta \tan \theta)$

$\qquad = 6 \sec^2 \theta \tan \theta$

65. $\qquad y = 2 \sin x + 3 \cos x$

$\qquad y' = 2 \cos x - 3 \sin x$

$\qquad y'' = -2 \sin x - 3 \cos x$

$y'' + y = -(2 \sin x + 3 \cos x) + (2 \sin x + 3 \cos x)$

$\qquad = 0$

67. $h(x) = \left(\dfrac{x - 3}{x^2 + 1}\right)^2$

$h'(x) = 2\left(\dfrac{x - 3}{x^2 + 1}\right)\left(\dfrac{(x^2 + 1)(1) - (x - 3)(2x)}{(x^2 + 1)^2}\right)$

$\qquad = \dfrac{2(x - 3)(-x^2 + 6x + 1)}{(x^2 + 1)^3}$

69. $f(s) = (s^2 - 1)^{5/2}(s^3 + 5)$

$f'(s) = (s^2 - 1)^{5/2}(3s^2) + (s^3 + 5)\left(\dfrac{5}{2}\right)(s^2 - 1)^{3/2}(2s)$

$\qquad = s(s^2 - 1)^{3/2}[3s(s^2 - 1) + 5(s^3 + 5)]$

$\qquad = s(s^2 - 1)^{3/2}(8s^3 - 3s + 25)$

71. $y = 3\cos(3x + 1)$

$y' = -3\sin(3x + 1)(3)$

$y' = -9\sin(3x + 1)$

73. $y = \dfrac{x}{2} - \dfrac{\sin 2x}{4}$

$y' = \dfrac{1}{2} - \dfrac{1}{4}\cos 2x(2) = \dfrac{1}{2}(1 - \cos 2x) = \sin^2 x$

75. $y = \dfrac{2}{3}\sin^{3/2} x - \dfrac{2}{7}\sin^{7/2} x$

$y' = \sin^{1/2} x \cos x - \sin^{5/2} x \cos x$

$\qquad = (\cos x)\sqrt{\sin x}(1 - \sin^2 x)$

$\qquad = (\cos^3 x)\sqrt{\sin x}$

77. $y = \dfrac{\sin \pi x}{x + 2}$

$y' = \dfrac{(x + 2)\pi \cos \pi x - \sin \pi x}{(x + 2)^2}$

79. $f(x) = \sqrt{1 - x^3}$

$f'(x) = \dfrac{1}{2}(1 - x^3)^{-1/2}(-3x^2) = \dfrac{-3x^2}{2\sqrt{1 - x^3}}$

$f'(-2) = \dfrac{-12}{2(3)} = -2$

81. $\qquad y = \dfrac{1}{2}\csc 2x$

$\qquad y' = -\csc 2x \cot 2x$

$\qquad y'\left(\dfrac{\pi}{4}\right) = 0$

83. $g(x) = 2x(x + 1)^{-1/2}$

$g'(x) = \dfrac{x + 2}{(x + 1)^{3/2}}$

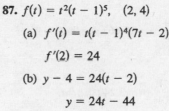

g' does not equal zero for any value of x in the domain. The graph of g has no horizontal tangent lines.

85. $f(t) = \sqrt{t + 1}\sqrt[3]{t + 1}$

$f(t) = (t + 1)^{1/2}(t + 1)^{1/3} = (t + 1)^{5/6}$

$f'(t) = \dfrac{5}{6(t + 1)^{1/6}}$

f' does not equal zero for any x in the domain. The graph of f has no horizontal tangent lines.

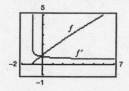

87. $f(t) = t^2(t - 1)^5$, $(2, 4)$

(a) $f'(t) = t(t - 1)^4(7t - 2)$

$\quad f'(2) = 24$

(b) $y - 4 = 24(t - 2)$

$\quad y = 24t - 44$

(c)

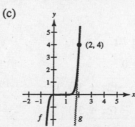

89. $y = \tan \sqrt{1 - x}, \quad \left(-2, \tan \sqrt{3}\right)$

(a) $y' = \dfrac{-1}{2\sqrt{1 - x}\,\cos^2 \sqrt{1 - x}}$

   $y'(-2) = \dfrac{-\sqrt{3}}{6\cos^2 \sqrt{3}} \approx -11.1983$

(b) $y - \tan \sqrt{3} = \dfrac{-\sqrt{3}}{6\cos^2 \sqrt{3}}(x + 2)$

   $y = \dfrac{-\sqrt{3}}{6\cos^2 \sqrt{3}}x + \tan \sqrt{3} - \dfrac{\sqrt{3}}{3\cos^2 \sqrt{3}}$

(c)

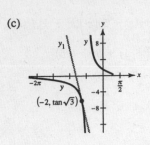

91. $y = 2x^2 + \sin 2x$

   $y' = 4x + 2\cos 2x$

   $y'' = 4 - 4\sin 2x$

93. $f(x) = \cot x$

   $f'(x) = -\csc^2 x$

   $f''(x) = -2\csc x(-\csc x \cdot \cot x)$

   $ = 2\csc^2 x \cot x$

95. $f(t) = \dfrac{t}{(1 - t)^2}$

   $f'(t) = \dfrac{t + 1}{(1 - t)^3}$

   $f''(t) = \dfrac{2(t + 2)}{(1 - t)^4}$

97. $g(\theta) = \tan 3\theta - \sin(\theta - 1)$

   $g'(\theta) = 3\sec^2 3\theta - \cos(\theta - 1)$

   $g''(\theta) = 18\sec^2 3\theta \tan 3\theta + \sin(\theta - 1)$

99. $T = \dfrac{700}{t^2 + 4t + 10}$

   $T = 700(t^2 + 4t + 10)^{-1}$

   $T' = \dfrac{-1400(t + 2)}{(t^2 + 4t + 10)^2}$

(a) When $t = 1$,

   $T' = \dfrac{-1400(1 + 2)}{(1 + 4 + 10)^2} \approx -18.667 \text{ deg/hr.}$

(b) When $t = 3$,

   $T' = \dfrac{-1400(3 + 2)}{(9 + 12 + 10)^2} \approx -7.284 \text{ deg/hr.}$

(c) When $t = 5$,

   $T' = \dfrac{-1400(5 + 2)}{(25 + 20 + 10)^2} \approx -3.240 \text{ deg/hr.}$

(d) When $t = 10$,

   $T' = \dfrac{-1400(10 + 2)}{(100 + 40 + 10)^2} \approx -0.747 \text{ deg/hr.}$

101. $x^2 + 3xy + y^3 = 10$

   $2x + 3xy' + 3y + 3y^2y' = 0$

   $3(x + y^2)y' = -(2x + 3y)$

   $y' = \dfrac{-(2x + 3y)}{3(x + y^2)}$

103.
$$y\sqrt{x} - x\sqrt{y} = 16$$

$$y\left(\frac{1}{2}x^{-1/2}\right) + x^{1/2}y' - x\left(\frac{1}{2}y^{-1/2}y'\right) - y^{1/2} = 0$$

$$\left(\sqrt{x} - \frac{x}{2\sqrt{y}}\right)y' = \sqrt{y} - \frac{y}{2\sqrt{x}}$$

$$\frac{2\sqrt{xy} - x}{2\sqrt{y}}y' = \frac{2\sqrt{xy} - y}{2\sqrt{x}}$$

$$y' = \frac{2\sqrt{xy} - y}{2\sqrt{x}} \cdot \frac{2\sqrt{y}}{2\sqrt{xy} - x} = \frac{2y\sqrt{x} - y\sqrt{y}}{2x\sqrt{y} - x\sqrt{x}}$$

105.
$$x \sin y = y \cos x$$

$$(x \cos y)y' + \sin y = -y \sin x + y' \cos x$$

$$y'(x \cos y - \cos x) = -y \sin x - \sin y$$

$$y' = \frac{y \sin x + \sin y}{\cos x - x \cos y}$$

107.
$$x^2 + y^2 = 20$$

$$2x + 2yy' = 0$$

$$y' = -\frac{x}{y}$$

At $(2, 4)$: $y' = -\frac{1}{2}$

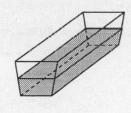

Tangent line: $y - 4 = -\frac{1}{2}(x - 2)$

$$x + 2y - 10 = 0$$

Normal line: $y - 4 = 2(x - 2)$

$$2x - y = 0$$

109. $y = \sqrt{x}$

$$\frac{dy}{dt} = 2 \text{ units/sec}$$

$$\frac{dy}{dt} = \frac{1}{2\sqrt{x}}\frac{dx}{dt} \implies \frac{dx}{dt} = 2\sqrt{x}\frac{dy}{dt} = 4\sqrt{x}$$

(a) When $x = \frac{1}{2}$, $\frac{dx}{dt} = 2\sqrt{2}$ units/sec.

(b) When $x = 1$, $\frac{dx}{dt} = 4$ units/sec.

(c) When $x = 4$, $\frac{dx}{dt} = 8$ units/sec.

111.
$$\frac{s}{h} = \frac{1/2}{2}$$

$$s = \frac{1}{4}h$$

$$\frac{dV}{dt} = 1$$

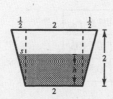

Width of water at depth h:

$$w = 2 + 2s = 2 + 2\left(\frac{1}{4}h\right) = \frac{4 + h}{2}$$

$$V = \frac{5}{2}\left(2 + \frac{4 + h}{2}\right)h = \frac{5}{4}(8 + h)h$$

$$\frac{dV}{dt} = \frac{5}{2}(4 + h)\frac{dh}{dt}$$

$$\frac{dh}{dt} = \frac{2(dV/dt)}{5(4 + h)}$$

When $h = 1$, $\frac{dh}{dt} = \frac{2}{25}$ m/min.

113.
$$s(t) = 60 - 4.9t^2$$

$$s'(t) = -9.8t$$

$$s = 35 = 60 - 4.9t^2$$

$$4.9t^2 = 25$$

$$t = \frac{5}{\sqrt{4.9}}$$

$$\tan 30 = \frac{1}{\sqrt{3}} = \frac{s(t)}{x(t)}$$

$$x(t) = \sqrt{3}\,s(t)$$

$$\frac{dx}{dt} = \sqrt{3}\frac{ds}{dt} = \sqrt{3}(-9.8)\frac{5}{\sqrt{4.9}} \approx -38.34 \text{ m/sec}$$

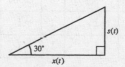

Problem Solving for Chapter 2

1. (a) $x^2 + (y - r)^2 = r^2$, Circle

$$x^2 = y, \quad \text{Parabola}$$

Substituting:

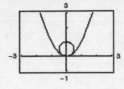

$$(y - r)^2 = r^2 - y$$

$$y^2 - 2ry + r^2 = r^2 - y$$

$$y^2 - 2ry + y = 0$$

$$y(y - 2r + 1) = 0$$

Since you want only one solution, let $1 - 2r = 0 \Rightarrow r = \frac{1}{2}$. Graph $y = x^2$ and $x^2 + \left(y - \frac{1}{2}\right)^2 = \frac{1}{4}$.

(b) Let (x, y) be a point of tangency:

$$x^2 + (y - b)^2 = 1 \Rightarrow 2x + 2(y - b)y' = 0 \Rightarrow y' = \frac{x}{b - y}, \quad \text{Circle}$$

$$y = x^2 \Rightarrow y' = 2x, \quad \text{Parabola}$$

Equating:

$$2x = \frac{x}{b - y}$$

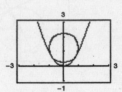

$$2(b - y) = 1$$

$$b - y = \frac{1}{2} \Rightarrow b = y + \frac{1}{2}$$

Also, $x^2 + (y - b)^2 = 1$ and $y = x^2$ imply:

$$y + (y - b)^2 = 1 \Rightarrow y + \left[y - \left(y + \frac{1}{2}\right)\right]^2 = 1 \Rightarrow y + \frac{1}{4} = 1 \Rightarrow y = \frac{3}{4} \text{ and } b = \frac{5}{4}$$

Center: $\left(0, \dfrac{5}{4}\right)$

Graph $y = x^2$ and $x^2 + \left(y - \frac{5}{4}\right)^2 = 1$.

3. (a)

$f(x) = \cos x$	$P_1(x) = a_0 + a_1 x$
$f(0) = 1$	$P_1(0) = a_0 \Rightarrow a_0 = 1$
$f'(0) = 0$	$P'_1(0) = a_1 \Rightarrow a_1 = 0$
$P_1(x) = 1$	

(b)

$f(x) = \cos x$	$P_2(x) = a_0 + a_1 x + a_2 x^2$
$f(0) = 1$	$P_2(0) = a_0 \Rightarrow a_0 = 1$
$f'(0) = 0$	$P'_2(0) = a_1 \Rightarrow a_1 = 0$
$f''(0) = -1$	$P''_2(0) = 2a_2 \Rightarrow a_2 = -\frac{1}{2}$
$P_2(x) = 1 - \frac{1}{2}x^2$	

(d)

$f(x) = \sin x$	$P_3(x) = a_0 + a_1 x + a_2 x^2 + a_3 x^3$
$f(0) = 0$	$P_3(0) = a_0 \Rightarrow a_0 = 0$
$f'(0) = 1$	$P'_3(0) = a_1 \Rightarrow a_1 = 1$
$f''(0) = 0$	$P''_3(0) = 2a_2 \Rightarrow a_2 = 0$
$f'''(0) = -1$	$P'''_3(0) = 6a_3 \Rightarrow a_3 = -\frac{1}{6}$
	$P_3(x) = x - \frac{1}{6}x^3$

(c)

x	-1.0	-0.1	-0.001	0	0.001	0.1	1.0
$\cos x$	0.5403	0.9950	≈ 1	1	≈ 1	0.9950	0.5403
$P_2(x)$	0.5	0.9950	≈ 1	1	≈ 1	0.9950	0.5

$P_2(x)$ is a good approximation of $f(x) = \cos x$ when x is near 0.

5. Let $p(x) = Ax^3 + Bx^2 + Cx + D$

$p'(x) = 3Ax^2 + 2Bx + C.$

At $(1, 1)$: $A + B + C + D = 1$ Equation 1   At $(-1, -3)$: $-A + B - C + D = -3$ Equation 3

$3A + 2B + C = 14$ Equation 2 $3A - 2B + C = -2$ Equation 4

Adding Equations 1 and 3: $2B + 2D = -2$

Subtracting Equations 1 and 3: $2A + 2C = 4$

Adding Equations 2 and 4: $6A + 2C = 12$

Subtracting Equations 2 and 4: $4B = 16$

Hence, $B = 4$ and $D = \frac{1}{2}(-2 - 2B) = -5$. Subtracting $2A + 2C = 4$ and $6A + 2C = 12$, you obtain $4A = 8 \Rightarrow A = 2$. Finally, $C = \frac{1}{2}(4 - 2A) = 0$. Thus, $p(x) = 2x^3 + 4x^2 - 5$.

7. (a) $x^4 = a^2x^2 - a^2y^2$

$a^2y^2 = a^2x^2 - x^4$

$y = \dfrac{\pm\sqrt{a^2x^2 - x^4}}{a}$

Graph: $y_1 = \dfrac{\sqrt{a^2x^2 - x^4}}{a}$ and $y_2 = -\dfrac{\sqrt{a^2x^2 - x^4}}{a}$.

(b)

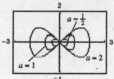

$(\pm a, 0)$ are the x-intercepts, along with $(0, 0)$.

(c) Differentiating implicitly:

$4x^3 = 2a^2x - 2a^2yy'$

$y' = \dfrac{2a^2x - 4x^3}{2a^2y}$

$= \dfrac{x(a^2 - 2x^2)}{a^2y} = 0 \Rightarrow 2x^2 = a^2 \Rightarrow x = \dfrac{\pm a}{\sqrt{2}}$

$\left(\dfrac{a^2}{2}\right)^2 = a^2\left(\dfrac{a^2}{2}\right) - a^2y^2$

$\dfrac{a^4}{4} = \dfrac{a^4}{2} - a^2y^2$

$a^2y^2 = \dfrac{a^4}{4}$

$y^2 = \dfrac{a^2}{4}$

$y = \pm\dfrac{a}{2}$

Four points:

$\left(\dfrac{a}{\sqrt{2}}, \dfrac{a}{2}\right), \left(\dfrac{a}{\sqrt{2}}, -\dfrac{a}{2}\right), \left(-\dfrac{a}{\sqrt{2}}, \dfrac{a}{2}\right), \left(\dfrac{-a}{\sqrt{2}}, -\dfrac{a}{2}\right)$

9. (a)

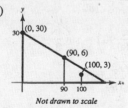

Not drawn to scale

Line determined by $(0, 30)$ and $(90, 6)$:

$y - 30 = \dfrac{30 - 6}{0 - 90}(x - 0)$

$= -\dfrac{24}{90}x = -\dfrac{4}{15}x \Rightarrow y = -\dfrac{4}{15}x + 30$

When $x = 100$:

$y = \dfrac{-4}{15}(100) + 30 = \dfrac{10}{3} > 3 \Rightarrow$ Shadow determined by man

(b)

Not drawn to scale

Line determined by $(0, 30)$ and $(60, 6)$:

$y - 30 = \dfrac{30 - 6}{0 - 60}(x - 0) = -\dfrac{2}{5}x \Rightarrow y = -\dfrac{2}{5}x + 30$

When $x = 70$:

$y = \dfrac{-2}{5}(70) + 30$

$= 2 < 3 \Rightarrow$ Shadow determined by child

—CONTINUED—

9. —CONTINUED—

(c) Need $(0, 30)$, $(d, 6)$, $(d + 10, 3)$ collinear.

$$\frac{30 - 6}{0 - d} = \frac{6 - 3}{d - (d + 10)} \Longrightarrow \frac{24}{d} = \frac{3}{10} \Longrightarrow d = 80 \text{ feet}$$

(d) Let y be the distance from the base of the street light to the tip of the shadow. We know that $dx/dt = -5$.

For $x > 80$, the shadow is determined by the man.

$$\frac{y}{30} = \frac{y - x}{6} \Longrightarrow y = \frac{5}{4}x \text{ and } \frac{dy}{dt} = \frac{5}{4}\frac{dx}{dt} = \frac{-25}{4}$$

For $x < 80$, the shadow is determined by the child.

$$\frac{y}{30} = \frac{y - x - 10}{3} \Longrightarrow y = \frac{10}{9}x + \frac{100}{9} \text{ and } \frac{dy}{dt} = \frac{10}{9}\frac{dx}{dt} = \frac{-50}{9}$$

Therefore:

$$\frac{dy}{dt} = \begin{cases} -\frac{25}{4}, & x > 80 \\ -\frac{50}{9}, & 0 < x < 80 \end{cases}$$

dy/dt is not continuous at $x = 80$.

11. $L'(x) = \displaystyle\lim_{\Delta x \to 0} \frac{L(x + \Delta x) - L(x)}{\Delta x}$

$= \displaystyle\lim_{\Delta x \to 0} \frac{L(x) + L(\Delta x) - L(x)}{\Delta x}$

$= \displaystyle\lim_{\Delta x \to 0} \frac{L(\Delta x)}{\Delta x}$

Also, $L'(0) = \displaystyle\lim_{\Delta x \to 0} \frac{L(\Delta x) - L(0)}{\Delta x}$.

But, $L(0) = 0$ because

$L(0) = L(0 + 0) = L(0) + L(0) \Longrightarrow L(0) = 0.$

Thus, $L'(x) = L'(0)$ for all x. The graph of L is a line through the origin of slope $L'(0)$.

13. (a)

z (degrees)	0.1	0.01	0.0001
$\dfrac{\sin z}{z}$	0.0174524	0.0174533	0.0174533

(b) $\displaystyle\lim_{z \to 0} \frac{\sin z}{z} \approx 0.0174533$

In fact, $\displaystyle\lim_{z \to 0} \frac{\sin z}{z} = \frac{\pi}{180}$.

(c) $\dfrac{d}{dz}(\sin z) = \displaystyle\lim_{\Delta z \to 0} \frac{\sin(z + \Delta z) - \sin z}{\Delta z}$

$= \displaystyle\lim_{\Delta z \to 0} \frac{\sin z \cdot \cos \Delta z + \sin \Delta z \cdot \cos z - \sin z}{\Delta z}$

$= \displaystyle\lim_{\Delta z \to 0} \left[\sin z\left(\frac{\cos \Delta z - 1}{\Delta z}\right)\right] + \displaystyle\lim_{\Delta z \to 0} \left[\cos z\left(\frac{\sin \Delta z}{\Delta z}\right)\right]$

$= (\sin z)(0) + (\cos z)\left(\frac{\pi}{180}\right) = \frac{\pi}{180}\cos z$

(d) $S(90) = \sin\left(\frac{\pi}{180}90\right) = \sin\frac{\pi}{2} = 1$

$C(180) = \cos\left(\frac{\pi}{180}180\right) = -1$

$\dfrac{d}{dz}S(z) = \dfrac{d}{dz}\sin(cz) = c \cdot \cos(cz) = \frac{\pi}{180}C(z)$

(e) The formulas for the derivatives are more complicated in degrees.

15. $j(t) = a'(t)$ (a) $j(t)$ is the rate of change of acceleration.

(b) $s(t) = -8.25t^2 + 66t$

$v(t) = -16.5t + 66$

$a(t) = -16.5$

$a'(t) = j(t) = 0$

(c) a is position.

b is acceleration.

c is jerk.

d is velocity.

CHAPTER 3
Applications of Differentiation

Section 3.1   Extrema on an Interval **104**

Section 3.2   Rolle's Theorem and the Mean Value Theorem **108**

Section 3.3   Increasing and Decreasing Functions and
the First Derivative Test **114**

Section 3.4   Concavity and the Second Derivative Test **124**

Section 3.5   Limits at Infinity **132**

Section 3.6   A Summary of Curve Sketching **140**

Section 3.7   Optimization Problems **149**

Section 3.8   Newton's Method **157**

Section 3.9   Differentials **162**

Review Exercises **165**

Problem Solving **174**

CHAPTER 3
Applications of Differentiation

Section 3.1   Extrema on an Interval

1. A: neither
   B: absolute maximum (and relative maximum)
   C: neither
   D: neither
   E: relative maximum
   F: relative minimum
   G: neither

3. $f(x) = \dfrac{x^2}{x^2 + 4}$

   $f'(x) = \dfrac{(x^2 + 4)(2x) - (x^2)(2x)}{(x^2 + 4)^2} = \dfrac{8x}{(x^2 + 4)^2}$

   $f'(0) = 0$

5. $f(x) = x + \dfrac{27}{2x^2} = x + \dfrac{27}{2}x^{-2}$

   $f'(x) = 1 - 27x^{-3} = 1 - \dfrac{27}{x^3}$

   $f'(3) = 1 - \dfrac{27}{3^3} = 1 - 1 = 0$

7. $f(x) = (x + 2)^{2/3}$

   $f'(x) = \dfrac{2}{3}(x + 2)^{-1/3}$

   $f'(-2)$ is undefined.

9. Critical number: $x = 2$

   $x = 2$: absolute maximum

11. Critical numbers: $x = 1, 2, 3$

    $x = 1, 3$: absolute maximum

    $x = 2$: absolute minimum

13. $f(x) = x^2(x - 3) = x^3 - 3x^2$

    $f'(x) = 3x^2 - 6x = 3x(x - 2)$

    Critical numbers: $x = 0, x = 2$

15. $g(t) = t\sqrt{4 - t},\ t < 3$

    $g'(t) = t\left[\dfrac{1}{2}(4 - t)^{-1/2}(-1)\right] + (4 - t)^{1/2}$

    $= \dfrac{1}{2}(4 - t)^{-1/2}[-t + 2(4 - t)]$

    $= \dfrac{8 - 3t}{2\sqrt{4 - t}}$

    Critical number: $t = \dfrac{8}{3}$.

17. $h(x) = \sin^2 x + \cos x,\ 0 < x < 2\pi$

    $h'(x) = 2\sin x \cos x - \sin x = \sin x(2\cos x - 1)$

    On $(0, 2\pi)$, critical numbers: $x = \dfrac{\pi}{3}, x = \pi, x = \dfrac{5\pi}{3}$

19. $f(x) = 2(3 - x),\ [-1, 2]$

    $f'(x) = -2 \Rightarrow$ No critical numbers

    Left endpoint: $(-1, 8)$ Maximum

    Right endpoint: $(2, 2)$ Minimum

21. $f(x) = -x^2 + 3x,\ [0, 3]$

    $f'(x) = -2x + 3$

    Left endpoint: $(0, 0)$ Minimum

    Critical number: $\left(\dfrac{3}{2}, \dfrac{9}{4}\right)$ Maximum

    Right endpoint: $(3, 0)$ Minimum

23. $f(x) = x^3 - \frac{3}{2}x^2,\ [-1, 2]$

$f'(x) = 3x^2 - 3x = 3x(x - 1)$

Left endpoint: $\left(-1, -\frac{5}{2}\right)$ Minimum

Right endpoint: $(2, 2)$ Maximum

Critical number: $(0, 0)$

Critical number: $\left(1, -\frac{1}{2}\right)$

25. $f(x) = 3x^{2/3} - 2x,\ [-1, 1]$

$f'(x) = 2x^{-1/3} - 2 = \dfrac{2\left(1 - \sqrt[3]{x}\right)}{\sqrt[3]{x}}$

Left endpoint: $(-1, 5)$ Maximum

Critical number: $(0, 0)$ Minimum

Right endpoint: $(1, 1)$

27. $g(t) = \dfrac{t^2}{t^2 + 3},\ [-1, 1]$

$g'(t) = \dfrac{6t}{(t^2 + 3)^2}$

Left endpoint: $\left(-1, \frac{1}{4}\right)$ Maximum

Critical number: $(0, 0)$ Minimum

Right endpoint: $\left(1, \frac{1}{4}\right)$ Maximum

29. $h(s) = \dfrac{1}{s - 2},\ [0, 1]$

$h'(s) = \dfrac{-1}{(s - 2)^2}$

Left endpoint: $\left(0, -\frac{1}{2}\right)$ Maximum

Right endpoint: $(1, -1)$ Minimum

31. $y = 3 - |t - 3|,\ [-1, 5]$

From the graph, you see that $t - 3$ is a critical number.

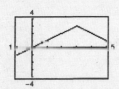

Left endpoint: $(-1, -1)$ Minimum

Right endpoint: $(5, 1)$

Critical number: $(3, 3)$ Maximum

33. $f(x) = \cos \pi x,\ \left[0, \frac{1}{6}\right]$

$f'(x) = -\pi \sin \pi x$

Left endpoint: $(0, 1)$ Maximum

Right endpoint: $\left(\dfrac{1}{6}, \dfrac{\sqrt{3}}{2}\right)$ Minimum

35. $y = \dfrac{4}{x} + \tan\left(\dfrac{\pi x}{8}\right),\ [1, 2]$

$y' = \dfrac{-4}{x^2} + \dfrac{\pi}{8}\sec^2\dfrac{\pi x}{8} = 0$

$\dfrac{\pi}{8}\sec^2\dfrac{\pi x}{8} = \dfrac{4}{x^2}$

On the interval $[1, 2]$, this equation has no solutions. Thus, there are no critical numbers.

Left endpoint: $\left(1, \sqrt{2} + 3\right) \approx (1, 4.4142)$ Maximum

Right endpoint: $(2, 3)$ Minimum

37. (a) Minimum: $(0, -3)$

     Maximum: $(2, 1)$

(b) Minimum: $(0, -3)$

(c) Maximum: $(2, 1)$

(d) No extrema

39. $f(x) = x^2 - 2x$

(a) Minimum: $(1, -1)$ (b) Maximum: $(3, 3)$ (c) Minimum: $(1, -1)$ (d) Minimum: $(1, -1)$

     Maximum: $(-1, 3)$

41. $f(x) = \begin{cases} 2x + 2, & 0 \le x \le 1 \\ 4x^2, & 1 < x \le 3 \end{cases}$

Left endpoint: $(0, 2)$ Minimum

Right endpoint: $(3, 36)$ Maximum

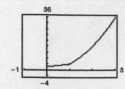

43. $f(x) = \dfrac{3}{x - 1}, (1, 4]$

Right endpoint: $(4, 1)$ Minimum

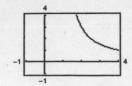

45. $f(x) = x^4 - 2x^3 + x + 1, \quad [-1, 3]$

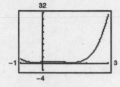

$f'(x) = 4x^3 - 6x^2 + 1 = (2x - 1)(2x^2 - 2x - 1) = 0$

$x = \dfrac{1}{2}, \dfrac{1 \pm \sqrt{3}}{2} \approx 0.5, -0.366, 1.366$

Maximum: $f(3) = 31$

Minimum: $f\left(\dfrac{1 \pm \sqrt{3}}{2}\right) = \dfrac{3}{4}$

47. (a)

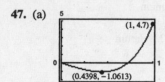

Maximum: $(1, 4.7)$ (endpoint)

Minimum: $(0.4398, -1.0613)$

(b)

$$f(x) = 3.2x^5 + 5x^3 - 3.5x, [0, 1]$$

$$f'(x) = 16x^4 + 15x^2 - 3.5$$

$$16x^4 + 15x^2 - 3.5 = 0$$

$$x^2 = \dfrac{-15 \pm \sqrt{(15)^2 - 4(16)(-3.5)}}{2(16)}$$

$$= \dfrac{-15 \pm \sqrt{449}}{32}$$

$$x = \sqrt{\dfrac{-15 + \sqrt{449}}{32}} \approx 0.4398$$

$$f(0) = 0$$

$$f(1) = 4.7 \text{ Maximum (endpoint)}$$

$$f\left(\sqrt{\dfrac{-15 + \sqrt{449}}{32}}\right) \approx -1.0613$$

Minimum: $(0.4398, -1.0613)$

49. $f(x) = (1 + x^3)^{1/2}, [0, 2]$

$f'(x) = \dfrac{3}{2}x^2(1 + x^3)^{-1/2}$

$f''(x) = \dfrac{3}{4}(x^4 + 4x)(1 + x^3)^{-3/2}$

$f'''(x) = -\dfrac{3}{8}(x^6 + 20x^3 - 8)(1 + x^3)^{-5/2}$

Setting $f''' = 0$, we have $x^6 + 20x^3 - 8 = 0$.

$x^3 = \dfrac{-20 \pm \sqrt{400 - 4(1)(-8)}}{2}$

$x = \sqrt[3]{-10 \pm \sqrt{108}} = \sqrt{3} - 1$

In the interval $[0, 2]$, choose

$x = \sqrt[3]{-10 + \sqrt{108}} = \sqrt{3} - 1 \approx 0.732.$

$\left|f''\left(\sqrt[3]{-10 + \sqrt{108}}\right)\right| \approx 1.47$ is the maximum value.

51. $f(x) = (x + 1)^{2/3}, [0, 2]$

$f'(x) = \dfrac{2}{3}(x + 1)^{-1/3}$

$f''(x) = -\dfrac{2}{9}(x + 1)^{-4/3}$

$f'''(x) = \dfrac{8}{27}(x + 1)^{-7/3}$

$f^{(4)}(x) = -\dfrac{56}{81}(x + 1)^{-10/3}$

$f^{(5)}(x) = \dfrac{560}{243}(x + 1)^{-13/3}$

$|f^{(4)}(0)| = \dfrac{56}{81}$ is the maximum value.

53.

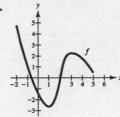

55. (a) Yes   (b) No

57. (a) No   (b) Yes

59. $P = VI - RI^2 = 12I - 0.5I^2, 0 \le I \le 15$

$P = 0$ when $I = 0$.

$P = 67.5$ when $I = 15$.

$P' = 12 - I = 0$

Critical number: $I = 12$ amps

When $I = 12$ amps, $P = 72$, the maximum output.

No, a 20-amp fuse would not increase the power output.
P is decreasing for $I > 12$.

61. $x = \dfrac{v^2 \sin 2\theta}{32}, \dfrac{\pi}{4} \le \theta \le \dfrac{3\pi}{4}$

$\dfrac{d\theta}{dt}$ is constant.

$\dfrac{dx}{dt} = \dfrac{dx}{d\theta} \dfrac{d\theta}{dt}$ (by the Chain Rule)

$= \dfrac{v^2 \cos 2\theta}{16} \dfrac{d\theta}{dt}$

In the interval $[\pi/4, 3\pi/4]$, $\theta = \pi/4, 3\pi/4$ indicate minimums for dx/dt and $\theta = \pi/2$ indicates a maximum for dx/dt. This implies that the sprinkler waters longest when $\theta = \pi/4$ and $3\pi/4$. Thus, the lawn farthest from the sprinkler gets the most water.

63. True. See Exercise 27.

65. True

67. If f has a maximum value at $x = c$, then $f(c) \ge f(x)$ for all x in I. Hence, $-f(c) \le -f(x)$ for all x in I. Thus, $-f$ has a minimum value at $x = c$.

69. (a) $y = ax^2 + bx + c$

$y' = 2ax + b$

The coordinates of B are $(500, 30)$, and those of A are $(-500, 45)$. From the slopes at A and B,

$-1000a + b = -0.09$

$1000a + b = 0.06.$

Solving these two equations, you obtain $a = 3/40,000$ and $b = -3/200$. From the points $(500, 30)$ and $(-500, 45)$, you obtain

$30 = \dfrac{3}{40,000} 500^2 + 500 \left(\dfrac{-3}{200} \right) + c$

$45 = \dfrac{3}{40,000} 500^2 - 500 \left(\dfrac{-3}{200} \right) + c.$

In both cases, $c = 18.75 = \dfrac{75}{4}$. Thus, $y = \dfrac{3}{40,000} x^2 - \dfrac{3}{200} x + \dfrac{75}{4}.$

—CONTINUED—

69. —CONTINUED—

(b)

x	-500	-400	-300	-200	-100	0	100	200	300	400	500
d	0	0.75	3	6.75	12	18.75	12	6.75	3	0.75	0

For $-500 \le x \le 0, d = (ax^2 + bx + c) - (-0.09x).$

For $0 \le x \le 500, d = (ax^2 + bx + c) - (0.06x).$

(c) The lowest point on the highway is $(100, 18)$, which is not directly over the point where the two hillsides come together.

Section 3.2   Rolle's Theorem and the Mean Value Theorem

1. Rolle's Theorem does not apply to $f(x) = 1 - |x - 1|$ over $[0, 2]$ since f is not differentiable at $x = 1$.

3. $f(x) = \left| \dfrac{1}{x} \right|$

$f(-1) = f(1) = 1$. But, f is not continuous on $[-1, 1]$.

5. $f(x) = x^2 - x - 2 = (x - 2)(x + 1)$

x-intercepts: $(-1, 0), (2, 0)$

$f'(x) = 2x - 1 = 0$ at $x = \dfrac{1}{2}.$

7. $f(x) = x\sqrt{x + 4}$

x-intercepts: $(-4, 0), (0, 0)$

$f'(x) = x\dfrac{1}{2}(x + 4)^{-1/2} + (x + 4)^{1/2}$

$\quad = (x + 4)^{-1/2}\left(\dfrac{x}{2} + (x + 4) \right)$

$f'(x) = \left(\dfrac{3}{2}x + 4 \right)(x + 4)^{-1/2} = 0$ at $x = -\dfrac{8}{3}$

9. $f(x) = x^2 + 3x - 4$

$f(-4) = f(1) = 0$

$f'(x) = 2x + 3 = 0$ for $x = \dfrac{-3}{2}$

$c = \dfrac{-3}{2}$ and $f'\left(\dfrac{-3}{2} \right) = 0$

11. $f(x) = x^2 - 2x, [0, 2]$

$f(0) = f(2) = 0$

f is continuous on $[0, 2]$. f is differentiable on $(0, 2)$. Rolle's Theorem applies.

$\qquad f'(x) = 2x - 2$

$\quad 2x - 2 = 0 \Rightarrow x = 1$

c-value: 1

13. $f(x) = (x - 1)(x - 2)(x - 3), [1, 3]$

$f(1) = f(3) = 0$

f is continuous on $[1, 3]$. f is differentiable on $(1, 3)$. Rolle's Theorem applies.

$\qquad\qquad f(x) = x^3 - 6x^2 + 11x - 6$

$\qquad\qquad f'(x) = 3x^2 - 12x + 11$

$\quad 3x^2 - 12x + 11 = 0 \Rightarrow x = \dfrac{6 \pm \sqrt{3}}{3}$

$c = \dfrac{6 - \sqrt{3}}{3}, c = \dfrac{6 + \sqrt{3}}{3}$

15. $f(x) = x^{2/3} - 1, [-8, 8]$

$f(-8) = f(8) = 3$

f is continuous on $[-8, 8]$. f is not differentiable on $(-8, 8)$ since $f'(0)$ does not exist. Rolle's Theorem does not apply.

17. $f(x) = \dfrac{x^2 - 2x - 3}{x + 2}$, $[-1, 3]$

$f(-1) = f(3) = 0$

f is continuous on $[-1, 3]$. (**Note:** The discontinuity, $x = -2$, is not in the interval.) f is differentiable on $(-1, 3)$. Rolle's Theorem applies.

$$f'(x) = \frac{(x + 2)(2x - 2) - (x^2 - 2x - 3)(1)}{(x + 2)^2} = 0$$

$$\frac{x^2 + 4x - 1}{(x + 2)^2} = 0$$

$$x = \frac{-4 \pm 2\sqrt{5}}{2} = -2 \pm \sqrt{5}$$

c-value: $-2 + \sqrt{5}$

19. $f(x) = \sin x$, $[0, 2\pi]$

$f(0) = f(2\pi) = 0$

f is continuous on $[0, 2\pi]$. f is differentiable on $(0, 2\pi)$. Rolle's Theorem applies.

$$f'(x) = \cos x$$

c-values: $\dfrac{\pi}{2}, \dfrac{3\pi}{2}$

21. $f(x) = \dfrac{6x}{\pi} - 4\sin^2 x$, $\left[0, \dfrac{\pi}{6}\right]$

$f(0) = f\left(\dfrac{\pi}{6}\right) = 0$

f is continuous on $[0, \pi/6]$. f is differentiable on $(0, \pi/6)$. Rolle's Theorem applies.

$$f'(x) = \frac{6}{\pi} - 8\sin x \cos x = 0$$

$$\frac{6}{\pi} = 8\sin x \cos x$$

$$\frac{3}{4\pi} = \frac{1}{2}\sin 2x$$

$$\frac{3}{2\pi} = \sin 2x$$

$$\frac{1}{2}\arcsin\left(\frac{3}{2\pi}\right) = x$$

$$x \approx 0.2489$$

c-value: 0.2489

23. $f(x) = \tan x$, $[0, \pi]$

$f(0) = f(\pi) = 0$

f is not continuous on $[0, \pi]$ since $f(\pi/2)$ does not exist. Rolle's Theorem does not apply.

25. $f(x) = |x| - 1$, $[-1, 1]$

$f(-1) = f(1) = 0$

f is continuous on $[-1, 1]$. f is not differentiable on $(-1, 1)$ since $f'(0)$ does not exist. Rolle's Theorem does not apply.

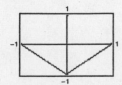

27. $f(x) = 4x - \tan \pi x, \left[-\frac{1}{4}, \frac{1}{4} \right]$

$$f\left(-\frac{1}{4} \right) = f\left(\frac{1}{4} \right) = 0$$

f is continuous on $[-1/4, 1/4]$. f is differentiable on $(-1/4, 1/4)$. Rolle's Theorem applies.

$$f'(x) = 4 - \pi \sec^2 \pi x = 0$$

$$\sec^2 \pi x = \frac{4}{\pi}$$

$$\sec \pi x = \pm \frac{2}{\sqrt{\pi}}$$

$$x = \pm \frac{1}{\pi} \operatorname{arcsec} \frac{2}{\sqrt{\pi}} = \pm \frac{1}{\pi} \arccos \frac{\sqrt{\pi}}{2}$$

$$\approx \pm 0.1533 \text{ radian}$$

c-values: ± 0.1533 radian

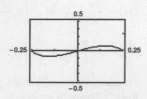

29. $f(t) = -16t^2 + 48t + 32$

(a) $f(1) = f(2) = 64$

(b) $v = f'(t)$ must be 0 at some time in $(1, 2)$.

$$f'(t) = -32t + 48 = 0$$

$$t = \frac{3}{2} \text{ seconds}$$

31.

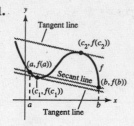

33. f is not continuous on the interval $[0, 6]$. (f is not continuous at $x = 2$.)

35. $f(x) = \frac{1}{x - 3}, [0, 6]$

f has a discontinuity at $x = 3$.

37. $f(x) = x^2 + 1$

(a)   slope $= \frac{5 - 2}{2 + 1} = 1$

secant line: $y - 2 = 1(x + 1)$

$$y = x + 3$$

(b) $f'(x) = 2x = 1 \implies c = \frac{1}{2}$

$$f'\left(\frac{1}{2} \right) = 1, \quad f\left(\frac{1}{2} \right) = \frac{5}{4}$$

(c)   Tangent line: $y - \frac{5}{4} = 1\left(x - \frac{1}{2} \right)$

$$y = x + \frac{3}{4}$$

(d)

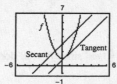

39. $f(x) = x^2$ is continuous on $[-2, 1]$ and differentiable on $(-2, 1)$.

$$\frac{f(1) - f(-2)}{1 - (-2)} = \frac{1 - 4}{3} = -1$$

$f'(x) = 2x = -1$ when $x = -\frac{1}{2}$. Therefore,

$$c = -\frac{1}{2}.$$

41. $f(x) = x^{2/3}$ is continuous on $[0, 1]$ and differentiable on $(0, 1)$.

$$\frac{f(1) - f(0)}{1 - 0} = 1$$

$$f'(x) = \frac{2}{3} x^{-1/3} = 1$$

$$x = \left(\frac{2}{3} \right)^3 = \frac{8}{27}$$

$$c = \frac{8}{27}$$

43. $f(x) = \sqrt{2-x}$ is continuous on $[-7, 2]$ and differentiable on $(-7, 2)$.

$$\frac{f(2) - f(-7)}{2 - (-7)} = \frac{0 - 3}{9} = -\frac{1}{3}$$

$$f'(x) = \frac{-1}{2\sqrt{2-x}} = -\frac{1}{3}$$

$$2\sqrt{2-x} = 3$$

$$\sqrt{2-x} = \frac{3}{2}$$

$$2 - x = \frac{9}{4}$$

$$x = -\frac{1}{4}$$

$$c = -\frac{1}{4}$$

45. $f(x) = \sin x$ is continuous on $[0, \pi]$ and differentiable on $(0, \pi)$.

$$\frac{f(\pi) - f(0)}{\pi - 0} = \frac{0 - 0}{\pi} = 0$$

$$f'(x) = \cos x = 0$$

$$c = \frac{\pi}{2}$$

47. $f(x) = \dfrac{x}{x + 1}, \quad \left[-\dfrac{1}{2}, 2\right]$

(a)

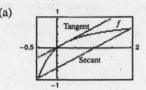

(b) Secant line:

$$\text{slope} = \frac{f(2) - f(-1/2)}{2 - (-1/2)} = \frac{2/3 - (-1)}{5/2} = \frac{2}{3}$$

$$y - \frac{2}{3} = \frac{2}{3}(x - 2)$$

$$3y - 2 = 2x - 4$$

$$3y - 2x + 2 = 0$$

(c) $f'(x) = \dfrac{1}{(x + 1)^2} = \dfrac{2}{3}$

$$(x + 1)^2 = \frac{3}{2}$$

$$x = -1 \pm \sqrt{\frac{3}{2}} = -1 \pm \frac{\sqrt{6}}{2}$$

In the interval $[-1/2, 2]$, $c = -1 + \left(\sqrt{6}/2\right)$.

$$f(c) = \frac{-1 + \left(\sqrt{6}/2\right)}{\left[-1 + \left(\sqrt{6}/2\right)\right] + 1} = \frac{-2 + \sqrt{6}}{\sqrt{6}} = \frac{-2}{\sqrt{6}} + 1$$

Tangent line: $y - 1 + \dfrac{2}{\sqrt{6}} = \dfrac{2}{3}\left(x - \dfrac{\sqrt{6}}{2} + 1\right)$

$$y - 1 + \frac{\sqrt{6}}{3} = \frac{2}{3}x - \frac{\sqrt{6}}{3} + \frac{2}{3}$$

$$3y - 2x - 5 + 2\sqrt{6} = 0$$

49. $f(x) = \sqrt{x}, \ [1, 9]$

$(1, 1), (9, 3)$

$$m = \frac{3 - 1}{9 - 1} = \frac{1}{4}$$

(a)

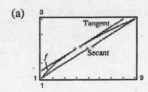

(b) Secant line: $y - 1 = \dfrac{1}{4}(x - 1)$

$$y = \frac{1}{4}x + \frac{3}{4}$$

$$0 = x - 4y + 3$$

(c) $f'(x) = \dfrac{1}{2\sqrt{x}}$

$$\frac{f(9) - f(1)}{9 - 1} = \frac{1}{4}$$

$$\frac{1}{2\sqrt{c}} = \frac{1}{4}$$

$$\sqrt{c} = 2$$

$$c = 4$$

$$(c, f(c)) = (4, 2)$$

$$m = f'(4) = \frac{1}{4}$$

Tangent line: $y - 2 = \dfrac{1}{4}(x - 4)$

$$y = \frac{1}{4}x + 1$$

$$0 = x - 4y + 4$$

51. $s(t) = -4.9t^2 + 500$

(a) $v_{avg} = \dfrac{s(3) - s(0)}{3 - 0} = \dfrac{455.9 - 500}{3} = -14.7$ m/sec

(b) $s(t)$ is continuous on $[0, 3]$ and differentiable on $(0, 3)$. Therefore, the Mean Value Theorem applies.

$$v(t) = s'(t) = -9.8t = -14.7 \text{ m/sec}$$

$$t = \frac{-14.7}{-9.8} = 1.5 \text{ seconds}$$

53. No. Let $f(x) = x^2$ on $[-1, 2]$.

$$f'(x) = 2x$$

$f'(0) = 0$ and zero is in the interval $(-1, 2)$ but
$f(-1) \neq f(2)$.

55. $f(x) = \begin{cases} 0, & x = 0 \\ 1 - x, & 0 < x \leq 1 \end{cases}$

No, this does not contradict Rolle's Theorem. f is not continuous on $[0, 1]$.

57. Let $S(t)$ be the position function of the plane. If $t = 0$ corresponds to 2 P.M., $S(0) = 0$, $S(5.5) = 2500$ and the Mean Value Theorem says that there exists a time t_0, $0 < t_0 < 5.5$, such that

$$S'(t_0) = v(t_0) = \frac{2500 - 0}{5.5 - 0} \approx 454.54.$$

Applying the Intermediate Value Theorem to the velocity function on the intervals $[0, t_0]$ and $[t_0, 5.5]$, you see that there are at least two times during the flight when the speed was 400 miles per hour. $(0 < 400 < 454.54)$

59. Let $S(t)$ be the difference in the positions of the 2 bicyclists,

$$S(t) = S_1(t) - S_2(t).$$

Since $S(0) = S(2.25) = 0$, there must exist a time $t_0 \in (0, 2.25)$ such that

$$S'(t_0) = v(t_0) = 0.$$

At this time, $v_1(t_0) = v_2(t_0)$.

61. (a) f is continuous on $[-10, 4]$ and changes sign, $(f(-8) > 0, f(3) < 0)$. By the Intermediate Value Theorem, there exists at least one value of x in $[-10, 4]$ satisfying $f(x) = 0$.

(b) There exist real numbers a and b such that $-10 < a < b < 4$ and $f(a) = f(b) = 2$. Therefore, by Rolle's Theorem there exists at least one number c in $(-10, 4)$ such that $f'(c) = 0$. This is called a critical number.

(c)

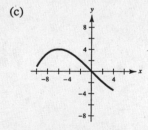

(d)

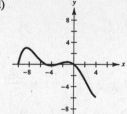

(e) No, f' did not have to be continuous on $[-10, 4]$.

63. f is continuous on $[-5, 5]$ and does not satisfy the conditions of the Mean Value Theorem.
$\Rightarrow f$ is not differentiable on $(-5, 5)$. Example: $f(x) = |x|$

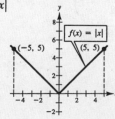

65. $f(x) = x^5 + x^3 + x + 1$

f is differentiable for all x.

$f(-1) = -2$ and $f(0) = 1$, so the Intermediate Value Theorem implies that f has at least one zero c in $[-1, 0]$, $f(c) = 0$.

Suppose f had 2 zeros, $f(c_1) = f(c_2) = 0$. Then Rolle's Theorem would guarantee the existence of a number a such that

$$f'(a) = f(c_2) - f(c_1) = 0.$$

But, $f'(x) = 5x^4 + 3x^2 + 1 > 0$ for all x. Hence, f has exactly one real zero.

67. f continuous at $x = 0$: $1 = b$
f continuous at $x = 1$: $a + 1 = 5 + c$
f differentiable at $x = 1$: $a = 2 + 4 = 6$ Hence, $c = 2$.

$$f(x) = \begin{cases} 1, & x = 0 \\ 6x + 1, & 0 < x \leq 1 \\ x^2 + 4x + 2, & 1 < x \leq 3 \end{cases}$$

$$= \begin{cases} 6x + 1, & 0 \leq x \leq 1 \\ x^2 + 4x + 2, & 1 < x \leq 3 \end{cases}$$

69. $f'(x) = 0$

$f(x) = c$

$f(2) = 5$

Hence, $f(x) = 5$.

71. $f'(x) = 2x$

$f(x) = x^2 + c$

$f(1) = 0 \implies 0 = 1 + c \implies c = -1$

Hence, $f(x) = x^2 - 1$.

73. False. $f(x) = 1/x$ has a discontinuity at $x = 0$.

75. True. A polynomial is continuous and differentiable everywhere.

77. Suppose that $p(x) = x^{2n+1} + ax + b$ has two real roots x_1 and x_2. Then by Rolle's Theorem, since $p(x_1) = p(x_2) = 0$, there exists c in (x_1, x_2) such that $p'(c) = 0$. But $p'(x) = (2n + 1)x^{2n} + a \neq 0$, since $n > 0, a > 0$. Therefore, $p(x)$ cannot have two real roots.

79. If $p(x) = Ax^2 + Bx + C$, then

$$p'(x) = 2Ax + B = \frac{f(b) - f(a)}{b - a} = \frac{(Ab^2 + Bb + C) - (Aa^2 + Ba + C)}{b - a}$$

$$= \frac{A(b^2 - a^2) + B(b - a)}{b - a}$$

$$= \frac{(b - a)[A(b + a) + B]}{b - a}$$

$$= A(b + a) + B.$$

Thus, $2Ax = A(b + a)$ and $x = (b + a)/2$ which is the midpoint of $[a, b]$.

81. Suppose $f(x)$ has two fixed points c_1 and c_2. Then, by the Mean Value Theorem, there exists c such that

$$f'(c) = \frac{f(c_2) - f(c_1)}{c_2 - c_1} = \frac{c_2 - c_1}{c_2 - c_1} = 1.$$

This contradicts the fact that $f'(x) < 1$ for all x.

83. Let $f(x) = \cos x$. f is continuous and differentiable for all real numbers. By the Mean Value Theorem, for any interval $[a, b]$, there exists c in (a, b) such that

$$\frac{f(b) - f(a)}{b - a} = f'(c)$$

$$\frac{\cos b - \cos a}{b - a} = -\sin c$$

$$\cos b - \cos a = (-\sin c)(b - a)$$

$$|\cos b - \cos a| = |-\sin c||b - a|$$

$$|\cos b - \cos a| \le |b - a| \text{ since } |-\sin c| \le 1.$$

85. Let $0 < a < b$. $f(x) = \sqrt{x}$ satisfies the hypotheses of the Mean Value Theorem on $[a, b]$. Hence, there exists c in (a, b) such that

$$f'(c) = \frac{1}{2\sqrt{c}} = \frac{f(b) - f(a)}{b - a} = \frac{\sqrt{b} - \sqrt{a}}{b - a}.$$

Thus $\sqrt{b} - \sqrt{a} = (b - a)\dfrac{1}{2\sqrt{c}} < \dfrac{b - a}{2\sqrt{a}}.$

Section 3.3 Increasing and Decreasing Functions and the First Derivative Test

1. (a) Increasing: $(0, 6)$ and $(8, 9)$. Largest: $(0, 6)$

(b) Decreasing: $(6, 8)$ and $(9, 10)$. Largest: $(6, 8)$

3. $f(x) = x^2 - 6x + 8$

Increasing on: $(3, \infty)$

Decreasing on: $(-\infty, 3)$

5. $y = \dfrac{x^3}{4} - 3x$

Increasing on: $(-\infty, -2), (2, \infty)$

Decreasing on: $(-2, 2)$

7. $f(x) = \sin x + 2, \quad 0 < x < 2\pi$

$f'(x) = \cos x$

Critical numbers: $x = \dfrac{\pi}{2}, \dfrac{3\pi}{2}$

Test intervals:	$0 < x < \dfrac{\pi}{2}$	$\dfrac{\pi}{2} < x < \dfrac{3\pi}{2}$	$\dfrac{3\pi}{2} < x < 2\pi$
Sign of $f'(x)$:	$f' > 0$	$f' < 0$	$f' > 0$
Conclusion:	Increasing	Decreasing	Increasing

Increasing on: $\left(0, \dfrac{\pi}{2}\right), \left(\dfrac{3\pi}{2}, 2\pi\right)$

Decreasing on: $\left(\dfrac{\pi}{2}, \dfrac{3\pi}{2}\right)$

9. $f(x) = \dfrac{1}{x^2} = x^{-2}$

$f'(x) = \dfrac{-2}{x^3}$

Discontinuity: $x = 0$

Test intervals:	$-\infty < x < 0$	$0 < x < \infty$
Sign of $f'(x)$:	$f' > 0$	$f' < 0$
Conclusion:	Increasing	Decreasing

Increasing on $(-\infty, 0)$

Decreasing on $(0, \infty)$

11. $g(x) = x^2 - 2x - 8$

$g'(x) = 2x - 2$

Critical number: $x = 1$

Test intervals:	$-\infty < x < 1$	$1 < x < \infty$
Sign of $g'(x)$:	$g' < 0$	$g' > 0$
Conclusion:	Decreasing	Increasing

Increasing on: $(1, \infty)$

Decreasing on: $(-\infty, 1)$

13. $y = x\sqrt{16 - x^2}$ Domain: $[-4, 4]$

$$y' = \frac{-2(x^2 - 8)}{\sqrt{16 - x^2}} = \frac{-2}{\sqrt{16 - x^2}}(x - 2\sqrt{2})(x + 2\sqrt{2})$$

Critical numbers: $x = \pm 2\sqrt{2}$

Test intervals:	$-4 < x < -2\sqrt{2}$	$-2\sqrt{2} < x < 2\sqrt{2}$	$2\sqrt{2} < x < 4$
Sign of y':	$y' < 0$	$y' > 0$	$y' < 0$
Conclusion:	Decreasing	Increasing	Decreasing

Increasing on $\left(-2\sqrt{2}, 2\sqrt{2}\right)$

Decreasing on $\left(-4, -2\sqrt{2}\right), \left(2\sqrt{2}, 4\right)$

15. $y = x - 2\cos x, \quad 0 < x < 2\pi$

$y' = 1 + 2\sin x$

$y' = 0: \ \sin x = -\dfrac{1}{2}$

Critical numbers: $x = \dfrac{7\pi}{6}, \ \dfrac{11\pi}{6}$

Test intervals:	$0 < x < \dfrac{7\pi}{6}$	$\dfrac{7\pi}{6} < x < \dfrac{11\pi}{6}$	$\dfrac{11\pi}{6} < x < 2\pi$
Sign of y'	$y' > 0$	$y' < 0$	$y' > 0$
Conclusion:	Increasing	Decreasing	Increasing

Increasing on: $\left(0, \dfrac{7\pi}{6}\right), \left(\dfrac{11\pi}{6}, 2\pi\right)$

Decreasing on: $\left(\dfrac{7\pi}{6}, \dfrac{11\pi}{6}\right)$

17. $f(x) = x^2 - 6x$

$f'(x) = 2x - 6 = 0$

Critical number: $x = 3$

Test intervals:	$-\infty < x < 3$	$3 < x < \infty$
Sign of $f'(x)$:	$f' < 0$	$f' > 0$
Conclusion:	Decreasing	Increasing

Increasing on: $(3, \infty)$

Decreasing on: $(-\infty, 3)$

Relative minimum: $(3, -9)$

19. $f(x) = -2x^2 + 4x + 3$

$f'(x) = -4x + 4 = 0$

Critical number: $x = 1$

Test intervals:	$-\infty < x < 1$	$1 < x < \infty$
Sign of $f'(x)$:	$f' > 0$	$f' < 0$
Conclusion:	Increasing	Decreasing

Increasing on: $(-\infty, 1)$

Decreasing on: $(1, \infty)$

Relative maximum: $(1, 5)$

21. $f(x) = 2x^3 + 3x^2 - 12x$

$f'(x) = 6x^2 + 6x - 12 = 6(x + 2)(x - 1) = 0$

Critical numbers: $x = -2, 1$

Test intervals:	$-\infty < x < -2$	$-2 < x < 1$	$1 < x < \infty$
Sign of $f'(x)$:	$f' > 0$	$f' < 0$	$f' > 0$
Conclusion:	Increasing	Decreasing	Increasing

Increasing on: $(-\infty, -2), (1, \infty)$

Decreasing on: $(-2, 1)$

Relative maximum: $(-2, 20)$

Relative minimum: $(1, -7)$

23. $f(x) = x^2(3 - x) = 3x^2 - x^3$

$f'(x) = 6x - 3x^2 = 3x(2 - x)$

Critical numbers: $x = 0, 2$

Test intervals:	$-\infty < x < 0$	$0 < x < 2$	$2 < x < \infty$
Sign of $f'(x)$:	$f' < 0$	$f' > 0$	$f' < 0$
Conclusion:	Decreasing	Increasing	Decreasing

Increasing on: $(0, 2)$

Decreasing on: $(-\infty, 0), (2, \infty)$

Relative maximum: $(2, 4)$

Relative minimum: $(0, 0)$

25. $f(x) = \dfrac{x^5 - 5x}{5}$

$f'(x) = x^4 - 1$

Critical numbers: $x = -1, 1$

Test intervals:	$-\infty < x < -1$	$-1 < x < 1$	$1 < x < \infty$
Sign of $f'(x)$:	$f' > 0$	$f' < 0$	$f' > 0$
Conclusion:	Increasing	Decreasing	Increasing

Increasing on: $(-\infty, -1), (1, \infty)$

Decreasing on: $(-1, 1)$

Relative maximum: $\left(-1, \frac{4}{5}\right)$

Relative minimum: $\left(1, -\frac{4}{5}\right)$

27. $f(x) = x^{1/3} + 1$

$f'(x) = \frac{1}{3}x^{-2/3} = \frac{1}{3x^{2/3}}$

Critical number: $x = 0$

Test intervals:	$-\infty < x < 0$	$0 < x < \infty$
Sign of $f'(x)$:	$f' > 0$	$f' > 0$
Conclusion:	Increasing	Increasing

Increasing on: $(-\infty, \infty)$

No relative extrema

29. $f(x) = (x - 1)^{2/3}$

$f'(x) = \frac{2}{3(x - 1)^{1/3}}$

Critical number: $x = 1$

Test intervals:	$-\infty < x < 1$	$1 < x < \infty$
Sign of $f'(x)$:	$f' < 0$	$f' > 0$
Conclusion:	Decreasing	Increasing

Increasing on: $(1, \infty)$

Decreasing on: $(-\infty, 1)$

Relative minimum: $(1, 0)$

31. $f(x) = 5 - |x - 5|$

$f'(x) = -\frac{x - 5}{|x - 5|} = \begin{cases} 1, & x < 5 \\ -1, & x > 5 \end{cases}$

Critical number: $x = 5$

Test intervals:	$-\infty < x < 5$	$5 < x < \infty$
Sign of $f'(x)$:	$f' > 0$	$f' < 0$
Conclusion:	Increasing	Decreasing

Increasing on: $(-\infty, 5)$

Decreasing on: $(5, \infty)$

Relative maximum: $(5, 5)$

33. $f(x) = x + \frac{1}{x}$

$f'(x) = 1 - \frac{1}{x^2} = \frac{x^2 - 1}{x^2}$

Critical numbers: $x = -1, 1$

Discontinuity: $x = 0$

Test intervals:	$-\infty < x < -1$	$-1 < x < 0$	$0 < x < 1$	$1 < x < \infty$
Sign of $f'(x)$:	$f' > 0$	$f' < 0$	$f' < 0$	$f' > 0$
Conclusion:	Increasing	Decreasing	Decreasing	Increasing

Increasing on: $(-\infty, -1), (1, \infty)$

Decreasing on: $(-1, 0), (0, 1)$

Relative maximum: $(-1, -2)$

Relative minimum: $(1, 2)$

35. $f(x) = \dfrac{x^2}{x^2 - 9}$

$f'(x) = \dfrac{(x^2 - 9)(2x) - (x^2)(2x)}{(x^2 - 9)^2} = \dfrac{-18x}{(x^2 - 9)^2}$

Critical number: $x = 0$

Discontinuities: $x = -3, 3$

Test intervals:	$-\infty < x < -3$	$-3 < x < 0$	$0 < x < 3$	$3 < x < \infty$
Sign of $f'(x)$:	$f' > 0$	$f' > 0$	$f' < 0$	$f' < 0$
Conclusion:	Increasing	Increasing	Decreasing	Decreasing

Increasing on: $(-\infty, -3), (-3, 0)$

Decreasing on: $(0, 3), (3, \infty)$

Relative maximum: $(0, 0)$

37. $f(x) = \dfrac{x^2 - 2x + 1}{x + 1}$

$f'(x) = \dfrac{(x + 1)(2x - 2) - (x^2 - 2x + 1)(1)}{(x + 1)^2} = \dfrac{x^2 + 2x - 3}{(x + 1)^2} = \dfrac{(x + 3)(x - 1)}{(x + 1)^2}$

Critical numbers: $x = -3, 1$

Discontinuity: $x = -1$

Test intervals:	$-\infty < x < -3$	$-3 < x < -1$	$-1 < x < 1$	$1 < x < \infty$
Sign of $f'(x)$:	$f' > 0$	$f' < 0$	$f' < 0$	$f' > 0$
Conclusion:	Increasing	Decreasing	Decreasing	Increasing

Increasing on: $(-\infty, -3), (1, \infty)$

Decreasing on: $(-3, -1), (-1, 1)$

Relative maximum: $(-3, -8)$

Relative minimum: $(1, 0)$

39. (a) $f(x) = \dfrac{x}{2} + \cos x, 0 < x < 2\pi$

(b) Relative maximum: $\left(\dfrac{\pi}{6}, \dfrac{\pi + 6\sqrt{3}}{12}\right)$

$\quad f'(x) = \dfrac{1}{2} - \sin x = 0$

Relative minimum: $\left(\dfrac{5\pi}{6}, \dfrac{5\pi - 6\sqrt{3}}{12}\right)$

Critical numbers: $x = \dfrac{\pi}{6}, \dfrac{5\pi}{6}$

Test intervals:	$0 < x < \dfrac{\pi}{6}$	$\dfrac{\pi}{6} < x < \dfrac{5\pi}{6}$	$\dfrac{5\pi}{6} < x < 2\pi$
Sign of $f'(x)$:	$f' > 0$	$f' < 0$	$f' > 0$
Conclusion:	Increasing	Decreasing	Increasing

Increasing on: $\left(0, \dfrac{\pi}{6}\right), \left(\dfrac{5\pi}{6}, 2\pi\right)$

Decreasing on: $\left(\dfrac{\pi}{6}, \dfrac{5\pi}{6}\right)$

41. (a) $f(x) = \sin x + \cos x, \qquad 0 < x < 2\pi$

$f'(x) = \cos x - \sin x = 0 \implies \sin x = \cos x$

Critical numbers: $x = \dfrac{\pi}{4}, \dfrac{5\pi}{4}$

Test intervals:	$0 < x < \dfrac{\pi}{4}$	$\dfrac{\pi}{4} < x < \dfrac{5\pi}{4}$	$\dfrac{5\pi}{4} < x < 2\pi$
Sign of $f'(x)$	$f' > 0$	$f' < 0$	$f' > 0$
Conclusion:	Increasing	Decreasing	Increasing

Increasing on: $\left(0, \dfrac{\pi}{4}\right), \left(\dfrac{5\pi}{4}, 2\pi\right)$

Decreasing on: $\left(\dfrac{\pi}{4}, \dfrac{5\pi}{4}\right)$

(b) Relative maximum: $\left(\dfrac{\pi}{4}, \sqrt{2}\right)$

     Relative minimum: $\left(\dfrac{5\pi}{4}, -\sqrt{2}\right)$

(c)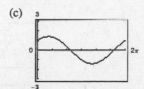

43. (a) $f(x) = \cos^2(2x), \qquad 0 < x < 2\pi$

$f'(x) = -2\cos 2x \sin 2x = 0 \implies \cos 2x = 0$ or $\sin 2x = 0$

Critical numbers: $x = \dfrac{\pi}{4}, \dfrac{3\pi}{4}, \dfrac{5\pi}{4}, \dfrac{7\pi}{4}, \dfrac{\pi}{2}, \pi, \dfrac{3\pi}{2}$

Test intervals:	$0 < x < \dfrac{\pi}{4}$	$\dfrac{\pi}{4} < x < \dfrac{\pi}{2}$	$\dfrac{\pi}{2} < x < \dfrac{3\pi}{4}$	$\dfrac{3\pi}{4} < x < \pi$
Sign of $f'(x)$	$f' < 0$	$f' > 0$	$f' < 0$	$f' > 0$
Conclusion:	Decreasing	Increasing	Decreasing	Increasing

Test intervals:	$\pi < x < \dfrac{5\pi}{4}$	$\dfrac{5\pi}{4} < x < \dfrac{3\pi}{2}$	$\dfrac{3\pi}{2} < x < \dfrac{7\pi}{4}$	$\dfrac{7\pi}{4} < x < 2\pi$
Sign of $f'(x)$	$f' < 0$	$f' > 0$	$f' < 0$	$f' > 0$
Conclusion:	Decreasing	Increasing	Decreasing	Increasing

Increasing on: $\left(\dfrac{\pi}{4}, \dfrac{\pi}{2}\right), \left(\dfrac{3\pi}{4}, \pi\right), \left(\dfrac{5\pi}{4}, \dfrac{3\pi}{2}\right), \left(\dfrac{7\pi}{4}, 2\pi\right)$

Decreasing on: $\left(0, \dfrac{\pi}{4}\right), \left(\dfrac{\pi}{2}, \dfrac{3\pi}{4}\right), \left(\pi, \dfrac{5\pi}{4}\right), \left(\dfrac{3\pi}{2}, \dfrac{7\pi}{4}\right)$

(b) Relative maxima: $\left(\dfrac{\pi}{2}, 1\right), (\pi, 1), \left(\dfrac{3\pi}{2}, 1\right)$

     Relative minima: $\left(\dfrac{\pi}{4}, 0\right), \left(\dfrac{3\pi}{4}, 0\right), \left(\dfrac{5\pi}{4}, 0\right), \left(\dfrac{7\pi}{4}, 0\right)$

(c)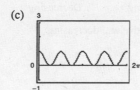

45. (a) $f(x) = \sin^2 x + \sin x, \quad 0 < x < 2\pi$

$f'(x) = 2\sin x \cos x + \cos x = \cos x (2\sin x + 1) = 0$

Critical numbers: $x = \dfrac{\pi}{2}, \dfrac{7\pi}{6}, \dfrac{3\pi}{2}, \dfrac{11\pi}{6}$

Test intervals:	$0 < x < \dfrac{\pi}{2}$	$\dfrac{\pi}{2} < x < \dfrac{7\pi}{6}$	$\dfrac{7\pi}{6} < x < \dfrac{3\pi}{2}$	$\dfrac{3\pi}{2} < x < \dfrac{11\pi}{6}$	$\dfrac{11\pi}{6} < x < 2\pi$
Sign of $f'(x)$:	$f' > 0$	$f' < 0$	$f' > 0$	$f' < 0$	$f' > 0$
Conclusion:	Increasing	Decreasing	Increasing	Decreasing	Increasing

Increasing on: $\left(0, \dfrac{\pi}{2}\right), \left(\dfrac{7\pi}{6}, \dfrac{3\pi}{2}\right), \left(\dfrac{11\pi}{6}, 2\pi\right)$

Decreasing on: $\left(\dfrac{\pi}{2}, \dfrac{7\pi}{6}\right), \left(\dfrac{3\pi}{2}, \dfrac{11\pi}{6}\right)$

(b) Relative minima: $\left(\dfrac{7\pi}{6}, -\dfrac{1}{4}\right), \left(\dfrac{11\pi}{6}, -\dfrac{1}{4}\right)$

Relative maxima: $\left(\dfrac{\pi}{2}, 2\right), \left(\dfrac{3\pi}{2}, 0\right)$

(c)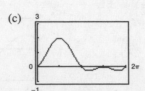

47. $f(x) = 2x\sqrt{9 - x^2}, [-3, 3]$

(a) $f'(x) = \dfrac{2(9 - 2x^2)}{\sqrt{9 - x^2}}$

(b)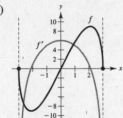

(c) $\dfrac{2(9 - 2x^2)}{\sqrt{9 - x^2}} = 0$

Critical numbers: $x = \pm\dfrac{3}{\sqrt{2}} = \pm\dfrac{3\sqrt{2}}{2}$

(d) Intervals:

$\left(-3, -\dfrac{3\sqrt{2}}{2}\right)$	$\left(-\dfrac{3\sqrt{2}}{2}, \dfrac{3\sqrt{2}}{2}\right)$	$\left(\dfrac{3\sqrt{2}}{2}, 3\right)$
$f'(x) < 0$	$f'(x) > 0$	$f'(x) < 0$
Decreasing	Increasing	Decreasing

f is increasing when f' is positive and decreasing when f' is negative.

49. $f(t) = t^2 \sin t, [0, 2\pi]$

(a) $f'(t) = t^2 \cos t + 2t \sin t$

$= t(t\cos t + 2\sin t)$

(b)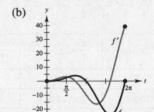

(c) $t(t\cos t + 2\sin t) = 0$

$t = 0$ or $t = -2\tan t$

$t\cot t = -2$

$t \approx 2.2889, 5.0870$ (graphing utility)

Critical numbers: $t = 2.2889, t = 5.0870$

(d) Intervals:

$(0, 2.2889)$	$(2.2889, 5.0870)$	$(5.0870, 2\pi)$
$f'(t) > 0$	$f'(t) < 0$	$f'(t) > 0$
Increasing	Decreasing	Increasing

f is increasing when f' is positive and decreasing when f' is negative.

51. (a) $f(x) = -3 \sin \frac{x}{3}$, $[0, 6\pi]$

$f'(x) = -\cos \frac{x}{3}$

(b)

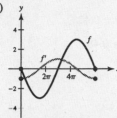

(c) Critical numbers: $x = \frac{3\pi}{2}, \frac{9\pi}{2}$

(d) Intervals:

$\left(0, \frac{3\pi}{2}\right)$	$\left(\frac{3\pi}{2}, \frac{9\pi}{2}\right)$	$\left(\frac{9\pi}{2}, 6\pi\right)$
$f' < 0$	$f' > 0$	$f' < 0$
Decreasing	Increasing	Decreasing

f is increasing when f' is positive and decreasing when f' is negative.

53. $f(x) = \dfrac{x^5 - 4x^3 + 3x}{x^2 - 1} = \dfrac{(x^2 - 1)(x^3 - 3x)}{x^2 - 1} = x^3 - 3x$, $x \neq \pm 1$

$f(x) = g(x) = x^3 - 3x$ for all $x \neq \pm 1$.

$f'(x) = 3x^2 - 3 = 3(x^2 - 1)$, $x \neq \pm 1 \implies f'(x) \neq 0$

f symmetric about origin

zeros of f: $(0, 0), \left(\pm\sqrt{3}, 0\right)$

No relative extrema

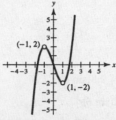

Holes at $(-1, 2)$ and $(1, -2)$

55. $f(x) = c$ is constant $\implies f'(x) = 0$.

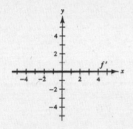

57. f is quadratic $\implies f'$ is a line.

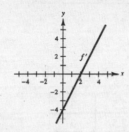

59. f has positive, but decreasing slope

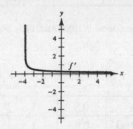

61. (a) f increasing on $(2, \infty)$ because $f' > 0$ on $(2, \infty)$

f decreasing on $(-\infty, 2)$ because $f' < 0$ on $(-\infty, 2)$

(b) f has a relative minimum at $x = 2$.

63. (a) f increasing on $(-\infty, 0)$ and $(1, \infty)$ because $f' > 0$ there

f decreasing on $(0, 1)$ because $f' < 0$ there

(b) f has a relative maximum at $x = 0$, and a relative minimum at $x = 1$.

In Exercises 65–70, $f'(x) > 0$ on $(-\infty, -4)$, $f'(x) < 0$ on $(-4, 6)$ and $f'(x) > 0$ on $(6, \infty)$.

65. $g(x) = f(x) + 5$

$g'(x) = f'(x)$

$g'(0) = f'(0) < 0$

67. $g(x) = -f(x)$

$g'(x) = -f'(x)$

$g'(-6) = -f'(-6) < 0$

69. $g(x) = f(x - 10)$

$g'(x) = f'(x - 10)$

$g'(0) = f'(-10) > 0$

71. $f'(x) \begin{cases} > 0, & x < 4 \Rightarrow f \text{ is increasing on } (-\infty, 4). \\ \text{undefined}, & x = 4 \\ < 0, & x > 4 \Rightarrow f \text{ is decreasing on } (4, \infty). \end{cases}$

Two possibilities for $f(x)$ are given below.

(a) 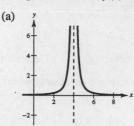   (b)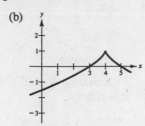

73. The critical numbers are in intervals $(-0.50, -0.25)$ and $(0.25, 0.50)$ since the sign of f' changes in these intervals. f is decreasing on approximately $(-1, -0.40)$, $(0.48, 1)$, and increasing on $(-0.40, 0.48)$.

Relative minimum when $x \approx -0.40$.

Relative maximum when $x \approx 0.48$.

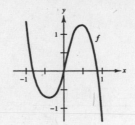

75. $s(t) = 4.9(\sin \theta)t^2$

(a) $s'(t) = 4.9(\sin \theta)(2t) = 9.8(\sin \theta)t$

speed $= |s'(t)| = |9.8(\sin \theta)t|$

(b)

θ	0	$\dfrac{\pi}{4}$	$\dfrac{\pi}{3}$	$\dfrac{\pi}{2}$	$\dfrac{2\pi}{3}$	$\dfrac{3\pi}{4}$	π
$|s'(t)|$	0	$4.9\sqrt{2}\,t$	$4.9\sqrt{3}\,t$	$9.8t$	$4.9\sqrt{3}\,t$	$4.9\sqrt{2}\,t$	0

The speed is maximum for $\theta = \dfrac{\pi}{2}$.

77. $f(x) = x$, $g(x) = \sin x$, $0 < x < \pi$

(a)

x	0.5	1	1.5	2	2.5	3
$f(x)$	0.5	1	1.5	2	2.5	3
$g(x)$	0.479	0.841	0.997	0.909	0.598	0.141

$f(x)$ seems greater than $g(x)$ on $(0, \pi)$.

(b)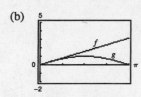

$x > \sin x$ on $(0, \pi)$

(c) Let $h(x) = f(x) - g(x) = x - \sin x$

$h'(x) = 1 - \cos x > 0$ on $(0, \pi)$.

Therefore, $h(x)$ is increasing on $(0, \pi)$. Since $h(0) = 0$, $h(x) > 0$ on $(0, \pi)$. Thus,

$$x - \sin x > 0$$
$$x > \sin x$$
$$f(x) > g(x) \text{ on } (0, \pi).$$

79. $v = k(R - r)r^2 = k(Rr^2 - r^3)$

$v' = k(2Rr - 3r^2)$

$\quad = kr(2R - 3r) = 0$

$r = 0$ or $\dfrac{2}{3}R$

Maximum when $r = \dfrac{2}{3}R$.

81. $P = \dfrac{vR_1R_2}{(R_1 + R_2)^2}$, v and R_1 are constant

$\dfrac{dP}{dR_2} = \dfrac{(R_1 + R_2)^2(vR_1) - vR_1R_2[2(R_1 + R_2)(1)]}{(R_1 + R_2)^4}$

$\quad = \dfrac{vR_1(R_1 - R_2)}{(R_1 + R_2)^3} = 0 \Rightarrow R_2 = R_1$

Maximum when $R_1 = R_2$.

83. (a) Using a graphing utility, ($t = 5$ represents 1995)
$$M = 5.267t^2 - 71.19t + 356.9.$$

(b)
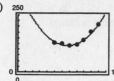

(c) $M'(t) = 10.534t - 71.19 = 0 \Rightarrow t \approx 6.8$

By the First Derivative Test, $t \approx 6.8$ is a minimum.

$M(6.8) \approx 116.3$

The minimum of the actual data is 115.6 at $t = 7$.

85. (a) $s(t) = 6t - t^2, t \geq 0$

$v(t) = 6 - 2t$

(b) $v(t) = 0$ when $t = 3$.

Moving in positive direction for $0 < t < 3$ because $v(t) > 0$ on $0 < t < 3$.

(c) Moving in negative direction when $t > 3$.

(d) The particle changes direction at $t = 3$.

87. (a) $s(t) = t^3 - 5t^2 + 4t, t \geq 0$

$v(t) = 3t^2 - 10t + 4$

(b) $v(t) = 0$ for $t = \dfrac{10 \pm \sqrt{100 - 48}}{6} = \dfrac{5 \pm \sqrt{13}}{3}$

Particle is moving in a positive direction on $\left(0, \dfrac{5 - \sqrt{13}}{3}\right) \approx (0, 0.4648)$ and $\left(\dfrac{5 + \sqrt{13}}{3}, \infty\right) \approx (2.8685, \infty)$ because $v > 0$ on these intervals.

(c) Particle is moving in a negative direction on $\left(\dfrac{5 - \sqrt{13}}{3}, \dfrac{5 + \sqrt{13}}{3}\right) \sim (0.4648, 2.8685)$

(d) The particle changes direction at $t = \dfrac{5 \pm \sqrt{13}}{3}$.

89. Answers will vary.

91. (a) Use a cubic polynomial
$$f(x) = a_3x^3 + a_2x^2 + a_1x + a_0.$$

(b) $f'(x) = 3a_3x^2 + 2a_2x + a_1.$

$(0, 0)$: $0 = a_0$ $(f(0) = 0)$

           $0 = a_1$ $(f'(0) = 0)$

$(2, 2)$: $2 = 8a_3 + 4a_2$ $(f(2) = 2)$

           $0 = 12a_3 + 4a_2$ $(f'(2) = 0)$

(c) The solution is $a_0 = a_1 = 0, a_2 = \dfrac{3}{2}, a_3 = -\dfrac{1}{2}$:

$$f(x) = -\dfrac{1}{2}x^3 + \dfrac{3}{2}x^2.$$

(d)
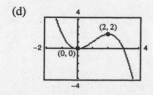

93. (a) Use a fourth degree polynomial
$$f(x) = a_4x^4 + a_3x^3 + a_2x^2 + a_1x + a_0.$$

(b) $f'(x) = 4a_4x^3 + 3a_3x^2 + 2a_2x + a_1$

$(0, 0)$: $0 = a_0$ $(f(0) = 0)$

           $0 = a_1$ $(f'(0) = 0)$

$(4, 0)$: $0 = 256a_4 + 64a_3 + 16a_2$ $(f(4) = 0)$

           $0 = 256a_4 + 48a_3 + 8a_2$ $(f'(4) = 0)$

$(2, 4)$: $4 = 16a_4 + 8a_3 + 4a_2$ $(f(2) = 4)$

           $0 = 32a_4 + 12a_3 + 4a_2$ $(f'(2) = 0)$

(c) The solution is $a_0 = a_1 = 0, a_2 = 4, a_3 = -2, a_4 = \dfrac{1}{4}$:

$$f(x) = \dfrac{1}{4}x^4 - 2x^3 + 4x^2$$

(d)
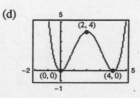

95. True.

Let $h(x) = f(x) + g(x)$ where f and g are increasing. Then $h'(x) = f'(x) + g'(x) > 0$ since $f'(x) > 0$ and $g'(x) > 0$.

97. False.

Let $f(x) = x^3$, then $f'(x) = 3x^2$ and f only has one critical number. Or, let $f(x) = x^3 + 3x + 1$, then $f'(x) = 3(x^2 + 1)$ has no critical numbers.

99. False. For example, $f(x) = x^3$ does not have a relative extrema at the critical number $x = 0$.

101. Assume that $f'(x) < 0$ for all x in the interval (a, b) and let $x_1 < x_2$ be any two points in the interval. By the Mean Value Theorem, we know there exists a number c such that $x_1 < c < x_2$, and

$$f'(c) = \frac{f(x_2) - f(x_1)}{x_2 - x_1}.$$

Since $f'(c) < 0$ and $x_2 - x_1 > 0$, then $f(x_2) - f(x_1) < 0$, which implies that $f(x_2) < f(x_1)$. Thus, f is decreasing on the interval.

103. Let $f(x) = (1 + x)^n - nx - 1$. Then

$$f'(x) = n(1 + x)^{n-1} - n$$

$$= n[(1 + x)^{n-1} - 1] > 0 \text{ since } x > 0 \text{ and } n > 1.$$

Thus, $f(x)$ is increasing on $(0, \infty)$. Since $f(0) = 0 \Rightarrow f(x) > 0$ on $(0, \infty)$

$$(1 + x)^n - nx - 1 > 0 \Rightarrow (1 + x)^n > 1 + nx.$$

105. Let x_1 and x_2 be two positive real numbers, $0 < x_1 < x_2$. Then

$$\frac{1}{x_1} > \frac{1}{x_2}$$

$$f(x_1) > f(x_2)$$

Thus, f is decreasing on $(0, \infty)$.

Section 3.4   Concavity and the Second Derivative Test

1. $y = x^2 - x - 2, y'' = 2$

Concave upward: $(-\infty, \infty)$

3. $f(x) = \dfrac{24}{x^2 + 12}, y'' = \dfrac{-144(4 - x^2)}{(x^2 + 12)^3}$

Concave upward: $(-\infty, -2), (2, \infty)$

Concave downward: $(-2, 2)$

5. $f(x) = \dfrac{x^2 + 1}{x^2 - 1}, y'' = \dfrac{4(3x^2 + 1)}{(x^2 - 1)^3}$

Concave upward: $(-\infty, -1), (1, \infty)$

Concave downward: $(-1, 1)$

7. $g(x) = 3x^2 - x^3$

$g'(x) = 6x - 3x^2$

$g''(x) = 6 - 6x$

Concave upward: $(-\infty, 1)$

Concave downward: $(1, \infty)$

9. $y = 2x - \tan x, \left(-\dfrac{\pi}{2}, \dfrac{\pi}{2}\right)$

$y' = 2 - \sec^2 x$

$y'' = -2 \sec^2 x \tan x$

Concave upward: $\left(-\dfrac{\pi}{2}, 0\right)$

Concave downward: $\left(0, \dfrac{\pi}{2}\right)$

11. $f(x) = x^3 - 6x^2 + 12x$

$f'(x) = 3x^2 - 12x + 12$

$f''(x) = 6(x - 2) = 0$ when $x = 2$.

The concavity changes at $x = 2$. $(2, 8)$ is a point of inflection.

Concave upward: $(2, \infty)$

Concave downward: $(-\infty, 2)$

13. $f(x) = \dfrac{1}{4}x^4 - 2x^2$

$f'(x) = x^3 - 4x$

$f''(x) = 3x^2 - 4$

$f''(x) = 3x^2 - 4 = 0$ when $x = \pm\dfrac{2}{\sqrt{3}}$.

Test interval:	$-\infty < x < -\dfrac{2}{\sqrt{3}}$	$-\dfrac{2}{\sqrt{3}} < x < \dfrac{2}{\sqrt{3}}$	$\dfrac{2}{\sqrt{3}} < x < \infty$
Sign of $f''(x)$:	$f''(x) > 0$	$f''(x) < 0$	$f''(x) > 0$
Conclusion:	Concave upward	Concave downward	Concave upward

Points of inflection: $\left(\pm\dfrac{2}{\sqrt{3}}, -\dfrac{20}{9}\right)$

15. $f(x) = x(x - 4)^3$

$f'(x) = x[3(x - 4)^2] + (x - 4)^3$

$\quad = (x - 4)^2(4x - 4)$

$f''(x) = 4(x - 1)[2(x - 4)] + 4(x - 4)^2$

$\quad = 4(x - 4)[2(x - 1) + (x - 4)]$

$\quad = 4(x - 4)(3x - 6) = 12(x - 4)(x - 2)$

$f''(x) = 12(x - 4)(x - 2) = 0$ when $x = 2, 4$.

Test interval:	$-\infty < x < 2$	$2 < x < 4$	$4 < x < \infty$
Sign of $f''(x)$:	$f''(x) > 0$	$f''(x) < 0$	$f''(x) > 0$
Conclusion:	Concave upward	Concave downward	Concave upward

Points of inflection: $(2, -16), (4, 0)$

17. $f(x) = x\sqrt{x + 3}$, Domain: $[-3, \infty)$

$f'(x) = x\left(\dfrac{1}{2}\right)(x + 3)^{-1/2} + \sqrt{x + 3} = \dfrac{3(x + 2)}{2\sqrt{x + 3}}$

$f''(x) = \dfrac{6\sqrt{x + 3} - 3(x + 2)(x + 3)^{-1/2}}{4(x + 3)} = \dfrac{3(x + 4)}{4(x + 3)^{3/2}}$

$f''(x) > 0$ on the entire domain of f (except for $x = -3$, for which $f''(x)$ is undefined). There are no points of inflection.

Concave upward on $(-3, \infty)$

19. $f(x) = \dfrac{x}{x^2 + 1}$

$f'(x) = \dfrac{1 - x^2}{(x^2 + 1)^2}$

$f''(x) = \dfrac{2x(x^2 - 3)}{(x^2 + 1)^3} = 0$ when $x = 0, \pm\sqrt{3}$.

Test intervals:	$-\infty < x < -\sqrt{3}$	$-\sqrt{3} < x < 0$	$0 < x < \sqrt{3}$	$\sqrt{3} < x < \infty$
Sign of $f'(x)$:	$f'' < 0$	$f'' > 0$	$f'' < 0$	$f'' > 0$
Conclusion:	Concave downward	Concave upward	Concave downward	Concave upward

Points of inflection: $\left(-\sqrt{3}, -\dfrac{\sqrt{3}}{4}\right), (0, 0), \left(\sqrt{3}, \dfrac{\sqrt{3}}{4}\right)$

21. $f(x) = \sin\dfrac{x}{2}, 0 \le x \le 4\pi$

$f'(x) = \dfrac{1}{2}\cos\left(\dfrac{x}{2}\right)$

$f''(x) = -\dfrac{1}{4}\sin\left(\dfrac{x}{2}\right)$

$f''(x) = 0$ when $x = 0, 2\pi, 4\pi$.

Test interval:	$0 < x < 2\pi$	$2\pi < x < 4\pi$
Sign of $f''(x)$:	$f'' < 0$	$f'' > 0$
Conclusion:	Concave downward	Concave upward

Point of inflection: $(2\pi, 0)$

23. $f(x) = \sec\left(x - \dfrac{\pi}{2}\right), 0 < x < 4\pi$

$f'(x) = \sec\left(x - \dfrac{\pi}{2}\right)\tan\left(x - \dfrac{\pi}{2}\right)$

$f''(x) = \sec^3\left(x - \dfrac{\pi}{2}\right) + \sec\left(x - \dfrac{\pi}{2}\right)\tan^2\left(x - \dfrac{\pi}{2}\right) \ne 0$ for any x in the domain of f.

Concave upward: $(0, \pi), (2\pi, 3\pi)$

Concave downward: $(\pi, 2\pi), (3\pi, 4\pi)$

No points of inflection

25. $f(x) = 2\sin x + \sin 2x, 0 \le x \le 2\pi$

$f'(x) = 2\cos x + 2\cos 2x$

$f''(x) = -2\sin x - 4\sin 2x = -2\sin x(1 + 4\cos x)$

$f''(x) = 0$ when $x = 0, 1.823, \pi, 4.460$.

Test interval:	$0 < x < 1.823$	$1.823 < x < \pi$	$\pi < x < 4.460$	$4.460 < x < 2\pi$
Sign of $f''(x)$:	$f'' < 0$	$f'' > 0$	$f'' < 0$	$f'' > 0$
Conclusion:	Concave downward	Concave upward	Concave downward	Concave upward

Points of inflection: $(1.823, 1.452), (\pi, 0), (4.46, -1.452)$

27. $f(x) = x^4 - 4x^3 + 2$

$f'(x) = 4x^3 - 12x^2 = 4x^2(x - 3)$

$f''(x) = 12x^2 - 24x = 12x(x - 2)$

Critical numbers: $x = 0, x = 3$

However, $f''(0) = 0$, so we must use the First Derivative Test. $f'(x) < 0$ on the intervals $(-\infty, 0)$ and $(0, 3)$; hence, $(0, 2)$ is not an extremum. $f''(3) > 0$ so $(3, -25)$ is a relative minimum.

29. $f(x) = (x - 5)^2$

$f'(x) = 2(x - 5)$

$f''(x) = 2$

Critical number: $x = 5$

$f''(5) > 0$

Therefore, $(5, 0)$ is a relative minimum.

31. $f(x) = x^3 - 3x^2 + 3$

$f'(x) = 3x^2 - 6x = 3x(x - 2)$

$f''(x) = 6x - 6 = 6(x - 1)$

Critical numbers: $x = 0, x = 2$

$f''(0) = -6 < 0$

Therefore, $(0, 3)$ is a relative maximum.

$f''(2) = 6 > 0$

Therefore, $(2, -1)$ is a relative minimum.

33. $g(x) = x^2(6 - x)^3$

$g'(x) = x(x - 6)^2(12 - 5x)$

$g''(x) = 4(6 - x)(5x^2 - 24x + 18)$

Critical numbers: $x = 0, \frac{12}{5}, 6$

$g''(0) = 432 > 0$

Therefore, $(0, 0)$ is a relative minimum.

$g''\left(\frac{12}{5}\right) = -155.52 < 0$

Therefore, $\left(\frac{12}{5}, 268.7\right)$ is a relative maximum.

$g''(6) = 0$

Test fails. By the First Derivative Test, $(6, 0)$ is not an extremum.

35. $f(x) = x^{2/3} - 3$

$f'(x) = \frac{2}{3x^{1/3}}$

$f''(x) = \frac{-2}{9x^{4/3}}$

Critical number: $x = 0$

However, $f''(0)$ is undefined, so we must use the First Derivative Test. Since $f'(x) < 0$ on $(-\infty, 0)$ and $f'(x) > 0$ on $(0, \infty)$, $(0, -3)$ is a relative minimum.

37. $f(x) = x + \frac{4}{x}$

$f'(x) = 1 - \frac{4}{x^2} = \frac{x^2 - 4}{x^2}$

$f''(x) = \frac{8}{x^3}$

Critical numbers: $x = \pm 2$

$f''(-2) < 0$

Therefore, $(-2, -4)$ is a relative maximum.

$f''(2) > 0$

Therefore, $(2, 4)$ is a relative minimum.

39. $f(x) = \cos x - x, \ 0 \le x \le 4\pi$

$f'(x) = -\sin x - 1 \le 0$

Therefore, f is non-increasing and there are no relative extrema.

41. $f(x) = 0.2x^2(x - 3)^3, [-1, 4]$

(a) $f'(x) = 0.2x(5x - 6)(x - 3)^2$

$f''(x) = (x - 3)(4x^2 - 9.6x + 3.6)$

$= 0.4(x - 3)(10x^2 - 24x + 9)$

(b) $f''(0) < 0 \Rightarrow (0, 0)$ is a relative maximum.

$f''\left(\frac{6}{5}\right) > 0 \Rightarrow (1.2, -1.6796)$ is a relative minimum.

Points of inflection:

$(3, 0), (0.4652, -0.7049), (1.9348, -0.9049)$

(c)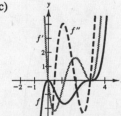

f is increasing when $f' > 0$ and decreasing when $f' < 0$. f is concave upward when $f'' > 0$ and concave downward when $f'' < 0$.

43. $f(x) = \sin x - \dfrac{1}{3}\sin 3x + \dfrac{1}{5}\sin 5x,\ [0,\ \pi]$

    (a) $f'(x) = \cos x - \cos 3x + \cos 5x$

         $f'(x) = 0$ when $x = \dfrac{\pi}{6},\ x = \dfrac{\pi}{2},\ x = \dfrac{5\pi}{6}.$

         $f''(x) = -\sin x + 3\sin 3x - 5\sin 5x$

         $f''(x) = 0$ when $x = \dfrac{\pi}{6},\ x = \dfrac{5\pi}{6},\ x \approx 1.1731,\ x \approx 1.9685$

    (b) $f''\!\left(\dfrac{\pi}{2}\right) < 0 \Longrightarrow \left(\dfrac{\pi}{2},\ 1.53333\right)$ is a relative maximum.

         Points of inflection: $\left(\dfrac{\pi}{6},\ 0.2667\right)$, $(1.1731,\ 0.9638)$,

                  $(1.9685,\ 0.9637),\ \left(\dfrac{5\pi}{6},\ 0.2667\right)$

         Note: $(0,\ 0)$ and $(\pi,\ 0)$ are not points of inflection since they are endpoints.

    (c)

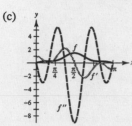

     The graph of f is increasing when $f' > 0$ and decreasing when $f' < 0$. f is concave upward when $f'' > 0$ and concave downward when $f'' < 0$.

45. (a)

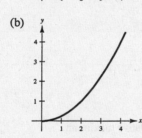

     $f' < 0$ means f decreasing

     f' increasing means concave upward

    (b)

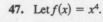

     $f' > 0$ means f increasing

     f' increasing means concave upward

47.   Let $f(x) = x^4$.

     $f''(x) = 12x^2$

     $f''(0) = 0$, but $(0,\ 0)$ is not a point of inflection.

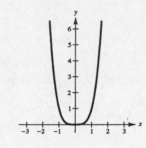

49.

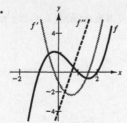

51.

53.

55.

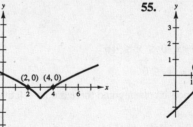

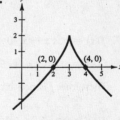

57. f'' is linear.

     f' is quadratic.

     f is cubic.

     f concave upwards on $(-\infty,\ 3)$, downward on $(3,\ \infty)$.

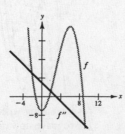

59. (a)

$n = 1:$

$f(x) = x - 2$

$f'(x) = 1$

$f''(x) = 0$

No inflection points

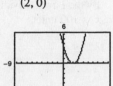

$n = 2:$

$f(x) = (x - 2)^2$

$f'(x) = 2(x - 2)$

$f''(x) = 2$

No inflection points

Relative minimum:
$(2, 0)$

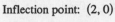

$n = 3:$

$f(x) = (x - 2)^3$

$f'(x) = 3(x - 2)^2$

$f''(x) = 6(x - 2)$

Inflection point: $(2, 0)$

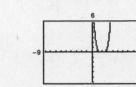

$n = 4:$

$f(x) = (x - 2)^4$

$f'(x) = 4(x - 2)^3$

$f''(x) = 12(x - 2)^2$

No inflection points:

Relative minimum:
$(2, 0)$

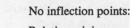

Conclusion: If $n \geq 3$ and n is odd, then $(2, 0)$ is an inflection point. If $n \geq 2$ and n is even, then $(2, 0)$ is a relative minimum.

(b) Let $f(x) = (x - 2)^n$, $f'(x) = n(x - 2)^{n-1}$, $f''(x) = n(n - 1)(x - 2)^{n-2}$.

For $n \geq 3$ and odd, $n - 2$ is also odd and the concavity changes at $x = 2$.

For $n \geq 4$ and even, $n - 2$ is also even and the concavity does not change at $x = 2$.

Thus, $x = 2$ is an inflection point if and only if $n \geq 3$ is odd.

61. $f(x) = ax^3 + bx^2 + cx + d$

Relative maximum: $(3, 3)$

Relative minimum: $(5, 1)$

Point of inflection: $(4, 2)$

$f'(x) = 3ax^2 + 2bx + c, f''(x) = 6ax + 2b$

$\left. \begin{array}{l} f(3) = 27a + 9b + 3c + d = 3 \\ f(5) = 125a + 25b + 5c + d = 1 \end{array} \right\} 98a + 16b + 2c = -2 \Rightarrow 49a + 8b + c = -1$

$f'(3) = 27a + 6b + c = 0, f''(4) = 24a + 2b = 0$

$\begin{array}{ll} 49a + 8b + c = -1 & 24a + 2b = \quad 0 \\ 27a + 6b + c = \quad 0 & 22a + 2b = -1 \\ \hline 22a + 2b \quad\quad = -1 & \overline{2a \quad\quad = \quad 1} \end{array}$

$a = \frac{1}{2}, b = -6, c = \frac{45}{2}, d = -24$

$f(x) = \frac{1}{2}x^3 - 6x^2 + \frac{45}{2}x - 24$

63. $f(x) = ax^3 + bx^2 + cx + d$

Maximum: $(-4, 1)$

Minimum: $(0, 0)$

(a) $f'(x) = 3ax^2 + 2bx + c, \quad f''(x) = 6ax + 2b$

$f(0) = 0 \Rightarrow d = 0$

$f(-4) = 1 \Rightarrow -64a + 16b - 4c = 1$

$f'(-4) = 0 \Rightarrow \quad 48a - 8b + c = 0$

$f'(0) = 0 \Rightarrow \quad\quad\quad\quad c = 0$

Solving this system yields $a = \frac{1}{32}$ and $b = 6a = \frac{3}{16}$.

$f(x) = \frac{1}{32}x^3 + \frac{3}{16}x^2$

(b) The plane would be descending at the greatest rate at the point of inflection.

$f''(x) = 6ax + 2b = \frac{3}{16}x + \frac{3}{8} = 0 \Rightarrow x = -2.$

Two miles from touchdown.

65. $D = 2x^4 - 5Lx^3 + 3L^2x^2$

$D' = 8x^3 - 15Lx^2 + 6L^2x = x(8x^2 - 15Lx + 6L^2) = 0$

$x = 0 \text{ or } x = \dfrac{15L \pm \sqrt{33}L}{16} = \left(\dfrac{15 \pm \sqrt{33}}{16}\right)L$

By the Second Derivative Test, the deflection is maximum when

$x = \left(\dfrac{15 - \sqrt{33}}{16}\right)L \approx 0.578L.$

67. $C = 0.5x^2 + 15x + 5000$

$\overline{C} = \dfrac{C}{x} = 0.5x + 15 + \dfrac{5000}{x}$

$\overline{C} = $ average cost per unit

$\dfrac{d\overline{C}}{dx} = 0.5 - \dfrac{5000}{x^2} = 0 \text{ when } x = 100$

By the First Derivative Test, $\overline{C}$ is minimized when $x = 100$ units.

69. $S = \dfrac{5000t^2}{8 + t^2}, \, 0 \leq t \leq 3$

(a)

t	0.5	1	1.5	2	2.5	3
S	151.5	555.6	1097.6	1666.7	2193.0	2647.1

Increasing at greatest rate when $t \approx 1.5$.

(b)

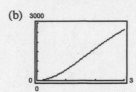

Increasing at greatest rate when $t \approx 1.5$.

(c) $S = \dfrac{5000t^2}{8 + t^2}$

$S'(t) = \dfrac{80,000t}{(8 + t^2)^2}$

$S''(t) = \dfrac{80,000(8 - 3t^2)}{(8 + t^2)^3}$

$S''(t) = 0$ for $t = \pm\sqrt{\dfrac{8}{3}}$. Hence, $t = \dfrac{2\sqrt{6}}{3} \approx 1.633$ yrs.

71. $f(x) = 2(\sin x + \cos x), \qquad f\left(\dfrac{\pi}{4}\right) = 2\sqrt{2}$

$f'(x) = 2(\cos x - \sin x), \qquad f'\left(\dfrac{\pi}{4}\right) = 0$

$f''(x) = 2(-\sin x - \cos x), \qquad f''\left(\dfrac{\pi}{4}\right) = -2\sqrt{2}$

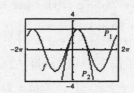

$P_1(x) = 2\sqrt{2} + 0\left(x - \dfrac{\pi}{4}\right) = 2\sqrt{2}$

$P_1'(x) = 0$

$P_2(x) = 2\sqrt{2} + 0\left(x - \dfrac{\pi}{4}\right) + \dfrac{1}{2}(-2\sqrt{2})\left(x - \dfrac{\pi}{4}\right)^2 = 2\sqrt{2} - \sqrt{2}\left(x - \dfrac{\pi}{4}\right)^2$

$P_2'(x) = -2\sqrt{2}\left(x - \dfrac{\pi}{4}\right)$

$P_2''(x) = -2\sqrt{2}$

The values of f, P_1, P_2, and their first derivatives are equal at $x = \pi/4$. The values of the second derivatives of f and P_2 are equal at $x = \pi/4$. The approximations worsen as you move away from $x = \pi/4$.

73. $f(x) = \sqrt{1-x}$, $\qquad$ $f(0) = 1$

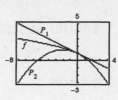

$f'(x) = -\dfrac{1}{2\sqrt{1-x}}$, $\qquad$ $f'(0) = -\dfrac{1}{2}$

$f''(x) = -\dfrac{1}{4(1-x)^{3/2}}$, $\qquad$ $f''(0) = -\dfrac{1}{4}$

$P_1(x) = 1 + \left(-\dfrac{1}{2}\right)(x-0) = 1 - \dfrac{x}{2}$

$P_1'(x) = -\dfrac{1}{2}$

$P_2(x) = 1 + \left(-\dfrac{1}{2}\right)(x-0) + \dfrac{1}{2}\left(-\dfrac{1}{4}\right)(x-0)^2 = 1 - \dfrac{x}{2} - \dfrac{x^2}{8}$

$P_2'(x) = -\dfrac{1}{2} - \dfrac{x}{4}$

$P_2''(x) = -\dfrac{1}{4}$

The values of f, P_1, P_2, and their first derivatives are equal at $x = 0$. The values of the second derivatives of f and P_2 are equal at $x = 0$. The approximations worsen as you move away from $x = 0$.

75. $f(x) = x \sin\left(\dfrac{1}{x}\right)$

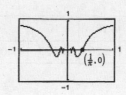

$f'(x) = x\left[-\dfrac{1}{x^2}\cos\left(\dfrac{1}{x}\right)\right] + \sin\left(\dfrac{1}{x}\right) = -\dfrac{1}{x}\cos\left(\dfrac{1}{x}\right) + \sin\left(\dfrac{1}{x}\right)$

$f''(x) = -\dfrac{1}{x}\left[\dfrac{1}{x^2}\sin\left(\dfrac{1}{x}\right)\right] + \dfrac{1}{x^2}\cos\left(\dfrac{1}{x}\right) - \dfrac{1}{x^2}\cos\left(\dfrac{1}{x}\right) = -\dfrac{1}{x^3}\sin\left(\dfrac{1}{x}\right) = 0$

$x = \dfrac{1}{\pi}$

Point of inflection: $\left(\dfrac{1}{\pi}, 0\right)$

When $x > 1/\pi$, $f'' < 0$, so the graph is concave downward.

77. Assume the zeros of f are all real. Then express the function as $f(x) = a(x - r_1)(x - r_2)(x - r_3)$ where r_1, r_2, and r_3 are the distinct zeros of f. From the Product Rule for a function involving three factors, we have

$f'(x) = a[(x - r_1)(x - r_2) + (x - r_1)(x - r_3) + (x - r_2)(x - r_3)]$

$f''(x) = a[(x - r_1) + (x - r_2) + (x - r_1) + (x - r_3) + (x - r_2) + (x - r_3)]$

$\qquad = a[6x - 2(r_1 + r_2 + r_3)]$.

Consequently, $f''(x) = 0$ if

$x = \dfrac{2(r_1 + r_2 + r_3)}{6} = \dfrac{r_1 + r_2 + r_3}{3} =$ (Average of r_1, r_2, and r_3).

79. True. Let $y = ax^3 + bx^2 + cx + d$, $a \neq 0$. Then $y'' = 6ax + 2b = 0$ when $x = -(b/3a)$, and the concavity changes at this point.

81. False. Concavity is determined by f''. For example, let $f(x) = x$ and $c = 2$. $f'(c) = f'(2) > 0$, but f is not concave upward at $c = 2$.

83. f and g are concave upward on (a, b) implies that f' and g' are increasing on (a, b), and hence $f'' > 0$ and $g'' > 0$. Thus, $(f + g)'' > 0 \implies f + g$ is concave upward on (a, b) by Theorem 3.7.

Section 3.5 Limits at Infinity

1. $\lim\limits_{x \to \infty} f(x) = 4$ means that $f(x)$ approaches 4 as x becomes large.

3. $f(x) = \dfrac{3x^2}{x^2 + 2}$

No vertical asymptotes

Horizontal asymptote: $y = 3$

Matches (f)

5. $f(x) = \dfrac{x}{x^2 + 2}$

No vertical asymptotes

Horizontal asymptote: $y = 0$

$f(1) < 1$

Matches (d)

7. $f(x) = \dfrac{4 \sin x}{x^2 + 1}$

No vertical asymptotes

Horizontal asymptote: $y = 0$

$f(1) > 1$

Matches (b)

9. $f(x) = \dfrac{4x + 3}{2x - 1}$

x	10^0	10^1	10^2	10^3	10^4	10^5	10^6
$f(x)$	7	2.26	2.025	2.0025	2.0003	2	2

$\lim\limits_{x \to \infty} f(x) = 2$

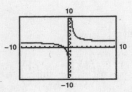

11. $f(x) = \dfrac{-6x}{\sqrt{4x^2 + 5}}$

x	10^0	10^1	10^2	10^3	10^4	10^5	10^6
$f(x)$	-2	-2.98	-2.9998	-3	-3	-3	-3

$\lim\limits_{x \to \infty} f(x) = -3$

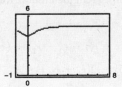

13. $f(x) = 5 - \dfrac{1}{x^2 + 1}$

x	10^0	10^1	10^2	10^3	10^4	10^5	10^6
$f(x)$	4.5	4.99	4.9999	4.999999	5	5	5

$\lim\limits_{x \to \infty} f(x) = 5$

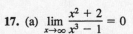

15. (a) $h(x) = \dfrac{f(x)}{x^2} = \dfrac{5x^3 - 3x^2 + 10}{x^2} = 5x - 3 + \dfrac{10}{x^2}$

$\lim\limits_{x \to \infty} h(x) = \infty$ (Limit does not exist)

(b) $h(x) = \dfrac{f(x)}{x^3} = \dfrac{5x^3 - 3x^2 + 10}{x^3} = 5 - \dfrac{3}{x} + \dfrac{10}{x^3}$

$\lim\limits_{x \to \infty} h(x) = 5$

(c) $h(x) = \dfrac{f(x)}{x^4} = \dfrac{5x^3 - 3x^2 + 10}{x^4} = \dfrac{5}{x} - \dfrac{3}{x^2} + \dfrac{10}{x^4}$

$\lim\limits_{x \to \infty} h(x) = 0$

17. (a) $\lim\limits_{x \to \infty} \dfrac{x^2 + 2}{x^3 - 1} = 0$

(b) $\lim\limits_{x \to \infty} \dfrac{x^2 + 2}{x^2 - 1} = 1$

(c) $\lim\limits_{x \to \infty} \dfrac{x^2 + 2}{x - 1} = \infty$ (Limit does not exist)

19. (a) $\lim\limits_{x\to\infty}\dfrac{5-2x^{3/2}}{3x^2-4}=0$ (b) $\lim\limits_{x\to\infty}\dfrac{5-2x^{3/2}}{3x^{3/2}-4}=-\dfrac{2}{3}$ (c) $\lim\limits_{x\to\infty}\dfrac{5-2x^{3/2}}{3x-4}=-\infty$ (Limit does not exist)

21. $\lim\limits_{x\to\infty}\dfrac{2x-1}{3x+2}=\lim\limits_{x\to\infty}\dfrac{2-(1/x)}{3+(2/x)}=\dfrac{2-0}{3+0}=\dfrac{2}{3}$

23. $\lim\limits_{x\to\infty}\dfrac{x}{x^2-1}=\lim\limits_{x\to\infty}\dfrac{1/x}{1-(1/x^2)}=\dfrac{0}{1}=0$

25. $\lim\limits_{x\to-\infty}\dfrac{5x^2}{x+3}=\lim\limits_{x\to-\infty}\dfrac{5x}{1+(3/x)}=-\infty$

Limit does not exist

27. $\lim\limits_{x\to-\infty}\dfrac{x}{\sqrt{x^2-x}}=\lim\limits_{x\to-\infty}\dfrac{1}{\dfrac{\sqrt{x^2-x}}{-\sqrt{x^2}}}$, $\left(\text{for } x<0 \text{ we have } x=-\sqrt{x^2}\right)$

$$=\lim\limits_{x\to-\infty}\dfrac{-1}{\sqrt{1-(1/x)}}=-1$$

29. $\lim\limits_{x\to-\infty}\dfrac{2x+1}{\sqrt{x^2-x}}=\lim\limits_{x\to-\infty}\dfrac{2+\dfrac{1}{x}}{\left(\dfrac{\sqrt{x^2-x}}{-\sqrt{x^2}}\right)}$ $\left(\text{for } x<0,\, x=-\sqrt{x^2}\right)$

$$=\lim\limits_{x\to-\infty}\dfrac{-2-\left(\dfrac{1}{x}\right)}{\sqrt{1-\dfrac{1}{x}}}=-2$$

31. Since $(-1/x)\le(\sin 2x)/x\le(1/x)$ for all $x\ne 0$, we have by the Squeeze Theorem,

$$\lim\limits_{x\to\infty}-\dfrac{1}{x}\le\lim\limits_{x\to\infty}\dfrac{\sin 2x}{x}\le\lim\limits_{x\to\infty}\dfrac{1}{x}$$

$$0\le\lim\limits_{x\to\infty}\dfrac{\sin 2x}{x}\le 0.$$

Therefore, $\lim\limits_{x\to\infty}\dfrac{\sin 2x}{x}=0.$

33. $\lim\limits_{x\to\infty}\dfrac{1}{2x+\sin x}=0$

35. $f(x)=\dfrac{|x|}{x+1}$

$$\lim\limits_{x\to\infty}\dfrac{|x|}{x+1}=1$$

$$\lim\limits_{x\to-\infty}\dfrac{|x|}{x+1}=-1$$

Therefore, $y=1$ and $y=-1$ are both horizontal asymptotes.

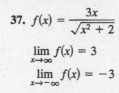

37. $f(x)=\dfrac{3x}{\sqrt{x^2+2}}$

$$\lim\limits_{x\to\infty}f(x)=3$$

$$\lim\limits_{x\to-\infty}f(x)=-3$$

Therefore, $y=3$ and $y=-3$ are both horizontal asymptotes.

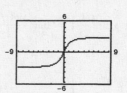

39. $\lim\limits_{x\to\infty}x\sin\dfrac{1}{x}=\lim\limits_{t\to 0^+}\dfrac{\sin t}{t}=1$

(Let $x=1/t$.)

41. $\lim\limits_{x \to -\infty} \left(x + \sqrt{x^2 + 3}\right) = \lim\limits_{x \to -\infty} \left[\left(x + \sqrt{x^2 + 3}\right) \cdot \dfrac{x - \sqrt{x^2 + 3}}{x - \sqrt{x^2 + 3}}\right] = \lim\limits_{x \to -\infty} \dfrac{-3}{x - \sqrt{x^2 + 3}} = 0$

43. $\lim\limits_{x \to \infty} \left(x - \sqrt{x^2 + x}\right) = \lim\limits_{x \to \infty} \left[\left(x - \sqrt{x^2 + x}\right) \cdot \dfrac{x + \sqrt{x^2 + x}}{x + \sqrt{x^2 + x}}\right]$

$\qquad = \lim\limits_{x \to \infty} \dfrac{-x}{x + \sqrt{x^2 + x}} = \lim\limits_{x \to \infty} \dfrac{-1}{1 + \sqrt{1 + (1/x)}} = -\dfrac{1}{2}$

45. $\lim\limits_{x \to \infty} \left(4x - \sqrt{16x^2 - x}\right) \dfrac{4x + \sqrt{16x^2 - x}}{4x + \sqrt{16x^2 - x}} = \lim\limits_{x \to \infty} \dfrac{16x^2 - (16x^2 - x)}{4x + \sqrt{(16x^2 - x)}}$

$\qquad = \lim\limits_{x \to \infty} \dfrac{x}{4x + \sqrt{16x^2 - x}}$

$\qquad = \lim\limits_{x \to \infty} \dfrac{1}{4 + \sqrt{16 - 1/x}}$

$\qquad = \dfrac{1}{4 + 4} = \dfrac{1}{8}$

47.

x	10^0	10^1	10^2	10^3	10^4	10^5	10^6
$f(x)$	1	0.513	0.501	0.500	0.500	0.500	0.500

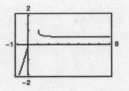

$\lim\limits_{x \to \infty} \left(x - \sqrt{x(x-1)}\right) = \lim\limits_{x \to \infty} \dfrac{x - \sqrt{x^2 - x}}{1} \cdot \dfrac{x + \sqrt{x^2 - x}}{x + \sqrt{x^2 - x}}$

$\qquad = \lim\limits_{x \to \infty} \dfrac{x}{x + \sqrt{x^2 - x}}$

$\qquad = \lim\limits_{x \to \infty} \dfrac{1}{1 + \sqrt{1 - (1/x)}}$

$\qquad = \dfrac{1}{2}$

49.

x	10^0	10^1	10^2	10^3	10^4	10^5	10^6
$f(x)$	0.479	0.500	0.500	0.500	0.500	0.500	0.500

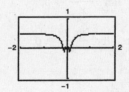

Let $x = 1/t$.

$\lim\limits_{x \to \infty} x \sin\left(\dfrac{1}{2x}\right) = \lim\limits_{t \to 0^+} \dfrac{\sin(t/2)}{t} = \lim\limits_{t \to 0^+} \dfrac{1}{2} \dfrac{\sin(t/2)}{t/2} = \dfrac{1}{2}$

51. (a)

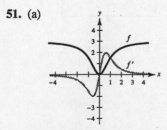

(b) $\lim\limits_{x \to \infty} f(x) = 3$ $\qquad$ $\lim\limits_{x \to \infty} f'(x) = 0$

(c) Since $\lim\limits_{x \to \infty} f(x) = 3$, the graph approaches that of a horizontal line, $\lim\limits_{x \to \infty} f'(x) = 0$.

53. Yes. For example, let $f(x) = \dfrac{6|x - 2|}{\sqrt{(x-2)^2 + 1}}$.

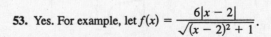

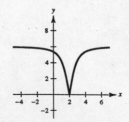

55. $y = \dfrac{2 + x}{1 - x}$

Intercepts: $(-2, 0), (0, 2)$

Symmetry: none

Horizontal asymptote: $y = -1$ since

$$\lim_{x \to -\infty} \frac{2 + x}{1 - x} = -1 = \lim_{x \to \infty} \frac{2 + x}{1 - x}.$$

Discontinuity: $x = 1$ (Vertical asymptote)

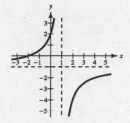

57. $y = \dfrac{x}{x^2 - 4}$

Intercept: $(0, 0)$

Symmetry: origin

Horizontal asymptote: $y = 0$

Vertical asymptote: $x = \pm 2$

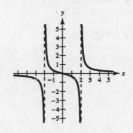

59. $y = \dfrac{x^2}{x^2 + 9}$

Intercept: $(0, 0)$

Symmetry: y-axis

Horizontal asymptote: $y = 1$ since

$$\lim_{x \to -\infty} \frac{x^2}{x^2 + 9} = 1 = \lim_{x \to \infty} \frac{x^2}{x^2 + 9}.$$

Relative minimum: $(0, 0)$

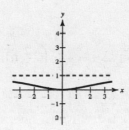

61. $y = \dfrac{2x^2}{x^2 - 4}$

Intercept: $(0, 0)$

Symmetry: y-axis

Horizontal asymptote: $y = 2$

Vertical asymptotes: $x = \pm 2$

Relative maximum: $(0, 0)$

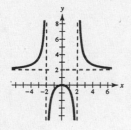

63. $xy^2 = 4$

Domain: $x > 0$

Intercepts: none

Symmetry: x-axis

Horizontal asymptote: $y = 0$ since

$$\lim_{x \to \infty} \frac{2}{\sqrt{x}} = 0 = \lim_{x \to \infty} -\frac{2}{\sqrt{x}}.$$

Discontinuity: $x = 0$ (Vertical asymptote)

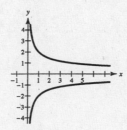

65. $y = \dfrac{2x}{1-x}$

Intercept: $(0, 0)$

Symmetry: none

Horizontal asymptote: $y = -2$ since

$$\lim_{x \to -\infty} \frac{2x}{1-x} = -2 = \lim_{x \to \infty} \frac{2x}{1-x}.$$

Discontinuity: $x = 1$ (Vertical asymptote)

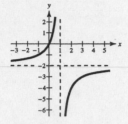

67. $y = 2 - \dfrac{3}{x^2}$

Intercepts: $\left(\pm\sqrt{3/2}, 0 \right)$

Symmetry: y-axis

Horizontal asymptote: $y = 2$ since

$$\lim_{x \to -\infty} \left(2 - \frac{3}{x^2} \right) = 2 = \lim_{x \to \infty} \left(2 - \frac{3}{x^2} \right).$$

Discontinuity: $x = 0$ (Vertical asymptote)

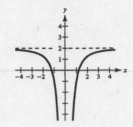

69. $y = 3 + \dfrac{2}{x}$

Intercept: $y = 0 = 3 + \dfrac{2}{x} \Longrightarrow \dfrac{2}{x} = -3 \Longrightarrow x = -\dfrac{2}{3}; \left(-\dfrac{2}{3}, 0 \right)$

Symmetry: none

Horizontal asymptote: $y = 3$

Vertical asymptote: $x = 0$

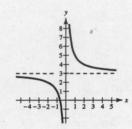

71. $y = \dfrac{x^3}{\sqrt{x^2 - 4}}$

Domain: $(-\infty, -2), (2, \infty)$

Intercepts: none

Symmetry: origin

Horizontal asymptote: none

Vertical asymptotes: $x = \pm 2$ (discontinuities)

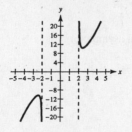

73. $f(x) = 5 - \dfrac{1}{x^2} = \dfrac{5x^2 - 1}{x^2}$

Domain: $(-\infty, 0), (0, \infty)$

$f'(x) = \dfrac{2}{x^3} \Longrightarrow$ No relative extrema

$f''(x) = -\dfrac{6}{x^4} \Longrightarrow$ No points of inflection

Vertical asymptote: $x = 0$

Horizontal asymptote: $y = 5$

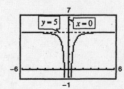

75. $f(x) = \dfrac{x}{x^2 - 4}$

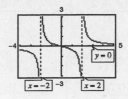

$$f'(x) = \frac{(x^2 - 4) - x(2x)}{(x^2 - 4)^2}$$

$$= \frac{-(x^2 + 4)}{(x^2 - 4)^2} \neq 0 \text{ for any } x \text{ in the domain of } f.$$

$$f''(x) = \frac{(x^2 - 4)^2(-2x) + (x^2 + 4)(2)(x^2 - 4)(2x)}{(x^2 - 4)^2}$$

$$= \frac{2x(x^2 + 12)}{(x^2 - 4)^3} = 0 \text{ when } x = 0.$$

Since $f''(x) > 0$ on $(-2, 0)$ and $f''(x) < 0$ on $(0, 2)$, then $(0, 0)$ is a point of inflection.

Vertical asymptotes: $x = \pm 2$

Horizontal asymptote: $y = 0$

77. $f(x) = \dfrac{x - 2}{x^2 - 4x + 3} = \dfrac{x - 2}{(x - 1)(x - 3)}$

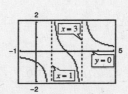

$$f'(x) = \frac{(x^2 - 4x + 3) - (x - 2)(2x - 4)}{(x^2 - 4x + 3)^2} = \frac{-x^2 + 4x - 5}{(x^2 - 4x + 3)^2} \neq 0$$

$$f''(x) = \frac{(x^2 - 4x + 3)^2(-2x + 4) - (-x^2 + 4x - 5)(2)(x^2 - 4x + 3)(2x - 4)}{(x^2 - 4x + 3)^4}$$

$$= \frac{2(x^3 - 6x^2 + 15x - 14)}{(x^2 - 4x + 3)^3} = \frac{2(x - 2)(x^2 - 4x + 7)}{(x^2 - 4x + 3)^3} = 0 \text{ when } x = 2.$$

Since $f''(x) > 0$ on $(1, 2)$ and $f''(x) < 0$ on $(2, 3)$, then $(2, 0)$ is a point of inflection.

Vertical asymptotes: $x = 1, x = 3$

Horizontal asymptote: $y = 0$

79. $f(x) = \dfrac{3x}{\sqrt{4x^2 + 1}}$

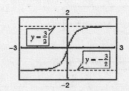

$$f'(x) = \frac{3}{(4x^2 + 1)^{3/2}} \Longrightarrow \text{No relative extrema}$$

$$f''(x) = \frac{-36x}{(4x^2 + 1)^{5/2}} = 0 \text{ when } x = 0.$$

Point of inflection: $(0, 0)$

Horizontal asymptotes: $y = \pm\dfrac{3}{2}$

No vertical asymptotes

81. $g(x) = \sin\left(\dfrac{x}{x - 2}\right), \; 3 < x < \infty$

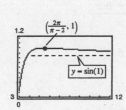

$$g'(x) = \frac{-2 \cos\left(\dfrac{x}{x - 2}\right)}{(x - 2)^2}$$

Horizontal asymptote: $y = \sin(1)$

Relative maximum: $\dfrac{x}{x - 2} = \dfrac{\pi}{2} \Longrightarrow x = \dfrac{2\pi}{\pi - 2} \approx 5.5039$

No vertical asymptotes

83. $f(x) = \dfrac{x^3 - 3x^2 + 2}{x(x - 3)}$, $g(x) = x + \dfrac{2}{x(x - 3)}$

(a)
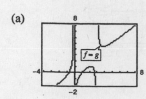

(b) $f(x) = \dfrac{x^3 - 3x^2 + 2}{x(x - 3)}$

$= \dfrac{x^2(x - 3)}{x(x - 3)} + \dfrac{2}{x(x - 3)}$

$= x + \dfrac{2}{x(x - 3)} = g(x)$

(c)
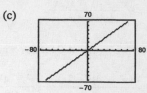

The graph appears as the slant asymptote $y = x$.

85. $C = 0.5x + 500$

$\overline{C} = \dfrac{C}{x}$

$\overline{C} = 0.5 + \dfrac{500}{x}$

$\displaystyle\lim_{x \to \infty}\left(0.5 + \dfrac{500}{x}\right) = 0.5$

87. $\displaystyle\lim_{t \to \infty} N(t) = \infty$

$\displaystyle\lim_{t \to \infty} E(t) = c$

89. $y = \dfrac{3.351t^2 + 42.461t - 543.730}{t^2}$

(a)
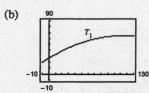

(b) Yes. $\displaystyle\lim_{t \to \infty} y = 3.351$

91. (a) $T_1(t) = -0.003t^2 + 0.677t + 26.564$

(b)
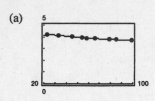

(c)
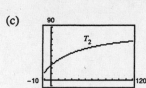

$T_2 = \dfrac{1451 + 86t}{58 + t}$

(d) $T_1(0) \approx 26.6$

$T_2(0) \approx 25.0$

(e) $\displaystyle\lim_{t \to \infty} T_2 = \dfrac{86}{1} = 86$

(f) No. The limiting temperature is 86. T_1 has no horizontal asymptote.

93. line: $mx - y + 4 = 0$

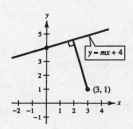
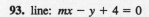

(a) $d = \dfrac{|Ax_1 + By_1 + C|}{\sqrt{A^2 + B^2}} = \dfrac{|m(3) - 1(1) + 4|}{\sqrt{m^2 + 1}}$

$= \dfrac{|3m + 3|}{\sqrt{m^2 + 1}}$

(b)
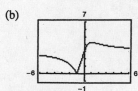

(c) $\displaystyle\lim_{m \to \infty} d(m) = 3 = \lim_{m \to -\infty} d(m)$

The line approaches the vertical line $x = 0$. Hence, the distance approaches 3.

95. $f(x) = \dfrac{2x^2}{x^2 + 2}$

(a) $\lim\limits_{x \to \infty} f(x) = 2 = L$

(b) $\qquad f(x_1) + \varepsilon = \dfrac{2x_1^2}{x_1^2 + 2} + \varepsilon = 2$

$2x_1^2 + \varepsilon x_1^2 + 2\varepsilon = 2x_1^2 + 4$

$x_1^2 \varepsilon = 4 - 2\varepsilon$

$x_1 = \sqrt{\dfrac{4 - 2\varepsilon}{\varepsilon}}$

$x_2 = -x_1$ by symmetry

(c) Let $M = \sqrt{\dfrac{4 - 2\varepsilon}{\varepsilon}} > 0$. For $x > M$:

$x > \sqrt{\dfrac{4 - 2\varepsilon}{\varepsilon}}$

$x^2 \varepsilon > 4 - 2\varepsilon$

$2x^2 + x^2 \varepsilon + 2\varepsilon > 2x^2 + 4$

$\dfrac{2x^2}{x^2 + 2} + \varepsilon > 2$

$\left| \dfrac{2x^2}{x^2 + 2} - 2 \right| > |-\varepsilon| = \varepsilon$

$|f(x) - L| > \varepsilon$

(d) Similarly, $N = -\sqrt{\dfrac{4 - 2\varepsilon}{\varepsilon}}$.

97. $\lim\limits_{x \to \infty} \dfrac{3x}{\sqrt{x^2 + 3}} = 3$

(a) For $\varepsilon = 0.5$, we need $M > 0$ such that

$|f(x) - L| = \left| \dfrac{3x}{\sqrt{x^2 + 3}} - 3 \right| < \varepsilon = 0.5$ whenever

$x > M \Rightarrow 2.5 < \dfrac{3x}{\sqrt{x^2 + 3}} < 3.5.$

By graphing $y_1 = \dfrac{3x}{\sqrt{x^2 + 3}}$, $y_2 = 2.5$ and $y_3 = 3.5$,

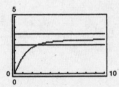

you see that the inequality is satisfied for $x > 2.7$, so let $M = 3$.

(b) For $\varepsilon = 0.1$, the inequality becomes

$2.9 < \dfrac{3x}{\sqrt{x^2 + 3}} < 3.1.$

From the corresponding graph you obtain $M = 7$.

99. $\lim\limits_{x \to \infty} \dfrac{1}{x^2} = 0$. Let $\varepsilon > 0$ be given. We need $M > 0$ such that

$|f(x) - L| = \left| \dfrac{1}{x^2} - 0 \right| = \dfrac{1}{x^2} < \varepsilon$ whenever $x > M.$

$x^2 > \dfrac{1}{\varepsilon}, x > \dfrac{1}{\sqrt{\varepsilon}}$, let $M = \dfrac{1}{\sqrt{\varepsilon}}.$

Hence, for $x > M$, we have

$x > \dfrac{1}{\sqrt{\varepsilon}} \Rightarrow x^2 > \dfrac{1}{\varepsilon} \Rightarrow \dfrac{1}{x^2} < \varepsilon \Rightarrow |f(x) - L| < \varepsilon.$

101. $\lim\limits_{x \to -\infty} \dfrac{1}{x^3} = 0$. Let $\varepsilon > 0$. We need to find $N < 0$ such that

$|f(x) - L| = \left| \dfrac{1}{x^3} - 0 \right| = \dfrac{1}{x^3} < \varepsilon$ whenever $x < N.$

$\dfrac{-1}{x^3} < \varepsilon \Rightarrow -x^3 > \dfrac{1}{\varepsilon} \Rightarrow x < \dfrac{-1}{\varepsilon^{1/3}}.$

Let $N = \dfrac{-1}{\sqrt[3]{\varepsilon}}.$

Hence, for $x < N < \dfrac{-1}{\sqrt[3]{\varepsilon}}$,

$\dfrac{1}{x} > -\sqrt[3]{\varepsilon}$

$-\dfrac{1}{x} < \sqrt[3]{\varepsilon}$

$-\dfrac{1}{x^3} < \varepsilon$

$\Rightarrow |f(x) - L| < \varepsilon.$

103. $\lim\limits_{x \to \infty} \dfrac{p(x)}{q(x)} = \lim\limits_{x \to \infty} \dfrac{a_n x^n + \cdots + a_1 x + a_0}{b_m x^m + \cdots + b_1 x + b_0}$

Divide $p(x)$ and $q(x)$ by x^m.

Case 1: If $n < m$: $\lim\limits_{x \to \infty} \dfrac{p(x)}{q(x)} = \lim\limits_{x \to \infty} \dfrac{\dfrac{a_n}{x^{m-n}} + \cdots + \dfrac{a_1}{x^{m-1}} + \dfrac{a_0}{x^m}}{b_m + \cdots + \dfrac{b_1}{x^{m-1}} + \dfrac{b_0}{x^m}} = \dfrac{0 + \cdots + 0 + 0}{b_m + \cdots + 0 + 0} = \dfrac{0}{b_m} = 0.$

Case 2: If $m = n$: $\lim\limits_{x \to \infty} \dfrac{p(x)}{q(x)} = \lim\limits_{x \to \infty} \dfrac{a_n + \cdots + \dfrac{a_1}{x^{m-1}} + \dfrac{a_0}{x^m}}{b_m + \cdots + \dfrac{b_1}{x^{m-1}} + \dfrac{b_0}{x^m}} = \dfrac{a_n + \cdots + 0 + 0}{b_m + \cdots + 0 + 0} = \dfrac{a_n}{b_m}.$

Case 3: If $n > m$: $\lim\limits_{x \to \infty} \dfrac{p(x)}{q(x)} = \lim\limits_{x \to \infty} \dfrac{a_n x^{n-m} + \cdots + \dfrac{a_1}{x^{m-1}} + \dfrac{a_0}{x^m}}{b_m + \cdots + \dfrac{b_1}{x^{m-1}} + \dfrac{b_0}{x^m}} = \dfrac{\pm \infty + \cdots + 0}{b_m + \cdots + 0} = \pm \infty.$

105. False. Let $f(x) = \dfrac{2x}{\sqrt{x^2 + 2}}$. (See Exercise 2.)

Section 3.6   A Summary of Curve Sketching

1. f has constant negative slope. Matches (d)

3. The slope is periodic, and zero at $x = 0$. Matches (a)

5. (a) $f'(x) = 0$ for $x = -2$ and $x = 2$

f' is negative for $-2 < x < 2$ (decreasing function).

f' is positive for $x > 2$ and $x < -2$ (increasing function).

(b) $f''(x) = 0$ at $x = 0$ (Inflection point).

f'' is positive for $x > 0$ (Concave upwards).

f'' is negative for $x < 0$ (Concave downward).

(c) f' is increasing on $(0, \infty)$. $(f'' > 0)$

(d) $f'(x)$ is minimum at $x = 0$. The rate of change of f at $x = 0$ is less than the rate of change of f for all other values of x.

7. $y = \dfrac{x^2}{x^2 + 3}$

$y' = \dfrac{6x}{(x^2 + 3)^2} = 0$ when $x = 0$.

$y'' = \dfrac{18(1 - x^2)}{(x^2 + 3)^3} = 0$ when $x = \pm 1$.

Horizontal asymptote: $y = 1$

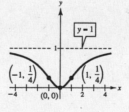

	y	y'	y''	Conclusion
$-\infty < x < -1$		$-$	$-$	Decreasing, concave down
$x = -1$	$\frac{1}{4}$	$-$	0	Point of inflection
$-1 < x < 0$		$-$	$+$	Decreasing, concave up
$x = 0$	0	0	$+$	Relative minimum
$0 < x < 1$		$+$	$+$	Increasing, concave up
$x = 1$	$\frac{1}{4}$	$+$	0	Point of inflection
$1 < x < \infty$		$+$	$-$	Increasing, concave down

9. $y = \dfrac{1}{x-2} - 3$

$y' = -\dfrac{1}{(x-2)^2} < 0$ when $x \neq 2$.

$y'' = \dfrac{2}{(x-2)^3}$

No relative extrema, no points of inflection

Intercepts: $\left(\dfrac{7}{3}, 0\right), \left(0, -\dfrac{7}{2}\right)$

Vertical asymptote: $x = 2$

Horizontal asymptote: $y = -3$

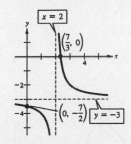

11. $y = \dfrac{2x}{x^2 - 1}$

$y' = \dfrac{-2(x^2 + 1)}{(x^2 - 1)^2} < 0$ if $x \neq \pm 1$.

$y'' = \dfrac{4x(x^2 + 3)}{(x^2 - 1)^3} = 0$ if $x = 0$.

Inflection point: $(0, 0)$

Intercept: $(0, 0)$

Vertical asymptotes: $x = \pm 1$

Horizontal asymptote: $y = 0$

Symmetry with respect to the origin

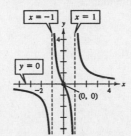

13. $g(x) = x + \dfrac{4}{x^2 + 1}$

$g'(x) = 1 - \dfrac{8x}{(x^2 + 1)^2} = \dfrac{x^4 + 2x^2 - 8x + 1}{(x^2 + 1)^2} = 0$ when $x \approx 0.1292, 1.6085$

$g''(x) = \dfrac{8(3x^2 - 1)}{(x^2 + 1)^3} = 0$ when $x = \pm\dfrac{\sqrt{3}}{3}$

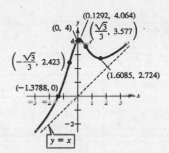

$g''(0.1292) < 0$, therefore, $(0.1292, 4.064)$ is relative maximum.

$g''(1.6085) > 0$, therefore, $(1.6085, 2.724)$ is a relative minimum.

Points of inflection: $\left(-\dfrac{\sqrt{3}}{3}, 2.423\right), \left(\dfrac{\sqrt{3}}{3}, 3.577\right)$

Intercepts: $(0, 4), (-1.3788, 0)$

Slant asymptote: $y = x$

15. $f(x) = \dfrac{x^2 + 1}{x} = x + \dfrac{1}{x}$

$f'(x) = 1 - \dfrac{1}{x^2} = 0$ when $x = \pm 1$.

$f''(x) = \dfrac{2}{x^3} \neq 0$

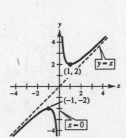

Relative maximum: $(-1, -2)$

Relative minimum: $(1, 2)$

Vertical asymptote: $x = 0$

Slant asymptote: $y = x$

17. $y = \dfrac{x^2 - 6x + 12}{x - 4} = x - 2 + \dfrac{4}{x - 4}$

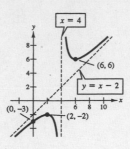

$$y' = 1 - \dfrac{4}{(x - 4)^2}$$

$$= \dfrac{(x - 2)(x - 6)}{(x - 4)^2} = 0 \text{ when } x = 2, 6.$$

$$y'' = \dfrac{8}{(x - 4)^3}$$

$y'' < 0$ when $x = 2$.

Therefore, $(2, -2)$ is a relative maximum.

$y'' > 0$ when $x = 6$.

Therefore, $(6, 6)$ is a relative minimum.

Vertical asymptote: $x = 4$

Slant asymptote: $y = x - 2$

19. $y = x\sqrt{4 - x}$,

Domain: $(-\infty, 4]$

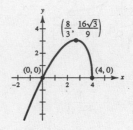

$$y' = \dfrac{8 - 3x}{2\sqrt{4 - x}} = 0 \text{ when } x = \dfrac{8}{3} \text{ and undefined when } x = 4.$$

$$y'' = \dfrac{3x - 16}{4(4 - x)^{3/2}} = 0 \text{ when } x = \dfrac{16}{3} \text{ and undefined when } x = 4.$$

Note: $x = \dfrac{16}{3}$ is not in the domain.

	y	y'	y''	Conclusion
$-\infty < x < \dfrac{8}{3}$		$+$	$-$	Increasing, concave down
$x = \dfrac{8}{3}$	$\dfrac{16}{3\sqrt{3}}$	0	$-$	Relative maximum
$\dfrac{8}{3} < x < 4$		$-$	$-$	Decreasing, concave down
$x = 4$	0	Undefined	Undefined	Endpoint

21. $h(x) = x\sqrt{9 - x^2}$ Domain: $-3 \le x \le 3$

$$h'(x) = \dfrac{9 - 2x^2}{\sqrt{9 - x^2}} = 0 \text{ when } x = \pm\dfrac{3}{\sqrt{2}} = \pm\dfrac{3\sqrt{2}}{2}.$$

$$h''(x) = \dfrac{x(2x^2 - 27)}{(9 - x^2)^{3/2}} = 0 \text{ when } x = 0.$$

Relative maximum: $\left(\dfrac{3\sqrt{2}}{2}, \dfrac{9}{2}\right)$

Relative minimum: $\left(-\dfrac{3\sqrt{2}}{2}, -\dfrac{9}{2}\right)$

Intercepts: $(0, 0), (\pm 3, 0)$

Symmetric with respect to the origin

Point of inflection: $(0, 0)$

23. $y = 3x^{2/3} - 2x$

$$y' = 2x^{-1/3} - 2 = \frac{2(1 - x^{1/3})}{x^{1/3}}$$

   $= 0$ when $x = 1$ and undefined when $x = 0$.

$$y'' = \frac{-2}{3x^{4/3}} < 0 \text{ when } x \neq 0.$$

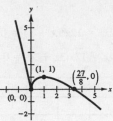

	y	y'	y''	Conclusion
$-\infty < x < 0$		$-$	$-$	Decreasing, concave down
$x = 0$	0	Undefined	Undefined	Relative minimum
$0 < x < 1$		$+$	$-$	Increasing, concave down
$x = 1$	1	0	$-$	Relative maximum
$1 < x < \infty$		$-$	$-$	Decreasing, concave down

25. $y = x^3 - 3x^2 + 3$

$y' = 3x^2 - 6x = 3x(x - 2) = 0$ when $x = 0, x = 2$.

$y'' = 6x - 6 = 6(x - 1) = 0$ when $x = 1$.

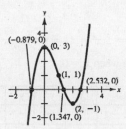

	y	y'	y''	Conclusion
$-\infty < x < 0$		$+$	$-$	Increasing, concave down
$x = 0$	3	0	$-$	Relative maximum
$0 < x < 1$		$-$	$-$	Decreasing, concave down
$x = 1$	1	$-$	0	Point of inflection
$1 < x < 2$		$-$	$+$	Decreasing, concave up
$x = 2$	-1	0	$+$	Relative minimum
$2 < x < \infty$		$+$	$+$	Increasing, concave up

27. $y = 2 - x - x^3$

$y' = 1 \quad 3x^2$

No critical numbers

$y'' = -6x = 0$ when $x = 0$.

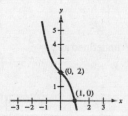

	y	y'	y''	Conclusion
$-\infty < x < 0$		$-$	$+$	Decreasing, concave up
$x = 0$	2	$-$	0	Point of inflection
$0 < x < \infty$		$-$	$-$	Decreasing, concave down

29. $y = 3x^4 + 4x^3$

$y' = 12x^3 + 12x^2 = 12x^2(x + 1) = 0$ when $x = 0, x = -1$.

$y'' = 36x^2 + 24x = 12x(3x + 2) = 0$ when $x = 0, x = -\frac{2}{3}$.

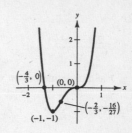

	y	y'	y''	Conclusion
$-\infty < x < -1$		$-$	$+$	Decreasing, concave up
$x = -1$	-1	0	$+$	Relative minimum
$-1 < x < -\frac{2}{3}$		$+$	$+$	Increasing, concave up
$x = -\frac{2}{3}$	$-\frac{16}{27}$	$+$	0	Point of inflection
$-\frac{2}{3} < x < 0$		$+$	$-$	Increasing, concave down
$x = 0$	0	0	0	Point of inflection
$0 < x < \infty$		$+$	$+$	Increasing, concave up

31. $y = x^5 - 5x$

$y' = 5x^4 - 5 = 5(x^4 - 1) = 0$ when $x = \pm 1$.

$y'' = 20x^3 = 0$ when $x = 0$.

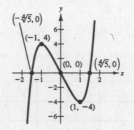

	y	y'	y''	Conclusion
$-\infty < x < -1$		$+$	$-$	Increasing, concave down
$x = -1$	4	0	$-$	Relative maximum
$-1 < x < 0$		$-$	$-$	Decreasing, concave down
$x = 0$	0	$-$	0	Point of inflection
$0 < x < 1$		$-$	$+$	Decreasing, concave up
$x = 1$	-4	0	$+$	Relative minimum
$1 < x < \infty$		$+$	$+$	Increasing, concave up

33. $y = |2x - 3|$

$y' = \dfrac{2(2x - 3)}{|2x - 3|}$ undefined at $x = \dfrac{3}{2}$.

$y'' = 0$

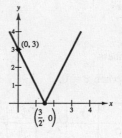

	y	y'	Conclusion
$-\infty < x < \frac{3}{2}$		$-$	Decreasing
$x = \frac{3}{2}$	0	Undefined	Relative minimum
$\frac{3}{2} < x < \infty$		$+$	Increasing

35. $f(x) = \dfrac{20x}{x^2 + 1} - \dfrac{1}{x} = \dfrac{19x^2 - 1}{x(x^2 + 1)}$

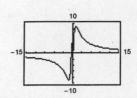

$x = 0$ vertical asymptote

$y = 0$ horizontal asymptote

Minimum: $(-1.10, -9.05)$

Maximum: $(1.10, 9.05)$

Points of inflection: $(-1.84, -7.86), (1.84, 7.86)$

37. $y = \dfrac{x}{\sqrt{x^2 + 7}}$

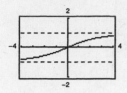

$(0, 0)$ point of inflection

$y = \pm 1$ horizontal asymptotes

39. $y = \sin x - \dfrac{1}{18} \sin 3x,\ 0 \le x \le 2\pi$

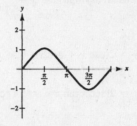

$$y' = \cos x - \frac{1}{6} \cos 3x$$

$$= \cos x - \frac{1}{6}[\cos 2x \cos x - \sin 2x \sin x]$$

$$= \cos x - \frac{1}{6}[(1 - 2\sin^2 x)\cos x - 2\sin^2 x \cos x]$$

$$= \cos x\left[1 - \frac{1}{6}(1 - 2\sin^2 x - 2\sin^2 x)\right]$$

$$= \cos x\left[\frac{5}{6} + \frac{2}{3}\sin^2 x\right]$$

$y' = 0$: $\cos x = 0 \Rightarrow x = \pi/2,\ 3\pi/2$

$\qquad \dfrac{5}{6} + \dfrac{2}{3}\sin^2 x = 0 \Rightarrow \sin^2 x = -5/4$, impossible

$y'' = -\sin x + \dfrac{1}{2}\sin 3x = 0 \Rightarrow 2\sin x = \sin 3x$

$$= \sin 2x \cos x + \cos 2x \sin x$$

$$= 2\sin x \cos^2 x + (2\cos^2 x - 1)\sin x$$

$$= \sin x(2\cos^2 x + 2\cos^2 x - 1)$$

$$= \sin x(4\cos^2 x - 1)$$

$\sin x = 0 \Rightarrow x = 0,\ \pi,\ 2\pi$

$2 = 4\cos^2 x - 1 \Rightarrow \cos x = \pm\sqrt{3}/2 \Rightarrow x = \dfrac{\pi}{6}, \dfrac{5\pi}{6}, \dfrac{7\pi}{6}, \dfrac{11\pi}{6}$

Relative maximum: $\left(\dfrac{\pi}{2}, \dfrac{19}{18}\right)$

Relative minimum: $\left(\dfrac{3\pi}{2}, -\dfrac{19}{18}\right)$

Inflection points: $\left(\dfrac{\pi}{6}, \dfrac{4}{9}\right), \left(\dfrac{5\pi}{6}, \dfrac{4}{9}\right), (\pi, 0), \left(\dfrac{7\pi}{6}, -\dfrac{4}{9}\right), \left(\dfrac{11\pi}{6}, -\dfrac{4}{9}\right)$

41. $y = 2x - \tan x, \ -\dfrac{\pi}{2} < x < \dfrac{\pi}{2}$

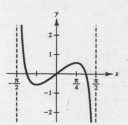

$y' = 2 - \sec^2 x = 0$ when $x = \pm\dfrac{\pi}{4}$.

$y'' = -2\sec^2 x \tan x = 0$ when $x = 0$.

Relative maximum: $\left(\dfrac{\pi}{4}, \dfrac{\pi}{2} - 1\right)$

Relative minimum: $\left(-\dfrac{\pi}{4}, 1 - \dfrac{\pi}{2}\right)$

Inflection point: $(0, 0)$

Vertical asymptotes: $x = \pm\dfrac{\pi}{2}$

43. $y = 2(\csc x + \sec x), \ 0 < x < \dfrac{\pi}{2}$

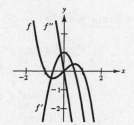

$y' = 2(\sec x \tan x - \csc x \cot x) = 0 \implies x = \dfrac{\pi}{4}$

Relative minimum: $\left(\dfrac{\pi}{4}, 4\sqrt{2}\right)$

Vertical asymptotes: $x = 0, \ x = \dfrac{\pi}{2}$

45. $g(x) = x \tan x, \ -\dfrac{3\pi}{2} < x < \dfrac{3\pi}{2}$

$g'(x) = \dfrac{x + \sin x \cos x}{\cos^2 x} = 0$ when $x = 0$.

$g''(x) = \dfrac{2(\cos x + x \sin x)}{\cos^3 x}$

Vertical asymptotes: $x = -\dfrac{3\pi}{2}, -\dfrac{\pi}{2}, \dfrac{\pi}{2}, \dfrac{3\pi}{2}$

Intercepts: $(-\pi, 0), (0, 0), (\pi, 0)$

Symmetric with respect to y-axis.

Increasing on $\left(0, \dfrac{\pi}{2}\right)$ and $\left(\dfrac{\pi}{2}, \dfrac{3\pi}{2}\right)$

Points of inflection: $(\pm 2.80, -1)$

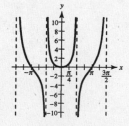

47. f is cubic.

f' is quadratic.

f'' is linear.

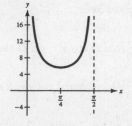

49.

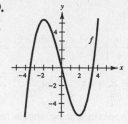

(any vertical translate of f will do)

51.

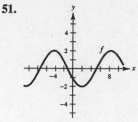

(any vertical translate of f will do)

53. Since the slope is negative, the function is decreasing on $(2, 8)$, and hence $f(3) > f(5)$.

55. $f(x) = \dfrac{4(x - 1)^2}{x^2 - 4x + 5}$

Vertical asymptote: none

Horizontal asymptote: $y = 4$

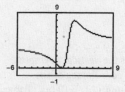

The graph crosses the horizontal asymptote $y = 4$. If a function has a vertical asymptote at $x = c$, the graph would not cross it since $f(c)$ is undefined.

57. $h(x) = \dfrac{\sin 2x}{x}$

Vertical asymptote: none

Horizontal asymptote: $y = 0$

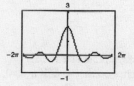

Yes, it is possible for a graph to cross its horizontal asymptote.

It is not possible to cross a vertical asymptote because the function is not continuous there.

59. $h(x) = \dfrac{6 - 2x}{3 - x}$

$= \dfrac{2(3 - x)}{3 - x} = \begin{cases} 2, & \text{if } x \neq 3 \\ \text{Undefined,} & \text{if } x = 3 \end{cases}$

The rational function is not reduced to lowest terms.

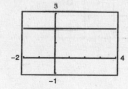

hole at $(3, 2)$

61. $f(x) = -\dfrac{x^2}{x - 2} - \dfrac{3x}{x-2} - \dfrac{1}{x-2} = -x + 1 + \dfrac{3}{x - 2}$

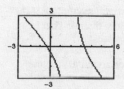

The graph appears to approach the slant asymptote $y = -x + 1$.

63. $f(x) = \dfrac{x^3}{x^2 + 1} = x - \dfrac{x}{x^2 + 1}$

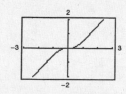

The graph appears to approach the slant asymptote $y = x$.

65. $f(x) = \dfrac{\cos^2 \pi x}{\sqrt{x^2 + 1}}$, $(0, 4)$

(a)

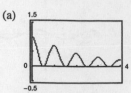

On $(0, 4)$ there seem to be 7 critical numbers:

0.5, 1.0, 1.5, 2.0, 2.5, 3.0, 3.5

(b) $f'(x) = \dfrac{-\cos \pi x(x \cos \pi x + 2\pi(x^2 + 1) \sin \pi x)}{(x^2 + 1)^{3/2}} = 0$

Critical numbers $\approx \dfrac{1}{2}, 0.97, \dfrac{3}{2}, 1.98, \dfrac{5}{2}, 2.98, \dfrac{7}{2}$.

The critical numbers where maxima occur appear to be integers in part (a), but approximating them using f' shows that they are not integers.

67. Vertical asymptote: $x = 5$

Horizontal asymptote: $y = 0$

$$y = \frac{1}{x - 5}$$

69. Vertical asymptote: $x = 5$

Slant asymptote: $y = 3x + 2$

$$y = 3x + 2 + \frac{1}{x - 5} = \frac{3x^2 - 13x - 9}{x - 5}$$

71. $f(x) = \dfrac{ax}{(x - b)^2}$

(a) The graph has a vertical asymptote at $x = b$. If $a > 0$, the graph approaches ∞ as $x \to b$. If $a < 0$, the graph approaches $-\infty$ as $x \to b$. The graph approaches its vertical asymptote faster as $|a| \to 0$.

(b) As b varies, the position of the vertical asymptote changes: $x = b$. Also, the coordinates of the minimum $(a > 0)$ or maximum $(a < 0)$ are changed.

73. $f(x) = \dfrac{3x^n}{x^4 + 1}$

(a) For n even, f is symmetric about the y-axis. For n odd, f is symmetric about the origin.

(b) The x-axis will be the horizontal asymptote if the degree of the numerator is less than 4. That is, $n = 0, 1, 2, 3$.

(c) $n = 4$ gives $y = 3$ as the horizontal asymptote.

(d) There is a slant asymptote $y = 3x$ if $n = 5$:

$$\frac{3x^5}{x^4 + 1} = 3x - \frac{3x}{x^4 + 1}.$$

(e)

n	0	1	2	3	4	5
M	1	2	3	2	1	0
N	2	3	4	5	2	3

75. (a)

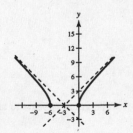

(b) When $t = 10$, $N = 2434$ bacteria.

(c) N is greatest (≈ 2518) at $t \approx 7.2$.

(d) $N'(t)$ is greatest when $t \approx 3.2$.

(Find the t-value of the point of inflection.)

(e) $\displaystyle \lim_{t \to \infty} N(t) = \frac{13,250}{7} \approx 1893 \, \text{bacteria}$

77. $y = \sqrt{x^2 + 6x} = \sqrt{(x + 3)^2 - 9}$

$y \to x + 3$ as $x \to \infty$, and $y \to -x - 3$ as $x \to -\infty$.

Section 3.7   Optimization Problems

1. (a)

First Number, x	Second Number	Product, P
10	$110 - 10$	$10(110 - 10) = 1000$
20	$110 - 20$	$20(110 - 20) = 1800$
30	$110 - 30$	$30(110 - 30) = 2400$
40	$110 - 40$	$40(110 - 40) = 2800$
50	$110 - 50$	$50(110 - 50) = 3000$
60	$110 - 60$	$60(110 - 60) = 3000$

(b)

First Number, x	Second Number	Product, P
10	$110 - 10$	$10(110 - 10) = 1000$
20	$110 - 20$	$20(110 - 20) = 1800$
30	$110 - 30$	$30(110 - 30) = 2400$
40	$110 \quad 40$	$40(110 - 40) = 2800$
50	$110 - 50$	$50(110 - 50) = 3000$
60	$110 - 60$	$60(110 - 60) = 3000$
70	$110 - 70$	$70(110 - 70) = 2800$
80	$110 - 80$	$80(110 - 80) = 2400$
90	$110 - 90$	$90(110 - 90) = 1800$
100	$110 - 100$	$100(110 - 100) = 1000$

The maximum is attained near $x = 50$ and 60.

(c) $P = x(110 - x) = 110x - x^2$

(d)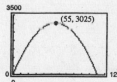

The solution appears to be $x = 55$.

(e) $\dfrac{dP}{dx} = 110 - 2x = 0$ when $x = 55$.

$\dfrac{d^2P}{dx^2} = -2 < 0$

P is a maximum when $x = 110 - x = 55$.
The two numbers are 55 and 55.

3. Let x and y be two positive numbers such that $x + y = S$.

$$P = xy = x(S - x) = Sx - x^2$$

$$\frac{dP}{dx} = S - 2x = 0 \text{ when } x = \frac{S}{2}.$$

$$\frac{d^2P}{dx^2} = -2 < 0 \text{ when } x = \frac{S}{2}.$$

P is a maximum when $x = y = S/2$.

5. Let x and y be two positive numbers such that $xy = 192$.

$$S = x + 3y = \frac{192}{y} + 3y$$

$$\frac{dS}{dy} = 3 - \frac{192}{y^2} = 0 \text{ when } y = 8.$$

$$\frac{d^2S}{dy^2} = \frac{384}{y^3} > 0 \text{ when } y = 8.$$

S is minimum when $y = 8$ and $x = 24$.

7. Let x and y be two positive numbers such that $x + 2y = 100$.

$$P = xy = y(100 - 2y) = 100y - 2y^2$$

$$\frac{dP}{dy} = 100 - 4y = 0 \text{ when } y = 25.$$

$$\frac{d^2P}{dy^2} = -4 < 0 \text{ when } y = 25.$$

P is a maximum when $x = 50$ and $y = 25$.

11. Let x be the length and y the width of the rectangle.

$$xy = 64$$

$$y = \frac{64}{x}$$

$$P = 2x + 2y = 2x + 2\left(\frac{64}{x}\right) = 2x + \frac{128}{x}$$

$$\frac{dP}{dx} = 2 - \frac{128}{x^2} = 0 \text{ when } x = 8.$$

$$\frac{d^2P}{dx^2} = \frac{256}{x^3} > 0 \text{ when } x = 8.$$

P is minimum when $x = y = 8$ feet.

9. Let x be the length and y the width of the rectangle.

$$2x + 2y = 100$$

$$y = 50 - x$$

$$A = xy = x(50 - x)$$

$$\frac{dA}{dx} = 50 - 2x = 0 \text{ when } x = 25.$$

$$\frac{d^2A}{dx^2} = -2 < 0 \text{ when } x = 25.$$

A is maximum when $x = y = 25$ meters.

13. $d = \sqrt{(x - 4)^2 + \left(\sqrt{x} - 0\right)^2}$

$ = \sqrt{x^2 - 7x + 16}$

Since d is smallest when the expression inside the radical is smallest, you need only find the critical numbers of

$$f(x) = x^2 - 7x + 16.$$

$$f'(x) = 2x - 7 = 0$$

$$x = \frac{7}{2}$$

By the First Derivative Test, the point nearest to $(4, 0)$ is $\left(7/2, \sqrt{7/2}\right)$.

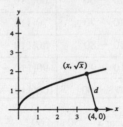

15. $d = \sqrt{(x - 2)^2 + [x^2 - (1/2)]^2}$

$ = \sqrt{x^4 - 4x + (17/4)}$

Since d is smallest when the expression inside the radical is smallest, you need only find the critical numbers of

$$f(x) = x^4 - 4x + \frac{17}{4}.$$

$$f'(x) = 4x^3 - 4 = 0$$

$$x = 1$$

By the First Derivative Test, the point nearest to $\left(2, \frac{1}{2}\right)$ is $(1, 1)$.

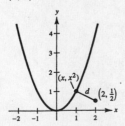

17. $\dfrac{dQ}{dx} = kx(Q_0 - x) = kQ_0 x - kx^2$

$$\frac{d^2Q}{dx^2} = kQ_0 - 2kx$$

$$\phantom{\frac{d^2Q}{dx^2}} = k(Q_0 - 2x) = 0 \text{ when } x = \frac{Q_0}{2}.$$

$$\frac{d^3Q}{dx^3} = -2k < 0 \text{ when } x = \frac{Q_0}{2}.$$

dQ/dx is maximum when $x = Q_0/2$.

19. $xy = 180,000$ (see figure)

$S = x + 2y = \left(x + \dfrac{360,000}{x}\right)$ where S is the length of fence needed.

$\dfrac{dS}{dx} = 1 - \dfrac{360,000}{x^2} = 0$ when $x = 600$.

$\dfrac{d^2S}{dx^2} = \dfrac{720,000}{x^3} > 0$ when $x = 600$.

S is a minimum when $x = 600$ meters and $y = 300$ meters.

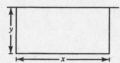

21. (a) $A = 4(\text{area of side}) + 2(\text{area of Top})$

    (1) $A = 4(3)(11) + 2(3)(3) = 150$ square inches

    (2) $A = 4(5)(5) + 2(5)(5) = 150$ square inches

    (3) $A = 4(3.25)(6) + 2(6)(6) = 150$ square inches

(b) $V = (\text{length})(\text{width})(\text{height})$

    (1) $V = (3)(3)(11) = 99$ cubic inches

    (2) $V = (5)(5)(5) = 125$ cubic inches

    (3) $V = (6)(6)(3.25) = 117$ cubic inches

(c) $S = 4xy + 2x^2 = 150 \implies y = \dfrac{150 - 2x^2}{4x}$

$V = x^2y = x^2\left(\dfrac{150 - 2x^2}{4x}\right) = \dfrac{75}{2}x - \dfrac{1}{2}x^3$

$V' = \dfrac{75}{2} - \dfrac{3}{2}x^2 = 0 \implies x = \pm 5$

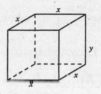

By the First Derivative Test, $x = 5$ yields the maximum volume. Dimensions: $5 \times 5 \times 5$. (A cube!)

23. $16 = 2y + x + \pi\left(\dfrac{x}{2}\right)$

$32 = 4y + 2x + \pi x$

$y = \dfrac{32 - 2x - \pi x}{4}$

$A = xy + \dfrac{\pi}{2}\left(\dfrac{x}{2}\right)^2 = \left(\dfrac{32 - 2x - \pi x}{4}\right)x + \dfrac{\pi x^2}{8}$

$\quad = 8x - \dfrac{1}{2}x^2 - \dfrac{\pi}{4}x^2 + \dfrac{\pi}{8}x^2$

$\dfrac{dA}{dx} = 8 - x - \dfrac{\pi}{2}x + \dfrac{\pi}{4}x = 8 - x\left(1 + \dfrac{\pi}{4}\right)$

$\quad = 0$ when $x = \dfrac{8}{1 + (\pi/4)} = \dfrac{32}{4 + \pi}$.

$\dfrac{d^2A}{dx^2} = -\left(1 + \dfrac{\pi}{4}\right) < 0$ when $x = \dfrac{32}{4 + \pi}$.

$y = \dfrac{32 - 2[32/(4 + \pi)] - \pi[32/(4 + \pi)]}{4} = \dfrac{16}{4 + \pi}$

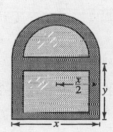

The area is maximum when $y = \dfrac{16}{4 + \pi}$ feet and $x = \dfrac{32}{4 + \pi}$ feet.

25. (a) $\dfrac{y-2}{0-1} = \dfrac{0-2}{x-1}$

$$y = 2 + \frac{2}{x-1}$$

$$L = \sqrt{x^2 + y^2} = \sqrt{x^2 + \left(2 + \frac{2}{x-1}\right)^2}$$

$$= \sqrt{x^2 + 4 + \frac{8}{x-1} + \frac{4}{(x-1)^2}}, \quad x > 1$$

(b)
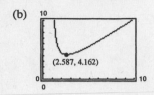

(2.587, 4.162)

L is minimum when $x \approx 2.587$ and $L \approx 4.162$.

(c) Area $= A(x) = \dfrac{1}{2}xy = \dfrac{1}{2}x\left(2 + \dfrac{2}{x-1}\right) = x + \dfrac{x}{x-1}$

$$A'(x) = 1 + \frac{(x-1)-x}{(x-1)^2} = 1 - \frac{1}{(x-1)^2} = 0$$

$$(x-1)^2 = 1$$

$$x - 1 = \pm 1$$

$$x = 0, 2 \ \text{(select } x = 2)$$

Then $y = 4$ and $A = 4$.

Vertices: $(0, 0), (2, 0), (0, 4)$

27. $A = 2xy = 2x\sqrt{25 - x^2}$ (see figure)

$$\frac{dA}{dx} = 2x\left(\frac{1}{2}\right)\left(\frac{-2x}{\sqrt{25-x^2}}\right) + 2\sqrt{25-x^2}$$

$$= 2\left(\frac{25 - 2x^2}{\sqrt{25-x^2}}\right) = 0 \text{ when } x = y = \frac{5\sqrt{2}}{2} \approx 3.54.$$

By the First Derivative Test, the inscribed rectangle of maximum area has vertices

$$\left(\pm\frac{5\sqrt{2}}{2}, 0\right), \left(\pm\frac{5\sqrt{2}}{2}, \frac{5\sqrt{2}}{2}\right).$$

Width: $\dfrac{5\sqrt{2}}{2}$; Length: $5\sqrt{2}$

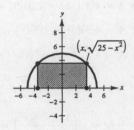

$\left(x, \sqrt{25 - x^2}\right)$

29. $xy = 30 \Rightarrow y = \dfrac{30}{x}$

$$A = (x+2)\left(\frac{30}{x} + 2\right) \quad \text{(see figure)}$$

$$\frac{dA}{dx} = (x+2)\left(\frac{-30}{x^2}\right) + \left(\frac{30}{x} + 2\right) = \frac{2(x^2 - 30)}{x^2} = 0 \text{ when } x = \sqrt{30}.$$

$$y = \frac{30}{\sqrt{30}} = \sqrt{30}$$

By the First Derivative Test, the dimensions $(x + 2)$ by $(y + 2)$ are $\left(2 + \sqrt{30}\right)$ by $\left(2 + \sqrt{30}\right)$ (approximately 7.477 by 7.477). These dimensions yield a minimum area.

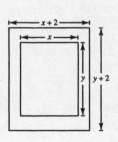

31. (a) $P = 2x + 2\pi r$

$$= 2x + 2\pi\left(\frac{y}{2}\right)$$

$$= 2x + \pi y = 200$$

$$\Rightarrow y = \frac{200 - 2x}{\pi} = \frac{2}{\pi}(100 - x)$$

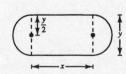

—CONTINUED—

31. —CONTINUED—

(b)

Length, x	Width, y	Area, xy
10	$\frac{2}{\pi}(100 - 10)$	$(10)\frac{2}{\pi}(100 - 10) \approx 573$
20	$\frac{2}{\pi}(100 - 20)$	$(20)\frac{2}{\pi}(100 - 20) \approx 1019$
30	$\frac{2}{\pi}(100 - 30)$	$(30)\frac{2}{\pi}(100 - 30) \approx 1337$
40	$\frac{2}{\pi}(100 - 40)$	$(40)\frac{2}{\pi}(100 - 40) \approx 1528$
50	$\frac{2}{\pi}(100 - 50)$	$(50)\frac{2}{\pi}(100 - 50) \approx 1592$
60	$\frac{2}{\pi}(100 - 60)$	$(60)\frac{2}{\pi}(100 - 60) = 1528$

The maximum area of the rectangle is approximately 1592 m².

(c) $A = xy = x\frac{2}{\pi}(100 - x) = \frac{2}{\pi}(100x - x^2)$

(d) $A' = \frac{2}{\pi}(100 - 2x)$. $A' = 0$ when $x = 50$.

Maximum when length = 50 m and width = $\frac{100}{\pi}$.

(e)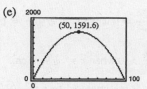

Maximum area is approximately 1591.55 m² ($x = 50$ m).

33. Let x be the sides of the square ends and y the length of the package.

$P = 4x + y = 108 \rightarrow y = 108 - 4x$

$V = x^2y = x^2(108 - 4x) = 108x^2 - 4x^3$

$\frac{dV}{dx} = 216x - 12x^2$

$= 12x(18 - x) = 0$ when $x = 18$.

$\frac{d^2V}{dx^2} = 216 - 24x = -216 < 0$ when $x = 18$.

The volume is maximum when $x = 18$ inches and $y = 108 - 4(18) = 36$ inches.

35. $V = \frac{1}{3}\pi x^2 h = \frac{1}{3}\pi x^2\left(r + \sqrt{r^2 - x^2}\right)$ (see figure)

$\frac{dV}{dx} = \frac{1}{3}\pi\left[\frac{-x^3}{\sqrt{r^2 - x^2}} + 2x\left(r + \sqrt{r^2 - x^2}\right)\right] = \frac{\pi x}{3\sqrt{r^2 - x^2}}\left(2r^2 + 2r\sqrt{r^2 - x^2} - 3x^2\right) = 0$

$2r^2 + 2r\sqrt{r^2 - x^2} - 3x^2 = 0$

$2r\sqrt{r^2 - x^2} = 3x^2 - 2r^2$

$4r^2(r^2 - x^2) = 9x^4 - 12x^2r^2 + 4r^4$

$0 = 9x^4 - 8x^2r^2 = x^2(9x^2 - 8r^2)$

$x = 0, \frac{2\sqrt{2}r}{3}$

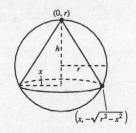

By the First Derivative Test, the volume is a maximum when

$x = \frac{2\sqrt{2}r}{3}$ and $h = r + \sqrt{r^2 - x^2} = \frac{4r}{3}$.

Thus, the maximum volume is

$V = \frac{1}{3}\pi\left(\frac{8r^2}{9}\right)\left(\frac{4r}{3}\right) = \frac{32\pi r^3}{81}$ cubic units.

37. No, there is no minimum area. If the sides are x and y, then $2x + 2y = 20 \implies y = 10 - x$.
The area is $A(x) = x(10 - x) = 10x - x^2$. This can be made arbitrarily small by selecting $x \approx 0$.

39. $V = 12 = \dfrac{4}{3}\pi r^3 + \pi r^2 h$

$$h = \frac{12 - (4/3)\pi r^3}{\pi r^2} = \frac{12}{\pi r^2} - \frac{4}{3}r$$

$$S = 4\pi r^2 + 2\pi rh = 4\pi r^2 + 2\pi r\left(\frac{12}{\pi r^2} - \frac{4}{3}r\right) = 4\pi r^2 + \frac{24}{r} - \frac{8}{3}\pi r^2 = \frac{4}{3}\pi r^2 + \frac{24}{r}$$

$$\frac{dS}{dr} = \frac{8}{3}\pi r - \frac{24}{r^2} = 0 \text{ when } r = \sqrt[3]{9/\pi} \approx 1.42 \text{ cm.}$$

$$\frac{d^2S}{dr^2} = \frac{8}{3}\pi + \frac{48}{r^3} > 0 \text{ when } r = \sqrt[3]{9/\pi} \text{ cm.}$$

The surface area is minimum when $r = \sqrt[3]{9/\pi}$ cm and $h = 0$. The resulting solid is a sphere of radius $r \approx 1.42$ cm.

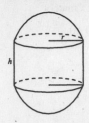

41. Let x be the length of a side of the square and y the length of a side of the triangle.

$$4x + 3y = 10$$

$$A = x^2 + \frac{1}{2}y\left(\frac{\sqrt{3}}{2}y\right)$$

$$= \frac{(10 - 3y)^2}{16} + \frac{\sqrt{3}}{4}y^2$$

$$\frac{dA}{dy} = \frac{1}{8}(10 - 3y)(-3) + \frac{\sqrt{3}}{2}y = 0$$

$$-30 + 9y + 4\sqrt{3}y = 0$$

$$y = \frac{30}{9 + 4\sqrt{3}}$$

$$\frac{d^2A}{dy^2} = \frac{9 + 4\sqrt{3}}{8} > 0$$

A is minimum when

$$y = \frac{30}{9 + 4\sqrt{3}} \text{ and } x = \frac{10\sqrt{3}}{9 + 4\sqrt{3}}.$$

43. Let S be the strength and k the constant of proportionality. Given $h^2 + w^2 = 24^2$, $h^2 = 24^2 - w^2$,

$$S = kwh^2$$

$$S = kw(576 - w^2) = k(576w - w^3)$$

$$\frac{dS}{dw} = k(576 - 3w^2) = 0 \text{ when } w = 8\sqrt{3}, h = 8\sqrt{6}.$$

$$\frac{d^2S}{dw^2} = -6kw < 0 \text{ when } w = 8\sqrt{3}.$$

These values yield a maximum.

45. $R = \dfrac{v_0{}^2}{g}\sin 2\theta$

$$\frac{dR}{d\theta} = \frac{2v_0{}^2}{g}\cos 2\theta = 0 \text{ when } \theta = \frac{\pi}{4}, \frac{3\pi}{4}.$$

$$\frac{d^2R}{d\theta^2} = -\frac{4v_0{}^2}{g}\sin 2\theta < 0 \text{ when } \theta = \frac{\pi}{4}.$$

By the Second Derivative Test, R is maximum when $\theta = \pi/4$.

47. $\sin \alpha = \dfrac{h}{s} \Rightarrow s = \dfrac{h}{\sin \alpha}, \; 0 < \alpha < \dfrac{\pi}{2}$

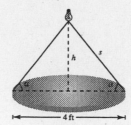

$\tan \alpha = \dfrac{h}{2} \Rightarrow h = 2 \tan \alpha \Rightarrow s = \dfrac{2 \tan \alpha}{\sin \alpha} = 2 \sec \alpha$

$I = \dfrac{k \sin \alpha}{s^2} = \dfrac{k \sin \alpha}{4 \sec^2 \alpha} = \dfrac{k}{4} \sin \alpha \cos^2 \alpha$

$\dfrac{dI}{d\alpha} = \dfrac{k}{4} \left[\sin \alpha (-2 \sin \alpha \cos \alpha) + \cos^2 \alpha (\cos \alpha) \right]$

$\qquad = \dfrac{k}{4} \cos \alpha [\cos^2 \alpha - 2 \sin^2 \alpha]$

$\qquad = \dfrac{k}{4} \cos \alpha [1 - 3 \sin^2 \alpha]$

$\qquad = 0$ when $\alpha = \dfrac{\pi}{2}, \dfrac{3\pi}{2}$, or when $\sin \alpha = \pm \dfrac{1}{\sqrt{3}}$.

Since α is acute, we have

$\sin \alpha = \dfrac{1}{\sqrt{3}} \Rightarrow h = 2 \tan \alpha = 2 \left(\dfrac{1}{\sqrt{2}} \right) = \sqrt{2}$ feet.

Since $(d^2 I)/(d\alpha^2) = (k/4) \sin \alpha (9 \sin^2 \alpha - 7) < 0$ when $\sin \alpha = 1/\sqrt{3}$, this yields a maximum.

49.

$$S = \sqrt{x^2 + 4}, \, L = \sqrt{1 + (3 - x)^2}$$

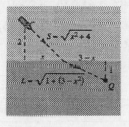

$$\text{Time} = T = \frac{\sqrt{x^2 + 4}}{2} + \frac{\sqrt{x^2 - 6x + 10}}{4}$$

$$\frac{dT}{dx} = \frac{x}{2\sqrt{x^2 + 4}} + \frac{x - 3}{4\sqrt{x^2 - 6x + 10}} = 0$$

$$\frac{x^2}{x^2 + 4} = \frac{9 - 6x + x^2}{4(x^2 - 6x + 10)}$$

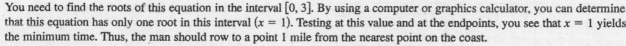

$$x^4 - 6x^3 + 9x^2 + 8x - 12 = 0$$

You need to find the roots of this equation in the interval $[0, 3]$. By using a computer or graphics calculator, you can determine that this equation has only one root in this interval ($x = 1$). Testing at this value and at the endpoints, you see that $x = 1$ yields the minimum time. Thus, the man should row to a point 1 mile from the nearest point on the coast.

51. $T = \dfrac{\sqrt{x^2 + 4}}{v_1} + \dfrac{\sqrt{x^2 - 6x + 10}}{v_2}$

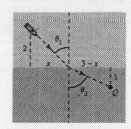

$\dfrac{dT}{dx} = \dfrac{x}{v_1 \sqrt{x^2 + 4}} + \dfrac{x - 3}{v_2 \sqrt{x^2 - 6x + 10}} = 0$

Since

$$\frac{x}{\sqrt{x^2 + 4}} = \sin \theta_1 \text{ and } \frac{x - 3}{\sqrt{x^2 - 6x + 10}} = -\sin \theta_2$$

we have

$$\frac{\sin \theta_1}{v_1} - \frac{\sin \theta_2}{v_2} = 0 \Rightarrow \frac{\sin \theta_1}{v_1} = \frac{\sin \theta_2}{v_2}.$$

Since

$$\frac{d^2 T}{dx^2} = \frac{4}{v_1 (x^2 + 4)^{3/2}} + \frac{1}{v_2 (x^2 - 6x + 10)^{3/2}} > 0$$

this condition yields a minimum time.

53. $f(x) = 2 - 2 \sin x$

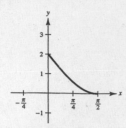

(a) Distance from origin to y-intercept is 2.
   Distance from origin to x-intercept is $\pi/2 \approx 1.57$.

(b) $d = \sqrt{x^2 + y^2} = \sqrt{x^2 + (2 - 2 \sin x)^2}$

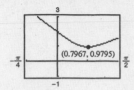

(0.7967, 0.9795)

Minimum distance $= 0.9795$ at $x = 0.7967$.

(c) Let $f(x) = d^2(x) = x^2 + (2 - 2 \sin x)^2$.

$f'(x) = 2x + 2(2 - 2 \sin x)(-2 \cos x)$

Setting $f'(x) = 0$, you obtain $x \approx 0.7967$, which corresponds to $d = 0.9795$.

55. $F \cos \theta = k(W - F \sin \theta)$

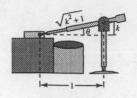

$$F = \frac{kW}{\cos \theta + k \sin \theta}$$

$$\frac{dF}{d\theta} = \frac{-kW(k \cos \theta - \sin \theta)}{(\cos \theta + k \sin \theta)^2} = 0$$

$k \cos \theta = \sin \theta \Rightarrow k = \tan \theta \Rightarrow \theta = \arctan k$

Since

$$\cos \theta + k \sin \theta = \frac{1}{\sqrt{k^2 + 1}} + \frac{k^2}{\sqrt{k^2 + 1}} = \sqrt{k^2 + 1},$$

the minimum force is

$$F = \frac{kW}{\cos \theta + k \sin \theta} = \frac{kW}{\sqrt{k^2 + 1}}.$$

57. (a)

Base 1	Base 2	Altitude	Area
8	$8 + 16 \cos 10°$	$8 \sin 10°$	≈ 22.1
8	$8 + 16 \cos 20°$	$8 \sin 20°$	≈ 42.5
8	$8 + 16 \cos 30°$	$8 \sin 30°$	≈ 59.7
8	$8 + 16 \cos 40°$	$8 \sin 40°$	≈ 72.7
8	$8 + 16 \cos 50°$	$8 \sin 50°$	≈ 80.5
8	$8 + 16 \cos 60°$	$8 \sin 60°$	≈ 83.1

(b)

Base 1	Base 2	Altitude	Area
8	$8 + 16 \cos 10°$	$8 \sin 10°$	≈ 22.1
8	$8 + 16 \cos 20°$	$8 \sin 20°$	≈ 42.5
8	$8 + 16 \cos 30°$	$8 \sin 30°$	≈ 59.7
8	$8 + 16 \cos 40°$	$8 \sin 40°$	≈ 72.7
8	$8 + 16 \cos 50°$	$8 \sin 50°$	≈ 80.5
8	$8 + 16 \cos 60°$	$8 \sin 60°$	≈ 83.1
8	$8 + 16 \cos 70°$	$8 \sin 70°$	≈ 80.7
8	$8 + 16 \cos 80°$	$8 \sin 80°$	≈ 74.0
8	$8 + 16 \cos 90°$	$8 \sin 90°$	$= 64.0$

The maximum cross-sectional area is approximately 83.1 square feet.

(c) $A = (a + b)\dfrac{h}{2}$

$= [8 + (8 + 16 \cos \theta)]\dfrac{8 \sin \theta}{2}$

$= 64(1 + \cos \theta) \sin \theta, \, 0° < \theta < 90°$

(e)

(60°, 83.1)

(d) $\dfrac{dA}{d\theta} = 64(1 + \cos \theta) \cos \theta + (-64 \sin \theta) \sin \theta$

$= 64(\cos \theta + \cos^2 \theta - \sin^2 \theta)$

$= 64(2 \cos^2 \theta + \cos \theta - 1)$

$= 64(2 \cos \theta - 1)(\cos \theta + 1)$

$= 0$ when $\theta = 60°, 180°, 300°$.

The maximum occurs when $\theta = 60°$.

59. $C = 100\left(\dfrac{200}{x^2} + \dfrac{x}{x+30}\right), 1 \le x$

$C' = 100\left(-\dfrac{400}{x^3} + \dfrac{30}{(x+30)^2}\right)$

Approximation: $x \approx 40.45$ units, or 4045 units

61. $S_1 = (4m - 1)^2 + (5m - 6)^2 + (10m - 3)^2$

$\dfrac{dS_1}{dm} = 2(4m - 1)(4) + 2(5m - 6)(5) + 2(10m - 3)(10) = 282m - 128 = 0$ when $m = \dfrac{64}{141}$.

Line: $y = \dfrac{64}{141}x$

$S = \left|4\left(\dfrac{64}{141}\right) - 1\right| + \left|5\left(\dfrac{64}{141}\right) - 6\right| + \left|10\left(\dfrac{64}{141}\right) - 3\right|$

$= \left|\dfrac{256}{141} - 1\right| + \left|\dfrac{320}{141} - 6\right| + \left|\dfrac{640}{141} - 3\right| = \dfrac{858}{141} \approx 6.1$ mi

63. $S_3 = \dfrac{|4m - 1|}{\sqrt{m^2 + 1}} + \dfrac{|5m - 6|}{\sqrt{m^2 + 1}} + \dfrac{|10m - 3|}{\sqrt{m^2 + 1}}$

Using a graphing utility, you can see that the minimum occurs when $x \approx 0.3$.

Line: $y \approx 0.3x$

$S_3 = \dfrac{|4(0.3) - 1| + |5(0.3) - 6| + |10(0.3) - 3|}{\sqrt{(0.3)^2 + 1}} \approx 4.5$ mi.

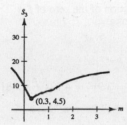

65. $f(x) = x^3 - 3x;\ x^4 + 36 \le 13x^2$

$x^4 - 13x^2 + 36 = (x^2 - 9)(x^2 - 4)$

$\qquad\qquad = (x - 3)(x - 2)(x + 2)(x + 3) \le 0$

Hence, $-3 \le x \le -2$ or $2 \le x \le 3$.

$f'(x) = 3x^2 - 3 = 3(x + 1)(x - 1)$

f is increasing on $(-\infty, -1)$ and $(1, \infty)$.

Hence, f is increasing on $[-3, -2]$ and $[2, 3]$.

$f(-2) = -2, f(3) = 18$. The maximum value of f is 18.

Section 3.8   Newton's Method

1. $f(x) = x^2 - 3$

$f'(x) = 2x$

$x_1 = 1.7$

n	x_n	$f(x_n)$	$f'(x_n)$	$\dfrac{f(x_n)}{f'(x_n)}$	$x_n - \dfrac{f(x_n)}{f'(x_n)}$
1	1.7000	-0.1100	3.4000	-0.0324	1.7324
2	1.7324	0.0012	3.4648	0.0003	1.7321

3. $f(x) = \sin x$

$f'(x) = \cos x$

$x_1 = 3$

n	x_n	$f(x_n)$	$f'(x_n)$	$\dfrac{f(x_n)}{f'(x_n)}$	$x_n - \dfrac{f(x_n)}{f'(x_n)}$
1	3.0000	0.1411	-0.9900	-0.1425	3.1425
2	3.1425	-0.0009	-1.0000	0.0009	3.1416

5. $f(x) = x^3 + x - 1$

$f'(x) = 3x^2 + 1$

Approximation of the zero of f is 0.682.

n	x_n	$f(x_n)$	$f'(x_n)$	$\dfrac{f(x_n)}{f'(x_n)}$	$x_n - \dfrac{f(x_n)}{f'(x_n)}$
1	0.5000	-0.3750	1.7500	-0.2143	0.7143
2	0.7143	0.0788	2.5307	0.0311	0.6832
3	0.6832	0.0021	2.4003	0.0009	0.6823

7. $f(x) = 3\sqrt{x-1} - x$

$f'(x) = \dfrac{3}{2\sqrt{x-1}} - 1$

Approximation of the zero of f is 1.146.

Similarly, the other zero is approximately 7.854.

n	x_n	$f(x_n)$	$f'(x_n)$	$\dfrac{f(x_n)}{f'(x_n)}$	$x_n - \dfrac{f(x_n)}{f'(x_n)}$
1	1.2000	0.1416	2.3541	0.0602	1.1398
2	1.1398	-0.0181	3.0118	-0.0060	1.1458
3	1.1458	-0.0003	2.9284	-0.0001	1.1459

9. $f(x) = x^3 + 3$

$f'(x) = 3x^2$

Approximation of the zero of f is -1.44224957.

n	x_n	$f(x_n)$	$f'(x_n)$	$\dfrac{f(x_n)}{f'(x_n)}$	$x_n - \dfrac{f(x_n)}{f'(x_n)}$
1	-1.5000	-0.3750	6.7500	-0.0556	-1.4444
2	-1.4444	-0.0134	6.2589	-0.0021	-1.4423
3	-1.4423	-0.00003	6.2407	-0.0000033	-1.4422

11. $f(x) = x^3 - 3.9x^2 + 4.79x - 1.881$

$f'(x) = 3x^2 - 7.8x + 4.79$

n	x_n	$f(x_n)$	$f'(x_n)$	$\dfrac{f(x_n)}{f'(x_n)}$	$x_n - \dfrac{f(x_n)}{f'(x_n)}$
1	0.5000	-0.3360	1.6400	-0.2049	0.7049
2	0.7049	-0.0921	0.7824	-0.1177	0.8226
3	0.8226	-0.0231	0.4037	-0.0573	0.8799
4	0.8799	-0.0045	0.2495	-0.0181	0.8980
5	0.8980.	-0.0004	0.2048	-0.0020	0.9000
6	0.9000	0.0000	0.2000	0.0000	0.9000

Approximation of the zero of f is 0.900.

n	x_n	$f(x_n)$	$f'(x_n)$	$\dfrac{f(x_n)}{f'(x_n)}$	$x_n - \dfrac{f(x_n)}{f'(x_n)}$
1	1.1	0.0000	-0.1600	-0.0000	1.1000

Approximation of the zero of f is 1.100.

n	x_n	$f(x_n)$	$f'(x_n)$	$\dfrac{f(x_n)}{f'(x_n)}$	$x_n - \dfrac{f(x_n)}{f'(x_n)}$
1	1.9	0.0000	0.8000	0.0000	1.9000

Approximation of the zero of f is 1.900.

13. $f(x) = x + \sin(x + 1)$

$f'(x) = 1 + \cos(x + 1)$

Approximation of the zero of f is -0.489.

n	x_n	$f(x_n)$	$f'(x_n)$	$\dfrac{f(x_n)}{f'(x_n)}$	$x_n - \dfrac{f(x_n)}{f'(x_n)}$
1	-0.5000	-0.0206	1.8776	-0.0110	-0.4890
2	-0.4890	0.0000	1.8723	0.0000	-0.4890

15. $h(x) = f(x) - g(x) = 2x + 1 - \sqrt{x+4}$

$h'(x) = 2 - \dfrac{1}{2\sqrt{x+4}}$

Point of intersection of the graphs of f and g occurs when $x \approx 0.569$.

n	x_n	$h(x_n)$	$h'(x_n)$	$\dfrac{h(x_n)}{h'(x_n)}$	$x_n - \dfrac{h(x_n)}{h'(x_n)}$
1	0.6000	0.0552	1.7669	0.0313	0.5687
2	0.5687	0.0000	1.7661	0.0000	0.5687

17. $h(x) = f(x) - g(x) = x - \tan x$

$h'(x) = 1 - \sec^2 x$

Point of intersection of the graphs of f and g occurs when $x \approx 4.493$.

Note: $f(x) = x$ and $g(x) - \tan x$ intersect infinitely often.

n	x_n	$h(x_n)$	$h'(x_n)$	$\dfrac{h(x_n)}{h'(x_n)}$	$x_n - \dfrac{h(x_n)}{h'(x_n)}$
1	4.5000	-0.1373	-21.5048	0.0064	4.4936
2	4.4936	-0.0039	-20.2271	0.0002	4.4934

19. (a) $f(x) = x^2 - a,\ a > 0$

$f'(x) = 2x$

$x_{n+1} = x_n - \dfrac{f(x_n)}{f'(x_n)}$

$= x_n - \dfrac{x_n^2 - a}{2x_n}$

$= \dfrac{1}{2}\left(x_n + \dfrac{a}{x_n}\right)$

(b) $\sqrt{5}$: $x_{n+1} = \dfrac{1}{2}\left(x_n + \dfrac{5}{x_n}\right)$. $x_1 = 2$

n	1	2	3	4
x_n	2	2.25	2.2361	2.2361

For example, given $x_1 = 2$,

$x_2 = \dfrac{1}{2}\left(2 + \dfrac{5}{2}\right) = \dfrac{9}{4} = 2.25.$

$\sqrt{5} \approx 2.236$

$\sqrt{7}$: $x_{n+1} = \dfrac{1}{2}\left(x_n + \dfrac{7}{x_n}\right)$, $x_1 = 2$

n	1	2	3	4	5
x_n	2	2.75	2.6477	2.6458	2.6458

$\sqrt{7} \approx 2.646$

21. $y - 2x^3 - 6x^2 + 6x - 1 = f(x)$

$y' - 6x^2 - 12x + 6 = f'(x)$

$x_1 = 1$

$f'(x) = 0$; therefore, the method fails.

n	x_n	$f(x_n)$	$f'(x_n)$
1	1	1	0

23. $y = -x^3 + 6x^2 - 10x + 6 = f(x)$

$y' = -3x^2 + 12x - 10 = f'(x)$

$x_1 = 2$

$x_2 = 1$

$x_3 = 2$

$x_4 = 1$ and so on.

Fails to converge

25. Answers will vary. See page 222.

Newton's Method uses tangent lines to approximate c such that $f(c) = 0$.

First, estimate an initial x_1 close to c (see graph).

Then determine x_2 by $x_2 = x_1 - \dfrac{f(x_1)}{f'(x_1)}$.

Calculate a third estimate by $x_3 = x_2 - \dfrac{f(x_2)}{f'(x_2)}$.

Continue this process until $|x_n - x_{n+1}|$ is within the desired accuracy.

Let x_{n+1} be the final approximation of c.

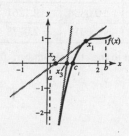

27. Let $g(x) = f(x) - x = \cos x - x$

$g'(x) = -\sin x - 1.$

The fixed point is approximately 0.74.

n	x_n	$g(x_n)$	$g'(x_n)$	$\dfrac{g(x_n)}{g'(x_n)}$	$x_n - \dfrac{g(x_n)}{g'(x_n)}$
1	1.0000	-0.4597	-1.8415	0.2496	0.7504
2	0.7504	-0.0190	-1.6819	0.0113	0.7391
3	0.7391	0.0000	-1.6736	0.0000	0.7391

29. $f(x) = x^3 - 3x^2 + 3,\ \ f'(x) = 3x^2 - 6x$

(a)

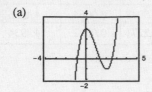

(b) $x_1 = 1$

$x_2 = x_1 - \dfrac{f(x_1)}{f'(x_1)} \approx 1.333$

Continuing, the zero is 1.347.

(c) $x_1 = \dfrac{1}{4}$

$x_2 = x_1 - \dfrac{f(x_1)}{f'(x_1)} \approx 2.405$

Continuing, the zero is 2.532.

(d)

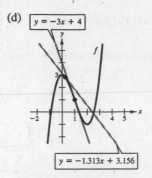

$y = -3x + 4$

$y = -1.313x + 3.156$

The x-intercepts correspond to the values resulting from the first iteration of Newton's Method.

(e) If the initial guess x_1 is not "close to" the desired zero of the function, the x-intercept of the tangent line may approximate another zero of the function.

31. $f(x) = \dfrac{1}{x} - a = 0$

$f'(x) = -\dfrac{1}{x^2}$

$x_{n+1} = x_n - \dfrac{(1/x_n) - a}{-1/x_n^2} = x_n + x_n^2\left(\dfrac{1}{x_n} - a\right) = x_n + x_n - x_n^2 a = 2x_n - x_n^2 a = x_n(2 - ax_n)$

33. $f(x) = x \cos x,\ \ (0, \pi)$

$f'(x) = -x \sin x + \cos x = 0$

Letting $F(x) = f'(x)$, we can use Newton's Method as follows.

$[F'(x) = -2 \sin x + x \cos x]$

n	x_n	$F(x_n)$	$F'(x_n)$	$\dfrac{F(x_n)}{F'(x_n)}$	$x_n - \dfrac{F(x_n)}{F'(x_n)}$
1	0.9000	-0.0834	-2.1261	0.0392	0.8608
2	0.8608	-0.0010	-2.0778	0.0005	0.8603

Approximation to the critical number: 0.860

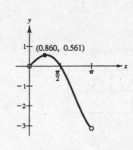

(0.860, 0.561)

35. $y = f(x) = 4 - x^2, (1, 0)$

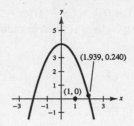

$$d = \sqrt{(x - 1)^2 + (y - 0)^2} = \sqrt{(x - 1)^2 + (4 - x^2)^2} = \sqrt{x^4 - 7x^2 - 2x + 17}$$

d is minimized when $D = x^4 - 7x^2 - 2x + 17$ is a minimum.

$$g(x) = D' = 4x^3 - 14x - 2$$
$$g'(x) = 12x^2 - 14$$

n	x_n	$g(x_n)$	$g'(x_n)$	$\dfrac{g(x_n)}{g'(x_n)}$	$x_n - \dfrac{g(x_n)}{g'(x_n)}$
1	2.0000	2.0000	34.0000	0.0588	1.9412
2	1.9412	0.0830	31.2191	0.0027	1.9385
3	1.9385	−0.0012	31.0934	0.0000	1.9385

$x \approx 1.939$

Point closest to $(1, 0)$ is $\approx (1.939, 0.240)$.

37.
$$\text{Minimize: } T = \frac{\text{Distance rowed}}{\text{Rate rowed}} + \frac{\text{Distance walked}}{\text{Rate walked}}$$

$$T = \frac{\sqrt{x^2 + 4}}{3} + \frac{\sqrt{x^2 - 6x + 10}}{4}$$

$$T' = \frac{x}{3\sqrt{x^2 + 4}} + \frac{x - 3}{4\sqrt{x^2 - 6x + 10}} = 0$$

$$4x\sqrt{x^2 - 6x + 10} = -3(x - 3)\sqrt{x^2 + 4}$$

$$16x^2(x^2 - 6x + 10) = 9(x - 3)^2(x^2 + 4)$$

$$7x^4 - 42x^3 + 43x^2 + 216x - 324 = 0$$

Let $f(x) = 7x^4 - 42x^3 + 43x^2 + 216x - 324$ and $f'(x) = 28x^3 - 126x^2 + 86x + 216$. Since $f(1) = -100$ and $f(2) = 56$, the solution is in the interval $(1, 2)$.

n	x_n	$f(x_n)$	$f'(x_n)$	$\dfrac{f(x_n)}{f'(x_n)}$	$x_n - \dfrac{f(x_n)}{f'(x_n)}$
1	1.7000	19.5887	135.6240	0.1444	1.5556
2	1.5556	−1.0480	150.2780	−0.0070	1.5626
3	1.5626	0.0014	49.5591	0.0000	1.5626

Approximation: $x \approx 1.563$ miles

39.
$$2,500,000 = -76x^3 + 4830x^2 - 320,000$$

$$76x^3 - 4830x^2 + 2,820,000 = 0$$

Let $f(x) = 76x^3 - 4830x^2 + 2,820,000$

$$f'(x) = 228x^2 - 9660x.$$

From the graph, choose $x_1 = 40$.

n	x_n	$f(x_n)$	$f'(x_n)$	$\dfrac{f(x_n)}{f'(x_n)}$	$x_n - \dfrac{f(x_n)}{f'(x_n)}$
1	40.0000	−44000.0000	−21600.0000	2.0370	37.9630
2	37.9630	17157.6209	−38131.4039	−0.4500	38.4130
3	38.4130	780.0914	−34642.2263	−0.0225	38.4355
4	38.4355	2.6308	−34465.3435	−0.0001	38.4356

The zero occurs when $x \approx 38.4356$ which corresponds to $384,356. (The larger zero is $x \approx 46.071$)

41. False. Let $f(x) = (x^2 - 1)/(x - 1)$. $x = 1$ is a discontinuity. It is not a zero of $f(x)$. This statement would be true if $f(x) = p(x)/q(x)$ is given in **reduced** form.

43. True

45. $f(x) = -\sin x$

$f'(x) = -\cos x$

Let $(x_0, y_1) = (x_0, -\sin(x_0))$ be a point on the graph of f. If (x_0, y_0) is a point of tangency, then

$$-\cos(x_0) = \frac{y_0 - 0}{x_0 - 0} = \frac{y_0}{x_0} = \frac{-\sin(x_0)}{x_0}.$$

Thus, $x_0 = \tan(x_0)$.

$x_0 \approx 4.4934$

Slope $= -\cos(x_0) \approx 0.217$

You can verify this answer by graphing $y_1 = -\sin x$ and the tangent line $y_2 = 0.217x$.

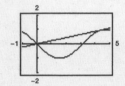

Section 3.9   Differentials

1. $f(x) = x^2$

$f'(x) = 2x$

Tangent line at $(2, 4)$: $y - f(2) = f'(2)(x - 2)$

$y - 4 = 4(x - 2)$

$y = 4x - 4$

x	1.9	1.99	2	2.01	2.1
$f(x) = x^2$	3.6100	3.9601	4	4.0401	4.4100
$T(x) = 4x - 4$	3.6000	3.9600	4	4.0400	4.4000

3. $f(x) = x^5$

$f'(x) = 5x^4$

Tangent line at $(2, 32)$: $y - f(2) = f'(2)(x - 2)$

$y - 32 = 80(x - 2)$

$y = 80x - 128$

x	1.9	1.99	2	2.01	2.1
$f(x) = x^5$	24.7610	31.2080	32	32.8080	40.8410
$T(x) = 80x - 128$	24.0000	31.2000	32	32.8000	40.0000

5. $f(x) = \sin x$

$f'(x) = \cos x$

Tangent line at $(2, \sin 2)$:

$y - f(2) = f'(2)(x - 2)$

$y - \sin 2 = (\cos 2)(x - 2)$

$y = (\cos 2)(x - 2) + \sin 2$

x	1.9	1.99	2	2.01	2.1
$f(x) = \sin x$	0.9463	0.9134	0.9093	0.9051	0.8632
$T(x) = (\cos 2)(x - 2) + \sin 2$	0.9509	0.9135	0.9093	0.9051	0.8677

7. $y = f(x) = \frac{1}{2}x^3, f'(x) = \frac{3}{2}x^2, x = 2, \Delta x = dx = 0.1$

$\Delta y = f(x + \Delta x) - f(x)$

$\quad = f(2.1) - f(2)$

$\quad = 0.6305$

$dy = f'(x)\,dx$

$\quad = f'(2)(0.1)$

$\quad = 6(0.1) = 0.6$

9. $y = f(x) = x^4 + 1, f'(x) = 4x^3, x = -1, \Delta x = dx = 0.01$

$\Delta y = f(x + \Delta x) - f(x)$

$\quad = f(-0.99) - f(-1)$

$\quad = [(-0.99)^4 + 1] - [(-1)^4 + 1] \approx -0.0394$

$dy = f'(x)\,dx$

$\quad = f'(-1)(0.01)$

$\quad = (-4)(0.01) = -0.04$

11. $y = 3x^2 - 4$

$dy = 6x\,dx$

13. $y = \dfrac{x + 1}{2x - 1}$

$dy = \dfrac{-3}{(2x - 1)^2}\,dx$

15. $y = x\sqrt{1 - x^2}$

$dy = \left(x\dfrac{-x}{\sqrt{1 - x^2}} + \sqrt{1 - x^2}\right)dx = \dfrac{1 - 2x^2}{\sqrt{1 - x^2}}\,dx$

17. $y = 2x - \cot^2 x$

$dy = (2 + 2\cot x \csc^2 x)\,dx$

$\quad = (2 + 2\cot x + 2\cot^3 x)\,dx$

19. $y = \dfrac{1}{3}\cos\left(\dfrac{6\pi x - 1}{2}\right)$

$dy = -\pi \sin\left(\dfrac{6\pi x - 1}{2}\right)dx$

21. (a) $f(1.9) = f(2 - 0.1) \approx f(2) + f'(2)(-0.1)$

$\quad \approx 1 + (1)(-0.1) = 0.9$

(b) $f(2.04) = f(2 + 0.04) \approx f(2) + f'(2)(0.04)$

$\quad \approx 1 + (1)(0.04) = 1.04$

23. (a) $f(1.9) = f(2 - 0.1) \approx f(2) + f'(2)(-0.1)$

$\quad \approx 1 + \left(-\frac{1}{2}\right)(-0.1) = 1.05$

(b) $f(2.04) = f(2 + 0.04) \approx f(2) + f'(2)(0.04)$

$\quad \approx 1 + \left(-\frac{1}{2}\right)(0.04) = 0.98$

25. (a) $g(2.93) = g(3 - 0.07) \approx g(3) + g'(3)(-0.07)$

$\quad \approx 8 + \left(-\frac{1}{2}\right)(-0.07) = 8.035$

(b) $g(3.1) = g(3 + 0.1) \approx g(3) + g'(3)(0.1)$

$\quad \approx 8 + \left(-\frac{1}{2}\right)(0.1) = 7.95$

27. $A = x^2$

$x = 12$

$\Delta x = dx = \pm\frac{1}{64}$

$dA = 2x\,dx$

$\Delta A \approx dA = 2(12)\left(\pm\frac{1}{64}\right)$

$\quad = \pm\frac{3}{8}$ square inches

29. $A = \pi r^2$

$r = 14$

$\Delta r = dr = \pm\frac{1}{4}$

$\Delta A \approx dA = 2\pi r\,dr = \pi(28)\left(\pm\frac{1}{4}\right)$

$\quad = \pm 7\pi$ square inches

31. (a) $x = 15$ centimeters

$\Delta x = dx = \pm 0.05$ centimeters

$A = x^2$

$dA = 2x\,dx = 2(15)(\pm 0.05)$

$\quad = \pm 1.5$ square centimeters

Percentage error:

$\dfrac{dA}{A} = \dfrac{1.5}{(15)^2} = 0.00666\ldots = \dfrac{2}{3}\%$

(b) $\dfrac{dA}{A} = \dfrac{2x\,dx}{x^2} = \dfrac{2\,dx}{x} \le 0.025$

$\dfrac{dx}{x} \le \dfrac{0.025}{2} = 0.0125 = 1.25\%$

33. $r = 6$ inches

$\Delta r = dr = \pm 0.02$ inches

(a) $V = \dfrac{4}{3}\pi r^3$

$dV = 4\pi r^2\, dr = 4\pi(6)^2(\pm 0.02) = \pm 2.88\pi$ cubic inches

(b) $S = 4\pi r^2$

$dS = 8\pi r\, dr = 8\pi(6)(\pm 0.02) = \pm 0.96\pi$ square inches

(c) Relative error: $\dfrac{dV}{V} = \dfrac{4\pi r^2\, dr}{(4/3)\pi r^3} = \dfrac{3\,dr}{r}$

$= \dfrac{3}{6}(0.02) = 0.01 = 1\%$

Relative error: $\dfrac{dS}{S} = \dfrac{8\pi r\, dr}{4\pi r^2} = \dfrac{2\,dr}{r}$

$= \dfrac{2(0.02)}{6} = 0.00666\ldots = \dfrac{2}{3}\%$

35. $V = \pi r^2 h = 40\pi r^2,\, r = 5$ cm, $h = 40$ cm, $dr = 0.2$ cm

$\Delta V \approx dV = 80\pi r\, dr = 80\pi(5)(0.2) = 80\pi$ cm^3

37. (a) $T = 2\pi\sqrt{L/g}$

$dT = \dfrac{\pi}{g\sqrt{L/g}}\, dL$

Relative error:

$\dfrac{dT}{T} = \dfrac{(\pi\, dL)/\left(g\sqrt{L/g}\right)}{2\pi\sqrt{L/g}}$

$= \dfrac{dL}{2L}$

$= \dfrac{1}{2}$ (relative error in L)

$= \dfrac{1}{2}(0.005) = 0.0025$

Percentage error: $\dfrac{dT}{T}(100) = 0.25\% = \dfrac{1}{4}\%$

(b) $(0.0025)(3600)(24) = 216$ seconds

$= 3.6$ minutes

39. $\theta = 26°45' = 26.75°$

$d\theta = \pm 15' = \pm 0.25°$

(a) $h = 9.5\csc\theta$

$dh = -9.5\csc\theta\cot\theta\, d\theta$

$\dfrac{dh}{h} = -\cot\theta\, d\theta$

$\left|\dfrac{dh}{h}\right| = (\cot 26.75°)(0.25°)$

Converting to radians,
$(\cot 0.4669)(0.0044) \approx 0.0087 = 0.87\%$ (in radians).

(b) $\left|\dfrac{dh}{h}\right| = \cot\theta\, d\theta \le 0.02$

$\dfrac{d\theta}{\theta} \le \dfrac{0.02}{\theta(\cot\theta)} = \dfrac{0.02\tan\theta}{\theta}$

$\dfrac{d\theta}{\theta} \le \dfrac{0.02\tan 26.75°}{26.75°} \approx \dfrac{0.02\tan 0.4669}{0.4669}$

$\approx 0.0216 = 2.16\%$ (in radians)

41. $r = \dfrac{v_0{}^2}{32}(\sin 2\theta)$

$v_0 = 2200$ ft/sec

θ changes from $10°$ to $11°$.

$dr = \dfrac{(2200)^2}{16}(\cos 2\theta)\, d\theta$

$\theta = 10\left(\dfrac{\pi}{180}\right)$

$d\theta = (11 - 10)\dfrac{\pi}{180}$

$\Delta r \approx dr$

$= \dfrac{(2200)^2}{16}\cos\left(\dfrac{20\pi}{180}\right)\left(\dfrac{\pi}{180}\right)$

≈ 4961 feet

43. Let $f(x) = \sqrt{x},\, x = 100,\, dx = -0.6$.

$f(x + \Delta x) \approx f(x) + f'(x)\, dx$

$= \sqrt{x} + \dfrac{1}{2\sqrt{x}}\, dx$

$f(x + \Delta x) = \sqrt{99.4}$

$\approx \sqrt{100} + \dfrac{1}{2\sqrt{100}}(-0.6) = 9.97$

Using a calculator: $\sqrt{99.4} \approx 9.96995$

45. Let $f(x) = \sqrt[4]{x}$, $x = 625$, $dx = -1$.

$$f(x + \Delta x) \approx f(x) + f'(x)\, dx = \sqrt[4]{x} + \frac{1}{4\sqrt[4]{x^3}}\, dx$$

$$f(x + \Delta x) = \sqrt[4]{624} \approx \sqrt[4]{625} + \frac{1}{4(\sqrt[4]{625})^3}(-1)$$

$$= 5 - \frac{1}{500} = 4.998$$

Using a calculator, $\sqrt[4]{624} \approx 4.9980$.

47. Let $f(x) = \sqrt{x}$, $x = 4$, $dx = 0.02$, $f'(x) = 1/(2\sqrt{x})$.

Then

$$f(4.02) \approx f(4) + f'(4)\, dx$$

$$\sqrt{4.02} \approx \sqrt{4} + \frac{1}{2\sqrt{4}}(0.02) = 2 + \frac{1}{4}(0.02).$$

49. $f(x) = \sqrt{x + 4}$

$$f'(x) = \frac{1}{2\sqrt{x + 4}}$$

At $(0, 2)$, $f(0) = 2$, $f'(0) = \frac{1}{4}$.

Tangent line: $y - 2 = \frac{1}{4}(x - 0)$

$$y = \frac{1}{4}x + 2$$

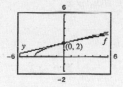

51. $f(x) = \tan x$

$$f'(x) = \sec^2 x$$

$$f(0) = 0$$

$$f'(0) = 1$$

Tangent line at $(0, 0)$: $y - 0 = (x - 0)$

$$y = x$$

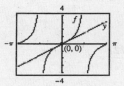

53. In general, when $\Delta x \to 0$, dy approaches Δy.

55. True

57. True

Review Exercises for Chapter 3

1. A number c in the domain of f is a critical number if $f'(c) = 0$ or f' is undefined at c.

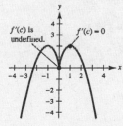

3. $g(x) = 2x + 5 \cos x$, $[0, 2\pi]$

$$g'(x) = 2 - 5 \sin x$$

$$= 0 \text{ when } \sin x = \tfrac{2}{5}.$$

Critical numbers: $x \approx 0.41$, $x \approx 2.73$

Left endpoint: $(0, 5)$

Critical number: $(0.41, 5.41)$

Critical number: $(2.73, 0.88)$ Minimum

Right endpoint: $(2\pi, 17.57)$ Maximum

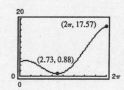

5. Yes. $f(-3) = f(2) = 0$. f is continuous on $[-3, 2]$, differentiable on $(-3, 2)$.

$f'(x) = (x + 3)(3x - 1) = 0$ for $x = \frac{1}{3}$.

$c = \frac{1}{3}$ satisfies $f'(c) = 0$.

7. $f(x) = 3 - |x - 4|$

(a)

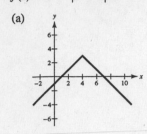

$f(1) = f(7) = 0$

(b) f is not differentiable at $x = 4$.

9. $f(x) = x^{2/3}, 1 \le x \le 8$

$f'(x) = \frac{2}{3}x^{-1/3}$

$\dfrac{f(b) - f(a)}{b - a} = \dfrac{4 - 1}{8 - 1} = \dfrac{3}{7}$

$f'(c) = \dfrac{2}{3}c^{-1/3} = \dfrac{3}{7}$

$c = \left(\dfrac{14}{9}\right)^3 = \dfrac{2744}{729} \approx 3.764$

11. $f(x) = x - \cos x, -\dfrac{\pi}{2} \le x \le \dfrac{\pi}{2}$

$f'(x) = 1 + \sin x$

$\dfrac{f(b) - f(a)}{b - a} = \dfrac{(\pi/2) - (-\pi/2)}{(\pi/2) - (-\pi/2)} = 1$

$f'(c) = 1 + \sin c = 1$

$c = 0$

13. $f(x) = Ax^2 + Bx + C$

$f'(x) = 2Ax + B$

$\dfrac{f(x_2) - f(x_1)}{x_2 - x_1} = \dfrac{A(x_2{}^2 - x_1{}^2) + B(x_2 - x_1)}{x_2 - x_1}$

$= A(x_1 + x_2) + B$

$f'(c) = 2Ac + B = A(x_1 + x_2) + B$

$2Ac = A(x_1 + x_2)$

$c = \dfrac{x_1 + x_2}{2} = \text{Midpoint of } [x_1, x_2]$

15. $f(x) = (x - 1)^2(x - 3)$

$f'(x) = (x - 1)^2(1) + (x - 3)(2)(x - 1)$

$= (x - 1)(3x - 7)$

Critical numbers: $x = 1$ and $x = \frac{7}{3}$

Interval:	$-\infty < x < 1$	$1 < x < \frac{7}{3}$	$\frac{7}{3} < x < \infty$
Sign of $f'(x)$:	$f'(x) > 0$	$f'(x) < 0$	$f'(x) > 0$
Conclusion:	Increasing	Decreasing	Increasing

17. $h(x) = \sqrt{x}(x - 3) = x^{3/2} - 3x^{1/2}$

Domain: $(0, \infty)$

$h'(x) = \dfrac{3}{2}x^{1/2} - \dfrac{3}{2}x^{-1/2}$

$= \dfrac{3}{2}x^{-1/2}(x - 1) = \dfrac{3(x - 1)}{2\sqrt{x}}$

Critical number: $x = 1$

Interval:	$0 < x < 1$	$1 < x < \infty$
Sign of $h'(x)$:	$h'(x) < 0$	$h'(x) > 0$
Conclusion:	Decreasing	Increasing

19. $h(t) = \frac{1}{4}t^4 - 8t$

$h'(t) = t^3 - 8 = 0$ when $t = 2$.

Relative minimum: $(2, -12)$

Test Interval:	$-\infty < t < 2$	$2 < t < \infty$
Sign of $h'(t)$:	$h'(t) < 0$	$h'(t) > 0$
Conclusion:	Decreasing	Increasing

21. $y = \frac{1}{3}\cos(12t) - \frac{1}{4}\sin(12t)$

$v = y' = -4\sin(12t) - 3\cos(12t)$

(a) When $t = \frac{\pi}{8}$, $y = \frac{1}{4}$ inch and $v = y' = 4$ inches/second.

(b) $y' = -4\sin(12t) - 3\cos(12t) = 0$ when $\dfrac{\sin(12t)}{\cos(12t)} = -\dfrac{3}{4} \Rightarrow \tan(12t) = -\dfrac{3}{4}.$

Therefore, $\sin(12t) = -\dfrac{3}{5}$ and $\cos(12t) = \dfrac{4}{5}$. The maximum displacement is

$$y = \left(\frac{1}{3}\right)\left(\frac{4}{5}\right) - \frac{1}{4}\left(-\frac{3}{5}\right) = \frac{5}{12} \text{ inch.}$$

(c) Period: $\dfrac{2\pi}{12} = \dfrac{\pi}{6}$

Frequency: $\dfrac{1}{\pi/6} = \dfrac{6}{\pi}$

23. $f(x) = x + \cos x,\ 0 \le x \le 2\pi$

$f'(x) = 1 - \sin x$

$f''(x) = -\cos x = 0$ when $x = \dfrac{\pi}{2}, \dfrac{3\pi}{2}.$

Points of inflection: $\left(\dfrac{\pi}{2}, \dfrac{\pi}{2}\right), \left(\dfrac{3\pi}{2}, \dfrac{3\pi}{2}\right)$

Test Interval:	$0 < x < \dfrac{\pi}{2}$	$\dfrac{\pi}{2} < x < \dfrac{3\pi}{2}$	$\dfrac{3\pi}{2} < x < 2\pi$
Sign of $f''(x)$:	$f''(x) < 0$	$f''(x) > 0$	$f''(x) < 0$
Conclusion:	Concave downward	Concave upward	Concave downward

25. $g(x) = 2x^2(1 - x^2)$

$g'(x) = -4x(2x^2 - 1)$ Critical numbers: $x = 0, +\dfrac{1}{\sqrt{2}}$

$g''(x) = 4 - 24x^2$

$g''(0) = 4 > 0$ Relative minimum at $(0, 0)$

$g''\left(\pm\dfrac{1}{\sqrt{2}}\right) = -8 < 0$ Relative maximums at $\left(\pm\dfrac{1}{\sqrt{2}}, \dfrac{1}{2}\right)$

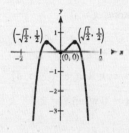

27.

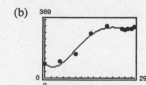

29. The first derivative is positive and the second derivative is negative. The graph is increasing and is concave down.

31. (a) $D = 0.0034t^4 - 0.2352t^3 + 4.9423t^2 - 20.8641t + 94.4025$

(b)

(c) Maximum at $(21.9, 319.5)$ (≈ 1992)

Minimum at $(2.6, 69.6)$ (≈ 1972)

(d) Outlays increasing at greatest rate at the point of inflection $(9.8, 173.7)$ (≈ 1979)

33. $\lim\limits_{x \to \infty} \dfrac{2x^2}{3x^2 + 5} = \lim\limits_{x \to \infty} \dfrac{2}{3 + 5/x^2} = \dfrac{2}{3}$

35. $\lim\limits_{x \to -\infty} \dfrac{3x^2}{x + 5} = -\infty$

37. $\lim\limits_{x \to \infty} \dfrac{5\cos x}{x} = 0$, since $|5\cos x| \le 5$.

39. $\lim\limits_{x \to -\infty} \dfrac{6x}{x + \cos x} = 6$

41. $h(x) = \dfrac{2x + 3}{x - 4}$

Discontinuity: $x = 4$

$$\lim\limits_{x \to \infty} \dfrac{2x + 3}{x - 4} = \lim\limits_{x \to \infty} \dfrac{2 + (3/x)}{1 - (4/x)} = 2$$

Vertical asymptote: $x = 4$

Horizontal asymptote: $y = 2$

43. $f(x) = \dfrac{3}{x} - 2$

Discontinuity: $x = 0$

$$\lim\limits_{x \to \infty} \left(\dfrac{3}{x} - 2\right) = -2$$

Vertical asymptote: $x = 0$

Horizontal asymptote: $y = -2$

45. $f(x) = x^3 + \dfrac{243}{x}$

Relative minimum: $(3, 108)$

Relative maximum: $(-3, -108)$

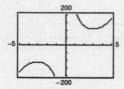

Vertical asymptote: $x = 0$

47. $f(x) = \dfrac{x - 1}{1 + 3x^2}$

Relative minimum: $(-0.155, -1.077)$

Relative maximum: $(2.155, 0.077)$

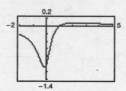

Horizontal asymptote: $y = 0$

49. $f(x) = 4x - x^2 = x(4 - x)$

Domain: $(-\infty, \infty)$; Range: $(-\infty, 4]$

$f'(x) = 4 - 2x = 0$ when $x = 2$.

$f''(x) = -2$

Therefore, $(2, 4)$ is a relative maximum.

Intercepts: $(0, 0), (4, 0)$

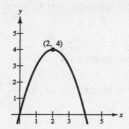

51. $f(x) = x\sqrt{16 - x^2}$

Domain: $[-4, 4]$; Range: $[-8, 8]$

$f'(x) = \dfrac{16 - 2x^2}{\sqrt{16 - x^2}} = 0$ when $x = \pm 2\sqrt{2}$ and undefined when $x = \pm 4$.

$f''(x) = \dfrac{2x(x^2 - 24)}{(16 - x^2)^{3/2}}$

$f''(-2\sqrt{2}) > 0$

Therefore, $(-2\sqrt{2}, -8)$ is a relative minimum.

$f''(2\sqrt{2}) < 0$

Therefore, $(2\sqrt{2}, 8)$ is a relative maximum.

Point of inflection: $(0, 0)$

Intercepts: $(-4, 0), (0, 0), (4, 0)$

Symmetry with respect to origin

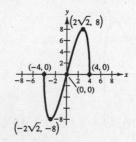

53. $f(x) = (x - 1)^3(x - 3)^2$

Domain: $(-\infty, \infty)$; Range: $(-\infty, \infty)$

$f'(x) = (x - 1)^2(x - 3)(5x - 11) = 0$ when $x = 1, \dfrac{11}{5}, 3$.

$f''(x) = 4(x - 1)(5x^2 - 22x + 23) = 0$ when $x = 1, \dfrac{11 \pm \sqrt{6}}{5}$.

$f''(3) > 0$

Therefore, $(3, 0)$ is a relative minimum.

$f''\left(\dfrac{11}{5}\right) < 0$

Therefore, $\left(\dfrac{11}{5}, \dfrac{3456}{3125}\right)$ is a relative maximum.

Points of inflection: $(1, 0), \left(\dfrac{11 - \sqrt{6}}{5}, 0.60\right), \left(\dfrac{11 + \sqrt{6}}{5}, 0.46\right)$

Intercepts: $(0, -9), (1, 0), (3, 0)$

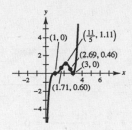

55. $f(x) = x^{1/3}(x + 3)^{2/3}$

Domain: $(-\infty, \infty)$; Range: $(-\infty, \infty)$

$f'(x) = \dfrac{x + 1}{(x + 3)^{1/3}x^{2/3}} = 0$ when $x = -1$ and undefined when $x = -3, 0$.

$f''(x) = \dfrac{-2}{x^{5/3}(x + 3)^{4/3}}$ is undefined when $x = 0, -3$.

By the First Derivative Test $(-3, 0)$ is a relative maximum and $\left(-1, -\sqrt[3]{4}\right)$ is a relative minimum. $(0, 0)$ is a point of inflection.

Intercepts: $(-3, 0), (0, 0)$

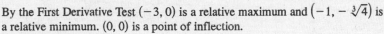

57. $f(x) = \dfrac{x + 1}{x - 1}$

Domain: $(-\infty, 1), (1, \infty)$; Range: $(-\infty, 1), (1, \infty)$

$f'(x) = \dfrac{-2}{(x - 1)^2} < 0$ if $x \neq 1$

$f''(x) = \dfrac{4}{(x - 1)^3}$

Horizontal asymptote: $y = 1$

Vertical asymptote: $x = 1$

Intercepts: $(-1, 0), (0, -1)$

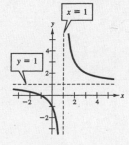

59. $f(x) = \dfrac{4}{1 + x^2}$

Domain: $(-\infty, \infty)$; Range: $(0, 4]$

$f'(x) = \dfrac{-8x}{(1 + x^2)^2} = 0$ when $x = 0$.

$f''(x) = \dfrac{-8(1 - 3x^2)}{(1 + x^2)^3} = 0$ when $x = \pm\dfrac{\sqrt{3}}{3}$.

$f''(0) < 0$

Therefore, $(0, 4)$ is a relative maximum.

Points of inflection: $\left(\pm\sqrt{3}/3, 3\right)$

Intercept: $(0, 4)$

Symmetric to the y-axis

Horizontal asymptote: $y = 0$

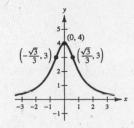

61. $f(x) = x^3 + x + \dfrac{4}{x}$

Domain: $(-\infty, 0), (0, \infty)$;   Range: $(-\infty, -6], [6, \infty)$

$$f'(x) = 3x^2 + 1 - \frac{4}{x^2} = \frac{3x^4 + x^2 - 4}{x^2} = \frac{(3x^2 + 4)(x^2 - 1)}{x^2} = 0 \text{ when } x = \pm 1.$$

$$f''(x) = 6x + \frac{8}{x^3} = \frac{6x^4 + 8}{x^3} \neq 0$$

$$f''(-1) < 0$$

Therefore, $(-1, -6)$ is a relative maximum.

$$f''(1) > 0$$

Therefore, $(1, 6)$ is a relative minimum.

Vertical asymptote: $x = 0$

Symmetric with respect to origin

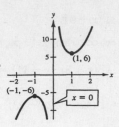

63. $f(x) = |x^2 - 9|$

Domain: $(-\infty, \infty)$; Range: $[0, \infty)$

$$f'(x) = \frac{2x(x^2 - 9)}{|x^2 - 9|} = 0 \text{ when } x = 0 \text{ and is undefined when } x = \pm 3.$$

$$f''(x) = \frac{2(x^2 - 9)}{|x^2 - 9|} \text{ is undefined at } x = \pm 3.$$

$$f''(0) < 0$$

Therefore, $(0, 9)$ is a relative maximum.

Relative minima: $(\pm 3, 0)$

Intercepts: $(\pm 3, 0), (0, 9)$

Symmetric to the y-axis

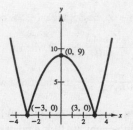

65. $f(x) = x + \cos x$

Domain: $[0, 2\pi]$; Range: $[1, 1 + 2\pi]$

$f'(x) = 1 - \sin x \geq 0$, f is increasing.

$f''(x) = -\cos x = 0$ when $x = \dfrac{\pi}{2}, \dfrac{3\pi}{2}$.

Points of inflection: $\left(\dfrac{\pi}{2}, \dfrac{\pi}{2} \right), \left(\dfrac{3\pi}{2}, \dfrac{3\pi}{2} \right)$

Intercept: $(0, 1)$

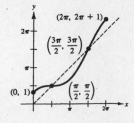

67. $x^2 + 4y^2 - 2x - 16y + 13 = 0$

(a) $(x^2 - 2x + 1) + 4(y^2 - 4y + 4) = -13 + 1 + 16$

$$(x - 1)^2 + 4(y - 2)^2 = 4$$

$$\frac{(x - 1)^2}{4} + \frac{(y - 2)^2}{1} = 1$$

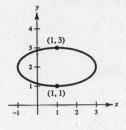

The graph is an ellipse:

Maximum: $(1, 3)$

Minimum: $(1, 1)$

(b) $x^2 + 4y^2 - 2x - 16y + 13 = 0$

$$2x + 8y\frac{dy}{dx} - 2 - 16\frac{dy}{dx} = 0$$

$$\frac{dy}{dx}(8y - 16) = 2 - 2x$$

$$\frac{dy}{dx} = \frac{2 - 2x}{8y - 16} = \frac{1 - x}{4y - 8}$$

The critical numbers are $x = 1$ and $y = 2$. These correspond to the points $(1, 1)$, $(1, 3)$, $(2, -1)$, and $(2, 3)$. Hence, the maximum is $(1, 3)$ and the minimum is $(1, 1)$.

69. Let $t = 0$ at noon.

$$L = d^2 = (100 - 12t)^2 + (-10t)^2 = 10,000 - 2400t + 244t^2$$

$$\frac{dL}{dt} = -2400 + 488t = 0 \text{ when } t = \frac{300}{61} \approx 4.92 \text{ hr.}$$

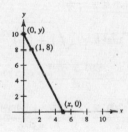

Ship A at $(40.98, 0)$; Ship B at $(0, -49.18)$

$$d^2 = 10,000 - 2400t + 244t^2$$

$$\approx 4098.36 \text{ when } t \approx 4.92 \approx 4:55 \text{ P.M.}$$

$$d \approx 64 \text{ km}$$

71. We have points $(0, y)$, $(x, 0)$, and $(1, 8)$. Thus,

$$m = \frac{y - 8}{0 - 1} = \frac{0 - 8}{x - 1} \text{ or } y = \frac{8x}{x - 1}.$$

Let $f(x) = L^2 = x^2 + \left(\frac{8x}{x - 1}\right)^2$.

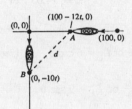

$$f'(x) = 2x + 128\left(\frac{x}{x - 1}\right)\left[\frac{(x - 1) - x}{(x - 1)^2}\right] = 0$$

$$x - \frac{64x}{(x - 1)^3} = 0$$

$$x[(x - 1)^3 - 64] = 0 \text{ when } x = 0, 5 \text{ (minimum)}.$$

Vertices of triangle: $(0, 0)$, $(5, 0)$, $(0, 10)$

73. $A = $ (Average of bases)(Height)

$$= \left(\frac{x + s}{2}\right)\frac{\sqrt{3s^2 + 2sx - x^2}}{2} \text{ (see figure)}$$

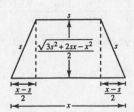

$$\frac{dA}{dx} = \frac{1}{4}\left[\frac{(s - x)(s + x)}{\sqrt{3s^2 + 2sx - x^2}} + \sqrt{3s^2 + 2sx - x^2}\right]$$

$$= \frac{2(2s - x)(s + x)}{4\sqrt{3s^2 + 2sx - x^2}} = 0 \text{ when } x = 2s.$$

A is a maximum when $x = 2s$.

75. You can form a right triangle with vertices $(0, 0)$, $(x, 0)$ and $(0, y)$. Assume that the hypotenuse of length L passes through $(4, 6)$.

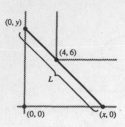

$$m = \frac{y - 6}{0 - 4} = \frac{6 - 0}{4 - x} \text{ or } y = \frac{6x}{x - 4}$$

Let $f(x) = L^2 = x^2 + y^2 = x^2 + \left(\frac{6x}{x - 4}\right)^2$.

$$f'(x) = 2x + 72\left(\frac{x}{x - 4}\right)\left[\frac{-4}{(x - 4)^2}\right] = 0$$

$x[(x - 4)^3 - 144] = 0$ when $x = 0$ or $x = 4 + \sqrt[3]{144}$.

$$L \approx 14.05 \text{ feet}$$

77. $\csc \theta = \dfrac{L_1}{6}$ or $L_1 = 6 \csc \theta$ (see figure)

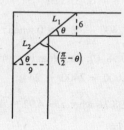

$\sec \theta = \dfrac{L_2}{9}$ or $L_2 = 9 \sec \theta$

$L = L_1 + L_2 = 6 \csc \theta + 9 \sec \theta$

$\dfrac{dL}{d\theta} = -6 \csc \theta \cot \theta + 9 \sec \theta \tan \theta = 0$

$\tan^3 \theta = \dfrac{2}{3} \Longrightarrow \tan \theta = \dfrac{\sqrt[3]{2}}{\sqrt[3]{3}}$

$\sec \theta = \sqrt{1 + \tan^2 \theta} = \sqrt{1 + \left(\dfrac{2}{3}\right)^{2/3}} = \dfrac{\sqrt{3^{2/3} + 2^{2/3}}}{3^{1/3}}$

$\csc \theta = \dfrac{\sec \theta}{\tan \theta} = \dfrac{\sqrt{3^{2/3} + 2^{2/3}}}{2^{1/3}}$

$L = 6\dfrac{(3^{2/3} + 2^{2/3})^{1/2}}{2^{1/3}} + 9\dfrac{(3^{2/3} + 2^{2/3})^{1/2}}{3^{1/3}} = 3(3^{2/3} + 2^{2/3})^{3/2} \text{ ft} \approx 21.07 \text{ ft}$ (Compare to Exercise 72 using $a = 9$ and $b = 6$.)

79. Total cost $=$ (Cost per hour)(Number of hours)

$$T = \left(\frac{v^2}{600} + 5\right)\left(\frac{110}{v}\right) = \frac{11v}{60} + \frac{550}{v}$$

$\dfrac{dT}{dv} = \dfrac{11}{60} - \dfrac{550}{v^2} = \dfrac{11v^2 - 33,000}{60v^2}$

$= 0$ when $v = \sqrt{3000} = 10\sqrt{30} \approx 54.8$ mph.

$\dfrac{d^2T}{dv^2} = \dfrac{1100}{v^3} > 0$ when $v = 10\sqrt{30}$ so this value yields a minimum.

81. $f(x) = x^3 - 3x - 1$

From the graph you can see that $f(x)$ has three real zeros.

$f'(x) = 3x^2 - 3$

n	x_n	$f(x_n)$	$f'(x_n)$	$\dfrac{f(x_n)}{f'(x_n)}$	$x_n - \dfrac{f(x_n)}{f'(x_n)}$
1	−1.5000	0.1250	3.7500	0.0333	−1.5333
2	−1.5333	−0.0049	4.0530	−0.0012	−1.5321

n	x_n	$f(x_n)$	$f'(x_n)$	$\dfrac{f(x_n)}{f'(x_n)}$	$x_n - \dfrac{f(x_n)}{f'(x_n)}$
1	−0.5000	0.3750	−2.2500	−0.1667	−0.3333
2	−0.3333	−0.0371	−2.6667	0.0139	−0.3472
3	−0.3472	−0.0003	−2.6384	0.0001	−0.3473

n	x_n	$f(x_n)$	$f'(x_n)$	$\dfrac{f(x_n)}{f'(x_n)}$	$x_n - \dfrac{f(x_n)}{f'(x_n)}$
1	1.9000	0.1590	7.8300	0.0203	1.8797
2	1.8797	0.0024	7.5998	0.0003	1.8794

The three real zeros of $f(x)$ are $x \approx -1.532$, $x \approx -0.347$, and $x \approx 1.879$.

83. Find the zeros of $f(x) = x^4 - x - 3$.

$f'(x) = 4x^3 - 1$

From the graph you can see that $f(x)$ has two real zeros.

f changes sign in $[-2, -1]$.

n	x_n	$f(x_n)$	$f'(x_n)$	$\dfrac{f(x_n)}{f'(x_n)}$	$x_n - \dfrac{f(x_n)}{f'(x_n)}$
1	−1.2000	0.2736	−7.9120	−0.0346	−1.1654
2	−1.1654	0.0100	−7.3312	−0.0014	−1.1640

On the interval $[-2, -1]$: $x \approx -1.164$.

f changes sign in $[1, 2]$.

n	x_n	$f(x_n)$	$f'(x_n)$	$\dfrac{f(x_n)}{f'(x_n)}$	$x_n - \dfrac{f(x_n)}{f'(x_n)}$
1	1.5000	0.5625	12.5000	0.0450	1.4550
2	1.4550	0.0268	11.3211	0.0024	1.4526
3	1.4526	−0.0003	11.2602	0.0000	1.4526

On the interval $[1, 2]$: $x \approx 1.453$.

85. $y = x(1 - \cos x) = x - x \cos x$

$\dfrac{dy}{dx} = 1 + x \sin x - \cos x$

$dy = (1 + x \sin x - \cos x)\, dx$

87. $\qquad S = 4\pi r^2 \quad dr = \Delta r = \pm 0.025$

$\qquad\quad dS = 8\pi r\, dr = 8\pi(9)(\pm 0.025)$

$\qquad\qquad\quad = \pm 1.8\pi$ square cm

$\dfrac{dS}{S}(100) = \dfrac{8\pi r\, dr}{4\pi r^2}(100) = \dfrac{2\, dr}{r}(100)$

$\qquad\qquad = \dfrac{2(\pm 0.025)}{9}(100) \approx \pm 0.56\%$

$\qquad V = \dfrac{4}{3}\pi r^3$

$\qquad\quad dV = 4\pi r^2\, dr = 4\pi(9)^2(\pm 0.025)$

$\qquad\qquad\quad = \pm 8.1\pi$ cubic cm

$\dfrac{dV}{V}(100) = \dfrac{4\pi r^2\, dr}{(4/3)\pi r^3}(100) = \dfrac{3\, dr}{r}(100)$

$\qquad\qquad = \dfrac{3(\pm 0.025)}{9}(100) \approx \pm 0.83\%$

Problem Solving for Chapter 3

1. $p(x) = x^4 + ax^2 + 1$

(a) $p'(x) = 4x^3 + 2ax = 2x(2x^2 + a)$

$\quad p''(x) = 12x^2 + 2a$

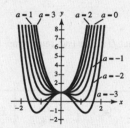

$\quad$ For $a \geq 0$, there is one relative minimum at $(0, 1)$.

(b) For $a < 0$, there is a relative maximum at $(0, 1)$.

(c) For $a < 0$, there are two relative minima at $x = \pm\sqrt{-\dfrac{a}{2}}$.

(d) There are either 1 or 3 critical points. The above analysis shows that there cannot be exactly two relative extrema.

3. $f(x) = \dfrac{c}{x} + x^2$

$f'(x) = -\dfrac{c}{x^2} + 2x = 0 \implies \dfrac{c}{x^2} = 2x \implies x^3 = \dfrac{c}{2} \implies x = \sqrt[3]{\dfrac{c}{2}}$

$f''(x) = \dfrac{2c}{x^3} + 2$

If $c = 0, f(x) = x^2$ has a relative minimum, but no relative maximum.

If $c > 0, x = \sqrt[3]{\dfrac{c}{2}}$ is a relative minimum, because $f''\!\left(\sqrt[3]{\dfrac{c}{2}}\right) > 0$.

If $c < 0, x = \sqrt[3]{\dfrac{c}{2}}$ is a relative minimum too.

Answer: all c.

5. Assume $y_1 < d < y_2$. Let $g(x) = f(x) - d(x - a)$. g is continuous on $[a, b]$ and therefore has a minimum $(c, g(c))$ on $[a, b]$. The point c cannot be an endpoint of $[a, b]$ because

$$g'(a) = f'(a) - d = y_1 - d < 0$$

$$g'(b) = f'(b) - d = y_2 - d > 0.$$

Hence, $a < c < b$ and $g'(c) = 0 \Rightarrow f'(c) = d$.

7. Set $\dfrac{f(b) - f(a) - f'(a)(b - a)}{(b - a)^2} = k.$

Define $F(x) = f(x) - f(a) - f'(a)(x - a) - k(x - a)^2$.

$F(a) = 0, F(b) = f(b) - f(a) - f'(a)(b - a) - k(b - a)^2 = 0$

F is continuous on $[a, b]$ and differentiable on (a, b).

There exists c_1, $a < c_1 < b$, satisfying $F'(c_1) = 0$.

$F'(x) = f'(x) - f'(a) - 2k(x - a)$ satisfies the hypothesis of Rolle's Theorem on $[a, c_1]$:

$$F'(a) = 0, F'(c_1) = 0.$$

There exists c_2, $a < c_2 < c_1$ satisfying $F''(c_2) = 0$.

Finally, $F''(x) = f''(x) - 2k$ and $F''(c_2) = 0$ implies that

$$k = \frac{f''(c_2)}{2}.$$

Thus, $k = \dfrac{f(b) - f(a) - f'(a)(b - a)}{(b - a)^2} = \dfrac{f''(c_2)}{2} \Rightarrow f(b) = f(a) + f'(a)(b - a) + \dfrac{1}{2}f''(c_2)(b - a)^2.$

9.

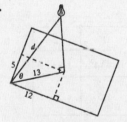

$d = \sqrt{13^2 + x^2}, \sin \theta = \dfrac{x}{d}.$

Let A be the amount of illumination at one of the corners, as indicated in the figure. Then

$$A = \frac{kI}{(13^2 + x^2)} \sin \theta = \frac{kIx}{(13^2 + x^2)^{3/2}}$$

$$A'(x) = kI\frac{(x^2 + 169)^{3/2}(1) - x\left(\dfrac{3}{2}\right)(x^2 + 169)^{1/2}(2x)}{(169 + x^2)^3} = 0$$

$$\Rightarrow (x^2 + 169)^{3/2} = 3x^2(x^2 + 169)^{1/2}$$

$$x^2 + 169 = 3x^2$$

$$2x^2 = 169$$

$$x = \frac{13}{\sqrt{2}} \approx 9.19 \text{ feet.}$$

By the First Derivative Test, this is a maximum.

11. Let T be the intersection of PQ and RS. Let MN be the perpendicular to SQ and PR passing through T.

Let $TM = x$ and $TN = b - x$.

$$\frac{SN}{b-x} = \frac{MR}{x} \implies SN = \frac{b-x}{x}MR$$

$$\frac{NQ}{b-x} = \frac{PM}{x} \implies NQ = \frac{b-x}{x}PM$$

$$SQ = \frac{b-x}{x}(MR + PM) = \frac{b-x}{x}d$$

$$A(x) = \text{Area} = \frac{1}{2}dx + \frac{1}{2}\left(\frac{b-x}{x}d\right)(b-x) = \frac{1}{2}d\left[x + \frac{(b-x)^2}{x}\right] = \frac{1}{2}d\left[\frac{2x^2 - 2bx + b^2}{x}\right]$$

$$A'(x) = \frac{1}{2}d\left[\frac{x(4x-2b) - (2x^2 - 2bx + b^2)}{x^2}\right]$$

$$A'(x) = 0 \implies 4x^2 - 2xb = 2x^2 - 2bx + b^2$$

$$2x^2 = b^2$$

$$x = \frac{b}{\sqrt{2}}$$

Hence, we have $SQ = \frac{b-x}{x}d = \frac{b - (b/\sqrt{2})}{b/\sqrt{2}}d = \left(\sqrt{2} - 1\right)d.$

Using the Second Derivative Test, this is a minimum. There is no maximum.

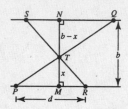

13. (a) Let $M > 0$ be given. Take $N = \sqrt{M}$.

Then whenever $x > N = \sqrt{M}$, you have

$$f(x) = x^2 > M.$$

(b) Let $\varepsilon > 0$ be given. Let $M = \sqrt{\dfrac{1}{\varepsilon}}$.

Then whenever $x > M = \sqrt{\dfrac{1}{\varepsilon}}$, you have

$$x^2 > \frac{1}{\varepsilon} \implies \frac{1}{x^2} < \varepsilon \implies \left|\frac{1}{x^2} - 0\right| < \varepsilon.$$

(c) Let $\varepsilon > 0$ be given. There exists $N > 0$ such that

$|f(x) - L| < \varepsilon$ whenever $x > N$.

Let $\delta = \dfrac{1}{N}$. Let $x = \dfrac{1}{y}$.

If $0 < y < \delta = \dfrac{1}{N}$, then $\dfrac{1}{x} < \dfrac{1}{N} \implies x > N$ and

$$|f(x) - L| = \left|f\left(\frac{1}{y}\right) - L\right| < \varepsilon.$$

15. (a)

x	0	0.5	1	2
$\sqrt{1+x}$	1	1.2247	1.4142	1.7321
$\frac{1}{2}x + 1$	1	1.25	1.5	2

(b) Let $f(x) = \sqrt{1+x}$. Using the Mean Value Theorem on the interval $[0, x]$, there exists c, $0 < c < x$, satisfying

$$f'(c) = \frac{1}{2\sqrt{1+c}} = \frac{f(x) - f(0)}{x - 0} = \frac{\sqrt{1+x} - 1}{x}.$$

Thus $\sqrt{1+x} = \dfrac{x}{2\sqrt{1+c}} + 1 < \dfrac{x}{2} + 1$

$\left(\text{because } \sqrt{1+c} > 1\right).$

17. (a) $s = \dfrac{v \frac{km}{hr}\left(1000 \frac{m}{km}\right)}{\left(3600 \frac{sec}{hr}\right)} = \dfrac{5}{18}v$

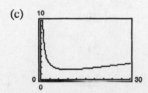

v	20	40	60	80	100
s	5.56	11.11	16.67	22.22	27.78
d	5.1	13.7	27.2	44.2	66.4

$d(t) = 0.071s^2 + 0.389s + 0.727$

(b) The distance between the back of the first vehicle and the front of the second vehicle is $d(t)$, the safe stopping distance. The first vehicle passes the given point in $5.5/s$ seconds, and the second vehicle takes $d(s)/s$ more seconds. Hence,

$$T = \frac{d(s)}{s} + \frac{5.5}{s}.$$

(c)

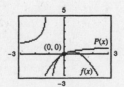

$$T = \frac{1}{s}(0.071s^2 + 0.389s + 0.727) + \frac{5.5}{s}$$

The minimum is attained when $s \approx 9.365$ m/sec.

(d) $T(s) = 0.071s + 0.389 + \dfrac{6.227}{s}$

$T'(s) = 0.071 - \dfrac{6.227}{s^2} \Rightarrow s^2 = \dfrac{6.227}{0.071}$

$\Rightarrow s \approx 9.365$ m/sec

$T(9.365) \approx 1.719$ seconds

$9.365 \text{ m/sec} \cdot \dfrac{3600}{1000} = 33.7$ km/hr

(e) $d(9.365) = 10.597$ m

19. $f(x) = \dfrac{x}{x + 1}, f'(x) = \dfrac{1}{(x + 1)^2}, f''(x) = \dfrac{-2}{(x + 1)^3}$

$P(0) = f(0): c_0 = 0$

$P'(0) = f'(0): c_1 = 1$

$P''(0) = f''(0): 2c_2 = -2 \Rightarrow c_2 = -1$

$P(x) = x - x^2$

C H A P T E R 4
Integration

Section 4.1   Antiderivatives and Indefinite Integration 179

Section 4.2   Area 184

Section 4.3   Riemann Sums and Definite Integrals 191

Section 4.4   The Fundamental Theorem of Calculus 195

Section 4.5   Integration by Substitution 200

Section 4.6   Numerical Integration 209

Review Exercises 214

Problem Solving 220

CHAPTER 4
Integration

Section 4.1   Antiderivatives and Indefinite Integration

1. $\dfrac{d}{dx}\left(\dfrac{3}{x^3} + C\right) = \dfrac{d}{dx}(3x^{-3} + C) = -9x^{-4} = \dfrac{-9}{x^4}$

3. $\dfrac{d}{dx}\left(\dfrac{1}{3}x^3 - 4x + C\right) = x^2 - 4 = (x-2)(x+2)$

5. $\dfrac{dy}{dt} = 3t^2$

$y = t^3 + C$

Check: $\dfrac{d}{dt}[t^3 + C] = 3t^2$

7. $\dfrac{dy}{dx} = x^{3/2}$

$y = \dfrac{2}{5}x^{5/2} + C$

Check: $\dfrac{d}{dx}\left[\dfrac{2}{5}x^{5/2} + C\right] = x^{3/2}$

	Given	_Rewrite_	_Integrate_	_Simplify_
9.	$\displaystyle\int \sqrt[3]{x}\,dx$	$\displaystyle\int x^{1/3}\,dx$	$\dfrac{x^{4/3}}{4/3} + C$	$\dfrac{3}{4}x^{4/3} + C$
11.	$\displaystyle\int \dfrac{1}{x\sqrt{x}}\,dx$	$\displaystyle\int x^{-3/2}\,dx$	$\dfrac{x^{-1/2}}{-1/2} + C$	$-\dfrac{2}{\sqrt{x}} + C$
13.	$\displaystyle\int \dfrac{1}{2x^3}\,dx$	$\dfrac{1}{2}\displaystyle\int x^{-3}\,dx$	$\dfrac{1}{2}\left(\dfrac{x^{-2}}{-2}\right) + C$	$-\dfrac{1}{4x^2} + C$

15. $\displaystyle\int (x+3)\,dx = \dfrac{x^2}{2} + 3x + C$

Check: $\dfrac{d}{dx}\left[\dfrac{x^2}{2} + 3x + C\right] = x + 3$

17. $\displaystyle\int (2x - 3x^2)\,dx = x^2 - x^3 + C$

Check: $\dfrac{d}{dx}[x^2 - x^3 + C] = 2x - 3x^2$

19. $\displaystyle\int (x^3 + 2)\,dx = \dfrac{1}{4}x^4 + 2x + C$

Check: $\dfrac{d}{dx}\left(\dfrac{1}{4}x^4 + 2x + C\right) = x^3 + 2$

21. $\displaystyle\int (x^{3/2} + 2x + 1)\,dx = \dfrac{2}{5}x^{5/2} + x^2 + x + C$

Check: $\dfrac{d}{dx}\left(\dfrac{2}{5}x^{5/2} + x^2 + x + C\right) = x^{3/2} + 2x + 1$

23. $\displaystyle\int \sqrt[3]{x^2}\,dx = \int x^{2/3}\,dx = \dfrac{x^{5/3}}{5/3} + C = \dfrac{3}{5}x^{5/3} + C$

Check: $\dfrac{d}{dx}\left(\dfrac{3}{5}x^{5/3} + C\right) = x^{2/3} = \sqrt[3]{x^2}$

25. $\displaystyle\int \dfrac{1}{x^3}\,dx = \int x^{-3}\,dx = \dfrac{x^{-2}}{-2} + C = -\dfrac{1}{2x^2} + C$

Check: $\dfrac{d}{dx}\left(-\dfrac{1}{2x^2} + C\right) = \dfrac{1}{x^3}$

27. $\displaystyle\int \dfrac{x^2 + x + 1}{\sqrt{x}}\,dx = \int (x^{3/2} + x^{1/2} + x^{-1/2})\,dx = \dfrac{2}{5}x^{5/2} + \dfrac{2}{3}x^{3/2} + 2x^{1/2} + C = \dfrac{2}{15}x^{1/2}(3x^2 + 5x + 15) + C$

Check: $\dfrac{d}{dx}\left(\dfrac{2}{5}x^{5/2} + \dfrac{2}{3}x^{3/2} + 2x^{1/2} + C\right) = x^{3/2} + x^{1/2} + x^{-1/2} = \dfrac{x^2 + x + 1}{\sqrt{x}}$

29. $\int (x + 1)(3x - 2)\, dx = \int (3x^2 + x - 2)\, dx$

$$= x^3 + \frac{1}{2}x^2 - 2x + C$$

Check: $\frac{d}{dx}\left(x^3 + \frac{1}{2}x^2 - 2x + C\right) = 3x^2 + x - 2$

$$= (x + 1)(3x - 2)$$

31. $\int y^2 \sqrt{y}\, dy = \int y^{5/2}\, dy = \frac{2}{7}y^{7/2} + C$

Check: $\frac{d}{dy}\left(\frac{2}{7}y^{7/2} + C\right) = y^{5/2} = y^2 \sqrt{y}$

33. $\int dx = \int 1\, dx = x + C$

Check: $\frac{d}{dx}(x + C) = 1$

35. $\int (2 \sin x + 3 \cos x)\, dx = -2 \cos x + 3 \sin x + C$

Check: $\frac{d}{dx}(-2 \cos x + 3 \sin x + C) = 2 \sin x + 3 \cos x$

37. $\int (1 - \csc t \cot t)\, dt = t + \csc t + C$

Check: $\frac{d}{dt}(t + \csc t + C) = 1 - \csc t \cot t$

39. $\int (\sec^2 \theta - \sin \theta)\, d\theta = \tan \theta + \cos \theta + C$

Check: $\frac{d}{d\theta}(\tan \theta + \cos \theta + C) = \sec^2 \theta - \sin \theta$

41. $\int (\tan^2 y + 1)\, dy = \int \sec^2 y\, dy = \tan y + C$

Check: $\frac{d}{dy}(\tan y + C) = \sec^2 y = \tan^2 y + 1$

43. $f'(x) = 2$

$f(x) = 2x + C$

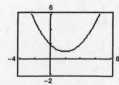

Answers will vary.

45. $f'(x) = 1 - x^2$

$f(x) = x - \frac{x^3}{3} + C$

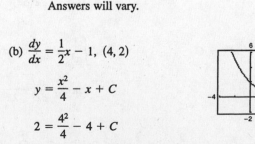

Answers will vary.

47. $\frac{dy}{dx} = 2x - 1, \ (1, 1)$

$y = \int (2x - 1)\, dx = x^2 - x + C$

$1 = (1)^2 - (1) + C \implies C = 1$

$y = x^2 - x + 1$

49. (a) Answers will vary.

(b) $\frac{dy}{dx} = \frac{1}{2}x - 1, \ (4, 2)$

$y = \frac{x^2}{4} - x + C$

$2 = \frac{4^2}{4} - 4 + C$

$2 = C$

$y = \frac{x^2}{4} - x + 2$

51. (a)

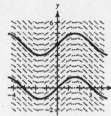

(b) $\dfrac{dy}{dx} = \cos x, \quad (0, 4)$

$y = \displaystyle\int \cos x \, dx = \sin x + C$

$4 = \sin(0) + C \Longrightarrow C = 4$

$y = \sin x + 4$

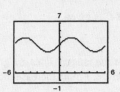

53. (a)

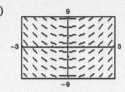

(b) $\dfrac{dy}{dx} = 2x, \quad (-2, -2)$

$y = \displaystyle\int 2x \, dx = x^2 + C$

$-2 = (-2)^2 + C = 4 + C \Longrightarrow C = -6$

$y = x^2 - 6$

(c)

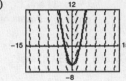

55. $f'(x) = 4x, f(0) = 6$

$f(x) = \displaystyle\int 4x \, dx = 2x^2 + C$

$f(0) = 6 = 2(0)^2 + C \Longrightarrow C = 6$

$f(x) = 2x^2 + 6$

57. $h'(t) = 8t^3 + 5, h(1) = -4$

$h(t) = \displaystyle\int (8t^3 + 5) \, dt = 2t^4 + 5t + C$

$h(1) = -4 = 2 + 5 + C \Longrightarrow C = -11$

$h(t) = 2t^4 + 5t - 11$

59. $f''(x) = 2$

$f'(2) = 5$

$f(2) = 10$

$f'(x) = \displaystyle\int 2 \, dx = 2x + C_1$

$f'(2) = 4 + C_1 = 5 \Longrightarrow C_1 = 1$

$f'(x) = 2x + 1$

$f(x) = \displaystyle\int (2x + 1) \, dx = x^2 + x + C_2$

$f(2) = 6 + C_2 = 10 \Longrightarrow C_2 = 4$

$f(x) = x^2 + x + 4$

61. $f''(x) = x^{-3/2}$

$f'(4) = 2$

$f(0) = 0$

$f'(x) = \displaystyle\int x^{-3/2} \, dx = -2x^{-1/2} + C_1 = -\dfrac{2}{\sqrt{x}} + C_1$

$f'(4) = -\dfrac{2}{2} + C_1 = 2 \Longrightarrow C_1 = 3$

$f'(x) = -\dfrac{2}{\sqrt{x}} + 3$

$f(x) = \displaystyle\int (-2x^{-1/2} + 3) \, dx = -4x^{1/2} + 3x + C_2$

$f(0) = 0 + 0 + C_2 = 0 \Longrightarrow C_2 = 0$

$f(x) = -4x^{1/2} + 3x = -4\sqrt{x} + 3x$

63. (a) $h(t) = \displaystyle\int (1.5t + 5) \, dt = 0.75t^2 + 5t + C$

$h(0) = 0 + 0 + C = 12 \Longrightarrow C = 12$

$h(t) = 0.75t^2 + 5t + 12$

(b) $h(6) = 0.75(6)^2 + 5(6) + 12 = 69$ cm

65. $f(0) = -4.$ Graph of f' is given.

(a) $f'(4) \approx -1.0$

(b) No. The slopes of the tangent lines are greater than 2 on $[0, 2]$. Therefore, f must increase more than 4 units on $[0, 4]$.

(c) No, $f(5) < f(4)$ because f is decreasing on $[4, 5]$.

(d) f is a maximum at $x = 3.5$ because $f'(3.5) \approx 0$ and the First Derivative Test.

(e) f is concave upward when f' is increasing on $(-\infty, 1)$ and $(5, \infty)$. f is concave downward on $(1, 5)$. Points of inflection at $x = 1, 5$.

(f) f'' is a minimum at $x = 3$.

(g)

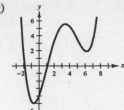

67. $a(t) = -32$ ft/sec^2

$$v(t) = \int -32 \, dt = -32t + C_1$$

$$v(0) = 60 = C_1$$

$$s(t) = \int (-32t + 60) \, dt = -16t^2 + 60t + C_2$$

$$s(0) = 6 = C_2$$

$$s(t) = -16t^2 + 60t + 6, \text{ Position function}$$

The ball reaches its maximum height when

$$v(t) = -32t + 60 = 0$$

$$32t = 60$$

$$t = \tfrac{15}{8} \text{ seconds.}$$

$$s\left(\tfrac{15}{8}\right) = -16\left(\tfrac{15}{8}\right)^2 + 60\left(\tfrac{15}{8}\right) + 6 = 62.25 \text{ feet}$$

69. From Exercise 68, we have:

$$s(t) = -16t^2 + v_0 t$$

$$s'(t) = -32t + v_0 = 0 \text{ when } t = \frac{v_0}{32} = \text{ time to reach}$$
maximum height.

$$s\left(\frac{v_0}{32}\right) = -16\left(\frac{v_0}{32}\right)^2 + v_0\left(\frac{v_0}{32}\right) = 550$$

$$-\frac{v_0^2}{64} + \frac{v_0^2}{32} = 550$$

$$v_0^2 = 35,200$$

$$v_0 \approx 187.617 \text{ ft/sec}$$

71. $a(t) = -9.8$

$$v(t) = \int -9.8 \, dt = -9.8t + C_1$$

$$v(0) = v_0 = C_1 \implies v(t) = -9.8t + v_0$$

$$f(t) = \int (-9.8t + v_0) \, dt = -4.9t^2 + v_0 t + C_2$$

$$f(0) = s_0 = C_2 \implies f(t) = -4.9t^2 + v_0 t + s_0$$

73. From Exercise 71, $f(t) = -4.9t^2 + 10t + 2.$

$$v(t) = -9.8t + 10 = 0 \text{ (Maximum height when } v = 0.)$$

$$9.8t = 10$$

$$t = \frac{10}{9.8}$$

$$f\left(\frac{10}{9.8}\right) \approx 7.1 \text{ m}$$

75. $a = -1.6$

$$v(t) = \int -1.6 \, dt = -1.6t + v_0 = -1.6t, \text{ since the stone was dropped, } v_0 = 0.$$

$$s(t) = \int (-1.6t) \, dt = -0.8t^2 + s_0$$

$$s(20) = 0 \implies -0.8(20)^2 + s_0 = 0$$

$$s_0 = 320$$

Thus, the height of the cliff is 320 meters.

$$v(t) = -1.6t$$

$$v(20) = -32 \text{ m/sec}$$

77. $x(t) = t^3 - 6t^2 + 9t - 2 \qquad 0 \le t \le 5$

(a) $v(t) = x'(t) = 3t^2 - 12t + 9$

$\qquad = 3(t^2 - 4t + 3) = 3(t - 1)(t - 3)$

$a(t) = v'(t) = 6t - 12 = 6(t - 2)$

(b) $v(t) > 0$ when $0 < t < 1$ or $3 < t < 5$.

(c) $a(t) = 6(t - 2) = 0$ when $t = 2$.

$v(2) = 3(1)(-1) = -3$

79. $v(t) = \dfrac{1}{\sqrt{t}} = t^{-1/2} \qquad t > 0$

$x(t) = \displaystyle\int v(t)\, dt = 2t^{1/2} + C$

$x(1) = 4 = 2(1) + C \implies C = 2$

$x(t) = 2t^{1/2} + 2$ position function

$a(t) = v'(t) = -\dfrac{1}{2}t^{-3/2} = \dfrac{-1}{2t^{3/2}}$ acceleration

81. (a) $v(0) - 25$ km/hr $= 25 \cdot \dfrac{1000}{3600} - \dfrac{250}{36}$ m/sec

$v(13) = 80$ km/hr $= 80 \cdot \dfrac{1000}{3600} = \dfrac{800}{36}$ m/sec

$a(t) = a$ (constant acceleration)

$v(t) = at + C$

$v(0) = \dfrac{250}{36} \implies v(t) = at + \dfrac{250}{36}$

$v(13) = \dfrac{800}{36} = 13a + \dfrac{250}{36}$

$\dfrac{550}{36} = 13a$

$a = \dfrac{550}{468} = \dfrac{275}{234} \approx 1.175$ m/sec^2

(b) $s(t) = a\dfrac{t^2}{2} + \dfrac{250}{36}t \quad (s(0) = 0)$

$s(13) = \dfrac{275}{234}\dfrac{(13)^2}{2} + \dfrac{250}{36}(13) \approx 189.58$ m

83. Truck: $v(t) = 30$

$\qquad s(t) = 30t$ (Let $s(0) = 0$.)

Automobile: $a(t) = 6$

$\qquad v(t) = 6t$ (Let $v(0) = 0$.)

$\qquad s(t) = 3t^2$ (Let $s(0) = 0$.)

At the point where the automobile overtakes the truck:

$30t = 3t^2$

$0 = 3t^2 - 30t$

$0 = 3t(t - 10)$ when $t = 10$ sec.

(a) $s(10) = 3(10)^2 = 300$ ft

(b) $v(10) = 6(10) = 60$ ft/sec ≈ 41 mph

85. $a(t) = k$

$v(t) = kt$

$s(t) = \dfrac{k}{2}t^2$ since $v(0) - s(0) = 0$.

At the time of lift-off, $kt = 160$ and $(k/2)t^2 = 0.7$. Since $(k/2)t^2 = 0.7$,

$t = \sqrt{\dfrac{1.4}{k}}$

$v\left(\sqrt{\dfrac{1.4}{k}}\right) = k\sqrt{\dfrac{1.4}{k}} = 160$

$1.4k = 160^2 \implies k = \dfrac{160^2}{1.4}$

$\approx 18{,}285.714$ mi/hr^2

≈ 7.45 ft/sec^2.

87. True

89. True

91. False. For example, $\displaystyle\int x \cdot x\, dx \neq \int x\, dx \cdot \int x\, dx$ because $\dfrac{x^3}{3} + C \neq \left(\dfrac{x^2}{2} + C_1\right)\left(\dfrac{x^2}{2} + C_2\right)$.

93. $f''(x) = 2x$

$f'(x) = x^2 + C$

$f'(2) = 0 \implies 4 + C = 0 \implies C = -4$

$f(x) = \dfrac{x^3}{3} - 4x + C_1$

$f(2) = 0 \implies \dfrac{8}{3} - 8 + C_1 = 0 \implies C_1 = \dfrac{16}{3}$

Answer: $f(x) = \dfrac{x^3}{3} - 4x + \dfrac{16}{3}$

95. $f'(x) = \begin{cases} 1, & 0 \le x < 2 \\ 3x, & 2 \le x \le 5 \end{cases}$

$f(x) = \begin{cases} x + C_1, & 0 \le x < 2 \\ \dfrac{3x^2}{2} + C_2, & 2 \le x \le 5 \end{cases}$

$f(1) = 3 \implies 1 + C_1 = 3 \implies C_1 = 2$

f is continuous: Values must agree at $x = 2$:

$4 = 6 + C_2 \implies C_2 = -2$

$f(x) = \begin{cases} x + 2, & 0 \le x < 2 \\ \dfrac{3x^2}{2} - 2, & 2 \le x \le 5 \end{cases}$

The left and right hand derivatives at $x = 2$ do not agree. Hence f is not differentiable at $x = 2$.

97. $f(x + y) = f(x)f(y) - g(x)g(y)$

$g(x + y) = f(x)g(y) + g(x)f(y)$

$f'(0) = 0$

[Note: $f(x) = \cos x$ and $g(x) = \sin x$ satisfy these conditions]

$f'(x + y) = f(x)f'(y) - g(x)g'(y)$ (Differentiate with respect to y)

$g'(x + y) = f(x)g'(y) + g(x)f'(y)$ (Differentiate with respect to y)

Letting $y = 0$, $f'(x) = f(x)f'(0) - g(x)g'(0) = -g(x)g'(0)$

$\qquad\qquad g'(x) = f(x)g'(0) + g(x)f'(0) = f(x)g'(0)$

Hence, $2f(x)f'(x) = -2f(x)g(x)g'(0)$

$\qquad 2g(x)g'(x) = 2g(x)f(x)g'(0)$.

Adding, $2f(x)f'(x) + 2g(x)g'(x) = 0$.

Integrating, $f(x)^2 + g(x)^2 = C$.

Clearly $C \ne 0$, for if $C = 0$, then $f(x)^2 = -g(x)^2 \implies f(x) = g(x) = 0$, which contradicts that f, g are nonconstant.

Now, $C = f(x + y)^2 + g(x + y)^2 = (f(x)f(y) - g(x)g(y))^2 + (f(x)g(y) + g(x)f(y))^2$

$\qquad\qquad\qquad\qquad = f(x)^2 f(y)^2 + g(x)^2 g(y)^2 + f(x)^2 g(y)^2 + g(x)^2 f(y)^2$

$\qquad\qquad\qquad\qquad = [f(x)^2 + g(x)^2][f(y)^2 + g(y)^2]$

$\qquad\qquad\qquad\qquad = C^2$

Thus, $C = 1$ and we have $f(x)^2 + g(x)^2 = 1$.

Section 4.2   Area

1. $\displaystyle\sum_{i=1}^{5} (2i + 1) = 2\sum_{i=1}^{5} i + \sum_{i=1}^{5} 1 = 2(1 + 2 + 3 + 4 + 5) + 5 = 35$

3. $\displaystyle\sum_{k=0}^{4} \dfrac{1}{k^2 + 1} = 1 + \dfrac{1}{2} + \dfrac{1}{5} + \dfrac{1}{10} + \dfrac{1}{17} = \dfrac{158}{85}$

5. $\displaystyle\sum_{k=1}^{4} c = c + c + c + c = 4c$

7. $\displaystyle\sum_{i=1}^{9} \dfrac{1}{3i}$

9. $\displaystyle\sum_{j=1}^{8} \left[5\left(\dfrac{j}{8}\right) + 3 \right]$

11. $\dfrac{2}{n}\displaystyle\sum_{i=1}^{n} \left[\left(\dfrac{2i}{n}\right)^3 - \left(\dfrac{2i}{n}\right) \right]$

13. $\dfrac{3}{n}\displaystyle\sum_{i=1}^{n} \left[2\left(1 + \dfrac{3i}{n}\right)^2 \right]$

15. $\displaystyle\sum_{i=1}^{20} 2i = 2\sum_{i=1}^{20} i = 2\left[\frac{20(21)}{2}\right] = 420$

17. $\displaystyle\sum_{i=1}^{20}(i-1)^2 = \sum_{i=1}^{19} i^2$

$$= \left[\frac{19(20)(39)}{6}\right] = 2470$$

19. $\displaystyle\sum_{i=1}^{15} i(i-1)^2 = \sum_{i=1}^{15} i^3 - 2\sum_{i=1}^{15} i^2 + \sum_{i=1}^{15} i$

$$= \frac{15^2(16)^2}{4} - 2\frac{15(16)(31)}{6} + \frac{15(16)}{2}$$

$$= 14{,}400 - 2480 + 120$$

$$= 12{,}040$$

21. sum seq$(x \boxed{\wedge} 2 + 3, x, 1, 20, 1) = 2930$ (*TI-82*)

$$\sum_{i=1}^{20}(i^2 + 3) = \frac{20(20 + 1)(2(20) + 1)}{6} + 3(20)$$

$$= \frac{(20)(21)(41)}{6} + 60 = 2930$$

23. $S = \left[3 + 4 + \frac{9}{2} + 5\right](1) = \frac{33}{2} = 16.5$

$s = \left[1 + 3 + 4 + \frac{9}{2}\right](1) = \frac{25}{2} = 12.5$

25. $S = [3 + 3 + 5](1) = 11$

$s = [2 + 2 + 3](1) = 7$

27. $S(4) = \sqrt{\frac{1}{4}}\left(\frac{1}{4}\right) + \sqrt{\frac{1}{2}}\left(\frac{1}{4}\right) + \sqrt{\frac{3}{4}}\left(\frac{1}{4}\right) + \sqrt{1}\left(\frac{1}{4}\right) = \frac{1 + \sqrt{2} + \sqrt{3} + 2}{8} \approx 0.768$

$s(4) = 0\left(\frac{1}{4}\right) + \sqrt{\frac{1}{4}}\left(\frac{1}{4}\right) + \sqrt{\frac{1}{2}}\left(\frac{1}{4}\right) + \sqrt{\frac{3}{4}}\left(\frac{1}{4}\right) = \frac{1 + \sqrt{2} + \sqrt{3}}{8} \approx 0.518$

29. $S(5) = 1\left(\frac{1}{5}\right) + \frac{1}{6/5}\left(\frac{1}{5}\right) + \frac{1}{7/5}\left(\frac{1}{5}\right) + \frac{1}{8/5}\left(\frac{1}{5}\right) + \frac{1}{9/5}\left(\frac{1}{5}\right) = \frac{1}{5} + \frac{1}{6} + \frac{1}{7} + \frac{1}{8} + \frac{1}{9} \approx 0.746$

$s(5) = \frac{1}{6/5}\left(\frac{1}{5}\right) + \frac{1}{7/5}\left(\frac{1}{5}\right) + \frac{1}{8/5}\left(\frac{1}{5}\right) + \frac{1}{9/5}\left(\frac{1}{5}\right) + \frac{1}{2}\left(\frac{1}{5}\right) = \frac{1}{6} + \frac{1}{7} + \frac{1}{8} + \frac{1}{9} + \frac{1}{10} \approx 0.646$

31. $\displaystyle\lim_{n\to\infty}\left[\left(\frac{81}{n^4}\right)\frac{n^2(n+1)^2}{4}\right] - \frac{81}{4}\lim_{n\to\infty}\left[\frac{n^4 + 2n^3 + n^2}{n^4}\right] - \frac{81}{4}(1) = \frac{81}{4}$

33. $\displaystyle\lim_{n\to\infty}\left[\left(\frac{18}{n^2}\right)\frac{n(n+1)}{2}\right] = \frac{18}{2}\lim_{n\to\infty}\left[\frac{n^2 + n}{n^2}\right] = \frac{18}{2}(1) = 9$

35. $\displaystyle\sum_{i=1}^{n}\frac{2i+1}{n^2} = \frac{1}{n^2}\sum_{i=1}^{n}(2i+1) = \frac{1}{n^2}\left[2\frac{n(n+1)}{2} + n\right] = \frac{n+2}{n} = 1 + \frac{2}{n} = S(n)$

$$S(10) = \frac{12}{10} = 1.2$$

$$S(100) = 1.02$$

$$S(1000) = 1.002$$

$$S(10{,}000) = 1.0002$$

37. $\displaystyle\sum_{k=1}^{n}\frac{6k(k-1)}{n^3} = \frac{6}{n^3}\sum_{k=1}^{n}(k^2 - k) = \frac{6}{n^3}\left[\frac{n(n+1)(2n+1)}{6} - \frac{n(n+1)}{2}\right]$

$$= \frac{6}{n^2}\left[\frac{2n^2 + 3n + 1 - 3n - 3}{6}\right] = \frac{1}{n^2}[2n^2 - 2] = 2 - \frac{2}{n^2} = S(n)$$

$$S(10) = 1.98$$

$$S(100) = 1.9998$$

$$S(1000) = 1.999998$$

$$S(10{,}000) = 1.99999998$$

39. $\lim\limits_{n\to\infty} \sum\limits_{i=1}^{n} \left(\dfrac{16i}{n^2}\right) = \lim\limits_{n\to\infty} \dfrac{16}{n^2} \sum\limits_{i=1}^{n} i = \lim\limits_{n\to\infty} \dfrac{16}{n^2}\left(\dfrac{n(n+1)}{2}\right) = \lim\limits_{n\to\infty} \left[8\left(\dfrac{n^2+n}{n^2}\right)\right] = 8 \lim\limits_{n\to\infty}\left(1 + \dfrac{1}{n}\right) = 8$

41. $\lim\limits_{n\to\infty} \sum\limits_{i=1}^{n} \dfrac{1}{n^3}(i-1)^2 = \lim\limits_{n\to\infty} \dfrac{1}{n^3}\sum\limits_{i=1}^{n-1} i^2 = \lim\limits_{n\to\infty} \dfrac{1}{n^3}\left[\dfrac{(n-1)(n)(2n-1)}{6}\right]$

$$= \lim\limits_{n\to\infty} \dfrac{1}{6}\left[\dfrac{2n^3 - 3n^2 + n}{n^3}\right] = \lim\limits_{n\to\infty}\left[\dfrac{1}{6}\left(\dfrac{2 - (3/n) + (1/n^2)}{1}\right)\right] = \dfrac{1}{3}$$

43. $\lim\limits_{n\to\infty} \sum\limits_{i=1}^{n}\left(1 + \dfrac{i}{n}\right)\left(\dfrac{2}{n}\right) = 2 \lim\limits_{n\to\infty} \dfrac{1}{n}\left[\sum\limits_{i=1}^{n} 1 + \dfrac{1}{n}\sum\limits_{i=1}^{n} i\right] = 2 \lim\limits_{n\to\infty} \dfrac{1}{n}\left[n + \dfrac{1}{n}\left(\dfrac{n(n+1)}{2}\right)\right] = 2 \lim\limits_{n\to\infty}\left[1 + \dfrac{n^2+n}{2n^2}\right] = 2\left(1 + \dfrac{1}{2}\right) = 3$

45. (a)

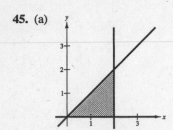

(b) $\Delta x = \dfrac{2-0}{n} = \dfrac{2}{n}$

Endpoints:

$$0 < 1\left(\dfrac{2}{n}\right) < 2\left(\dfrac{2}{n}\right) < \cdots < (n-1)\left(\dfrac{2}{n}\right) < n\left(\dfrac{2}{n}\right) = 2$$

(c) Since $y = x$ is increasing, $f(m_i) = f(x_{i-1})$ on $[x_{i-1}, x_i]$.

$$s(n) = \sum\limits_{i=1}^{n} f(x_{i-1})\,\Delta x$$

$$= \sum\limits_{i=1}^{n} f\left(\dfrac{2i-2}{n}\right)\left(\dfrac{2}{n}\right) = \sum\limits_{i=1}^{n}\left[(i-1)\left(\dfrac{2}{n}\right)\right]\left(\dfrac{2}{n}\right)$$

(d) $f(M_i) = f(x_i)$ on $[x_{i-1}, x_i]$

$$S(n) = \sum\limits_{i=1}^{n} f(x_i)\,\Delta x = \sum\limits_{i=1}^{n} f\left(\dfrac{2i}{n}\right)\dfrac{2}{n} = \sum\limits_{i=1}^{n}\left[i\left(\dfrac{2}{n}\right)\right]\left(\dfrac{2}{n}\right)$$

(e)

x	5	10	50	100
$s(n)$	1.6	1.8	1.96	1.98
$S(n)$	2.4	2.2	2.04	2.02

(f) $\lim\limits_{n\to\infty} \sum\limits_{i=1}^{n}\left[(i-1)\left(\dfrac{2}{n}\right)\right]\left(\dfrac{2}{n}\right) = \lim\limits_{n\to\infty} \dfrac{4}{n^2}\sum\limits_{i=1}^{n}(i-1)$

$$= \lim\limits_{n\to\infty} \dfrac{4}{n^2}\left[\dfrac{n(n+1)}{2} - n\right]$$

$$= \lim\limits_{n\to\infty}\left[\dfrac{2(n+1)}{n} - \dfrac{4}{n}\right] = 2$$

$\lim\limits_{n\to\infty} \sum\limits_{i=1}^{n}\left[i\left(\dfrac{2}{n}\right)\right]\left(\dfrac{2}{n}\right) = \lim\limits_{n\to\infty} \dfrac{4}{n^2}\sum\limits_{i=1}^{n} i$

$$= \lim\limits_{n\to\infty}\left(\dfrac{4}{n^2}\right)\dfrac{n(n+1)}{2}$$

$$= \lim\limits_{n\to\infty} \dfrac{2(n+1)}{n} = 2$$

47. $y = -2x + 3$ on $[0, 1]$. $\left(\text{Note: } \Delta x = \dfrac{1-0}{n} = \dfrac{1}{n}\right)$

$$s(n) = \sum\limits_{i=1}^{n} f\left(\dfrac{i}{n}\right)\left(\dfrac{1}{n}\right) = \sum\limits_{i=1}^{n}\left[-2\left(\dfrac{i}{n}\right) + 3\right]\left(\dfrac{1}{n}\right)$$

$$= 3 - \dfrac{2}{n^2}\sum\limits_{i=1}^{n} i = 3 - \dfrac{2(n+1)n}{2n^2} = 2 - \dfrac{1}{n}$$

Area $= \lim\limits_{n\to\infty} s(n) = 2$

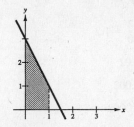

49. $y = x^2 + 2$ on $[0, 1]$. $\left(\text{Note: } \Delta x = \dfrac{1}{n}\right)$

$$S(n) = \sum\limits_{i=1}^{n} f\left(\dfrac{i}{n}\right)\left(\dfrac{1}{n}\right) = \sum\limits_{i=1}^{n}\left[\left(\dfrac{i}{n}\right)^2 + 2\right]\left(\dfrac{1}{n}\right)$$

$$= \left[\dfrac{1}{n^3}\sum\limits_{i=1}^{n} i^2\right] + 2 = \dfrac{n(n+1)(2n+1)}{6n^3} + 2 = \dfrac{1}{6}\left(2 + \dfrac{3}{n} + \dfrac{1}{n^2}\right) + 2$$

Area $= \lim\limits_{n\to\infty} S(n) = \dfrac{7}{3}$

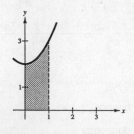

51. $y = 16 - x^2$ on $[1, 3]$. $\left(\textbf{Note: } \Delta x = \dfrac{2}{n}\right)$

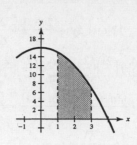

$$s(n) = \sum_{i=1}^{n} f\left(1 + \frac{2i}{n}\right)\left(\frac{2}{n}\right) = \sum_{i=1}^{n}\left[16 - \left(1 + \frac{2i}{n}\right)^2\right]\left(\frac{2}{n}\right)$$

$$= \frac{2}{n}\sum_{i=1}^{n}\left[15 - \frac{4i^2}{n^2} - \frac{4i}{n}\right]$$

$$= \frac{2}{n}\left[15n - \frac{4}{n^2}\frac{n(n+1)(2n+1)}{6} - \frac{4}{n}\frac{n(n+1)}{2}\right]$$

$$= 30 - \frac{8}{6n^2}(n+1)(2n+1) - \frac{4}{n}(n+1)$$

$$\text{Area} = \lim_{n \to \infty} s(n) = 30 - \frac{8}{3} - 4 = \frac{70}{3} = 23\frac{1}{3}$$

53. $y = 64 - x^3$ on $[1, 4]$. $\left(\textbf{Note: } \Delta x = \dfrac{4-1}{n} = \dfrac{3}{n}\right)$

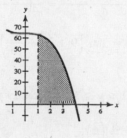

$$s(n) = \sum_{i=1}^{n} f\left(1 + \frac{3i}{n}\right)\left(\frac{3}{n}\right) = \sum_{i=1}^{n}\left[64 - \left(1 + \frac{3i}{n}\right)^3\right]\left(\frac{3}{n}\right)$$

$$= \frac{3}{n}\sum_{i=1}^{n}\left[63 - \frac{27i^3}{n^3} - \frac{27i^2}{n^2} - \frac{9i}{n}\right]$$

$$= \frac{3}{n}\left[63n - \frac{27}{n^3}\frac{n^2(n+1)^2}{4} - \frac{27}{n^2}\frac{n(n+1)(2n+1)}{6} - \frac{9}{n}\frac{n(n+1)}{2}\right]$$

$$= 189 - \frac{81}{4n^2}(n+1)^2 - \frac{81}{6n^2}(n+1)(2n+1) - \frac{27}{2}\frac{n+1}{n}$$

$$\text{Area} = \lim_{n \to \infty} s(n) = 189 - \frac{81}{4} - 27 - \frac{27}{2} = \frac{513}{4} = 128.25$$

55. $y = x^2 - x^3$ on $[-1, 1]$. $\left(\textbf{Note: } \Delta x = \dfrac{1-(-1)}{n} = \dfrac{2}{n}\right)$

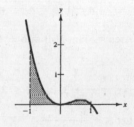

Again, $T(n)$ is neither an upper nor a lower sum.

$$T(n) = \sum_{i=1}^{n} f\left(-1 + \frac{2i}{n}\right)\left(\frac{2}{n}\right) = \sum_{i=1}^{n}\left[\left(-1 + \frac{2i}{n}\right)^2 - \left(-1 + \frac{2i}{n}\right)^3\right]\left(\frac{2}{n}\right)$$

$$= \sum_{i=1}^{n}\left[\left(1 - \frac{4i}{n} + \frac{4i^2}{n^2}\right) - \left(-1 + \frac{6i}{n} - \frac{12i^2}{n^2} + \frac{8i^3}{n^3}\right)\right]\left(\frac{2}{n}\right)$$

$$= \sum_{i=1}^{n}\left[2 - \frac{10i}{n} + \frac{16i^2}{n^2} - \frac{8i^3}{n^3}\right]\left(\frac{2}{n}\right) = \frac{4}{n}\sum_{i=1}^{n}1 - \frac{20}{n^2}\sum_{i=1}^{n}i + \frac{32}{n^3}\sum_{i=1}^{n}i^2 - \frac{16}{n^4}\sum_{i=1}^{n}i^3$$

$$= \frac{4}{n}(n) - \frac{20}{n^2}\cdot\frac{n(n+1)}{2} + \frac{32}{n^3}\cdot\frac{n(n+1)(2n+1)}{6} - \frac{16}{n^4}\cdot\frac{n^2(n+1)^2}{4}$$

$$= 4 - 10\left(1 + \frac{1}{n}\right) + \frac{16}{3}\left(2 + \frac{3}{n} + \frac{1}{n^2}\right) - 4\left(1 + \frac{2}{n} + \frac{1}{n^2}\right)$$

$$\text{Area} = \lim_{n \to \infty} T(n) = 4 - 10 + \frac{32}{3} - 4 = \frac{2}{3}$$

57. $f(y) = 3y, 0 \le y \le 2$ $\left(\text{Note: } \Delta y = \dfrac{2 - 0}{n} = \dfrac{2}{n}\right)$

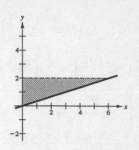

$$S(n) = \sum_{i=1}^{n} f(m_i)\,\Delta y = \sum_{i=1}^{n} f\left(\frac{2i}{n}\right)\left(\frac{2}{n}\right) = \sum_{i=1}^{n} 3\left(\frac{2i}{n}\right)\left(\frac{2}{n}\right)$$

$$= \frac{12}{n^2}\sum_{i=1}^{n} i = \left(\frac{12}{n^2}\right)\cdot\frac{n(n+1)}{2} = \frac{6(n+1)}{n} = 6 + \frac{6}{n}$$

$$\text{Area} = \lim_{n\to\infty} S(n) = \lim_{n\to\infty}\left(6 + \frac{6}{n}\right) = 6$$

59. $f(y) = y^2, 0 \le y \le 3$ $\left(\text{Note: } \Delta y = \dfrac{3 - 0}{n} = \dfrac{3}{n}\right)$

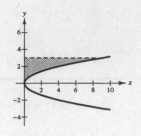

$$S(n) = \sum_{i=1}^{n} f\left(\frac{3i}{n}\right)\left(\frac{3}{n}\right) = \sum_{i=1}^{n}\left(\frac{3i}{n}\right)^2\left(\frac{3}{n}\right) = \frac{27}{n^3}\sum_{i=1}^{n} i^2$$

$$= \frac{27}{n^3}\cdot\frac{n(n+1)(2n+1)}{6} = \frac{9}{n^2}\left(\frac{2n^2 + 3n + 1}{2}\right) = 9 + \frac{27}{2n} + \frac{9}{2n^2}$$

$$\text{Area} = \lim_{n\to\infty} S(n) = \lim_{n\to\infty}\left(9 + \frac{27}{2n} + \frac{9}{2n^2}\right) = 9$$

61. $g(y) = 4y^2 - y^3, 1 \le y \le 3.$ $\left(\text{Note: } \Delta y = \dfrac{3 - 1}{n} = \dfrac{2}{n}\right)$

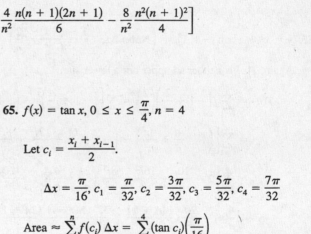

$$S(n) = \sum_{i=1}^{n} g\left(1 + \frac{2i}{n}\right)\left(\frac{2}{n}\right)$$

$$= \sum_{i=1}^{n}\left[4\left(1 + \frac{2i}{n}\right)^2 - \left(1 + \frac{2i}{n}\right)^3\right]\frac{2}{n}$$

$$= \frac{2}{n}\sum_{i=1}^{n} 4\left[1 + \frac{4i}{n} + \frac{4i^2}{n^2}\right] - \left[1 + \frac{6i}{n} + \frac{12i^2}{n^2} + \frac{8i^3}{n^3}\right]$$

$$= \frac{2}{n}\sum_{i=1}^{n}\left[3 + \frac{10i}{n} + \frac{4i^2}{n^2} - \frac{8i^3}{n^3}\right] = \frac{2}{n}\left[3n + \frac{10}{n}\frac{n(n+1)}{2} + \frac{4}{n^2}\frac{n(n+1)(2n+1)}{6} - \frac{8}{n^2}\frac{n^2(n+1)^2}{4}\right]$$

$$\text{Area} = \lim_{n\to\infty} S(n) = 6 + 10 + \frac{8}{3} - 4 = \frac{44}{3}$$

63. $f(x) = x^2 + 3, 0 \le x \le 2, n = 4$

Let $c_i = \dfrac{x_i + x_{i-1}}{2}$.

$\Delta x = \dfrac{1}{2}, c_1 = \dfrac{1}{4}, c_2 = \dfrac{3}{4}, c_3 = \dfrac{5}{4}, c_4 = \dfrac{7}{4}$

$$\text{Area} \approx \sum_{i=1}^{n} f(c_i)\,\Delta x = \sum_{i=1}^{4}[c_i^2 + 3]\left(\frac{1}{2}\right)$$

$$= \frac{1}{2}\left[\left(\frac{1}{16} + 3\right) + \left(\frac{9}{16} + 3\right) + \left(\frac{25}{16} + 3\right) + \left(\frac{49}{16} + 3\right)\right]$$

$$= \frac{69}{8}$$

65. $f(x) = \tan x, 0 \le x \le \dfrac{\pi}{4}, n = 4$

Let $c_i = \dfrac{x_i + x_{i-1}}{2}$.

$\Delta x = \dfrac{\pi}{16}, c_1 = \dfrac{\pi}{32}, c_2 = \dfrac{3\pi}{32}, c_3 = \dfrac{5\pi}{32}, c_4 = \dfrac{7\pi}{32}$

$$\text{Area} \approx \sum_{i=1}^{n} f(c_i)\,\Delta x = \sum_{i=1}^{4}(\tan c_i)\left(\frac{\pi}{16}\right)$$

$$= \frac{\pi}{16}\left(\tan\frac{\pi}{32} + \tan\frac{3\pi}{32} + \tan\frac{5\pi}{32} + \tan\frac{7\pi}{32}\right) \approx 0.345$$

67. $f(x) = \sqrt{x}$ on $[0, 4]$.

n	4	8	12	16	20
Approximate area	5.3838	5.3523	5.3439	5.3403	5.3384

(Exact value is 16/3.)

69. $f(x) = \tan\left(\dfrac{\pi x}{8}\right)$ on $[1, 3]$.

n	4	8	12	16	20
Approximate area	2.2223	2.2387	2.2418	2.2430	2.2435

71.

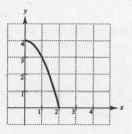

(b) $A \approx 6$ square units

73. We can use the line $y = x$ bounded by $x = a$ and $x = b$. The sum of the areas of these inscribed rectangles is the lower sum.

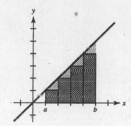

The sum of the areas of these circumscribed rectangles is the upper sum.

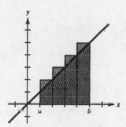

We can see that the rectangles do not contain all of the area in the first graph and the rectangles in the second graph cover more than the area of the region.

The exact value of the area lies between these two sums.

75. (a)

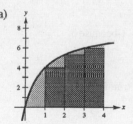

Lower sum:

$s(4) = 0 + 4 + 5\frac{1}{3} + 6 = 15\frac{1}{3} = \frac{46}{3} \approx 15.333$

(b)

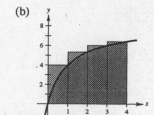

Upper sum:

$S(4) = 4 + 5\frac{1}{3} + 6 + 6\frac{2}{5} = 21\frac{11}{15} = \frac{326}{15} \approx 21.733$

(c)

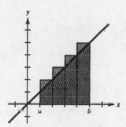

Midpoint Rule:

$M(4) = 2\frac{2}{3} + 4\frac{4}{5} + 5\frac{5}{7} + 6\frac{2}{9} = \frac{6112}{315} \approx 19.403$

(d) In each case, $\Delta x = 4/n$. The lower sum uses left endpoints, $(i - 1)(4/n)$. The upper sum uses right endpoints, $(i)(4/n)$. The Midpoint Rule uses midpoints, $\left(i - \frac{1}{2}\right)(4/n)$.

—CONTINUED—

75. —CONTINUED—

(e)

n	4	8	20	100	200
$s(n)$	15.333	17.368	18.459	18.995	19.06
$S(n)$	21.733	20.568	19.739	19.251	19.188
$M(n)$	19.403	19.201	19.137	19.125	19.125

(f) $s(n)$ increases because the lower sum approaches the exact value as n increases. $S(n)$ decreases because the upper sum approaches the exact value as n increases. Because of the shape of the graph, the lower sum is always smaller than the exact value, whereas the upper sum is always larger.

77. True. (Theorem 4.2 (2))

79. Suppose there are n rows and $n + 1$ columns in the figure. The stars on the left total $1 + 2 + \cdots + n$, as do the stars on the right. There are $n(n + 1)$ stars in total, hence

$$2[1 + 2 + \cdots + n] = n(n + 1)$$

$$1 + 2 + \cdots + n = \tfrac{1}{2}(n)(n + 1).$$

81. (a) $y = (-4.09 \times 10^{-5})x^3 + 0.016x^2 - 2.67x + 452.9$

(b)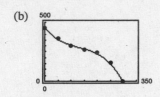

(c) Using the integration capability of a graphing utility, you obtain

$$A \approx 76,897.5 \text{ ft}^2.$$

83. (a) $\displaystyle\sum_{i=1}^{n} 2i = n(n + 1)$

The formula is true for $n = 1$: $2 = 1(1 + 1) = 2$.

Assume that the formula is true for $n = k$:

$$\sum_{i=1}^{k} 2i = k(k + 1).$$

Then we have $\displaystyle\sum_{i=1}^{k+1} 2i = \sum_{i=1}^{k} 2i + 2(k + 1)$

$$= k(k + 1) + 2(k + 1)$$

$$= (k + 1)(k + 2)$$

which shows that the formula is true for $n = k + 1$.

(b) $\displaystyle\sum_{i=1}^{n} i^3 = \frac{n^2(n + 1)^2}{4}$

The formula is true for $n = 1$ because

$$1^3 = \frac{1^2(1 + 1)^2}{4} = \frac{4}{4} = 1.$$

Assume that the formula is true for $n = k$:

$$\sum_{i=1}^{k} i^3 = \frac{k^2(k + 1)^2}{4}.$$

Then we have $\displaystyle\sum_{i=1}^{k+1} i^3 = \sum_{i=1}^{k} i^3 + (k + 1)^3$

$$= \frac{k^2(k + 1)^2}{4} + (k + 1)^3$$

$$= \frac{(k + 1)^2}{4}[k^2 + 4(k + 1)]$$

$$= \frac{(k + 1)^2}{4}(k + 2)^2$$

which shows that the formula is true for $n = k + 1$.

Section 4.3   Riemann Sums and Definite Integrals

1. $f(x) = \sqrt{x}, y = 0, x = 0, x = 3, c_i = \dfrac{3i^2}{n^2}$

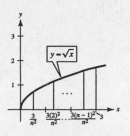

$$\Delta x_i = \frac{3i^2}{n^2} - \frac{3(i-1)^2}{n^2} = \frac{3}{n^2}(2i-1)$$

$$\lim_{n \to \infty} \sum_{i=1}^{n} f(c_i)\,\Delta x_i = \lim_{n \to \infty} \sum_{i=1}^{n} \sqrt{\frac{3i^2}{n^2}}\,\frac{3}{n^2}(2i-1)$$

$$= \lim_{n \to \infty} \frac{3\sqrt{3}}{n^3} \sum_{i=1}^{n} (2i^2 - i)$$

$$= \lim_{n \to \infty} \frac{3\sqrt{3}}{n^3}\left[2\frac{n(n+1)(2n+1)}{6} - \frac{n(n+1)}{2}\right]$$

$$= \lim_{n \to \infty} 3\sqrt{3}\left[\frac{(n+1)(2n+1)}{3n^2} - \frac{n+1}{2n^2}\right]$$

$$= 3\sqrt{3}\left[\frac{2}{3} - 0\right] = 2\sqrt{3} \approx 3.464$$

3. $y = 6$ on $[4, 10]$. $\left(\text{Note: } \Delta x = \dfrac{10-4}{n} = \dfrac{6}{n}, \|\Delta\| \to 0 \text{ as } n \to \infty\right)$

$$\sum_{i=1}^{n} f(c_i)\,\Delta x_i = \sum_{i=1}^{n} f\left(4 + \frac{6i}{n}\right)\left(\frac{6}{n}\right) = \sum_{i=1}^{n} 6\left(\frac{6}{n}\right) = \sum_{i=1}^{n} \frac{36}{n} = \frac{1}{n}\sum_{i=1}^{n} 36 = \frac{1}{n}(36n) = 36$$

$$\int_{4}^{10} 6\,dx = \lim_{n \to \infty} 36 = 36$$

5. $y = x^3$ on $[-1, 1]$. $\left(\text{Note: } \Delta x = \dfrac{1-(-1)}{n} = \dfrac{2}{n}, \|\Delta\| \to 0 \text{ as } n \to \infty\right)$

$$\sum_{i=1}^{n} f(c_i)\,\Delta x_i = \sum_{i=1}^{n} f\left(-1 + \frac{2i}{n}\right)\left(\frac{2}{n}\right) = \sum_{i=1}^{n}\left(-1 + \frac{2i}{n}\right)^3\left(\frac{2}{n}\right) = \sum_{i=1}^{n}\left[-1 + \frac{6i}{n} - \frac{12i^2}{n^2} + \frac{8i^3}{n^3}\right]\left(\frac{2}{n}\right)$$

$$= -2 + \frac{12}{n^2}\sum_{i=1}^{n} i \quad \frac{24}{n^3}\sum_{i=1}^{n} i^2 + \frac{16}{n^4}\sum_{i=1}^{n} i^3$$

$$= -2 + 6\left(1 + \frac{1}{n}\right) - 4\left(2 + \frac{3}{n} + \frac{1}{n^2}\right) + 4\left(1 + \frac{2}{n} + \frac{1}{n^2}\right) = \frac{2}{n}$$

$$\int_{-1}^{1} x^3\,dx = \lim_{n \to \infty} \frac{2}{n} = 0$$

7. $y = x^2 + 1$ on $[1, 2]$. $\left(\text{Note: } \Delta x = \dfrac{2-1}{n} = \dfrac{1}{n}, \|\Delta\| \to 0 \text{ as } n \to \infty\right)$

$$\sum_{i=1}^{n} f(c_i)\,\Delta x_i = \sum_{i=1}^{n} f\left(1 + \frac{i}{n}\right)\left(\frac{1}{n}\right) = \sum_{i=1}^{n}\left[\left(1 + \frac{i}{n}\right)^2 + 1\right]\left(\frac{1}{n}\right) = \sum_{i=1}^{n}\left[1 + \frac{2i}{n} + \frac{i^2}{n^2} + 1\right]\left(\frac{1}{n}\right)$$

$$= 2 + \frac{2}{n^2}\sum_{i=1}^{n} i + \frac{1}{n^3}\sum_{i=1}^{n} i^2 = 2 + \left(1 + \frac{1}{n}\right) + \frac{1}{6}\left(2 + \frac{3}{n} + \frac{1}{n^2}\right) = \frac{10}{3} + \frac{3}{2n} + \frac{1}{6n^2}$$

$$\int_{1}^{2} (x^2 + 1)\,dx = \lim_{n \to \infty}\left(\frac{10}{3} + \frac{3}{2n} + \frac{1}{6n^2}\right) = \frac{10}{3}$$

9. $\lim\limits_{\|\Delta I\|\to 0} \sum\limits_{i=1}^{n} (3c_i + 10)\,\Delta x_i = \int_{-1}^{5}(3x+10)\,dx$

on the interval $[-1, 5]$.

11. $\lim\limits_{\|\Delta I\|\to 0} \sum\limits_{i=1}^{n} \sqrt{c_i^2 + 4}\,\Delta x_i = \int_{0}^{3}\sqrt{x^2+4}\,dx$

on the interval $[0, 3]$.

13. $\int_{0}^{5} 3\,dx$

15. $\int_{-4}^{4}(4 - |x|)\,dx$

17. $\int_{-2}^{2}(4 - x^2)\,dx$

19. $\int_{0}^{\pi} \sin x\,dx$

21. $\int_{0}^{2} y^3\,dy$

23. Rectangle

$A = bh = 3(4)$

$A = \int_{0}^{3} 4\,dx = 12$

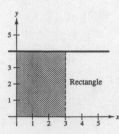

25. Triangle

$A = \dfrac{1}{2}bh = \dfrac{1}{2}(4)(4)$

$A = \int_{0}^{4} x\,dx = 8$

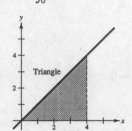

27. Trapezoid

$A = \dfrac{b_1 + b_2}{2}h = \left(\dfrac{5+9}{2}\right)2$

$A = \int_{0}^{2}(2x+5)\,dx = 14$

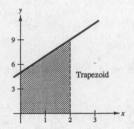

29. Triangle

$A = \dfrac{1}{2}bh = \dfrac{1}{2}(2)(1)$

$A = \int_{-1}^{1}(1 - |x|)\,dx = 1$

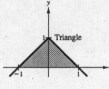

31. Semicircle

$A = \dfrac{1}{2}\pi r^2 = \dfrac{1}{2}\pi(3)^2$

$A = \int_{-3}^{3}\sqrt{9 - x^2}\,dx = \dfrac{9\pi}{2}$

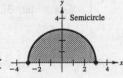

In Exercises 33–40, $\int_{2}^{4} x^3\,dx = 60,\ \int_{2}^{4} x\,dx = 6,\ \int_{2}^{4} dx = 2$

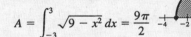

33. $\int_{4}^{2} x\,dx = -\int_{2}^{4} x\,dx = -6$

35. $\int_{2}^{4} 4x\,dx = 4\int_{2}^{4} x\,dx = 4(6) = 24$

37. $\int_{2}^{4}(x-8)\,dx = \int_{2}^{4} x\,dx - 8\int_{2}^{4} dx = 6 - 8(2) = -10$

39. $\int_{2}^{4}\left(\dfrac{1}{2}x^3 - 3x + 2\right)dx = \dfrac{1}{2}\int_{2}^{4} x^3\,dx - 3\int_{2}^{4} x\,dx + 2\int_{2}^{4} dx$

$= \dfrac{1}{2}(60) - 3(6) + 2(2) = 16$

41. (a) $\int_{0}^{7} f(x)\,dx = \int_{0}^{5} f(x)\,dx + \int_{5}^{7} f(x)\,dx = 10 + 3 = 13$

(b) $\int_{5}^{0} f(x)\,dx = -\int_{0}^{5} f(x)\,dx = -10$

(c) $\int_{5}^{5} f(x)\,dx = 0$

(d) $\int_{0}^{5} 3f(x)\,dx = 3\int_{0}^{5} f(x)\,dx = 3(10) = 30$

43. (a) $\int_2^6 [f(x) + g(x)] \, dx = \int_2^6 f(x) \, dx + \int_2^6 g(x) \, dx$

$$= 10 + (-2) = 8$$

(b) $\int_2^6 [g(x) - f(x)] \, dx = \int_2^6 g(x) \, dx - \int_2^6 f(x) \, dx$

$$= -2 - 10 = -12$$

(c) $\int_2^6 2g(x) \, dx = 2 \int_2^6 g(x) \, dx = 2(-2) = -4$

(d) $\int_2^6 3f(x) \, dx = 3 \int_2^6 f(x) \, dx = 3(10) = 30$

45. Lower estimate: $[24 + 12 - 4 - 20 - 36](2) = -48$

Upper estimate: $[32 + 24 + 12 - 4 - 20](2) = 88$

47. (a) Quarter circle below x-axis: $-\frac{1}{4}\pi r^2 = -\frac{1}{4}\pi(2)^2 = -\pi$

(b) Triangle: $\frac{1}{2}bh = \frac{1}{2}(4)(2) = 4$

(c) Triangle + Semicircle below x-axis: $-\frac{1}{2}(2)(1) - \frac{1}{2}\pi(2)^2 = -(1 + 2\pi)$

(d) Sum of parts (b) and (c): $4 - (1 + 2\pi) = 3 - 2\pi$

(e) Sum of absolute values of (b) and (c): $4 + (1 + 2\pi) = 5 + 2\pi$

(f) Answer to (d) plus $2(10) = 20$: $(3 - 2\pi) + 20 = 23 - 2\pi$

49. (a) $\int_0^5 [f(x) + 2] \, dx = \int_0^5 f(x) \, dx + \int_0^5 2 \, dx = 4 + 10 = 14$

(b) $\int_{-2}^3 f(x + 2) \, dx = \int_0^5 f(x) \, dx = 4$ (Let $u = x + 2$.)

(c) $\int_{-5}^5 f(x) \, dx = 2 \int_0^5 f(x) \, dx = 2(4) = 8$ (f even)

(d) $\int_{-5}^5 f(x) \, dx = 0$ (f odd)

51. The left endpoint approximation will be greater than the actual area: >

53. $f(x) = \dfrac{1}{x - 4}$

is not integrable on the interval $[3, 5]$ because f has a discontinuity at $x = 4$.

55.

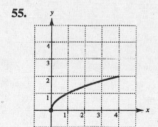

(a) $A \approx 5$ square units

57.

(d) $\int_0^1 2 \sin \pi x \, dx \approx \frac{1}{2}(1)(2) \approx 1$

59. $\int_0^3 x\sqrt{3 - x} \, dx$

n	4	8	12	16	20
$L(n)$	3.6830	3.9956	4.0707	4.1016	4.1177
$M(n)$	4.3082	4.2076	4.1838	4.1740	4.1690
$R(n)$	3.6830	3.9956	4.0707	4.1016	4.1177

61. $\int_0^{\pi/2} \sin^2 x \, dx$

n	4	8	12	16	20
$L(n)$	0.5890	0.6872	0.7199	0.7363	0.7461
$M(n)$	0.7854	0.7854	0.7854	0.7854	0.7854
$R(n)$	0.9817	0.8836	0.8508	0.8345	0.8247

63. True

65. True

67. False

$$\int_0^2 (-x) \, dx = -2$$

69. $f(x) = x^2 + 3x, \, [0, 8]$

$x_0 = 0, x_1 = 1, x_2 = 3, x_3 = 7, x_4 = 8$

$\Delta x_1 = 1, \Delta x_2 = 2, \Delta x_3 = 4, \Delta x_4 = 1$

$c_1 = 1, c_2 = 2, c_3 = 5, c_4 = 8$

$$\sum_{i=1}^4 f(c_i) \, \Delta x = f(1) \, \Delta x_1 + f(2) \, \Delta x_2 + f(5) \, \Delta x_3 + f(8) \, \Delta x_4$$

$$= (4)(1) + (10)(2) + (40)(4) + (88)(1) = 272$$

71. $\Delta x = \dfrac{b-a}{n}, c_i = a + i(\Delta x) = a + i\left(\dfrac{b-a}{n}\right)$

$$\int_a^b x \, dx = \lim_{\|\Delta\| \to 0} \sum_{i=1}^n f(c_i) \, \Delta x$$

$$= \lim_{n \to \infty} \sum_{i=1}^n \left[a + i\left(\frac{b-a}{n}\right)\right]\left(\frac{b-a}{n}\right)$$

$$= \lim_{n \to \infty} \left[\left(\frac{b-a}{n}\right)\sum_{i=1}^n a + \left(\frac{b-a}{n}\right)^2 \sum_{i=1}^n i\right]$$

$$= \lim_{n \to \infty} \left[\frac{b-a}{n}(an) + \left(\frac{b-a}{n}\right)^2 \frac{n(n+1)}{2}\right]$$

$$= \lim_{n \to \infty} \left[a(b-a) + \frac{(b-a)^2}{n}\frac{n+1}{2}\right]$$

$$= a(b-a) + \frac{(b-a)^2}{2}$$

$$= (b-a)\left[a + \frac{b-a}{2}\right]$$

$$= \frac{(b-a)(a+b)}{2} = \frac{b^2 - a^2}{2}$$

73. $f(x) = \begin{cases} 1, & x \text{ is rational} \\ 0, & x \text{ is irrational} \end{cases}$

is not integrable on the interval $[0, 1]$. As $\|\Delta\| \to 0$, $f(c_i) = 1$ or $f(c_i) = 0$ in each subinterval since there are an infinite number of both rational and irrational numbers in any interval, no matter how small.

75. The function f is nonnegative between $x = -1$ and $x = 1$.

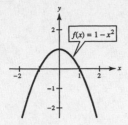

Hence,

$$\int_a^b (1 - x^2)\, dx$$

is a maximum for $a = -1$ and $b = 1$.

77. Let $f(x) = x^2, 0 \leq x \leq 1$, and $\Delta x_i = 1/n$. The appropriate Riemann Sum is

$$\sum_{i=1}^{n} f(c_i)\, \Delta x_i = \sum_{i=1}^{n} \left(\frac{i}{n}\right)^2 \frac{1}{n} = \frac{1}{n^3} \sum_{i=1}^{n} i^2.$$

$$\lim_{n \to \infty} \frac{1}{n^3}[1^2 + 2^2 + 3^2 + \cdots + n^2] = \lim_{n \to \infty} \frac{1}{n^3} \cdot \frac{n(2n + 1)(n + 1)}{6}$$

$$= \lim_{n \to \infty} \frac{2n^2 + 3n + 1}{6n^2} = \lim_{n \to \infty} \left(\frac{1}{3} + \frac{1}{2n} + \frac{1}{6n^2}\right) = \frac{1}{3}$$

Section 4.4   The Fundamental Theorem of Calculus

1. $f(x) = \dfrac{4}{x^2 + 1}$

$\displaystyle\int_0^\pi \frac{4}{x^2 + 1}\, dx$ is positive.

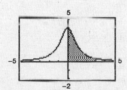

3. $f(x) = x\sqrt{x^2 + 1}$

$\displaystyle\int_{-2}^{2} x\sqrt{x^2 + 1}\, dx = 0$

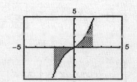

5. $\displaystyle\int_0^1 2x\, dx = \left[x^2\right]_0^1 = 1 - 0 = 1$

7. $\displaystyle\int_{-1}^0 (x - 2)\, dx = \left[\frac{x^2}{2} - 2x\right]_{-1}^0 = 0 - \left(\frac{1}{2} + 2\right) = -\frac{5}{2}$

9. $\displaystyle\int_{-1}^1 (t^2 - 2)\, dt = \left[\frac{t^3}{3} - 2t\right]_{-1}^1 = \left(\frac{1}{3} - 2\right) - \left(-\frac{1}{3} + 2\right) = -\frac{10}{3}$

11. $\displaystyle\int_0^1 (2t - 1)^2\, dt = \int_0^1 (4t^2 - 4t + 1)\, dt = \left[\frac{4}{3}t^3 - 2t^2 + t\right]_0^1 = \frac{4}{3} - 2 + 1 = \frac{1}{3}$

13. $\displaystyle\int_1^2 \left(\frac{3}{x^2} - 1\right)\, dx = \left[-\frac{3}{x} - x\right]_1^2 = \left(-\frac{3}{2} - 2\right) - (-3 - 1) = \frac{1}{2}$

15. $\displaystyle\int_1^4 \frac{u - 2}{\sqrt{u}}\, du = \int_1^4 (u^{1/2} - 2u^{-1/2})\, du = \left[\frac{2}{3}u^{3/2} - 4u^{1/2}\right]_1^4 = \left[\frac{2}{3}(\sqrt{4})^3 - 4\sqrt{4}\right] - \left[\frac{2}{3} - 4\right] = \frac{2}{3}$

17. $\displaystyle\int_{-1}^{1} \left(\sqrt[3]{t} - 2\right) dt = \left[\frac{3}{4}t^{4/3} - 2t\right]_{-1}^{1} = \left(\frac{3}{4} - 2\right) - \left(\frac{3}{4} + 2\right) = -4$

19. $\displaystyle\int_{0}^{1} \frac{x - \sqrt{x}}{3} \, dx = \frac{1}{3}\int_{0}^{1} \left(x - x^{1/2}\right) dx = \frac{1}{3}\left[\frac{x^2}{2} - \frac{2}{3}x^{3/2}\right]_{0}^{1} = \frac{1}{3}\left(\frac{1}{2} - \frac{2}{3}\right) = -\frac{1}{18}$

21. $\displaystyle\int_{-1}^{0} \left(t^{1/3} - t^{2/3}\right) dt = \left[\frac{3}{4}t^{4/3} - \frac{3}{5}t^{5/3}\right]_{-1}^{0} = 0 - \left(\frac{3}{4} + \frac{3}{5}\right) = -\frac{27}{20}$

23. $\displaystyle\int_{0}^{3} |2x - 3| \, dx = \int_{0}^{3/2} (3 - 2x) \, dx + \int_{3/2}^{3} (2x - 3) \, dx \;\left(\text{split up the integral at the zero } x = \frac{3}{2}\right)$

$$= \left[3x - x^2\right]_{0}^{3/2} + \left[x^2 - 3x\right]_{3/2}^{3} = \left(\frac{9}{2} - \frac{9}{4}\right) - 0 + (9 - 9) - \left(\frac{9}{4} - \frac{9}{2}\right) = 2\left(\frac{9}{2} - \frac{9}{4}\right) = \frac{9}{2}$$

25. $\displaystyle\int_{0}^{3} |x^2 - 4| \, dx = \int_{0}^{2} (4 - x^2) \, dx + \int_{2}^{3} (x^2 - 4) \, dx$

$$= \left[4x - \frac{x^3}{3}\right]_{0}^{2} + \left[\frac{x^3}{3} - 4x\right]_{2}^{3}$$

$$= \left(8 - \frac{8}{3}\right) + (9 - 12) - \left(\frac{8}{3} - 8\right)$$

$$= \frac{23}{3}$$

27. $\displaystyle\int_{0}^{\pi} (1 + \sin x) \, dx = \left[x - \cos x\right]_{0}^{\pi} = (\pi + 1) - (0 - 1) = 2 + \pi$

29. $\displaystyle\int_{-\pi/6}^{\pi/6} \sec^2 x \, dx = \left[\tan x\right]_{-\pi/6}^{\pi/6} = \frac{\sqrt{3}}{3} - \left(-\frac{\sqrt{3}}{3}\right) = \frac{2\sqrt{3}}{3}$

31. $\displaystyle\int_{-\pi/3}^{\pi/3} 4 \sec \theta \tan \theta \, d\theta = \left[4 \sec \theta\right]_{-\pi/3}^{\pi/3} = 4(2) - 4(2) = 0$ **33.** $A = \displaystyle\int_{0}^{1} (x - x^2) \, dx = \left[\frac{x^2}{2} - \frac{x^3}{3}\right]_{0}^{1} = \frac{1}{6}$

35. $A = \displaystyle\int_{0}^{3} (3 - x)\sqrt{x} \, dx = \int_{0}^{3} \left(3x^{1/2} - x^{3/2}\right) dx = \left[2x^{3/2} - \frac{2}{5}x^{5/2}\right]_{0}^{3} = \left[\frac{x\sqrt{x}}{5}(10 - 2x)\right]_{0}^{3} = \frac{12\sqrt{3}}{5}$

37. $A = \displaystyle\int_{0}^{\pi/2} \cos x \, dx = \left[\sin x\right]_{0}^{\pi/2} = 1$

39. Since $y \geq 0$ on $[0, 2]$,

$$A = \int_{0}^{2} (3x^2 + 1) \, dx = \left[x^3 + x\right]_{0}^{2} = 8 + 2 = 10.$$

41. Since $y \geq 0$ on $[0, 2]$,

$$A = \int_{0}^{2} (x^3 + x) \, dx = \left[\frac{x^4}{4} + \frac{x^2}{2}\right]_{0}^{2} = 4 + 2 = 6.$$

43. $\int_0^2 \left(x - 2\sqrt{x}\right) dx = \left[\dfrac{x^2}{2} - \dfrac{4x^{3/2}}{3}\right]_0^2 = 2 - \dfrac{8\sqrt{2}}{3}$

$$f(c)(2 - 0) = \dfrac{6 - 8\sqrt{2}}{3}$$

$$c - 2\sqrt{c} = \dfrac{3 - 4\sqrt{2}}{3}$$

$$c - 2\sqrt{c} + 1 = \dfrac{3 - 4\sqrt{2}}{3} + 1$$

$$\left(\sqrt{c} - 1\right)^2 = \dfrac{6 - 4\sqrt{2}}{3}$$

$$\sqrt{c} - 1 = \pm\sqrt{\dfrac{6 - 4\sqrt{2}}{3}}$$

$$c = \left[1 \pm \sqrt{\dfrac{6 - 4\sqrt{2}}{3}}\right]^2$$

$$c \approx 0.4380 \text{ or } c \approx 1.7908$$

45. $\int_{-\pi/4}^{\pi/4} 2 \sec^2 x \, dx = \left[2 \tan x\right]_{-\pi/4}^{\pi/4} = 2(1) - 2(-1) = 4$

$$f(c)\left[\dfrac{\pi}{4} - \left(-\dfrac{\pi}{4}\right)\right] = 4$$

$$2 \sec^2 c = \dfrac{8}{\pi}$$

$$\sec^2 c = \dfrac{4}{\pi}$$

$$\sec c = \pm\dfrac{2}{\sqrt{\pi}}$$

$$c = \pm \text{arcsec}\left(\dfrac{2}{\sqrt{\pi}}\right)$$

$$= \pm \arccos \dfrac{\sqrt{\pi}}{2} \approx \pm 0.4817$$

47. $\dfrac{1}{2 - (-2)}\int_{-2}^{2} (4 - x^2) \, dx = \dfrac{1}{4}\left[4x - \dfrac{1}{3}x^3\right]_{-2}^{2} = \dfrac{1}{4}\left[\left(8 - \dfrac{8}{3}\right) - \left(-8 + \dfrac{8}{3}\right)\right] = \dfrac{8}{3}$

Average value $= \dfrac{8}{3}$

$4 - x^2 = \dfrac{8}{3}$ when $x^2 = 4 - \dfrac{8}{3}$ or $x = +\dfrac{2\sqrt{3}}{3} \approx \pm 1.155$.

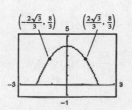

49. $\dfrac{1}{\pi - 0}\int_0^{\pi} \sin x \, dx = \left[-\dfrac{1}{\pi} \cos x\right]_0^{\pi} = \dfrac{2}{\pi}$

Average value $= \dfrac{2}{\pi}$

$\sin x = \dfrac{2}{\pi}$

$x \approx 0.690, 2.451$

51. The distance traveled is $\int_0^8 v(t) \, dt$. The area under the curve from $0 \le t \le 8$ is approximately $(18 \text{ squares})(30) \approx 540$ ft.

53. If f is continuous on $[a, b]$ and $F'(x) = f(x)$ on $[a, b]$,

then $\int_a^b f(x) \, dx = F(b) - F(a)$.

55. $\int_0^2 f(x) \, dx = -(\text{area of region } A) = -1.5$

57. $\int_0^6 |f(x)| \, dx = -\int_0^2 f(x) \, dx + \int_2^6 f(x) \, dx = 1.5 + 5.0 = 6.5$

59. $\int_0^6 [2 + f(x)] \, dx = \int_0^6 2 \, dx + \int_0^6 f(x) \, dx$

$$= 12 + 3.5 = 15.5$$

61. (a) $F(x) = k \sec^2 x$

$F(0) = k = 500$

$F(x) = 500 \sec^2 x$

(b) $\dfrac{1}{\pi/3 - 0}\int_0^{\pi/3} 500 \sec^2 x \, dx = \dfrac{1500}{\pi}\left[\tan x\right]_0^{\pi/3}$

$$= \dfrac{1500}{\pi}\left(\sqrt{3} - 0\right)$$

$$\approx 826.99 \text{ newtons}$$

$$\approx 827 \text{ newtons}$$

63. $\dfrac{1}{5-0}\displaystyle\int_0^5 (0.1729t + 0.1522t^2 - 0.0374t^3)\,dt \approx \dfrac{1}{5}\Big[0.08645t^2 + 0.05073t^3 - 0.00935t^4 \Big]_0^5 \approx 0.5318 \text{ liter}$

65. (a) $v = -8.61 \times 10^{-4}t^3 + 0.0782t^2 - 0.208t + 0.0952$

(b)

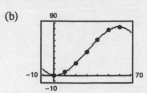

(c) $\displaystyle\int_0^{60} v(t)\,dt = \left[\dfrac{-8.61 \times 10^{-4}t^4}{4} + \dfrac{0.0782t^3}{3} - \dfrac{0.208t^2}{2} + 0.0952t \right]_0^{60} \approx 2476 \text{ meters}$

67. $F(x) = \displaystyle\int_0^x (t - 5)\,dt = \left[\dfrac{t^2}{2} - 5t \right]_0^x = \dfrac{x^2}{2} - 5x$

$F(2) = \dfrac{4}{2} - 5(2) = -8$

$F(5) = \dfrac{25}{2} - 5(5) = -\dfrac{25}{2}$

$F(8) = \dfrac{64}{2} - 5(8) = -8$

69. $F(x) = \displaystyle\int_1^x \dfrac{10}{v^2}\,dv = \int_1^x 10v^{-2}\,dv = \dfrac{-10}{v}\bigg]_1^x$

$\qquad = -\dfrac{10}{x} + 10 = 10\left(1 - \dfrac{1}{x} \right)$

$F(2) = 10\left(\dfrac{1}{2}\right) = 5$

$F(5) = 10\left(\dfrac{4}{5}\right) = 8$

$F(8) = 10\left(\dfrac{7}{8}\right) = \dfrac{35}{4}$

71. $F(x) = \displaystyle\int_1^x \cos\theta\,d\theta = \sin\theta\bigg]_1^x = \sin x - \sin 1$

$F(2) = \sin 2 - \sin 1 = 0.0678$

$F(5) = \sin 5 - \sin 1 \approx -1.8004$

$F(8) = \sin 8 - \sin 1 \approx 0.1479$

73. $g(x) = \displaystyle\int_0^x f(t)\,dt$

(a) $g(0) = \displaystyle\int_0^0 f(t)\,dt = 0$

$g(2) = \displaystyle\int_0^2 f(t)\,dt \approx 4 + 2 + 1 = 7$

$g(4) = \displaystyle\int_0^4 f(t)\,dt \approx 7 + 2 = 9$

$g(6) = \displaystyle\int_0^6 f(t)\,dt \approx 9 + (-1) = 8$

$g(8) = \displaystyle\int_0^8 f(t)\,dt \approx 8 - 3 = 5$

(b) g increasing on $(0, 4)$ and decreasing on $(4, 8)$

(c) g is a maximum of 9 at $x = 4$.

(d)

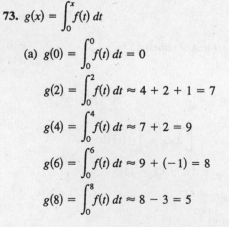

75. (a) $\displaystyle\int_0^x (t + 2)\,dt = \left[\dfrac{t^2}{2} + 2t \right]_0^x = \dfrac{1}{2}x^2 + 2x$

(b) $\dfrac{d}{dx}\left[\dfrac{1}{2}x^2 + 2x \right] = x + 2$

77. (a) $\displaystyle\int_8^x \sqrt[3]{t}\,dt = \left[\dfrac{3}{4}t^{4/3} \right]_8^x = \dfrac{3}{4}(x^{4/3} - 16) = \dfrac{3}{4}x^{4/3} - 12$

(b) $\dfrac{d}{dx}\left[\dfrac{3}{4}x^{4/3} - 12 \right] = x^{1/3} = \sqrt[3]{x}$

79. (a) $\int_{x/4}^{x} \sec^2 t \, dt = \Big[\tan t \Big]_{x/4}^{x} = \tan x - 1$

(b) $\dfrac{d}{dx}[\tan x - 1] = \sec^2 x$

81. $F(x) = \int_{-2}^{x} (t^2 - 2t) \, dt$

$F'(x) = x^2 - 2x$

83. $F(x) = \int_{-1}^{x} \sqrt{t^4 + 1} \, dt$

$F'(x) = \sqrt{x^4 + 1}$

85. $F(x) = \int_{0}^{x} t \cos t \, dt$

$F'(x) = x \cos x$

87. $F(x) = \int_{x}^{x+2} (4t + 1) \, dt$

$= \Big[2t^2 + t \Big]_{x}^{x+2}$

$= [2(x + 2)^2 + (x + 2)] - [2x^2 + x]$

$= 8x + 10$

$F'(x) = 8$

Alternate solution:

$F(x) = \int_{x}^{x+2} (4t + 1) \, dt$

$= \int_{x}^{0} (4t + 1) \, dt + \int_{0}^{x+2} (4t + 1) \, dt$

$= -\int_{0}^{x} (4t + 1) \, dt + \int_{0}^{x+2} (4t + 1) \, dt$

$F'(x) = -(4x + 1) + 4(x + 2) + 1 = 8$

89. $F(x) = \int_{0}^{\sin x} \sqrt{t} \, dt = \Big[\dfrac{2}{3} t^{3/2} \Big]_{0}^{\sin x} = \dfrac{2}{3}(\sin x)^{3/2}$

$F'(x) = (\sin x)^{1/2} \cos x = \cos x \sqrt{\sin x}$

Alternate solution:

$F(x) = \int_{0}^{\sin x} \sqrt{t} \, dt$

$F'(x) = \sqrt{\sin x} \, \dfrac{d}{dx}(\sin x) - \sqrt{\sin x}(\cos x)$

91. $F(x) = \int_{0}^{x^3} \sin t^2 \, dt$

$F'(x) = \sin(x^3)^2 \cdot 3x^2 = 3x^2 \sin x^6$

93. $g(x) = \int_{0}^{x} f(t) \, dt$

$g(0) = 0, \ g(1) \approx \dfrac{1}{2}, \ g(2) \approx 1, \ g(3) \approx \dfrac{1}{2}, \ g(4) = 0$

g has a relative maximum at $x = 2$.

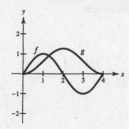

95. (a) $C(x) = 5000 \left(25 + 3 \int_{0}^{x} t^{1/4} \, dt \right)$

$= 5000 \left(25 + 3 \Big[\dfrac{4}{5} t^{5/4} \Big]_{0}^{x} \right)$

$= 5000 \left(25 + \dfrac{12}{5} x^{5/4} \right) = 1000(125 + 12x^{5/4})$

(b) $C(1) = 1000(125 + 12(1)) = \$137{,}000$

$C(5) = 1000(125 + 12(5)^{5/4}) \approx \$214{,}721$

$C(10) = 1000(125 + 12(10)^{5/4}) \approx \$338{,}394$

97. $x(t) = t^3 - 6t^2 + 9t - 2$

$x'(t) = 3t^2 - 12t + 9$

$\quad = 3(t^2 - 4t + 3)$

$\quad = 3(t - 3)(t - 1)$

Total distance $= \displaystyle\int_0^5 |x'(t)|\, dt$

$\quad = \displaystyle\int_0^5 3|(t - 3)(t - 1)|\, dt$

$\quad = 3\displaystyle\int_0^1 (t^2 - 4t + 3)\, dt - 3\int_1^3 (t^2 - 4t + 3)\, dt + 3\int_3^5 (t^2 - 4t + 3)\, dt$

$\quad = 4 + 4 + 20$

$\quad = 28 \text{ units}$

99. Total distance $= \displaystyle\int_1^4 |x'(t)|\, dt$

$\quad = \displaystyle\int_1^4 |v(t)|\, dt$

$\quad = \displaystyle\int_1^4 \frac{1}{\sqrt{t}}\, dt$

$\quad = 2t^{1/2}\Big]_1^4$

$\quad = 2(2 - 1) = 2 \text{ units}$

101. True

103. The function $f(x) = x^{-2}$ is not continuous on $[-1, 1]$.

$\displaystyle\int_{-1}^1 x^{-2}\, dx = \int_{-1}^0 x^{-2}\, dx + \int_0^1 x^{-2}\, dx$

Each of these integrals is infinite. $f(x) = x^{-2}$ has a nonremovable discontinuity at $x = 0$.

105. $f(x) = \displaystyle\int_0^{1/x} \frac{1}{t^2 + 1}\, dt + \int_0^x \frac{1}{t^2 + 1}\, dt$

By the Second Fundamental Theorem of Calculus, we have

$f'(x) = \dfrac{1}{(1/x)^2 + 1}\left(-\dfrac{1}{x^2}\right) + \dfrac{1}{x^2 + 1}$

$\quad = -\dfrac{1}{1 + x^2} + \dfrac{1}{x^2 + 1} = 0.$

Since $f'(x) = 0$, $f(x)$ must be constant.

Section 4.5   Integration by Substitution

$\displaystyle\int f(g(x))g'(x)\, dx$	$u = g(x)$	$du = g'(x)\, dx$
1. $\displaystyle\int (5x^2 + 1)^2(10x)\, dx$	$5x^2 + 1$	$10x\, dx$
3. $\displaystyle\int \frac{x}{\sqrt{x^2 + 1}}\, dx$	$x^2 + 1$	$2x\, dx$
5. $\displaystyle\int \tan^2 x \sec^2 x\, dx$	$\tan x$	$\sec^2 x\, dx$

7. $\displaystyle\int (1 + 2x)^4(2)\,dx = \frac{(1 + 2x)^5}{5} + C$

     Check: $\dfrac{d}{dx}\left[\dfrac{(1 + 2x)^5}{5} + C\right] = 2(1 + 2x)^4$

9. $\displaystyle\int (9 - x^2)^{1/2}(-2x)\,dx = \frac{(9 - x^2)^{3/2}}{3/2} + C = \frac{2}{3}(9 - x^2)^{3/2} + C$

     Check: $\dfrac{d}{dx}\left[\dfrac{2}{3}(9 - x^2)^{3/2} + C\right] = \dfrac{2}{3}\cdot\dfrac{3}{2}(9 - x^2)^{1/2}(-2x) = \sqrt{9 - x^2}(-2x)$

11. $\displaystyle\int x^3(x^4 + 3)^2\,dx = \frac{1}{4}\int (x^4 + 3)^2(4x^3)\,dx = \frac{1}{4}\frac{(x^4 + 3)^3}{3} + C = \frac{(x^4 + 3)^3}{12} + C$

     Check: $\dfrac{d}{dx}\left[\dfrac{(x^4 + 3)^3}{12} + C\right] = \dfrac{3(x^4 + 3)^2}{12}(4x^3) = (x^4 + 3)^2(x^3)$

13. $\displaystyle\int x^2(x^3 - 1)^4\,dx = \frac{1}{3}\int (x^3 - 1)^4(3x^2)\,dx = \frac{1}{3}\left[\frac{(x^3 - 1)^5}{5}\right] + C = \frac{(x^3 - 1)^5}{15} + C$

     Check: $\dfrac{d}{dx}\left[\dfrac{(x^3 - 1)^5}{15} + C\right] = \dfrac{5(x^3 - 1)^4(3x^2)}{15} = x^2(x^3 - 1)^4$

15. $\displaystyle\int t\sqrt{t^2 + 2}\,dt = \frac{1}{2}\int (t^2 + 2)^{1/2}(2t)\,dt = \frac{1}{2}\frac{(t^2 + 2)^{3/2}}{3/2} + C = \frac{(t^2 + 2)^{3/2}}{3} + C$

     Check: $\dfrac{d}{dt}\left[\dfrac{(t^2 + 2)^{3/2}}{3} + C\right] = \dfrac{3/2(t^2 + 2)^{1/2}(2t)}{3} = (t^2 + 2)^{1/2}t$

17. $\displaystyle\int 5x(1 - x^2)^{1/3}\,dx = -\frac{5}{2}\int (1 - x^2)^{1/3}(-2x)\,dx = -\frac{5}{2}\cdot\frac{(1 - x^2)^{4/3}}{4/3} + C = -\frac{15}{8}(1 - x^2)^{4/3} + C$

     Check: $\dfrac{d}{dx}\left[-\dfrac{15}{8}(1 - x^2)^{4/3} + C\right] = -\dfrac{15}{8}\cdot\dfrac{4}{3}(1 - x^2)^{1/3}(-2x) = 5x(1 - x^2)^{1/3} = 5x\sqrt[3]{1 - x^2}$

19. $\displaystyle\int \frac{x}{(1 - x^2)^3}\,dx = -\frac{1}{2}\int (1 - x^2)^{-3}(-2x)\,dx = -\frac{1}{2}\frac{(1 - x^2)^{-2}}{-2} + C = \frac{1}{4(1 - x^2)^2} + C$

     Check: $\dfrac{d}{dx}\left[\dfrac{1}{4(1 - x^2)^2} + C\right] = \dfrac{1}{4}(-2)(1 - x^2)^{-3}(-2x) = \dfrac{x}{(1 - x^2)^3}$

21. $\displaystyle\int \frac{x^2}{(1 + x^3)^2}\,dx = \frac{1}{3}\int (1 + x^3)^{-2}(3x^2)\,dx = \frac{1}{3}\left[\frac{(1 + x^3)^{-1}}{-1}\right] + C = -\frac{1}{3(1 + x^3)} + C$

     Check: $\dfrac{d}{dx}\left[-\dfrac{1}{3(1 + x^3)} + C\right] = -\dfrac{1}{3}(-1)(1 + x^3)^{-2}(3x^2) = \dfrac{x^2}{(1 + x^3)^2}$

23. $\displaystyle\int \frac{x}{\sqrt{1 - x^2}}\,dx = -\frac{1}{2}\int (1 - x^2)^{-1/2}(-2x)\,dx = -\frac{1}{2}\frac{(1 - x^2)^{1/2}}{1/2} + C = -\sqrt{1 - x^2} + C$

     Check: $\dfrac{d}{dx}\left[-(1 - x^2)^{1/2} + C\right] = -\dfrac{1}{2}(1 - x^2)^{-1/2}(-2x) = \dfrac{x}{\sqrt{1 - x^2}}$

25. $\displaystyle\int\left(1 + \frac{1}{t}\right)^3\left(\frac{1}{t^2}\right) dt = -\int\left(1 + \frac{1}{t}\right)^3\left(-\frac{1}{t^2}\right) dt = -\frac{[1 + (1/t)]^4}{4} + C$

    Check: $\displaystyle\frac{d}{dt}\left[-\frac{[1 + (1/t)]^4}{4} + C\right] = -\frac{1}{4}(4)\left(1 + \frac{1}{t}\right)^3\left(-\frac{1}{t^2}\right) = \frac{1}{t^2}\left(1 + \frac{1}{t}\right)^3$

27. $\displaystyle\int\frac{1}{\sqrt{2x}}\,dx = \frac{1}{2}\int (2x)^{-1/2}\,2\,dx = \frac{1}{2}\left[\frac{(2x)^{1/2}}{1/2}\right] + C = \sqrt{2x} + C$

    Alternate Solution: $\displaystyle\int\frac{1}{\sqrt{2x}}\,dx = \frac{1}{\sqrt{2}}\int x^{-1/2}\,dx = \frac{1}{\sqrt{2}}\frac{x^{1/2}}{(1/2)} + C = \sqrt{2x} + C$

    Check: $\displaystyle\frac{d}{dx}\left[\sqrt{2x} + C\right] = \frac{1}{2}(2x)^{-1/2}(2) = \frac{1}{\sqrt{2x}}$

29. $\displaystyle\int\frac{x^2 + 3x + 7}{\sqrt{x}}\,dx = \int (x^{3/2} + 3x^{1/2} + 7x^{-1/2})\,dx = \frac{2}{5}x^{5/2} + 2x^{3/2} + 14x^{1/2} + C = \frac{2}{5}\sqrt{x}(x^2 + 5x + 35) + C$

    Check: $\displaystyle\frac{d}{dx}\left[\frac{2}{5}x^{5/2} + 2x^{3/2} + 14x^{1/2} + C\right] = \frac{x^2 + 3x + 7}{\sqrt{x}}$

31. $\displaystyle\int t^2\left(t - \frac{2}{t}\right) dt = \int (t^3 - 2t)\,dt = \frac{1}{4}t^4 - t^2 + C$

    Check: $\displaystyle\frac{d}{dt}\left[\frac{1}{4}t^4 - t^2 + C\right] = t^3 - 2t = t^2\left(t - \frac{2}{t}\right)$

33. $\displaystyle\int (9 - y)\sqrt{y}\,dy = \int (9y^{1/2} - y^{3/2})\,dy = 9\left(\frac{2}{3}y^{3/2}\right) - \frac{2}{5}y^{5/2} + C = \frac{2}{5}y^{3/2}(15 - y) + C$

    Check: $\displaystyle\frac{d}{dy}\left[\frac{2}{5}y^{3/2}(15 - y) + C\right] = \frac{d}{dy}\left[6y^{3/2} - \frac{2}{5}y^{5/2} + C\right] = 9y^{1/2} - y^{3/2} = (9 - y)\sqrt{y}$

35. $\displaystyle y = \int\left[4x + \frac{4x}{\sqrt{16 - x^2}}\right] dx$

    $\displaystyle = 4\int x\,dx - 2\int (16 - x^2)^{-1/2}(-2x)\,dx$

    $\displaystyle = 4\left(\frac{x^2}{2}\right) - 2\left[\frac{(16 - x^2)^{1/2}}{1/2}\right] + C$

    $\displaystyle = 2x^2 - 4\sqrt{16 - x^2} + C$

37. $\displaystyle y = \int\frac{x + 1}{(x^2 + 2x - 3)^2}\,dx$

    $\displaystyle = \frac{1}{2}\int (x^2 + 2x - 3)^{-2}(2x + 2)\,dx$

    $\displaystyle = \frac{1}{2}\left[\frac{(x^2 + 2x - 3)^{-1}}{-1}\right] + C$

    $\displaystyle = -\frac{1}{2(x^2 + 2x - 3)} + C$

39. (a)

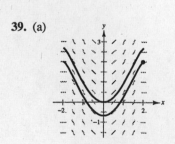

    (b) $\displaystyle\frac{dy}{dx} = x\sqrt{4 - x^2},\ (2, 2)$

    $\displaystyle y = \int x\sqrt{4 - x^2}\,dx = -\frac{1}{2}\int (4 - x^2)^{1/2}(-2x\,dx)$

    $\displaystyle = -\frac{1}{2}\cdot\frac{2}{3}(4 - x^2)^{3/2} + C = -\frac{1}{3}(4 - x^2)^{3/2} + C$

    $(2, 2):\ 2 = -\frac{1}{3}(4 - 2^2)^{3/2} + C \Rightarrow C = 2$

    $\displaystyle y = -\frac{1}{3}(4 - x^2)^{3/2} + 2$

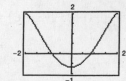

41. (a)

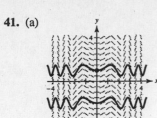

(b) $\dfrac{dy}{dx} = x \cos x^2$, $(0, 1)$

$$y = \int x \cos x^2\, dx = \frac{1}{2}\int \cos(x^2)2x\, dx$$

$$= \frac{1}{2}\sin(x^2) + C$$

$(0, 1)$: $1 = \frac{1}{2}\sin(0) + C \Rightarrow C = 1$

$$y = \frac{1}{2}\sin(x^2) + 1$$

43. $\displaystyle \int \pi \sin \pi x\, dx = -\cos \pi x + C$

45. $\displaystyle \int \sin 2x\, dx = \frac{1}{2}\int (\sin 2x)(2x)\, dx = -\frac{1}{2}\cos 2x + C$

47. $\displaystyle \int \frac{1}{\theta^2}\cos\frac{1}{\theta}\, d\theta = -\int \cos\frac{1}{\theta}\left(-\frac{1}{\theta^2}\right) d\theta = -\sin\frac{1}{\theta} + C$

49. $\displaystyle \int \sin 2x \cos 2x\, dx = \frac{1}{2}\int (\sin 2x)(2\cos 2x)\, dx = \frac{1}{2}\frac{(\sin 2x)^2}{2} + C = \frac{1}{4}\sin^2 2x + C$ OR

$\displaystyle \int \sin 2x \cos 2x\, dx = -\frac{1}{2}\int (\cos 2x)(-2\sin 2x)\, dx = -\frac{1}{2}\frac{(\cos 2x)^2}{2} + C_1 = -\frac{1}{4}\cos^2 2x + C_1$ OR

$\displaystyle \int \sin 2x \cos 2x\, dx = \frac{1}{2}\int 2\sin 2x \cos 2x\, dx = \frac{1}{2}\int \sin 4x\, dx = -\frac{1}{8}\cos 4x + C_2$

51. $\displaystyle \int \tan^4 x \sec^2 x\, dx = \frac{\tan^5 x}{5} + C = \frac{1}{5}\tan^5 x + C$

53. $\displaystyle \int \frac{\csc^2 x}{\cot^3 x}\, dx = -\int (\cot x)^{-3}(-\csc^2 x)\, dx$

$$= -\frac{(\cot x)^{-2}}{-2} + C = \frac{1}{2\cot^2 x} + C = \frac{1}{2}\tan^2 x + C = \frac{1}{2}(\sec^2 x - 1) + C = \frac{1}{2}\sec^2 x + C_1$$

55. $\displaystyle \int \cot^2 x\, dx = \int (\csc^2 x - 1)\, dx = -\cot x - x + C$

57. $f(x) = \displaystyle \int \cos\frac{x}{2}\, dx = 2\sin\frac{x}{2} + C$

Since $f(0) = 3 = 2\sin 0 + C$, $C = 3$. Thus,

$$f(x) = 2\sin\frac{x}{2} + 3.$$

59. $f'(x) = \sin 4x$, $\left(\dfrac{\pi}{4}, \dfrac{-3}{4}\right)$

$$f(x) = \frac{-1}{4}\cos 4x + C$$

$$f\left(\frac{\pi}{4}\right) = \frac{-1}{4}\cos\left(4\left(\frac{\pi}{4}\right)\right) + C = \frac{-3}{4}$$

$$-\frac{1}{4}(-1) + C = \frac{-3}{4}$$

$$C = -1$$

$$f(x) = -\frac{1}{4}\cos 4x - 1$$

61. $f'(x) = 2x(4x^2 - 10)^2$, $(2, 10)$

$$f(x) = \frac{(4x^2 - 10)^3}{12} + C = \frac{2(2x^2 - 5)^3}{3} + C$$

$$f(2) = \frac{2(8 - 5)^3}{3} + C = 18 + C = 10 \Rightarrow C = -8$$

$$f(x) = \frac{2}{3}(2x^2 - 5)^3 - 8$$

63. $u = x + 2, x = u - 2, dx = du$

$$\int x\sqrt{x + 2}\, dx = \int (u - 2)\sqrt{u}\, du$$

$$= \int (u^{3/2} - 2u^{1/2})\, du$$

$$= \frac{2}{5}u^{5/2} - \frac{4}{3}u^{3/2} + C$$

$$= \frac{2u^{3/2}}{15}(3u - 10) + C$$

$$= \frac{2}{15}(x + 2)^{3/2}[3(x + 2) - 10] + C$$

$$= \frac{2}{15}(x + 2)^{3/2}(3x - 4) + C$$

65. $u = 1 - x, x = 1 - u, dx = -du$

$$\int x^2\sqrt{1 - x}\, dx = -\int (1 - u)^2\sqrt{u}\, du$$

$$= -\int (u^{1/2} - 2u^{3/2} + u^{5/2})\, du$$

$$= -\left(\frac{2}{3}u^{3/2} - \frac{4}{5}u^{5/2} + \frac{2}{7}u^{7/2}\right) + C$$

$$= -\frac{2u^{3/2}}{105}(35 - 42u + 15u^2) + C$$

$$= -\frac{2}{105}(1 - x)^{3/2}[35 - 42(1 - x) + 15(1 - x)^2] + C$$

$$= -\frac{2}{105}(1 - x)^{3/2}(15x^2 + 12x + 8) + C$$

67. $u = 2x - 1, x = \frac{1}{2}(u + 1), dx = \frac{1}{2}du$

$$\int \frac{x^2 - 1}{\sqrt{2x - 1}}\, dx = \int \frac{[(1/2)(u + 1)]^2 - 1}{\sqrt{u}}\frac{1}{2}\, du$$

$$= \frac{1}{8}\int u^{-1/2}[(u^2 + 2u + 1) - 4]\, du$$

$$= \frac{1}{8}\int (u^{3/2} + 2u^{1/2} - 3u^{-1/2})\, du$$

$$= \frac{1}{8}\left(\frac{2}{5}u^{5/2} + \frac{4}{3}u^{3/2} - 6u^{1/2}\right) + C$$

$$= \frac{u^{1/2}}{60}(3u^2 + 10u - 45) + C$$

$$= \frac{\sqrt{2x - 1}}{60}[3(2x - 1)^2 + 10(2x - 1) - 45] + C$$

$$= \frac{1}{60}\sqrt{2x - 1}(12x^2 + 8x - 52) + C$$

$$= \frac{1}{15}\sqrt{2x - 1}(3x^2 + 2x - 13) + C$$

69. $u = x + 1, x = u - 1, dx = du$

$$\int \frac{-x}{(x+1) - \sqrt{x+1}}\, dx = \int \frac{-(u-1)}{u - \sqrt{u}}\, du$$

$$= -\int \frac{(\sqrt{u}+1)(\sqrt{u}-1)}{\sqrt{u}(\sqrt{u}-1)}\, du$$

$$= -\int (1 + u^{-1/2})\, du$$

$$= -(u + 2u^{1/2}) + C$$

$$= -u - 2\sqrt{u} + C$$

$$= -(x+1) - 2\sqrt{x+1} + C$$

$$= -x - 2\sqrt{x+1} - 1 + C$$

$$= -(x + 2\sqrt{x+1}) + C_1$$

where $C_1 = -1 + C$.

71. Let $u = x^2 + 1, du = 2x\, dx$.

$$\int_{-1}^{1} x(x^2+1)^3\, dx = \frac{1}{2}\int_{-1}^{1} (x^2+1)^3(2x)\, dx = \left[\frac{1}{8}(x^2+1)^4\right]_{-1}^{1} = 0$$

73. Let $u = x^3 + 1, du = 3x^2\, dx$.

$$\int_{1}^{2} 2x^2\sqrt{x^3+1}\, dx = 2 \cdot \frac{1}{3}\int_{1}^{2} (x^3+1)^{1/2}(3x^2)\, dx$$

$$= \left[\frac{2}{3}\frac{(x^3+1)^{3/2}}{3/2}\right]_{1}^{2}$$

$$= \frac{4}{9}\left[(x^3+1)^{3/2}\right]_{1}^{2}$$

$$= \frac{4}{9}\left[27 - 2\sqrt{2}\right] = 12 - \frac{8}{9}\sqrt{2}$$

75. Let $u = 2x + 1, du = 2\, dx$.

$$\int_{0}^{4} \frac{1}{\sqrt{2x+1}}\, dx = \frac{1}{2}\int_{0}^{4} (2x+1)^{-1/2}(2)\, dx = \left[\sqrt{2x+1}\right]_{0}^{4} = \sqrt{9} - \sqrt{1} = 2$$

77. Let $u = 1 + \sqrt{x}, du = \frac{1}{2\sqrt{x}}\, dx$.

$$\int_{1}^{9} \frac{1}{\sqrt{x}(1 + \sqrt{x})^2}\, dx = 2\int_{1}^{9} (1 + \sqrt{x})^{-2}\left(\frac{1}{2\sqrt{x}}\right) dx = \left[-\frac{2}{1 + \sqrt{x}}\right]_{1}^{9} = -\frac{1}{2} + 1 = \frac{1}{2}$$

79. $u = 2 - x, x = 2 - u, dx = -du$

When $x = 1, u = 1$. When $x = 2, u = 0$.

$$\int_{1}^{2} (x-1)\sqrt{2-x}\, dx = \int_{1}^{0} -[(2-u)-1]\sqrt{u}\, du = \int_{1}^{0} (u^{3/2} - u^{1/2})\, du = \left[\frac{2}{5}u^{5/2} - \frac{2}{3}u^{3/2}\right]_{1}^{0} = -\left[\frac{2}{5} - \frac{2}{3}\right] = \frac{4}{15}$$

81. $\displaystyle\int_{0}^{\pi/2} \cos\left(\frac{2}{3}x\right) dx = \left[\frac{3}{2}\sin\left(\frac{2}{3}x\right)\right]_{0}^{\pi/2} = \frac{3}{2}\left(\frac{\sqrt{3}}{2}\right) = \frac{3\sqrt{3}}{4}$

83. $\dfrac{dy}{dx} = 18x^2(2x^3 + 1)^2, \ (0, 4)$

$\qquad y = 3\displaystyle\int (2x^3 + 1)^2 (6x^2) \, dx \quad (u = 2x^3 + 1)$

$\qquad y = 3\dfrac{(2x^3 + 1)^3}{3} + C = (2x^3 + 1)^3 + C$

$\qquad 4 = 1^3 + C \Longrightarrow C = 3$

$\qquad y = (2x^3 + 1)^3 + 3$

85. $\dfrac{dy}{dx} = \dfrac{2x}{\sqrt{2x^2 - 1}}, \ (5, 4)$

$\qquad y = \dfrac{1}{2}\displaystyle\int (2x^2 - 1)^{-1/2} \, (4x \, dx) \quad (u = 2x^2 - 1)$

$\qquad y = \dfrac{1}{2} \dfrac{(2x^2 - 1)^{1/2}}{1/2} + C = \sqrt{2x^2 - 1} + C$

$\qquad 4 = \sqrt{49} + C = 7 + C \Longrightarrow C = -3$

$\qquad y = \sqrt{2x^2 - 1} - 3$

87. $u = x + 1, \ x = u - 1, \ dx = du$

$\qquad$ When $x = 0, \ u = 1.$ When $x = 7, \ u = 8.$

$\qquad \text{Area} = \displaystyle\int_0^7 x\sqrt[3]{x + 1} \, dx = \int_1^8 (u - 1)\sqrt[3]{u} \, du$

$\qquad\qquad = \displaystyle\int_1^8 (u^{4/3} - u^{1/3}) \, du = \left[\dfrac{3}{7}u^{7/3} - \dfrac{3}{4}u^{4/3}\right]_1^8 = \left(\dfrac{384}{7} - 12\right) - \left(\dfrac{3}{7} - \dfrac{3}{4}\right) = \dfrac{1209}{28}$

89. $A = \displaystyle\int_0^\pi (2 \sin x + \sin 2x) \, dx = -\left[2 \cos x + \dfrac{1}{2} \cos 2x\right]_0^\pi = 4$

91. $\text{Area} = \displaystyle\int_{\pi/2}^{2\pi/3} \sec^2\left(\dfrac{x}{2}\right) dx = 2\int_{\pi/2}^{2\pi/3} \sec^2\left(\dfrac{x}{2}\right)\left(\dfrac{1}{2}\right) dx = \left[2 \tan\left(\dfrac{x}{2}\right)\right]_{\pi/2}^{2\pi/3} = 2\left(\sqrt{3} - 1\right)$

93. $\displaystyle\int_0^4 \dfrac{x}{\sqrt{2x + 1}} \, dx \approx 3.333 = \dfrac{10}{3}$

95. $\displaystyle\int_3^7 x\sqrt{x - 3} \, dx \approx 28.8 = \dfrac{144}{5}$

97. $\displaystyle\int_0^3 \left(\theta + \cos\dfrac{\theta}{6}\right) d\theta \approx 7.377$

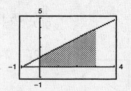

99. $\displaystyle\int (2x - 1)^2 \, dx = \dfrac{1}{2}\int (2x - 1)^2 \, 2 \, dx = \dfrac{1}{6}(2x - 1)^3 + C_1 = \dfrac{4}{3}x^3 - 2x^2 + x - \dfrac{1}{6} + C_1$

$\qquad \displaystyle\int (2x - 1)^2 \, dx = \int (4x^2 - 4x + 1) \, dx = \dfrac{4}{3}x^3 - 2x^2 + x + C_2$

$\qquad$ They differ by a constant: $C_2 = C_1 - \dfrac{1}{6}.$

101. $f(x) = x^2(x^2 + 1)$ is even.

$\qquad \displaystyle\int_{-2}^2 x^2(x^2 + 1) \, dx = 2\int_0^2 (x^4 + x^2) \, dx = 2\left[\dfrac{x^5}{5} + \dfrac{x^3}{3}\right]_0^2$

$\qquad\qquad = 2\left[\dfrac{32}{5} + \dfrac{8}{3}\right] = \dfrac{272}{15}$

103. $f(x) = \sin^2 x \cos x$ is even.

$\qquad \displaystyle\int_{-\pi/2}^{\pi/2} \sin^2 x \cos x \, dx = \int_0^{\pi/2} \sin^2 x(\cos x) \, dx$

$\qquad\qquad = 2\left[\dfrac{\sin^3 x}{3}\right]_0^{\pi/2}$

$\qquad\qquad = \dfrac{2}{3}$

105. $\int_0^2 x^2 \, dx = \left[\dfrac{x^3}{3}\right]_0^2 = \dfrac{8}{3}$; the function x^2 is an even function.

(a) $\int_{-2}^0 x^2 \, dx = \int_0^2 x^2 \, dx = \dfrac{8}{3}$

(b) $\int_{-2}^2 x^2 \, dx = 2\int_0^2 x^2 \, dx = \dfrac{16}{3}$

(c) $\int_0^2 (-x^2) \, dx = -\int_0^2 x^2 \, dx = -\dfrac{8}{3}$

(d) $\int_{-2}^0 3x^2 \, dx = 3\int_0^2 x^2 \, dx = 8$

107. $\int_{-4}^4 (x^3 + 6x^2 - 2x - 3) \, dx = \int_{-4}^4 (x^3 - 2x) \, dx + \int_{-4}^4 (6x^2 - 3) \, dx = 0 + 2\int_0^4 (6x^2 - 3) \, dx = 2\left[2x^3 - 3x\right]_0^4 = 232$

109. If $u = 5 - x^2$, then $du = -2x \, dx$ and $\int x(5 - x^2)^3 \, dx = -\dfrac{1}{2}\int (5 - x^2)^3(-2x) \, dx = -\dfrac{1}{2}\int u^3 \, du$.

111. $\dfrac{dQ}{dt} = k(100 - t)^2$

$Q(t) = \int k(100 - t)^2 \, dt = -\dfrac{k}{3}(100 - t)^3 + C$

$Q(100) = C = 0$

$Q(t) = -\dfrac{k}{3}(100 - t)^3$

$Q(0) - \dfrac{k}{3}(100)^3 = 2{,}000{,}000 \Rightarrow k = -6$

Thus, $Q(t) = 2(100 - t)^3$. When $t = 50$, $Q(50) = \$250{,}000$.

113. $R = 3.121 + 2.399 \sin(0.524t + 1.377)$

(a)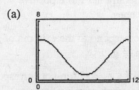

Relative minimum: $(6.4, 0.7)$ or June

Relative maximum: $(0.4, 5.5)$ or January

(b) $\int_0^{12} R(t) \, dt \approx 37.47$ inches

(c) $\dfrac{1}{3}\int_9^{12} R(t) \, dt \approx \dfrac{1}{3}(13) = 4.33$ inches

115. (a)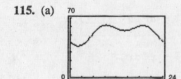

Maximum flow: $R \approx 61.713$ at $t = 9.36$.

$[(18.861, 61.178)$ is a relative maximum.$]$

(b) Volume $= \int_0^{24} R(t) \, dt \approx 1272$ (5 thousands of gallons)

117. $u = 1 - x, \, x = 1 - u, \, dx = -du$

When $x = a$, $u = 1 - a$. When $x = b$, $u = 1 - b$.

$P_{a,b} = \int_a^b \dfrac{15}{4}x\sqrt{1 - x} \, dx = \dfrac{15}{4}\int_{1-a}^{1-b} -(1 - u)\sqrt{u} \, du$

$= \dfrac{15}{4}\int_{1-a}^{1-b} (u^{3/2} - u^{1/2}) \, du = \dfrac{15}{4}\left[\dfrac{2}{5}u^{5/2} - \dfrac{2}{3}u^{3/2}\right]_{1-a}^{1-b} = \dfrac{15}{4}\left[\dfrac{2u^{3/2}}{15}(3u - 5)\right]_{1-a}^{1-b} = \left[-\dfrac{(1 - x)^{3/2}}{2}(3x + 2)\right]_a^b$

(a) $P_{0.50,\,0.75} = \left[-\dfrac{(1 - x)^{3/2}}{2}(3x + 2)\right]_{0.50}^{0.75} = 0.353 = 35.3\%$

(b) $P_{0,\,b} = \left[-\dfrac{(1 - x)^{3/2}}{2}(3x + 2)\right]_0^b = -\dfrac{(1 - b)^{3/2}}{2}(3b + 2) + 1 = 0.5$

$(1 - b)^{3/2}(3b + 2) = 1$

$b \approx 0.586 = 58.6\%$

119. (a) $C = 0.1 \int_8^{20} \left[12 \sin \frac{\pi(t-8)}{12} \right] dt = \left[-\frac{14.4}{\pi} \cos \frac{\pi(t-8)}{12} \right]_8^{20} = \frac{-14.4}{\pi}(-1 - 1) \approx \$9.17$

   (b) $C = 0.1 \int_{10}^{18} \left[12 \sin \frac{\pi(t-8)}{12} - 6 \right] dt = \left[-\frac{14.4}{\pi} \cos \frac{\pi(t-8)}{12} - 0.6t \right]_{10}^{18}$

$$= \left[\frac{-14.4}{\pi}\left(\frac{-\sqrt{3}}{2} \right) - 10.8 \right] - \left[\frac{-14.4}{\pi}\left(\frac{\sqrt{3}}{2} \right) - 6 \right] \approx \$3.14$$

Savings $\approx 9.17 - 3.14 = \$6.03$.

121. (a)

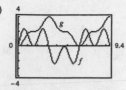

(e) The graph of h is that of g shifted 2 units downward.

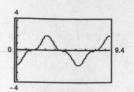

(b) g is nonnegative because the graph of f is positive at the beginning, and generally has more positive sections than negative ones.

(c) The points on g that correspond to the extrema of f are points of inflection of g.

(d) No, some zeros of f, like $x = \pi/2$, do not correspond to an extrema of g. The graph of g continues to increase after $x = \pi/2$ because f remains above the x-axis.

$g(t) = \int_0^t f(x)\, dx$

$= \int_0^{\pi/2} f(x)\, dx + \int_{\pi/2}^t f(x)\, dx = 2 + h(t)$.

123. (a) Let $u = 1 - x, du = -dx, x = 1 - u$

$x = 0 \Rightarrow u = 1, x = 1 \Rightarrow u = 0$

$\int_0^1 x^2(1 - x)^5\, dx = \int_1^0 (1 - u)^2 u^5(-du)$

$= \int_0^1 u^5(1 - u)^2\, du$

$= \int_0^1 x^5(1 - x)^2\, dx$

(b) Let $u = 1 - x, du = -dx, x = 1 - u$

$x = 0 \Rightarrow u = 1, x = 1 \Rightarrow u = 0$

$\int_0^1 x^a(1 - x)^b\, dx = \int_1^0 (1 - u)^a u^b(-du)$

$= \int_0^1 u^b(1 - u)^a\, du$

$= \int_0^1 x^b(1 - x)^a\, dx$

125. False

$$\int (2x + 1)^2\, dx = \frac{1}{2}\int (2x + 1)^2\, 2\, dx = \frac{1}{6}(2x + 1)^3 + C$$

127. True

$$\int_{-10}^{10} (ax^3 + bx^2 + cx + d)\, dx = \underbrace{\int_{-10}^{10} (ax^3 + cx)\, dx}_{\text{Odd}} + \underbrace{\int_{-10}^{10} (bx^2 + d)\, dx}_{\text{Even}} = 0 + 2\int_0^{10} (bx^2 + d)\, dx$$

129. True

$$4\int \sin x \cos x\, dx = 2\int \sin 2x\, dx = -\cos 2x + C$$

131. Let $u = cx$, $du = c\,dx$:

$$c\int_a^b f(cx)\,dx = c\int_{ca}^{cb} f(u)\,\frac{du}{c}$$

$$= \int_{ca}^{cb} f(u)\,du$$

$$= \int_{ca}^{cb} f(x)\,dx$$

133. Because f is odd, $f(-x) = -f(x)$. Then

$$\int_{-a}^{a} f(x)\,dx = \int_{-a}^{0} f(x)\,dx + \int_{0}^{a} f(x)\,dx$$

$$= -\int_{0}^{-a} f(x)\,dx + \int_{0}^{a} f(x)\,dx.$$

Let $x = -u$, $dx = -du$ in the first integral.

When $x = 0$, $u = 0$. When $x = -a$, $u = a$.

$$\int_{-a}^{a} f(x)\,dx = -\int_{0}^{a} f(-u)(-du) + \int_{0}^{a} f(x)\,dx$$

$$= -\int_{0}^{a} f(u)\,du + \int_{0}^{a} f(x)\,dx = 0$$

135. Let $f(x) = a_0 + a_1 x + a_2 x^2 + \cdots + a_n x^n$.

$$\int_{0}^{1} f(x)\,dx = \left[a_0 x + a_1 \frac{x^2}{2} + a_2 \frac{x^3}{3} + \cdots + a_n \frac{x^{n+1}}{n+1} \right]_0^1$$

$$= a_0 + \frac{a_1}{2} + \frac{a_2}{3} + \cdots + \frac{a_n}{n+1} = 0 \quad \text{(Given)}$$

By the Mean Value Theorem for Integrals, there exists c in $[0, 1]$ such that

$$\int_{0}^{1} f(x)\,dx = f(c)(1 - 0)$$

$$0 = f(c).$$

Thus the equation has at least one real zero.

Section 4.6   Numerical Integration

1. Exact: $\displaystyle\int_{0}^{2} x^2\,dx = \left[\frac{1}{3}x^3\right]_0^2 = \frac{8}{3} \approx 2.6667$

Trapezoidal: $\displaystyle\int_{0}^{2} x^2\,dx \approx \frac{1}{4}\left[0 + 2\left(\frac{1}{2}\right)^2 + 2(1)^2 + 2\left(\frac{3}{2}\right)^2 + (2)^2\right] = \frac{11}{4} = 2.7500$

Simpson's: $\displaystyle\int_{0}^{2} x^2\,dx \approx \frac{1}{6}\left[0 + 4\left(\frac{1}{2}\right)^2 + 2(1)^2 + 4\left(\frac{3}{2}\right)^2 + (2)^2\right] = \frac{8}{3} \approx 2.6667$

3. Exact: $\displaystyle\int_{0}^{2} x^3\,dx = \left[\frac{x^4}{4}\right]_0^2 = 4.0000$

Trapezoidal: $\displaystyle\int_{0}^{2} x^3\,dx \approx \frac{1}{4}\left[0 + 2\left(\frac{1}{2}\right)^3 + 2(1)^3 + 2\left(\frac{3}{2}\right)^3 + (2)^3\right] = \frac{17}{4} = 4.2500$

Simpson's: $\displaystyle\int_{0}^{2} x^3\,dx \approx \frac{1}{6}\left[0 + 4\left(\frac{1}{2}\right)^3 + 2(1)^3 + 4\left(\frac{3}{2}\right)^3 + (2)^3\right] = \frac{24}{6} = 4.0000$

5. Exact: $\displaystyle\int_{0}^{2} x^3\,dx = \left[\frac{1}{4}x^4\right]_0^2 = 4.0000$

Trapezoidal: $\displaystyle\int_{0}^{2} x^3\,dx \approx \frac{1}{8}\left[0 + 2\left(\frac{1}{4}\right)^3 + 2\left(\frac{2}{4}\right)^3 + 2\left(\frac{3}{4}\right)^3 + 2(1)^3 + 2\left(\frac{5}{4}\right)^3 + 2\left(\frac{6}{4}\right)^3 + 2\left(\frac{7}{4}\right)^3 + 8\right] = 4.0625$

Simpson's: $\displaystyle\int_{0}^{2} x^3\,dx \approx \frac{1}{12}\left[0 + 4\left(\frac{1}{4}\right)^3 + 2\left(\frac{2}{4}\right)^3 + 4\left(\frac{3}{4}\right)^3 + 2(1)^3 + 4\left(\frac{5}{4}\right)^3 + 2\left(\frac{6}{4}\right)^3 + 4\left(\frac{7}{4}\right)^3 + 8\right] = 4.0000$

7. Exact: $\displaystyle\int_4^9 \sqrt{x}\,dx = \left[\frac{2}{3}x^{3/2}\right]_4^9 = 18 - \frac{16}{3} = \frac{38}{3} \approx 12.6667$

Trapezoidal: $\displaystyle\int_4^9 \sqrt{x}\,dx \approx \frac{5}{16}\left[2 + 2\sqrt{\frac{37}{8}} + 2\sqrt{\frac{21}{4}} + 2\sqrt{\frac{47}{8}} + 2\sqrt{\frac{26}{4}} + 2\sqrt{\frac{57}{8}} + 2\sqrt{\frac{31}{4}} + 2\sqrt{\frac{67}{8}} + 3\right]$

≈ 12.6640

Simpson's: $\displaystyle\int_4^9 \sqrt{x}\,dx \approx \frac{5}{24}\left[2 + 4\sqrt{\frac{37}{8}} + \sqrt{21} + 4\sqrt{\frac{47}{8}} + \sqrt{26} + 4\sqrt{\frac{57}{8}} + \sqrt{31} + 4\sqrt{\frac{67}{8}} + 3\right] \approx 12.6667$

9. Exact: $\displaystyle\int_1^2 \frac{1}{(x+1)^2}\,dx = \left[-\frac{1}{x+1}\right]_1^2 = -\frac{1}{3} + \frac{1}{2} = \frac{1}{6} \approx 0.1667$

Trapezoidal: $\displaystyle\int_1^2 \frac{1}{(x+1)^2}\,dx \approx \frac{1}{8}\left[\frac{1}{4} + 2\left(\frac{1}{((5/4)+1)^2}\right) + 2\left(\frac{1}{((3/2)+1)^2}\right) + 2\left(\frac{1}{((7/4)+1)^2}\right) + \frac{1}{9}\right]$

$= \frac{1}{8}\left(\frac{1}{4} + \frac{32}{81} + \frac{8}{25} + \frac{32}{121} + \frac{1}{9}\right) \approx 0.1676$

Simpson's: $\displaystyle\int_1^2 \frac{1}{(x+1)^2}\,dx \approx \frac{1}{12}\left[\frac{1}{4} + 4\left(\frac{1}{((5/4)+1)^2}\right) + 2\left(\frac{1}{((3/2)+1)^2}\right) + 4\left(\frac{1}{((7/4)+1)^2}\right) + \frac{1}{9}\right]$

$= \frac{1}{12}\left(\frac{1}{4} + \frac{64}{81} + \frac{8}{25} + \frac{64}{121} + \frac{1}{9}\right) \approx 0.1667$

11. Trapezoidal: $\displaystyle\int_0^2 \sqrt{1+x^3}\,dx \approx \frac{1}{4}\left[1 + 2\sqrt{1+(1/8)} + 2\sqrt{2} + 2\sqrt{1+(27/8)} + 3\right] \approx 3.283$

Simpson's: $\displaystyle\int_0^2 \sqrt{1+x^3}\,dx \approx \frac{1}{6}\left[1 + 4\sqrt{1+(1/8)} + 2\sqrt{2} + 4\sqrt{1+(27/8)} + 3\right] \approx 3.240$

Graphing utility: 3.241

13. $\displaystyle\int_0^1 \sqrt{x}\sqrt{1-x}\,dx = \int_0^1 \sqrt{x(1-x)}\,dx$

Trapezoidal: $\displaystyle\int_0^1 \sqrt{x(1-x)}\,dx \approx \frac{1}{8}\left[0 + 2\sqrt{\frac{1}{4}\left(1-\frac{1}{4}\right)} + 2\sqrt{\frac{1}{2}\left(1-\frac{1}{2}\right)} + 2\sqrt{\frac{3}{4}\left(1-\frac{3}{4}\right)}\right] \approx 0.342$

Simpson's: $\displaystyle\int_0^1 \sqrt{x(1-x)}\,dx \approx \frac{1}{12}\left[0 + 4\sqrt{\frac{1}{4}\left(1-\frac{1}{4}\right)} + 2\sqrt{\frac{1}{2}\left(1-\frac{1}{2}\right)} + 4\sqrt{\frac{3}{4}\left(1-\frac{3}{4}\right)}\right] \approx 0.372$

Graphing utility: 0.393

15. Trapezoidal: $\displaystyle\int_0^{\sqrt{\pi/2}} \cos(x^2)\,dx \approx \frac{\sqrt{\pi/2}}{8}\left[\cos 0 + 2\cos\left(\frac{\sqrt{\pi/2}}{4}\right)^2 + 2\cos\left(\frac{\sqrt{\pi/2}}{2}\right)^2 + 2\cos\left(\frac{3\sqrt{\pi/2}}{4}\right)^2 + \cos\left(\sqrt{\frac{\pi}{2}}\right)^2\right]$

≈ 0.957

Simpson's: $\displaystyle\int_0^{\sqrt{\pi/2}} \cos(x^2)\,dx \approx \frac{\sqrt{\pi/2}}{12}\left[\cos 0 + 4\cos\left(\frac{\sqrt{\pi/2}}{4}\right)^2 + 2\cos\left(\frac{\sqrt{\pi/2}}{2}\right)^2 + 4\cos\left(\frac{3\sqrt{\pi/2}}{4}\right)^2 + \cos\left(\sqrt{\frac{\pi}{2}}\right)^2\right]$

≈ 0.978

Graphing utility: 0.977

17. Trapezoidal: $\displaystyle\int_{1}^{1.1} \sin x^2\,dx \approx \frac{1}{80}[\sin(1) + 2\sin(1.025)^2 + 2\sin(1.05)^2 + 2\sin(1.075)^2 + \sin(1.1)^2] \approx 0.089$

Simpson's: $\displaystyle\int_{1}^{1.1} \sin x^2\,dx \approx \frac{1}{120}[\sin(1) + 4\sin(1.025)^2 + 2\sin(1.05)^2 + 4\sin(1.075)^2 + \sin(1.1)^2] \approx 0.089$

Graphing utility: 0.089

19. Trapezoidal: $\displaystyle\int_{0}^{\pi/4} x\tan x\,dx \approx \frac{\pi}{32}\left[0 + 2\left(\frac{\pi}{16}\right)\tan\left(\frac{\pi}{16}\right) + 2\left(\frac{2\pi}{16}\right)\tan\left(\frac{2\pi}{16}\right) + 2\left(\frac{3\pi}{16}\right)\tan\left(\frac{3\pi}{16}\right) + \frac{\pi}{4}\right] \approx 0.194$

Simpson's: $\displaystyle\int_{0}^{\pi/4} x\tan x\,dx \approx \frac{\pi}{48}\left[0 + 4\left(\frac{\pi}{16}\right)\tan\left(\frac{\pi}{16}\right) + 2\left(\frac{2\pi}{16}\right)\tan\left(\frac{2\pi}{16}\right) + 4\left(\frac{3\pi}{16}\right)\tan\left(\frac{3\pi}{16}\right) + \frac{\pi}{4}\right] \approx 0.186$

Graphing utility: 0.186

21.

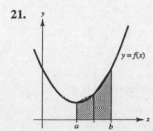

The Trapezoidal Rule overestimates the area if the graph of the integrand is concave up.

23. $f(x) = x^3$

$f'(x) = 3x^2$

$f''(x) = 6x$

$f'''(x) = 6$

$f^{(4)}(x) = 0$

(a) Trapezoidal: Error $\leq \dfrac{(2-0)^3}{12(4^2)}(12) = 0.5$ since

$|f''(x)|$ is maximum in $[0, 2]$ when $x = 2$.

(b) Simpson's: Error $\leq \dfrac{(2-0)^5}{180(4^4)}(0) = 0$ since

$f^{(4)}(x) = 0.$

25. $f(x) = \dfrac{1}{x+1}$

$f'(x) = \dfrac{-1}{(x+1)^2}$

$f''(x) = \dfrac{2}{(x+1)^3}$

$f'''(x) = \dfrac{-6}{(x+1)^4}$

$f^{(4)}(x) = \dfrac{24}{(x+1)^5}$

(a) Trapezoidal: Error $\leq \dfrac{(1-0)^2}{12(4^2)}(2) = \dfrac{1}{96} \approx 0.01$ since

$f''(x)$ is maximum in $[0, 1]$ when $x = 0$.

(b) Simpson's: Error $\leq \dfrac{(1-0)^5}{180(4^4)}(24) = \dfrac{1}{1920} \approx 0.0005$

since $f^{(4)}(x)$ is maximum in $[0, 1]$ when $x = 0$.

27. $f(x) = \cos x$

$f'(x) = -\sin x$

$f''(x) = -\cos x$

$f'''(x) = \sin x$

$f^{(4)}(x) = \cos x$

(a) Trapezoidal: Error $\leq \dfrac{(\pi-0)^3}{12(4^2)}(1) = \dfrac{\pi^3}{192} \approx 0.1615$

because $|f''(x)|$ is at most 1 on $[0, \pi]$.

(b) Simpson's:

Error $\leq \dfrac{(\pi-0)^5}{180(4^4)}(1) = \dfrac{\pi^5}{46{,}080} \approx 0.006641$

because $|f^{(4)}(x)|$ is at most 1 on $[0, \pi]$.

29. $f''(x) = \dfrac{2}{x^3}$ in $[1, 3]$.

(a) $|f''(x)|$ is maximum when $x = 1$ and $|f''(1)| = 2$.

Trapezoidal: Error $\leq \dfrac{2^3}{12n^2}(2) < 0.00001$, $n^2 > 133{,}333.33$, $n > 365.15$; let $n = 366$.

$f^{(4)}(x) = \dfrac{24}{x^5}$ in $[1, 3]$.

(b) $|f^{(4)}(x)|$ is maximum when $x = 1$ and $|f^{(4)}(1)| = 24$.

Simpson's: Error $\leq \dfrac{2^5}{180n^4}(24) < 0.00001$, $n^4 > 426{,}666.67$, $n > 25.56$; let $n = 26$.

31. $f(x) = (x + 2)^{1/2}, \quad 0 \leq x \leq 2$

$f'(x) = \dfrac{1}{2}(x + 2)^{-1/2}$

$f''(x) = -\dfrac{1}{4}(x + 2)^{-3/2}$

$f'''(x) = \dfrac{3}{8}(x + 2)^{-5/2}$

$f^{(4)}(x) = \dfrac{-15}{16}(x + 2)^{-7/2}$

(a) Maximum of $|f''(x)| = \left| \dfrac{-1}{4(x + 2)^{3/2}} \right|$ is $\dfrac{\sqrt{2}}{16} \approx 0.0884$.

Trapezoidal: Error $\leq \dfrac{(2 - 0)^3}{12n^2}\left(\dfrac{\sqrt{2}}{16} \right) \leq 0.00001$

$n^2 \geq \dfrac{8\sqrt{2}}{12(16)}10^5 = \dfrac{\sqrt{2}}{24}10^5$

$n \geq 76.8$. Let $n = 77$.

(b) Maximum of $|f^{(4)}(x)| = \left| \dfrac{-15}{16(x + 2)^{7/2}} \right|$ is

$\dfrac{15\sqrt{2}}{256} \approx 0.0829$.

Simpson's: Error $\leq \dfrac{2^5}{180n^4}\left(\dfrac{15\sqrt{2}}{256} \right) \leq 0.00001$

$n^4 \geq \dfrac{32(15)\sqrt{2}}{180(256)}10^5 = \dfrac{\sqrt{2}}{96}10^5$

$n \geq 6.2$. Let $n = 8$ (even).

33. $f(x) = \cos(\pi x), \quad 0 \leq x \leq 1$

$f'(x) = -\pi \sin(\pi x)$

$f''(x) = -\pi^2 \cos(\pi x)$

$f'''(x) = \pi^3 \sin(\pi x)$

$f^{(4)}(x) = \pi^4 \cos(\pi x)$

(a) Maximum of $|f''(x)| = |-\pi^2 \cos(\pi x)|$ is π^2.

Trapezoidal: Error $\leq \dfrac{(1 - 0)^3}{12n^2}\pi^2 \leq 0.00001$

$n^2 \geq \dfrac{\pi^2}{12} \cdot 10^5$

$n \geq 286.8$. Let $n = 287$.

(b) Maximum of $|f^{(4)}(x)| = |\pi^4 \cos(\pi x)|$ is π^4.

Simpson's: Error $\leq \dfrac{1}{180n^4}\pi^4 \leq 0.00001$

$n^4 \geq \dfrac{\pi^4}{180} \cdot 10^5$

$n \geq 15.3$. Let $n = 16$.

35. $f(x) = \sqrt{1 + x}$

(a) $f''(x) = -\dfrac{1}{4(1 + x)^{3/2}}$ in $[0, 2]$.

$|f''(x)|$ is maximum when $x = 0$ and $|f''(0)| = \dfrac{1}{4}$.

Trapezoidal: Error $\leq \dfrac{8}{12n^2}\left(\dfrac{1}{4} \right) < 0.00001$, $n^2 > 16{,}666.67$, $n > 129.10$; let $n = 130$.

(b) $f^{(4)}(x) = \dfrac{-15}{16(1 + x)^{7/2}}$ in $[0, 2]$

$|f^{(4)}(x)|$ is maximum when $x = 0$ and $|f^{(4)}(0)| = \dfrac{15}{16}$.

Simpson's: Error $\leq \dfrac{32}{180n^4}\left(\dfrac{15}{16} \right) < 0.00001$, $n^4 > 16{,}666.67$, $n > 11.36$; let $n = 12$.

37. $f(x) = \tan(x^2)$

(a) $f''(x) = 2\sec^2(x^2)[1 + 4x^2 \tan(x^2)]$ in $[0, 1]$.

$|f''(x)|$ is maximum when $x = 1$ and $|f''(1)| \approx 49.5305$.

Trapezoidal: Error $\leq \dfrac{(1 - 0)^3}{12n^2}(49.5305) < 0.00001$, $n^2 > 412{,}754.17$, $n > 642.46$; let $n = 643$.

(b) $f^{(4)}(x) = 8\sec^2(x^2)[12x^2 + (3 + 32x^4)\tan(x^2) + 36x^2\tan^2(x^2) + 48x^4\tan^3(x^2)]$ in $[0, 1]$

$|f^{(4)}(x)|$ is maximum when $x = 1$ and $|f^{(4)}(1)| \approx 9184.4734$.

Simpson's: Error $\leq \dfrac{(1 - 0)^5}{180n^4}(9184.4734) < 0.00001$, $n^4 > 5{,}102{,}485.22$, $n > 47.53$; let $n = 48$.

39. (a) $b - a = 4 - 0 = 4$, $n = 4$

$$\int_0^4 f(x)\, dx \approx \frac{4}{8}[3 + 2(7) + 2(9) + 2(7) + 0]$$

$$= \frac{1}{2}(49) = \frac{49}{2} = 24.5$$

(b) $\displaystyle\int_0^4 f(x)\, dx \approx \frac{4}{12}[3 + 4(7) + 2(9) + 4(7) + 0]$

$$= \frac{77}{3} \approx 25.67$$

41. The program will vary depending upon the computer or programmable calculator that you use.

43. $f(x) = \sqrt{1 - x^2}$ on $[0, 1]$.

n	$L(n)$	$M(n)$	$R(n)$	$T(n)$	$S(n)$
4	0.8739	0.7960	0.6239	0.7489	0.7709
8	0.8350	0.7892	0.7100	0.7725	0.7803
10	0.8261	0.7881	0.7261	0.7761	0.7818
12	0.8200	0.7875	0.7367	0.7783	0.7826
16	0.8121	0.7867	0.7496	0.7808	0.7836
20	0.8071	0.7864	0.7571	0.7821	0.7841

45. $A = \displaystyle\int_0^{\pi/2} \sqrt{x} \cos x \, dx$

Simpson's Rule: $n = 14$

$$\int_0^{\pi/2} \sqrt{x} \cos x \, dx \approx \frac{\pi}{84}\left[\sqrt{0}\cos 0 + 4\sqrt{\frac{\pi}{28}}\cos\frac{\pi}{28} + 2\sqrt{\frac{\pi}{14}}\cos\frac{\pi}{14} + 4\sqrt{\frac{3\pi}{28}}\cos\frac{3\pi}{28} + \cdots + \sqrt{\frac{\pi}{2}}\cos\frac{\pi}{2} \right]$$

$$\approx 0.701$$

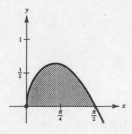

47. $W = \displaystyle\int_0^5 100x\sqrt{125 - x^3}\, dx$

Simpson's Rule: $n = 12$

$$\int_0^5 100x\sqrt{125 - x^3}\, dx \approx \frac{5}{3(12)}\left[0 + 400\left(\frac{5}{12}\right)\sqrt{125 - \left(\frac{5}{12}\right)^3} + 200\left(\frac{10}{12}\right)\sqrt{125 - \left(\frac{10}{12}\right)^3}\right.$$

$$\left. + 400\left(\frac{15}{12}\right)\sqrt{125 - \left(\frac{15}{12}\right)^3} + \cdots + 0\right] \approx 10{,}233.58 \text{ ft} \cdot \text{lb}$$

49. $\displaystyle\int_0^{1/2} \frac{6}{\sqrt{1 - x^2}}\, dx$ Simpson's Rule, $n = 6$

$$\pi \approx \frac{\left(\dfrac{1}{2} - 0\right)}{3(6)}[6 + 4(6.0209) + 2(6.0851) + 4(6.1968) + 2(6.3640) + 4(6.6002) + 6.9282]$$

$$\approx \frac{1}{36}[113.098] \approx 3.1416$$

51. Area $\approx \dfrac{1000}{2(10)}[125 + 2(125) + 2(120) + 2(112) + 2(90) + 2(90) + 2(95) + 2(88) + 2(75) + 2(35)] = 89{,}250 \text{ sq m}$

53. Let $f(x) = Ax^3 + Bx^2 + Cx + D$. Then $f^{(4)}(x) = 0$.

Simpson's: Error $\leq \dfrac{(b - a)^5}{180n^4}(0) = 0$

Therefore, Simpson's Rule is exact when approximating the integral of a cubic polynomial.

Example: $\displaystyle\int_0^1 x^3\, dx = \frac{1}{6}\left[0 + 4\left(\frac{1}{2}\right)^3 + 1\right] = \frac{1}{4}$

This is the exact value of the integral.

55. The quadratic polynomial

$$p(x) = \frac{(x - x_2)(x - x_3)}{(x_1 - x_2)(x_1 - x_3)}y_1 + \frac{(x - x_1)(x - x_3)}{(x_2 - x_1)(x_2 - x_3)}y_2 + \frac{(x - x_1)(x - x_2)}{(x_3 - x_1)(x_3 - x_2)}y_3$$

passes through the three points.

Review Exercises for Chapter 4

1.

3. $\displaystyle\int (2x^2 + x - 1)\, dx = \frac{2}{3}x^3 + \frac{1}{2}x^2 - x + C$

5. $\displaystyle\int \frac{x^3 + 1}{x^2}\, dx = \int\left(x + \frac{1}{x^2}\right)dx = \frac{1}{2}x^2 - \frac{1}{x} + C$

7. $\displaystyle\int (4x - 3\sin x)\, dx = 2x^2 + 3\cos x + C$

9. $f'(x) = -2x$, $(-1, 1)$

$$f(x) = \int -2x \, dx = -x^2 + C$$

When $x = -1$:

$$y = -1 + C = 1.$$

$$C = 2$$

$$y = 2 - x^2$$

11. (a)

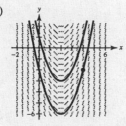

(b) $\dfrac{dy}{dx} = 2x - 4$, $(4, -2)$

$$y = \int (2x - 4) \, dx = x^2 - 4x + C$$

$$-2 = 16 - 16 + C \implies C = -2$$

$$y = x^2 - 4x - 2$$

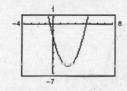

13. $a(t) = a$

$$v(t) = \int a \, dt = at + C_1$$

$$v(0) = 0 + C_1 = 0 \text{ when } C_1 = 0.$$

$$v(t) = at$$

$$s(t) = \int at \, dt = \frac{a}{2}t^2 + C_2$$

$$s(0) = 0 + C_2 = 0 \text{ when } C_2 = 0.$$

$$s(t) = \frac{a}{2}t^2$$

$$s(30) = \frac{a}{2}(30)^2 = 3600 \text{ or}$$

$$a - \frac{2(3600)}{(30)^2} = 8 \text{ ft/sec}^2.$$

$$v(30) = 8(30) = 240 \text{ ft/sec}$$

15. $a(t) = -32$

$$v(t) = -32t + 96$$

$$s(t) = -16t^2 + 96t$$

(a) $v(t) = -32t + 96 = 0$ when $t = 3$ sec.

(b) $s(3) = -144 + 288 = 144$ ft

(c) $v(t) = -32t + 96 = \dfrac{96}{2}$ when $t = \dfrac{3}{2}$ sec.

(d) $s\left(\dfrac{3}{2}\right) = -16\left(\dfrac{9}{4}\right) + 96\left(\dfrac{3}{2}\right) = 108$ ft

17. $\displaystyle\sum_{i=1}^{8} \frac{1}{4i} = \frac{1}{4(1)} + \frac{1}{4(2)} + \cdots + \frac{1}{4(8)}$

19. $\displaystyle\sum_{i=1}^{n} \left(\frac{3}{n}\right)\left(\frac{i+1}{n}\right)^2 = \frac{3}{n}\left(\frac{1+1}{n}\right)^2 + \frac{3}{n}\left(\frac{2+1}{n}\right)^2 + \cdots + \frac{3}{n}\left(\frac{n+1}{n}\right)^2$

21. $\displaystyle\sum_{i=1}^{10} 3i = 3\left(\frac{10(11)}{2}\right) = 165$

23. $\displaystyle\sum_{i=1}^{20} (i+1)^2 = \sum_{i=1}^{20} (i^2 + 2i + 1)$

$$= \frac{20(21)(41)}{6} + 2\frac{20(21)}{2} + 20$$

$$= 2870 + 420 + 20 = 3310$$

25. (a) $\displaystyle\sum_{i=1}^{10} (2i - 1)$

(b) $\displaystyle\sum_{i=1}^{n} i^3$

(c) $\displaystyle\sum_{i=1}^{10} (4i + 2)$

27. $y = \dfrac{10}{x^2 + 1}, \Delta x = \dfrac{1}{2}, n = 4$

$$S(n) = S(4) = \frac{1}{2}\left[\frac{10}{1} + \frac{10}{(1/2)^2 + 1} + \frac{10}{(1)^2 + 1} + \frac{10}{(3/2)^2 + 1}\right]$$

$$\approx 13.0385$$

$$s(n) = s(4) = \frac{1}{2}\left[\frac{10}{(1/2)^2 + 1} + \frac{10}{1 + 1} + \frac{10}{(3/2)^2 + 1} + \frac{10}{2^2 + 1}\right]$$

$$\approx 9.0385$$

$9.0385 < \text{Area of Region} < 13.0385$

29. $y = 6 - x, \Delta x = \dfrac{4}{n}$, right endpoints

$$\text{Area} = \lim_{n \to \infty} \sum_{i=1}^{n} f(c_i)\, \Delta x$$

$$= \lim_{n \to \infty} \sum_{i=1}^{n} \left(6 - \frac{4i}{n}\right)\frac{4}{n}$$

$$= \lim_{n \to \infty} \frac{4}{n}\left[6n - \frac{4}{n}\frac{n(n + 1)}{2}\right]$$

$$= \lim_{n \to \infty} \left[24 - 8\frac{n + 1}{n}\right] = 24 - 8 = 16$$

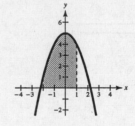

31. $y = 5 - x^2, \Delta x = \dfrac{3}{n}$

$$\text{Area} = \lim_{n \to \infty} \sum_{i=1}^{n} f(c_i)\, \Delta x$$

$$= \lim_{n \to \infty} \sum_{i=1}^{n} \left[5 - \left(-2 + \frac{3i}{n}\right)^2\right]\left(\frac{3}{n}\right)$$

$$= \lim_{n \to \infty} \frac{3}{n} \sum_{i=1}^{n} \left[1 + \frac{12i}{n} - \frac{9i^2}{n^2}\right]$$

$$= \lim_{n \to \infty} \frac{3}{n}\left[n + \frac{12}{n}\frac{n(n + 1)}{2} - \frac{9}{n^2}\frac{n(n + 1)(2n + 1)}{6}\right]$$

$$= \lim_{n \to \infty} \left[3 + 18\frac{n + 1}{n} - \frac{9}{2}\frac{(n + 1)(2n + 1)}{n^2}\right]$$

$$= 3 + 18 - 9 = 12$$

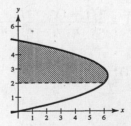

33. $x = 5y - y^2, 2 \le y \le 5, \Delta y = \dfrac{3}{n}$

$$\text{Area} = \lim_{n \to \infty} \sum_{i=1}^{n} \left[5\left(2 + \frac{3i}{n}\right) - \left(2 + \frac{3i}{n}\right)^2\right]\left(\frac{3}{n}\right)$$

$$= \lim_{n \to \infty} \frac{3}{n} \sum_{i=1}^{n} \left[10 + \frac{15i}{n} - 4 - 12\frac{i}{n} - \frac{9i^2}{n^2}\right]$$

$$= \lim_{n \to \infty} \frac{3}{n} \sum_{i=1}^{n} \left[6 + \frac{3i}{n} - \frac{9i^2}{n^2}\right]$$

$$= \lim_{n \to \infty} \frac{3}{n}\left[6n + \frac{3}{n}\frac{n(n + 1)}{2} - \frac{9}{n^2}\frac{n(n + 1)(2n + 1)}{6}\right]$$

$$= \left[18 + \frac{9}{2} - 9\right] = \frac{27}{2}$$

35. $\lim\limits_{\|\Delta\|\to 0} \sum\limits_{i=1}^{n}(2c_i - 3)\,\Delta x_i = \int_4^6 (2x - 3)\,dx$

37. $\int_{-2}^{0} (3x + 6)\,dx$

39.

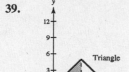

$$\int_0^5 (5 - |x - 5|)\,dx = \int_0^5 (5 - (5 - x))\,dx = \int_0^5 x\,dx = \frac{25}{2} \quad (\text{triangle})$$

41. (a) $\int_2^6 [f(x) + g(x)]\,dx = \int_2^6 f(x)\,dx + \int_2^6 g(x)\,dx = 10 + 3 = 13$

(b) $\int_2^6 [f(x) - g(x)]\,dx = \int_2^6 f(x)\,dx - \int_2^6 g(x)\,dx = 10 - 3 = 7$

(c) $\int_2^6 [2f(x) - 3g(x)]\,dx = 2\int_2^6 f(x)\,dx - 3\int_2^6 g(x)\,dx = 2(10) - 3(3) = 11$

(d) $\int_2^6 5f(x)\,dx = 5\int_2^6 f(x)\,dx = 5(10) = 50$

43. $\int_0^4 (2 + x)\,dx = \left[2x + \frac{x^2}{2}\right]_0^4 = 8 + \frac{16}{2} = 16$

45. $\int_{-1}^{1} (4t^3 - 2t)\,dt = \left[t^4 - t^2\right]_{-1}^{1} - 0$

47. $\int_4^9 x\sqrt{x}\,dx = \int_4^9 x^{3/2}\,dx = \left[\frac{2}{5}x^{5/2}\right]_4^9 = \frac{2}{5}\left[(\sqrt{9})^5 - (\sqrt{4})^5\right] = \frac{2}{5}(243 - 32) = \frac{422}{5}$

49. $\int_0^{3\pi/4} \sin\theta\,d\theta = \left[-\cos\theta\right]_0^{3\pi/4} = -\left(-\frac{\sqrt{2}}{2}\right) + 1 = 1 + \frac{\sqrt{2}}{2} = \frac{\sqrt{2} + 2}{2}$

51. $\int_1^3 (2x - 1)\,dx = \left[x^2 - x\right]_1^3 = 6$

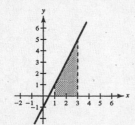

53. $\int_3^4 (x^2 - 9)\,dx = \left[\frac{x^3}{3} - 9x\right]_3^4$

$$= \left(\frac{64}{3} - 36\right) - (9 - 27)$$

$$= \frac{64}{3} - \frac{54}{3} = \frac{10}{3}$$

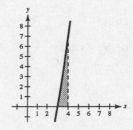

55. $\int_0^1 (x - x^3)\, dx = \left[\dfrac{x^2}{2} - \dfrac{x^4}{4}\right]_0^1 = \dfrac{1}{2} - \dfrac{1}{4} = \dfrac{1}{4}$

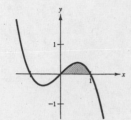

57. Area $= \int_0^1 \sin x\, dx$

$= -\cos x\Big]_0^1$

$= -\cos(1) + 1$

$= 1 - \cos(1)$

≈ 0.460

59. Area $= \int_1^9 \dfrac{4}{\sqrt{x}}\, dx = \left[\dfrac{4x^{1/2}}{(1/2)}\right]_1^9 = 8(3 - 1) = 16$

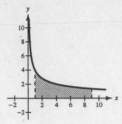

61. $\dfrac{1}{9 - 4}\int_4^9 \dfrac{1}{\sqrt{x}}\, dx = \left[\dfrac{1}{5} 2\sqrt{x}\right]_4^9 = \dfrac{2}{5}(3 - 2) = \dfrac{2}{5}$ Average value

$\dfrac{2}{5} = \dfrac{1}{\sqrt{x}}$

$\sqrt{x} = \dfrac{5}{2}$

$x = \dfrac{25}{4}$

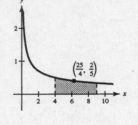

63. $F'(x) = x^2\sqrt{1 + x^3}$

65. $F'(x) = x^2 + 3x + 2$

67. $\int (x^2 + 1)^3\, dx = \int (x^6 + 3x^4 + 3x^2 + 1)\, dx = \dfrac{x^7}{7} + \dfrac{3}{5}x^5 + x^3 + x + C$

69. $u = x^3 + 3,\ du = 3x^2\, dx$

$\int \dfrac{x^2}{\sqrt{x^3 + 3}}\, dx = \int (x^3 + 3)^{-1/2} x^2\, dx = \dfrac{1}{3}\int (x^3 + 3)^{-1/2}\, 3x^2\, dx = \dfrac{2}{3}(x^3 + 3)^{1/2} + C$

71. $u = 1 - 3x^2,\ du = -6x\, dx$

$\int x(1 - 3x^2)^4\, dx = -\dfrac{1}{6}\int (1 - 3x^2)^4(-6x\, dx) = -\dfrac{1}{30}(1 - 3x^2)^5 + C = \dfrac{1}{30}(3x^2 - 1)^5 + C$

73. $\int \sin^3 x \cos x\, dx = \dfrac{1}{4}\sin^4 x + C$

75. $\int \dfrac{\sin \theta}{\sqrt{1 - \cos \theta}}\, d\theta = \int (1 - \cos \theta)^{-1/2} \sin \theta\, d\theta = 2(1 - \cos \theta)^{1/2} + C = 2\sqrt{1 - \cos \theta} + C$

77. $\displaystyle\int \tan^n x \sec^2 x\, dx = \frac{\tan^{n+1} x}{n+1} + C, n \neq -1$

79. $\displaystyle\int (1 + \sec \pi x)^2 \sec \pi x \tan \pi x\, dx = \frac{1}{\pi}\int (1 + \sec \pi x)^2 (\pi \sec \pi x \tan \pi x)\, dx = \frac{1}{3\pi}(1 + \sec \pi x)^3 + C$

81. $\displaystyle\int_{-1}^{2} x(x^2 - 4)\, dx = \frac{1}{2}\int_{-1}^{2}(x^2 - 4)(2x)\, dx = \frac{1}{2}\frac{(x^2-4)^2}{2}\Big]_{-1}^{2} = \frac{1}{4}[0 - 9] = -\frac{9}{4}$

83. $\displaystyle\int_{0}^{3}\frac{1}{\sqrt{1+x}}\, dx = \int_{0}^{3}(1+x)^{-1/2}\, dx = \Big[2(1+x)^{1/2}\Big]_{0}^{3} = 4 - 2 = 2$

85. $u = 1 - y, y = 1 - u, dy = -du$

When $y = 0$, $u = 1$. When $y = 1$, $u = 0$.

$$2\pi \int_{0}^{1}(y+1)\sqrt{1-y}\, dy = 2\pi \int_{1}^{0} -[(1-u)+1]\sqrt{u}\, du$$

$$= 2\pi \int_{1}^{0}(u^{3/2} - 2u^{1/2})\, du = 2\pi\left[\frac{2}{5}u^{5/2} - \frac{4}{3}u^{3/2}\right]_{1}^{0} = \frac{28\pi}{15}$$

87. $\displaystyle\int_{0}^{\pi}\cos\left(\frac{x}{2}\right)dx = 2\int_{0}^{\pi}\cos\left(\frac{x}{2}\right)\frac{1}{2}\, dx = \left[2\sin\left(\frac{x}{2}\right)\right]_{0}^{\pi} = 2$

89. (a)

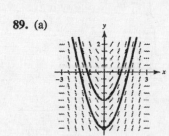

(b) $\dfrac{dy}{dx} = x\sqrt{9 - x^2}, \quad (0, -4)$

$$y = \int (9 - x^2)^{1/2} x\, dx = \frac{-1}{2}\frac{(9-x^2)^{3/2}}{3/2} + C = -\frac{1}{3}(9 - x^2)^{3/2} + C$$

$$-4 = -\frac{1}{3}(9 - 0)^{3/2} + C = -\frac{1}{3}(27) + C \implies C = 5$$

$$y = -\frac{1}{3}(9 - x^2)^{3/2} + 5$$

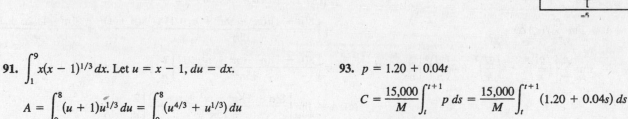

91. $\displaystyle\int_{1}^{9} x(x-1)^{1/3}\, dx$. Let $u = x - 1, du = dx$.

$$A = \int_{0}^{8}(u+1)u^{1/3}\, du = \int_{0}^{8}(u^{4/3} + u^{1/3})\, du$$

$$= \left[\frac{3u^{7/3}}{7} + \frac{3u^{4/3}}{4}\right]_{0}^{8}$$

$$= \frac{3}{7}(128) + \frac{3}{4}(16) = \frac{468}{7}$$

93. $p = 1.20 + 0.04t$

$$C = \frac{15,000}{M}\int_{t}^{t+1} p\, ds = \frac{15,000}{M}\int_{t}^{t+1}(1.20 + 0.04s)\, ds$$

(a) 2000 corresponds to $t = 10$.

$$C = \frac{15,000}{M}\int_{10}^{11}[1.20 + 0.04t]\, dt$$

$$= \frac{15,000}{M}\Big[1.20t + 0.02t^2\Big]_{10}^{11} = \frac{24,300}{M}$$

(b) 2005 corresponds to $t = 15$.

$$C = \frac{15,000}{M}\Big[1.20t + 0.02t^2\Big]_{15}^{16} = \frac{27,300}{M}$$

95. Trapezoidal Rule $(n = 4)$: $\displaystyle\int_1^2 \frac{1}{1 + x^3}\,dx \approx \frac{1}{8}\left[\frac{1}{1 + 1^3} + \frac{2}{1 + (1.25)^3} + \frac{2}{1 + (1.5)^3} + \frac{2}{1 + (1.75)^3} + \frac{1}{1 + 2^3}\right] \approx 0.257$

Simpson's Rule $(n = 4)$: $\displaystyle\int_1^2 \frac{1}{1 + x^3}\,dx \approx \frac{1}{12}\left[\frac{1}{1 + 1^3} + \frac{4}{1 + (1.25)^3} + \frac{2}{1 + (1.5)^3} + \frac{4}{1 + (1.75)^3} + \frac{1}{1 + 2^3}\right] \approx 0.254$

Graphing utility: 0.254

97. Trapezoidal Rule $(n = 4)$: $\displaystyle\int_0^{\pi/2} \sqrt{x}\cos x\,dx \approx 0.637$

Simpson's Rule $(n = 4)$: 0.685

Graphing Utility: 0.704

Problem Solving for Chapter 4

1. (a) $L(1) = \displaystyle\int_1^1 \frac{1}{t}\,dt = 0$

(b) $L'(x) = \dfrac{1}{x}$ by the Second Fundamental Theorem of Calculus.

$L'(1) = 1$

(c) $L(x) = 1 = \displaystyle\int_1^x \frac{1}{t}\,dt$ for $x \approx 2.718$

$\displaystyle\int_1^{2.718} \frac{1}{t}\,dt = 0.999896$

(**Note:** The exact value of x is e, the base of the natural logarithm function.)

(d) We first show that $\displaystyle\int_1^{x_1} \frac{1}{t}\,dt = \int_{1/x_1}^1 \frac{1}{t}\,dt$.

To see this, let $u = \dfrac{t}{x_1}$ and $du = \dfrac{1}{x_1}\,dt$.

Then $\displaystyle\int_1^{x_1} \frac{1}{t}\,dt = \int_{1/x_1}^1 \frac{1}{ux_1}(x_1\,du) = \int_{1/x_1}^1 \frac{1}{u}\,du = \int_{1/x_1}^1 \frac{1}{t}\,dt.$

Now, $L(x_1 x_2) = \displaystyle\int_1^{x_1 x_2} \frac{1}{t}\,dt = \int_{1/x_1}^{x_2} \frac{1}{u}\,du\left(\text{using } u = \frac{t}{x_1}\right)$

$= \displaystyle\int_{1/x_1}^1 \frac{1}{u}\,du + \int_1^{x_2} \frac{1}{u}\,du$

$= \displaystyle\int_1^{x_1} \frac{1}{u}\,du + \int_1^{x_2} \frac{1}{u}\,du$

$= L(x_1) + L(x_2).$

3. $y = x^4 - 4x^3 + 4x^2$, $[0, 2]$, $c_i = \dfrac{2i}{n}$

(a) $\Delta x = \dfrac{2}{n}$, $f(x) = x^4 - 4x^3 + 4x^2$

$A = \displaystyle\lim_{n\to\infty}\sum_{i=1}^n f(c_i)\,\Delta x$

$= \displaystyle\lim_{n\to\infty}\sum_{i=1}^n\left[\left(\frac{2i}{n}\right)^4 - 4\left(\frac{2i}{n}\right)^3 + 4\left(\frac{2i}{n}\right)^2\right]\frac{2}{n}$

(b) $\displaystyle\sum_{i=1}^n\left[\left(\frac{2i}{n}\right)^4 - 4\left(\frac{2i}{n}\right)^3 + 4\left(\frac{2i}{n}\right)^2\right]\frac{2}{n}$

$= \left[\dfrac{8(n + 1)(6n^3 + 9n^2 + n - 1)}{15n^3} - \dfrac{8(n + 1)^2}{n} + \dfrac{8(n + 1)(2n + 1)}{3n}\right]\dfrac{2}{n}$

$= \left[\dfrac{8(n + 1)(n^3 - n^2 + n - 1)}{15n^3}\right]\dfrac{2}{n}$

(c) $A = \displaystyle\lim_{n\to\infty}\left[\dfrac{8(n + 1)(n^3 - n^2 + n - 1)}{15n^3}\right]\dfrac{2}{n} = \dfrac{16}{15}$

5. $S(x) = \displaystyle\int_0^x \sin\left(\frac{\pi t^2}{2}\right)dt$ **(a)**

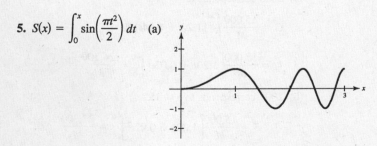

(b)

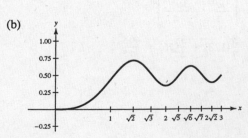

The zeros of $y = \sin\dfrac{\pi x^2}{2}$ correspond to the relative extrema of $S(x)$.

—CONTINUED—

5. —CONTINUED—

(c) $S'(x) = \sin\dfrac{\pi x^2}{2} = 0 \Rightarrow \dfrac{\pi x^2}{2} = n\pi \Rightarrow x^2 = 2n \Rightarrow x = \sqrt{2n}$, n integer.

Relative maximum at $x = \sqrt{2} \approx 1.4142$ and $x = \sqrt{6} \approx 2.4495$

Relative minimum at $x = 2$ and $x = \sqrt{8} \approx 2.8284$

(d) $S''(x) = \cos\left(\dfrac{\pi x^2}{2}\right)(\pi x) = 0 \Rightarrow \dfrac{\pi x^2}{2} = \dfrac{\pi}{2} + n\pi \Rightarrow x^2 = 1 + 2n \Rightarrow x = \sqrt{1 + 2n}$, n integer

Points of inflection at $x = 1$, $\sqrt{3}$, $\sqrt{5}$, and $\sqrt{7}$.

7. (a) Area $= \displaystyle\int_{-3}^{3} (9 - x^2)\, dx = 2\int_{0}^{3} (9 - x^2)\, dx$

$= 2\left[9x - \dfrac{x^3}{3}\right]_0^3$

$= 2[27 - 9] = 36$

(b) Base $= 6$, height $= 9$, Area $= \dfrac{2}{3} bh = \dfrac{2}{3}(6)(9) = 36$

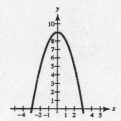

(c) Let the parabola be given by $y = b^2 - a^2 x^2$, $a, b > 0$.

Area $= 2\displaystyle\int_{0}^{b/a} (b^2 - a^2 x^2)\, dx$

$= 2\left[b^2 x - a^2 \dfrac{x^3}{3}\right]_0^{b/a}$

$= 2\left[b^2\left(\dfrac{b}{a}\right) - \dfrac{a^2}{3}\left(\dfrac{b}{a}\right)^3\right]$

$= 2\left[\dfrac{b^3}{a} - \dfrac{1}{3}\dfrac{b^3}{a}\right] = \dfrac{4}{3}\dfrac{b^3}{a}$

Base $= \dfrac{2b}{a}$, height $= b^2$

Archimedes' Formula: Area $= \dfrac{2}{3}\left(\dfrac{2b}{a}\right)(b^2) = \dfrac{4}{3}\dfrac{b^3}{a}$

9. (a)

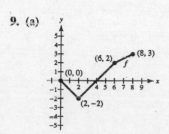

(b)

x	0	1	2	3	4	5	6	7	8
$F(x)$	0	$-\dfrac{1}{2}$	-2	$-\dfrac{7}{2}$	-4	$-\dfrac{7}{2}$	-2	$\dfrac{1}{4}$	3

(c) $f(x) = \begin{cases} -x, & 0 \le x < 2 \\ x - 4, & 2 \le x < 6 \\ \dfrac{1}{2}x - 1, & 6 \le x \le 8 \end{cases}$

$F(x) = \displaystyle\int_{0}^{x} f(t)\, dt = \begin{cases} (-x^2/2), & 0 \le x < 2 \\ (x^2/2) - 4x + 4, & 2 \le x < 6 \\ (1/4)x^2 - x - 5, & 6 \le x \le 8 \end{cases}$

$F'(x) = f(x)$. F is decreasing on $(0, 4)$ and increasing on $(4, 8)$. Therefore, the minimum is -4 at $x = 4$, and the maximum is 3 at $x = 8$.

(d) $F''(x) = f'(x) = \begin{cases} -1, & 0 < x < 2 \\ 1, & 2 < x < 6 \\ \dfrac{1}{2}, & 6 < x < 8 \end{cases}$

$x = 2$ is a point of inflection, whereas $x = 6$ is not.

11. $\displaystyle\int_{0}^{x} f(t)(x - t)\, dt = \int_{0}^{x} xf(t)\, dt - \int_{0}^{x} tf(t)\, dt = x\int_{0}^{x} f(t)\, dt - \int_{0}^{x} tf(t)\, dt$

Thus, $\dfrac{d}{dx}\displaystyle\int_{0}^{x} f(t)(x - t)\, dt = xf(x) + \int_{0}^{x} f(t)\, dt - xf(x) = \int_{0}^{x} f(t)\, dt$

Differentiating the other integral,

$\dfrac{d}{dx}\displaystyle\int_{0}^{x}\left(\int_{0}^{x} f(v)\, dv\right) dt = \int_{0}^{x} f(v)\, dv.$

Thus, the two original integrals have equal derivatives,

$\displaystyle\int_{0}^{x} f(t)(x - t)\, dt = \int_{0}^{x}\left(\int_{0}^{t} f(v)\, dv\right) dt + C.$

Letting $x = 0$, we see that $C = 0$.

13. Consider $\int_0^1 \sqrt{x}\, dx = \frac{2}{3} x^{3/2} \Big]_0^1 = \frac{2}{3}$. The corresponding

Riemann Sum using right-hand endpoints is

$$S(n) = \frac{1}{n}\left[\sqrt{\frac{1}{n}} + \sqrt{\frac{2}{n}} + \cdots + \sqrt{\frac{n}{n}} \right]$$

$$= \frac{1}{n^{3/2}}\left[\sqrt{1} + \sqrt{2} + \cdots + \sqrt{n} \right].$$

Thus, $\displaystyle\lim_{n \to \infty} \frac{\sqrt{1} + \sqrt{2} + \cdots + \sqrt{n}}{n^{3/2}} = \frac{2}{3}$.

15. By Theorem 4.8, $0 < f(x) \le M \implies \int_a^b f(x)\, dx \le \int_a^b M\, dx = M(b-a)$.

Similarly, $m \le f(x) \implies m(b-a) = \int_a^b m\, dx \le \int_a^b f(x)\, dx$.

Thus, $m(b-a) \le \int_a^b f(x)\, dx \le M(b-a)$. On the interval $[0, 1]$, $1 \le \sqrt{1 + x^4} \le \sqrt{2}$ and $b - a = 1$.

Thus, $1 \le \int_0^1 \sqrt{1 + x^4}\, dx \le \sqrt{2}$. $\left(\textbf{Note: } \int_0^1 \sqrt{1 + x^4}\, dx \approx 1.0894\right)$

17. (a) $(1 + i)^3 = 1 + 3i + 3i^2 + i^3 \implies (1 + i)^3 - i^3 = 3i^2 + 3i + 1$

(b) $3i^2 + 3i + 1 = (i + 1)^3 - i^3$

$$\sum_{i=1}^n (3i^2 + 3i + 1) = \sum_{i=1}^n [(i + 1)^3 - i^3]$$

$$= (2^3 - 1^3) + (3^3 - 2^3) + \cdots + [((n + 1)^3 - n^3)]$$

$$= (n + 1)^3 - 1$$

Hence, $(n + 1)^3 = \displaystyle\sum_{i=1}^n (3i^2 + 3i + 1) + 1$.

(c) $(n + 1)^3 - 1 = \displaystyle\sum_{i=1}^n (3i^2 + 3i + 1) = \sum_{i=1}^n 3i^2 + \frac{3(n)(n + 1)}{2} + n$

$$\implies \sum_{i=1}^n 3i^2 = n^3 + 3n^2 + 3n - \frac{3n(n + 1)}{2} - n$$

$$= \frac{2n^3 + 6n^2 + 6n - 3n^2 - 3n - 2n}{2}$$

$$= \frac{2n^3 + 3n^2 + n}{2}$$

$$= \frac{n(n + 1)(2n + 1)}{2}$$

$$\implies \sum_{i=1}^n i^2 = \frac{n(n + 1)(2n + 1)}{6}$$

19. (a) $R < I < T < L$

(b) $S(4) = \dfrac{4 - 0}{3(4)}[f(0) + 4f(1) + 2f(2) + 4f(3) + f(4)]$

$$\approx \frac{1}{3}\left[4 + 4(2) + 2(1) + 4\left(\frac{1}{2}\right) + \frac{1}{4} \right] \approx 5.417$$

C H A P T E R 5
Logarithmic, Exponential, and Other Transcendental Functions

Section 5.1   The Natural Logarithmic Function: Differentiation 224

Section 5.2   The Natural Logarithmic Function: Integration 230

Section 5.3   Inverse Functions 236

Section 5.4   Exponential Functions: Differentiation and Integration . . 242

Section 5.5   Bases Other than *e* and Applications 249

Section 5.6   Inverse Trigonometric Functions: Differentiation 255

Section 5.7   Inverse Trigonometric Functions: Integration 261

Section 5.8   Hyperbolic Functions 266

Review Exercises   272

Problem Solving 277

C H A P T E R 5
Logarithmic, Exponential, and Other Transcendental Functions

Section 5.1   The Natural Logarithmic Function: Differentiation

1. Simpson's Rule: $n = 10$

x	0.5	1.5	2	2.5	3	3.5	4
$\displaystyle\int_{1}^{x}\frac{1}{t}\,dt$	-0.6932	0.4055	0.6932	0.9163	1.0987	1.2529	1.3865

Note: $\displaystyle\int_{1}^{0.5}\frac{1}{t}\,dt = -\int_{0.5}^{1}\frac{1}{t}\,dt$

3. (a) $\ln 45 \approx 3.8067$

    (b) $\displaystyle\int_{1}^{45}\frac{1}{t}\,dt \approx 3.8067$

5. (a) $\ln 0.8 \approx -0.2231$

    (b) $\displaystyle\int_{1}^{0.8}\frac{1}{t}\,dt \approx -0.2231$

7. $f(x) = \ln x + 2$

Vertical shift 2 units upward

Matches (b)

9. $f(x) = \ln(x - 1)$

Horizontal shift 1 unit to the right

Matches (a)

11. $f(x) = 3 \ln x$

Domain: $x > 0$

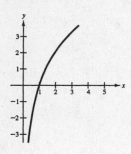

13. $f(x) = \ln 2x$

Domain: $x > 0$

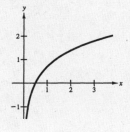

15. $f(x) = \ln(x - 1)$

Domain: $x > 1$

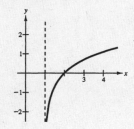

17. (a) $\ln 6 = \ln 2 + \ln 3 \approx 1.7917$

    (b) $\ln\frac{2}{3} = \ln 2 - \ln 3 \approx -0.4055$

    (c) $\ln 81 = \ln 3^4 = 4 \ln 3 \approx 4.3944$

    (d) $\ln\sqrt{3} = \ln 3^{1/2} = \frac{1}{2}\ln 3 \approx 0.5493$

19. $\ln\dfrac{2}{3} = \ln 2 - \ln 3$

21. $\ln\dfrac{xy}{z} = \ln x + \ln y - \ln z$

23. $\ln\sqrt[3]{a^2 + 1} = \ln(a^2 + 1)^{1/3} = \frac{1}{3}\ln(a^2 + 1)$

25. $\ln\left(\dfrac{x^2 - 1}{x^3}\right)^3 = 3[\ln(x^2 - 1) - \ln x^3]$

$\qquad\qquad\qquad\qquad = 3[\ln(x + 1) + \ln(x - 1) - 3\ln x]$

27. $\ln z(z-1)^2 = \ln z + \ln(z-1)^2$

$\qquad = \ln z + 2\ln(z-1)$

29. $\ln(x-2) - \ln(x+2) = \ln\dfrac{x-2}{x+2}$

31. $\dfrac{1}{3}[2\ln(x+3) + \ln x - \ln(x^2-1)] = \dfrac{1}{3}\ln\dfrac{x(x+3)^2}{x^2-1}$

$\qquad\qquad\qquad = \ln\sqrt[3]{\dfrac{x(x+3)^2}{x^2-1}}$

33. $2\ln 3 - \dfrac{1}{2}\ln(x^2+1) = \ln 9 - \ln\sqrt{x^2+1} = \ln\dfrac{9}{\sqrt{x^2+1}}$

35. (a)

(b) $f(x) = \ln\dfrac{x^2}{4} = \ln x^2 - \ln 4 = 2\ln x - \ln 4 = g(x)$

since $x > 0$.

37. $\displaystyle\lim_{x\to 3^+} \ln(x-3) = -\infty$

39. $\displaystyle\lim_{x\to 2^-} \ln[x^2(3-x)] = \ln 4$

$\qquad\qquad\qquad\qquad \approx 1.3863$

41. $y = \ln x^3 = 3\ln x$

$y' = \dfrac{3}{x}$

Slope at $(1,0)$ is $\dfrac{3}{1} = 3$.

$y - 0 = 3(x-1)$

$\qquad y = 3x - 3$ Tangent line

43. $y = \ln x^2 = 2\ln x$

$y' = \dfrac{2}{x}$

Slope at $(1,0)$ is 2.

$y - 0 = 2(x-1)$

$\qquad y = 2x - 2$ Tangent line

45. $g(x) = \ln x^2 = 2\ln x$

$g'(x) = \dfrac{2}{x}$

47. $y = (\ln x)^4$

$\dfrac{dy}{dx} = 4(\ln x)^3\left(\dfrac{1}{x}\right) = \dfrac{4(\ln x)^3}{x}$

49. $y = \ln\left[x\sqrt{x^2-1}\right] = \ln x + \dfrac{1}{2}\ln(x^2-1)$

$\dfrac{dy}{dx} = \dfrac{1}{x} + \dfrac{1}{2}\left(\dfrac{2x}{x^2-1}\right) = \dfrac{2x^2-1}{x(x^2-1)}$

51. $f(x) = \ln\dfrac{x}{x^2+1} = \ln x - \ln(x^2+1)$

$f'(x) = \dfrac{1}{x} - \dfrac{2x}{x^2+1} = \dfrac{1-x^2}{x(x^2+1)}$

53. $g(t) = \dfrac{\ln t}{t^2}$

$g'(t) = \dfrac{t^2(1/t) - 2t\ln t}{t^4} = \dfrac{1-2\ln t}{t^3}$

55. $y = \ln(\ln x^2)$

$\dfrac{dy}{dx} = \dfrac{1}{\ln x^2}\dfrac{d}{dx}(\ln x^2) = \dfrac{(2x/x^2)}{\ln x^2} = \dfrac{2}{x\ln x^2} = \dfrac{1}{x\ln x}$

57. $y = \ln\sqrt{\dfrac{x+1}{x-1}} = \dfrac{1}{2}[\ln(x+1) - \ln(x-1)]$

$\dfrac{dy}{dx} = \dfrac{1}{2}\left[\dfrac{1}{x+1} - \dfrac{1}{x-1}\right] = \dfrac{1}{1-x^2}$

59. $f(x) = \ln\dfrac{\sqrt{4+x^2}}{x} = \dfrac{1}{2}\ln(4+x^2) - \ln x$

$f'(x) = \dfrac{x}{4+x^2} - \dfrac{1}{x} = \dfrac{-4}{x(x^2+4)}$

61. $y = \dfrac{-\sqrt{x^2 + 1}}{x} + \ln\left(x + \sqrt{x^2 + 1}\right)$

$\dfrac{dy}{dx} = \dfrac{-x\left(x/\sqrt{x^2+1}\right) + \sqrt{x^2+1}}{x^2} + \left(\dfrac{1}{x + \sqrt{x^2+1}}\right)\left(1 + \dfrac{x}{\sqrt{x^2+1}}\right)$

$= \dfrac{1}{x^2\sqrt{x^2+1}} + \left(\dfrac{1}{x + \sqrt{x^2+1}}\right)\left(\dfrac{\sqrt{x^2+1} + x}{\sqrt{x^2+1}}\right)$

$= \dfrac{1}{x^2\sqrt{x^2+1}} + \dfrac{1}{\sqrt{x^2+1}} = \dfrac{1 + x^2}{x^2\sqrt{x^2+1}} = \dfrac{\sqrt{x^2+1}}{x^2}$

63. $y = \ln|\sin x|$

$\dfrac{dy}{dx} = \dfrac{\cos x}{\sin x} = \cot x$

65. $y = \ln\left|\dfrac{\cos x}{\cos x - 1}\right|$

$= \ln|\cos x| - \ln|\cos x - 1|$

$\dfrac{dy}{dx} = \dfrac{-\sin x}{\cos x} - \dfrac{-\sin x}{\cos x - 1}$

$= -\tan x + \dfrac{\sin x}{\cos x - 1}$

67. $y = \ln\left|\dfrac{-1 + \sin x}{2 + \sin x}\right|$

$= \ln|-1 + \sin x| - \ln|2 + \sin x|$

$\dfrac{dy}{dx} = \dfrac{\cos x}{-1 + \sin x} - \dfrac{\cos x}{2 + \sin x}$

$= \dfrac{3\cos x}{(\sin x - 1)(\sin x + 2)}$

69. $f(x) = \displaystyle\int_2^{\ln(2x)} (t + 1)\, dt$

$f'(x) = [\ln(2x) + 1]\left(\dfrac{1}{x}\right) = \dfrac{\ln(2x) + 1}{x}$

Second solution:

$f(x) = \displaystyle\int_2^{\ln(2x)} (t + 1)\, dt$

$= \left[\dfrac{t^2}{2} + t\right]_2^{\ln 2x} = \left[\dfrac{[\ln(2x)]^2}{2} + \ln(2x)\right] - [2 + 2]$

$f'(x) = \dfrac{1}{2}\,2\ln(2x)\dfrac{1}{x} + \dfrac{1}{2x}(2) = \dfrac{\ln(2x) + 1}{x}$

71. (a) $y = 3x^2 - \ln x, \quad (1, 3)$

$\dfrac{dy}{dx} = 6x - \dfrac{1}{x}$

When $x = 1, \dfrac{dy}{dx} = 5.$

Tangent line: $y - 3 = 5(x - 1)$

$y = 5x - 2$

$0 = 5x - y - 2$

(b)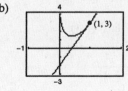

73. (a) $f(x) = \ln\sqrt{1 + \sin^2 x}$

$= \dfrac{1}{2}\ln(1 + \sin^2 x), \quad \left(\dfrac{\pi}{4}, \ln\sqrt{\dfrac{3}{2}}\right)$

$f'(x) = \dfrac{2\sin x \cos x}{2(1 + \sin^2 x)} = \dfrac{\sin x \cos x}{1 + \sin^2 x}$

$f'\left(\dfrac{\pi}{4}\right) = \dfrac{\left(\sqrt{2}/2\right)\left(\sqrt{2}/2\right)}{(3/2)} = \dfrac{1}{3}$

Tangent line: $y - \ln\sqrt{\dfrac{3}{2}} = \dfrac{1}{3}\left(x - \dfrac{\pi}{4}\right)$

$y = \dfrac{1}{3}x + \dfrac{1}{2}\ln\left(\dfrac{3}{2}\right) - \dfrac{\pi}{12}$

(b)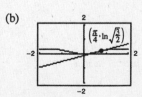

75. (a) $f(x) = x^3 \ln x$, $(1, 0)$

$f'(x) = 3x^2 \ln x + x^2$

$f'(1) = 1$

Tangent line: $y - 0 = 1(x - 1)$

$y = x - 1$

(b)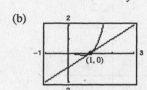

77. $x^2 - 3 \ln y + y^2 = 10$

$2x - \dfrac{3}{y}\dfrac{dy}{dx} + 2y\dfrac{dy}{dx} = 0$

$2x = \dfrac{dy}{dx}\left(\dfrac{3}{y} - 2y\right)$

$\dfrac{dy}{dx} = \dfrac{2x}{(3/y) - 2y} = \dfrac{2xy}{3 - 2y^2}$

79. $x + y - 1 = \ln(x^2 + y^2)$, $(1, 0)$

$1 + y' = \dfrac{2x + 2yy'}{x^2 + y^2}$

$x^2 + y^2 + (x^2 + y^2)y' = 2x + 2yy'$

At $(1, 0)$: $1 + y' = 2$

$y' = 1$

Tangent line: $y = x - 1$

81. $y = 2(\ln x) + 3$

$y' = \dfrac{2}{x}$

$y'' = -\dfrac{2}{x^2}$

$xy'' + y' = x\left(-\dfrac{2}{x^2}\right) + \dfrac{2}{x} = 0$

83. $y = \dfrac{x^2}{2} - \ln x$

Domain: $x > 0$

$y' = x - \dfrac{1}{x}$

$= \dfrac{(x + 1)(x - 1)}{x}$

$= 0$ when $x = 1$.

$y'' = 1 + \dfrac{1}{x^2} > 0$

Relative minimum: $\left(1, \dfrac{1}{2}\right)$

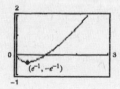

85. $y = x \ln x$

Domain: $x > 0$

$y' = x\left(\dfrac{1}{x}\right) + \ln x = 1 + \ln x = 0$ when $x = e^{-1}$.

$y'' = \dfrac{1}{x} > 0$

Relative minimum: $(e^{-1}, -e^{-1})$

87. $y = \dfrac{x}{\ln x}$

Domain: $0 < x < 1, x > 1$

$y' = \dfrac{(\ln x)(1) - (x)(1/x)}{(\ln x)^2} = \dfrac{\ln x - 1}{(\ln x)^2} = 0$ when $x = e$.

$y'' = \dfrac{(\ln x)^2(1/x) - (\ln x - 1)(2/x)\ln x}{(\ln x)^4}$

$= \dfrac{2 - \ln x}{x(\ln x)^3} = 0$ when $x = e^2$.

Relative minimum: (e, e)

Point of inflection: $\left(e^2, \dfrac{e^2}{2}\right)$

89. $f(x) = \ln x,$ $\qquad\qquad\qquad$ $f(1) = 0$

$f'(x) = \dfrac{1}{x},$ $\qquad\qquad\qquad$ $f'(1) = 1$

$f''(x) = -\dfrac{1}{x^2},$ $\qquad\qquad\quad$ $f''(1) = -1$

$P_1(x) = f(1) + f'(1)(x - 1) = x - 1,$ $\quad$ $P_1(1) = 0$

$P_2(x) = f(1) + f'(1)(x - 1) + \dfrac{1}{2}f''(1)(x - 1)^2$

$\qquad = (x - 1) - \dfrac{1}{2}(x - 1)^2,$ $\qquad$ $P_2(1) = 0$

$P_1'(x) = 1,$ $\qquad\qquad\qquad\qquad$ $P_1'(1) = 1$

$P_2'(x) = 1 - (x - 1) = 2 - x,$ $\qquad$ $P_2'(1) = 1$

$P_2''(x) = -1,$ $\qquad\qquad\qquad\qquad$ $P_2''(1) = -1$

The values of $f, P_1, P_2,$ and their first derivatives agree at $x = 1$. The values of the second derivatives of f and P_2 agree at $x = 1$.

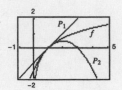

91. Find x such that $\ln x = -x$.

$f(x) = \ln x + x = 0$

$f'(x) = \dfrac{1}{x} + 1$

$x_{n+1} = x_n - \dfrac{f(x_n)}{f'(x_n)} = x_n\left[\dfrac{1 - \ln x_n}{1 + x_n}\right]$

n	1	2	3
x_n	0.5	0.5644	0.5671
$f(x_n)$	-0.1931	-0.0076	-0.0001

Approximate root: $x = 0.567$

93. $\qquad y = x\sqrt{x^2 - 1}$

$\ln y = \ln x + \dfrac{1}{2}\ln(x^2 - 1)$

$\dfrac{1}{y}\left(\dfrac{dy}{dx}\right) = \dfrac{1}{x} + \dfrac{x}{x^2 - 1}$

$\dfrac{dy}{dx} = y\left[\dfrac{2x^2 - 1}{x(x^2 - 1)}\right] = \dfrac{2x^2 - 1}{\sqrt{x^2 - 1}}$

95. $\qquad y = \dfrac{x^2\sqrt{3x - 2}}{(x - 1)^2}$

$\ln y = 2\ln x + \dfrac{1}{2}\ln(3x - 2) - 2\ln(x - 1)$

$\dfrac{1}{y}\left(\dfrac{dy}{dx}\right) = \dfrac{2}{x} + \dfrac{3}{2(3x - 2)} - \dfrac{2}{x - 1}$

$\dfrac{dy}{dx} = y\left[\dfrac{3x^2 - 15x + 8}{2x(3x - 2)(x - 1)}\right]$

$\qquad\quad = \dfrac{3x^3 - 15x^2 + 8x}{2(x - 1)^3\sqrt{3x - 2}}$

97. $\qquad y = \dfrac{x(x - 1)^{3/2}}{\sqrt{x + 1}}$

$\ln y = \ln x + \dfrac{3}{2}\ln(x - 1) - \dfrac{1}{2}\ln(x + 1)$

$\dfrac{1}{y}\left(\dfrac{dy}{dx}\right) = \dfrac{1}{x} + \dfrac{3}{2}\left(\dfrac{1}{x - 1}\right) - \dfrac{1}{2}\left(\dfrac{1}{x + 1}\right)$

$\dfrac{dy}{dx} = \dfrac{y}{2}\left[\dfrac{2}{x} + \dfrac{3}{x - 1} - \dfrac{1}{x + 1}\right]$

$\qquad\quad = \dfrac{y}{2}\left[\dfrac{4x^2 + 4x - 2}{x(x^2 - 1)}\right] = \dfrac{(2x^2 + 2x - 1)\sqrt{x - 1}}{(x + 1)^{3/2}}$

99. Answers will vary. See Theorems 5.1 and 5.2.

101. $g(x) = \ln f(x)$, $f(x) > 0$

$$g'(x) = \frac{f'(x)}{f(x)}$$

(a) Yes. If the graph of g is increasing, then $g'(x) > 0$. Since $f(x) > 0$, you know that $f'(x) = g'(x)f(x)$ and thus, $f'(x) > 0$. Therefore, the graph of f is increasing.

(b) No. Let $f(x) = x^2 + 1$ (positive and concave up). $g(x) = \ln(x^2 + 1)$ is not concave up.

103. False

$$\ln x + \ln 25 = \ln(25x)$$

$$\neq \ln(x + 25)$$

105. $t = \dfrac{5.315}{-6.7968 + \ln x}$, $\quad 1000 < x$

(a)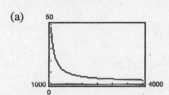

(b) $t(1167.41) \approx 20$ years

$$T = (1167.41)(20)(12) = \$280,178.40$$

(c) $t(1068.45) \approx 30$ years

$$T = (1068.45)(30)(12) = \$384,642.00$$

(d) $\dfrac{dt}{dx} = -5.315(-6.7968 + \ln x)^{-2}\left(\dfrac{1}{x}\right)$

$$= -\frac{5.315}{x(-6.7968 + \ln x)^2}$$

When $x = 1167.41$, $dt/dx \approx -0.0645$. When $x = 1068.45$, $dt/dx \approx -0.1585$.

(e) There are two obvious benefits to paying a higher monthly payment:

1. The term is lower
2. The total amount paid is lower.

107. (a)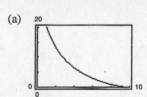

(b) $T'(p) = \dfrac{34.96}{p} + \dfrac{3.955}{\sqrt{p}}$

$T'(10) \approx 4.75$ deg/lb/in^2

$T'(70) \approx 0.97$ deg/lb/in^2

(c)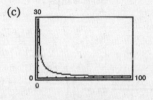

$$\lim_{p \to \infty} T'(p) = 0$$

109. $y = 10 \ln\left(\dfrac{10 + \sqrt{100 - x^2}}{x}\right) - \sqrt{100 - x^2} = 10\left[\ln\left(10 + \sqrt{100 - x^2}\right) - \ln x\right] - \sqrt{100 - x^2}$

(a) [graph, 0 to 20 vertical, 0 to 10 horizontal]

(c) $\displaystyle\lim_{x \to 10^-} \dfrac{dy}{dx} = 0$

(b) $\dfrac{dy}{dx} = 10\left[\dfrac{-x}{\sqrt{100 - x^2}(10 + \sqrt{100 - x^2})} - \dfrac{1}{x}\right] + \dfrac{x}{\sqrt{100 - x^2}}$

$$= \frac{x}{\sqrt{100 - x^2}}\left[\frac{-10}{10 + \sqrt{100 - x^2}}\right] - \frac{10}{x} + \frac{x}{\sqrt{100 - x^2}}$$

$$= \frac{x}{\sqrt{100 - x^2}}\left[\frac{-10}{10 + \sqrt{100 - x^2}} + 1\right] - \frac{10}{x}$$

$$= \frac{x}{\sqrt{100 - x^2}}\left[\frac{\sqrt{100 - x^2}}{10 + \sqrt{100 - x^2}}\right] - \frac{10}{x}$$

$$= \frac{x}{10 + \sqrt{100 - x^2}} - \frac{10}{x}$$

$$= \frac{x\left(10 - \sqrt{100 - x^2}\right)}{x^2} - \frac{10}{x} = -\frac{\sqrt{100 - x^2}}{x}$$

When $x = 5$, $dy/dx = -\sqrt{3}$. When $x = 9$, $dy/dx = -\sqrt{19}/9$.

111. $y = \ln x$

$y' = \dfrac{1}{x} > 0$ for $x > 0$.

Since $\ln x$ is increasing on its entire domain $(0, \infty)$, it is a strictly monotonic function and therefore, is one-to-one.

Section 5.2   The Natural Logarithmic Function: Integration

1. $\displaystyle\int \frac{5}{x}\,dx = 5\int \frac{1}{x}\,dx = 5\ln|x| + C$

3. $u = x + 1,\ du = dx$

$\displaystyle\int \frac{1}{x+1}\,dx = \ln|x+1| + C$

5. $u = 3 - 2x,\ du = -2\,dx$

$\displaystyle\int \frac{1}{3-2x}\,dx = -\frac{1}{2}\int \frac{1}{3-2x}(-2)\,dx$

$\qquad = -\dfrac{1}{2}\ln|3 - 2x| + C$

7. $u = x^2 + 1,\ du = 2x\,dx$

$\displaystyle\int \frac{x}{x^2+1}\,dx = \frac{1}{2}\int \frac{1}{x^2+1}(2x)\,dx$

$\qquad = \dfrac{1}{2}\ln(x^2 + 1) + C$

$\qquad = \ln\sqrt{x^2 + 1} + C$

9. $\displaystyle\int \frac{x^2 - 4}{x}\,dx = \int\left(x - \frac{4}{x}\right)dx$

$\qquad = \dfrac{x^2}{2} - 4\ln|x| + C$

11. $u = x^3 + 3x^2 + 9x,\ du = 3(x^2 + 2x + 3)\,dx$

$\displaystyle\int \frac{x^2 + 2x + 3}{x^3 + 3x^2 + 9x}\,dx = \frac{1}{3}\int \frac{3(x^2 + 2x + 3)}{x^3 + 3x^2 + 9x}\,dx$

$\qquad = \dfrac{1}{3}\ln|x^3 + 3x^2 + 9x| + C$

13. $\displaystyle\int \frac{x^2 - 3x + 2}{x + 1}\,dx = \int\left(x - 4 + \frac{6}{x+1}\right)dx$

$\qquad = \dfrac{x^2}{2} - 4x + 6\ln|x + 1| + C$

15. $\displaystyle\int \frac{x^3 - 3x^2 + 5}{x - 3}\,dx = \int\left(x^2 + \frac{5}{x-3}\right)dx$

$\qquad = \dfrac{x^3}{3} + 5\ln|x - 3| + C$

17. $\displaystyle\int \frac{x^4 + x - 4}{x^2 + 2}\,dx = \int\left(x^2 - 2 + \frac{x}{x^2+2}\right)dx$

$\qquad = \dfrac{x^3}{3} - 2x + \dfrac{1}{2}\ln(x^2 + 2) + C$

19. $u = \ln x,\ du = \dfrac{1}{x}\,dx$

$\displaystyle\int \frac{(\ln x)^2}{x}\,dx = \frac{1}{3}(\ln x)^3 + C$

21. $u = x + 1,\ du = dx$

$\displaystyle\int \frac{1}{\sqrt{x+1}}\,dx = \int (x + 1)^{-1/2}\,dx$

$\qquad = 2(x + 1)^{1/2} + C$

$\qquad = 2\sqrt{x + 1} + C$

23. $\displaystyle\int \frac{2x}{(x-1)^2}\,dx = \int \frac{2x - 2 + 2}{(x-1)^2}\,dx$

$\qquad = \displaystyle\int \frac{2(x-1)}{(x-1)^2}\,dx + 2\int \frac{1}{(x-1)^2}\,dx$

$\qquad = 2\displaystyle\int \frac{1}{x-1}\,dx + 2\int \frac{1}{(x-1)^2}\,dx$

$\qquad = 2\ln|x - 1| - \dfrac{2}{(x-1)} + C$

25. $u = 1 + \sqrt{2x}$, $du = \dfrac{1}{\sqrt{2x}}\,dx \implies (u - 1)\,du = dx$

$$\int \frac{1}{1 + \sqrt{2x}}\,dx = \int \frac{(u - 1)}{u}\,du = \int \left(1 - \frac{1}{u}\right) du$$

$$= u - \ln|u| + C_1$$

$$= \left(1 + \sqrt{2x}\right) - \ln\left|1 + \sqrt{2x}\right| + C_1$$

$$= \sqrt{2x} - \ln\left(1 + \sqrt{2x}\right) + C$$

where $C = C_1 + 1$.

27. $u = \sqrt{x} - 3$, $du = \dfrac{1}{2\sqrt{x}}\,dx \implies 2(u + 3)\,du = dx$

$$\int \frac{\sqrt{x}}{\sqrt{x} - 3}\,dx = 2\int \frac{(u + 3)^2}{u}\,du$$

$$= 2\int \frac{u^2 + 6u + 9}{u}\,du = 2\int \left(u + 6 + \frac{9}{u}\right) du$$

$$= 2\left[\frac{u^2}{2} + 6u + 9\ln|u|\right] + C_1$$

$$= u^2 + 12u + 18\ln|u| + C_1$$

$$= \left(\sqrt{x} - 3\right)^2 + 12\left(\sqrt{x} - 3\right) + 18\ln\left|\sqrt{x} - 3\right| + C_1$$

$$= x + 6\sqrt{x} + 18\ln\left|\sqrt{x} - 3\right| + C$$

where $C = C_1 - 27$.

29. $\displaystyle\int \frac{\cos\theta}{\sin\theta}\,d\theta = \ln|\sin\theta| + C$

$(u = \sin\theta, du = \cos\theta\,d\theta)$

31. $\displaystyle\int \csc 2x\,dx = \frac{1}{2}\int (\csc 2x)(2)\,dx$

$$= -\frac{1}{2}\ln|\csc 2x + \cot 2x| + C$$

33. $\displaystyle\int \frac{\cos t}{1 + \sin t}\,dt = \ln|1 + \sin t| + C$

35. $\displaystyle\int \frac{\sec x \tan x}{\sec x - 1}\,dx = \ln|\sec x - 1| + C$

37. $y = \displaystyle\int \frac{3}{2 - x}\,dx$

$$= -3\int \frac{1}{x - 2}\,dx$$

$$= -3\ln|x - 2| + C$$

$(1, 0):\ 0 = -3\ln|1 - 2| + C \implies C = 0$

$y = -3\ln|x - 2|$

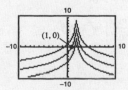

39. $s = \displaystyle\int \tan(2\theta)\,d\theta$

$$= \frac{1}{2}\int \tan(2\theta)(2\,d\theta)$$

$$= -\frac{1}{2}\ln|\cos 2\theta| + C$$

$(0, 2):\ 2 = -\dfrac{1}{2}\ln|\cos(0)| + C \implies C = 2$

$$s = -\frac{1}{2}\ln|\cos 2\theta| + 2$$

41. $f''(x) = \dfrac{2}{x^2} = 2x^{-2}, \; x > 0$

$f'(x) = \dfrac{-2}{x} + C$

$f'(1) = 1 = -2 + C \Rightarrow C = 3$

$f'(x) = \dfrac{-2}{x} + 3$

$f(x) = -2\ln x + 3x + C_1$

$f(1) = 1 = -2(0) + 3 + C_1 \Rightarrow C_1 = -2$

$f(x) = -2\ln x + 3x - 2$

43. $\dfrac{dy}{dx} = \dfrac{1}{x+2}, (0, 1)$

(a)

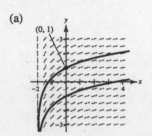

(b) $y = \displaystyle\int \dfrac{1}{x+2}\, dx = \ln|x+2| + C$

$y(0) = 1 \Rightarrow 1 = \ln 2 + C \Rightarrow C = 1 - \ln 2$

Hence, $y = \ln|x+2| + 1 - \ln 2 = \ln\left|\dfrac{x+2}{2}\right| + 1.$

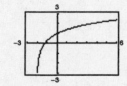

45. (a)

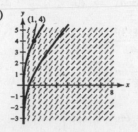

(b) $\dfrac{dy}{dx} = 1 + \dfrac{1}{x}, \quad (1, 4)$

$y = x + \ln x + C$

$4 = 1 + 0 + C \Rightarrow C = 3$

$y = x + \ln x + 3$

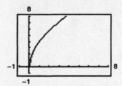

47. $\displaystyle\int_0^4 \dfrac{5}{3x+1}\, dx = \left[\dfrac{5}{3}\ln|3x+1|\right]_0^4$

$= \dfrac{5}{3}\ln 13 \approx 4.275$

49. $u = 1 + \ln x, \; du = \dfrac{1}{x}\, dx$

$\displaystyle\int_1^e \dfrac{(1 + \ln x)^2}{x}\, dx = \left[\dfrac{1}{3}(1 + \ln x)^3\right]_1^e$

$= \dfrac{7}{3}$

51. $\displaystyle\int_0^2 \dfrac{x^2 - 2}{x+1}\, dx = \int_0^2 \left(x - 1 - \dfrac{1}{x+1}\right) dx$

$= \left[\dfrac{1}{2}x^2 - x - \ln|x+1|\right]_0^2 = -\ln 3$

53. $\displaystyle\int_1^2 \dfrac{1 - \cos\theta}{\theta - \sin\theta}\, d\theta = \Big[\ln|\theta - \sin\theta|\Big]_1^2$

$= \ln\left|\dfrac{2 - \sin 2}{1 - \sin 1}\right| \approx 1.929$

55. $\displaystyle\int \dfrac{1}{1 + \sqrt{x}}\, dx = 2\left(1 + \sqrt{x}\right) - 2\ln\left(1 + \sqrt{x}\right) + C_1$

$= 2\left[\sqrt{x} - \ln\left(1 + \sqrt{x}\right)\right] + C$ where $C = C_1 + 2.$

57. $\int \dfrac{\sqrt{x}}{x-1}\,dx = \ln\left(\dfrac{\sqrt{x}-1}{\sqrt{x}+1}\right) + 2\sqrt{x} + C$

59. $\displaystyle\int_{\pi/4}^{\pi/2}(\csc x - \sin x)\,dx = \left[-\ln|\csc x + \cot x| + \cos x\right]_{\pi/4}^{\pi/2}$

$$= \ln\left(\sqrt{2}+1\right) - \dfrac{\sqrt{2}}{2} \approx 0.174$$

Note: In Exercises 61–63, you can use the Second Fundamental Theorem of Calculus or integrate the function.

61. $F(x) = \displaystyle\int_{1}^{x} \dfrac{1}{t}\,dt$

$F'(x) = \dfrac{1}{x}$

63. $F(x) = \displaystyle\int_{1}^{3x} \dfrac{1}{t}\,dt$

$F'(x) = \dfrac{1}{3x}(3) = \dfrac{1}{x}$

(by Second Fundamental Theorem of Calculus)

Alternate Solution:

$F(x) = \displaystyle\int_{1}^{3x} \dfrac{1}{t}\,dt = \left[\ln|t|\right]_{1}^{3x} = \ln|3x|$

$F'(x) = \dfrac{1}{3x}(3) = \dfrac{1}{x}$

65

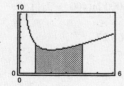

$A \approx 1.25$; Matches (d)

67. $A = \displaystyle\int_{1}^{3} \dfrac{4}{x}\,dx = 4\ln|x|\Big]_{1}^{3} = 4\ln 3$

69. $A = \displaystyle\int_{0}^{\pi/4} \tan x\,dx = -\ln|\cos x|\Big]_{0}^{\pi/4}$

$$= -\ln\dfrac{\sqrt{2}}{2} + 0$$

$$= \ln\sqrt{2} = \dfrac{\ln 2}{2} \approx 0.3466$$

71. $A = \displaystyle\int_{1}^{4} \dfrac{x^2+4}{x}\,dx = \int_{1}^{4}\left(x + \dfrac{4}{x}\right)dx$

$$= \left[\dfrac{x^2}{2} + 4\ln x\right]_{1}^{4} = (8 + 4\ln 4) - \dfrac{1}{2}$$

$$= \dfrac{15}{2} + 8\ln 2 \approx 13.045 \text{ square units}$$

73. $\displaystyle\int_{0}^{2} 2\sec\dfrac{\pi x}{6}\,dx = \dfrac{12}{\pi}\int_{0}^{2}\sec\left(\dfrac{\pi x}{6}\right)\dfrac{\pi}{6}\,dx$

$$= \left[\dfrac{12}{\pi}\ln\left|\sec\dfrac{\pi x}{6} + \tan\dfrac{\pi x}{6}\right|\right]_{0}^{2}$$

$$= \dfrac{12}{\pi}\ln\left|\sec\dfrac{\pi}{3} + \tan\dfrac{\pi}{3}\right| - \dfrac{12}{\pi}\ln|1 + 0|$$

$$= \dfrac{12}{\pi}\ln\left(2 + \sqrt{3}\right) \approx 5.03041$$

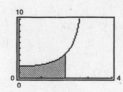

75. $f(x) = \dfrac{12}{x}$, $b - a = 5 - 1 = 4$, $n = 4$

Trapezoid: $\dfrac{4}{2(4)}[f(1) + 2f(2) + 2f(3) + 2f(4) + f(5)] = \dfrac{1}{2}[12 + 12 + 8 + 6 + 2.4] = 20.2$

Simpson: $\dfrac{4}{3(4)}[f(1) + 4f(2) + 2f(3) + 4f(4) + f(5)] = \dfrac{1}{3}[12 + 24 + 8 + 12 + 2.4] \approx 19.4667$

Calculator: $\displaystyle\int_1^5 \dfrac{12}{x}\,dx \approx 19.3133$

Exact: $12 \ln 5$

77. $f(x) = \ln x$, $b - a = 6 - 2 = 4$, $n = 4$

Trapezoid: $\dfrac{4}{2(4)}[f(2) + 2f(3) + 2f(4) + 2f(5) + f(6)] = \dfrac{1}{2}[0.6931 + 2.1972 + 2.7726 + 3.2189 + 1.7918] \approx 5.3368$

Simpson: $\dfrac{4}{3(4)}[f(2) + 4f(3) + 2f(4) + 4f(5) + f(6)] \approx 5.3632$

Calculator: $\displaystyle\int_2^6 \ln x\,dx \approx 5.3643$

79. Power Rule

81. Substitution: $(u = x^2 + 4)$ and Log Rule

83. $-\ln|\cos x| + C = \ln\left|\dfrac{1}{\cos x}\right| + C$

$= \ln|\sec x| + C$

85. $\ln|\sec x + \tan x| + C = \ln\left|\dfrac{(\sec x + \tan x)(\sec x - \tan x)}{(\sec x - \tan x)}\right| + C = \ln\left|\dfrac{\sec^2 x - \tan^2 x}{\sec x - \tan x}\right| + C$

$= \ln\left|\dfrac{1}{\sec x - \tan x}\right| + C = -\ln|\sec x - \tan x| + C$

87. Average value $= \dfrac{1}{4 - 2}\displaystyle\int_2^4 \dfrac{8}{x^2}\,dx$

$= 4\displaystyle\int_2^4 x^{-2}\,dx$

$= \left[-4\dfrac{1}{x}\right]_2^4$

$= -4\left(\dfrac{1}{4} - \dfrac{1}{2}\right) = 1$

89. Average value $= \dfrac{1}{e - 1}\displaystyle\int_1^e \dfrac{\ln x}{x}\,dx$

$= \dfrac{1}{e - 1}\left[\dfrac{(\ln x)^2}{2}\right]_1^e$

$= \dfrac{1}{e - 1}\left(\dfrac{1}{2}\right)$

$= \dfrac{1}{2e - 2} \approx 0.291$

91. $P(t) = \displaystyle\int \dfrac{3000}{1 + 0.25t}\,dt = (3000)(4)\int \dfrac{0.25}{1 + 0.25t}\,dt$

$= 12{,}000 \ln|1 + 0.25t| + C$

$P(0) = 12{,}000 \ln|1 + 0.25(0)| + C = 1000$

$C = 1000$

$P(t) = 12{,}000 \ln|1 + 0.25t| + 1000$

$= 1000[12 \ln|1 + 0.25t| + 1]$

$P(3) = 1000[12(\ln 1.75) + 1] \approx 7715$

93. $\dfrac{1}{50 - 40}\displaystyle\int_{40}^{50} \dfrac{90{,}000}{400 + 3x}\,dx = \left[3000 \ln|400 + 3x|\right]_{40}^{50}$

$\approx \$168.27$

95. (a) $2x^2 - y^2 = 8$

$$y^2 = 2x^2 - 8$$

$$y_1 = \sqrt{2x^2 - 8}$$

$$y_2 = -\sqrt{2x^2 - 8}$$

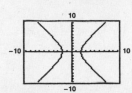

(b) $y^2 = e^{-\int(1/x)\,dx} = e^{-\ln x + C} = e^{\ln(1/x)}(e^C) = \dfrac{1}{x}k$

Let $k = 4$ and graph $y^2 = \dfrac{4}{x}$. $\left(y_1 = \dfrac{2}{\sqrt{x}}, y_2 = -\dfrac{2}{\sqrt{x}}\right)$

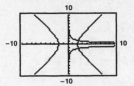

(c) In part (a): $2x^2 - y^2 = 8$ 　　　In part (b): $y^2 = \dfrac{4}{x} = 4x^{-1}$

$$4x - 2yy' = 0 \qquad\qquad\qquad 2yy' = \dfrac{-4}{x^2}$$

$$y' = \dfrac{2x}{y} \qquad\qquad\qquad y' = \dfrac{-2}{yx^2} = \dfrac{-2y}{y^2x^2} = \dfrac{-2y}{4x} = \dfrac{-y}{2x}$$

Using a graphing utility the graphs intersect at (2.214, 1.344). The slopes are 3.295 and $-0.304 = (-1)/3.295$, respectively.

97. False 　　　　　　　　　　　　　　**99. True**

$$\frac{1}{2}(\ln x) = \ln(x^{1/2}) \qquad\qquad \int \frac{1}{x}\,dx = \ln|x| + C_1$$

$$\neq (\ln x)^{1/2} \qquad\qquad\qquad = \ln|x| + \ln|C|$$

$$\qquad\qquad\qquad\qquad\qquad = \ln|Cx|, \ C \neq 0$$

101. $f(x) = \dfrac{x}{1 + x^2}$

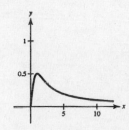

(a) $y = \dfrac{1}{2}x$ intersects $f(x) = \dfrac{x}{1+x^2}$:

$$\frac{1}{2}x = \frac{x}{1+x^2}$$

$$1 + x^2 = 2$$

$$x = 1$$

$$A = \int_0^1 \left(\left[\frac{x}{1+x^2}\right] - \frac{1}{2}x\right) dx$$

$$= \left[\frac{1}{2}\ln(x^2 + 1) - \frac{x^2}{4}\right]_0^1$$

$$= \frac{1}{2}\ln 2 - \frac{1}{4}$$

(b) $f'(x) = \dfrac{(1+x^2) - x(2x)}{(1+x^2)^2} = \dfrac{1 - x^2}{(1+x^2)^2}$

$$f'(0) = 1$$

Hence, for $0 < m < 1$, the graphs of f and $y = mx$ enclose a finite region.

(c)

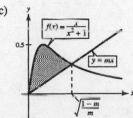

$$\frac{x}{1+x^2} = mx$$

$$1 = m + mx^2$$

$$x^2 = \frac{1-m}{m}$$

$$x = \sqrt{\frac{1-m}{m}}, \quad \text{intersection point}$$

$$A = \int_0^{\sqrt{(1-m)/m}} \left(\frac{x}{1+x^2} - mx\right) dx, \quad 0 < m < 1$$

$$= \left[\frac{1}{2}\ln(1+x^2) - \frac{mx^2}{2}\right]_0^{\sqrt{(1-m)/m}}$$

$$= \frac{1}{2}\ln\left(1 + \frac{1-m}{m}\right) - \frac{1}{2}m\left(\frac{1-m}{m}\right)$$

$$= \frac{1}{2}\ln\left(\frac{1}{m}\right) - \frac{1}{2}(1-m)$$

$$= \frac{1}{2}[m - \ln(m) - 1]$$

Section 5.3   Inverse Functions

1. (a) $f(x) = 5x + 1$

$$g(x) = \frac{x - 1}{5}$$

$$f(g(x)) = f\left(\frac{x - 1}{5}\right) = 5\left(\frac{x - 1}{5}\right) + 1 = x$$

$$g(f(x)) = g(5x + 1) = \frac{(5x - 1) - 1}{5} = x$$

(b)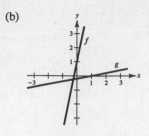

3. (a) $f(x) = x^3$

$$g(x) = \sqrt[3]{x}$$

$$f(g(x)) = f(\sqrt[3]{x}) = (\sqrt[3]{x})^3 = x$$

$$g(f(x)) = g(x^3) = \sqrt[3]{x^3} = x$$

(b)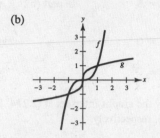

5. (a) $f(x) = \sqrt{x - 4}$

$$g(x) = x^2 + 4, \quad x \geq 0$$

$$f(g(x)) = f(x^2 + 4)$$

$$= \sqrt{(x^2 + 4) - 4} = \sqrt{x^2} = x$$

$$g(f(x)) = g(\sqrt{x - 4})$$

$$= (\sqrt{x - 4})^2 + 4 = x - 4 + 4 = x$$

(b)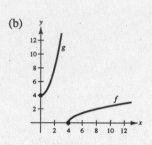

7. (a) $f(x) = \dfrac{1}{x}$

$$g(x) = \frac{1}{x}$$

$$f(g(x)) = \frac{1}{1/x} = x$$

$$g(f(x)) = \frac{1}{1/x} = x$$

(b)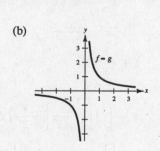

9. Matches (c)

11. Matches (a)

13. $f(x) = \frac{3}{4}x + 6$

One-to-one; has an inverse

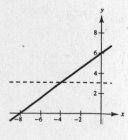

15. $f(\theta) = \sin \theta$

Not one-to-one; does not have an inverse

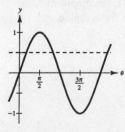

17. $h(s) = \dfrac{1}{s - 2} - 3$

One-to-one; has an inverse

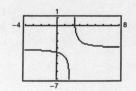

19. $f(x) = \ln x$

One-to-one; has an inverse

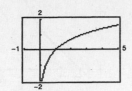

21. $g(x) = (x + 5)^3$

One-to-one; has an inverse

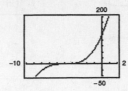

23. $f(x) = \ln(x - 3), \quad x > 3$

$$f'(x) = \frac{1}{x - 3} > 0 \text{ for } x > 3$$

f is increasing on $(3, \infty)$.
Therefore, f is strictly monotonic
and has an inverse.

25. $f(x) = \dfrac{x^4}{4} - 2x^2$

$f'(x) = x^3 - 4x = 0$ when $x = 0, 2, -2$

f is not strictly monotonic on $(-\infty, \infty)$. Therefore,
f does not have an inverse.

27. $f(x) = 2 - x - x^3$

$f'(x) = -1 - 3x^2 < 0$ for all x

f is decreasing on $(-\infty, \infty)$. Therefore, f is strictly
monotonic and has an inverse.

29. $f(x) = 2x - 3 = y$

$$x = \frac{y + 3}{2}$$

$$y = \frac{x + 3}{2}$$

$$f^{-1}(x) = \frac{x + 3}{2}$$

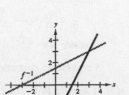

31. $f(x) = x^5 = y$

$$x = \sqrt[5]{y}$$

$$y = \sqrt[5]{x}$$

$$f^{-1}(x) = \sqrt[5]{x} = x^{1/5}$$

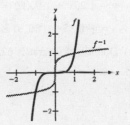

33. $f(x) = \sqrt{x} = y$

$$x = y^2$$

$$y = x^2$$

$$f^{-1}(x) = x^2, \quad x \geq 0$$

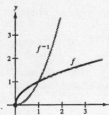

35. $f(x) = \sqrt{4 - x^2} = y, \quad 0 \leq x \leq 2$

$$x = \sqrt{4 - y^2}$$

$$y = \sqrt{4 - x^2}$$

$$f^{-1}(x) = \sqrt{4 - x^2}, \quad 0 \leq x \leq 2$$

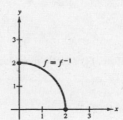

37. $f(x) = \sqrt[3]{x - 1} = y$

$$x = y^3 + 1$$

$$y = x^3 + 1$$

$$f^{-1}(x) = x^3 + 1$$

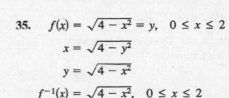

The graphs of f and f^{-1} are reflections of each other
across the line $y = x$.

39. $f(x) = x^{2/3} = y, \quad x \geq 0$

$$x = y^{3/2}$$

$$y = x^{3/2}$$

$$f^{-1}(x) = x^{3/2}, \quad x \geq 0$$

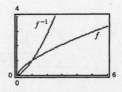

The graphs of f and f^{-1} are reflections of each other
across the line $y = x$.

41. $f(x) = \dfrac{x}{\sqrt{x^2 + 7}} = y$

$x = \dfrac{\sqrt{7}y}{\sqrt{1 - y^2}}$

$y = \dfrac{\sqrt{7}x}{\sqrt{1 - x^2}}$

$f^{-1}(x) = \dfrac{\sqrt{7}x}{\sqrt{1 - x^2}}, \quad -1 < x < 1$

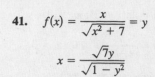 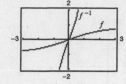

The graphs of f and f^{-1} are reflections of each other across the line $y = x$.

43.

x	1	2	3	4
$f^{-1}(x)$	0	1	2	4

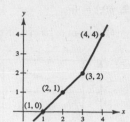

45. (a) Let x be the number of pounds of the commodity costing 1.25 per pound. Since there are 50 pounds total, the amount of the second commodity is $50 - x$. The total cost is

$y = 1.25x + 1.60(50 - x)$

$= -0.35x + 80, \ 0 \le x \le 50.$

(c) Domain of inverse is $62.5 \le x \le 80$.

(b) We find the inverse of the original function:

$y = -0.35x + 80$

$0.35x = 80 - y$

$x = \frac{100}{35}(80 - y)$

Inverse: $y = \frac{100}{35}(80 - x) = \frac{20}{7}(80 - x)$

x represents cost and y represents pounds.

(d) If $x = 73$ in the inverse function,
$y = \frac{100}{35}(80 - 73) = \frac{100}{5} = 20$ pounds.

47. $f(x) = (x - 4)^2$ on $[4, \infty)$

$f'(x) = 2(x - 4) > 0$ on $(4, \infty)$

f is increasing on $[4, \infty)$. Therefore, f is strictly monotonic and has an inverse.

49. $f(x) = \dfrac{4}{x^2}$ on $(0, \infty)$

$f'(x) = -\dfrac{8}{x^3} < 0$ on $(0, \infty)$

f is decreasing on $(0, \infty)$. Therefore, f is strictly monotonic and has an inverse.

51. $f(x) = \cos x$ on $[0, \pi]$

$f'(x) = -\sin x < 0$ on $(0, \pi)$

f is decreasing on $[0, \pi]$. Therefore, f is strictly monotonic and has an inverse.

53. $f(x) = \dfrac{x}{x^2 - 4} = y$ on $(-2, 2)$

$x^2y - 4y = x$

$x^2y - x - 4y = 0$

$a = y, b = -1, c = -4y$

$x = \dfrac{1 \pm \sqrt{1 - 4(y)(-4y)}}{2y} = \dfrac{1 \pm \sqrt{1 + 16y^2}}{2y}$

$y = f^{-1}(x) = \begin{cases} (1 - \sqrt{1 + 16x^2})/2x, & \text{if } x \neq 0 \\ 0, & \text{if } x = 0 \end{cases}$

Domain of f^{-1}: all x

Range of f^{-1}: $-2 < y < 2$

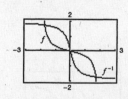

The graphs of f and f^{-1} are reflections of each other across the line $y = x$.

55. (a), (b)

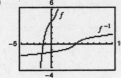

(c) Yes, f is one-to-one and has an inverse. The inverse relation is an inverse function.

57. (a), (b)

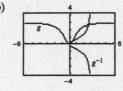

(c) g is not one-to-one and does not have an inverse. The inverse relation is not an inverse function.

59. $f(x) = \sqrt{x - 2}$, Domain: $x \geq 2$

$$f'(x) = \frac{1}{2\sqrt{x - 2}} > 0 \text{ for } x > 2$$

f is one-to-one; has an inverse

$$\sqrt{x - 2} = y$$

$$x - 2 = y^2$$

$$x = y^2 + 2$$

$$y = x^2 + 2$$

$$f^{-1}(x) = x^2 + 2, \quad x \geq 0$$

61. $f(x) = |x - 2|, \quad x \leq 2$

$$= -(x - 2)$$

$$= 2 - x$$

f is one-to-one; has an inverse

$$2 - x = y$$

$$2 - y = x$$

$$f^{-1}(x) = 2 - x, \quad x \geq 0$$

63. $f(x) = (x - 3)^2$ is one-to-one for $x \geq 3$.

$$(x - 3)^2 = y$$

$$x - 3 = \sqrt{y}$$

$$x = \sqrt{y} + 3$$

$$y = \sqrt{x} + 3$$

$$f^{-1}(x) = \sqrt{x} + 3, \quad x \geq 0$$

(Answer is not unique.)

65. $f(x) = |x + 3|$ is one-to-one for $x \geq -3$.

$$x + 3 = y$$

$$x = y - 3$$

$$y = x - 3$$

$$f^{-1}(x) = x - 3, \quad x \geq 0$$

(Answer is not unique.)

67. Yes, the volume is an increasing function, and hence one-to-one. The inverse function gives the time t corresponding to the volume V.

69. No, $C(t)$ is not one-to-one because long distance costs are step functions. A call lasting 2.1 minutes costs the same as one lasting 2.2 minutes.

71.
$$f(x) = x^3 + 2x - 1, \quad f(1) = 2 = a$$

$$f'(x) = 3x^2 + 2$$

$$(f^{-1})'(2) = \frac{1}{f'(f^{-1}(2))} = \frac{1}{f'(1)} = \frac{1}{3(1)^2 + 2} = \frac{1}{5}$$

73.
$$f(x) = \sin x, \quad f\left(\frac{\pi}{6}\right) = \frac{1}{2} = a$$

$$f'(x) = \cos x$$

$$(f^{-1})'\left(\frac{1}{2}\right) = \frac{1}{f'(f^{-1}(1/2))} = \frac{1}{f'(\pi/6)} = \frac{1}{\cos(\pi/6)}$$

$$= \frac{1}{\sqrt{3}/2} = \frac{2\sqrt{3}}{3}$$

75.
$$f(x) = x^3 - \frac{4}{x}, \quad f(2) = 6 = a$$

$$f'(x) = 3x^2 + \frac{4}{x^2}$$

$$(f^{-1})'(6) = \frac{1}{f'(f^{-1}(6))} = \frac{1}{f'(2)} = \frac{1}{3(2)^2 + (4/2^2)} = \frac{1}{13}$$

77. (a) Domain f = Domain f^{-1} = $(-\infty, \infty)$

(b) Range f = Range f^{-1} = $(-\infty, \infty)$

(c)

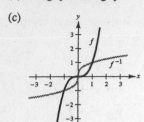

(d) $\quad f(x) = x^3, \ \left(\dfrac{1}{2}, \dfrac{1}{8}\right)$

$$f'(x) = 3x^2$$

$$f'\left(\frac{1}{2}\right) = \frac{3}{4}$$

$$f^{-1}(x) = \sqrt[3]{x}, \ \left(\frac{1}{8}, \frac{1}{2}\right)$$

$$(f^{-1})'(x) = \frac{1}{3\sqrt[3]{x^2}}$$

$$(f^{-1})'\left(\frac{1}{8}\right) = \frac{4}{3}$$

79. (a) Domain f = $[4, \infty)$, Domain f^{-1} = $[0, \infty)$

(b) Range f = $[0, \infty)$, Range f^{-1} = $[4, \infty)$

(c)

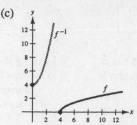

(d) $\quad f(x) = \sqrt{x-4}, \ (5, 1)$

$$f'(x) = \frac{1}{2\sqrt{x-4}}$$

$$f'(5) = \frac{1}{2}$$

$$f^{-1}(x) = x^2 + 4, \ (1, 5)$$

$$(f^{-1})'(x) = 2x$$

$$(f^{-1})'(1) = 2$$

81. $\quad x = y^3 - 7y^2 + 2$

$$1 = 3y^2 \frac{dy}{dx} - 14y \frac{dy}{dx}$$

$$\frac{dy}{dx} = \frac{1}{3y^2 - 14y}$$

At $(-4, 1), \dfrac{dy}{dx} = \dfrac{1}{3 - 14} = \dfrac{-1}{11}.$

Alternate Solution:

Let $f(x) = x^3 - 7x^2 + 2$. Then $f'(x) = 3x^2 - 14x$ and $f'(1) = -11$. Hence,

$$\frac{dy}{dx} = \frac{1}{-11} = \frac{-1}{11}.$$

In Exercises 83–85, use the following.

$f(x) = \frac{1}{8}x - 3$ and $g(x) = x^3$

$f^{-1}(x) = 8(x + 3)$ and $g^{-1}(x) = \sqrt[3]{x}$

83. $(f^{-1} \circ g^{-1})(1) = f^{-1}(g^{-1}(1)) = f^{-1}(1) = 32$

85. $(f^{-1} \circ f^{-1})(6) = f^{-1}(f^{-1}(6)) = f^{-1}(72) = 600$

In Exercises 87–89, use the following.

$f(x) = x + 4$ and $g(x) = 2x - 5$

$f^{-1}(x) = x - 4$ and $g^{-1}(x) = \dfrac{x + 5}{2}$

87. $(g^{-1} \circ f^{-1})(x) = g^{-1}(f^{-1}(x))$

$$= g^{-1}(x - 4)$$

$$= \frac{(x - 4) + 5}{2}$$

$$= \frac{x + 1}{2}$$

89. $(f \circ g)(x) = f(g(x))$

$$= f(2x - 5)$$

$$= (2x - 5) + 4$$

$$= 2x - 1$$

Hence, $(f \circ g)^{-1}(x) = \dfrac{x + 1}{2}.$

Note: $(f \circ g)^{-1} = g^{-1} \circ f^{-1}$

91. Answers will vary. See page 343 and Example 3.

93. f is not one-to-one because many different x-values yield the same y-value.

Example: $f(0) = f(\pi) = 0$

Not continuous at $\dfrac{(2n - 1)\pi}{2}$, where n is an integer.

95. $f(x) = k(2 - x - x^3)$ is one-to-one. Since $f^{-1}(3) = -2$,

$$f(-2) = 3 = k(2 - (-2) - (-2)^3) = 12k \implies k = \tfrac{1}{4}.$$

97. Let f and g be one-to-one functions.

(a) Let $(f \circ g)(x_1) = (f \circ g)(x_2)$

$$f(g(x_1)) = f(g(x_2))$$

$$g(x_1) = g(x_2) \qquad \text{(Because } f \text{ is one-to-one.)}$$

$$x_1 = x_2 \qquad \text{(Because } g \text{ is one-to-one.)}$$

Thus, $f \circ g$ is one-to-one.

(b) Let $(f \circ g)(x) = y$, then $x = (f \circ g)^{-1}(y)$. Also:

$$(f \circ g)(x) = y$$

$$f(g(x)) = y$$

$$g(x) = f^{-1}(y)$$

$$x = g^{-1}(f^{-1}(y))$$

$$x = (g^{-1} \circ f^{-1})(y)$$

Thus, $(f \circ g)^{-1}(y) = (g^{-1} \circ f^{-1})(y)$ and $(f \circ g)^{-1} = g^{-1} \circ f^{-1}$.

99. Suppose $g(x)$ and $h(x)$ are both inverses of $f(x)$. Then the graph of $f(x)$ contains the point (a, b) if and only if the graphs of $g(x)$ and $h(x)$ contain the point (b, a). Since the graphs of $g(x)$ and $h(x)$ are the same, $g(x) = h(x)$. Therefore, the inverse of $f(x)$ is unique.

101. False. Let $f(x) = x^2$.

103. True

105. Not true.

Let $f(x) = \begin{cases} x, & 0 \le x \le 1 \\ 1 - x, & 1 < x \le 2 \end{cases}$.

f is one-to-one, but not strictly monotonic.

107.
$$f(x) = \int_2^x \frac{dt}{\sqrt{1 + t^4}}, \quad f(2) = 0$$

$$f'(x) = \frac{1}{\sqrt{1 + x^4}}$$

$$(f^{-1})'(0) = \frac{1}{f'(2)} = \frac{1}{1/\sqrt{17}} = \sqrt{17}$$

109. (a)
$$y = \frac{x - 2}{x - 1}$$

$$x = \frac{y - 2}{y - 1}$$

$$xy - x = y - 2$$

$$xy - y = x - 2$$

$$y = \frac{x - 2}{x - 1}$$

Hence, if $f(x) = \dfrac{x - 2}{x - 1}$, then $f^{-1}(x) = f(x)$.

(b) The graph of f is symmetric about the line $y = x$.

Section 5.4 Exponential Functions: Differentiation and Integration

1. $e^{\ln x} = 4$

$x = 4$

3. $e^x = 12$

$x = \ln 12 \approx 2.485$

5. $9 - 2e^x = 7$

$2e^x = 2$

$e^x = 1$

$x = 0$

7. $50e^{-x} = 30$

$e^{-x} = \frac{3}{5}$

$-x = \ln\left(\frac{3}{5}\right)$

$x = \ln\left(\frac{5}{3}\right)$

≈ 0.511

9. $\ln x = 2$

$x = e^2 \approx 7.389$

11. $\ln(x - 3) = 2$

$x - 3 = e^2$

$x = 3 + e^2 \approx 10.389$

13. $\ln\sqrt{x + 2} = 1$

$\sqrt{x + 2} = e^1 = e$

$x + 2 = e^2$

$x = e^2 - 2 \approx 5.389$

15. $y = e^{-x}$

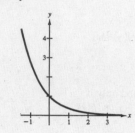

17. $y = e^{-x^2}$

Symmetric with respect to the y-axis

Horizontal asymptote: $y = 0$

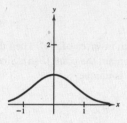

19. (a)

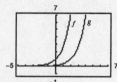

Horizontal shift 2 units to the right

(b)

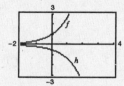

A reflection in the x-axis and a vertical shrink

(c)

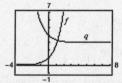

Vertical shift 3 units upward and a reflection in the y-axis

21. $y = Ce^{ax}$

Horizontal asymptote: $y = 0$

Matches (c)

23. $y = C(1 - e^{-ax})$

Vertical shift C units

Reflection in both the x- and y-axes

Matches (a)

25. $f(x) = e^{2x}$

$g(x) = \ln\sqrt{x} = \frac{1}{2}\ln x$

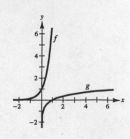

27. $f(x) = e^x - 1$

$g(x) = \ln(x + 1)$

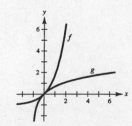

29.

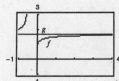

As $x \to \infty$, the graph of f approaches the graph of g.

$$\lim_{x \to \infty}\left(1 + \frac{0.5}{x}\right)^x = e^{0.5}$$

31. $\left(1 + \dfrac{1}{1,000,000}\right)^{1,000,000} \approx 2.718280469$

$$e \approx 2.718281828$$

$$e > \left(1 + \dfrac{1}{1,000,000}\right)^{1,000,000}$$

33. (a) $y = e^{3x}$

$y' = 3e^{3x}$

$y'(0) = 3$

$y - 1 = 3(x - 0)$

$y = 3x + 1$ Tangent line

(b) $y = e^{-3x}$

$y' = -3e^{-3x}$

$y'(0) = -3$

$y - 1 = -3(x - 0)$

$y = -3x + 1$ Tangent line

35. $f(x) = e^{2x}$

$f'(x) = 2e^{2x}$

37. $y = e^{\sqrt{x}}$

$\dfrac{dy}{dx} = \dfrac{e^{\sqrt{x}}}{2\sqrt{x}}$

39. $g(t) = (e^{-t} + e^{t})^3$

$g'(t) = 3(e^{-t} + e^{t})^2(e^{t} - e^{-t})$

41. $y = \ln(1 + e^{2x})$

$\dfrac{dy}{dx} = \dfrac{2e^{2x}}{1 + e^{2x}}$

43. $y = \dfrac{2}{e^x + e^{-x}} = 2(e^x + e^{-x})^{-1}$

$\dfrac{dy}{dx} = -2(e^x + e^{-x})^{-2}(e^x - e^{-x}) = \dfrac{-2(e^x - e^{-x})}{(e^x + e^{-x})^2}$

45. $y = e^x(\sin x + \cos x)$

$\dfrac{dy}{dx} = e^x(\cos x - \sin x) + (\sin x + \cos x)(e^x)$

$= e^x(2 \cos x) - 2e^x \cos x$

47. $F(x) = \displaystyle\int_{\pi}^{\ln x} \cos e^t \, dt$

$F'(x) = \cos(e^{\ln x}) \cdot \dfrac{1}{x} = \dfrac{\cos(x)}{x}$

49. $f(x) = e^{1-x}, \quad (1, 1)$

$f'(x) = -e^{1-x}, \quad f'(1) = -1$

$y - 1 = -1(x - 1)$

$y = -x + 2$ Tangent line

51. $y = \ln(e^{x^2}) = x^2, \quad (-2, 4)$

$y' = 2x, \quad y'(-2) = -4$

$y - 4 = -4(x + 2)$

$y = -4x - 4$ Tangent line

53. $y = x^2e^x - 2xe^x + 2e^x, \quad (1, e)$

$y' = x^2e^x + 2xe^x - 2xe^x - 2e^x + 2e^x = x^2e^x$

$y'(1) = e$

$y - e = e(x - 1)$

$y = ex$ Tangent line

55. $f(x) = e^{-x} \ln x, \quad (1, 0)$

$f'(x) = e^{-x}\left(\dfrac{1}{x}\right) - e^{-x} \ln x = e^{-x}\left(\dfrac{1}{x} - \ln x\right)$

$f'(1) = e^{-1}$

$y - 0 = e^{-1}(x - 1)$

$y = \dfrac{1}{e}x - \dfrac{1}{e}$ Tangent line

57. $xe^y - 10x + 3y = 0$

$xe^y \dfrac{dy}{dx} + e^y - 10 + 3\dfrac{dy}{dx} = 0$

$\dfrac{dy}{dx}(xe^y + 3) = 10 - e^y$

$\dfrac{dy}{dx} = \dfrac{10 - e^y}{xe^y + 3}$

59. $xe^y + ye^x = 1, \quad (0, 1)$

$xe^yy' + e^y + ye^x + y'e^x = 0$

At $(0, 1)$: $e + 1 + y' = 0$

$y' = -e - 1$

Tangent line: $y - 1 = (-e - 1)(x - 0)$

$y = (-e - 1)x + 1$

61. $f(x) = (3 + 2x)e^{-3x}$

$\quad f'(x) = (3 + 2x)(-3e^{-3x}) + 2e^{-3x}$

$\qquad = (-7 - 6x)e^{-3x}$

$\quad f''(x) = (-7 - 6x)(-3e^{-3x}) - 6e^{-3x}$

$\qquad = 3(6x + 5)e^{-3x}$

63. $\qquad y = e^x\left(\cos \sqrt{2}x + \sin \sqrt{2}x\right)$

$\qquad y' = e^x\left(-\sqrt{2}\sin \sqrt{2}x + \sqrt{2}\cos \sqrt{2}x\right) + e^x\left(\cos \sqrt{2}x + \sin \sqrt{2}x\right)$

$\qquad\quad = e^x\left[\left(1 + \sqrt{2}\right)\cos\sqrt{2}x + \left(1 - \sqrt{2}\right)\sin\sqrt{2}x\right]$

$\qquad y'' = e^x\left[-\left(\sqrt{2} + 2\right)\sin \sqrt{2}x + \left(\sqrt{2} - 2\right)\cos \sqrt{2}x\right] + e^x\left[\left(1 + \sqrt{2}\right)\cos\sqrt{2}x + \left(1 - \sqrt{2}\right)\sin \sqrt{2}x\right]$

$\qquad\quad = e^x\left[\left(-1 - 2\sqrt{2}\right)\sin \sqrt{2}x + \left(-1 + 2\sqrt{2}\right)\cos \sqrt{2}x\right]$

$\quad -2y' + 3y = -2e^x\left[\left(1 + \sqrt{2}\right)\cos \sqrt{2}x + \left(1 - \sqrt{2}\right)\sin \sqrt{2}x\right] + 3e^x\left[\cos \sqrt{2}x + \sin \sqrt{2}x\right]$

$\qquad\qquad\quad = e^x\left[\left(1 - 2\sqrt{2}\right)\cos \sqrt{2}x + \left(1 + 2\sqrt{2}\right)\sin \sqrt{2}x\right] = -y''$

Therefore, $-2y' + 3y = -y'' \implies y'' - 2y' + 3y = 0.$

65. $f(x) = \dfrac{e^x + e^{-x}}{2}$

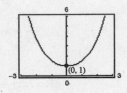

$\quad f'(x) = \dfrac{e^x - e^{-x}}{2} = 0$ when $x = 0.$

$\quad f''(x) = \dfrac{e^x + e^{-x}}{2} > 0$

Relative minimum: $(0, 1)$

67. $g(x) = \dfrac{1}{\sqrt{2\pi}}e^{-(x-2)^2/2}$

$\quad g'(x) = \dfrac{-1}{\sqrt{2\pi}}(x - 2)e^{-(x-2)^2/2}$

$\quad g''(x) = \dfrac{1}{\sqrt{2\pi}}(x - 1)(x - 3)e^{-(x-2)^2/2}$

Relative maximum: $\left(2, \dfrac{1}{\sqrt{2\pi}}\right) \approx (2, 0.399)$

Points of inflection: $\left(1, \dfrac{1}{\sqrt{2\pi}}e^{-1/2}\right), \left(3, \dfrac{1}{\sqrt{2\pi}}e^{-1/2}\right) \approx (1, 0.242), (3, 0.242)$

69. $f(x) = x^2 e^{-x}$

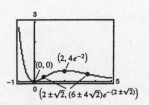

$\quad f'(x) = -x^2 e^{-x} + 2xe^{-x} = xe^{-x}(2 - x) = 0$ when $x = 0, 2.$

$\quad f''(x) = -e^{-x}(2x - x^2) + e^{-x}(2 - 2x)$

$\qquad = e^{-x}(x^2 - 4x + 2) = 0$ when $x = 2 \pm \sqrt{2}.$

Relative minimum: $(0, 0)$

Relative maximum: $(2, 4e^{-2})$

$x = 2 \pm \sqrt{2}$

$y = \left(2 \pm \sqrt{2}\right)^2 e^{-(2 \pm \sqrt{2})}$

Points of inflection: $(3.414, 0.384), (0.586, 0.191)$

71. $g(t) = 1 + (2 + t)e^{-t}$

$g'(t) = -(1 + t)e^{-t}$

$g''(t) = te^{-t}$

Relative maximum: $(-1, 1 + e) \approx (-1, 3.718)$

Point of inflection: $(0, 3)$

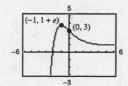

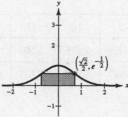

73. $A = (\text{base})(\text{height}) = 2xe^{-x^2}$

$\dfrac{dA}{dx} = -4x^2e^{-x^2} + 2e^{-x^2}$

$= 2e^{-x^2}(1 - 2x^2) = 0$ when $x = \dfrac{\sqrt{2}}{2}$.

$A = \sqrt{2}e^{-1/2}$

75. $y = \dfrac{L}{1 + ae^{-x/b}}, \quad a > 0, b > 0, L > 0$

$y' = \dfrac{-L\left(-\dfrac{a}{b}e^{-x/b}\right)}{(1 + ae^{-x/b})^2} = \dfrac{\dfrac{aL}{b}e^{-x/b}}{(1 + ae^{-x/b})^2}$

$y'' = \dfrac{(1 + ae^{-x/b})^2\left(\dfrac{-aL}{b^2}e^{-x/b}\right) - \left(\dfrac{aL}{b}e^{-x/b}\right)2(1 + ae^{-x/b})\left(\dfrac{-a}{b}e^{-x/b}\right)}{(1 + ae^{-x/b})^4}$

$= \dfrac{(1 + ae^{-x/b})\left(\dfrac{-aL}{b^2}e^{-x/b}\right) + 2\left(\dfrac{aL}{b}e^{-x/b}\right)\left(\dfrac{a}{b}e^{-x/b}\right)}{(1 + ae^{-x/b})^3}$

$= \dfrac{Lae^{-x/b}[ae^{-x/b} - 1]}{(1 + ae^{-x/b})^3 b^2}$

$y'' = 0$ if $ae^{-x/b} = 1 \Rightarrow \dfrac{-x}{b} = \ln\left(\dfrac{1}{a}\right) \Rightarrow x = b \ln a$

$y(b \ln a) = \dfrac{L}{1 + ae^{-(b \ln a)/b}} = \dfrac{L}{1 + a(1/a)} = \dfrac{L}{2}$

Therefore, the y-coordinate of the inflection point is $L/2$.

77. $f(x) = e^{2x}$

$f'(x) = 2e^{2x}$

Let $(x, y) = (x, e^{2x})$ be the point on the graph where the tangent line passes through the origin. Equating slopes,

$2e^{2x} = \dfrac{e^{2x} - 0}{x - 0}$

$2 = \dfrac{1}{x}$

$x = \dfrac{1}{2}, y = e, y' = 2e$.

Point: $\left(\dfrac{1}{2}, e\right)$

Tangent line: $y - e = 2e\left(x - \dfrac{1}{2}\right)$

$y = 2ex$

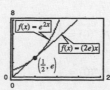

79. $V = 15,000e^{-0.6286t}, \quad 0 \leq t \leq 10$

(a)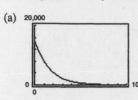

(b) $\dfrac{dV}{dt} = -9429e^{-0.6286t}$

When $t = 1, \dfrac{dV}{dt} \approx -5028.84$.

When $t = 5, \dfrac{dV}{dt} \approx -406.89$.

(c)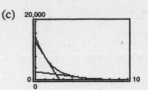

81.

h	0	5	10	15	20
P	10,332	5583	2376	1240	517
$\ln P$	9.243	8.627	7.773	7.123	6.248

(a)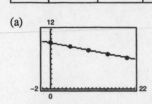

$y = -0.1499h + 9.3018$ is the regression line for data $(h, \ln P)$.

(c)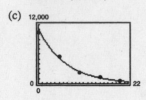

(b) $\ln P = ah + b$

$P = e^{ah+b} = e^b e^{ah}$

$P = Ce^{ah}, C = e^b$

For our data, $a = -0.1499$ and $C = e^{9.3018} = 10,957.7$.

$P = 10,957.7e^{-0.1499h}$

(d) $\dfrac{dP}{dh} = (10,957.71)(-0.1499)e^{-0.1499h}$

$= -1642.56e^{-0.1499h}$

For $h = 5, \dfrac{dP}{dh} = -776.3$. For $h = 18, \dfrac{dP}{dh} \approx -110.6$.

83.

$f(x) = e^{x/2}$,	$f(0) = 1$
$f'(x) = \dfrac{1}{2}e^{x/2}$,	$f'(0) = \dfrac{1}{2}$
$f''(x) = \dfrac{1}{4}e^{x/2}$,	$f''(0) = \dfrac{1}{4}$
$P_1(x) = 1 + \dfrac{1}{2}(x - 0) = \dfrac{x}{2} + 1$,	$P_1(0) = 1$
$P_1'(x) = \dfrac{1}{2}$,	$P_1'(0) = \dfrac{1}{2}$
$P_2(x) = 1 + \dfrac{1}{2}(x - 0) + \dfrac{1}{8}(x - 0)^2$	$P_2(0) = 1$
$\quad = \dfrac{x^2}{8} + \dfrac{x}{2} + 1$	
$P_2'(x) = \dfrac{1}{4}x + \dfrac{1}{2}$,	$P_2'(0) = \dfrac{1}{2}$
$P_2''(x) = \dfrac{1}{4}$,	$P_2''(0) = \dfrac{1}{4}$

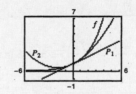

The values of f, P_1, P_2 and their first derivatives agree at $x = 0$. The values of the second derivatives of f and P_2 agree at $x = 0$.

85. Let $u = 5x, du = 5\, dx$.

$\displaystyle\int e^{5x}(5)\, dx = e^{5x} + C$

87. $\displaystyle\int \dfrac{e^{\sqrt{x}}}{\sqrt{x}}\, dx = 2\int e^{\sqrt{x}}\left(\dfrac{1}{2\sqrt{x}}\right) dx = 2e^{\sqrt{x}} + C$

89. Let $u = 1 + e^{-x}$, $du = -e^{-x} dx$.

$$\int \frac{e^{-x}}{1 + e^{-x}} dx = -\int \frac{-e^{-x}}{1 + e^{-x}} dx = -\ln(1 + e^{-x}) + C = \ln\left(\frac{e^x}{e^x + 1}\right) + C = x - \ln(e^x + 1) + C$$

91. Let $u = 1 - e^x$, $du = -e^x dx$.

$$\int e^x \sqrt{1 - e^x} \, dx = -\int (1 - e^x)^{1/2}(-e^x) \, dx$$

$$= -\frac{2}{3}(1 - e^x)^{3/2} + C$$

93. Let $u = e^x - e^{-x}$, $du = (e^x + e^{-x}) \, dx$.

$$\int \frac{e^x + e^{-x}}{e^x - e^{-x}} dx = \ln|e^x - e^{-x}| + C$$

95. $\displaystyle \int \frac{5 - e^x}{e^{2x}} dx = \int 5e^{-2x} dx - \int e^{-x} dx$

$$= -\frac{5}{2}e^{-2x} + e^{-x} + C$$

97. $\displaystyle \int e^{-x} \tan(e^{-x}) \, dx = -\int [\tan(e^{-x})](-e^{-x}) \, dx$

$$= \ln|\cos(e^{-x})| + C$$

99. Let $u = -2x$, $du = -2 \, dx$.

$$\int_0^1 e^{-2x} \, dx = -\frac{1}{2}\int_0^1 e^{-2x}(-2) \, dx = \left[-\frac{1}{2}e^{-2x}\right]_0^1$$

$$= \frac{1}{2}(1 - e^{-2}) = \frac{e^2 - 1}{2e^2}$$

101. $\displaystyle \int_0^1 xe^{-x^2} \, dx = -\frac{1}{2}\int_0^1 e^{-x^2}(-2x) \, dx$

$$= -\frac{1}{2}\left[e^{-x^2}\right]_0^1$$

$$= -\frac{1}{2}[e^{-1} - 1]$$

$$= \frac{1 - (1/e)}{2} = \frac{e - 1}{2e}$$

103. Let $u = \dfrac{3}{x}$, $du = -\dfrac{3}{x^2} \, dx$.

$$\int_1^3 \frac{e^{3/x}}{x^2} dx = -\frac{1}{3}\int_1^3 e^{3/x}\left(-\frac{3}{x^2}\right) dx$$

$$= \left[-\frac{1}{3}e^{3/x}\right]_1^3 = \frac{e}{3}(e^2 - 1)$$

105. $\displaystyle \int_0^{\pi/2} e^{\sin \pi x} \cos \pi x \, dx = \frac{1}{\pi}\int_0^{\pi/2} e^{\sin \pi x}(\pi \cos \pi x) \, dx$

$$= \frac{1}{\pi}\left[e^{\sin \pi x}\right]_0^{\pi/2}$$

$$= \frac{1}{\pi}[e^{\sin(\pi^2/2)} - 1]$$

107. Let $u = ax^2$, $du = 2ax \, dx$. (Assume $a \neq 0$.)

$$y = \int xe^{ax^2} \, dx$$

$$= \frac{1}{2a}\int e^{ax^2}(2ax) \, dx = \frac{1}{2a}e^{ax^2} + C$$

109. $f'(x) = \displaystyle \int \frac{1}{2}(e^x + e^{-x}) \, dx = \frac{1}{2}(e^x - e^{-x}) + C_1$

$f'(0) = C_1 = 0$

$f(x) = \displaystyle \int \frac{1}{2}(e^x - e^{-x}) \, dx = \frac{1}{2}(e^x + e^{-x}) + C_2$

$f(0) = 1 + C_2 = 1 \Rightarrow C_2 = 0$

$f(x) = \dfrac{1}{2}(e^x + e^{-x})$

111. (a)

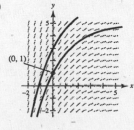

(b) $\dfrac{dy}{dx} = 2e^{-x/2}$, $(0, 1)$

$$y = \int 2e^{-x/2} \, dx = -4\int e^{-x/2}\left(-\frac{1}{2} \, dx\right)$$

$$= -4e^{-x/2} + C$$

$(0, 1)$: $1 = -4e^0 + C = -4 + C \Rightarrow C = 5$

$y = -4e^{-x/2} + 5$

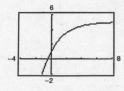

113. $\int_0^5 e^x \, dx = \left[e^x \right]_0^5 = e^5 - 1 \approx 147.413$

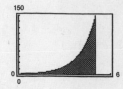

115. $\int_0^{\sqrt{6}} xe^{-x^2/4} \, dx = \left[-2e^{-x^2/4} \right]_0^{\sqrt{6}}$

$$= -2e^{-3/2} + 2 \approx 1.554$$

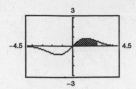

117. $\int_0^4 \sqrt{x} \, e^x \, dx, \ n = 12$

Midpoint Rule: 92.1898

Trapezoidal Rule: 93.8371

Simpson's Rule: 92.7385

Graphing utility: 92.7437

119. $0.0665 \int_{48}^{60} e^{-0.0139(t-48)^2} \, dt$

Graphing utility: $0.4772 = 47.72\%$

121. $\int_0^x e^t \, dt \geq \int_0^x 1 \, dt$

$$\left[e^t \right]_0^x \geq \left[t \right]_0^x$$

$e^x - 1 \geq x \implies e^x \geq 1 + x \text{ for } x \geq 0$

123. $f(x) = e^x$. Domain is $(-\infty, \infty)$ and range is $(0, \infty)$. f is continuous, increasing, one-to-one, and concave upwards on its entire domain.

$$\lim_{x \to -\infty} e^x = 0 \text{ and } \lim_{x \to \infty} e^x = \infty.$$

125. Yes. $f(x) = Ce^x$, C a constant.

127. $e^{-x} = x \implies f(x) = x - e^{-x}$

$f'(x) = 1 + e^{-x}$

$$x_{n+1} = x_n - \frac{f(x_n)}{f'(x_n)} = x_n - \frac{x_n - e^{-x_n}}{1 + e^{-x_n}}$$

$x_1 = 1$

$$x_2 = x_1 - \frac{f(x_1)}{f'(x_1)} \approx 0.5379$$

$$x_3 = x_2 - \frac{f(x_2)}{f'(x_2)} \approx 0.5670$$

$$x_4 = x_3 - \frac{f(x_3)}{f'(x_3)} \approx 0.5671$$

We approximate the root of f to be $x = 0.567$.

129. $\ln \dfrac{e^a}{e^b} = \ln e^a - \ln e^b = a - b$

$\ln e^{a-b} = a - b$

Therefore, $\ln \dfrac{e^a}{e^b} = \ln e^{a-b}$ and since $y = \ln x$ is one-to-one, we have $\dfrac{e^a}{e^b} = e^{a-b}$.

Section 5.5   Bases Other than *e* and Applications

1. $\log_2 \frac{1}{8} = \log_2 2^{-3} = -3$

3. $\log_7 1 = 0$

5. (a) $2^3 = 8$

       $\log_2 8 = 3$

(b) $3^{-1} = \frac{1}{3}$

      $\log_3 \frac{1}{3} = -1$

7. (a) $\log_{10} 0.01 = -2$

       $10^{-2} = 0.01$

(b) $\log_{0.5} 8 = -3$

     $0.5^{-3} = 8$

     $\left(\frac{1}{2}\right)^{-3} = 8$

9. $y = 3^x$

x	-2	-1	0	1	2
y	$\frac{1}{9}$	$\frac{1}{3}$	1	3	9

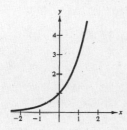

11. $y = \left(\frac{1}{3}\right)^x = 3^{-x}$

x	-2	-1	0	1	2
y	9	3	1	$\frac{1}{3}$	$\frac{1}{9}$

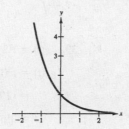

13. $h(x) = 5^{x-2}$

x	-1	0	1	2	3
y	$\frac{1}{125}$	$\frac{1}{25}$	$\frac{1}{5}$	1	5

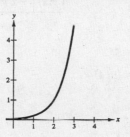

15. (a) $\log_{10} 1000 = x$

      $10^x = 1000$

        $x = 3$

(b) $\log_{10} 0.1 = x$

     $10^x = 0.1$

       $x = -1$

17. (a) $\log_3 x = -1$

      $3^{-1} = x$

       $x = \frac{1}{3}$

(b) $\log_2 x = -4$

     $2^{-4} = x$

      $x = \frac{1}{16}$

19. (a) $x^2 - x = \log_5 25$

       $x^2 - x = \log_5 5^2 = 2$

     $x^2 - x - 2 = 0$

    $(x + 1)(x - 2) = 0$

        $x = -1$ OR $x = 2$

(b) $3x + 5 = \log_2 64$

    $3x + 5 = \log_2 2^6 = 6$

        $3x = 1$

         $x = \frac{1}{3}$

21. $3^{2x} = 75$

  $2x \ln 3 = \ln 75$

       $x = \left(\frac{1}{2}\right)\frac{\ln 75}{\ln 3} \approx 1.965$

23. $2^{3-z} = 625$

  $(3 - z)\ln 2 = \ln 625$

      $3 - z = \dfrac{\ln 625}{\ln 2}$

         $z = 3 - \dfrac{\ln 625}{\ln 2} \approx -6.288$

25. $\left(1 + \dfrac{0.09}{12}\right)^{12t} = 3$

  $12t \ln\left(1 + \dfrac{0.09}{12}\right) = \ln 3$

        $t = \left(\dfrac{1}{12}\right)\dfrac{\ln 3}{\ln\left(1 + \dfrac{0.09}{12}\right)} \approx 12.253$

27. $\log_2(x - 1) = 5$

    $x - 1 = 2^5 = 32$

        $x = 33$

29. $\log_3 x^2 = 4.5$

$\quad x^2 = 3^{4.5}$

$\quad x = \pm\sqrt{3^{4.5}} \approx \pm 11.845$

31. $g(x) = 6(2^{1-x}) - 25$

Zero: $x \approx -1.059$

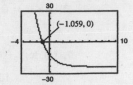

33. $h(s) = 32\log_{10}(s - 2) + 15$

Zero: $s \approx 2.340$

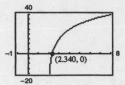

35. $f(x) = 4^x$

$\quad g(x) = \log_4 x$

x	-2	-1	0	$\frac{1}{2}$	1
$f(x)$	$\frac{1}{16}$	$\frac{1}{4}$	1	2	4

x	$\frac{1}{16}$	$\frac{1}{4}$	1	2	4
$g(x)$	-2	-1	0	$\frac{1}{2}$	1

37. $f(x) = 4^x$

$\quad f'(x) = (\ln 4)\,4^x$

39. $g(t) = t^2 2^t$

$\quad g'(t) = t^2(\ln 2)2^t + (2t)2^t$

$\qquad = t2^t(t\ln 2 + 2)$

$\qquad = 2^t t(2 + t\ln 2)$

41. $h(\theta) = 2^{-\theta}\cos \pi\theta$

$\quad h'(\theta) = 2^{-\theta}(-\pi\sin \pi\theta) - (\ln 2)2^{-\theta}\cos \pi\theta$

$\qquad = -2^{-\theta}[(\ln 2)\cos \pi\theta + \pi\sin \pi\theta]$

43. $f(x) = \log_2 \dfrac{x^2}{x - 1}$

$\qquad = 2\log_2 x - \log_2(x - 1)$

$\quad f'(x) = \dfrac{2}{x\ln 2} - \dfrac{1}{(x - 1)\ln 2}$

$\qquad = \dfrac{x - 2}{(\ln 2)x(x - 1)}$

45. $y = \log_5 \sqrt{x^2 - 1} = \dfrac{1}{2}\log_5(x^2 - 1)$

$\quad \dfrac{dy}{dx} = \dfrac{1}{2}\cdot\dfrac{2x}{(x^2 - 1)\ln 5} = \dfrac{x}{(x^2 - 1)\ln 5}$

47. $g(t) = \dfrac{10\log_4 t}{t} = \dfrac{10}{\ln 4}\left(\dfrac{\ln t}{t}\right)$

$\quad g'(t) = \dfrac{10}{\ln 4}\left[\dfrac{t(1/t) - \ln t}{t^2}\right]$

$\qquad = \dfrac{10}{t^2\ln 4}[1 - \ln t]$

$\qquad = \dfrac{5}{t^2\ln 2}(1 - \ln t)$

49. $y = 2^{-x}, \quad (-1, 2)$

$\quad y' = -2^{-x}\ln(2)$

$\quad$ At $(-1, 2)$, $y' = -2\ln(2)$.

$\quad$ Tangent line: $y - 2 = -2\ln(2)(x + 1)$

$\qquad\qquad\qquad y = [-2\ln(2)]x + 2 - 2\ln(2)$

51. $y = \log_3 x, \quad (27, 3)$

$\quad y' = \dfrac{1}{x\ln 3}$

$\quad$ At $(27, 3)$, $y' = \dfrac{1}{27\ln 3}$.

$\quad$ Tangent line: $y - 3 = \dfrac{1}{27\ln 3}(x - 27)$

$\qquad\qquad\qquad y = \dfrac{1}{27\ln 3}x + 3 - \dfrac{1}{\ln 3}$

53. $y = x^{2/x}$

$\ln y = \dfrac{2}{x} \ln x$

$\dfrac{1}{y}\left(\dfrac{dy}{dx}\right) = \dfrac{2}{x}\left(\dfrac{1}{x}\right) + \ln x\left(-\dfrac{2}{x^2}\right) = \dfrac{2}{x^2}(1 - \ln x)$

$\dfrac{dy}{dx} = \dfrac{2y}{x^2}(1 - \ln x) = 2x^{(2/x)-2}(1 - \ln x)$

55. $y = (x - 2)^{x+1}$

$\ln y = (x + 1)\ln(x - 2)$

$\dfrac{1}{y}\left(\dfrac{dy}{dx}\right) = (x + 1)\left(\dfrac{1}{x - 2}\right) + \ln(x - 2)$

$\dfrac{dy}{dx} = y\left[\dfrac{x + 1}{x - 2} + \ln(x - 2)\right]$

$= (x - 2)^{x+1}\left[\dfrac{x + 1}{x - 2} + \ln(x - 2)\right]$

57. $y = x^{\sin x},\quad \left(\dfrac{\pi}{2}, \dfrac{\pi}{2}\right)$

$\ln y = \sin x \ln x$

$\dfrac{y'}{y} = \dfrac{\sin x}{x} + \cos x \ln x$

At $\left(\dfrac{\pi}{2}, \dfrac{\pi}{2}\right)$: $\dfrac{y'}{(\pi/2)} = \dfrac{1}{(\pi/2)} + 0$

$y' = 1$

Tangent line: $y - \dfrac{\pi}{2} = 1\left(x - \dfrac{\pi}{2}\right)$

$y = x$

59. $y = (\ln x)^{\cos x},\quad (e, 1)$

$\ln y = \cos x \cdot \ln(\ln x)$

$\dfrac{y'}{y} = \cos x \cdot \dfrac{1}{x \ln x} - \sin x \cdot \ln(\ln x)$

At $(e, 1)$, $y' = \cos(e)\dfrac{1}{e} - 0.$

Tangent line: $y - 1 = \dfrac{\cos(e)}{e}(x - e)$

$y = \dfrac{\cos(e)}{e}x + 1 - \cos(e)$

61. $\displaystyle\int 3^x \, dx = \dfrac{3^x}{\ln 3} + C$

63. $\displaystyle\int x\left(5^{-x^2}\right) dx = -\dfrac{1}{2}\int 5^{-x^2}(-2x)\, dx$

$= -\left(\dfrac{1}{2}\right)\dfrac{5^{-x^2}}{\ln 5} + C$

$= \dfrac{-1}{2\ln 5}\left(5^{-x^2}\right) + C$

65. $\displaystyle\int \dfrac{3^{2x}}{1 + 3^{2x}}\, dx, \; u = 1 + 3^{2x}, \; du = 2(\ln 3)3^{2x}\, dx$

$\dfrac{1}{2\ln 3}\int \dfrac{(2\ln 3)3^{2x}}{1 + 3^{2x}}\, dx = \dfrac{1}{2\ln 3}\ln(1 + 3^{2x}) + C$

67. $\displaystyle\int_{-1}^{2} 2^x \, dx = \left[\dfrac{2^x}{\ln 2}\right]_{-1}^{2}$

$= \dfrac{1}{\ln 2}\left[4 - \dfrac{1}{2}\right] = \dfrac{7}{2\ln 2} = \dfrac{7}{\ln 4}$

69. $\displaystyle\int_{0}^{1} (5^x - 3^x)\, dx = \left[\dfrac{5^x}{\ln 5} - \dfrac{3^x}{\ln 3}\right]_{0}^{1}$

$= \left(\dfrac{5}{\ln 5} - \dfrac{3}{\ln 3}\right) - \left(\dfrac{1}{\ln 5} - \dfrac{1}{\ln 3}\right)$

$= \dfrac{4}{\ln 5} - \dfrac{2}{\ln 3}$

71.   Area $= \displaystyle\int_{0}^{3} 3^x \, dx$

$= \left[\dfrac{1}{\ln 3}3^x\right]_{0}^{3}$

$= \dfrac{1}{\ln 3}(27 - 1) = \dfrac{26}{\ln 3}$

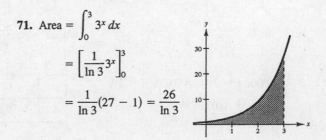

73. $\dfrac{dy}{dx} = 0.4^{x/3},\quad \left(0, \dfrac{1}{2}\right)$

(a)

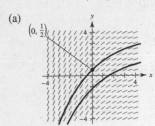

(b) $y = \displaystyle\int 0.4^{x/3}\,dx = 3\int 0.4^{x/3}\left(\dfrac{1}{3}\,dx\right)$

$\qquad = \dfrac{3}{\ln 0.4}\, 0.4^{x/3} + C$

$\qquad \dfrac{1}{2} = \dfrac{3}{\ln 0.4} + C \Rightarrow C = \dfrac{1}{2} - \dfrac{3}{\ln 0.4}$

$\qquad y = \dfrac{3}{\ln 0.4}(0.4)^{x/3} + \dfrac{1}{2} - \dfrac{3}{\ln 0.4}$

$\qquad = \dfrac{3}{\ln 0.4}(0.4^{x/3} - 1) + \dfrac{1}{2}$

$\qquad = \dfrac{3(1 - 0.4^{x/3})}{\ln 2.5} + \dfrac{1}{2}$

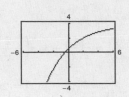

75.

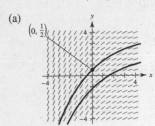

x	1	2	8
y	0	1	3

(a) y is an exponential function of x: False

(b) y is a logarithmic function of x: True; $y = \log_2 x$

(c) x is an exponential function of y: True, $2^y = x$

(d) y is a linear function of x: False

77. $f(x) = \log_2 x \Rightarrow f'(x) = \dfrac{1}{x \ln 2}$

$\qquad g(x) = x^x \Rightarrow g'(x) = x^x(1 + \ln x)$

Note: Let $y = g(x)$. Then:

$\qquad \ln y = \ln x^x = x \ln x$

$\qquad \dfrac{1}{y}\,y' = x \cdot \dfrac{1}{x} + \ln x$

$\qquad y' = y(1 + \ln x)$

$\qquad y' = x^x(1 + \ln x) = g'(x)$

$\qquad h(x) = x^2 \Rightarrow h'(x) = 2x$

$\qquad k(x) = 2^x \Rightarrow k'(x) = (\ln 2)2^x$

From greatest to smallest rate of growth:

$\qquad g(x),\ k(x),\ h(x),\ f(x)$

79. $C(t) = P(1.05)^t$

(a) $C(10) = 24.95(1.05)^{10}$

$\qquad \approx \$40.64$

(b) $\dfrac{dC}{dt} = P(\ln 1.05)(1.05)^t$

When $t = 1$, $\dfrac{dC}{dt} \approx 0.051P$.

When $t = 8$, $\dfrac{dC}{dt} \approx 0.072P$.

(c) $\dfrac{dC}{dt} = (\ln 1.05)[P(1.05)^t]$

$\qquad = (\ln 1.05)C(t)$

The constant of proportionality is $\ln 1.05$.

81. $P = \$1000,\ r = 3\tfrac{1}{2}\% = 0.035,\ t = 10$

$\qquad A = 1000\left(1 + \dfrac{0.035}{n}\right)^{10n}$

$\qquad A = 1000e^{(0.035)(10)} = 1419.07$

n	1	2	4	12	365	Continuous
A	1410.60	1414.78	1416.91	1418.34	1419.04	1419.07

83. $P = \$1000,\ r = 5\% = 0.05,\ t = 30$

$\qquad A = 1000\left(1 + \dfrac{0.05}{n}\right)^{30n}$

$\qquad A = 1000e^{(0.05)30} = 4481.69$

n	1	2	4	12	365	Continuous
A	4321.94	4399.79	4440.21	4467.74	4481.23	4481.69

85. $100,000 = Pe^{0.05t} \implies P = 100,000e^{-0.05t}$

t	1	10	20	30	40	50
P	95,122.94	60,653.07	36,787.94	22,313.02	13,533.53	8208.50

87. $100,000 = P\left(1 + \dfrac{0.05}{12}\right)^{12t} \implies P = 100,000\left(1 + \dfrac{0.05}{12}\right)^{-12t}$

t	1	10	20	30	40	50
P	95,132.82	60,716.10	36,864.45	22,382.66	13,589.88	8251.24

89. (a) $A = 20,000\left(1 + \dfrac{0.06}{365}\right)^{(365)(8)} \approx \$32,320.21$

(b) $A = \$30,000$

(c) $A = 8000\left(1 + \dfrac{0.06}{365}\right)^{(365)(8)} + 20,000\left(1 + \dfrac{0.06}{365}\right)^{(365)(4)}$

   $\approx \$12,928.09 + 25,424.48 = \$38,352.57$

(d) $A = 9000\left[\left(1 + \dfrac{0.06}{365}\right)^{(365)(8)} + \left(1 + \dfrac{0.06}{365}\right)^{(365)(4)} + 1\right]$

   $\approx \$34,985.11$

Take option (c).

91. (a) $\lim\limits_{t \to \infty} 6.7e^{(-48.1)/t} = 6.7e^0 = 6.7$ million ft^3

(b) $V' = \dfrac{322.27}{t^2}e^{-(48.1)/t}$

$V'(20) \approx 0.073$ million ft^3/yr

$V'(60) \approx 0.040$ million ft^3/yr

93. $y = \dfrac{300}{3 + 17e^{-0.0625x}}$

(a)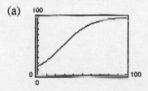

(b) If $x = 2$ (2000 egg masses), $y \approx 16.67 \approx 16.7\%$.

(c) If $y = 66.67\%$, then $x \approx 38.8$ or 38,800 egg masses.

(d) $y = 300(3 + 17e^{-0.0625x})^{-1}$

$y' = \dfrac{318.75e^{-0.0625x}}{(3 + 17e^{-0.0625x})^2}$

$y'' = \dfrac{19.921875e^{-0.0625x}(17e^{-0.0625x} - 3)}{(3 + 17e^{-0.0625x})^3}$

$17e^{-0.0625x} - 3 = 0 \implies$

   $x \approx 27.8$ or 27,800 egg masses.

95. (a) $B = 4.7539(6.7744)^d = 4.7539e^{1.9132d}$

(b)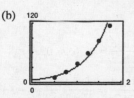

(c) $B'(d) = 9.0952e^{1.9132d}$

   $B'(0.8) \approx 42.03$ tons/inch

   $B'(1.5) \approx 160.38$ tons/inch

97. (a) $\displaystyle\int_0^4 f(t)\, dt \approx 5.67$ (b)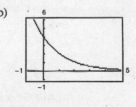

$\displaystyle\int_0^4 g(t)\, dt \approx 5.67$

$\displaystyle\int_0^4 h(t)\, dt \approx 5.67$

(c) The functions appear to be equal: $f(t) = g(t) = h(t)$.
Analytically,

$$f(t) = 4\left(\frac{3}{8}\right)^{2t/3} = 4\left[\left(\frac{3}{8}\right)^{2/3}\right]^t = 4\left(\frac{9^{1/3}}{4}\right)^t = g(t)$$

$$h(t) = 4e^{-0.653886t} = 4[e^{-0.653886}]^t \approx 4(0.52002)^t$$

$$g(t) = 4\left(\frac{9^{1/3}}{4}\right)^t \approx 4(0.52002)^t$$

No. The definite integrals over a given interval may be
equal when the functions are not equal.

99.

t	0	1	2	3	4
y	1200	720	432	259.20	155.52

$y = C(k^t)$

When $t = 0$, $y = 1200 \implies C = 1200$.

$y = 1200(k^t)$

$\dfrac{720}{1200} = 0.6, \dfrac{432}{720} = 0.6, \dfrac{259.20}{432} = 0.6, \dfrac{155.52}{259.20} = 0.6$

Let $k = 0.6$.

$y = 1200(0.6)^t$

101. False. e is an irrational number.

103. True

$\begin{aligned} f(g(x)) &= 2 + e^{\ln(x-2)} \\ &= 2 + x - 2 = x \\ g(f(x)) &= \ln(2 + e^x - 2) \\ &= \ln e^x = x \end{aligned}$

105. True

$\dfrac{d}{dx}[e^x] = e^x$ and $\dfrac{d}{dx}[e^{-x}] = -e^{-x}$

$e^x = e^{-x}$ when $x = 0$.

$(e^0)(-e^{-0}) = -1$

107. $\dfrac{dy}{dt} = \dfrac{8}{25}y\left(\dfrac{5}{4} - y\right)$, $y(0) = 1$

$\dfrac{dy}{y[(5/4) - y]} = \dfrac{8}{25}dt \implies \dfrac{4}{5}\int\left(\dfrac{1}{y} + \dfrac{1}{(5/4) - y}\right)dy = \int \dfrac{8}{25}dt \implies$

$\ln y - \ln\left(\dfrac{5}{4} - y\right) = \dfrac{2}{5}t + C$

$\ln\left(\dfrac{y}{(5/4) - y}\right) = \dfrac{2}{5}t + C$

$\dfrac{y}{(5/4) - y} = e^{(2/5)t+C} = C_1 e^{(2/5)t}$

$y(0) = 1 \implies C_1 = 4 \implies 4e^{(2/5)t} = \dfrac{y}{(5/4) - y}$

$\implies 4e^{(2/5)t}\left(\dfrac{5}{4} - y\right) = y \implies 5e^{(2/5)t} = 4e^{(2/5)t}y + y = (4e^{(2/5)t} + 1)y$

$\implies y = \dfrac{5e^{(2/5)t}}{4e^{(2/5)t} + 1} = \dfrac{5}{4 + e^{-0.4t}} = \dfrac{1.25}{1 + 0.25e^{-0.4t}}$

109. (a)

$y^x = x^y$

$x \ln y = y \ln x$

$x\dfrac{y'}{y} + \ln y = \dfrac{y}{x} + y' \ln x$

$y'\left[\dfrac{x}{y} - \ln x\right] = \dfrac{y}{x} - \ln y$

$y' = \dfrac{(y/x) - \ln y}{(x/y) - \ln x}$

$y' = \dfrac{y^2 - xy \ln y}{x^2 - xy \ln x}$

(b) (i) At (c, c): $y' = \dfrac{c^2 - c^2 \ln c}{c^2 - c^2 \ln c} = 1$, $(c \neq 0, e)$

(ii) At $(2, 4)$: $y' = \dfrac{16 - 8\ln 4}{4 - 8\ln 2} = \dfrac{4 - 4\ln 2}{1 - 2\ln 2} \approx -3.1774$

(iii) At $(4, 2)$: $y' = \dfrac{4 - 8\ln 2}{16 - 8\ln 4} = \dfrac{1 - 2\ln 2}{4 - 4\ln 2} \approx -0.3147$

(c) y' is undefined for

$x^2 = xy \ln x \implies x = y \ln x = \ln x^y \implies e^x = x^y.$

At (e, e), y' is undefined.

111. Let $f(x) = \dfrac{\ln x}{x}$, $x > 0$.

$f'(x) = \dfrac{1 - \ln x}{x^2} < 0$ for $x > e \Rightarrow f$ is decreasing for $x \geq e$. Hence, for $e \leq x < y$:

$$f(x) > f(y)$$

$$\frac{\ln x}{x} > \frac{\ln y}{y}$$

$$(xy)\frac{\ln x}{x} > (xy)\frac{\ln y}{y}$$

$$\ln x^y > \ln y^x$$

$$x^y > y^x$$

For $n \geq 8$, $e < \sqrt{n} < \sqrt{n+1}$, $\left(\sqrt{8} \approx 2.828\right)$ and so letting $x = \sqrt{n}$, $y = \sqrt{n+1}$, we have

$$\left(\sqrt{n}\right)^{\sqrt{n+1}} > \left(\sqrt{n+1}\right)^{\sqrt{n}}.$$

Note: $\sqrt{8}^{\sqrt{9}} \approx 22.6$ and $\sqrt{9}^{\sqrt{8}} \approx 22.4$.

Note: This same argument shows $e^{\pi} > \pi^e$.

Section 5.6   Inverse Trigonometric Functions: Differentiation

1. $y = \arcsin x$

(a)

x	-1	-0.8	-0.6	-0.4	-0.2	0	0.2	0.4	0.6	0.8	1
y	-1.571	-0.927	-0.644	-0.412	-0.201	0	0.201	0.412	0.644	0.927	1.571

(b)
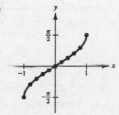

(c)
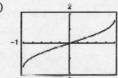

(d) Symmetric about origin:

$$\arcsin(-x) = -\arcsin x$$

Intercept: $(0, 0)$

3. $y = \arccos x$

$\left(-\dfrac{\sqrt{2}}{2}, \dfrac{3\pi}{4}\right)$ because $\cos\left(\dfrac{3\pi}{4}\right) = -\dfrac{\sqrt{2}}{2}$

$\left(\dfrac{1}{2}, \dfrac{\pi}{3}\right)$ because $\cos\left(\dfrac{\pi}{3}\right) = \dfrac{1}{2}$

$\left(\dfrac{\sqrt{3}}{2}, \dfrac{\pi}{6}\right)$ because $\cos\left(\dfrac{\pi}{6}\right) = \dfrac{\sqrt{3}}{2}$

5. $\arcsin \dfrac{1}{2} = \dfrac{\pi}{6}$

7. $\arccos \dfrac{1}{2} = \dfrac{\pi}{3}$

9. $\arctan \dfrac{\sqrt{3}}{3} = \dfrac{\pi}{6}$

11. $\operatorname{arccsc}\left(-\sqrt{2}\right) = -\dfrac{\pi}{4}$

13. $\arccos(-0.8) \approx 2.50$

15. $\operatorname{arcsec}(1.269) = \arccos\left(\dfrac{1}{1.269}\right)$

$$\approx 0.66$$

17. (a) $\sin\left(\arctan\dfrac{3}{4}\right) = \dfrac{3}{5}$

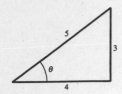

(b) $\sec\left(\arcsin\dfrac{4}{5}\right) = \dfrac{5}{3}$

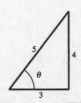

19. (a) $\cot\left[\arcsin\left(-\dfrac{1}{2}\right)\right] = \cot\left(-\dfrac{\pi}{6}\right) = -\sqrt{3}$

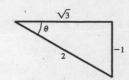

(b) $\csc\left[\arctan\left(-\dfrac{5}{12}\right)\right] = -\dfrac{13}{5}$

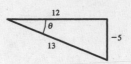

21. $y = \cos(\arcsin 2x)$

$\theta = \arcsin 2x$

$y = \cos\theta = \sqrt{1 - 4x^2}$

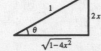

23. $y = \sin(\text{arcsec } x)$

$\theta = \text{arcsec } x,\ 0 \le \theta \le \pi,\ \theta \ne \dfrac{\pi}{2}$

$y = \sin\theta = \dfrac{\sqrt{x^2 - 1}}{|x|}$

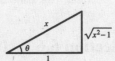

The absolute value bars on x
are necessary because of the
restriction $0 \le \theta \le \pi,\ \theta \ne \pi/2$, and $\sin\theta$
for this domain must always be nonnegative.

25. $y = \tan\left(\text{arcsec }\dfrac{x}{3}\right)$

$\theta = \text{arcsec }\dfrac{x}{3}$

$y = \tan\theta = \dfrac{\sqrt{x^2 - 9}}{3}$

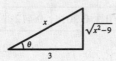

27. $y = \csc\left(\arctan\dfrac{x}{\sqrt{2}}\right)$

$\theta = \arctan\dfrac{x}{\sqrt{2}}$

$y = \csc\theta = \dfrac{\sqrt{x^2 + 2}}{x}$

29. (a)

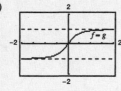

(b) Let $y = \arctan(2x)$

$\tan y = 2x$

$\sin y = \sin(\arctan(2x))$

$= \dfrac{2x}{\sqrt{1 + 4x^2}}.$

(c) Asymptotes: $y = \pm 1$

31. $\arcsin(3x - \pi) = \dfrac{1}{2}$

$3x - \pi = \sin\left(\dfrac{1}{2}\right)$

$x = \dfrac{1}{3}\left[\sin\left(\dfrac{1}{2}\right) + \pi\right] \approx 1.207$

33. $\arcsin\sqrt{2x} = \arccos\sqrt{x}$

$\sqrt{2x} = \sin\left(\arccos\sqrt{x}\right)$

$\sqrt{2x} = \sqrt{1 - x},\quad 0 \le x \le 1$

$2x = 1 - x$

$3x = 1$

$x = \dfrac{1}{3}$

35. (a) $\operatorname{arccsc} x = \arcsin \dfrac{1}{x}, \quad |x| \geq 1$

Let $y = \operatorname{arccsc} x$. Then for

$$-\frac{\pi}{2} \leq y < 0 \text{ and } 0 < y \leq \frac{\pi}{2},$$

$\csc y = x \implies \sin y = 1/x$. Thus, $y = \arcsin(1/x)$.
Therefore, $\operatorname{arccsc} x = \arcsin(1/x)$.

(b) $\arctan x + \arctan \dfrac{1}{x} = \dfrac{\pi}{2}, \quad x > 0$

Let $y = \arctan x + \arctan(1/x)$. Then,

$$\tan y = \frac{\tan(\arctan x) + \tan[\arctan(1/x)]}{1 - \tan(\arctan x)\tan[\arctan(1/x)]}$$

$$= \frac{x + (1/x)}{1 - x(1/x)}$$

$$= \frac{x + (1/x)}{0} \text{ (which is undefined).}$$

Thus, $y = \pi/2$. Therefore, $\arctan x + \arctan(1/x) = \pi/2$.

37. $f(x) = \arcsin(x - 1)$

$x - 1 = \sin y$

$x = 1 + \sin y$

Domain: $[0, 2]$

Range: $\left[-\dfrac{\pi}{2}, \dfrac{\pi}{2}\right]$

$f(x)$ is the graph of $\arcsin x$ shifted 1 unit to the right.

39. $f(x) = \operatorname{arcsec} 2x$

$2x = \sec y$

$x = \dfrac{1}{2}\sec y$

Domain: $\left(-\infty, -\dfrac{1}{2}\right], \left[\dfrac{1}{2}, \infty\right)$

Range: $\left[0, \dfrac{\pi}{2}\right), \left(\dfrac{\pi}{2}, \pi\right]$

41. $f(x) = 2\arcsin(x - 1)$

$$f'(x) = \frac{2}{\sqrt{1 - (x-1)^2}}$$

$$= \frac{2}{\sqrt{2x - x^2}}$$

43. $g(x) = 3\arccos \dfrac{x}{2}$

$$g'(x) = \frac{-3(1/2)}{\sqrt{1 - (x^2/4)}} = \frac{-3}{\sqrt{4 - x^2}}$$

45. $f(x) = \arctan \dfrac{x}{a}$

$$f'(x) = \frac{1/a}{1 + (x^2/a^2)} = \frac{a}{a^2 + x^2}$$

47. $g(x) = \dfrac{\arcsin 3x}{x}$

$$g'(x) = \frac{x\left(3/\sqrt{1 - 9x^2}\right) - \arcsin 3x}{x^2}$$

$$= \frac{3x - \sqrt{1 - 9x^2}\,\arcsin 3x}{x^2\sqrt{1 - 9x^2}}$$

49. $h(t) = \sin(\arccos t) = \sqrt{1 - t^2}$

$$h'(t) = \frac{1}{2}(1 - t^2)^{-1/2}(-2t)$$

$$= \frac{-t}{\sqrt{1 - t^2}}$$

51. $y = x\arccos x - \sqrt{1 - x^2}$

$$y' = \arccos x - \frac{x}{\sqrt{1 - x^2}} - \frac{1}{2}(1 - x^2)^{-1/2}(-2x)$$

$$= \arccos x$$

53. $y = \dfrac{1}{2}\left(\dfrac{1}{2}\ln\dfrac{x + 1}{x - 1} + \arctan x\right)$

$$= \frac{1}{4}[\ln(x + 1) - \ln(x - 1)] + \frac{1}{2}\arctan x$$

$$\frac{dy}{dx} = \frac{1}{4}\left(\frac{1}{x + 1} - \frac{1}{x - 1}\right) + \frac{1/2}{1 + x^2} = \frac{1}{1 - x^4}$$

55. $y = x\arcsin x + \sqrt{1 - x^2}$

$$\frac{dy}{dx} = x\left(\frac{1}{\sqrt{1 - x^2}}\right) + \arcsin x - \frac{x}{\sqrt{1 - x^2}} = \arcsin x$$

57. $y = 8 \arcsin \dfrac{x}{4} - \dfrac{x\sqrt{16 - x^2}}{2}$

$y' = 2\dfrac{1}{\sqrt{1 - (x/4)^2}} - \dfrac{\sqrt{16 - x^2}}{2} - \dfrac{x}{4}(16 - x^2)^{-1/2}(-2x)$

$\quad = \dfrac{8}{\sqrt{16 - x^2}} - \dfrac{\sqrt{16 - x^2}}{2} + \dfrac{x^2}{2\sqrt{16 - x^2}}$

$\quad = \dfrac{16 - (16 - x^2) + x^2}{2\sqrt{16 - x^2}} = \dfrac{x^2}{\sqrt{16 - x^2}}$

59. $y = \arctan x + \dfrac{x}{1 + x^2}$

$y' = \dfrac{1}{1 + x^2} + \dfrac{(1 + x^2) - x(2x)}{(1 + x^2)^2}$

$\quad = \dfrac{(1 + x^2) + (1 - x^2)}{(1 + x^2)^2}$

$\quad = \dfrac{2}{(1 + x^2)^2}$

61. $y = 2 \arcsin x, \quad \left(\dfrac{1}{2}, \dfrac{\pi}{3}\right)$

$y' = \dfrac{2}{\sqrt{1 - x^2}}$

At $\left(\dfrac{1}{2}, \dfrac{\pi}{3}\right), y' = \dfrac{2}{\sqrt{1 - (1/4)}} = \dfrac{4}{\sqrt{3}}$.

Tangent line: $y - \dfrac{\pi}{3} = \dfrac{4}{\sqrt{3}}\left(x - \dfrac{1}{2}\right)$

$\qquad y = \dfrac{4}{\sqrt{3}}x + \dfrac{\pi}{3} - \dfrac{2}{\sqrt{3}}$

$\qquad y = \dfrac{4\sqrt{3}}{3}x + \dfrac{\pi}{3} - \dfrac{2\sqrt{3}}{3}$

63. $y = \arctan\left(\dfrac{x}{2}\right), \quad \left(2, \dfrac{\pi}{4}\right)$

$y' = \dfrac{1}{1 + (x^2/4)}\left(\dfrac{1}{2}\right) = \dfrac{2}{4 + x^2}$

At $\left(2, \dfrac{\pi}{4}\right), y' = \dfrac{2}{4 + 4} = \dfrac{1}{4}$.

Tangent line: $y - \dfrac{\pi}{4} = \dfrac{1}{4}(x - 2)$

$\qquad y = \dfrac{1}{4}x + \dfrac{\pi}{4} - \dfrac{1}{2}$

65. $y = 4x \arccos(x - 1), \quad (1, 2\pi)$

$y' = 4x\dfrac{-1}{\sqrt{1 - (x - 1)^2}} + 4\arccos(x - 1)$

At $(1, 2\pi), y' = -4 + 2\pi$.

Tangent line: $y - 2\pi = (2\pi - 4)(x - 1)$

$\qquad y = (2\pi - 4)x + 4$

67. $f(x) = \arctan x, \quad a = 0$

$f(0) = 0$

$f'(x) = \dfrac{1}{1 + x^2}, \quad f'(0) = 1$

$f''(x) = \dfrac{-2x}{(1 + x^2)^2}, \quad f''(0) = 0$

$P_1(x) = f(0) + f'(0)x = x$

$P_2(x) = f(0) + f'(0)x + \dfrac{1}{2}f''(0)x^2 = x$

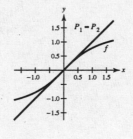

69. $f(x) = \arcsin x, \quad a = \dfrac{1}{2}$

$f'(x) = \dfrac{1}{\sqrt{1 - x^2}}$

$f''(x) = \dfrac{x}{(1 - x^2)^{3/2}}$

$P_1(x) = f\left(\dfrac{1}{2}\right) + f'\left(\dfrac{1}{2}\right)\left(x - \dfrac{1}{2}\right) = \dfrac{\pi}{6} + \dfrac{2\sqrt{3}}{3}\left(x - \dfrac{1}{2}\right)$

$P_2(x) = f\left(\dfrac{1}{2}\right) + f'\left(\dfrac{1}{2}\right)\left(x - \dfrac{1}{2}\right) + \dfrac{1}{2}f''\left(\dfrac{1}{2}\right)\left(x - \dfrac{1}{2}\right)^2 = \dfrac{\pi}{6} + \dfrac{2\sqrt{3}}{3}\left(x - \dfrac{1}{2}\right) + \dfrac{2\sqrt{3}}{9}\left(x - \dfrac{1}{2}\right)^2$

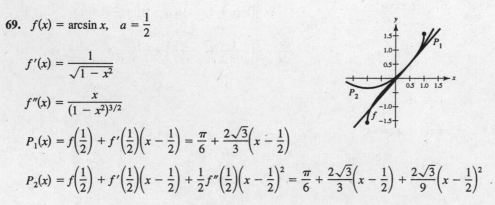

71. $f(x) = \text{arcsec } x - x$

$$f'(x) = \frac{1}{|x|\sqrt{x^2-1}} - 1$$

$$= 0 \text{ when } |x|\sqrt{x^2-1} = 1$$

$$x^2(x^2-1) = 1$$

$$x^4 - x^2 - 1 = 0 \text{ when } x^2 = \frac{1+\sqrt{5}}{2} \text{ or }$$

$$x = \pm\sqrt{\frac{1+\sqrt{5}}{2}} = \pm 1.272$$

Relative maximum: $(1.272, -0.606)$

Relative minimum: $(-1.272, 3.747)$

73. $f(x) = \arctan x - \arctan(x-4)$

$$f'(x) = \frac{1}{1+x^2} - \frac{1}{1+(x-4)^2} = 0$$

$$1 + x^2 = 1 + (x-4)^2$$

$$0 = -8x + 16$$

$$x = 2$$

By the First Derivative Test, $(2, 2.214)$ is a relative maximum.

75. $x^2 + x \arctan y = y - 1, \quad \left(-\frac{\pi}{4}, 1\right)$

$$2x + \arctan y + \frac{x}{1+y^2}y' = y'$$

$$\left(1 - \frac{x}{1+y^2}\right)y' = 2x + \arctan y$$

$$y' = \frac{2x + \arctan y}{1 - \dfrac{x}{1+y^2}}$$

At $\left(-\frac{\pi}{4}, 1\right)$, $y' = \dfrac{-\dfrac{\pi}{2} + \dfrac{\pi}{4}}{1 - \dfrac{-\pi/4}{2}} = \dfrac{-\dfrac{\pi}{2}}{2 + \dfrac{\pi}{4}} = \dfrac{-2\pi}{8+\pi}$.

Tangent line: $y - 1 = \dfrac{-2\pi}{8+\pi}\left(x + \dfrac{\pi}{4}\right)$

$$y = \frac{-2\pi}{8+\pi}x + 1 - \frac{\pi^2}{16+2\pi}$$

77. $\arcsin x + \arcsin y = \dfrac{\pi}{2}, \quad \left(\dfrac{\sqrt{2}}{2}, \dfrac{\sqrt{2}}{2}\right)$

$$\frac{1}{\sqrt{1-x^2}} + \frac{1}{\sqrt{1-y^2}}y' = 0$$

$$\frac{1}{\sqrt{1-y^2}}y' = \frac{-1}{\sqrt{1-x^2}}$$

At $\left(\dfrac{\sqrt{2}}{2}, \dfrac{\sqrt{2}}{2}\right)$, $y' = -1$

Tangent line: $y - \dfrac{\sqrt{2}}{2} = -1\left(x - \dfrac{\sqrt{2}}{2}\right)$

$$y = -x + \sqrt{2}$$

79. The trigonometric functions are not one-to-one on $(-\infty, \infty)$, so their domains must be restricted to intervals on which they are one-to-one.

81. $y = \text{arccot } x, \quad 0 < y < \pi$

$$x = \cot y$$

$$\tan y = \frac{1}{x}$$

So, graph the function
$y = \arctan(1/x)$ for $x > 0$ and
$y = \arctan(1/x) + \pi$ for $x < 0$.

83. False

$$\arccos \frac{1}{2} = \frac{\pi}{3}$$

since the range is $[0, \pi]$.

85. True

$$\frac{d}{dx}[\arctan x] = \frac{1}{1+x^2} > 0 \text{ for all } x.$$

87. True

$$\frac{d}{dx}[\arctan(\tan x)] = \frac{\sec^2 x}{1+\tan^2 x}$$

$$= \frac{\sec^2 x}{\sec^2 x} = 1$$

89. (a) $\cot \theta = \dfrac{x}{5}$

$$\theta = \operatorname{arccot}\left(\dfrac{x}{5}\right)$$

(b) $\dfrac{d\theta}{dt} = \dfrac{-1/5}{1 + (x/5)^2} \dfrac{dx}{dt} = \dfrac{-5}{x^2 + 25} \dfrac{dx}{dt}$

If $\dfrac{dx}{dt} = -400$ and $x = 10$, $\dfrac{d\theta}{dt} = 16$ rad/hr.

If $\dfrac{dx}{dt} = -400$ and $x = 3$, $\dfrac{d\theta}{dt} \approx 58.824$ rad/hr.

91. (a) $\qquad h(t) = -16t^2 + 256$

$-16t^2 + 256 = 0$ when $t = 4$ sec

(b) $\tan \theta = \dfrac{h}{500} = \dfrac{-16t^2 + 256}{500}$

$$\theta = \arctan\left[\dfrac{16}{500}(-t^2 + 16)\right]$$

$$\dfrac{d\theta}{dt} = \dfrac{-8t/125}{1 + [(4/125)(-t^2 + 16)]^2}$$

$$= \dfrac{-1000t}{15{,}625 + 16(16 - t^2)^2}$$

When $t = 1$, $d\theta/dt \approx -0.0520$ rad/sec.

When $t = 2$, $d\theta/dt \approx -0.1116$ rad/sec.

93. (a) $\tan(\arctan x + \arctan y) = \dfrac{\tan(\arctan x) + \tan(\arctan y)}{1 - \tan(\arctan x)\tan(\arctan y)}$

$$= \dfrac{x + y}{1 - xy}, \quad xy \neq 1$$

Therefore,

$$\arctan x + \arctan y = \arctan\left(\dfrac{x + y}{1 - xy}\right), xy \neq 1.$$

(b) Let $x = \frac{1}{2}$ and $y = \frac{1}{3}$.

$$\arctan\left(\dfrac{1}{2}\right) + \arctan\left(\dfrac{1}{3}\right) = \arctan\dfrac{(1/2) + (1/3)}{1 - [(1/2) \cdot (1/3)]}$$

$$= \arctan\dfrac{5/6}{1 - (1/6)}$$

$$= \arctan\dfrac{5/6}{5/6} = \arctan 1 = \dfrac{\pi}{4}$$

95. $f(x) = kx + \sin x$

$f'(x) = k + \cos x \geq 0$ for $k \geq 1$

$f'(x) = k + \cos x \leq 0$ for $k \leq -1$

Therefore, $f(x) = kx + \sin x$ is strictly monotonic and has an inverse for $k \leq -1$ or $k \geq 1$.

97. (a) $f(x) = \arccos x + \arcsin x$

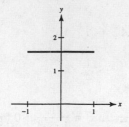

(b) The graph of f is the constant function $y = \pi/2$.

(c) Let $\quad u = \arccos x \quad$ and $\quad v = \arcsin x$

$\qquad \cos u = x \qquad$ and $\quad \sin v = x.$

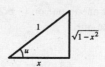

$\sin(u + v) = \sin u \cos v + \sin v \cos u$

$$= \sqrt{1 - x^2}\sqrt{1 - x^2} + x \cdot x$$

$$= 1 - x^2 + x^2 = 1$$

Hence, $u + v = \pi/2$.

Thus, $\arccos x + \arcsin x = \pi/2$.

99.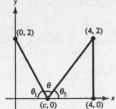

$\tan \theta_1 = \dfrac{2}{c}, \ \tan \theta_2 = \dfrac{2}{4-c}, \ 0 < c < 4$

To maximize θ, we minimize $f(c) = \theta_1 + \theta_2$.

$$f(c) = \arctan\left(\frac{2}{c}\right) + \arctan\left(\frac{2}{4-c}\right)$$

$$f'(c) = \frac{-2}{c^2 + 4} + \frac{2}{(4-c)^2 + 4} = 0$$

$$\frac{1}{c^2 + 4} = \frac{1}{(4-c)^2 + 4}$$

$$c^2 + 4 = c^2 - 8c + 16 + 4$$

$$8c = 16$$

$$c = 2$$

By the First Derivative Test, $c = 2$ is a minimum. Hence, $(c, f(c)) = (2, \pi/2)$ is a relative maximum for the angle θ. Checking the endpoints:

$c = 0$: $\tan \theta = \dfrac{4}{2} = 2 \implies \theta \approx 1.107$

$c = 4$: $\tan \theta = \dfrac{4}{2} = 2 \implies \theta \approx 1.107$

$c = 2$: $\theta = \pi - \theta_1 - \theta_2 = \dfrac{\pi}{2} \approx 1.5708$

Thus, $(2, \pi/2)$ is the absolute maximum.

101. $f(x) = \sec x, \ \ 0 \le x < \dfrac{\pi}{2}, \ \pi \le x < \dfrac{3\pi}{2}$

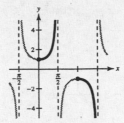

(a) $y = \text{arcsec } x, \ \ x \le -1 \ \ \ \text{ or } \ \ x \ge 1$

$$0 \le y < \frac{\pi}{2} \ \ \text{ or } \ \ \pi \le y < \frac{3\pi}{2}$$

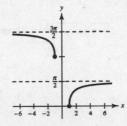

(b) $y = \text{arcsec } x$

$x = \sec y$

$1 = \sec y \tan y \cdot y'$

$y' = \dfrac{1}{\sec y \tan y}$

$\ \ \ = \dfrac{1}{x\sqrt{x^2 - 1}}$

$\tan^2 y + 1 = \sec^2 y$

$\ \ \ \ \tan y = +\sqrt{\sec^2 y - 1}$

On $0 \le y < \pi/2$ and $\pi \le y < 3\pi/2$, $\tan y \ge 0$.

Section 5.7   Inverse Trigonometric Functions: Integration

1. $\displaystyle\int \frac{5}{\sqrt{9 - x^2}}\, dx = 5 \arcsin\left(\frac{x}{3}\right) + C$

3. $\displaystyle\int \frac{7}{16 + x^2}\, dx = \frac{7}{4} \arctan\left(\frac{x}{4}\right) + C$

5. $\displaystyle\int \frac{1}{x\sqrt{4x^2 - 1}}\, dx = \int \frac{2}{2x\sqrt{(2x)^2 - 1}}\, dx = \text{arcsec}|2x| + C$

7. $\displaystyle\int \frac{x^3}{x^2 + 1}\, dx = \int \left[x - \frac{x}{x^2 + 1} \right] dx = \int x\, dx - \frac{1}{2}\int \frac{2x}{x^2 + 1}\, dx = \frac{1}{2}x^2 - \frac{1}{2}\ln(x^2 + 1) + C$ (Use long division.)

9. $\displaystyle\int \frac{1}{\sqrt{1 - (x + 1)^2}}\, dx = \arcsin(x + 1) + C$

11. Let $u = t^2, du = 2t\, dt$.

$$\int \frac{t}{\sqrt{1 - t^4}}\, dt = \frac{1}{2}\int \frac{1}{\sqrt{1 - (t^2)^2}}(2t)\, dt = \frac{1}{2}\arcsin(t^2) + C$$

13. Let $u = e^{2x}$, $du = 2e^{2x}\,dx$.

$$\int \frac{e^{2x}}{4 + e^{4x}}\,dx = \frac{1}{2}\int \frac{2e^{2x}}{4 + (e^{2x})^2}\,dx = \frac{1}{4}\arctan\frac{e^{2x}}{2} + C$$

15. $\displaystyle \int \frac{1}{\sqrt{x}\,\sqrt{1-x}}\,dx$, $u = \sqrt{x}$, $x = u^2$, $dx = 2u\,du$

$$\int \frac{1}{u\sqrt{1-u^2}}(2u\,du) = 2\int \frac{du}{\sqrt{1-u^2}} = 2\arcsin u + C$$
$$= 2\arcsin\sqrt{x} + C$$

17. $\displaystyle \int \frac{x-3}{x^2+1}\,dx = \frac{1}{2}\int \frac{2x}{x^2+1}\,dx - 3\int \frac{1}{x^2+1}\,dx$

$$= \frac{1}{2}\ln(x^2+1) - 3\arctan x + C$$

19. $\displaystyle \int \frac{x+5}{\sqrt{9-(x-3)^2}}\,dx = \int \frac{(x-3)}{\sqrt{9-(x-3)^2}}\,dx + \int \frac{8}{\sqrt{9-(x-3)^2}}\,dx$

$$= -\sqrt{9-(x-3)^2} - 8\arcsin\left(\frac{x-3}{3}\right) + C = -\sqrt{6x-x^2} + 8\arcsin\left(\frac{x}{3}-1\right) + C$$

21. Let $u = 3x$, $du = 3\,dx$.

$$\int_0^{1/6} \frac{1}{\sqrt{1-9x^2}}\,dx = \frac{1}{3}\int_0^{1/6} \frac{1}{\sqrt{1-(3x)^2}}(3)\,dx$$
$$= \left[\frac{1}{3}\arcsin(3x)\right]_0^{1/6} = \frac{\pi}{18}$$

23. Let $u = 2x$, $du = 2\,dx$.

$$\int_0^{\sqrt{3}/2} \frac{1}{1+4x^2}\,dx = \frac{1}{2}\int_0^{\sqrt{3}/2} \frac{2}{1+(2x)^2}\,dx$$
$$= \left[\frac{1}{2}\arctan(2x)\right]_0^{\sqrt{3}/2} = \frac{\pi}{6}$$

25. Let $u = \arcsin x$, $du = \dfrac{1}{\sqrt{1-x^2}}\,dx$.

$$\int_0^{1/\sqrt{2}} \frac{\arcsin x}{\sqrt{1-x^2}}\,dx = \left[\frac{1}{2}\arcsin^2 x\right]_0^{1/\sqrt{2}} = \frac{\pi^2}{32} \approx 0.308$$

27. Let $u = 1 - x^2$, $du = -2x\,dx$.

$$\int_{-1/2}^0 \frac{x}{\sqrt{1-x^2}}\,dx = -\frac{1}{2}\int_{-1/2}^0 (1-x^2)^{-1/2}(-2x)\,dx$$
$$= \left[-\sqrt{1-x^2}\right]_{-1/2}^0 = \frac{\sqrt{3}-2}{2}$$
$$\approx -0.134$$

29. Let $u = \cos x$, $du = -\sin x\,dx$.

$$\int_{\pi/2}^{\pi} \frac{\sin x}{1+\cos^2 x}\,dx = -\int_{\pi/2}^{\pi} \frac{-\sin x}{1+\cos^2 x}\,dx$$
$$= \left[-\arctan(\cos x)\right]_{\pi/2}^{\pi} = \frac{\pi}{4}$$

31. $\displaystyle \int_0^2 \frac{dx}{x^2-2x+2} = \int_0^2 \frac{1}{1+(x-1)^2}\,dx$

$$= \left[\arctan(x-1)\right]_0^2 = \frac{\pi}{2}$$

33. $\displaystyle \int \frac{2x}{x^2+6x+13}\,dx = \int \frac{2x+6}{x^2+6x+13}\,dx - 6\int \frac{1}{x^2+6x+13}\,dx = \int \frac{2x+6}{x^2+6x+13}\,dx - 6\int \frac{1}{4+(x+3)^2}\,dx$

$$= \ln|x^2+6x+13| - 3\arctan\left(\frac{x+3}{2}\right) + C$$

35. $\displaystyle \int \frac{1}{\sqrt{-x^2-4x}}\,dx = \int \frac{1}{\sqrt{4-(x+2)^2}}\,dx$

$$= \arcsin\left(\frac{x+2}{2}\right) + C$$

37. Let $u = -x^2 - 4x$, $du = (-2x-4)\,dx$.

$$\int \frac{x+2}{\sqrt{-x^2-4x}}\,dx = -\frac{1}{2}\int (-x^2-4x)^{-1/2}(-2x-4)\,dx$$
$$= -\sqrt{-x^2-4x} + C$$

39. $\displaystyle\int_2^3 \frac{2x-3}{\sqrt{4x-x^2}}\,dx = \int_2^3 \frac{2x-4}{\sqrt{4x-x^2}}\,dx + \int_2^3 \frac{1}{\sqrt{4x-x^2}}\,dx = -\int_2^3 (4x-x^2)^{-1/2}(4-2x)\,dx + \int_2^3 \frac{1}{\sqrt{4-(x-2)^2}}\,dx$

$$= \left[-2\sqrt{4x-x^2} + \arcsin\!\left(\frac{x-2}{2}\right) \right]_2^3 = 4 - 2\sqrt{3} + \frac{\pi}{6} \approx 1.059$$

41. Let $u = x^2 + 1$, $du = 2x\,dx$.

$$\int \frac{x}{x^4 + 2x^2 + 2}\,dx = \frac{1}{2}\int \frac{2x}{(x^2+1)^2+1}\,dx$$

$$= \frac{1}{2}\arctan(x^2+1) + C$$

43. Let $u = \sqrt{e^t - 3}$. Then $u^2 + 3 = e^t$, $2u\,du = e^t\,dt$, and $\dfrac{2u\,du}{u^2+3} = dt$.

$$\int \sqrt{e^t - 3}\,dt = \int \frac{2u^2}{u^2+3}\,du = \int 2\,du - \int 6\,\frac{1}{u^2+3}\,du$$

$$= 2u - 2\sqrt{3}\arctan\frac{u}{\sqrt{3}} + C = 2\sqrt{e^t-3} - 2\sqrt{3}\arctan\sqrt{\frac{e^t-3}{3}} + C$$

45. $\displaystyle\int_1^3 \frac{dx}{\sqrt{x}(1+x)}$

Let $u = \sqrt{x}$, $u^2 = x$, $2u\,du = dx$, $1+x = 1+u^2$.

$$\int_1^{\sqrt{3}} \frac{2u\,du}{u(1+u^2)} = \int_1^{\sqrt{3}} \frac{2}{1+u^2}\,du$$

$$= 2\arctan(u)\Big]_1^{\sqrt{3}}$$

$$= 2\left(\frac{\pi}{3} - \frac{\pi}{4}\right) = \frac{\pi}{6}$$

47. (a) $\displaystyle\int \frac{1}{\sqrt{1-x^2}}\,dx = \arcsin x + C, \quad u = x$

(b) $\displaystyle\int \frac{x}{\sqrt{1-x^2}}\,dx = -\sqrt{1-x^2} + C, \quad u = 1 - x^2$

(c) $\displaystyle\int \frac{1}{x\sqrt{1-x^2}}\,dx$ cannot be evaluated using the basic integration rules.

49. (a) $\displaystyle\int \sqrt{x-1}\,dx = \frac{2}{3}(x-1)^{3/2} + C, \quad u = x - 1$

(b) Let $u = \sqrt{x-1}$. Then $x = u^2 + 1$ and $dx = 2u\,du$.

$$\int x\sqrt{x-1}\,dx = \int (u^2+1)(u)(2u)\,du = 2\int (u^4+u^2)\,du = 2\left(\frac{u^5}{5} + \frac{u^3}{3}\right) + C$$

$$= \frac{2}{15}u^3(3u^2+5) + C = \frac{2}{15}(x-1)^{3/2}[3(x-1)+5] + C = \frac{2}{15}(x-1)^{3/2}(3x+2) + C$$

(c) Let $u = \sqrt{x-1}$. Then $x = u^2 + 1$ and $dx = 2u\,du$.

$$\int \frac{x}{\sqrt{x-1}}\,dx = \int \frac{u^2+1}{u}(2u)\,du = 2\int (u^2+1)\,du = 2\left(\frac{u^3}{3} + u\right) + C = \frac{2}{3}u(u^2+3) + C = \frac{2}{3}\sqrt{x-1}(x+2) + C$$

Note: In (b) and (c), substitution was necessary *before* the basic integration rules could be used.

51. Area $\approx (1)(1) = 1$

Matches (c)

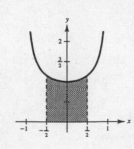

53. $y' = \dfrac{1}{\sqrt{4-x^2}}, \quad (0, \pi)$

$$y = \int \frac{1}{\sqrt{4-x^2}}\,dx = \arcsin\!\left(\frac{x}{2}\right) + C$$

$$y(0) = \pi = C$$

$$y = \arcsin\!\left(\frac{x}{2}\right) + \pi$$

55. (a)

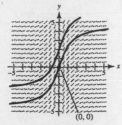

(b) $\dfrac{dy}{dx} = \dfrac{3}{1+x^2}$, $\quad (0, 0)$

$y = 3\displaystyle\int \dfrac{dx}{1+x^2} = 3\arctan x + C$

$(0, 0): 0 = 3\arctan(0) + C \implies C = 0$

$y = 3\arctan x$

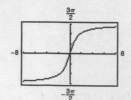

57. (a)

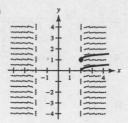

(b) $y' = \dfrac{1}{x\sqrt{x^2-4}}$, $\quad (2, 1)$

$y = \displaystyle\int \dfrac{1}{x\sqrt{x^2-4}}\,dx = \dfrac{1}{2}\,\text{arcsec}\,\dfrac{|x|}{2} + C$

$1 = \dfrac{1}{2}\,\text{arcsec}(1) + C = C$

$y = \dfrac{1}{2}\,\text{arcsec}\,\dfrac{x}{2} + 1$, $\quad x \geq 2$

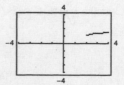

59. $\dfrac{dy}{dx} = \dfrac{10}{x\sqrt{x^2-1}}$, $\quad y(3) = 0$

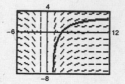

61. $\dfrac{dy}{dx} = \dfrac{2y}{\sqrt{16-x^2}}$, $\quad y(0) = 2$

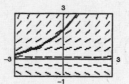

63. $A = \displaystyle\int_1^3 \dfrac{1}{x^2-2x+5}\,dx = \int_1^3 \dfrac{1}{(x-1)^2+4}\,dx$

$= \dfrac{1}{2}\arctan\left(\dfrac{x-1}{2}\right)\Big]_1^3$

$= \dfrac{1}{2}\arctan(1) - \dfrac{1}{2}\arctan(0)$

$= \dfrac{\pi}{8}$

65. Area $= \displaystyle\int_0^1 \dfrac{1}{\sqrt{4-x^2}}\,dx$

$= \arcsin\left(\dfrac{x}{2}\right)\Big]_0^1$

$= \arcsin\left(\dfrac{1}{2}\right) - \arcsin(0)$

$= \dfrac{\pi}{6}$

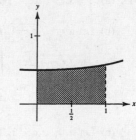

67. Area $= \displaystyle\int_{-\pi/2}^{\pi/2} \dfrac{3\cos x}{1+\sin^2 x}\,dx = 3\int_{-\pi/2}^{\pi/2} \dfrac{1}{1+\sin^2 x}(\cos x\,dx)$

$= 3\arctan(\sin x)\Big]_{-\pi/2}^{\pi/2}$

$= 3\arctan(1) - 3\arctan(-1)$

$= \dfrac{3\pi}{4} + \dfrac{3\pi}{4} = \dfrac{3\pi}{2}$

69. (a) $\dfrac{d}{dx}\left[\ln x - \dfrac{1}{2}\ln(1 + x^2) - \dfrac{\arctan x}{x} + C\right] = \dfrac{1}{x} - \dfrac{x}{1 + x^2} - \left(\dfrac{x[1/(1 + x^2)] - \arctan x}{x^2}\right)$

$$= \dfrac{1 + x^2 - x^2}{x(1 + x^2)} - \dfrac{1}{x(1 + x^2)} + \dfrac{\arctan x}{x^2} = \dfrac{\arctan x}{x^2}$$

Thus, $\displaystyle\int \dfrac{\arctan x}{x^2}\,dx = \ln x - \dfrac{1}{2}\ln(1 + x^2) - \dfrac{\arctan x}{x} + C.$

(b) $A = \displaystyle\int_1^{\sqrt{3}} \dfrac{\arctan x}{x^2}\,dx$

$= \left[\ln x - \dfrac{1}{2}\ln(1 + x^2) - \dfrac{\arctan x}{x}\right]_1^{\sqrt{3}}$

$= \left(\ln \sqrt{3} - \dfrac{1}{2}\ln(4) - \dfrac{\arctan \sqrt{3}}{\sqrt{3}}\right) - \left(\dfrac{-1}{2}\ln 2 - \arctan(1)\right)$

$= \dfrac{1}{2}\ln 3 - \dfrac{1}{2}\ln 2 - \dfrac{\pi\sqrt{3}}{9} + \dfrac{\pi}{4} \approx 0.3835$

71. (a)

Shaded area is given by $\displaystyle\int_0^1 \arcsin x\,dx.$

(b) $\displaystyle\int_0^1 \arcsin x\,dx \approx 0.5708$

(c) Divide the rectangle into two regions.

Area rectangle $= (\text{base})(\text{height}) = 1\left(\dfrac{\pi}{2}\right) = \dfrac{\pi}{2}$

Area rectangle $= \displaystyle\int_0^1 \arcsin x\,dx + \int_0^{\pi/2} \sin y\,dy$

$\dfrac{\pi}{2} = \displaystyle\int_0^1 \arcsin x\,dx + (-\cos y)\Big]_0^{\pi/2}$

$= \displaystyle\int_0^1 \arcsin x\,dx + 1$

Hence, $\displaystyle\int_0^1 \arcsin x\,dx = \dfrac{\pi}{2} - 1, \quad (\approx 0.5708).$

73. $F(x) = \dfrac{1}{2}\displaystyle\int_x^{x+2} \dfrac{2}{t^2 + 1}\,dt$

(a) $F(x)$ represents the average value of $f(x)$ over the interval $[x, x + 2]$. Maximum at $x = -1$, since the graph is greatest on $[-1, 1]$.

(b) $F(x) = \left[\arctan t\right]_x^{x+2} = \arctan(x + 2) - \arctan x$

$F'(x) = \dfrac{1}{1 + (x + 2)^2} - \dfrac{1}{1 + x^2} = \dfrac{(1 + x^2) - (x^2 + 4x + 5)}{(x^2 + 1)(x^2 + 4x + 5)} = \dfrac{-4(x + 1)}{(x^2 + 1)(x^2 + 4x + 5)} = 0$ when $x = -1.$

75. False, $\displaystyle\int \dfrac{dx}{3x\sqrt{9x^2 - 16}} = \dfrac{1}{12}\text{arcsec}\,\dfrac{|3x|}{4} + C$

77. True

$\dfrac{d}{dx}\left[-\arccos\dfrac{x}{2} + C\right] = \dfrac{1/2}{\sqrt{1 - (x/2)^2}} = \dfrac{1}{\sqrt{4 - x^2}}$

79. $\dfrac{d}{dx}\left[\arcsin\left(\dfrac{u}{a}\right) + C\right] = \dfrac{1}{\sqrt{1 - (u^2/a^2)}}\left(\dfrac{u'}{a}\right) = \dfrac{u'}{\sqrt{a^2 - u^2}}$

Thus, $\displaystyle\int \dfrac{du}{\sqrt{a^2 - u^2}} = \arcsin\left(\dfrac{u}{a}\right) + C.$

81. Assume $u > 0$.

$$\frac{d}{dx}\left[\frac{1}{a}\operatorname{arcsec}\frac{u}{a} + C\right] = \frac{1}{a}\left[\frac{u'/a}{(u/a)\sqrt{(u/a)^2 - 1}}\right] = \frac{1}{a}\left[\frac{u'}{u\sqrt{(u^2 - a^2)/a^2}}\right] = \frac{u'}{u\sqrt{u^2 - a^2}}.$$

The case $u < 0$ is handled in a similar manner. Thus,

$$\int\frac{du}{u\sqrt{u^2 - a^2}} = \int\frac{u'}{u\sqrt{u^2 - a^2}}\,dx = \frac{1}{a}\operatorname{arcsec}\frac{|u|}{a} + C.$$

83. (a) $v(t) = -32t + 500$

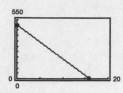

(c)

$$\int\frac{1}{32 + kv^2}\,dv = -\int dt$$

$$\frac{1}{\sqrt{32k}}\arctan\left(\sqrt{\frac{k}{32}}v\right) = -t + C_1$$

$$\arctan\left(\sqrt{\frac{k}{32}}v\right) = -\sqrt{32k}\,t + C$$

$$\sqrt{\frac{k}{32}}v = \tan\left(C - \sqrt{32k}\,t\right)$$

$$v = \sqrt{\frac{32}{k}}\tan\left(C - \sqrt{32k}\,t\right)$$

When $t = 0$, $v = 500$, $C = \arctan\left(500\sqrt{k/32}\right)$, and we have

$$v(t) = \sqrt{\frac{32}{k}}\tan\left[\arctan\left(500\sqrt{\frac{k}{32}}\right) - \sqrt{32k}\,t\right].$$

(e) $h = \displaystyle\int_0^{6.86}\sqrt{32{,}000}\,\tan\left[\arctan\left(500\sqrt{0.00003125}\right) - \sqrt{0.032}\,t\right]dt$

Simpson's Rule: $n = 10$; $h \approx 1088$ feet

(f) Air resistance lowers the maximum height.

(b) $s(t) = \displaystyle\int v(t)\,dt = \int(-32t + 500)\,dt$

$$= -16t^2 + 500t + C$$

$$s(0) = -16(0) + 500(0) + C = 0 \implies C = 0$$

$$s(t) = -16t^2 + 500t$$

When the object reaches its maximum height, $v(t) = 0$.

$$v(t) = -32t + 500 = 0$$

$$-32t = -500$$

$$t = 15.625$$

$$s(15.625) = -16(15.625)^2 + 500(15.625)$$

$$= 3906.25 \text{ ft (Maximum height)}$$

(d) When $k = 0.001$:

$$v(t) = \sqrt{32{,}000}\,\tan\left[\arctan\left(500\sqrt{0.00003125}\right) - \sqrt{0.032}\,t\right]$$

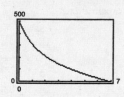

$v(t) = 0$ when $t_0 \approx 6.86$ sec.

Section 5.8　Hyperbolic Functions

1. (a) $\sinh 3 = \dfrac{e^3 - e^{-3}}{2} \approx 10.018$

(b) $\tanh(-2) = \dfrac{\sinh(-2)}{\cosh(-2)} = \dfrac{e^{-2} - e^2}{e^{-2} + e^2} \approx -0.964$

3. (a) $\operatorname{csch}(\ln 2) = \dfrac{2}{e^{\ln 2} - e^{-\ln 2}} = \dfrac{2}{2 - (1/2)} = \dfrac{4}{3}$

(b) $\coth(\ln 5) = \dfrac{\cosh(\ln 5)}{\sinh(\ln 5)} = \dfrac{e^{\ln 5} + e^{-\ln 5}}{e^{\ln 5} - e^{-\ln 5}}$

$$= \dfrac{5 + (1/5)}{5 - (1/5)} = \dfrac{13}{12}$$

5. (a) $\cosh^{-1} 2 = \ln\left(2 + \sqrt{3}\right) \approx 1.317$

(b) $\text{sech}^{-1} \dfrac{2}{3} = \ln\left(\dfrac{1 + \sqrt{1 - (4/9)}}{2/3}\right) \approx 0.962$

7. $\tanh^2 x + \text{sech}^2 x = \left(\dfrac{e^x - e^{-x}}{e^x + e^{-x}}\right)^2 + \left(\dfrac{2}{e^x + e^{-x}}\right)^2 = \dfrac{e^{2x} - 2 + e^{-2x} + 4}{(e^x + e^{-x})^2} = \dfrac{e^{2x} + 2 + e^{-2x}}{e^{2x} + 2 + e^{-2x}} = 1$

9. $\sinh x \cosh y + \cosh x \sinh y = \left(\dfrac{e^x - e^{-x}}{2}\right)\left(\dfrac{e^y + e^{-y}}{2}\right) + \left(\dfrac{e^x + e^{-x}}{2}\right)\left(\dfrac{e^y - e^{-y}}{2}\right)$

$$= \frac{1}{4}\left[e^{x+y} - e^{-x+y} + e^{x-y} - e^{-(x+y)} + e^{x+y} + e^{-x+y} - e^{x-y} - e^{-(x+y)}\right]$$

$$= \frac{1}{4}\left[2(e^{x+y} - e^{-(x+y)})\right] = \frac{e^{(x+y)} - e^{-(x+y)}}{2} = \sinh(x + y)$$

11. $3 \sinh x + 4 \sinh^3 x = \sinh x (3 + 4 \sinh^2 x) = \left(\dfrac{e^x - e^{-x}}{2}\right)\left[3 + 4\left(\dfrac{e^x - e^{-x}}{2}\right)^2\right]$

$$= \left(\frac{e^x - e^{-x}}{2}\right)[3 + e^{2x} - 2 + e^{-2x}] = \frac{1}{2}(e^x - e^{-x})(e^{2x} + e^{-2x} + 1)$$

$$= \frac{1}{2}\left[e^{3x} + e^{-x} + e^x - e^x - e^{-3x} - e^{-x}\right] = \frac{e^{3x} - e^{-3x}}{2} = \sinh(3x)$$

13. $\qquad \sinh x = \dfrac{3}{2}$

$\cosh^2 x \quad \left(\dfrac{3}{2}\right)^2 - 1 \rightarrow \cosh^2 x = \dfrac{13}{4} \Rightarrow \cosh x = \dfrac{\sqrt{13}}{2}$

$\tanh x = \dfrac{3/2}{\sqrt{13}/2} = \dfrac{3\sqrt{13}}{13}$

$\text{csch } x = \dfrac{1}{3/2} = \dfrac{2}{3}$

$\text{sech } x = \dfrac{1}{\sqrt{13}/2} = \dfrac{2\sqrt{13}}{13}$

$\coth x = \dfrac{1}{3/\sqrt{13}} = \dfrac{\sqrt{13}}{3}$

15. $y = \text{sech}(x + 1)$

$y' = -\text{sech}(x + 1) \tanh(x + 1)$

17. $f(x) = \ln(\sinh x)$

$f'(x) = \dfrac{1}{\sinh x}(\cosh x) = \coth x$

19. $y = \ln\left(\tanh \dfrac{x}{2}\right)$

$y' = \dfrac{1/2}{\tanh(x/2)} \text{sech}^2\left(\dfrac{x}{2}\right)$

$= \dfrac{1}{2 \sinh(x/2) \cosh(x/2)}$

$= \dfrac{1}{\sinh x} = \text{csch } x$

21. $h(x) = \dfrac{1}{4} \sinh 2x - \dfrac{x}{2}$

$h'(x) = \dfrac{1}{2} \cosh(2x) - \dfrac{1}{2} = \dfrac{\cosh(2x) - 1}{2} = \sinh^2 x$

23. $f(t) = \arctan(\sinh t)$

$f'(t) = \dfrac{1}{1 + \sinh^2 t}(\cosh t) = \dfrac{\cosh t}{\cosh^2 t} = \text{sech } t$

25. $y = \sinh(1 - x^2)$, $(1, 0)$

$y' = \cosh(1 - x^2)(-2x)$

$y'(1) = -2$

Tangent line: $y - 0 = -2(x - 1)$

$$y = -2x + 2$$

27. $y = (\cosh x - \sinh x)^2$, $(0, 1)$

$y' = 2(\cosh x - \sinh x)(\sinh x - \cosh x)$

At $(0, 1)$, $y' = 2(1)(-1) = -2$.

Tangent line: $y - 1 = -2(x - 0)$

$$y = -2x + 1$$

29. $f(x) = \sin x \sinh x - \cos x \cosh x$, $-4 \le x \le 4$

$f'(x) = \sin x \cosh x + \cos x \sinh x - \cos x \sinh x + \sin x \cosh x$

$\quad = 2 \sin x \cosh x = 0$ when $x = 0, \pm\pi$.

Relative maxima: $(\pm\pi, \cosh \pi)$

Relative minimum: $(0, -1)$

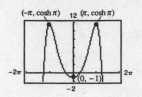

31. $g(x) = x \operatorname{sech} x$

$g'(x) = \operatorname{sech} x - x \operatorname{sech} x \tanh x$

$\quad = \operatorname{sech} x(1 - x \tanh x) = 0$

$x \tanh x = 1$

Using a graphing utility, $x \approx \pm 1.1997$.

By the First Derivative Test, $(1.1997, 0.6627)$ is a relative maximum and $(-1.1997, -0.6627)$ is a relative minimum.

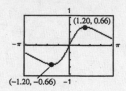

33. $y = a \sinh x$

$y' = a \cosh x$

$y'' = a \sinh x$

$y''' = a \cosh x$

Therefore, $y''' - y' = 0$.

35. $f(x) = \tanh x$, $\qquad f(0) = 0$

$f'(x) = \operatorname{sech}^2 x$, $\qquad f'(0) = 1$

$f''(x) = -2 \operatorname{sech}^2 x \tanh x$, $\quad f''(0) = 0$

$P_1(x) = f(0) + f'(0)(x - 0) = x$

$P_2(x) = f(0) + f'(0)(x - 0) + \frac{1}{2}f''(0)(x - 0)^2 = x$

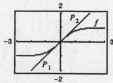

37. (a) $y = 10 + 15 \cosh \dfrac{x}{15}$, $-15 \le x \le 15$

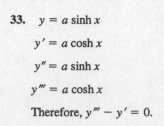

(b) At $x = \pm 15$, $y = 10 + 15 \cosh(1) \approx 33.146$.

At $x = 0$, $y = 10 + 15 \cosh(0) = 25$.

(c) $y' = \sinh \dfrac{x}{15}$. At $x = 15$, $y' = \sinh(1) \approx 1.175$.

39. Let $u = 1 - 2x$, $du = -2 \, dx$.

$$\int \sinh(1 - 2x) \, dx = -\frac{1}{2}\int \sinh(1 - 2x)(-2) \, dx$$

$$= -\frac{1}{2}\cosh(1 - 2x) + C$$

41. Let $u = \cosh(x - 1)$, $du = \sinh(x - 1) \, dx$.

$$\int \cosh^2(x - 1) \sinh(x - 1) \, dx = \frac{1}{3}\cosh^3(x - 1) + C$$

43. Let $u = \sinh x$, $du = \cosh x \, dx$.

$$\int \frac{\cosh x}{\sinh x} \, dx = \ln|\sinh x| + C$$

45. Let $u = \dfrac{x^2}{2}$, $du = x \, dx$.

$$\int x \, \mathrm{csch}^2 \frac{x^2}{2} \, dx = \int \left(\mathrm{csch}^2 \frac{x^2}{2} \right) x \, dx = -\coth \frac{x^2}{2} + C$$

47. Let $u = \dfrac{1}{x}$, $du = -\dfrac{1}{x^2} \, dx$.

$$\int \frac{\mathrm{csch}(1/x) \coth(1/x)}{x^2} \, dx = -\int \mathrm{csch} \frac{1}{x} \coth \frac{1}{x} \left(-\frac{1}{x^2} \right) dx$$

$$= \mathrm{csch} \frac{1}{x} + C$$

49. Let $u = x^2$, $du = 2x \, dx$.

$$\int \frac{x}{x^4 + 1} \, dx = \frac{1}{2} \int \frac{2x}{(x^2)^2 + 1} \, dx = \frac{1}{2} \arctan(x^2) + C$$

51. $\displaystyle\int_0^{\ln 2} \tanh x \, dx = \int_0^{\ln 2} \frac{\sinh x}{\cosh x} \, dx, \quad (u = \cosh x)$

$$= \ln(\cosh x) \Big]_0^{\ln 2}$$

$$= \ln(\cosh(\ln 2) - \ln(\cosh(0))$$

$$= \ln\left(\frac{5}{4} \right) - 0 = \ln\left(\frac{5}{4} \right)$$

Note: $\cosh(\ln 2) = \dfrac{e^{\ln 2} + e^{-\ln 2}}{2} = \dfrac{2 + (1/2)}{2} - \dfrac{5}{4}$

53. $\displaystyle\int_0^4 \frac{1}{25 - x^2} \, dx = \frac{1}{10} \int \frac{1}{5 - x} \, dx + \frac{1}{10} \int \frac{1}{5 + x} \, dx$

$$= \left[\frac{1}{10} \ln\left| \frac{5 + x}{5 - x} \right| \right]_0^4 = \frac{1}{10} \ln 9 = \frac{1}{5} \ln 3$$

55. Let $u = 2x$, $du = 2 \, dx$.

$$\int_0^{\sqrt{2}/4} \frac{2}{\sqrt{1 - 4x^2}} \, dx = \int_0^{\sqrt{2}/4} \frac{1}{\sqrt{1 - (2x)^2}} (2) \, dx$$

$$= \left[\arcsin(2x) \right]_0^{\sqrt{2}/4} = \frac{\pi}{4}$$

57. $y = \cosh^{-1}(3x)$

$$y' = \frac{3}{\sqrt{9x^2 - 1}}$$

59. $y = \sinh^{-1}(\tan x)$

$$y' = \frac{1}{\sqrt{\tan^2 x + 1}} (\sec^2 x) = |\sec x|$$

61. $y = \tanh^{-1}(\sin 2x)$

$$y' = \frac{1}{1 - \sin^2 2x} (2 \cos 2x) = 2 \sec 2x$$

63. $y = 2x \sinh^{-1}(2x) - \sqrt{1 + 4x^2}$

$$y' = 2x\left(\frac{2}{\sqrt{1 + 4x^2}} \right) + 2 \sinh^{-1}(2x) - \frac{4x}{\sqrt{1 + 4x^2}}$$

$$= 2 \sinh^{-1}(2x)$$

65. Answers will vary.

67. $\displaystyle\lim_{x \to \infty} \sinh x = \infty$

69. $\displaystyle\lim_{x \to \infty} \mathrm{sech} \, x = 0$

71. $\displaystyle\lim_{x \to 0} \frac{\sinh x}{x} = \lim_{x \to 0} \frac{e^x - e^{-x}}{2x} = 1$

73. $\displaystyle\int \frac{1}{\sqrt{1 + e^{2x}}} \, dx = \int \frac{e^x}{e^x \sqrt{1 + (e^x)^2}} \, dx = -\mathrm{csch}^{-1}(e^x) + C$

$$= -\ln\left(\frac{1 + \sqrt{1 + e^{2x}}}{e^x} \right) + C$$

75. Let $u = \sqrt{x}$, $du = \dfrac{1}{2\sqrt{x}}\,dx$.

$$\int \frac{1}{\sqrt{x}\sqrt{1+x}}\,dx = 2\int \frac{1}{\sqrt{1+(\sqrt{x})^2}}\left(\frac{1}{2\sqrt{x}}\right)dx = 2\sinh^{-1}\sqrt{x} + C = 2\ln\left(\sqrt{x} + \sqrt{1+x}\right) + C$$

77. $\displaystyle\int \frac{-1}{4x - x^2}\,dx = \int \frac{1}{(x-2)^2 - 4}\,dx = \frac{1}{4}\ln\left|\frac{(x-2)-2}{(x-2)+2}\right| = \frac{1}{4}\ln\left|\frac{x-4}{x}\right| + C$

79. $\displaystyle\int \frac{1}{1 - 4x - 2x^2}\,dx = \int \frac{1}{3 - 2(x+1)^2}\,dx$

$$= \frac{1}{\sqrt{2}}\int \frac{\sqrt{2}}{(\sqrt{3})^2 - [\sqrt{2}(x+1)]^2}\,dx$$

$$= \frac{1}{\sqrt{2}}\cdot\frac{1}{2\sqrt{3}}\ln\left|\frac{\sqrt{3} + \sqrt{2}(x+1)}{\sqrt{3} - \sqrt{2}(x+1)}\right| + C = \frac{1}{2\sqrt{6}}\ln\left|\frac{\sqrt{2}(x+1) + \sqrt{3}}{\sqrt{2}(x+1) - \sqrt{3}}\right| + C$$

81. Let $u = 4x - 1$, $du = 4\,dx$.

$$y = \int \frac{1}{\sqrt{80 + 8x - 16x^2}}\,dx$$

$$= \frac{1}{4}\int \frac{4}{\sqrt{81 - (4x-1)^2}}\,dx = \frac{1}{4}\arcsin\left(\frac{4x-1}{9}\right) + C$$

83. $\displaystyle y = \int \frac{x^3 - 21x}{5 + 4x - x^2}\,dx = \int\left(-x - 4 + \frac{20}{5 + 4x - x^2}\right)dx$

$$= \int(-x - 4)\,dx + 20\int \frac{1}{3^2 - (x-2)^2}\,dx$$

$$= -\frac{x^2}{2} - 4x + \frac{20}{6}\ln\left|\frac{3 + (x-2)}{3 - (x-2)}\right| + C$$

$$= -\frac{x^2}{2} - 4x + \frac{10}{3}\ln\left|\frac{1 + x}{5 - x}\right| + C$$

$$= \frac{-x^2}{2} - 4x - \frac{10}{3}\ln\left|\frac{5 - x}{x + 1}\right| + C$$

85. $\displaystyle A = 2\int_0^4 \operatorname{sech}\frac{x}{2}\,dx$

$$= 2\int_0^4 \frac{2}{e^{x/2} + e^{-x/2}}\,dx$$

$$= 4\int_0^4 \frac{e^{x/2}}{(e^{x/2})^2 + 1}\,dx$$

$$= \left[8\arctan(e^{x/2})\right]_0^4$$

$$= 8\arctan(e^2) - 2\pi \approx 5.207$$

87. $\displaystyle A = \int_0^2 \frac{5x}{\sqrt{x^4 + 1}}\,dx$

$$= \frac{5}{2}\int_0^2 \frac{2x}{\sqrt{(x^2)^2 + 1}}\,dx$$

$$= \left[\frac{5}{2}\ln\left(x^2 + \sqrt{x^4 + 1}\right)\right]_0^2$$

$$= \frac{5}{2}\ln\left(4 + \sqrt{17}\right) \approx 5.237$$

89. (a) $\displaystyle\int_0^{\sqrt{3}} \frac{dx}{\sqrt{x^2 + 1}} = \ln\left(x + \sqrt{x^2 + 1}\right)\Big]_0^{\sqrt{3}}$

$$= \ln\left(\sqrt{3} + 2\right) \approx 1.317$$

(b) $\displaystyle\int_0^{\sqrt{3}} \frac{dx}{\sqrt{x^2 + 1}} = \sinh^{-1}x\Big]_0^{\sqrt{3}} = \sinh^{-1}\left(\sqrt{3}\right) \approx 1.317$

91. $\int \dfrac{3k}{16} \, dt = \int \dfrac{1}{x^2 - 12x + 32} \, dx$

$\dfrac{3kt}{16} = \int \dfrac{1}{(x - 6)^2 - 4} \, dx = \dfrac{1}{2(2)} \ln \left| \dfrac{(x - 6) - 2}{(x - 6) + 2} \right| + C = \dfrac{1}{4} \ln \left| \dfrac{x - 8}{x - 4} \right| + C$

When $x = 0$:

$t = 0$

$C = -\dfrac{1}{4} \ln(2)$

When $x = 1$:

$t = 10$

$\dfrac{30k}{16} = \dfrac{1}{4} \ln \left| \dfrac{-7}{-3} \right| - \dfrac{1}{4} \ln(2) = \dfrac{1}{4} \ln \left(\dfrac{7}{6} \right)$

$k = \dfrac{2}{15} \ln \left(\dfrac{7}{6} \right)$

When $t = 20$:

$\left(\dfrac{3}{16} \right) \left(\dfrac{2}{15} \right) \ln \left(\dfrac{7}{6} \right) (20) = \dfrac{1}{4} \ln \dfrac{x - 8}{2x - 8}$

$\ln \left(\dfrac{7}{6} \right)^2 = \ln \dfrac{x - 8}{2x - 8}$

$\dfrac{49}{36} = \dfrac{x - 8}{2x - 8}$

$62x = 104$

$x = \dfrac{104}{62} = \dfrac{52}{31} \approx 1.677 \text{ kg}$

93. $y = a \operatorname{sech}^{-1} \dfrac{x}{a} - \sqrt{a^2 - x^2}, \quad a > 0$

$\dfrac{dy}{dx} = \dfrac{-1}{(x/a)\sqrt{1 - (x^2/a^2)}} + \dfrac{x}{\sqrt{a^2 - x^2}} = \dfrac{-a^2}{x\sqrt{a^2 - x^2}} + \dfrac{x}{\sqrt{a^2 - x^2}} = \dfrac{x^2 - a^2}{x\sqrt{a^2 - x^2}} = \dfrac{-\sqrt{a^2 - x^2}}{x}$

95. Let $\quad u = \tanh^{-1} x, \quad -1 < x < 1$

$\tanh u = x.$

$\dfrac{\sinh u}{\cosh u} = \dfrac{e^u - e^{-u}}{e^u + e^{-u}} = x$

$e^u - e^{-u} = xe^u + xe^{-u}$

$e^{2u} - 1 = xe^{2u} + x$

$e^{2u}(1 - x) = 1 + x$

$e^{2u} = \dfrac{1 + x}{1 - x}$

$2u = \ln \left(\dfrac{1 + x}{1 - x} \right)$

$u = \dfrac{1}{2} \ln \left(\dfrac{1 + x}{1 - x} \right), \quad -1 < x < 1$

97. $\displaystyle\int_{-b}^{b} e^{xt} \, dt = \dfrac{e^{xt}}{x} \Bigg]_{-b}^{b}$

$= \dfrac{e^{xb}}{x} - \dfrac{e^{-xb}}{x}$

$= \dfrac{2}{x} \left[\dfrac{e^{xb} - e^{-xb}}{2} \right]$

$= \dfrac{2}{x} \sinh(xb)$

99. $\qquad y = \operatorname{sech}^{-1} x$

$\operatorname{sech} y = x$

$-(\operatorname{sech} y)(\tanh y)y' = 1$

$y' = \dfrac{-1}{(\operatorname{sech} y)(\tanh y)}$

$= \dfrac{-1}{(\operatorname{sech} y)\sqrt{1 - \operatorname{sech}^2 y}} = \dfrac{-1}{x\sqrt{1 - x^2}}$

101. $\qquad y = \sinh^{-1} x$

$\sinh y = x$

$(\cosh y)y' = 1$

$y' = \dfrac{1}{\cosh y} = \dfrac{1}{\sqrt{\sinh^2 y + 1}} = \dfrac{1}{\sqrt{x^2 + 1}}$

103. $y = c \cosh \dfrac{x}{c}$

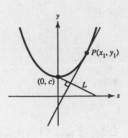

Let $P(x_1, y_1)$ be a point on the catenary.

$y' = \sinh \dfrac{x}{c}$

The slope at P is $\sinh(x_1/c)$. The equation of line L is

$$y - c = \frac{-1}{\sinh(x_1/c)}(x - 0).$$

When $y = 0$, $c = \dfrac{x}{\sinh(x_1/c)} \implies x = c \sinh\left(\dfrac{x_1}{c}\right)$. The length of L is

$$\sqrt{c^2 \sinh^2\left(\frac{x_1}{c}\right) + c^2} = c \cdot \cosh \frac{x_1}{c} = y_1,$$

the ordinate y_1 of the point P.

Review Exercises for Chapter 5

1. $f(x) = \ln x + 3$

Vertical shift 3 units upward

Vertical asymptote: $x = 0$

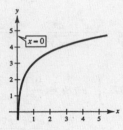

3. $\ln \sqrt[5]{\dfrac{4x^2 - 1}{4x^2 + 1}} = \dfrac{1}{5} \ln \dfrac{(2x - 1)(2x + 1)}{4x^2 + 1} = \dfrac{1}{5}[\ln(2x - 1) + \ln(2x + 1) - \ln(4x^2 + 1)]$

5. $\ln 3 + \dfrac{1}{3}\ln(4 - x^2) - \ln x = \ln 3 + \ln \sqrt[3]{4 - x^2} - \ln x$

$$= \ln\left(\frac{3\sqrt[3]{4 - x^2}}{x}\right)$$

7. $\ln \sqrt{x + 1} = 2$

$\quad \sqrt{x + 1} = e^2$

$\quad x + 1 = e^4$

$\quad x = e^4 - 1 \approx 53.598$

9. $g(x) = \ln \sqrt{x} = \dfrac{1}{2}\ln x$

$g'(x) = \dfrac{1}{2x}$

11. $f(x) = x\sqrt{\ln x}$

$f'(x) = \left(\dfrac{x}{2}\right)(\ln x)^{-1/2}\left(\dfrac{1}{x}\right) + \sqrt{\ln x}$

$\quad = \dfrac{1}{2\sqrt{\ln x}} + \sqrt{\ln x} = \dfrac{1 + 2\ln x}{2\sqrt{\ln x}}$

13. $y = \dfrac{1}{b^2}[a + bx - a\ln(a + bx)]$

$\dfrac{dy}{dx} = \dfrac{1}{b^2}\left(b - \dfrac{ab}{a + bx}\right) = \dfrac{x}{a + bx}$

15. $y = \ln(2 + x) + \dfrac{2}{2 + x}, \; (-1, 2)$

$y' = \dfrac{1}{2 + x} - \dfrac{2}{(2 + x)^2}$

$y'(-1) = 1 - 2 = -1$

Tangent line: $y - 2 = -1(x + 1)$

$\qquad\qquad\qquad y = -x + 1$

17. $u = 7x - 2, du = 7 \, dx$

$$\int \frac{1}{7x - 2} \, dx = \frac{1}{7} \int \frac{1}{7x - 2} (7) \, dx = \frac{1}{7} \ln|7x - 2| + C$$

19. $\displaystyle\int \frac{\sin x}{1 + \cos x} \, dx = -\int \frac{-\sin x}{1 + \cos x} \, dx$

$$= -\ln|1 + \cos x| + C$$

21. $\displaystyle\int_1^4 \frac{x + 1}{x} \, dx = \int_1^4 \left(1 + \frac{1}{x}\right) dx = \left[x + \ln|x|\right]_1^4$

$$= 3 + \ln 4$$

23. $\displaystyle\int_0^{\pi/3} \sec\theta \, d\theta = \left[\ln|\sec\theta + \tan\theta|\right]_0^{\pi/3} = \ln(2 + \sqrt{3})$

25. (a) $\quad f(x) = \frac{1}{2}x - 3$ (b)

$$y = \frac{1}{2}x - 3$$
$$2(y + 3) = x$$
$$2(x + 3) = y$$
$$f^{-1}(x) = 2x + 6$$

(c) $f^{-1}(f(x)) = f^{-1}\left(\frac{1}{2}x - 3\right) = 2\left(\frac{1}{2}x - 3\right) + 6 = x$

$\quad f(f^{-1}(x)) = f(2x + 6) = \frac{1}{2}(2x + 6) - 3 = x$

27. (a) $\quad f(x) = \sqrt{x + 1}$ (b)

$$y = \sqrt{x + 1}$$
$$y^2 - 1 = x$$
$$x^2 - 1 = y$$
$$f^{-1}(x) = x^2 - 1, \ x \geq 0$$

(c) $f^{-1}(f(x)) = f^{-1}\left(\sqrt{x + 1}\right) = \sqrt{(x^2 - 1)^2} - 1 = x$

$\quad f(f^{-1}(x)) = f(x^2 - 1) = \sqrt{(x^2 - 1) + 1}$

$$= \sqrt{x^2} = x \text{ for } x \geq 0.$$

29. (a) $\quad f(x) = \sqrt[3]{x + 1}$

$$y = \sqrt[3]{x + 1}$$
$$y^3 - 1 = x$$
$$x^3 - 1 = y$$
$$f^{-1}(x) = x^3 - 1$$

(b)

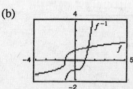

(c) $f^{-1}(f(x)) = f^{-1}\left(\sqrt[3]{x + 1}\right)$

$$= \left(\sqrt[3]{x + 1}\right)^3 - 1 = x$$

$\quad f(f^{-1}(x)) = f(x^3 - 1) = \sqrt[3]{(x^3 - 1) + 1} = x$

31. $\quad\quad f(x) = x^3 + 2$

$$f^{-1}(x) = (x - 2)^{1/3}$$
$$(f^{-1})'(x) = \frac{1}{3}(x - 2)^{-2/3}$$
$$(f^{-1})'(-1) = \frac{1}{3}(-1 - 2)^{-2/3} = \frac{1}{3(-3)^{2/3}}$$
$$= \frac{1}{3^{5/3}} \approx 0.160$$

or

$$f(-3^{1/3}) = -1$$
$$f'(x) = 3x^2$$
$$f'(-3^{1/3}) = 3(-3^{1/3})^2 = 3^{5/3}$$
$$(f^{-1})'(-1) = \frac{1}{f'(-3^{1/3})} = \frac{1}{3^{5/3}}$$

33. $\quad\quad f(x) = \tan x$

$$f\left(\frac{\pi}{6}\right) = \frac{\sqrt{3}}{3}$$
$$f'(x) = \sec^2 x$$
$$f'\left(\frac{\pi}{6}\right) = \frac{4}{3}$$
$$(f^{-1})'\left(\frac{\sqrt{3}}{3}\right) = \frac{1}{f'(\pi/6)} = \frac{3}{4}$$

35. (a) $\quad f(x) = \ln\sqrt{x}$ (b)

$$y = \ln\sqrt{x}$$
$$e^y = \sqrt{x}$$
$$e^{2y} = x$$
$$e^{2x} = y$$
$$f^{-1}(x) = e^{2x}$$

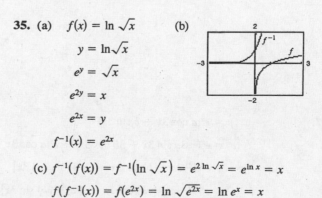

(c) $f^{-1}(f(x)) = f^{-1}\left(\ln\sqrt{x}\right) = e^{2\ln\sqrt{x}} = e^{\ln x} = x$

$\quad f(f^{-1}(x)) = f(e^{2x}) = \ln\sqrt{e^{2x}} = \ln e^x = x$

37. $y = e^{-x/2}$

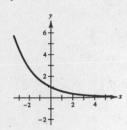

39. $g(t) = t^2 e^t$

$$g'(t) = t^2 e^t + 2te^t = te^t(t + 2)$$

41. $y = \sqrt{e^{2x} + e^{-2x}}$

$$y' = \frac{1}{2}(e^{2x} + e^{-2x})^{-1/2}(2e^{2x} - 2e^{-2x}) = \frac{e^{2x} - e^{-2x}}{\sqrt{e^{2x} + e^{-2x}}}$$

43. $g(x) = \dfrac{x^2}{e^x}$

$$g'(x) = \frac{e^x(2x) - x^2 e^x}{e^{2x}} = \frac{x(2 - x)}{e^x}$$

45. $f(x) = \ln(e^{-x^2}) = -x^2, \quad (2, -4)$

$f'(x) = -2x$

$f'(2) = -4$

Tangent line: $y + 4 = -4(x - 2)$

$$y = -4x + 4$$

47.
$$y(\ln x) + y^2 = 0$$

$$y\left(\frac{1}{x}\right) + (\ln x)\left(\frac{dy}{dx}\right) + 2y\left(\frac{dy}{dx}\right) = 0$$

$$(2y + \ln x)\frac{dy}{dx} = \frac{-y}{x}$$

$$\frac{dy}{dx} = \frac{-y}{x(2y + \ln x)}$$

49. $\displaystyle\int_0^1 xe^{-3x^2}\, dx = -\frac{1}{6}\int_0^1 e^{-3x^2}(-6x\, dx)$

$$= -\frac{1}{6}e^{-3x^2}\bigg]_0^1$$

$$= -\frac{1}{6}[e^{-3} - 1]$$

$$= \frac{1}{6}\left(1 - \frac{1}{e^3}\right)$$

51. $\displaystyle\int \frac{e^{4x} - e^{2x} + 1}{e^x}\, dx = \int (e^{3x} - e^x + e^{-x})\, dx$

$$= \frac{1}{3}e^{3x} - e^x - e^{-x} + C$$

$$= \frac{e^{4x} - 3e^{2x} - 3}{3e^x} + C$$

53. $\displaystyle\int xe^{1-x^2}\, dx = -\frac{1}{2}\int e^{1-x^2}(-2x)\, dx$

$$= -\frac{1}{2}e^{1-x^2} + C$$

55. $\displaystyle\int_1^3 \frac{e^x}{e^x - 1}\, dx$

Let $u = e^x - 1, du = e^x\, dx$.

$$\int_1^3 \frac{e^x}{e^x - 1}\, dx = \ln|e^x - 1|\bigg]_1^3$$

$$= \ln(e^3 - 1) - \ln(e - 1)$$

$$= \ln\left(\frac{e^3 - 1}{e - 1}\right)$$

$$= \ln(e^2 + e + 1)$$

57.
$$y = e^x(a\cos 3x + b\sin 3x)$$

$$y' = e^x(-3a\sin 3x + 3b\cos 3x) + e^x(a\cos 3x + b\sin 3x)$$

$$= e^x[(-3a + b)\sin 3x + (a + 3b)\cos 3x]$$

$$y'' = e^x[3(-3a + b)\cos 3x - 3(a + 3b)\sin 3x] + e^x[(-3a + b)\sin 3x + (a + 3b)\cos 3x]$$

$$= e^x[(-6a - 8b)\sin 3x + (-8a + 6b)\cos 3x]$$

$$y'' - 2y' + 10y = e^x\{[(-6a - 8b) - 2(-3a + b) + 10b]\sin 3x + [(-8a + 6b) - 2(a + 3b) + 10a]\cos 3x\} = 0$$

59. Area $= \displaystyle\int_0^4 xe^{-x^2}\,dx = \left[-\frac{1}{2}e^{-x^2} \right]_0^4 = -\frac{1}{2}(e^{-16} - 1) \approx 0.500$

61. $y = 3^{x/2}$

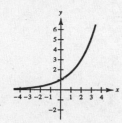

63. $y = \log_2(x - 1)$

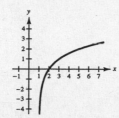

65. $f(x) = 3^{x-1}$

$f'(x) = 3^{x-1}\ln 3$

67. $y = x^{2x+1}$

$\ln y = (2x + 1)\ln x$

$\dfrac{y'}{y} = \dfrac{2x + 1}{x} + 2\ln x$

$y' = y\left(\dfrac{2x + 1}{x} + 2\ln x\right) = x^{2x+1}\left(\dfrac{2x + 1}{x} + 2\ln x\right)$

69. $g(x) = \log_3 \sqrt{1 - x} = \dfrac{1}{2}\log_3(1 - x)$

$g'(x) = \left(\dfrac{1}{2}\right)\dfrac{-1}{(1 - x)\ln 3} = \dfrac{1}{2(x - 1)\ln 3}$

71. $\displaystyle\int (x + 1)5^{(x+1)^2}\,dx = \left(\frac{1}{2}\right)\frac{1}{\ln 5}5^{(x+1)^2} + C$

73. $t = 50\log_{10}\left(\dfrac{18,000}{18,000 - h}\right)$

(a) Domain: $0 \le h < 18,000$

(b)

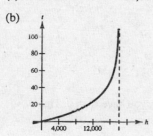

Vertical asymptote: $h = 18,000$

(c) $\qquad t = 50\log_{10}\left(\dfrac{18,000}{18,000 - h}\right)$

$10^{t/50} = \dfrac{18,000}{18,000 - h}$

$18,000 - h = 18,000(10^{-t/50})$

$h = 18,000(1 - 10^{-t/50})$

$\dfrac{dh}{dt} = 360\ln 10\left(\dfrac{1}{10}\right)^{t/50}$ is greatest when $t = 0$.

75. $f(x) = 2\arctan(x + 3)$

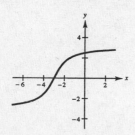

77. (a)

Let $\theta = \arcsin \dfrac{1}{2}$

$\sin \theta = \dfrac{1}{2}$

$\sin\left(\arcsin \dfrac{1}{2}\right) = \sin \theta = \dfrac{1}{2}.$

(b)

Let $\theta = \arcsin \dfrac{1}{2}$

$\sin \theta = \dfrac{1}{2}$

$\cos\left(\arcsin \dfrac{1}{2}\right) = \cos \theta = \dfrac{\sqrt{3}}{2}.$

79. $y = \tan(\arcsin x) = \dfrac{x}{\sqrt{1 - x^2}}$

$y' = \dfrac{(1 - x^2)^{1/2} + x^2(1 - x^2)^{-1/2}}{1 - x^2} = (1 - x^2)^{-3/2}$

81. $y = x \operatorname{arcsec} x$

$y' = \dfrac{x}{|x|\sqrt{x^2 - 1}} + \operatorname{arcsec} x$

83. $y = x(\arcsin x)^2 - 2x + 2\sqrt{1 - x^2}\arcsin x$

$y' = \dfrac{2x \arcsin x}{\sqrt{1 - x^2}} + (\arcsin x)^2 - 2 + \dfrac{2\sqrt{1 - x^2}}{\sqrt{1 - x^2}} - \dfrac{2x}{\sqrt{1 - x^2}}\arcsin x = (\arcsin x)^2$

85. Let $u = e^{2x}$, $du = 2e^{2x}\,dx$.

$\displaystyle\int \dfrac{1}{e^{2x} + e^{-2x}}\,dx = \int \dfrac{e^{2x}}{1 + e^{4x}}\,dx = \dfrac{1}{2}\int \dfrac{1}{1 + (e^{2x})^2}(2e^{2x})\,dx = \dfrac{1}{2}\arctan(e^{2x}) + C$

87. Let $u = x^2$, $du = 2x\,dx$.

$\displaystyle\int \dfrac{x}{\sqrt{1 - x^4}}\,dx = \dfrac{1}{2}\int \dfrac{1}{\sqrt{1 - (x^2)^2}}(2x)\,dx = \dfrac{1}{2}\arcsin x^2 + C$

89. Let $u = \arctan\left(\dfrac{x}{2}\right)$, $du = \dfrac{2}{4 + x^2}\,dx$.

$\displaystyle\int \dfrac{\arctan(x/2)}{4 + x^2}\,dx = \dfrac{1}{2}\int \left(\arctan \dfrac{x}{2}\right)\left(\dfrac{2}{4 + x^2}\right)dx$

$= \dfrac{1}{4}\left(\arctan \dfrac{x}{2}\right)^2 + C$

91. $A = \displaystyle\int_0^1 \dfrac{4 - x}{\sqrt{4 - x^2}}\,dx$

$= 4\displaystyle\int_0^1 \dfrac{1}{\sqrt{4 - x^2}}\,dx + \dfrac{1}{2}\int_0^1 (4 - x^2)^{-1/2}(-2x)\,dx$

$= \left[4\arcsin\left(\dfrac{x}{2}\right) + \sqrt{4 - x^2}\,\right]_0^1$

$= \left(4\arcsin\left(\dfrac{1}{2}\right) + \sqrt{3}\right) - 2$

$= \dfrac{2\pi}{3} + \sqrt{3} - 2 \approx 1.8264$

93. $\displaystyle\int \dfrac{dy}{\sqrt{A^2 - y^2}} = \int \sqrt{\dfrac{k}{m}}\,dt$

$\arcsin\left(\dfrac{y}{A}\right) = \sqrt{\dfrac{k}{m}}\,t + C$

Since $y = 0$ when $t = 0$, you have $C = 0$. Thus,

$\sin\left(\sqrt{\dfrac{k}{m}}\,t\right) = \dfrac{y}{A}$

$y = A\sin\left(\sqrt{\dfrac{k}{m}}\,t\right).$

95. $y = x\tanh^{-1} 2x$

$y' = x\left(\dfrac{2}{1 - 4x^2}\right) + \tanh^{-1} 2x = \dfrac{2x}{1 - 4x^2} + \tanh^{-1} 2x$

97. Let $u = x^3$, $du = 3x^2\,dx$.

$\displaystyle\int x^2(\operatorname{sech} x^3)^2\,dx = \dfrac{1}{3}\int (\operatorname{sech} x^3)^2(3x^2)\,dx = \dfrac{1}{3}\tanh x^3 + C$

Problem Solving for Chapter 5

1. $\tan \theta_1 = \dfrac{3}{x}$

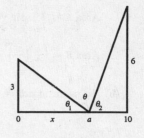

$\tan \theta_2 = \dfrac{6}{10 - x}$

Minimize $\theta_1 + \theta_2$:

$$f(x) = \theta_1 + \theta_2 = \arctan\left(\frac{3}{x}\right) + \arctan\left(\frac{6}{10 - x}\right)$$

$$f'(x) = \frac{1}{1 + \dfrac{9}{x^2}}\left(\frac{-3}{x^2}\right) + \frac{1}{1 + \dfrac{36}{(10 - x)^2}}\left(\frac{6}{(10 - x)^2}\right) = 0$$

$$\frac{3}{x^2 + 9} = \frac{6}{(10 - x)^2 + 36}$$

$$(10 - x)^2 + 36 = 2(x^2 + 9)$$

$$100 - 20x + x^2 + 36 = 2x^2 + 18$$

$$x^2 + 20x - 118 = 0$$

$$x = \frac{-20 \pm \sqrt{20^2 - 4(-118)}}{2} = -10 \pm \sqrt{218}$$

$a = -10 + \sqrt{218} \approx 4.7648 \quad f(a) \approx 1.4153$

$\theta = \pi - (\theta_1 + \theta_2) \approx 1.7263 \quad \text{or} \quad 98.9°$

Endpoints: $a = 0$: $\theta \approx 1.0304$

$\qquad\qquad\quad a = 10$: $\theta \approx 1.2793$

Maximum is 1.7263 at $a = -10 + \sqrt{218} \approx 4.7648$.

3. (a) $f(x) = \dfrac{\ln(x + 1)}{x}, \quad -1 \le x \le 1$

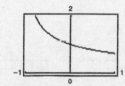

(b) $\displaystyle\lim_{x \to 0} f(x) = 1$

(c) Let $g(x) = \ln x$, $g'(x) = 1/x$, and $g'(1) = 1$. From the definition of derivative

$$g'(1) = \lim_{x \to 0} \frac{g(1 + x) - g(1)}{x} = \lim_{x \to 0} \frac{\ln(1 + x)}{x}.$$

Thus, $\displaystyle\lim_{x \to 0} f(x) = 1$.

5. $y = 0.5^x$ and $y = 1.2^x$ intersect $y = x$. $y = 2^x$ does not intersect $y = x$.
Suppose $y = x$ is tangent to $y = a^x$ at (x, y).

$a^x = x \implies a = x^{1/x}$.

$y' = a^x \ln a = 1 \implies x \ln x^{1/x} = 1 \implies \ln x = 1 \implies x = e, a = e^{1/e}$

For $0 < a \le e^{1/e} \approx 1.445$, the curve $y = a^x$ intersects $y = x$.

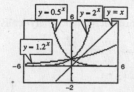

7. (a) $y = f(x) = \arcsin x$

$\sin y = x$

$$\text{Area } A = \int_{\pi/6}^{\pi/4} \sin y \cdot dy = -\cos y \Big]_{\pi/6}^{\pi/4} = -\frac{\sqrt{2}}{2} + \frac{\sqrt{3}}{2} = \frac{\sqrt{3} - \sqrt{2}}{2} \approx 0.1589$$

$$\text{Area } B = \left(\frac{1}{2}\right)\left(\frac{\pi}{6}\right) = \frac{\pi}{12} \approx 0.2618$$

(b) $$\int_{1/2}^{\sqrt{2}/2} \arcsin x \, dx = \text{Area}(C) = \left(\frac{\pi}{4}\right)\left(\frac{\sqrt{2}}{2}\right) - A - B$$

$$= \frac{\pi\sqrt{2}}{8} - \frac{\sqrt{3} - \sqrt{2}}{2} - \frac{\pi}{12} = \pi\left(\frac{\sqrt{2}}{8} - \frac{1}{12}\right) + \frac{\sqrt{2} - \sqrt{3}}{2} \approx 0.1346$$

(c) $$\text{Area } A = \int_0^{\ln 3} e^y \, dy$$

$$= e^y \Big]_0^{\ln 3} = 3 - 1 = 2$$

$$\text{Area } B = \int_1^3 \ln x \, dx = 3(\ln 3) - A = 3\ln 3 - 2 = \ln 27 - 2 \approx 1.2958$$

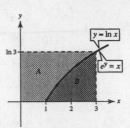

(d) $\tan y = x$

$$\text{Area } A = \int_{\pi/4}^{\pi/3} \tan y \, dy$$

$$= -\ln|\cos y| \Big]_{\pi/4}^{\pi/3}$$

$$= -\ln\frac{1}{2} + \ln\frac{\sqrt{2}}{2} = \ln\sqrt{2} = \frac{1}{2}\ln 2$$

$$\text{Area } C = \int_1^{\sqrt{3}} \arctan x \, dx = \left(\frac{\pi}{3}\right)(\sqrt{3}) - \frac{1}{2}\ln 2 - \left(\frac{\pi}{4}\right)(1)$$

$$= \frac{\pi}{12}(4\sqrt{3} - 3) - \frac{1}{2}\ln 2 \approx 0.6818$$

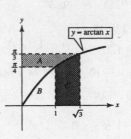

9. $y = e^x$

$y' = e^x$

$y - b = e^a(x - a)$

$y = e^a x - a e^a + b$ Tangent line

If $y = 0$: $e^a x = a e^a - b$

$bx = ab - b$ $(b = e^a)$

$x = a - 1$

$c = a - 1$

Thus, $a - c = a - (a - 1) = 1$.

11. Let $u = \tan x, du = \sec^2 x \, dx$.

$$\text{Area } = \int_0^{\pi/4} \frac{1}{\sin^2 x + 4\cos^2 x} dx = \int_0^{\pi/4} \frac{\sec^2 x}{\tan^2 x + 4} dx$$

$$= \int_0^1 \frac{du}{u^2 + 4}$$

$$= \left[\frac{1}{2}\arctan\left(\frac{u}{2}\right)\right]_0^1$$

$$= \frac{1}{2}\arctan\left(\frac{1}{2}\right)$$

13. (a) $u = 985.93 - \left(985.93 - \frac{(120{,}000)(0.095)}{12}\right)\left(1 + \frac{0.095}{12}\right)^{12t}$

$v = \left(985.93 - \frac{(120{,}000)(0.095)}{12}\right)\left(1 + \frac{0.095}{12}\right)^{12t}$

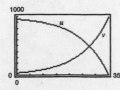

(b) The larger part goes for interest. The curves intersect when $t \approx 27.7$ years.

(c) The slopes are negatives of each other. Analytically,

$$u = 985.93 - v \implies \frac{du}{dt} = -\frac{dv}{dt}$$

$$u'(15) = -v'(15) = -14.06.$$

(d) $t = 12.7$ years

Again, the larger part goes for interest.

CHAPTER 6
Differential Equations

Section 6.1   Slope Fields and Euler's Method 280

Section 6.2   Differential Equations: Growth and Decay   286

Section 6.3   Separation of Variables and the Logistic Equation 292

Section 6.4   First-Order Linear Differential Equations   298

Review Exercises   304

Problem Solving 309

CHAPTER 6
Differential Equations

Section 6.1   Slope Fields and Euler's Method

1. Differential equation: $y' = 4y$

   Solution: $y = Ce^{4x}$

   Check: $y' = 4Ce^{4x} = 4y$

3. Differential equation: $y' = \dfrac{2xy}{x^2 - y^2}$

   Solution: $x^2 + y^2 = Cy$

   Check: $2x + 2yy' = Cy'$

   $$y' = \frac{-2x}{(2y - C)}$$

   $$y' = \frac{-2xy}{2y^2 - Cy}$$

   $$= \frac{-2xy}{2y^2 - (x^2 + y^2)}$$

   $$= \frac{-2xy}{y^2 - x^2}$$

   $$= \frac{2xy}{x^2 - y^2}$$

5. Differential equation: $y'' + y = 0$

   Solution: $y = C_1 \cos x + C_2 \sin x$

   Check: $y' = -C_1 \sin x + C_2 \cos x$

   $\qquad\quad y'' = -C_1 \cos x - C_2 \sin x$

   $\qquad y'' + y = -C_1 \cos x - C_2 \sin x + C_1 \cos x + C_2 \sin x = 0$

7. Differential Equation: $y'' + y = \tan x$

   $y = -\cos x \ln|\sec x + \tan x|$

   $y' = (-\cos x)\dfrac{1}{\sec x + \tan x}(\sec x \cdot \tan x + \sec^2 x) + \sin x \ln|\sec x + \tan x|$

   $\quad = \dfrac{(-\cos x)}{\sec x + \tan x}(\sec x)(\tan x + \sec x) + \sin x \ln|\sec x + \tan x|$

   $\quad = -1 + \sin x \ln|\sec x + \tan x|$

   $y'' = (\sin x)\dfrac{1}{\sec x + \tan x}(\sec x \cdot \tan x + \sec^2 x) + \cos x \ln|\sec x + \tan x|$

   $\quad = (\sin x)(\sec x) + \cos x \ln|\sec x + \tan x|$

   Substituting,

   $\qquad y'' + y = (\sin x)(\sec x) + \cos x \ln|\sec x + \tan x| - \cos x \ln|\sec x + \tan x|$

   $\qquad\qquad = \tan x.$

9. $y = \sin x \cos x - \cos^2 x$

$y' = -\sin^2 x + \cos^2 x + 2 \cos x \sin x$

$\quad = -1 + 2 \cos^2 x + \sin 2x$

Differential Equation:

$2y + y' = 2(\sin x \cos x - \cos^2 x) + (-1 + 2 \cos^2 x + \sin 2x)$

$\qquad = 2 \sin x \cos x - 1 + \sin 2x$

$\qquad = 2 \sin 2x - 1$

Initial condition:

$$y\left(\frac{\pi}{4}\right) = \sin\frac{\pi}{4}\cos\frac{\pi}{4} - \cos^2\frac{\pi}{4} = \frac{\sqrt{2}}{2}\cdot\frac{\sqrt{2}}{2} - \left(\frac{\sqrt{2}}{2}\right)^2 = 0$$

11. $y = 6e^{-2x^2}$

$y' = 6e^{-2x^2}(-4x) = -24xe^{-2x^2}$

Differential Equation:

$y' = -24xe^{-2x^2} = -4x(6e^{-2x^2}) = -4xy$

Initial condition:

$y(0) = 6e^{-2(0)^2} = 6e^0 = 6(1) = 6$

In Exercises 13–17, the differential equation is $y^{(4)} - 16y = 0$.

13. $y = 3 \cos x$

$\quad y^{(4)} = 3 \cos x$

$y^{(4)} - 16y = -45 \cos x \neq 0,$

No

15. $y = e^{-2x}$

$\quad y^{(4)} = 16e^{-2x}$

$y^{(4)} - 16y = 16e^{-2x} - 16e^{-2x} = 0,$

Yes

17. $y = C_1 e^{2x} + C_2 e^{-2x} + C_3 \sin 2x + C_4 \cos 2x$

$\quad y^{(4)} = 16C_1 e^{2x} + 16C_2 e^{-2x} + 16C_3 \sin 2x + 16C_4 \cos 2x$

$y^{(4)} - 16y = 0,$

Yes

In 19–23, the differential equation is $xy' - 2y = x^3 e^x$.

19. $y = x^2,\ y' = 2x$

$xy' - 2y = x(2x) - 2(x^2) = 0 \neq x^3 e^x,$

No

21. $y = x^2(2 + e^x),\ \ y' = x^2(e^x) + 2x(2 + e^x)$

$xy' - 2y = x[x^2 e^x + 2xe^x + 4x] - 2[x^2 e^x + 2x^2] = x^3 e^x,$

Yes

23. $y = \ln x,\ y' = \dfrac{1}{x}$

$xy' - 2y = x\left(\dfrac{1}{x}\right) - 2 \ln x \neq x^3 e^x,$

No

25. $y = Ce^{-x/2}$ passes through $(0, 3)$.

$3 = Ce^0 = C \implies C = 3$

Particular solution: $y = 3e^{-x/2}$

27. $y^2 = Cx^3$ passes through $(4, 4)$.

$16 = C(64) \implies C = \frac{1}{4}$

Particular solution: $y^2 = \frac{1}{4}x^3$ or $4y^2 = x^3$

29. Differential equation: $4yy' - x = 0$

General solution: $4y^2 - x^2 = C$

Particular solutions: $C = 0$, Two intersecting lines
$\qquad\qquad\qquad\quad C = \pm 1$, $C = \pm 4$, Hyperbolas

31. Differential equation: $y' + 2y = 0$

General solution: $y = Ce^{-2x}$

$y' + 2y = C(-2)e^{-2x} + 2(Ce^{-2x}) = 0$

Initial condition: $y(0) = 3$, $3 = Ce^0 = C$

Particular solution: $y = 3e^{-2x}$

33. Differential equation: $y'' + 9y = 0$

General solution: $y = C_1 \sin 3x + C_2 \cos 3x$

$y' = 3C_1 \cos 3x - 3C_2 \sin 3x,$

$y'' = -9C_1 \sin 3x - 9C_2 \cos 3x$

$y'' + 9y = (-9C_1 \sin 3x - 9C_2 \cos 3x)$
$\qquad\qquad + 9(C_1 \sin 3x + C_2 \cos 3x) = 0$

Initial conditions: $y\left(\dfrac{\pi}{6}\right) = 2$, $y'\left(\dfrac{\pi}{6}\right) = 1$

$2 = C_1 \sin\left(\dfrac{\pi}{2}\right) + C_2 \cos\left(\dfrac{\pi}{2}\right) \implies C_1 = 2$

$y' = 3C_1 \cos 3x - 3C_2 \sin 3x$

$1 = 3C_1 \cos\left(\dfrac{\pi}{2}\right) - 3C_2 \sin\left(\dfrac{\pi}{2}\right)$

$\quad = -3C_2 \implies C_2 = -\dfrac{1}{3}$

Particular solution: $y = 2 \sin 3x - \dfrac{1}{3} \cos 3x$

35. Differential equation: $x^2y'' - 3xy' + 3y = 0$

General solution: $y = C_1x + C_2x^3$

$y' = C_1 + 3C_2x^2, y'' = 6C_2x$

$x^2y'' - 3xy' + 3y = x^2(6C_2x) - 3x(C_1 + 3C_2x^2) + 3(C_1x + C_2x^3) = 0$

Initial conditions: $y(2) = 0$, $y'(2) = 4$

$\quad 0 = 2C_1 + 8C_2$

$\quad y' = C_1 + 3C_2x^2$

$\quad 4 = C_1 + 12C_2$

$\quad \left.\begin{array}{l} C_1 + 4C_2 = 0 \\ C_1 + 12C_2 = 4 \end{array}\right\}$ $C_2 = \frac{1}{2}$, $C_1 = -2$

Particular solution: $y = -2x + \frac{1}{2}x^3$

37. $\dfrac{dy}{dx} = 3x^2$

$y = \displaystyle\int 3x^2 \, dx = x^3 + C$

39. $\dfrac{dy}{dx} = \dfrac{x}{1 + x^2}$

$y = \displaystyle\int \dfrac{x}{1 + x^2} \, dx = \dfrac{1}{2} \ln(1 + x^2) + C$

$(u = 1 + x^2, du = 2x \, dx)$

41. $\dfrac{dy}{dx} = \dfrac{x - 2}{x} = 1 - \dfrac{2}{x}$

$y = \displaystyle\int \left[1 - \dfrac{2}{x}\right] dx$

$\quad = x - 2 \ln|x| + C = x - \ln x^2 + C$

43. $\dfrac{dy}{dx} = \sin 2x$

$y = \displaystyle\int \sin 2x \, dx = -\dfrac{1}{2} \cos 2x + C$

$(u = 2x, du = 2 \, dx)$

45. $\dfrac{dy}{dx} = x\sqrt{x-3}$

Let $u = \sqrt{x-3}$, then $x = u^2 + 3$ and $dx = 2u\,du$.

$$y = \int x\sqrt{x-3}\,dx = \int (u^2 + 3)(u)(2u)\,du$$

$$= 2\int (u^4 + 3u^2)\,du = 2\left(\frac{u^5}{5} + u^3\right) + C = \frac{2}{5}(x-3)^{5/2} + 2(x-3)^{3/2} + C$$

47. $\dfrac{dy}{dx} = xe^{x^2}$

$$y = \int xe^{x^2}\,dx = \frac{1}{2}e^{x^2} + C$$

$(u = x^2,\ du = 2x\,dx)$

49.

x	-4	-2	0	2	4	8
y	2	0	4	4	6	8
dy/dx	-2	Undef.	0	$\frac{1}{2}$	$\frac{2}{3}$	1

51.

x	-4	-2	0	2	4	8
y	2	0	4	4	6	8
dy/dx	$-2\sqrt{2}$	-2	0	0	$-2\sqrt{2}$	-8

53. $\dfrac{dy}{dx} = \cos(2x)$

For $x = \pi$, $\dfrac{dy}{dx} = 1$. Matches (b).

55. $\dfrac{dy}{dx} = e^{-2x}$

As $x \to \infty$, $\dfrac{dy}{dx} \to 0$. Matches (d).

57. (a), (b)

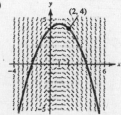

(c) As $x \to \infty$, $y \to -\infty$

As $x \to -\infty$, $y \to -\infty$

59. (a), (b)

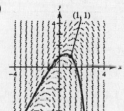

(c) As $x \to \infty$, $y \to -\infty$

As $x \to -\infty$, $y \to -\infty$

61. (a) $y' = \dfrac{1}{x}$, $y(1) = 0$

(b) $y' = \dfrac{1}{x}$, $y(2) = -1$

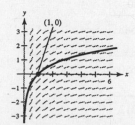

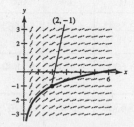

As $x \to \infty$, $y \to \infty$

As $x \to \infty$, $y \to \infty$

[Note: The solution is $y = \ln x$.]

63. $\dfrac{dy}{dx} = 0.5y;\ y(0) = 6$

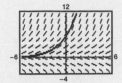

65. $\dfrac{dy}{dx} = 0.02y(10 - y),\ y(0) = 2$

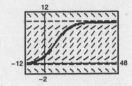

67. Slope field for $y' = 0.4y(3 - x)$ with solution passing through $(0, 1)$.

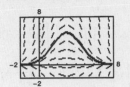

69. $y' = x + y,\quad y(0) = 2,\quad n = 10,\quad h = 0.1$

$y_1 = y_0 + hF(x_0, y_0) = 2 + (0.1)(0 + 2) = 2.2$

$y_2 = y_1 + hF(x_1, y_1) = 2.2 + (0.1)(0.1 + 2.2) = 2.43,$ etc.

n	0	1	2	3	4	5	6	7	8	9	10
x_n	0	0.1	0.2	0.3	0.4	0.5	0.6	0.7	0.8	0.9	1.0
y_n	2	2.2	2.43	2.693	2.992	3.332	3.715	4.146	4.631	5.174	5.781

71. $y' = 3x - 2y,\quad y(0) = 3,\quad n = 10,\quad h = 0.05$

$y_1 = y_0 + hF(x_0, y_0) = 3 + (0.05)(3(0) - 2(3)) = 2.7$

$y_2 = y_1 + hF(x_1, y_1) = 2.7 + (0.05)(3(0.05) - 2(2.7)) = 2.4375,$ etc.

n	0	1	2	3	4	5	6	7	8	9	10
x_n	0	0.05	0.1	0.15	0.2	0.25	0.3	0.35	0.4	0.45	0.5
y_n	3	2.7	2.438	2.209	2.010	1.839	1.693	1.569	1.464	1.378	1.308

73. $y' = e^{xy},\quad y(0) = 1,\quad n = 10,\quad h = 0.1$

$y_1 = y_0 + hF(x_0, y_0) = 1 + (0.1)e^{0(1)} = 1.1$

$y_2 = y_1 + hF(x_1, y_1) = 1.1 + (0.1)e^{(0.1)(1.1)} \approx 1.2116,$ etc.

n	0	1	2	3	4	5	6	7	8	9	10
x_n	0	0.1	0.2	0.3	0.4	0.5	0.6	0.7	0.8	0.9	1.0
y_n	1	1.1	1.212	1.339	1.488	1.670	1.900	2.213	2.684	3.540	5.958

75. $\dfrac{dy}{dx} = y,\ y = 3e^x,\ y(0) = 3$

x	0	0.2	0.4	0.6	0.8	1.0
$y(x)$ (exact)	3	3.6642	4.4755	5.4664	6.6766	8.1548
$y(x)$ ($h = 0.2$)	3	3.6000	4.3200	5.1840	6.2208	7.4650
$y(x)$ ($h = 0.1$)	3	3.6300	4.3923	5.3147	6.4308	7.7812

77. $\dfrac{dy}{dx} = y + \cos x$, $y = \dfrac{1}{2}(\sin x - \cos x + e^x)$, $y(0) = 0$

x	0	0.2	0.4	0.6	0.8	1.0
$y(x)$ (exact)	0	0.2200	0.4801	0.7807	0.1231	0.5097
$y(x)$ ($h = 0.2$)	0	0.2000	0.4360	0.7074	0.0140	0.3561
$y(x)$ ($h = 0.1$)	0	0.2095	0.4568	0.7418	0.0649	0.4273

79. $\dfrac{dy}{dt} = -\dfrac{1}{2}(y - 72)$, $y(0) = 140$, $h = 0.1$

(a)

t	0	1	2	3
Euler	140	112.7	96.4	86.6

(b) $y = 72 + 68e^{-t/2}$ exact

t	0	1	2	3
Exact	140	113.24	97.016	87.173

81. A general solution of order n has n arbitrary constants while in a particular solution initial conditions are given in order to solve for all these constants.

83. Consider $y' = F(x, y)$, $y(x_0) = y_0$. Begin with a point (x_0, y_0) that satisfies the initial condition, $y(x_0) = y_0$. Then using a step size of h, find the point $(x_1, y_1) = (x_0 + h, y_0 + hF(x_0, y_0))$. Continue generating the sequence of points $(x_{n+1}, y_{n+1}) = (x_n + h, y_n + hF(x_n, y_n))$.

85. False. Consider Example 2. $y = x^3$ is a solution to $xy' - 3y = 0$, but $y = x^3 + 1$ is not a solution.

87. True

89. $\dfrac{dy}{dx} = -2y$, $y(0) = 4$, $y = 4e^{-2x}$ solution

(a)

x	0	0.2	0.4	0.6	0.8	1.0
y	4	2.6813	1.7973	1.2048	0.8076	0.5413
y_1	4	2.5600	1.6384	1.0486	0.6711	0.4295
y_2	4	2.4000	1.4400	0.8640	0.5184	0.3110
e_1	0	0.1213	0.1589	0.1562	0.1365	0.1118
e_2	0	0.2813	0.3573	0.3408	0.2892	0.2303
r		0.4312	0.4447	0.4583	0.4720	0.4855

(b) If h is halved, then the error is approximately halved ($r \approx 0.5$).

(c) When $h = 0.05$, the errors will again be approximately halved.

91. (a) $L\dfrac{dI}{dt} + RI = E(t)$

$4\dfrac{dI}{dt} + 12I = 24$

$\dfrac{dI}{dt} = \dfrac{1}{4}(24 - 12I) = 6 - 3I$

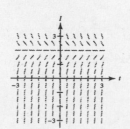

(b) As $t \to \infty$, $I \to 2$. That is, $\displaystyle\lim_{t \to \infty} I(t) = 2$.

In fact, $I = 2$ is a solution to the differential equation.

93. $y = A \sin \omega t$

$y' = A\omega \cos \omega t$

$y'' = -A\omega^2 \sin \omega t$

$y'' + 16y = 0$

$-A\omega^2 \sin \omega t + 16A \sin \omega t = 0$

$A \sin \omega t[16 - \omega^2] = 0$

If $A \neq 0$, then $\omega = \pm 4$ radians/sec.

95. Let the vertical line $x = k$ cut the graph of the solution $y = f(x)$ at (k, t).

The tangent line at (k, t) is

$y - t = f'(k)(x - k)$

Since $y' + p(x)y = q(x)$, we have

$y - t = [q(k) - p(k)t](x - k)$

For any value of t, this line passes through the point $\left(k + \dfrac{1}{p(k)}, \dfrac{q(k)}{p(k)}\right)$.

To see this, note that

$$\frac{q(k)}{p(k)} - t \overset{?}{=} [q(k) - p(k)t]\left(k + \frac{1}{p(k)} - k\right)$$

$$\overset{?}{=} q(k)k - p(k)tk + \frac{q(k)}{p(k)} - t - kq(k) + p(k)kt$$

$$= \frac{q(k)}{p(k)} - t.$$

Section 6.2   Differential Equations: Growth and Decay

1. $\dfrac{dy}{dx} = x + 2$

$y = \displaystyle\int (x + 2)\, dx = \frac{x^2}{2} + 2x + C$

3. $\dfrac{dy}{dx} = y + 2$

$\dfrac{dy}{y + 2} = dx$

$\displaystyle\int \frac{1}{y + 2}\, dy = \int dx$

$\ln|y + 2| = x + C_1$

$y + 2 = e^{x + C_1} = Ce^x$

$y = Ce^x - 2$

5. $y' = \dfrac{5x}{y}$

$yy' = 5x$

$\displaystyle\int yy'\, dx = \int 5x\, dx$

$\displaystyle\int y\, dy = \int 5x\, dx$

$\dfrac{1}{2}y^2 = \dfrac{5}{2}x^2 + C_1$

$y^2 - 5x^2 = C$

7. $y' = \sqrt{x}\, y$

$\dfrac{y'}{y} = \sqrt{x}$

$\displaystyle\int \frac{y'}{y}\, dx = \int \sqrt{x}\, dx$

$\displaystyle\int \frac{dy}{y} = \int \sqrt{x}\, dx$

$\ln|y| = \dfrac{2}{3}x^{3/2} + C_1$

$y = e^{(2/3)x^{3/2} + C_1}$

$\doteq e^{C_1}e^{(2/3)x^{3/2}}$

$= Ce^{(2/3)x^{3/2}}$

9. $(1 + x^2)y' - 2xy = 0$

$$y' = \frac{2xy}{1 + x^2}$$

$$\frac{y'}{y} = \frac{2x}{1 + x^2}$$

$$\int \frac{y'}{y}\,dx = \int \frac{2x}{1 + x^2}\,dx$$

$$\int \frac{dy}{y} = \int \frac{2x}{1 + x^2}\,dx$$

$$\ln|y| = \ln(1 + x^2) + C_1$$

$$\ln|y| = \ln(1 + x^2) + \ln C$$

$$\ln|y| = \ln[C(1 + x^2)]$$

$$y = C(1 + x^2)$$

11. $\dfrac{dQ}{dt} = \dfrac{k}{t^2}$

$$\int \frac{dQ}{dt}\,dt = \int \frac{k}{t^2}\,dt$$

$$\int dQ = -\frac{k}{t} + C$$

$$Q = -\frac{k}{t} + C$$

13. $\dfrac{dN}{ds} = k(250 - s)$

$$\int \frac{dN}{ds}\,ds = \int k(250 - s)\,ds$$

$$\int dN = -\frac{k}{2}(250 - s)^2 + C$$

$$N = -\frac{k}{2}(250 - s)^2 + C$$

15. (a)

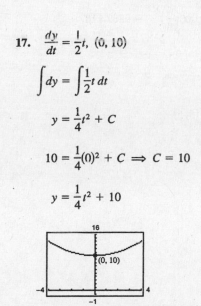

(b) $\dfrac{dy}{dx} = x(6 - y), \quad (0, 0)$

$$\frac{dy}{y - 6} = -x\,dx$$

$$\ln|y - 6| = \frac{-x^2}{2} + C$$

$$y - 6 = e^{-x^2/2 + C} = C_1 e^{-x^2/2}$$

$$y = 6 + C_1 e^{-x^2/2}$$

$(0, 0): 0 = 6 + C_1 \Rightarrow C_1 = -6 \Rightarrow y = 6 - 6e^{-x^2/2}$

17. $\dfrac{dy}{dt} = \dfrac{1}{2}t, \quad (0, 10)$

$$\int dy = \int \frac{1}{2}t\,dt$$

$$y = \frac{1}{4}t^2 + C$$

$$10 = \frac{1}{4}(0)^2 + C \Rightarrow C = 10$$

$$y = \frac{1}{4}t^2 + 10$$

19. $\dfrac{dy}{dt} = -\dfrac{1}{2}y,\ (0, 10)$

$$\int \dfrac{dy}{y} = \int -\dfrac{1}{2}\,dt$$

$$\ln|y| = -\dfrac{1}{2}t + C_1$$

$$y = e^{-(t/2)+C_1} = e^{C_1}\,e^{-t/2} = Ce^{-t/2}$$

$$10 = Ce^0 \Rightarrow C = 10$$

$$y = 10e^{-t/2}$$

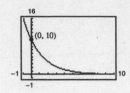

21. $\dfrac{dy}{dx} = ky$

$y = Ce^{kx}$ (Theorem 5.16)

$(0, 4):\ 4 = Ce^0 = C$

$(3, 10):\ 10 = 4e^{3k} \Rightarrow k = \dfrac{1}{3}\ln\!\left(\dfrac{5}{2}\right)$

When $x = 6$, $y = 4e^{1/3\,\ln(5/2)(6)} = 4e^{\ln(5/2)^2}$

$$= 4\left(\dfrac{5}{2}\right)^2 = 25.$$

23. $\dfrac{dV}{dt} = kV$

$V = Ce^{kt}$ (Theorem 5.16)

$(0, 20{,}000):\ C = 20{,}000$

$(4, 12{,}500):\ 12{,}500 = 20{,}000e^{4k} \Rightarrow k = \dfrac{1}{4}\ln\!\left(\dfrac{5}{8}\right)$

When $t = 6$, $V = 20{,}000e^{1/4\,\ln(5/8)(6)} = 20{,}000e^{\ln(5/8)^{3/2}}$

$$= 20{,}000\left(\dfrac{5}{8}\right)^{3/2} \approx 9882.118.$$

25. $y = Ce^{kt},\ \left(0, \dfrac{1}{2}\right),\ (5, 5)$

$C = \dfrac{1}{2}$

$y = \dfrac{1}{2}e^{kt}$

$5 = \dfrac{1}{2}e^{5k}$

$k = \dfrac{\ln 10}{5}$

$y = \dfrac{1}{2}e^{(\ln 10/5)t} = \dfrac{1}{2}(10^{t/5})$ or $y \approx \dfrac{1}{2}e^{0.4605t}$

27. $y = Ce^{kt},\ (1, 1),\ (5, 5)$

$1 = Ce^k$

$5 = Ce^{5k}$

$5Ce^k = Ce^{5k}$

$5e^k = e^{5k}$

$5 = e^{4k}$

$k = \dfrac{\ln 5}{4} \approx 0.4024$

$y = Ce^{0.4024t}$

$1 = Ce^{0.4024}$

$C \approx 0.6687$ $(C = 5^{-1/4})$

$y \approx 0.6687e^{0.4024t}$

29. In the model $y = Ce^{kt}$, C represents the initial value of y (when $t = 0$). k is the proportionality constant.

31. $\dfrac{dy}{dx} = \dfrac{1}{2}xy$

$\dfrac{dy}{dx} > 0$ when $xy > 0$. Quadrants I and III.

33. Since the initial quantity is 10 grams,

$$y = 10e^{kt}.$$

Since the half-life is 1599 years,

$$5 = 10e^{k(1599)}$$

$$k = \frac{1}{1599}\ln\left(\frac{1}{2}\right).$$

Thus, $y = 10e^{[\ln(1/2)/1599]t}$.

When $t = 1000$, $y = 10e^{[\ln(1/2)/1599](1000)} \approx 6.48$ g.

When $t = 10{,}000$, $y \approx 0.13$ g.

35. Since the half-life is 1599 years,

$$\frac{1}{2} = 1e^{k(1599)}$$

$$k = \frac{1}{1599}\ln\left(\frac{1}{2}\right).$$

Since there are 0.5 gram after 10,000 years,

$$0.5 = Ce^{[\ln(1/2)/1599](10{,}000)}$$

$$C \approx 38.158.$$

Hence, the initial quantity is approximately 38.158 g.

When $t = 1000$, $y = 38.158e^{[\ln(1/2)/1599](1000)}$

$$\approx 24.74 \text{ g.}$$

37. Since the initial quantity is 5 grams, $C = 5$.

Since the half-life is 5715 years,

$$2.5 = 5e^{k(5715)}$$

$$k = \frac{1}{5715}\ln\left(\frac{1}{2}\right).$$

When $t = 1000$ years, $y = 5e^{[\ln(1/2)/5715](1000)}$

$$\approx 4.43 \text{ g.}$$

When $t = 10{,}000$ years, $y = 5e^{[\ln(1/2)/5715](10{,}000)}$

$$\approx 1.49 \text{ g.}$$

39. Since the half-life is 24,100 years,

$$\frac{1}{2} = 1e^{k(24{,}100)}$$

$$k = \frac{1}{24{,}100}\ln\left(\frac{1}{2}\right).$$

Since there are 2.1 grams after 1000 years,

$$2.1 = Ce^{[\ln(1/2)/24{,}100](1000)}$$

$$C \approx 2.161.$$

Thus, the initial quantity is approximately 2.161 g.

When $t = 10{,}000$, $y = 2.161e^{[\ln(1/2)/24{,}100](10{,}000)}$

$$\approx 1.62 \text{ g.}$$

41. $y = Ce^{kt}$

$$\frac{1}{2}C = Ce^{k(1599)}$$

$$k = \frac{1}{1599}\ln\left(\frac{1}{2}\right)$$

When $t = 100$, $y = Ce^{[\ln(1/2)/1599](100)}$

$$\approx 0.9576\,C$$

Therefore, 95.76% remains after 100 years.

43. Since $A = 1000e^{0.06t}$, the time to double is given by $2000 = 1000e^{0.06t}$ and we have

$$2 = e^{0.06t}$$

$$\ln 2 = 0.06t$$

$$t = \frac{\ln 2}{0.06} \approx 11.55 \text{ years.}$$

Amount after 10 years: $A = 1000e^{(0.06)(10)} \approx \$1822.12$

45. Since $A = 750e^{rt}$ and $A = 1500$ when $t = 7.75$, we have the following.

$$1500 = 750e^{7.75r}$$

$$r = \frac{\ln 2}{7.75} \approx 0.0894 = 8.94\%$$

Amount after 10 years: $A = 750e^{0.0894(10)} \approx \$1833.67$

47. Since $A = 500e^{rt}$ and $A = 1292.85$ when $t = 10$, we have the following.

$$1292.85 = 500e^{10r}$$

$$r = \frac{\ln(1292.85/500)}{10} \approx 0.0950 = 9.50\%$$

The time to double is given by

$$1000 = 500e^{0.0950t}$$

$$t = \frac{\ln 2}{0.095} \approx 7.30 \text{ years.}$$

49. $500,000 = P\left(1 + \dfrac{0.075}{12}\right)^{(12)(20)}$

$\quad P = 500,000\left(1 + \dfrac{0.075}{12}\right)^{-240}$

$\quad\quad \approx \$112,087.09$

51. $500,000 = P\left(1 + \dfrac{0.08}{12}\right)^{(12)(35)}$

$\quad P = 500,000\left(1 + \dfrac{0.08}{12}\right)^{-420}$

$\quad\quad = \$30,688.87$

53. (a) $2000 = 1000(1 + 0.07)^t$

$\quad 2 = 1.07^t$

$\quad \ln 2 = t \ln 1.07$

$\quad t = \dfrac{\ln 2}{\ln 1.07} \approx 10.24 \text{ years}$

(b) $2000 = 1000\left(1 + \dfrac{0.07}{12}\right)^{12t}$

$\quad 2 = \left(1 + \dfrac{0.007}{12}\right)^{12t}$

$\quad \ln 2 = 12t \ln\left(1 + \dfrac{0.07}{12}\right)$

$\quad t = \dfrac{\ln 2}{12 \ln(1 + (0.07/12))} \approx 9.93 \text{ years}$

(c) $2000 = 1000\left(1 + \dfrac{0.07}{365}\right)^{365t}$

$\quad 2 = \left(1 + \dfrac{0.07}{365}\right)^{365t}$

$\quad \ln 2 = 365t \ln\left(1 + \dfrac{0.07}{365}\right)$

$\quad t = \dfrac{\ln 2}{365 \ln(1 + (0.07/365))} \approx 9.90 \text{ years}$

(d) $2000 = 1000e^{(0.07)t}$

$\quad 2 = e^{0.07t}$

$\quad \ln 2 = 0.07t$

$\quad t = \dfrac{\ln 2}{0.07} \approx 9.90 \text{ years}$

55. (a) $2000 = 1000(1 + 0.085)^t$

$\quad 2 = 1.085^t$

$\quad \ln 2 = t \ln 1.085$

$\quad t = \dfrac{\ln 2}{\ln 1.085} \approx 8.50 \text{ years}$

(b) $2000 = 1000\left(1 + \dfrac{0.085}{12}\right)^{12t}$

$\quad 2 = \left(1 + \dfrac{0.085}{12}\right)^{12t}$

$\quad \ln 2 = 12t \ln\left(1 + \dfrac{0.085}{12}\right)$

$\quad t = \dfrac{1}{12} \dfrac{\ln 2}{\ln\left(1 + \dfrac{0.085}{12}\right)} \approx 8.18 \text{ years}$

(c) $2000 = 1000\left(1 + \dfrac{0.085}{365}\right)^{365t}$

$\quad 2 = \left(1 + \dfrac{0.085}{365}\right)^{365t}$

$\quad \ln 2 = 365t \ln\left(1 + \dfrac{0.085}{365}\right)$

$\quad t = \dfrac{1}{365} \dfrac{\ln 2}{\ln\left(1 + \dfrac{0.085}{365}\right)} \approx 8.16 \text{ years}$

(d) $2000 = 1000e^{0.085t}$

$\quad 2 = e^{0.085t}$

$\quad \ln 2 = 0.085t$

$\quad t = \dfrac{\ln 2}{0.085} \approx 8.15 \text{ years}$

57. (a) $P = Ce^{kt} = Ce^{-0.009t}$

$\quad P(1) = 7.7 = Ce^{-0.009(1)}$

$\quad\quad C \approx 7.7696$

$\quad P = 7.7696e^{-0.009t}$

(b) For $t = 15$, $P = 7.7696e^{-0.009(15)} \approx 6.79$ million.

(c) If $k < 0$, the population is decreasing.

59. (a) $P = Ce^{kt} = Ce^{0.026t}$

$\quad 5.2 = P(1) = Ce^{0.026(1)}$

$\quad\quad C \approx 5.0665$

$\quad P = 5.0665e^{0.026t}$

(b) For $t = 15$, $P = 5.0665e^{0.026(15)} \approx 7.48$ million.

(c) For $k > 0$, the population is increasing.

61. (a) $N = 100.1596(1.2455)^t$

(b) $N = 400$ when $t = 6.3$ hours (graphing utility)

Analytically,

$$400 = 100.1596(1.2455)^t$$

$$1.2455^t = \frac{400}{100.1596} = 3.9936$$

$$t \ln 1.2455 = \ln 3.9936$$

$$t = \frac{\ln 3.9936}{\ln 1.2455} \approx 6.3 \text{ hours.}$$

63. (a) $19 = 30(1 - e^{20k})$

$$30e^{20k} = 11$$

$$k = \frac{\ln(11/30)}{20} \approx -0.0502$$

$$N \approx 30(1 - e^{-0.0502t})$$

(b) $25 = 30(1 - e^{-0.0502t})$

$$e^{-0.0502t} = \frac{1}{6}$$

$$t = \frac{-\ln 6}{-0.0502} \approx 36 \text{ days}$$

65. (a) $P_1 = Ce^{kt} = 181e^{kt}$

$$205 = 181e^{10k} \implies k = \frac{1}{10} \ln\left(\frac{205}{181}\right) \approx 0.01245$$

$$P_1 \approx 181e^{0.01245t} \approx 181(1.01253)^t$$

(c)

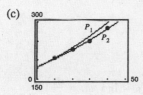

The model P_2 fits the data better.

(b) Using a graphing utility, $P_2 \approx 182.3248(1.01091)^t$

(d) Using the model P_2,

$$320 = 182.3248(1.01091)^t$$

$$\frac{320}{182.3248} = (1.01091)^t$$

$$t = \frac{\ln(320/182.3248)}{\ln(1.01091)} \approx 51.8 \text{ years, or 2011.}$$

67. $\beta(I) = 10 \log_{10} \frac{I}{I_0}, I_0 = 10^{-16}$

(a) $\beta(10^{-14}) = 10 \log_{10} \frac{10^{-14}}{10^{-16}} = 20$ decibels

(b) $\beta(10^{-9}) = 10 \log_{10} \frac{10^{-9}}{10^{-16}} = 70$ decibels

(c) $\beta(10^{-6.5}) = 10 \log_{10} \frac{10^{-6.5}}{10^{-16}} = 95$ decibels

(d) $\beta(10^{-4}) = 10 \log_{10} \frac{10^{-4}}{10^{-16}} = 120$ decibels

69. $A(t) = V(t)e^{-0.10t} = 100,000e^{0.8\sqrt{t}} e^{-0.10t} = 100,000e^{0.8\sqrt{t}-0.10t}$

$$\frac{dA}{dt} = 100,000\left(\frac{0.4}{\sqrt{t}} - 0.10\right)e^{0.8\sqrt{t}-0.10t} = 0 \text{ when } 16.$$

The timber should be harvested in the year 2014, (1998 + 16). **Note:** You could also use a graphing utility to graph $A(t)$ and find the maximum of $A(t)$. Use the viewing rectangle $0 \le x \le 30$ and $0 \le y \le 600,000$.

71.   Since $\frac{dy}{dt} = k(y - 80)$

$$\int \frac{1}{y - 80} \, dy = \int k \, dt$$

$$\ln(y - 80) = kt + C.$$

When $t = 0, y = 1500$. Thus, $C = \ln 1420$.

When $t = 1, y = 1120$. Thus,

$$k(1) + \ln 1420 = \ln(1120 - 80)$$

$$k = \ln 1040 - \ln 1420 = \ln \frac{104}{142}.$$

Thus, $y = 1420e^{[\ln(104/142)]t} + 80$.

When $t = 5, y \approx 379.2°$.

73. False. If $y = Ce^{kt}, y' = Cke^{kt} \ne$ constant.

75. True

Section 6.3 Separation of Variables and the Logistic Equation

1. $\dfrac{dy}{dx} = \dfrac{x}{y}$

$$\int y\,dy = \int x\,dx$$

$$\frac{y^2}{2} = \frac{x^2}{2} + C_1$$

$$y^2 - x^2 = C$$

3. $\dfrac{dr}{ds} = 0.05r$

$$\int \frac{dr}{r} = \int 0.05\,ds$$

$$\ln|r| = 0.05s + C_1$$

$$r = e^{0.05s + C_1} = Ce^{0.05s}$$

5. $(2 + x)y' = 3y$

$$\int \frac{dy}{y} = \int \frac{3}{2+x}\,dx$$

$$\ln|y| = 3\ln|2 + x| + \ln C = \ln|C(2+x)^3|$$

$$y = C(x+2)^3$$

7. $yy' = \sin x$

$$\int y\,dy = \int \sin x\,dx$$

$$\frac{y^2}{2} = -\cos x + C_1$$

$$y^2 = -2\cos x + C$$

9. $\sqrt{1 - 4x^2}\,y' = x$

$$dy = \frac{x}{\sqrt{1 - 4x^2}}\,dx$$

$$\int dy = \int \frac{x}{\sqrt{1 - 4x^2}}\,dx$$

$$= -\frac{1}{8}\int (1 - 4x^2)^{-1/2}(-8x\,dx)$$

$$y = -\frac{1}{4}(1 - 4x^2)^{1/2} + C$$

11. $y \ln x - xy' = 0$

$$\int \frac{dy}{y} = \int \frac{\ln x}{x}\,dx \quad \left(u = \ln x,\, du = \frac{dx}{x}\right)$$

$$\ln|y| = \frac{1}{2}(\ln x)^2 + C_1$$

$$y = e^{(1/2)(\ln x)^2 + C_1} = Ce^{(\ln x)^2/2}$$

13. $yy' - e^x = 0$

$$\int y\,dy = \int e^x\,dx$$

$$\frac{y^2}{2} = e^x + C_1$$

$$y^2 = 2e^x + C$$

Initial condition: $y(0) = 4,\ 16 = 2 + C,\ C = 14$

Particular solution: $y^2 = 2e^x + 14$

15. $y(x + 1) + y' = 0$

$$\int \frac{dy}{y} = -\int (x + 1)\,dx$$

$$\ln|y| = -\frac{(x+1)^2}{2} + C_1$$

$$y = Ce^{-(x+1)^2/2}$$

Initial condition: $y(-2) = 1,\ 1 = Ce^{-1/2},\ C = e^{1/2}$

Particular solution: $y = e^{[1-(x+1)^2]/2} = e^{-(x^2+2x)/2}$

17. $y(1 + x^2)y' = x(1 + y^2)$

$$\frac{y}{1 + y^2}\,dy = \frac{x}{1 + x^2}\,dx$$

$$\frac{1}{2}\ln(1 + y^2) = \frac{1}{2}\ln(1 + x^2) + C_1$$

$$\ln(1 + y^2) = \ln(1 + x^2) + \ln C = \ln[C(1 + x^2)]$$

$$1 + y^2 = C(1 + x^2)$$

$$y(0) = \sqrt{3}:\ 1 + 3 = C \Rightarrow C = 4$$

$$1 + y^2 = 4(1 + x^2)$$

$$y^2 = 3 + 4x^2$$

19. $\dfrac{du}{dv} = uv \sin v^2$

$$\int \frac{du}{u} = \int v \sin v^2\,dv$$

$$\ln|u| = -\frac{1}{2}\cos v^2 + C_1$$

$$u = Ce^{-(\cos v^2)/2}$$

Initial condition: $u(0) = 1,\ C = \dfrac{1}{e^{-1/2}} = e^{1/2}$

Particular solution: $u = e^{(1-\cos v^2)/2}$

21. $dP - kP\,dt = 0$

$$\int \frac{dP}{P} = k\int dt$$

$$\ln|P| = kt + C_1$$

$$P = Ce^{kt}$$

Initial condition: $P(0) = P_0$, $P_0 = Ce^0 = C$

Particular solution: $P = P_0 e^{kt}$

23. $\dfrac{dy}{dx} = \dfrac{-9x}{16y}$

$$\int 16y\,dy = -\int 9x\,dx$$

$$8y^2 = \frac{-9}{2}x^2 + C$$

Initial condition: $y(1) = 1$, $8 = -\dfrac{9}{2} + C$, $C = \dfrac{25}{2}$

Particular solution: $8y^2 = \dfrac{-9}{2}x^2 + \dfrac{25}{2}$

$$16y^2 + 9x^2 = 25$$

25. $m = \dfrac{dy}{dx} = \dfrac{0 - y}{(x + 2) - x} = -\dfrac{y}{2}$

$$\int \frac{dy}{y} = \int -\frac{1}{2}\,dx$$

$$\ln|y| = -\frac{1}{2}x + C_1$$

$$y = Ce^{-x/2}$$

27. $f(x, y) = x^3 - 4xy^2 + y^3$

$$f(tx, ty) = t^3x^3 - 4txt^2y^2 + t^3y^3$$

$$= t^3(x^3 - 4xy^2 + y^3)$$

Homogeneous of degree 3

29. $f(x, y) = \dfrac{x^2 y^2}{\sqrt{x^2 + y^2}}$

$$f(tx, ty) = \frac{t^4 x^2 y^2}{\sqrt{t^2 x^2 + t^2 y^2}} = t^3 \frac{x^2 y^2}{\sqrt{x^2 + y^2}}$$

Homogeneous of degree 3

31. $f(x, y) = 2\ln xy$

$$f(tx, ty) = 2\ln[txty]$$

$$= 2\ln[t^2 xy] = 2(\ln t^2 + \ln xy)$$

Not homogeneous

33. $f(x, y) = 2\ln \dfrac{x}{y}$

$$f(tx, ty) = 2\ln \frac{tx}{ty} = 2\ln \frac{x}{y}$$

Homogeneous degree 0

35. $y' = \dfrac{x + y}{2x}$, $y = vx$

$$v + x\frac{dv}{dx} = \frac{x + vx}{2x}$$

$$x\frac{dv}{dx} = \frac{1 + v}{2} - v = \frac{1 - v}{2}$$

$$2\int \frac{dv}{1 - v} = \int \frac{dx}{x}$$

$$-\ln(1 - v)^2 = \ln|x| + \ln C = \ln|Cx|$$

$$\frac{1}{(1 - v)^2} = |Cx|$$

$$\frac{1}{[1 - (y/x)]^2} = |Cx|$$

$$\frac{x^2}{(x - y)^2} = |Cx|$$

$$|x| = C(x - y)^2$$

37. $y' = \dfrac{x - y}{x + y}$, $y = vx$

$$v + x\frac{dv}{dx} = \frac{x - xv}{x + xv}$$

$$v\,dx + x\,dv = \frac{1 - v}{1 + v}\,dx$$

$$x\,dv = \left(\frac{1 - v}{1 + v} - v\right)dx = \frac{1 - 2v - v^2}{1 + v}\,dx$$

$$\int \frac{v + 1}{v^2 + 2v - 1}\,dv = -\int \frac{dx}{x}$$

$$\frac{1}{2}\ln|v^2 + 2v - 1| = -\ln|x| + \ln C_1 = \ln\left|\frac{C_1}{x}\right|$$

$$|v^2 + 2v - 1| = \frac{C}{x^2}$$

$$\left|\frac{y^2}{x^2} + 2\frac{y}{x} - 1\right| = \frac{C}{x^2}$$

$$|y^2 + 2xy - x^2| = C$$

39. $\quad y' = \dfrac{xy}{x^2 - y^2}, y = vx$

$$v + x\dfrac{dv}{dx} = \dfrac{x^2 v}{x^2 - x^2 v^2}$$

$$v\,dx + x\,dv = \dfrac{v}{1 - v^2}\,dx$$

$$x\,dv = \left(\dfrac{v}{1 - v^2} - v\right) dx = \left(\dfrac{v^3}{1 - v^2}\right) dx$$

$$\int \dfrac{1 - v^2}{v^3}\,dv = \int \dfrac{dx}{x}$$

$$-\dfrac{1}{2v^2} - \ln|v| = \ln|x| + \ln C_1 = \ln|C_1 x|$$

$$\dfrac{-1}{2v^2} = \ln|C_1 xv|$$

$$\dfrac{-x^2}{2y^2} = \ln|C_1 y|$$

$$y = Ce^{-x^2/2y^2}$$

41. $\quad x\,dy - (2xe^{-y/x} + y)\,dx = 0, y = vx$

$$x(v\,dx + x\,dv) - (2xe^{-v} + vx)\,dx = 0$$

$$\int e^v\,dv = \int \dfrac{2}{x}\,dx$$

$$e^v = \ln C_1 x^2$$

$$e^{y/x} = \ln C_1 + \ln x^2$$

$$e^{y/x} = C + \ln x^2$$

Initial condition: $y(1) = 0, 1 = C$

Particular solution: $e^{y/x} = 1 + \ln x^2$

43. $\quad \left(x \sec \dfrac{y}{x} + y\right) dx - x\,dy = 0, y = vx$

$$(x \sec v + xv)dx - x(v\,dx + x\,dv) = 0$$

$$(\sec v + v)\,dx = v\,dx + x\,dv$$

$$\int \cos v\,dv = \int \dfrac{dx}{x}$$

$$\sin v = \ln x + \ln C_1$$

$$x = Ce^{\sin v}$$

$$= Ce^{\sin(y/x)}$$

Initial condition: $y(1) = 0, 1 = Ce^0 = C$

Particular solution: $x = e^{\sin(y/x)}$

45. $\dfrac{dy}{dx} = x$

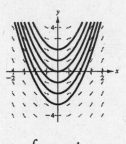

$$y = \int x\,dx = \dfrac{1}{2}x^2 + C$$

47. $\dfrac{dy}{dx} = 4 - y$

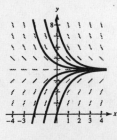

$$\int \dfrac{dy}{4 - y} = \int dx$$

$$\ln|4 - y| = -x + C_1$$

$$4 - y = e^{-x + C_1}$$

$$y = 4 + Ce^{-x}$$

49. (a) Euler's Method gives $y(1) \approx 0.1602$.

(b) $\quad \dfrac{dy}{dx} = -6xy$

$$\int \dfrac{dy}{y} = \int -6x$$

$$\ln|y| = -3x^2 + C_1$$

$$y = Ce^{-3x^2}$$

$$y(0) = 5 \Rightarrow C = 5$$

$$y = 5e^{-3x^2}$$

(c) At $x = 1, y = 5e^{-3(1)} \approx 0.2489$.

Error: $0.2489 - 0.1602 \approx 0.0887$

51. (a) Euler's Method gives $y(2) \approx 3.0318$.

(b) $\dfrac{dy}{dx} = \dfrac{2x + 12}{3y^2 - 4}$

$$\int (3y^2 - 4)\, dy = \int (2x + 12)\, dx$$

$$y^3 - 4y = x^2 + 12x + C$$

$$y(1) = 2: \ 2^3 - 4(2) = 1 + 12 + C \implies C = -13$$

$$y^3 - 4y = x^2 + 12x - 13$$

(c) For $x = 2$,

$$y^3 - 4y = 2^2 + 12(2) - 13 = 15$$

$$y^3 - 4y - 15 = 0$$

$$(y - 3)(y^2 + 3y + 5) = 0 \implies y = 3.$$

Error: $3.0318 - 3 = 0.0318$

53. $\dfrac{dy}{dt} = ky, \quad y = Ce^{kt}$

$y(0) = y_0 = C \qquad$ initial amount

$$\dfrac{y_0}{2} = y_0 e^{k(1599)}$$

$$k = \dfrac{1}{1599} \ln\!\left(\dfrac{1}{2}\right)$$

$$y = Ce^{[\ln(1/2)/1599]t}$$

When $t = 25$, $y = 0.989C$ or 98.9%.

55. $\dfrac{dy}{dx} = k(y - 4)$

The direction field satisfies $(dy/dx) = 0$ along $y = 4$; but not along $y = 0$. Matches (a).

57. $\dfrac{dy}{dx} = ky(y - 4)$

The direction field satisfies $(dy/dx) = 0$ along $y = 0$ and $y = 4$. Matches (c).

59. $\quad \dfrac{dw}{dt} = k(1200 - w)$

$$\int \dfrac{dw}{1200 - w} = \int k\, dt$$

$$\ln|1200 - w| = -kt + C_1$$

$$1200 - w = e^{-kt + C_1} = Ce^{-kt}$$

$$w = 1200 - Ce^{-kt}$$

$$w(0) = 60 = 1200 - C \implies C = 1200 - 60 = 1140$$

$$w = 1200 - 1140e^{-kt}$$

(a)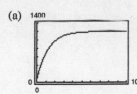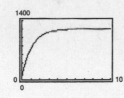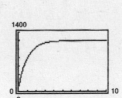

(b) $k = 0.8$: $t = 1.31$ years

$k = 0.9$: $t = 1.16$ years

$k = 1.0$: $t = 1.05$ years

(c) Maximum weight: 1200 pounds

$$\lim_{t \to \infty} w = 1200$$

61. Given family (circles): $x^2 + y^2 = C$

$$2x + 2yy' = 0$$

$$y' = -\frac{x}{y}$$

Orthogonal trajectory (lines): $y' = \frac{y}{x}$

$$\int \frac{dy}{y} = \int \frac{dx}{x}$$

$$\ln|y| = \ln|x| + \ln K$$

$$y = Kx$$

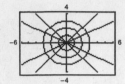

63. Given family (parabolas): $x^2 = Cy$

$$2x = Cy'$$

$$y' = \frac{2x}{C} = \frac{2x}{x^2/y} = \frac{2y}{x}$$

Orthogonal trajectory (ellipses): $y' = -\frac{x}{2y}$

$$2\int y \, dy = -\int x \, dx$$

$$y^2 = -\frac{x^2}{2} + K_1$$

$$x^2 + 2y^2 = K$$

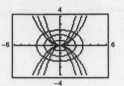

65. Given family: $y^2 = Cx^3$

$$2yy' = 3Cx^2$$

$$y' = \frac{3Cx^2}{2y} = \frac{3x^2}{2y}\left(\frac{y^2}{x^3}\right) = \frac{3y}{2x}$$

Orthogonal trajectory (ellipses): $y' = -\frac{2x}{3y}$

$$3\int y \, dy = -2\int x \, dx$$

$$\frac{3y^2}{2} = -x^2 + K_1$$

$$3y^2 + 2x^2 = K$$

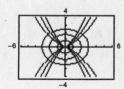

67. $y = \dfrac{12}{1 + e^{-x}}$

Since $y(0) = 6$, it matches (c) or (d).

Since (d) approaches its horizontal asymptote slower than (c), it matches (d).

69. $y = \dfrac{12}{1 + \dfrac{1}{2}e^{-x}}$

Since $y(0) = \dfrac{12}{\left(\dfrac{3}{2}\right)} = 8$, it matches (b).

71. $P(t) = \dfrac{1500}{1 + 24e^{-0.75t}}$

(a) $k = 0.75$

(b) $L = 1500$

(c) $P(0) = \dfrac{1500}{1 + 24} = 60$

(d) $750 = \dfrac{1500}{1 + 24e^{-0.75t}}$

$1 + 24e^{-0.75t} = 2$

$e^{-0.75t} = \dfrac{1}{24}$

$-0.75t = \ln\left(\dfrac{1}{24}\right) = -\ln 24$

$t = \dfrac{\ln 24}{0.75} \approx 4.2374$

(e) $\dfrac{dP}{dt} = 0.75P\left(1 - \dfrac{P}{1500}\right), \qquad P(0) = 60$

73. $\dfrac{dP}{dt} = 3P\left(1 - \dfrac{P}{100}\right)$

(a) $k = 3$

(b) $L = 100$

(c)

(d) $\dfrac{d^2P}{dt^2} = 3P'\left(1 - \dfrac{P}{100}\right) + 3P\left(\dfrac{-P'}{100}\right)$

$= 3\left[3P\left(1 - \dfrac{P}{100}\right)\right]\left(1 - \dfrac{P}{100}\right) - \dfrac{3P}{100}\left[3P\left(1 - \dfrac{P}{100}\right)\right]$

$= 9P\left(1 - \dfrac{P}{100}\right)\left(1 - \dfrac{P}{100} - \dfrac{P}{100}\right)$

$= 9P\left(1 - \dfrac{P}{100}\right)\left(1 - \dfrac{2P}{100}\right)$

$\dfrac{d^2P}{dt^2} = 0$ for $P = 50$, and by the first Derivative Test,

this is a maximum. $\left(\text{Note: } P = 50 = \dfrac{L}{2} = \dfrac{100}{2}\right)$

75. $\dfrac{dy}{dt} = y\left(1 - \dfrac{y}{40}\right), \qquad y(0) = 8$

$k = 1, \ L = 40$

$y = \dfrac{L}{1 + be^{-kt}} = \dfrac{40}{1 + be^{-t}}$

$y(0) = 8: \ 8 = \dfrac{40}{1 + b} \implies b = 4$

Solution: $y = \dfrac{40}{1 + 4e^{-t}}$

77. $\dfrac{dy}{dt} = \dfrac{4y}{5} - \dfrac{y^2}{150} = \dfrac{4}{5}y\left(1 - \dfrac{y}{120}\right), \qquad y(0) = 8$

$k = \dfrac{4}{5} = 0.8, \ L = 120$

$y = \dfrac{L}{1 + be^{-kt}} = \dfrac{120}{1 + be^{-0.8t}}$

$y(0) = 8: \ 8 = \dfrac{120}{1 + b} \implies b = 14$

Solution: $y = \dfrac{120}{1 + 14e^{-0.8t}}$

79. (a) $y = \dfrac{L}{1 + be^{-kt}}, \ L = 200, \ y(0) = 25$

$25 = \dfrac{200}{1 + b} \implies b = 7$

$39 = \dfrac{200}{1 + 7e^{-k(2)}}$

$1 + 7e^{-2k} = \dfrac{200}{39}$

$e^{-2k} = \dfrac{23}{39}$

$k = -\dfrac{1}{2}\ln\left(\dfrac{23}{39}\right) = \dfrac{1}{2}\ln\left(\dfrac{39}{23}\right) \approx 0.2640$

$y = \dfrac{200}{1 + 7e^{-0.2640t}}$

(b) For $t = 5$, $y \approx 70$ panthers.

(c) $100 = \dfrac{200}{1 + 7e^{-0.264t}}$

$1 + 7e^{-0.264t} = 2$

$-0.264t = \ln\left(\dfrac{1}{7}\right)$

$t \approx 7.37$ years

(d) $\dfrac{dy}{dt} = ky\left(1 - \dfrac{y}{L}\right) = 0.264y\left(1 - \dfrac{y}{200}\right), \qquad y(0) = 25$

Using Euler's Method, $y \approx 220.5$ when $t = 6$.

(e) y is increasing most rapidly where $y = 200/2 = 100$, corresponds to $t \approx 7.37$ years.

81. A differential equation can be solved by separation of variables if it can be written in the form

$$M(x) + N(y)\frac{dy}{dx} = 0.$$

To solve a separable equation, rewrite as,

$$M(x)\,dx = -N(y)\,dy$$

and integrate both sides.

83. Two families of curves are mutually orthogonal if each curve in the first family intersects each curve in the second family at right angles.

85. False. $\dfrac{dy}{dx} = \dfrac{x}{y}$ is separable, but $y = 0$ is not a solution.

87. False

$$f(tx, ty) = t^2x^2 + t^2\,xy + 2$$
$$\neq t^2 f(x, y)$$

89. $y = \dfrac{1}{1 + be^{-kt}}$

$$y' = \frac{-1}{(1 + be^{-kt})^2}(-bke^{-kt})$$

$$= \frac{k}{(1 + be^{-kt})} \cdot \frac{be^{-kt}}{(1 + be^{-kt})}$$

$$= \frac{k}{(1 + be^{-kt})} \cdot \frac{1 + be^{-kt} - 1}{(1 + be^{-kt})}$$

$$= \frac{k}{(1 + be^{-kt})} \cdot \left(1 - \frac{1}{1 + be^{-kt}}\right)$$

$$= ky(1 - y)$$

Section 6.4   First-Order Linear Differential Equations

1. $x^3 y' + xy = e^x + 1$

$$y' + \frac{1}{x^2}y = \frac{1}{x^3}(e^x + 1)$$

Linear

3. $y' + y\cos x = xy^2$

Not linear, because of the xy^2-term.

5. $\dfrac{dy}{dx} + \left(\dfrac{1}{x}\right)y = 3x + 4$

Integrating factor: $e^{\int(1/x)\,dx} = e^{\ln x} = x$

$$xy = \int x(3x + 4)\,dx = x^3 + 2x^2 + C$$

$$y = x^2 + 2x + \frac{C}{x}$$

7. $y - y = 10$

Integrating factor: $e^{\int -1\,dx} = e^{-x}$

$$e^{-x}y' - e^{-x}y = 10e^{-x}$$

$$ye^{-x} = \int 10e^{-x}\,dx = -10e^{-x} + C$$

$$y = -10 + Ce^x$$

9. $(y + 1)\cos x\,dx = dy$

$$y' = (y + 1)\cos x = y\cos x + \cos x$$

$$y' - (\cos x)y = \cos x$$

Integrating factor: $e^{\int -\cos x\,dx} = e^{-\sin x}$

$$y'e^{-\sin x} - (\cos x)e^{-\sin x}y = (\cos x)e^{-\sin x}$$

$$ye^{-\sin x} = \int (\cos x)e^{-\sin x}\,dx$$

$$= -e^{-\sin x} + C$$

$$y = -1 + Ce^{\sin x}$$

11. $(x - 1)y' + y = x^2 - 1$

$$y' + \left(\frac{1}{x - 1}\right)y = x + 1$$

Integrating factor: $e^{\int[1/(x-1)]\,dx} = e^{\ln|x-1|} = x - 1$

$$y(x - 1) = \int (x^2 - 1)\,dx = \frac{1}{3}x^3 - x + C_1$$

$$y = \frac{x^3 - 3x + C}{3(x - 1)}$$

13. $y' - 3x^2y = e^{x^3}$

Integrating factor: $e^{-\int 3x^2\,dx} = e^{-x^3}$

$$ye^{-x^3} = \int e^{x^3}e^{-x^3}\,dx = \int dx = x + C$$

$$y = (x + C)e^{x^3}$$

15. (a),(c)

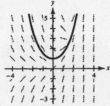

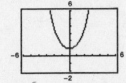

(b) $\dfrac{dy}{dx} = e^x - y$

$\dfrac{dy}{dx} + y = e^x$ Integrating factor: $e^{\int dx} = e^x$

$e^x y' + e^x y = e^{2x}$

$(ye^x) = \displaystyle\int e^{2x}\,dx$

$ye^x = \dfrac{1}{2}e^{2x} + C$

$y(0) = 1 \Rightarrow 1 = \dfrac{1}{2} + C \Rightarrow C = \dfrac{1}{2}$

$ye^x = \dfrac{1}{2}e^{2x} + \dfrac{1}{2}$

$y = \dfrac{1}{2}e^x + \dfrac{1}{2}e^{-x} = \dfrac{1}{2}(e^x + e^{-x})$

17. $y'\cos^2 x + y - 1 = 0$

$y' + (\sec^2 x)y = \sec^2 x$

Integrating factor: $e^{\int \sec^2 x\,dx} = e^{\tan x}$

$ye^{\tan x} = \displaystyle\int \sec^2 xe^{\tan x}\,dx = e^{\tan x} + C$

$y = 1 + Ce^{-\tan x}$

Initial condition: $y(0) = 5, C = 4$

Particular solution: $y = 1 + 4e^{\tan x}$

19. $y' + y\tan x = \sec x + \cos x$

Integrating factor: $e^{\int \tan x\,dx} = e^{\ln|\sec x|} = \sec x$

$y\sec x = \displaystyle\int \sec x(\sec x + \cos x)\,dx = \tan x + x + C$

$y = \sin x + x\cos x + C\cos x$

Initial condition: $y(0) = 1, 1 = C$

Particular solution: $y = \sin x + (x + 1)\cos x$

21. $y' + \left(\dfrac{1}{x}\right)y = 0$

Integrating factor: $e^{\int(1/x)\,dx} = e^{\ln|x|} = x$

Separation of variables:

$\dfrac{dy}{dx} = -\dfrac{y}{x}$

$\displaystyle\int \dfrac{1}{y}\,dy = \int -\dfrac{1}{x}\,dx$

$\ln y = -\ln x + \ln C$

$\ln xy = \ln C$

$xy = C$

Initial condition: $y(2) = 2, C = 4$

Particular solution: $xy = 4$

23. $x\,dy = (x + y + 2)\,dx$

$\dfrac{dy}{dx} = \dfrac{x + y + 2}{x} = \dfrac{y}{x} + 1 + \dfrac{2}{x}$

$\dfrac{dy}{dx} - \dfrac{1}{x}y = 1 + \dfrac{2}{x}$ Linear

$u(x) = e^{\int -(1/x)\,dx} = \dfrac{1}{x}$

$y = x\displaystyle\int \left(1 + \dfrac{2}{x}\right)\dfrac{1}{x}\,dx = x\int \left(\dfrac{1}{x} + \dfrac{2}{x^2}\right)dx$

$= x\left[\ln|x| + \dfrac{-2}{x} + C\right]$

$= -2 + x\ln|x| + Cx$

$y(1) = 10 = -2 + C \Rightarrow C = 12$

$y = -2 + x\ln|x| + 12x$

25. $y' + 3x^2y = x^2y^3$

$n = 3, Q = x^2, P = 3x^2$

$y^{-2}e^{\int(-2)3x^2\,dx} = \int(-2)x^2e^{\int(-2)3x^2\,dx}\,dx$

$y^{-2}e^{-2x^3} = -\int 2x^2e^{-2x^3}\,dx$

$y^{-2}e^{-2x^3} = \frac{1}{3}e^{-2x^3} + C$

$y^{-2} = \frac{1}{3} + Ce^{2x^3}$

$\frac{1}{y^2} = Ce^{2x^3} + \frac{1}{3}$

27. $y' + \left(\frac{1}{x}\right)y = xy^2$

$n = 2, Q = x, P = x^{-1}$

$e^{\int-(1/x)\,dx} = e^{-\ln|x|} = x^{-1}$

$y^{-1}x^{-1} = \int -x(x^{-1})\,dx = -x + C$

$\frac{1}{y} = -x^2 + Cx$

$y = \frac{1}{Cx - x^2}$

29. $y' - y = e^x\sqrt[3]{y}$, $n = \frac{1}{3}$, $Q = e^x$, $P = -1$

$e^{\int-(2/3)\,dx} = e^{-(2/3)x}$

$y^{2/3}e^{-(2/3)x} = \int \frac{2}{3}e^xe^{-(2/3)x}\,dx = \int \frac{2}{3}e^{(1/3)x}\,dx$

$y^{2/3}e^{-(2/3)x} = 2e^{(1/3)x} + C$

$y^{2/3} = 2e^x + Ce^{2x/3}$

31. (a)

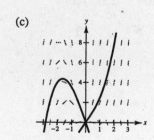

(c)

(b) $\dfrac{dy}{dx} - \dfrac{1}{x}y = x^2$

Integrating factor: $e^{-1/x\,dx} = e^{-\ln x} = \dfrac{1}{x}$

$\dfrac{1}{x}y' - \dfrac{1}{x^2}y = x$

$\left(\dfrac{1}{x}y\right) = \int x\,dx = \dfrac{x^2}{2} + C$

$y = \dfrac{x^3}{2} + Cx$

$(-2, 4): 4 = \dfrac{-8}{2} - 2C \Rightarrow C = -4 \Rightarrow y = \dfrac{x^3}{2} - 4x = \dfrac{1}{2}x(x^2 - 8)$

$(2, 8): 8 = \dfrac{8}{2} + 2C \Rightarrow C = 2 \Rightarrow y = \dfrac{x^3}{2} + 2x = \dfrac{1}{2}x(x^2 + 4)$

33. (a)

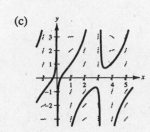

(b) $y' + (\cot x)y = 2$

Integrating factor: $e^{\int \cot x \, dx} = e^{\ln|\sin x|} = \sin x$

$$y' \sin x + (\cos x)y = 2 \sin x$$

$$y \sin x = \int 2 \sin x \, dx = -2 \cos x + C$$

$$y = -2 \cot x + C \csc x$$

$(1, 1)$: $1 = -2 \cot 1 + C \csc 1 \Rightarrow C = \dfrac{1 + 2 \cot 1}{\csc 1} = \sin 1 + 2 \cos 1$

$$y = -2 \cot x + (\sin 1 + 2 \cos 1) \csc x$$

$(3, -1)$: $-1 = -2 \cot 3 + C \csc 3 \Rightarrow C = \dfrac{2 \cot 3 - 1}{\csc 3} = 2 \cos 3 - \sin 3$

$$y = -2 \cot x + (2 \cos 3 - \sin 3) \csc x$$

(c)

35. $\dfrac{dP}{dt} = kP + N, N \text{ constant}$

$$\frac{dP}{kP + N} = dt$$

$$\int \frac{1}{kP + N} dP = \int dt$$

$$\frac{1}{k} \ln(kP + N) = t + C_1$$

$$\ln(kP + N) = kt + C_2$$

$$kP + N = e^{kt + C_2}$$

$$P = \frac{C_3 e^{kt} - N}{k}$$

$$P = Ce^{kt} - \frac{N}{k}$$

When $t = 0$: $P = P_0$

$$P_0 = C - \frac{N}{k} \Rightarrow C = P_0 + \frac{N}{k}$$

$$P = \left(P_0 + \frac{N}{k}\right)e^{kt} - \frac{N}{k}$$

37. (a) $A = \dfrac{P}{r}(e^{rt} - 1)$

$$A = \frac{100,000}{0.06}(e^{0.06(5)} - 1) \approx 583,098.01$$

(b) $A = \dfrac{250,000}{0.05}(e^{0.05(10)} - 1) \approx 3,243,606.35$

39. (a) $\dfrac{dQ}{dt} = q - kQ, q \text{ constant}$

(b) $Q' + kQ = q$

Let $P(t) = k$, $Q(t) = q$, then the integrating factor is $u(t) = e^{kt}$.

$$Q = e^{-kt} \int qe^{kt} \, dt = e^{-kt}\left(\frac{q}{k}e^{kt} + C\right) = \frac{q}{k} + Ce^{-kt}$$

When $t = 0$: $Q = Q_0$

$$Q_0 = \frac{q}{k} + C \Rightarrow C = Q_0 - \frac{q}{k}$$

$$Q = \frac{q}{k} + \left(Q_0 - \frac{q}{k}\right)e^{-kt}$$

(c) $\displaystyle\lim_{t \to \infty} Q = \dfrac{q}{k}$

41. Let Q be the number of pounds of concentrate in the solution at any time t. Since the number of gallons of solution in the tank at any time t is $v_0 + (r_1 - r_2)t$ and since the tank loses r_2 gallons of solution per minute, it must lose concentrate at the rate

$$\left[\frac{Q}{v_0 + (r_1 - r_2)t}\right]r_2.$$

The solution gains concentrate at the rate $r_1 q_1$. Therefore, the net rate of change is

$$\frac{dQ}{dt} = q_1 r_1 - \left[\frac{Q}{v_0 + (r_1 - r_2)t}\right]r_2 \quad \text{or} \quad \frac{dQ}{dt} + \frac{r_2 Q}{v_0 + (r_1 - r_2)t} = q_1 r_1.$$

43. (a) $Q' + \dfrac{r^2 Q}{v_0 + (r_1 - r_2)t} = q_1 r_1$

$Q(0) = q_0, q_0 = 25, q_1 = 0, v_0 = 200,$

$r_1 = 10, r_2 = 10, Q' + \dfrac{1}{20}Q = 0$

$\displaystyle\int \frac{1}{Q}\,dQ = \int -\frac{1}{20}\,dt$

$\ln Q = -\dfrac{1}{20}t + \ln C_1$

$Q = Ce^{-(1/20)t}$

Initial condition: $Q(0) = 25, C = 25$

Particular solution: $Q = 25e^{-(1/20)t}$

(b) $\quad 15 = 25e^{-(1/20)t}$

$\ln\left(\dfrac{3}{5}\right) = -\dfrac{1}{20}t$

$t = -20\ln\left(\dfrac{3}{5}\right) \approx 10.2 \text{ min}$

(c) $\displaystyle\lim_{t\to\infty} 25e^{-(1/20)t} = 0$

45. (a) The volume of the solution in the tank is given by $v_0 + (r_1 - r_2)t$. Therefore, $100 + (5 - 3)t = 200$ or $t = 50$ minutes.

(b) $Q' + \dfrac{r_2 Q}{v_0 + (r_1 - r_2)t} = q_1 r_1$

$Q(0) = q_0, q_0 = 0, q_1 = 0.5, v_0 = 100, r_1 = 5, r_2 = 3, Q' + \dfrac{3}{100 + 2t}Q = 2.5$

Integrating factor: $e^{\int [3/(100+2t)]\,dt} = (50 + t)^{3/2}$

$Q(50 + t)^{3/2} = \displaystyle\int 2.5(50 + t)^{3/2}\,dt = (50 + t)^{5/2} + C$

$Q = (50 + t) + C(50 + t)^{-3/2}$

Initial condition: $Q(0) = 0, 0 = 50 + C(50^{-3/2}), C = -50^{5/2}$

Particular solution: $\quad Q = (50 + t) - 50^{-5/2}(50 + t)^{-3/2}$

$Q(50) = 100 - 50^{5/2}(100)^{-3/2} = 100 - \dfrac{25}{\sqrt{2}} \approx 82.32 \text{ lbs}$

47. From Example 6,

$\dfrac{dv}{dt} + \dfrac{kv}{m} = g$

$v = \dfrac{mg}{k}(1 - e^{-kt/m}),$ Solution

$g = -32, mg = -8, v(5) = -101, m = \dfrac{-8}{g} = \dfrac{1}{4}$

implies that

$-101 = \dfrac{-8}{k}(1 - e^{-5k/(1/4)}).$

Using a graphing utility, $k \approx 0.050165$, and

$v = -159.47(1 - e^{-0.2007t}).$

As $t \to \infty$, $v \to -159.47$ ft/sec. The graph of v is shown below.

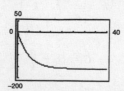

49. $L\dfrac{dI}{dt} + RI = E_0, I' + \dfrac{R}{L}I = \dfrac{E_0}{L}$

Integrating factor: $e^{\int (R/L)\,dt} = e^{Rt/L}$

$I\,e^{Rt/L} = \displaystyle\int \dfrac{E_0}{L}e^{Rt/L}\,dt = \dfrac{E_0}{R}e^{Rt/L} + C$

$\qquad I = \dfrac{E_0}{R} + Ce^{-Rt/L}$

51. $\dfrac{dy}{dx} + P(x)y = Q(x)$ Standard form

$\quad u(x) = e^{\int P(x)\,dx}$ Integrating factor

53. $y' - 2x = 0$

$\displaystyle\int dy = \int 2x\,dx$

$\qquad y = x^2 + C$

Matches c.

55. $y' - 2xy = 0$

$\displaystyle\int \dfrac{dy}{y} = \int 2x\,dx$

$\quad \ln y = x^2 + C_1$

$\qquad y = Ce^{x^2}$

Matches a.

57. $e^{2x+y}\,dx - e^{x-y}\,dy = 0$

Separation of variables:

$\quad e^{2x}e^y\,dx = e^x e^{-y}\,dy$

$\displaystyle\int e^x\,dx = \int e^{-2y}\,dy$

$\qquad e^x = -\dfrac{1}{2}e^{-2y} + C_1$

$\quad 2e^x + e^{-2y} = C$

59. $(y\cos x - \cos x)\,dx + dy = 0$

Separation of variables:

$\displaystyle\int \cos x\,dx = \int \dfrac{-1}{y-1}\,dy$

$\qquad \sin x = -\ln(y-1) + \ln C$

$\quad \ln(y-1) = -\sin x + \ln C$

$\qquad y = Ce^{-\sin x} + 1$

61. $(3y^2 + 4xy)\,dx + (2xy + x^2)\,dy = 0$

Homogeneous: $y = vx, dy = v\,dx + x\,dv$

$(3v^2x^2 + 4vx^2)\,dx + (2vx^2 + x^2)(v\,dx + x\,dv) = 0$

$\displaystyle\int \dfrac{5}{x}\,dx + \int \left(\dfrac{2v+1}{v^2+v}\right)dv = 0$

$\qquad \ln x^5 + \ln|v^2 + v| = \ln C$

$\qquad x^5(v^2 + v) = C$

$\qquad x^3y^2 + x^4y = C$

63. $(2y - e^x)\,dx + x\,dy = 0$

Linear: $y' + \left(\dfrac{2}{x}\right)y = \dfrac{1}{x}e^x$

Integrating factor: $e^{\int (2/x)\,dx} = e^{\ln x^2} = x^2$

$yx^2 = \displaystyle\int x^2 \dfrac{1}{x}e^x\,dx = e^x(x-1) + C$

$\quad y = \dfrac{e^x}{x^2}(x-1) + \dfrac{C}{x^2}$

65. $(x^2y^4 - 1)\,dx + x^3y^3\,dy = 0$

$y' + \left(\dfrac{1}{x}\right)y = x^{-3}y^{-3}$

Bernoulli: $n = -3, Q = x^{-3}, P = x^{-1},$

$e^{\int (4/x)\,dx} = e^{\ln x^4} = x^4$

$\quad y^4x^4 = \displaystyle\int 4(x^{-3})(x^4)\,dx = 2x^2 + C$

$x^4y^4 - 2x^2 = C$

67. $3(y - 4x^2)\,dx = -x\,dy$

$\qquad x\dfrac{dy}{dx} = -3y + 12x^2$

$\qquad y' + \dfrac{3}{x}y = 12x$

Integrating factor: $e^{\int (3/x)\,dx} = e^{3\ln x} = x^3$

$y'x^3 + \dfrac{3}{x}x^3y = 12x(x^3) = 12x^4$

$\qquad yx^3 = \displaystyle\int 12x^4\,dx = \dfrac{12}{5}x^5 + C$

$\qquad y = \dfrac{12}{5}x^2 + \dfrac{C}{x^3}$

69. False. The equation contains $\sqrt{y}$.

Review Exercises for Chapter 6

1. $y = x^3$, $y' = 3x^2$

$x^2 y' + 3y = x^2[3x^2] + 3[x^3] = 3(x^4 + x^3) \neq 6x^3$

Not a solution

3. $\dfrac{dy}{dx} = 2x^2 + 5$

$y = \displaystyle\int (2x^2 + 5)\, dx = \dfrac{2x^3}{3} + 5x + C$

5. $\dfrac{dy}{dx} = \cos 2x$

$y = \displaystyle\int \cos 2x\, dx = \dfrac{1}{2}\sin 2x + C$

7. $\dfrac{dy}{dx} = 2x\sqrt{x - 7}$

$y = \displaystyle\int 2x\sqrt{x - 7}\, dx$

Let $u = x - 7$, $du = dx$, $x = u + 7$:

$y = \displaystyle\int 2(u + 7)u^{1/2}\, du$

$= \dfrac{4}{5}u^{5/2} + \dfrac{28}{3}u^{3/2} + C$

$= \dfrac{4}{5}(x - 7)^{5/2} + \dfrac{28}{3}(x - 7)^{3/2} + C$

$= \dfrac{4}{15}(x - 7)^{3/2}(3x + 14) + C$

9. $\dfrac{dy}{dx} = \dfrac{2x}{y}$

x	-4	-2	0	2	4	8
y	2	0	4	4	6	8
dy/dx	-4	Undef.	0	1	$4/3$	2

11. $y' = -x - 2$, $\quad (-1, 1)$

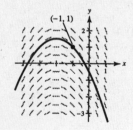

13. $y' = \dfrac{1}{4}x^2 - \dfrac{1}{3}x$, $\quad (0, 3)$

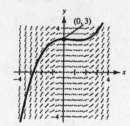

15. $y' = \dfrac{xy}{x^2 + 4}$, $\quad (0, 1)$

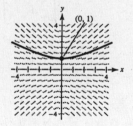

17. $\dfrac{dy}{dx} = 6 - x$

$y = \displaystyle\int (6 - x)\, dx = 6x - \dfrac{x^2}{2} + C$

19. $\quad\dfrac{dy}{dx} = (3 + y)^2$

$\displaystyle\int (3 + y)^{-2}\, dy = \int dx$

$-(3 + y)^{-1} = x + C$

$3 + y = \dfrac{-1}{x + C}$

$y = -3 - \dfrac{1}{x + C}$

21. $(2 + x)y' - xy = 0$

$$(2 + x)\frac{dy}{dx} = xy$$

$$\frac{1}{y}\, dy = \frac{x}{2 + x}\, dx$$

$$\frac{1}{y}\, dy = \left(1 - \frac{2}{2 + x}\right)dx$$

$$\ln|y| = x - 2\ln|2 + x| + C_1$$

$$y = Ce^x(2 + x)^{-2} = \frac{Ce^x}{(2 + x)^2}$$

23. $y = Ce^{kt}$

$$\left(0, \frac{3}{4}\right): \quad \frac{3}{4} = C$$

$$(5, 5): \quad 5 = \frac{3}{4}e^{k(5)}$$

$$\frac{20}{3} = e^{5k}$$

$$k = \frac{1}{5}\ln\left(\frac{20}{3}\right)$$

$$y = \frac{3}{4}e^{[\ln(20/3)/5]t} \approx \frac{3}{4}e^{0.379t}$$

25. $y = Ce^{kt}$

$(0, 5): \quad C = 5$

$$\left(5, \frac{1}{6}\right): \quad \frac{1}{6} = 5e^{5k}$$

$$k = \frac{1}{5}\ln\left(\frac{1}{30}\right) = \frac{-\ln 30}{5}$$

$$y = 5e^{[-t\ln 30]/5} \approx 5e^{-0.680t}$$

27. $\frac{dP}{dh} = kp$, $P(0) = 30$

$$P(h) = 30e^{kh}$$

$$P(18{,}000) = 30e^{18{,}000k} = 15$$

$$k = \frac{\ln(1/2)}{18{,}000} = \frac{-\ln 2}{18{,}000}$$

$$P(h) = 30e^{-(h\ln 2)/18{,}000}$$

$$P(35{,}000) = 30e^{-(35{,}000\ln 2)/18{,}000} \approx 7.79 \text{ inches}$$

29. $S = Ce^{k/t}$

(a) $\qquad S = 5$ when $t = 1$

$$5 = Ce^k$$

$$\lim_{t \to \infty} Ce^{k/t} = C = 30$$

$$5 = 30e^k$$

$$k = \ln\frac{1}{6} \approx -1.7918$$

$$S = 30\left(\frac{1}{6}\right) \approx 30e^{-1.7918/t}$$

(b) When $t = 5$, $S \approx 20.9646$ which is 20,965 units.

(c)

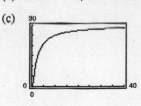

31. $P = Ce^{0.015t}$

$$2C = Ce^{0.015t}$$

$$2 = e^{0.015t}$$

$$\ln 2 = 0.015t$$

$$t = \frac{\ln 2}{0.015} \approx 46.21 \text{ years}$$

33. $\frac{dy}{dx} = \frac{x^2 + 3}{x}$

$$\int dy = \int\left(x + \frac{3}{x}\right)dx$$

$$y = \frac{x^2}{2} + 3\ln|x| + C$$

35. $y' - 2xy = 0$

$$\frac{dy}{dx} = 2xy$$

$$\int\frac{1}{y}\, dy = \int 2x\, dx$$

$$\ln|y| = x^2 + C_1$$

$$e^{x^2 + C_1} = y$$

$$y = Ce^{x^2}$$

37. $\dfrac{dy}{dx} = \dfrac{x^2 + y^2}{2xy}$ (homogeneous differential equation)

$(x^2 + y^2)\,dx - 2xy\,dy = 0$

Let $y = vx,\ dy = x\,dv + v\,dx$.

$(x^2 + v^2 x^2)\,dx - 2x(vx)(x\,dv + v\,dx) = 0$

$(x^2 + v^2 x^2 - 2x^2 v^2)\,dx - 2x^3 v\,dv = 0$

$(x^2 - x^2 v^2)\,dx = 2x^3 v\,dv$

$(1 - v^2)\,dx = 2xv\,dv$

$\displaystyle\int \frac{dx}{x} = \int \frac{2v}{1 - v^2}\,dv$

$\ln|x| = -\ln|1 - v^2| + C_1 = -\ln|1 - v^2| + \ln C$

$x = \dfrac{C}{1 - v^2} = \dfrac{C}{1 - (y/x)^2} = \dfrac{Cx^2}{x^2 - y^2}$

$1 = \dfrac{Cx}{x^2 - y^2}$ or $C_1 = \dfrac{x}{x^2 - y^2}$

39. $y = C_1 x + C_2 x^3$

$y' = C_1 + 3C_2 x^2$

$y'' = 6C_2 x$

$x^2 y'' - 3xy' + 3y = x^2(6C_2 x) - 3x(C_1 + 3C_2 x^2) + 3(C_1 x + C_2 x^3)$

$\qquad\qquad = 6C_2 x^3 - 3C_1 x - 9C_2 x^3 + 3C_1 x + 3C_2 x^3 = 0$

$x = 2, y = 0$: $0 = 2C_1 + 8C_2 \implies C_1 = -4C_2$

$x = 2, y' = 4$: $4 = C_1 + 12C_2$

$\qquad\qquad 4 = (-4C_2) + 12C_2 = 8C_2 \implies C_2 = \tfrac{1}{2}, C_1 = -2$

$y = -2x + \tfrac{1}{2}x^3$

41. $\dfrac{dy}{dx} = \dfrac{-4x}{y}$

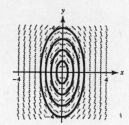

$\displaystyle\int y\,dy = \int -4x\,dx$

$\dfrac{y^2}{2} = -2x^2 + C_1$

$4x^2 + y^2 = C$ ellipses

43. $P(t) = \dfrac{7200}{1 + 44e^{-0.55t}}$

(a) $k = 0.55$

(b) $L = 7200$

(c) $P(0) = \dfrac{7200}{1 + 44} = 160$

(d) $\qquad 3600 = \dfrac{7200}{1 + 44e^{-0.55t}}$

$1 + 44e^{-0.55t} = 2$

$e^{-0.55t} = \dfrac{1}{44}$

$t = \dfrac{-1}{0.55}\ln\left(\dfrac{1}{44}\right) \approx 6.88$ yrs.

(e) $\dfrac{dP}{dt} = 0.55P\left(1 - \dfrac{P}{7200}\right)$

45. (a) $L = 20{,}400$, $y(0) = 1200$, $y(1) = 2000$

$$y = \frac{20{,}400}{1 + be^{-kt}}$$

$$y(0) = 1200 = \frac{20{,}400}{1 + b} \implies b = 16$$

$$y(1) = 2000 = \frac{20{,}400}{1 + 16e^{-k}}$$

$$16e^{-k} = \frac{46}{5}$$

$$k = -\ln\frac{23}{40} = \ln\frac{40}{23} \approx 0.553$$

$$y = \frac{20{,}400}{1 + 16e^{-0.553t}}$$

(b) $y(8) \approx 17{,}118$ trout

(c) $10{,}000 = \dfrac{20{,}400}{1 + 16e^{-0.553t}} \implies t \approx 4.94$ yrs.

47. $y' - y = 8$

$$P(x) = -1, \quad Q(x) = 8$$

$$u(x) = e^{\int -dx} = e^{-x}$$

$$y = \frac{1}{e^{-x}}\int 8e^{-x}\,dx$$

$$= e^{x}[-8e^{-x} + C]$$

$$= -8 + Ce^{x}$$

49. $4y' = e^{x/y} + y$

$$y' - \frac{1}{4}y = \frac{1}{4}e^{x/4}$$

$$P(x) - -\frac{1}{4}, \quad Q(x) = \frac{1}{4}e^{x/4}$$

$$u(x) = e^{\int -(1/4)\,dx} = e^{-(1/4)x}$$

$$y = \frac{1}{e^{-(1/4)x}}\int \frac{1}{4}e^{x/4}e^{-(1/4)x}\,dx$$

$$= e^{(1/4)x}\left[\frac{1}{4}x + C\right]$$

$$= \frac{1}{4}xe^{x/4} + Ce^{x/4}$$

51. $(x - 2)y' + y = 1$

$$\frac{dy}{dx} + \frac{1}{x - 2}y = \frac{1}{x - 2}$$

$$P(x) = \frac{1}{x - 2}, \quad Q(x) = \frac{1}{x - 2}$$

$$u(x) = e^{\int (1/x-2)\,dx} = e^{\ln|x-2|} = x - 2$$

$$y = \frac{1}{x - 2}\int \left(\frac{1}{x - 2}\right)(x - 2)\,dx$$

$$= \frac{1}{x - 2}[x + c]$$

53. $(3y + \sin 2x)\,dx - dy = 0$

$$y' - 3y = \sin 2x$$

Integrating factor: $e^{\int -3\,dx} = e^{-3x}$

$$ye^{-3x} = \int e^{-3x}\sin 2x\,dx$$

$$= \frac{1}{13}e^{-3x}(-3\sin 2x - 2\cos 2x) + C$$

$$y = -\frac{1}{13}(3\sin 2x + 2\cos 2x) + Ce^{3x}$$

55. $y' + 5y = e^{5x}$

Integrating factor: $e^{\int 5\,dx} = e^{5x}$

$$ye^{5x} = \int e^{10x}\,dx = \frac{1}{10}e^{10x} + C$$

$$y = \frac{1}{10}e^{5x} + Ce^{-5x}$$

57. $y' + y = xy^2$ Bernoulli equation

$n = 2$, let $z = y^{1-2} = y^{-1}$, $z' = -y^{-2}y'$.

$(-y^{-2})y' + (-y^{-2})y = -x$

$\qquad\qquad z' - z = -x$ Linear equation

$u(x) = e^{\int -dx} = e^{-x}$

$z = \dfrac{1}{e^{-x}} \displaystyle\int -xe^{-x}\,dx = e^x[xe^{-x} + e^{-x} + C]$

$y^{-1} = x + 1 + Ce^x$

$y = \dfrac{1}{x + 1 + Ce^x}$

59. $y' + \dfrac{1}{x}y = \dfrac{y^3}{x^2}$ Bernoulli equation

$n = 3$, let $z = y^{1-3} = y^{-2}$, $z' = -2y^{-3}y'$.

$(-2y^{-3})y' + \dfrac{1}{x}y(-2y^{-3}) = \dfrac{-2}{x^2}$

$\qquad\qquad z' - \dfrac{2}{x}z = \dfrac{-2}{x^2}$ Linear equation

$u(x) = e^{\int -(2/x)\,dx} = e^{-2\ln x} = x^{-2}$

$z = \dfrac{1}{x^{-2}} \displaystyle\int \dfrac{-2}{x^2}(x^{-2})\,dx = x^2\left[\dfrac{2x^{-3}}{3} + C\right]$

$\dfrac{1}{y^2} = \dfrac{2}{3x} + Cx^2$

61. Answers will vary. Sample Answer:

$(x^2 + 3y^2)\,dx - 2xy\,dy = 0$

Solution: Let $y = vx$, $dy = x\,dv + v\,dx$.

$(x^2 + 3v^2x^2)\,dx - 2x(vx)(x\,dv + v\,dx) = 0$

$(x^2 + v^2x^2)\,dx - 2x^3v\,dv = 0$

$(1 + v^2)\,dx = 2xv\,dv$

$\displaystyle\int \dfrac{dx}{x} = \int \dfrac{2v}{1 + v^2}\,dv$

$\ln|x| = \ln|1 + v^2| + C_1$

$x = C(1 + v^2) = C\left(1 + \dfrac{y^2}{x^2}\right)$

$x^3 = C(x^2 + y^2)$

63. Answers will vary.

Sample Answer: $x^3y' + 2x^2y = 1$

$\qquad\qquad y' + \dfrac{2}{x}y = \dfrac{1}{x^3}$

$u(x) = e^{\int (2/x)\,dx} = x^2$

$y = \dfrac{1}{x^2} \displaystyle\int \dfrac{1}{x^3}(x^2)\,dx = \dfrac{1}{x^2}[\ln|x| + C]$

Problem Solving for Chapter 6

1. (a) $\dfrac{dy}{dt} = y^{1.01}$

$$\int y^{-1.01}\, dy = \int dt$$

$$\frac{y^{-0.01}}{-0.01} = t + C_1$$

$$\frac{1}{y^{0.01}} = -0.01t + C$$

$$y^{0.01} = \frac{1}{C - 0.01t}$$

$$y = \frac{1}{(C - 0.01t)^{100}}$$

$y(0) = 1$: $1 = \dfrac{1}{C^{100}} \Rightarrow C = 1$

Hence, $y = \dfrac{1}{(1 - 0.01t)^{100}}.$

For $T = 100$, $\displaystyle\lim_{t \to T^-} y = \infty.$

(b) $\displaystyle\int y^{-(1+\varepsilon)}\, dy = \int k\, dt$

$$\frac{y^{-\varepsilon}}{-\varepsilon} = kt + C_1$$

$$y^{-\varepsilon} = -\varepsilon kt + C$$

$$y = \frac{1}{(C - \varepsilon kt)^{1/\varepsilon}}$$

$y(0) = y_0 = \dfrac{1}{C^{1/\varepsilon}} \Rightarrow C^{1/\varepsilon} = \dfrac{1}{y_0} \Rightarrow C = \left(\dfrac{1}{y_0}\right)^{\varepsilon}$

Hence, $y = \dfrac{1}{\left(\dfrac{1}{y_0^{\varepsilon}} - \varepsilon kt\right)^{1/\varepsilon}}.$

For $t \to \dfrac{1}{y_0^{\varepsilon}\varepsilon k}$, $y \to \infty.$

3. (a) $\dfrac{dS}{dt} = k_1 S(L - S)$

$S = \dfrac{L}{1 + Ce^{-kt}}$ is a solution because

$$\frac{dS}{dt} = -L(1 + Ce^{-kt})^{-2}(-Cke^{-kt})$$

$$= \frac{LC ke^{-kt}}{(1 + Ce^{-kt})^2}$$

$$= \left(\frac{k}{L}\right)\frac{L}{1 + Ce^{-kt}} \cdot \frac{C Le^{-kt}}{1 + Ce^{-kt}}$$

$$= \left(\frac{k}{L}\right)\frac{L}{1 + Ce^{-kt}} \cdot \left(L - \frac{L}{1 + Ce^{-kt}}\right)$$

$$= k_1 S(L - S), \text{ where } k_1 = \frac{k}{L}.$$

$L = 100$. Also, $S = 10$ when $t = 0 \Rightarrow C = 9$. And, $S = 20$ when $t = 1 \Rightarrow k = -\ln\frac{4}{9}.$

Particular Solution: $s = \dfrac{100}{1 + 9e^{\ln(4/9)t}}$

$$= \frac{100}{1 + 9e^{-0.8109t}}$$

(b) $\dfrac{dS}{dt} = k_1 S(100 - S)$

$$\frac{d^2S}{dt^2} = k_1\left[S\left(-\frac{dS}{dt}\right) + (100 - S)\frac{dS}{dt}\right]$$

$$= k_1(100 - 2S)\frac{dS}{dt}$$

$= 0$ when $S = 50$ or $\dfrac{dS}{dt} = 0.$

Choosing $S = 50$, we have:

$$50 = \frac{100}{1 + 9e^{\ln(4/9)t}}$$

$$2 = 1 + 9e^{\ln(4/9)t}$$

$$\frac{\ln(1/9)}{\ln(4/9)} = t$$

$t \approx 2.7$ months (This is the inflection point.)

(d)

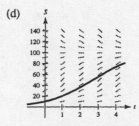

(e) Sales will decrease toward the line $S = L$.

(c)

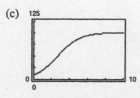

5. Let $u = \frac{1}{2} k \left(t - \frac{\ln b}{k} \right)$.

$$1 + \tanh u = 1 + \frac{e^4 - e^{-u}}{e^u + e^{-u}} = \frac{2}{1 + e^{-2u}}$$

$$e^{-2u} = e^{-k(t - (\ln b/k))} = e^{\ln b} e^{-kt} = b e^{-kt}$$

Finally,

$$\frac{1}{2} L \left[1 + \tanh\left(\frac{1}{2} k \left(t - \frac{\ln b}{k} \right) \right) \right] = \frac{L}{2} [1 + \tanh u]$$

$$= \frac{L}{2} \frac{2}{1 + b e^{-kt}}$$

$$= \frac{L}{1 + b e^{-kt}}.$$

The graph of the logistics function is just a shift of the graph of the hyperbolic tangent, as shown in Section 5.10.

7. (a) $A(h) \dfrac{dh}{dt} = -k\sqrt{2gh}$

$$\pi r^2 \frac{dh}{dt} = -k\sqrt{64h}$$

$$h^{-1/2} \, dh = \frac{-8k}{\pi r^2} \, dt = -C \, dt, \quad C = \frac{8k}{\pi r^2}$$

$$2\sqrt{h} = -Ct + C_1$$

$$2\sqrt{18} = C_1 \quad (\text{at } t = 0, h = 18)$$

Hence, $2\sqrt{h} = -Ct + 6\sqrt{2}$.

At $t = 30(60) = 1800$, $h = 12$:

$$2\sqrt{12} = -1800\,C + 6\sqrt{2}$$

$$\frac{6\sqrt{2} - 4\sqrt{3}}{1800} = C \approx 0.000865$$

Hence, $2\sqrt{h} = -0.000865t + 6\sqrt{2}$.

$$h = 0 \implies t = \frac{6\sqrt{2}}{0.000865} \approx 9809.1 \text{ seconds (2 hr, 43 min, 29 sec)}$$

(b) $t = 3600 \text{ sec} \implies 2\sqrt{h} = -0.000865(3600) + 6\sqrt{2}$

$$\implies h \approx 7.21 \text{ feet}$$

9. Let the radio receiver be located at $(x_0, 0)$. The tangent line to $y = x - x^2$ joins $(-1, 1)$ and $(x_0, 0)$.

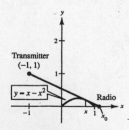

—**CONTINUED**—

9. —**CONTINUED**—

(a) If (x, y) is the point of tangency on $y = x - x^2$, then

$$1 - 2x = \frac{y - 1}{x + 1} = \frac{x - x^2 - 1}{x + 1}$$

$$x - 2x^2 + 1 - 2x = x - x^2 - 1$$

$$x^2 + 2x - 2 = 0$$

$$x = \left(\frac{-2 \pm \sqrt{4 + 8}}{2}\right) = -1 + \sqrt{3}$$

$$y = x - x^2 = 3\sqrt{3} - 5.$$

Then $\dfrac{1 - 0}{-1 - x_0} = \dfrac{1 - 3\sqrt{3} + 5}{-1 + 1 - \sqrt{3}} = \dfrac{6 - 3\sqrt{3}}{-\sqrt{3}}$

$$\sqrt{3} = (1 + x_0)(6 - 3\sqrt{3})$$

$$= 6 - 3\sqrt{3} + x_0(6 - 3\sqrt{3})$$

$$x_0 = \frac{4\sqrt{3} - 6}{6 - 3\sqrt{3}} \approx 1.155.$$

(c)

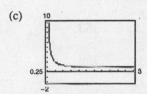

There is a vertical asymptote at $h = \frac{1}{4}$, which is the height of the mountain.

(b) Now let the transmitter be located at $(-1, h)$.

$$1 - 2x = \frac{y - h}{x + 1} = \frac{x - x^2 - h}{x + 1}$$

$$x - 2x^2 + 1 - 2x = x - x^2 - h$$

$$x^2 + 2x - h - 1 = 0$$

$$x = \frac{\left(-2 \pm \sqrt{4 + 4(h + 1)}\right)}{2}$$

$$= -1 + \sqrt{2 + h}$$

$$y = x - x^2$$

$$= 3\sqrt{2 + h} - h - 4$$

Then, $\dfrac{h - 0}{-1 - x_0} = \dfrac{h - \left(3\sqrt{2 + h} - h - 4\right)}{-1 - \left(-1 + \sqrt{2 + h}\right)}$

$$= \frac{2h + 4 - 3\sqrt{2 + h}}{-\sqrt{2 + h}}$$

$$\frac{x_0 + 1}{h} = \frac{\sqrt{2 + h}}{2h + 4 - 3\sqrt{2 + h}}$$

$$x_0 = \frac{h\sqrt{2 + h}}{2h + 4 - 3\sqrt{2 + h}} - 1.$$

11. (a) $\displaystyle\int \frac{dC}{C} = \int -\frac{R}{V}\, dt$

$$\ln|C| = -\frac{R}{V}t + K_1$$

$$C = Ke^{-Rt/V}$$

Since $C = C_0$ when $t = 0$, it follows that $K = C_0$ and the function is $C = C_0 e^{-Rt/V}$.

(b) Finally, as $t \to \infty$, we have

$$\lim_{t \to \infty} C = \lim_{t \to \infty} C_0 e^{-Rt/V} = 0.$$

13. (a) $\displaystyle\int \frac{1}{Q - RC}\, dC = \int \frac{1}{V}\, dt$

$$-\frac{1}{R}\ln|Q - RC| = \frac{t}{V} + K_1$$

$$Q - RC = e^{-R[(t/V) + K_1]}$$

$$C = \frac{1}{R}\left(Q - e^{-R[(t/V) + K_1]}\right)$$

$$= \frac{1}{R}\left(Q - Ke^{-Rt/V}\right)$$

Since $C = 0$ when $t = 0$, it follows that $K = Q$ and we have $C = \dfrac{Q}{R}(1 - e^{-Rt/V})$.

(b) As $t \to \infty$, the limit of C is Q/R.

C H A P T E R 7
Applications of Integration

Section 7.1   Area of a Region Between Two Curves 313

Section 7.2   Volume: The Disk Method 321

Section 7.3   Volume: The Shell Method 328

Section 7.4   Arc Length and Surfaces of Revolution 335

Section 7.5   Work 341

Section 7.6   Moments, Centers of Mass, and Centroids 344

Section 7.7   Fluid Pressure and Fluid Force 350

Review Exercises 353

Problem Solving 359

CHAPTER 7
Applications of Integration

Section 7.1   Area of a Region Between Two Curves

1. $A = \int_0^6 [0 - (x^2 - 6x)] \, dx = -\int_0^6 (x^2 - 6x) \, dx$

3. $A = \int_0^3 [(-x^2 + 2x + 3) - (x^2 - 4x + 3)] \, dx$

$= \int_0^3 (-2x^2 + 6x) \, dx$

5. $A = 2\int_{-1}^0 3(x^3 - x) \, dx = 6\int_{-1}^0 (x^3 - x) \, dx$

or $-6\int_0^1 (x^3 - x) \, dx$

7. $\int_0^4 \left[(x + 1) - \frac{x}{2}\right] dx$

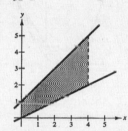

9. $\int_0^6 \left[4(2^{-x/3}) - \frac{x}{6}\right] dx$

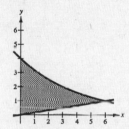

11. $\int_{-\pi/3}^{\pi/3} (2 - \sec x) \, dx$

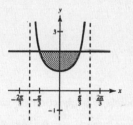

13. (a)

$x = 4 - y^2$

$x = y - 2$

$4 - y^2 = y - 2$

$y^2 + y - 6 = 0$

$(y + 3)(y - 2) = 0$

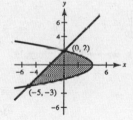

Intersection points: $(0, 2)$ and $(-5, -3)$

$A = \int_{-5}^0 \left[(x + 2) + \sqrt{4 - x}\right] dx + \int_0^4 2\sqrt{4 - x} \, dx = \frac{61}{6} + \frac{32}{3} = \frac{125}{6}$

(b) $A = \int_{-3}^2 [(4 - y^2) - (y - 2)] \, dy = \frac{125}{6}$

15. $f(x) = x + 1$

$g(x) = (x - 1)^2$

$A \approx 4$

Matches (d)

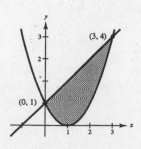

17. $A = \int_0^2 \left[\left(\frac{1}{2}x^3 + 2 \right) - (x + 1) \right] dx$

$= \int_0^2 \left(\frac{1}{2}x^3 - x + 1 \right) dx$

$= \left[\frac{x^4}{8} - \frac{x^2}{2} + x \right]_0^2$

$= \left(\frac{16}{8} - \frac{4}{2} + 2 \right) - 0 = 2$

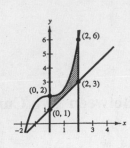

19. The points of intersection are given by:

$x^2 - 4x = 0$

$x(x - 4) = 0$ when $x = 0, 4$

$A = \int_0^4 [g(x) - f(x)] \, dx$

$= -\int_0^4 (x^2 - 4x) \, dx$

$= -\left[\frac{x^3}{3} - 2x^2 \right]_0^4$

$= \frac{32}{3}$

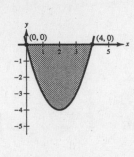

21. The points of intersection are given by:

$x^2 + 2x + 1 = 3x + 3$

$(x - 2)(x + 1) = 0$ when $x = -1, 2$

$A = \int_{-1}^2 [g(x) - f(x)] \, dx$

$= \int_{-1}^2 [(3x + 3) - (x^2 + 2x + 1)] \, dx$

$= \int_{-1}^2 (2 + x - x^2) \, dx$

$= \left[2x + \frac{x^2}{2} - \frac{x^3}{3} \right]_{-1}^2 = \frac{9}{2}$

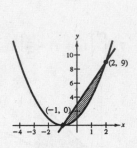

23. The points of intersection are given by:

$x = 2 - x$ and $x = 0$ and $2 - x = 0$

$x = 1$ $x = 0$ $x = 2$

$A = \int_0^1 [(2 - y) - (y)] \, dy = \left[2y - y^2 \right]_0^1 = 1$

Note that if we integrate with respect to x, we need two integrals. Also, note that the region is a triangle.

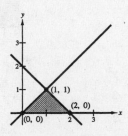

25. The points of intersection are given by:

$\sqrt{3x} + 1 = x + 1$

$\sqrt{3x} = x$ when $x = 0, 3$

$A = \int_0^3 [f(x) - g(x)] \, dx$

$= \int_0^3 \left[\left(\sqrt{3x} + 1 \right) - (x + 1) \right] dx$

$= \int_0^3 [(3x)^{1/2} - x] \, dx$

$= \left[\frac{2}{9}(3x)^{3/2} - \frac{x^2}{2} \right]_0^3 = \frac{3}{2}$

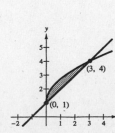

27. The points of intersection are given by:

$y^2 = y + 2$

$(y - 2)(y + 1) = 0$ when $y = -1, 2$

$A = \int_{-1}^2 [g(y) - f(y)] \, dy$

$= \int_{-1}^2 [(y + 2) - y^2] \, dy$

$= \left[2y + \frac{y^2}{2} - \frac{y^3}{3} \right]_{-1}^2 = \frac{9}{2}$

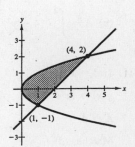

29. $A = \int_{-1}^{2} [f(y) - g(y)] \, dy$

$= \int_{-1}^{2} [(y^2 + 1) - 0] \, dy$

$= \left[\dfrac{y^3}{3} + y \right]_{-1}^{2} = 6$

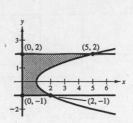

31. $y = \dfrac{10}{x} \implies x = \dfrac{10}{y}$

$A = \int_{2}^{10} \dfrac{10}{y} \, dy$

$= \left[10 \ln y \right]_{2}^{10}$

$= 10(\ln 10 - \ln 2)$

$= 10 \ln 5 \approx 16.0944$

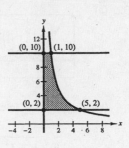

33. (a)

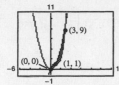

(c) Numerical approximation:
   $0.417 + 2.667 \approx 3.083$

(b) The points of intersection are given by:

$x^3 - 3x^2 + 3x = x^2$

$x(x - 1)(x - 3) = 0$ when $x = 0, 1, 3$

$A = \int_{0}^{1} [f(x) - g(x)] \, dx + \int_{1}^{3} [g(x) - f(x)] \, dx$

$= \int_{0}^{1} [(x^3 - 3x^2 + 3x) - x^2] \, dx + \int_{1}^{3} [x^2 - (x^3 - 3x^2 + 3x)] \, dx$

$= \int_{0}^{1} (x^3 - 4x^2 + 3x) \, dx + \int_{1}^{3} (-x^3 + 4x^2 - 3x) \, dx$

$= \left[\dfrac{x^4}{4} - \dfrac{4}{3}x^3 + \dfrac{3}{2}x^2 \right]_{0}^{1} + \left[\dfrac{-x^4}{4} + \dfrac{4}{3}x^3 - \dfrac{3}{2}x^2 \right]_{1}^{3} = \dfrac{5}{12} + \dfrac{8}{3} = \dfrac{37}{12}$

35. (a)

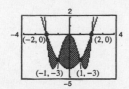

(c) Numerical approximation: 21.333

(b) The points of intersection are given by:

$x^2 - 4x + 3 = 3 + 4x - x^2$

$2x(x - 4) = 0$ when $x = 0, 4$

$A = \int_{0}^{4} [(3 + 4x - x^2) - (x^2 - 4x + 3)] \, dx$

$= \int_{0}^{4} (-2x^2 + 8x) \, dx$

$= \left[-\dfrac{2x^3}{3} + 4x^2 \right]_{0}^{4} = \dfrac{64}{3}$

37. (a) $f(x) = x^4 - 4x^2, \quad g(x) = x^2 - 4$

(c) Numerical approximation:
   $5.067 + 2.933 = 8.0$

(b) The points of intersection are given by:

$x^4 - 4x^2 = x^2 - 4$

$x^4 - 5x^2 + 4 = 0$

$(x^2 - 4)(x^2 - 1) = 0$ when $x = \pm 2, \pm 1$

By symmetry:

$A = 2 \int_{0}^{1} [(x^4 - 4x^2) - (x^2 - 4)] \, dx + 2 \int_{1}^{2} [(x^2 - 4) - (x^4 - 4x^2)] \, dx$

$= 2 \int_{0}^{1} (x^4 - 5x^2 + 4) \, dx + 2 \int_{1}^{2} (-x^4 + 5x^2 - 4) \, dx$

$= 2 \left[\dfrac{x^5}{5} - \dfrac{5x^3}{3} + 4x \right]_{0}^{1} + 2 \left[-\dfrac{x^5}{5} + \dfrac{5x^3}{3} - 4x \right]_{1}^{2}$

$= 2 \left[\dfrac{1}{5} - \dfrac{5}{3} + 4 \right] + 2 \left[\left(-\dfrac{32}{5} + \dfrac{40}{3} - 8 \right) - \left(-\dfrac{1}{5} + \dfrac{5}{3} - 4 \right) \right] = 8$

39. (a)

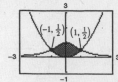

(b) The points of intersection are given by:

$$\frac{1}{1 + x^2} = \frac{x^2}{2}$$

$$x^4 + x^2 - 2 = 0$$

$$(x^2 + 2)(x^2 - 1) = 0$$

$$x = \pm 1$$

$$A = 2 \int_0^1 [f(x) - g(x)] \, dx$$

$$= 2 \int_0^1 \left[\frac{1}{1 + x^2} - \frac{x^2}{2} \right] dx$$

$$= 2 \left[\arctan x - \frac{x^3}{6} \right]_0^1$$

$$= 2 \left(\frac{\pi}{4} - \frac{1}{6} \right) = \frac{\pi}{2} - \frac{1}{3} \approx 1.237$$

(c) Numerical approximation: 1.237

41. (a)

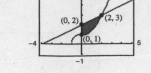

(b) and (c) $\sqrt{1 + x^3} \le \frac{1}{2}x + 2$ on $[0, 2]$

You must use numerical integration because $y = \sqrt{1 + x^3}$ does not have an elementary antiderivative.

$$A = \int_0^2 \left[\frac{1}{2}x + 2 - \sqrt{1 + x^3} \right] dx \approx 1.759$$

43. $A = 2 \int_0^{\pi/3} [f(x) - g(x)] \, dx$

$$= 2 \int_0^{\pi/3} (2 \sin x - \tan x) \, dx$$

$$= 2 \left[-2 \cos x + \ln|\cos x| \right]_0^{\pi/3}$$

$$= 2(1 - \ln 2) \approx 0.614$$

45. $A = \int_0^{2\pi} [(2 - \cos x) - \cos x] \, dx$

$$= 2 \int_0^{2\pi} (1 - \cos x) \, dx$$

$$= 2 \left[x - \sin x \right]_0^{2\pi} = 4\pi \approx 12.566$$

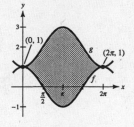

47. $A = \int_0^1 [x e^{-x^2} - 0] \, dx$

$$= \left[-\frac{1}{2} e^{-x^2} \right]_0^1 = \frac{1}{2} \left(1 - \frac{1}{e} \right) \approx 0.316$$

49. (a)

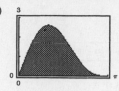

(b) $A = \int_0^{\pi} (2 \sin x + \sin 2x) \, dx$

$= \left[-2 \cos x - \frac{1}{2} \cos 2x \right]_0^{\pi}$

$= \left(2 - \frac{1}{2} \right) - \left(-2 - \frac{1}{2} \right) = 4$

(c) Numerical approximation: 4.0

51. (a)

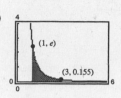

(b) $A = \int_1^3 \frac{1}{x^2} e^{1/x} \, dx$

$= \left[-e^{-1/x} \right]_1^3$

$= e - e^{1/3}$

(c) Numerical approximation: 1.323

53. (a)

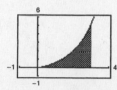

(b) The integral

$$A = \int_0^3 \sqrt{\frac{x^3}{4 - x}} \, dx$$

does not have an elementary antiderivative.

(c) $A \approx 4.7721$

55. (a)

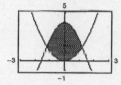

(b) The intersection points are difficult to determine by hand.

(c) Area $= \int_{-c}^{c} [4 \cos x - x^2] \, dx \approx 6.3043$ where $c \approx 1.201538$.

57. $F(x) = \int_0^x \left(\frac{1}{2} t + 1 \right) dt = \left[\frac{t^2}{4} + t \right]_0^x = \frac{x^2}{4} + x$

(a) $F(0) = 0$

(b) $F(2) = \frac{2^2}{4} + 2 = 3$

(c) $F(6) = \frac{6^2}{4} + 6 = 15$

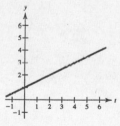

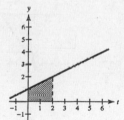

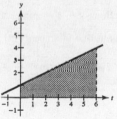

59. $F(\alpha) = \int_{-1}^{\alpha} \cos \frac{\pi \theta}{2} \, d\theta = \left[\frac{2}{\pi} \sin \frac{\pi \theta}{2} \right]_{-1}^{\alpha} = \frac{2}{\pi} \sin \frac{\pi \alpha}{2} + \frac{2}{\pi}$

(a) $F(-1) = 0$

(b) $F(0) = \frac{2}{\pi} \approx 0.6366$

(c) $F\left(\frac{1}{2} \right) = \frac{2 + \sqrt{2}}{\pi} \approx 1.0868$

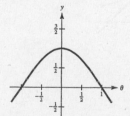

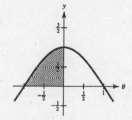

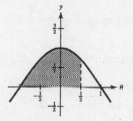

61. $A = \int_{2}^{4}\left[\left(\frac{9}{2}x - 12\right) - (x - 5)\right] dx + \int_{4}^{6}\left[\left(-\frac{5}{2}x + 16\right) - (x - 5)\right] dx$

$= \int_{2}^{4}\left(\frac{7}{2}x - 7\right) dx + \int_{4}^{6}\left(-\frac{7}{2}x + 21\right) dx$

$= \left[\frac{7}{4}x^2 - 7x\right]_{2}^{4} + \left[-\frac{7}{4}x^2 + 21x\right]_{4}^{6} = 7 + 7 = 14$

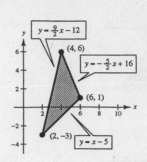

63. Left boundary line: $y = x + 2 \Longleftrightarrow x = y - 2$

Right boundary line: $y = x - 2 \Longleftrightarrow x = y + 2$

$A = \int_{-2}^{2} [(y + 2) - (y - 2)] \, dy$

$= \int_{-2}^{2} 4 \, dy = 4y \Big]_{-2}^{2} = 8 - (-8) = 16$

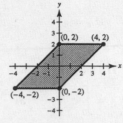

65. Answers will vary. If you let $\Delta x = 6$ and $n = 10$, $b - a = 10(6) = 60$.

(a) Area $\approx \dfrac{60}{2(10)}[0 + 2(14) + 2(14) + 2(12) + 2(12) + 2(15) + 2(20) + 2(23) + 2(25) + 2(26) + 0]$

$= 3[322] = 966$ sq ft

(b) Area $\approx \dfrac{60}{3(10)}[0 + 4(14) + 2(14) + 4(12) + 2(12) + 4(15) + 2(20) + 4(23) + 2(25) + 4(26) + 0]$

$= 2[502] = 1004$ sq ft

67. $f(x) = x^3$

$f'(x) = 3x^2$

At $(1, 1)$, $f'(1) = 3$.

Tangent line: $y - 1 = 3(x - 1)$ or $y = 3x - 2$

The tangent line intersects $f(x) = x^3$ at $x = -2$.

$A = \int_{-2}^{1} [x^3 - (3x - 2)] \, dx = \left[\frac{x^4}{4} - \frac{3x^2}{2} + 2x\right]_{-2}^{1} = \frac{27}{4}$

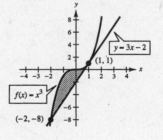

69. $f(x) = \dfrac{1}{x^2 + 1}$

$f'(x) = -\dfrac{2x}{(x^2 + 1)^2}$

At $\left(1, \dfrac{1}{2}\right)$, $f'(1) = -\dfrac{1}{2}$.

Tangent line: $y - \dfrac{1}{2} = -\dfrac{1}{2}(x - 1)$ or $y = -\dfrac{1}{2}x + 1$

The tangent line intersects $f(x) = \dfrac{1}{x^2 + 1}$ at $x = 0$.

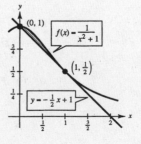

$A = \int_{0}^{1}\left[\frac{1}{x^2 + 1} - \left(-\frac{1}{2}x + 1\right)\right] dx = \left[\arctan x + \frac{x^2}{4} - x\right]_{0}^{1} = \frac{\pi - 3}{4} \approx 0.0354$

71. $x^4 - 2x^2 + 1 \leq 1 - x^2$ on $[-1, 1]$

$$A = \int_{-1}^{1} [(1 - x^2) - (x^4 - 2x^2 + 1)] \, dx$$

$$= \int_{-1}^{1} (x^2 - x^4) \, dx$$

$$= \left[\frac{x^3}{3} - \frac{x^5}{5} \right]_{-1}^{1} = \frac{4}{15}$$

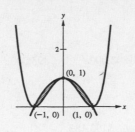

You can use a single integral because $x^4 - 2x^2 + 1 \leq 1 - x^2$ on $[-1, 1]$.

73. Offer 2 is better because the accumulated salary (area under the curve) is larger.

75.
$$A = \int_{-3}^{3} (9 - x^2) \, dx = 36$$

$$\int_{-\sqrt{9-b}}^{\sqrt{9-b}} [(9 - x^2) - b] \, dx = 18$$

$$\int_{0}^{\sqrt{9-b}} [(9 - b) - x^2] \, dx = 9$$

$$\left[(9 - b)x - \frac{x^3}{3} \right]_{0}^{\sqrt{9-b}} = 9$$

$$\frac{2}{3}(9 - b)^{3/2} = 9$$

$$(9 - b)^{3/2} = \frac{27}{2}$$

$$9 - b = \frac{9}{\sqrt[3]{4}}$$

$$b = 9 - \frac{9}{\sqrt[3]{4}} \approx 3.330$$

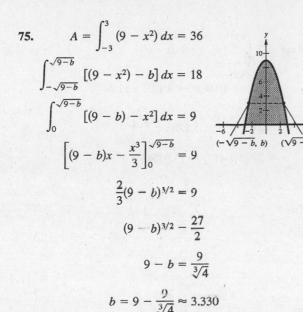

77. Area of triangle OAB is $\frac{1}{2}(4)(4) = 8$.

$$4 = \int_{0}^{a} (4 - x) \, dx = \left[4x - \frac{x^2}{2} \right]_{0}^{a} = 4a - \frac{a^2}{2}$$

$$a^2 - 8a + 8 = 0$$

$$a = 4 \pm 2\sqrt{2}$$

Since $0 < a < 4$, select $a = 4 - 2\sqrt{2} \approx 1.172$.

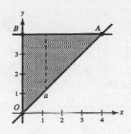

79. $\displaystyle \lim_{\|\Delta\| \to 0} \sum_{i=1}^{n} (x_i - x_i^2) \, \Delta x$

where $x_i = \dfrac{i}{n}$ and $\Delta x = \dfrac{1}{n}$ is the same as

$$\int_{0}^{1} (x - x^2) \, dx = \left[\frac{x^2}{2} - \frac{x^3}{3} \right]_{0}^{1} = \frac{1}{6}.$$

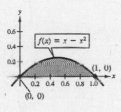

81. $\displaystyle \int_{0}^{5} [(7.21 + 0.58t) - (7.21 + 0.45t)] \, dt = \int_{0}^{5} 0.13t \, dt = \left[\frac{0.13t^2}{2} \right]_{0}^{5} = \$1.625$ billion

83. (a) $y_1 = (270.3151)(1.0586)^t = 270.3151e^{0.05695t}$

(b) $y_2 = (239.9704)(1.0416)^t = 239.9704e^{0.04074t}$

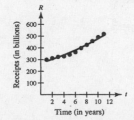

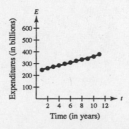

(c) Surplus $= \displaystyle \int_{12}^{17} (y_1 - y_2) \, dt \approx 926.4$ billion dollars

(Answers will vary.)

(d) No, $y_1 > y_2$ forever because $1.0586 > 1.0416$.
No, these models are not accurate for the future.
According to news, $E > R$ eventually.

85. 5%: $P_1 = 893,000e^{(0.05)t}$

$3\frac{1}{2}$%: $P_2 = 893,000e^{(0.035)t}$

Difference in profits over 5 years: $\displaystyle\int_0^5 [893,000e^{0.05t} - 893,000e^{0.035t}]\, dt = 893,000\left[\dfrac{e^{0.05t}}{0.05} - \dfrac{e^{0.035t}}{0.035}\right]_0^5$

$$\approx 893,000[(25.6805 - 34.0356) - (20 - 28.5714)]$$

$$\approx 893,000(0.2163) \approx \$193,156$$

Note: Using a graphing utility, you obtain $193,183.

87. The curves intersect at the point where the slope of y_2 equals that of y_1, 1.

$$y_2 = 0.08x^2 + k \implies y'_2 = 0.16x = 1 \implies x = \frac{1}{0.16} = 6.25$$

(a) The value of k is given by

$$y_1 = y_2$$
$$6.25 = (0.08)(6.25)^2 + k$$
$$k = 3.125.$$

(b) Area $= 2\displaystyle\int_0^{6.25} (y_2 - y_1)\, dx$

$$= 2\int_0^{6.25} (0.08x^2 + 3.125 - x)\, dx$$

$$= 2\left[\frac{0.08x^3}{3} + 3.125x - \frac{x^2}{2}\right]_0^{6.25}$$

$$= 2(6.510417) \approx 13.02083$$

89. (a) $A \approx 6.031 - 2\left[\pi\left(\dfrac{1}{16}\right)^2\right] - 2\left[\pi\left(\dfrac{1}{8}\right)^2\right] \approx 5.908$

(b) $V = 2A \approx 2(5.908) \approx 11.816 \text{ m}^3$

(c) $5000V \approx 5000(11.816) = 59,082$ pounds

91. True

93. Line: $y = \dfrac{-3}{7\pi}x$

$$A = \int_0^{7\pi/6}\left[\sin x + \frac{3x}{7\pi}\right] dx$$

$$= \left[-\cos x + \frac{3x^2}{14\pi}\right]_0^{7\pi/6}$$

$$= \frac{\sqrt{3}}{2} + \frac{7\pi}{24} + 1$$

$$\approx 2.7823$$

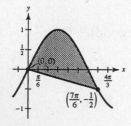

95. We want to find c such that:

$$\int_0^b [(2x - 3x^3) - c]\, dx = 0$$

$$\left[x^2 - \tfrac{3}{4}x^4 - cx\right]_0^b = 0$$

$$b^2 - \tfrac{3}{4}b^4 - cb = 0$$

But, $c = 2b - 3b^3$ because (b, c) is on the graph.

$$b^2 - \tfrac{3}{4}b^4 - (2b - 3b^3)b = 0$$

$$4 - 3b^2 - 8 + 12b^2 = 0$$

$$9b^2 = 4$$

$$b = \tfrac{2}{3}$$

$$c = \tfrac{4}{9}$$

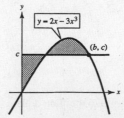

Section 7.2   Volume: The Disk Method

1. $V = \pi \int_0^1 (-x + 1)^2 \, dx = \pi \int_0^1 (x^2 - 2x + 1) \, dx = \pi \left[\dfrac{x^3}{3} - x^2 + x \right]_0^1 = \dfrac{\pi}{3}$

3. $V = \pi \int_1^4 (\sqrt{x})^2 \, dx = \pi \int_1^4 x \, dx = \pi \left[\dfrac{x^2}{2} \right]_1^4 = \dfrac{15\pi}{2}$

5. $V = \pi \int_0^1 [(x^2)^2 - (x^3)^2] \, dx = \pi \int_0^1 (x^4 - x^6) \, dx = \pi \left[\dfrac{x^5}{5} - \dfrac{x^7}{7} \right]_0^1 = \dfrac{2\pi}{35}$

7. $y = x^2 \implies x = \sqrt{y}$

$V = \pi \int_0^4 (\sqrt{y})^2 \, dy = \pi \int_0^4 y \, dy$

$= \pi \left[\dfrac{y^2}{2} \right]_0^4 = 8\pi$

9. $y = x^{2/3} \implies x = y^{3/2}$

$V = \pi \int_0^1 (y^{3/2})^2 \, dy = \pi \int_0^1 y^3 \, dy = \pi \left[\dfrac{y^4}{4} \right]_0^1 = \dfrac{\pi}{4}$

11. $y = \sqrt{x}, \; y = 0, \; x = 4$

(a) $R(x) = \sqrt{x}, \; r(x) = 0$

$V = \pi \int_0^4 (\sqrt{x})^2 \, dx$

$= \pi \int_0^4 x \, dx = \left[\dfrac{\pi}{2}x^2 \right]_0^4 = 8\pi$

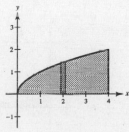

(b) $R(y) = 4, \; r(y) = y^2$

$V = \pi \int_0^2 (16 - y^4) \, dy$

$= \pi \left[16y - \dfrac{1}{5}y^5 \right]_0^2 = \dfrac{128\pi}{5}$

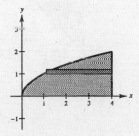

(c) $R(y) = 4 - y^2, \; r(y) = 0$

$V = \pi \int_0^2 (4 - y^2)^2 \, dy$

$= \pi \int_0^2 (16 - 8y^2 + y^4) \, dy$

$= \pi \left[16y - \dfrac{8}{3}y^3 + \dfrac{1}{5}y^5 \right]_0^2 = \dfrac{256\pi}{15}$

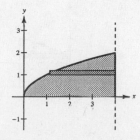

(d) $R(y) = 6 - y^2, \; r(y) = 2$

$V = \pi \int_0^2 [(6 - y^2)^2 - 4] \, dy$

$= \pi \int_0^2 (32 - 12y^2 + y^4) \, dy$

$= \pi \left[32y - 4y^3 + \dfrac{1}{5}y^5 \right]_0^2 = \dfrac{192\pi}{5}$

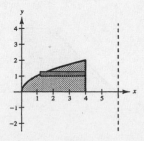

13. $y = x^2$, $y = 4x - x^2$ intersect at $(0, 0)$ and $(2, 4)$.

(a) $R(x) = 4x - x^2$, $r(x) = x^2$

$$V = \pi \int_0^2 [(4x - x^2)^2 - x^4]\, dx$$

$$= \pi \int_0^2 (16x^2 - 8x^3)\, dx$$

$$= \pi \left[\frac{16}{3}x^3 - 2x^4 \right]_0^2 = \frac{32\pi}{3}$$

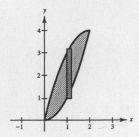

(b) $R(x) = 6 - x^2$, $r(x) = 6 - (4x - x^2)$

$$V = \pi \int_0^2 [(6 - x^2)^2 - (6 - 4x + x^2)^2]\, dx$$

$$= 8\pi \int_0^2 (x^3 - 5x^2 + 6x)\, dx$$

$$= 8\pi \left[\frac{x^4}{4} - \frac{5}{3}x^3 + 3x^2 \right]_0^2 = \frac{64\pi}{3}$$

15. $R(x) = 4 - x$, $r(x) = 1$

$$V = \pi \int_0^3 [(4 - x)^2 - (1)^2]\, dx$$

$$= \pi \int_0^3 (x^2 - 8x + 15)\, dx$$

$$= \pi \left[\frac{x^3}{3} - 4x^2 + 15x \right]_0^3$$

$$= 18\pi$$

17. $R(x) = 4$, $r(x) = 4 - \dfrac{1}{1 + x}$

$$V = \pi \int_0^3 \left[4^2 - \left(4 - \frac{1}{1 + x} \right)^2 \right] dx$$

$$= \pi \int_0^3 \left[\frac{8}{1 + x} - \frac{1}{(1 + x)^2} \right] dx$$

$$= \pi \left[8 \ln(1 + x) + \frac{1}{1 + x} \right]_0^3$$

$$= \pi \left[8 \ln 4 + \frac{1}{4} - 1 \right]$$

$$= \left(8 \ln 4 - \frac{3}{4} \right) \pi$$

$$\approx 32.485$$

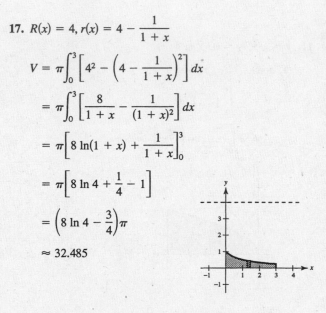

19. $R(y) = 6 - y$, $r(y) = 0$

$$V = \pi \int_0^4 (6 - y)^2\, dy$$

$$= \pi \int_0^4 (y^2 - 12y + 36)\, dy$$

$$= \pi \left[\frac{y^3}{3} - 6y^2 + 36y \right]_0^4$$

$$= \frac{208\pi}{3}$$

21. $R(y) = 6 - y^2$, $r(y) = 2$

$$V = \pi \int_{-2}^2 [(6 - y^2)^2 - (2)^2]\, dy$$

$$= 2\pi \int_0^2 (y^4 - 12y^2 + 32)\, dy$$

$$= 2\pi \left[\frac{y^5}{5} - 4y^3 + 32y \right]_0^2$$

$$= \frac{384\pi}{5}$$

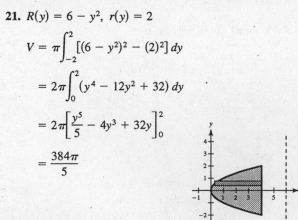

23. $R(x) = \dfrac{1}{\sqrt{x+1}},\ r(x) = 0$

$$V = \pi \int_0^3 \left(\frac{1}{\sqrt{x+1}}\right)^2 dx$$

$$= \pi \int_0^3 \frac{1}{x+1}\, dx$$

$$= \left[\, \pi \ln|x+1| \,\right]_0^3$$

$$= \pi \ln 4 \approx 4.355$$

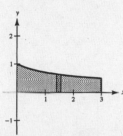

25. $R(x) = \dfrac{1}{x},\ r(x) = 0$

$$V = \pi \int_1^4 \left(\frac{1}{x}\right)^2 dx$$

$$= \pi \left[-\frac{1}{x}\right]_1^4$$

$$= \frac{3\pi}{4}$$

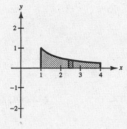

27. $R(x) = e^{-x},\ r(x) = 0$

$$V = \pi \int_0^1 (e^{-x})^2\, dx$$

$$= \pi \int_0^1 e^{-2x}\, dx$$

$$= \left[-\frac{\pi}{2} e^{-2x}\right]_0^1$$

$$= \frac{\pi}{2}(1 - e^{-2}) \approx 1.358$$

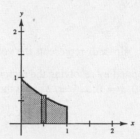

29.
$$x^2 + 1 = -x^2 + 2x + 5$$
$$2x^2 - 2x - 4 = 0$$
$$x^2 - x - 2 = 0$$
$$(x - 2)(x + 1) = 0$$

The curves intersect at $(-1, 2)$ and $(2, 5)$.

$$V = \pi \int_0^2 \left[(5 + 2x - x^2)^2 - (x^2 + 1)^2\right] dx + \pi \int_2^3 \left[(x^2 + 1)^2 - (5 + 2x - x^2)^2\right] dx$$

$$= \pi \int_0^2 (-4x^3 - 8x^2 + 20x + 24)\, dx + \pi \int_2^3 (4x^3 + 8x^2 - 20x - 24)\, dx$$

$$= \pi \left[-x^4 - \frac{8}{3}x^3 + 10x^2 + 24x\right]_0^2 + \pi \left[x^4 + \frac{8}{3}x^3 - 10x^2 - 24x\right]_2^3$$

$$= \pi \frac{152}{3} + \pi \frac{125}{3} = \frac{277\pi}{3}$$

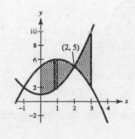

31. $y = 6 - 3x \implies x = \dfrac{1}{3}(6 - y)$

$$V = \pi \int_0^6 \left[\frac{1}{3}(6 - y)\right]^2 dy$$

$$= \frac{\pi}{9} \int_0^6 [36 - 12y + y^2]\, dy$$

$$= \frac{\pi}{9}\left[36y - 6y^2 + \frac{y^3}{3}\right]_0^6$$

$$= \frac{\pi}{9}\left[216 - 216 + \frac{216}{3}\right]$$

$$= 8\pi = \frac{1}{3}\pi r^2 h, \quad \text{Volume of cone}$$

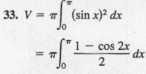

33. $V = \pi \displaystyle\int_0^\pi (\sin x)^2\, dx$

$$= \pi \int_0^\pi \frac{1 - \cos 2x}{2}\, dx$$

$$= \frac{\pi}{2}\left[x - \frac{1}{2}\sin 2x\right]_0^\pi$$

$$= \frac{\pi}{2}[\pi] = \frac{\pi^2}{2}$$

Numerical approximation: 4.9348

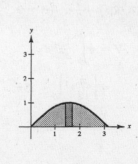

35. $V = \pi \int_1^2 (e^{x-1})^2 \, dx$

$= \pi \int_1^2 e^{2x-2} \, dx$

$= \frac{\pi}{2} e^{2x-2} \Big]_1^2$

$= \frac{\pi}{2}(e^2 - 1)$

Numerical approximation:
10.0359

37. $V = \pi \int_0^2 [e^{-x^2}]^2 \, dx \approx 1.9686$

39. $V = \pi \int_0^5 [2 \arctan(0.2x)]^2 \, dx$

≈ 15.4115

41. $\pi \int_0^{\pi/2} \sin^2 x \, dx$ represents the volume of the solid generated by revolving the region bounded by $y = \sin x$, $y = 0$, $x = 0$, $x = \pi/2$ about the x-axis.

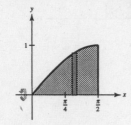

43. $A \approx 3$

Matches (a)

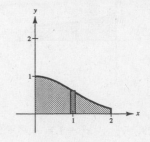

45.

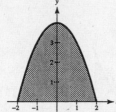   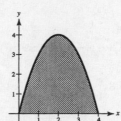

The volumes are the same because the solid has been translated horizontally. $(4x - x^2 = 4 - (x - 2)^2)$

47. $R(x) = \frac{1}{2}x, \ r(x) = 0$

$V = \pi \int_0^6 \frac{1}{4} x^2 \, dx$

$= \left[\frac{\pi}{12} x^3 \right]_0^6 = 18\pi$

Note: $V = \frac{1}{3} \pi r^2 h$

$= \frac{1}{3} \pi (3^2) 6$

$= 18\pi$

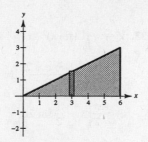

49. $R(x) = \sqrt{r^2 - x^2}, \ r(x) = 0$

$V = \pi \int_{-r}^{r} (r^2 - x^2) \, dx$

$= 2\pi \int_0^r (r^2 - x^2) \, dx$

$= 2\pi \left[r^2 x - \frac{1}{3} x^3 \right]_0^r$

$= 2\pi \left(r^3 - \frac{1}{3} r^3 \right) = \frac{4}{3} \pi r^3$

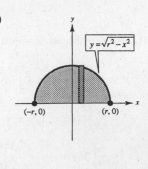

51. $x = r - \dfrac{r}{H}y = r\left(1 - \dfrac{y}{H}\right),\; R(y) = r\left(1 - \dfrac{y}{H}\right),\; r(y) = 0$

$$V = \pi\int_0^h \left[r\left(1 - \dfrac{y}{H}\right)\right]^2 dy = \pi r^2 \int_0^h \left(1 - \dfrac{2}{H}y + \dfrac{1}{H^2}y^2\right) dy$$

$$= \pi r^2\left[y - \dfrac{1}{H}y^2 + \dfrac{1}{3H^2}y^3\right]_0^h$$

$$= \pi r^2\left(h - \dfrac{h^2}{H} + \dfrac{h^3}{3H^2}\right)$$

$$= \pi r^2 h\left(1 - \dfrac{h}{H} + \dfrac{h^2}{3H^2}\right)$$

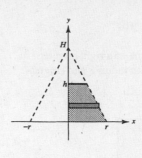

53. $V = \pi\int_0^2 \left(\dfrac{1}{8}x^2\sqrt{2-x}\right)^2 dx = \dfrac{\pi}{64}\int_0^2 x^4(2-x)\, dx = \dfrac{\pi}{64}\left[\dfrac{2x^5}{5} - \dfrac{x^6}{6}\right]_0^2 = \dfrac{\pi}{30}$

55. (a) $R(x) = \dfrac{3}{5}\sqrt{25 - x^2},\; r(x) = 0$

$$V = \dfrac{9\pi}{25}\int_{-5}^5 (25 - x^2)\, dx$$

$$= \dfrac{18\pi}{25}\int_0^5 (25 - x^2)\, dx$$

$$= \dfrac{18\pi}{25}\left[25x - \dfrac{x^3}{3}\right]_0^5$$

$$= 60\pi$$

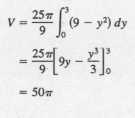

(b) $R(y) = \dfrac{5}{3}\sqrt{9 - y^2},\; r(y) = 0,\; x \geq 0$

$$V = \dfrac{25\pi}{9}\int_0^3 (9 - y^2)\, dy$$

$$= \dfrac{25\pi}{9}\left[9y - \dfrac{y^3}{3}\right]_0^3$$

$$= 50\pi$$

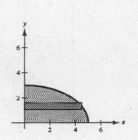

57. Total volume: $V = \dfrac{4\pi(50)^3}{3} = \dfrac{500,000\pi}{3}\; \text{ft}^3$

Volume of water in the tank:

$$\pi\int_{-50}^{y_0} \left(\sqrt{2500 - y^2}\right)^2 dy - \pi\int_{-50}^{y_0} (2500 - y^2)\, dy$$

$$= \pi\left[2500y - \dfrac{y^3}{3}\right]_{-50}^{y_0}$$

$$= \pi\left(2500y_0 - \dfrac{y_0^3}{3} + \dfrac{250,000}{3}\right)$$

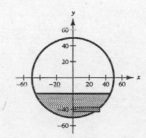

When the tank is one-fourth of its capacity:

$$\dfrac{1}{4}\left(\dfrac{500,000\pi}{3}\right) = \pi\left(2500y_0 - \dfrac{y_0^3}{3} + \dfrac{250,000}{3}\right)$$

$$125,000 = 7500y_0 - y_0^3 + 250,000$$

$$y_0^3 - 7500y_0 - 125,000 = 0$$

$$y_0 \approx -17.36$$

Depth: $-17.36 - (-50) = 32.64\ \text{feet}$

When the tank is three-fourths of its capacity the depth is $100 - 32.64 = 67.36$ feet.

59. (a) $\pi \displaystyle\int_0^h r^2 \, dx$ (ii)

is the volume of a right circular cylinder with radius r and height h.

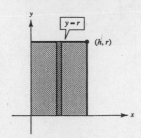

(b) $\pi \displaystyle\int_{-b}^b \left(a\sqrt{1 - \dfrac{x^2}{b^2}} \right)^2 dx$ (iv)

is the volume of an ellipsoid with axes $2a$ and $2b$.

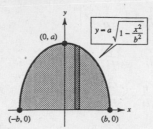

(c) $\pi \displaystyle\int_{-r}^r \left(\sqrt{r^2 - x^2} \right)^2 dx$ (iii)

is the volume of a sphere with radius r.

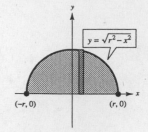

(d) $\pi \displaystyle\int_0^h \left(\dfrac{rx}{h} \right)^2 dx$ (i)

is the volume of a right circular cone with the radius of the base as r and height h.

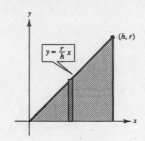

(e) $\pi \displaystyle\int_{-r}^r \left[\left(R + \sqrt{r^2 - x^2} \right)^2 - \left(R - \sqrt{r^2 - x^2} \right)^2 \right] dx$ (v)

is the volume of a torus with the radius of its circular cross section as r and the distance from the axis of the torus to the center of its cross section as R.

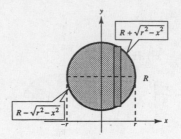

61.

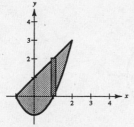

Base of cross section $= (x + 1) - (x^2 - 1) = 2 + x - x^2$

(a) $A(x) = b^2 = (2 + x - x^2)^2$

$\qquad = 4 + 4x - 3x^2 - 2x^3 + x^4$

$\quad V = \displaystyle\int_{-1}^2 (4 + 4x - 3x^2 - 2x^3 + x^4) \, dx$

$\qquad = \left[4x + 2x^2 - x^3 - \dfrac{1}{2}x^4 + \dfrac{1}{5}x^5 \right]_{-1}^2 = \dfrac{81}{10}$

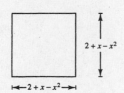

(b) $A(x) = bh = (2 + x - x^2)1$

$\quad V = \displaystyle\int_{-1}^2 (2 + x - x^2) \, dx = \left[2x + \dfrac{x^2}{2} - \dfrac{x^3}{3} \right]_{-1}^2 = \dfrac{9}{2}$

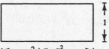

63.

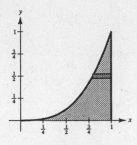

Base of cross section $= 1 - \sqrt[3]{y}$

(a) $A(y) = b^2 = \left(1 - \sqrt[3]{y}\right)^2$

$$V = \int_0^1 \left(1 - \sqrt[3]{y}\right)^2 dy$$

$$= \int_0^1 (1 - 2y^{1/3} + y^{2/3}) \, dy$$

$$= \left[y - \frac{3}{2}y^{4/3} + \frac{3}{5}y^{5/3}\right]_0^1 = \frac{1}{10}$$

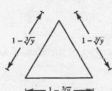

(c) $A(y) = \frac{1}{2}bh = \frac{1}{2}(1 - \sqrt[3]{y})\left(\frac{\sqrt{3}}{2}\right)(1 - \sqrt[3]{y})$

$$= \frac{\sqrt{3}}{4}\left(1 - \sqrt[3]{y}\right)^2$$

$$V = \frac{\sqrt{3}}{4}\int_0^1 \left(1 - \sqrt[3]{y}\right)^2 dy = \frac{\sqrt{3}}{4}\left(\frac{1}{10}\right) = \frac{\sqrt{3}}{40}$$

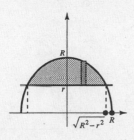

(b) $A(y) = \frac{1}{2}\pi r^2 = \frac{1}{2}\pi\left(\frac{1 - \sqrt[3]{y}}{2}\right)^2 = \frac{1}{8}\pi\left(1 - \sqrt[3]{y}\right)^2$

$$V = \frac{1}{8}\pi\int_0^1 \left(1 - \sqrt[3]{y}\right)^2 dy = \frac{\pi}{8}\left(\frac{1}{10}\right) = \frac{\pi}{80}$$

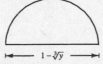

(d) $A(y) = \frac{1}{2}\pi ab = \frac{\pi}{2}(2)(1 - \sqrt[3]{y})\frac{1 - \sqrt[3]{y}}{2} = \frac{\pi}{2}\left(1 - \sqrt[3]{y}\right)^2$

$$V = \frac{\pi}{2}\int_0^1 \left(1 - \sqrt[3]{y}\right)^2 dy = \frac{\pi}{2}\left(\frac{1}{10}\right) = \frac{\pi}{20}$$

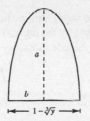

65. $V = \pi\displaystyle\int_{-\sqrt{R^2-r^2}}^{\sqrt{R^2-r^2}} \left[\left(\sqrt{R^2-x^2}\right)^2 - r^2\right] dx$

$$= 2\pi\int_0^{\sqrt{R^2-r^2}} (R^2 - r^2 - x^2) \, dx$$

$$= 2\pi\left[(R^2 - r^2)x - \frac{x^3}{3}\right]_0^{\sqrt{R^2-r^2}}$$

$$= 2\pi\left[(R^2 - r^2)^{3/2} - \frac{(R^2 - r^2)^{3/2}}{3}\right]$$

$$= \frac{4}{3}\pi(R^2 - r^2)^{3/2}$$

67. $V = \pi\displaystyle\int_0^1 y^2 \, dy = \pi\frac{y^3}{3}\bigg]_0^1 = \frac{\pi}{3}$

69. $V = \pi \displaystyle\int_0^1 (x^2 - x^4)\,dx$

$= \pi \left[\dfrac{x^3}{3} - \dfrac{x^5}{5} \right]_0^1$

$= \pi \left[\dfrac{1}{3} - \dfrac{1}{5} \right]$

$= \dfrac{2\pi}{15}$

71. $V = \pi \displaystyle\int_0^1 (1 - y)\,dy$

$= \pi \left[y - \dfrac{y^2}{2} \right]_0^1$

$= \pi \left[1 - \dfrac{1}{2} \right]$

$= \dfrac{\pi}{2}$

73. $V = \pi \displaystyle\int_0^1 (y - y^2)\,dy$

$= \pi \left[\dfrac{y^2}{2} - \dfrac{y^3}{3} \right]_0^1$

$= \pi \left[\dfrac{1}{2} - \dfrac{1}{3} \right]$

$= \dfrac{\pi}{6}$

75. (a) When $a = 1$: $|x| + |y| = 1$ represents a square.

When $a = 2$: $|x|^2 + |y|^2 = 1$ represents a circle.

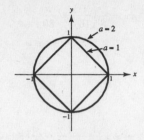

(b) $|y| = (1 - |x|^a)^{1/a}$

$A = 2 \displaystyle\int_{-1}^1 (1 - |x|^a)^{1/a}\,dx = 4 \int_0^1 (1 - x^a)^{1/a}\,dx$

To approximate the volume of the solid, form n slices, each of whose area is approximated by the integral above. Then sum the volumes of these n slices.

77. (a) $(x - R)^2 + y^2 = r^2$

$x = R \pm \sqrt{r^2 - y^2}$

$V = 2\pi \displaystyle\int_0^r \left(\left[R + \sqrt{r^2 - y^2} \right]^2 - \left[R - \sqrt{r^2 - y^2} \right]^2 \right) dy$

$= 2\pi \displaystyle\int_0^r 4R \sqrt{r^2 - y^2}\,dy$

$= 8\pi R \displaystyle\int_0^r \sqrt{r^2 - y^2}\,dy$

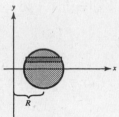

(b) $\displaystyle\int_0^r \sqrt{r^2 - y^2}\,dy$ is one-quarter of the area of a circle of radius r, $\tfrac{1}{4}\pi r^2$.

$V = 8\pi R \left(\tfrac{1}{4}\pi r^2 \right) = 2\pi^2 r^2 R$

Section 7.3   Volume: The Shell Method

1. $p(x) = x$, $h(x) = x$

$V = 2\pi \displaystyle\int_0^2 x(x)\,dx$

$= \left[\dfrac{2\pi x^3}{3} \right]_0^2 = \dfrac{16\pi}{3}$

3. $p(x) = x$, $h(x) = \sqrt{x}$

$V = 2\pi \displaystyle\int_0^4 x\sqrt{x}\,dx$

$= 2\pi \displaystyle\int_0^4 x^{3/2}\,dx$

$= \left[\dfrac{4\pi}{5} x^{5/2} \right]_0^4 = \dfrac{128\pi}{5}$

5. $p(x) = x$, $h(x) = x^2$

$V = 2\pi \displaystyle\int_0^2 x^3\,dx$

$= \left[\dfrac{\pi}{2} x^4 \right]_0^2 = 8\pi$

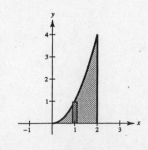

7. $p(x) = x$, $h(x) = (4x - x^2) - x^2 = 4x - 2x^2$

$V = 2\pi \displaystyle\int_0^2 x(4x - 2x^2)\,dx$

$= 4\pi \displaystyle\int_0^2 (2x^2 - x^3)\,dx$

$= 4\pi \left[\dfrac{2}{3} x^3 - \dfrac{1}{4} x^4 \right]_0^2 = \dfrac{16\pi}{3}$

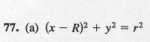

9. $p(x) = x$

$\quad h(x) = 4 - (4x - x^2)$

$\qquad = x^2 - 4x + 4$

$\quad V = 2\pi \int_0^2 (x^3 - 4x^2 + 4x)\, dx$

$\qquad = 2\pi \left[\dfrac{x^4}{4} - \dfrac{4}{3}x^3 + 2x^2 \right]_0^2$

$\qquad = \dfrac{8\pi}{3}$

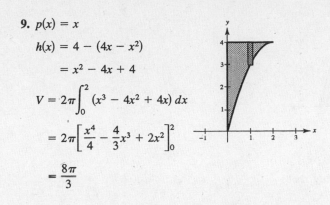

11. $p(x) = x,\ h(x) = \dfrac{1}{\sqrt{2\pi}} e^{-x^2/2}$

$\quad V = 2\pi \int_0^1 x \left(\dfrac{1}{\sqrt{2\pi}} e^{-x^2/2} \right) dx$

$\qquad = \sqrt{2\pi} \int_0^1 e^{-x^2/2} x\, dx$

$\qquad = \left[-\sqrt{2\pi}\, e^{-x^2/2} \right]_0^1$

$\qquad = \sqrt{2\pi} \left(1 - \dfrac{1}{\sqrt{e}} \right)$

$\qquad \approx 0.986$

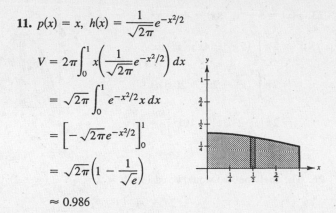

13. $p(y) = y,\ h(y) = 2 - y$

$\quad V = 2\pi \int_0^2 y(2 - y)\, dy$

$\qquad = 2\pi \int_0^2 (2y - y^2)\, dy$

$\qquad = 2\pi \left[y^2 - \dfrac{y^3}{3} \right]_0^2 = \dfrac{8\pi}{3}$

15. $p(y) = y$ and $h(y) = 1$ if $0 \le y < \dfrac{1}{2}$.

$\quad p(y) = y$ and $h(y) = \dfrac{1}{y} - 1$ if $\dfrac{1}{2} \le y \le 1$.

$\quad V = 2\pi \int_0^{1/2} y\, dy + 2\pi \int_{1/2}^1 (1 - y)\, dy$

$\qquad = 2\pi \left[\dfrac{y^2}{2} \right]_0^{1/2} + 2\pi \left[y - \dfrac{y^2}{2} \right]_{1/2}^1 = \dfrac{\pi}{4} + \dfrac{\pi}{4} = \dfrac{\pi}{2}$

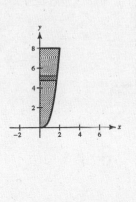

17. $p(y) = y,\ h(y) = \sqrt[3]{y}$

$\quad V = 2\pi \int_0^8 y \sqrt[3]{y}\, dy$

$\qquad = 2\pi \int_0^8 y^{4/3}\, dy$

$\qquad = \left[2\pi \left(\dfrac{3}{7} \right) y^{7/3} \right]_0^8$

$\qquad = \dfrac{6\pi}{7}(2^7) = \dfrac{768\pi}{7}$

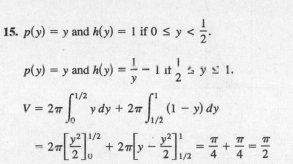

19. $p(y) = y,\ h(y) = (4 - y) - (y) = 4 - 2y$

$\quad V = 2\pi \int_0^2 y(4 - 2y)\, dy$

$\qquad = 2\pi \int_0^2 (4y - 2y^2)\, dy$

$\qquad = 2\pi \left[2y^2 - \dfrac{2}{3}y^3 \right]_0^2$

$\qquad = 2\pi \left[8 - \dfrac{16}{3} \right] = \dfrac{16\pi}{3}$

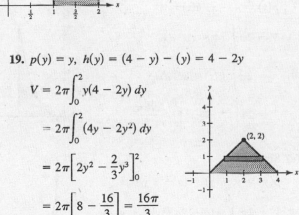

21. $p(x) = 4 - x,\ h(x) = 4x - x^2 - x^2 = 4x - 2x^2$

$\quad V = 2\pi \int_0^2 (4 - x)(4x - 2x^2)\, dx$

$\qquad = 2\pi(2) \int_0^2 (x^3 - 6x^2 + 8x)\, dx$

$\qquad = 4\pi \left[\dfrac{x^4}{4} - 2x^3 + 4x^2 \right]_0^2 = 16\pi$

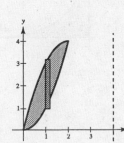

23. $p(x) = 5 - x, \; h(x) = 4x - x^2$

$$V = 2\pi \int_0^4 (5 - x)(4x - x^2)\, dx$$

$$= 2\pi \int_0^4 (x^3 - 9x^2 + 20x)\, dx$$

$$= 2\pi \left[\frac{x^4}{4} - 3x^3 + 10x^2 \right]_0^4 = 64\pi$$

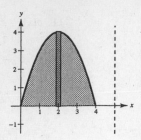

25. The shell method would be easier: $V = 2\pi \int_0^4 [4 - (y - 2)^2]\, y \, dy$ shells

Using the disk method: $V = \pi \int_0^4 \left[\left(2 + \sqrt{4 - x}\right)^2 - \left(2 - \sqrt{4 - x}\right)^2 \right] dx$ $\quad \left[\text{Note: } V = \dfrac{128\pi}{3} \right]$

27. (a) **Disk**

$R(x) = x^3, \; r(x) = 0$

$$V = \pi \int_0^2 x^6 \, dx = \pi \left[\frac{x^7}{7} \right]_0^2 = \frac{128\pi}{7}$$

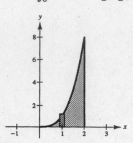

(b) **Shell**

$p(x) = x, \; h(x) = x^3$

$$V = 2\pi \int_0^2 x^4 \, dx = 2\pi \left[\frac{x^5}{5} \right]_0^2 = \frac{64\pi}{5}$$

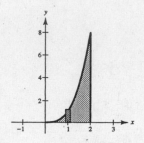

(c) **Shell**

$p(x) = 4 - x, \; h(x) = x^3$

$$V = 2\pi \int_0^2 (4 - x)x^3 \, dx$$

$$= 2\pi \int_0^2 (4x^3 - x^4)\, dx$$

$$= 2\pi \left[x^4 - \frac{1}{5}x^5 \right]_0^2 = \frac{96\pi}{5}$$

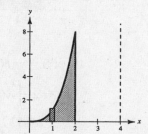

29. (a) **Shell**

$p(y) = y, \; h(y) = (a^{1/2} - y^{1/2})^2$

$$V = 2\pi \int_0^a y(a - 2a^{1/2}y^{1/2} + y)\, dy$$

$$= 2\pi \int_0^a (ay - 2a^{1/2}y^{3/2} + y^2)\, dy$$

$$= 2\pi \left[\frac{a}{2}y^2 - \frac{4a^{1/2}}{5}y^{5/2} + \frac{y^3}{3} \right]_0^a$$

$$= 2\pi \left[\frac{a^3}{2} - \frac{4a^3}{5} + \frac{a^3}{3} \right] = \frac{\pi a^3}{15}$$

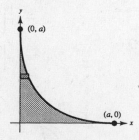

(b) Same as part (a) by symmetry

—**CONTINUED**—

29. —CONTINUED—

(c) **Shell**

$$p(x) = a - x, \ h(x) = (a^{1/2} - x^{1/2})^2$$

$$V = 2\pi \int_0^a (a - x)(a^{1/2} - x^{1/2})^2 \, dx$$

$$= 2\pi \int_0^a (a^2 - 2a^{3/2}x^{1/2} + 2a^{1/2}x^{3/2} - x^2) \, dx$$

$$= 2\pi \left[a^2x - \frac{4}{3}a^{3/2}x^{3/2} + \frac{4}{5}a^{1/2}x^{5/2} - \frac{1}{3}x^3 \right]_0^a = \frac{4\pi a^3}{15}$$

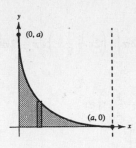

31. Answers will vary.   (a) The rectangles would be vertical.   (b) The rectangles would be horizontal.

33. $\pi \int_1^5 (x - 1) \, dx = \pi \int_1^5 \left(\sqrt{x - 1} \right)^2 dx$

This integral represents the volume of the solid generated by revolving the region bounded by $y = \sqrt{x - 1}, y = 0$, and $x = 5$ about the x-axis by using the disk method.

$$2\pi \int_0^2 y[5 - (y^2 + 1)] \, dy$$

represents this same volume by using the shell method.

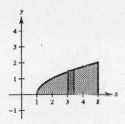

Disk method

35. (a)

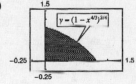

(b) $x^{4/3} + y^{4/3} = 1, x = 0, y = 0$

$y = (1 - x^{4/3})^{3/4}$

$V = 2\pi \int_0^1 x(1 - x^{4/3})^{3/4} \, dx \approx 1.5056$

37. (a)

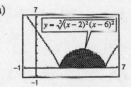

(b) $V = 2\pi \int_2^6 x\sqrt[3]{(x - 2)^2(x - 6)^2} \, dx \approx 187.240$

39. $y = 2e^{-x}, y = 0, x = 0, x = 2$

Volume ≈ 7.5

Matches (d)

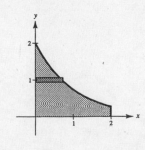

41. $p(x) = x, \ h(x) = 2 - \dfrac{1}{2}x^2$

$$V = 2\pi \int_0^2 x\left(2 - \frac{1}{2}x^2\right) dx = 2\pi \int_0^2 \left(2x - \frac{1}{2}x^3\right) dx = 2\pi\left[x^2 - \frac{1}{8}x^4\right]_0^2 = 4\pi \quad \text{(total volume)}$$

Now find x_0 such that:

$$\pi = 2\pi \int_0^{x_0} \left(2x - \frac{1}{2}x^3\right) dx$$

$$1 = 2\left[x^2 - \frac{1}{8}x^4\right]_0^{x_0}$$

$$1 = 2x_0^2 - \frac{1}{4}x_0^4$$

$x_0^4 - 8x_0^2 + 4 = 0$

$$x_0^2 = 4 \pm 2\sqrt{3} \qquad \text{(Quadratic Formula)}$$

Take $x_0 = \sqrt{4 - 2\sqrt{3}} \approx 0.73205$, since the other root is too large.

Diameter: $2\sqrt{4 - 2\sqrt{3}} \approx 1.464$

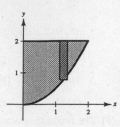

43. $V = 4\pi \displaystyle\int_{-1}^1 (2 - x)\sqrt{1 - x^2}\, dx$

$$= 8\pi \int_{-1}^1 \sqrt{1 - x^2}\, dx - 4\pi \int_{-1}^1 x\sqrt{1 - x^2}\, dx$$

$$= 8\pi\left(\frac{\pi}{2}\right) + 2\pi \int_{-1}^1 x(1 - x^2)^{1/2}(-2)\, dx$$

$$= 4\pi^2 + \left[2\pi\left(\frac{2}{3}\right)(1 - x^2)^{3/2}\right]_{-1}^1 = 4\pi^2$$

45. (a) $\dfrac{d}{dx}[\sin x - x\cos x + C] = \cos x + x\sin x - \cos x = x\sin x$

Hence, $\displaystyle\int x\sin x\, dx = \sin x - x\cos x + C.$

(b) (i) $p(x) = x, h(x) = \sin x$

$$V = 2\pi \int_0^{\pi/2} x\sin x\, dx$$

$$= 2\pi\left[\sin x - x\cos x\right]_0^{\pi/2}$$

$$= 2\pi[(1 - 0) - 0] = 2\pi$$

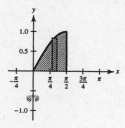

(ii) $p(x) = x, h(x) = 2\sin x - (-\sin x) = 3\sin x$

$$V = 2\pi \int_0^{\pi} x(3\sin x)\, dx$$

$$= 6\pi \int_0^{\pi} x\sin x\, dx$$

$$= 6\pi\left[\sin x - x\cos x\right]_0^{\pi}$$

$$= 6\pi[\pi] = 6\pi^2$$

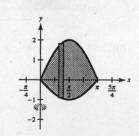

47. $2\pi \int_0^2 x^3 \, dx = 2\pi \int_0^2 x(x^2) \, dx$

    (a) Plane region bounded by $y = x^2, y = 0, x = 0, x = 2$

    (b) Revolved about the y-axis

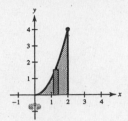

Other answers possible

49. $2\pi \int_0^6 (y + 2)\sqrt{6 - y} \, dy$

    (a) Plane region bounded by $x = \sqrt{6 - y}, x = 0, y = 0$

    (b) Revolved around line $y = -2$

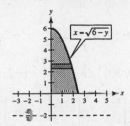

Other answers possible

51. Disk Method

$R(y) = \sqrt{r^2 - y^2}$

$r(y) = 0$

$V = \pi \int_{r-h}^r (r^2 - y^2) \, dy$

$\quad = \pi \left[r^2 y - \dfrac{y^3}{3} \right]_{r-h}^r = \dfrac{1}{3}\pi h^2 (3r - h)$

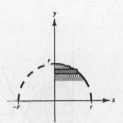

53. (a) Area region $= \displaystyle\int_0^b [ab^n - ax^n] \, dx$

$\quad = \left[ab^n x - a\dfrac{x^{n+1}}{n+1} \right]_0^b$

$\quad = ab^{n+1} - a\dfrac{b^{n+1}}{n+1}$

$\quad = ab^{n+1}\left(1 - \dfrac{1}{n+1} \right) = ab^{n+1}\left(\dfrac{n}{n+1} \right)$

$\quad R_1(n) = \dfrac{ab^{n+1}[n/(n+1)]}{(ab^n)b} = \dfrac{n}{n+1}$

(b) $\displaystyle\lim_{n \to \infty} R_1(n) = \lim_{n \to \infty} \dfrac{n}{n+1} = 1$

$\quad \displaystyle\lim_{n \to \infty} (ab^n)b = \infty$

(c) **Disk Method:**

$V = 2\pi \displaystyle\int_0^b x(ab^n - ax^n) \, dx$

$\quad = 2\pi a \displaystyle\int_0^b (xb^n - x^{n+1}) \, dx$

$\quad = 2\pi a \left[\dfrac{b^n}{2}x^2 - \dfrac{x^{n+2}}{n+2} \right]_0^b$

$\quad = 2\pi a \left[\dfrac{b^{n+2}}{2} - \dfrac{b^{n+2}}{n+2} \right] = \pi ab^{n+2}\left(\dfrac{n}{n+2} \right)$

$\quad R_2(n) = \dfrac{\pi ab^{n+2}[n/(n+2)]}{(\pi b^2)(ab^n)} = \left(\dfrac{n}{n+2} \right)$

(d) $\displaystyle\lim_{n \to \infty} R_2(n) = \lim_{n \to \infty} \left(\dfrac{n}{n+2} \right) = 1$

$\quad \displaystyle\lim_{n \to \infty} (\pi b^2)(ab^n) = \infty$

(e) As $n \to \infty$, the graph approaches the line $x = 1$.

55. (a) $V = 2\pi \displaystyle\int_0^4 xf(x) \, dx$

$\quad = \dfrac{2\pi(40)}{3(4)}[0 + 4(10)(45) + 2(20)(40) + 4(30)(20) + 0]$

$\quad = \dfrac{20\pi}{3}[5800] \approx 121{,}475 \text{ cubic feet}$

—CONTINUED—

55. —CONTINUED—

(b) Top line: $y - 50 = \dfrac{40 - 50}{20 - 0}(x - 0) = -\dfrac{1}{2}x \Rightarrow y = -\dfrac{1}{2}x + 50$

Bottom line: $y - 40 = \dfrac{0 - 40}{40 - 20}(x - 20) = -2(x - 20) \Rightarrow y = -2x + 80$

$$V = 2\pi \int_0^{20} x\left(-\dfrac{1}{2}x + 50\right) dx + 2\pi \int_{20}^{40} x(-2x + 80)\, dx$$

$$= 2\pi \int_0^{20} \left(-\dfrac{1}{2}x^2 + 50x\right) dx + 2\pi \int_{20}^{40} (-2x^2 + 80x)\, dx$$

$$= 2\pi\left[-\dfrac{x^3}{6} + 25x^2\right]_0^{20} + 2\pi\left[-\dfrac{2x^3}{3} + 40x^2\right]_{20}^{40}$$

$$= 2\pi\left[\dfrac{26{,}000}{3}\right] + 2\pi\left[\dfrac{32{,}000}{3}\right]$$

$$\approx 121{,}475 \text{ cubic feet}$$

(Note that Simpson's Rule is exact for this problem.)

57. $y^2 = x(4 - x)^2, \quad 0 \le x \le 4$

$y_1 = \sqrt{x(4 - x)^2} = (4 - x)\sqrt{x}$

$y_2 = -\sqrt{x(4 - x)^2} = -(4 - x)\sqrt{x}$

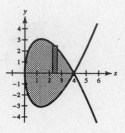

(a) $V = \pi \displaystyle\int_0^4 x(4 - x)^2\, dx$

$$= \pi \int_0^4 (x^3 - 8x^2 + 16x)\, dx$$

$$= \pi\left[\dfrac{x^4}{4} - \dfrac{8x^3}{3} + 8x^2\right]_0^4 = \dfrac{64\pi}{3}$$

(b) $V = 4\pi \displaystyle\int_0^4 x(4 - x)\sqrt{x}\, dx$

$$= 4\pi \int_0^4 (4x^{3/2} - x^{5/2})\, dx$$

$$= 4\pi\left[\dfrac{8}{5}x^{5/2} - \dfrac{2}{7}x^{7/2}\right]_0^4 = \dfrac{2048\pi}{35}$$

(c) $V = 4\pi \displaystyle\int_0^4 (4 - x)(4 - x)\sqrt{x}\, dx$

$$= 4\pi \int_0^4 \left(16\sqrt{x} - 8x^{3/2} + x^{5/2}\right) dx$$

$$= 4\pi\left[\dfrac{32}{3}x^{3/2} - \dfrac{16}{5}x^{5/2} + \dfrac{2}{7}x^{7/2}\right]_0^4 = \dfrac{8192\pi}{105}$$

59. $V_1 = \pi \displaystyle\int_{1/4}^c \dfrac{1}{x^2}\, dx = \pi\left[-\dfrac{1}{x}\right]_{1/4}^c = \pi\left[-\dfrac{1}{c} + 4\right] = \dfrac{4c - 1}{c}\pi$

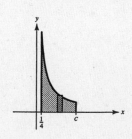

$V_2 = 2\pi \displaystyle\int_{1/4}^c x\left(\dfrac{1}{x}\right) dx = 2\pi x\Big]_{1/4}^c = 2\pi\left(c - \dfrac{1}{4}\right)$

$V_1 = V_2 \Rightarrow \dfrac{4c - 1}{c}\pi = 2\pi\left(c - \dfrac{1}{4}\right)$

$$4c - 1 = 2c\left(c - \dfrac{1}{4}\right)$$

$$4c^2 - 9c + 2 = 0$$

$$(4c - 1)(c - 2) = 0$$

$$c = 2 \quad \left(c = \dfrac{1}{4} \text{ yields no volume.}\right)$$

Section 7.4 Arc Length and Surfaces of Revolution

1. $(0, 0)$, $(5, 12)$

(a) $d = \sqrt{(5 - 0)^2 + (12 - 0)^2}$

$= 13$

(b) $y = \dfrac{12}{5}x$

$y' = \dfrac{12}{5}$

$s = \displaystyle\int_0^5 \sqrt{1 + \left(\dfrac{12}{5}\right)^2}\, dx$

$= \left[\dfrac{13}{5}x\right]_0^5 = 13$

3. $y = \dfrac{2}{3}x^{3/2} + 1$

$y' = x^{1/2}, \quad 0 \le x \le 1$

$s = \displaystyle\int_0^1 \sqrt{1 + x}\, dx$

$= \left[\dfrac{2}{3}(1 + x)^{3/2}\right]_0^1$

$= \dfrac{2}{3}\left(\sqrt{8} - 1\right) \approx 1.219$

5. $y = \dfrac{3}{2}x^{2/3}$

$y' = \dfrac{1}{x^{1/3}}, \quad 1 \le x \le 8$

$s = \displaystyle\int_1^8 \sqrt{1 + \left(\dfrac{1}{x^{1/3}}\right)^2}\, dx$

$= \displaystyle\int_1^8 \sqrt{\dfrac{x^{2/3} + 1}{x^{2/3}}}\, dx$

$= \dfrac{3}{2}\displaystyle\int_1^8 \sqrt{x^{2/3} + 1}\left(\dfrac{2}{3x^{1/3}}\right) dx$

$= \dfrac{3}{2}\left[\dfrac{2}{3}(x^{2/3} + 1)^{3/2}\right]_1^8$

$= 5\sqrt{5} - 2\sqrt{2} \approx 8.352$

7. $y = \dfrac{x^5}{10} + \dfrac{1}{6x^3}$

$y' = \dfrac{1}{2}x^4 - \dfrac{1}{2x^4}$

$1 + (y')^2 = \left(\dfrac{1}{2}x^4 + \dfrac{1}{2x^4}\right)^2, \quad 1 \le x \le 2$

$s = \displaystyle\int_a^b \sqrt{1 + (y')^2}\, dx$

$= \displaystyle\int_1^2 \sqrt{\left(\dfrac{1}{2}x^4 + \dfrac{1}{2x^4}\right)^2}\, dx$

$= \displaystyle\int_1^2 \left(\dfrac{1}{2}x^4 + \dfrac{1}{2x^4}\right) dx$

$= \left[\dfrac{1}{10}x^5 - \dfrac{1}{6x^3}\right]_1^2 = \dfrac{779}{240} \approx 3.2458$

9. $y = \ln(\sin x), \quad \left[\dfrac{\pi}{4}, \dfrac{3\pi}{4}\right]$

$y' = \dfrac{1}{\sin x}\cos x = \cot x$

$1 + (y')^2 = 1 + \cot^2 x = \csc^2 x$

$s = \displaystyle\int_{\pi/4}^{3\pi/4} \csc x\, dx$

$= \left[\ln|\csc x - \cot x|\right]_{\pi/4}^{3\pi/4}$

$= \ln\left(\sqrt{2} + 1\right) - \ln\left(\sqrt{2} - 1\right) \approx 1.763$

11. $y = \dfrac{1}{2}(e^x + e^{-x})$

$y' = \dfrac{1}{2}(e^x - e^{-x}), \quad [0, 2]$

$1 + (y')^2 = \left[\dfrac{1}{2}(e^x + e^{-x})\right]^2, \quad [0, 2]$

$s = \displaystyle\int_0^2 \sqrt{\left[\dfrac{1}{2}(e^x + e^{-x})\right]^2}\, dx$

$= \dfrac{1}{2}\displaystyle\int_0^2 (e^x + e^{-x})\, dx$

$= \dfrac{1}{2}\left[e^x - e^{-x}\right]_0^2 = \dfrac{1}{2}\left(e^2 - \dfrac{1}{e^2}\right) \approx 3.627$

13. $x = \dfrac{1}{3}(y^2 + 2)^{3/2}, \quad 0 \le y \le 4$

$\dfrac{dx}{dy} = y(y^2 + 2)^{1/2}$

$s = \displaystyle\int_0^4 \sqrt{1 + y^2(y^2 + 2)}\, dy$

$= \displaystyle\int_0^4 \sqrt{y^4 + 2y^2 + 1}\, dy$

$= \displaystyle\int_0^4 (y^2 + 1)\, dy$

$= \left[\dfrac{y^3}{3} + y\right]_0^4 = \dfrac{64}{3} + 4 = \dfrac{76}{3}$

15. (a) $y = 4 - x^2, \quad 0 \le x \le 2$

(b) $\quad y' = -2x$

$1 + (y')^2 = 1 + 4x^2$

$$L = \int_0^2 \sqrt{1 + 4x^2}\, dx$$

(c) $L \approx 4.647$

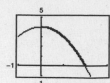

17. (a) $y = \dfrac{1}{x}, \quad 1 \le x \le 3$

(b) $\quad y' = -\dfrac{1}{x^2}$

$1 + (y')^2 = 1 + \dfrac{1}{x^4}$

$$L = \int_1^3 \sqrt{1 + \dfrac{1}{x^4}}\, dx$$

(c) $L \approx 2.147$

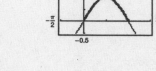

19. (a) $y = \sin x, \quad 0 \le x \le \pi$

(b) $\quad y' = \cos x$

$1 + (y')^2 = 1 + \cos^2 x$

$$L = \int_0^\pi \sqrt{1 + \cos^2 x}\, dx$$

(c) $L \approx 3.820$

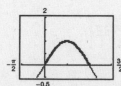

21. (a) $x = e^{-y}, \quad 0 \le y \le 2$

$y = -\ln x$

$1 \ge x \ge e^{-2} \approx 0.135$

(b) $\quad y' = -\dfrac{1}{x}$

$1 + (y')^2 = 1 + \dfrac{1}{x^2}$

$$L = \int_{e^{-2}}^1 \sqrt{1 + \dfrac{1}{x^2}}\, dx$$

(c) $L \approx 2.221$

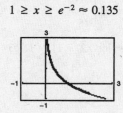

Alternatively, you can do all the computations with respect to y.

(a) $x = e^{-y}, \quad 0 \le y \le 2$

(b) $\quad \dfrac{dx}{dy} = -e^{-y}$

$1 + \left(\dfrac{dx}{dy}\right)^2 = 1 + e^{-2y}$

$$L = \int_0^2 \sqrt{1 + e^{-2y}}\, dy$$

(c) $L \approx 2.221$

23. (a) $y = 2 \arctan x, \quad 0 \le x \le 1$

(b) $y' = \dfrac{2}{1 + x^2}$

$$L = \int_0^1 \sqrt{1 + \dfrac{4}{(1 + x^2)^2}}\, dx$$

(c) $L \approx 1.871$

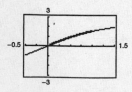

25. $\int_0^2 \sqrt{1 + \left[\dfrac{d}{dx}\left(\dfrac{5}{x^2 + 1}\right)\right]^2}\, dx$

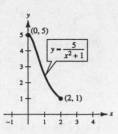

$s \approx 5$

Matches (b)

27. $y = x^3$, $[0, 4]$

(a) $d = \sqrt{(4 - 0)^2 + (64 - 0)^2} \approx 64.125$

(b) $d = \sqrt{(1 - 0)^2 + (1 - 0)^2} + \sqrt{(2 - 1)^2 + (8 - 1)^2} + \sqrt{(3 - 2)^2 + (27 - 8)^2} + \sqrt{(4 - 3)^2 + (64 - 27)^2}$

≈ 64.525

(c) $s = \int_0^4 \sqrt{1 + (3x^2)^2}\, dx = \int_0^4 \sqrt{1 + 9x^4}\, dx \approx 64.666$ (Simpson's Rule, $n = 10$)

(d) 64.672

29. (a) $f(x) = x^{2/3}$ 　　　　　　　　　　　　　　　(b) No, $f'(0)$ is not defined.

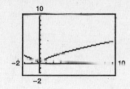

(c) 　　$f'(x) = \dfrac{2}{3}x^{-1/3}$

$1 + f'(x)^2 = 1 + \dfrac{4}{9x^{2/3}} = \dfrac{9x^{2/3} + 4}{9x^{2/3}}$

Divide $[-1, 8]$ into two intervals.

$[-1, 0]$: $s_1 = \int_{-1}^0 \sqrt{\dfrac{9x^{2/3} + 4}{9x^{2/3}}}\, dx$ 　　　　$[0, 8]$: $s_2 = \int_0^8 \sqrt{\dfrac{9x^{2/3} + 4}{9x^{2/3}}}\, dx$

$= \dfrac{-1}{3}\int_{-1}^0 \sqrt{9x^{2/3} + 4}\dfrac{1}{x^{1/3}}\, dx,$ $(x0)$ 　　$= \dfrac{1}{3}\int_0^8 \sqrt{9x^{2/3} + 4}\dfrac{1}{x^{1/3}}\, dx,$ $(x \geq 0)$

$= -\dfrac{1}{18}\int_{-1}^0 (9x^{2/3} + 4)^{1/2}\left(\dfrac{6}{x^{1/3}}\right) dx$ 　　$= \dfrac{1}{27}(9x^{2/3} + 4)^{3/2}\Big]_0^8$

$= -\dfrac{1}{27}(9x^{2/3} + 4)^{3/2}\Big]_{-1}^0$ 　　　　　$= \dfrac{1}{27}(40^{3/2} - 4^{3/2})$

$= -\dfrac{1}{27}(4^{3/2} - 13^{3/2})$ 　　　　　　　$= \dfrac{1}{27}(40^{3/2} - 8) \approx 9.0734$

$= -\dfrac{1}{27}(8 - 13^{3/2}) \approx 1.4397$

$s_1 + s_2 = \dfrac{1}{27}[40^{3/2} - 8 - 8 + 13^{3/2}]$

$= \dfrac{1}{27}[40^{3/2} + 13^{3/2} - 16] \approx 10.5131$

31. (a)

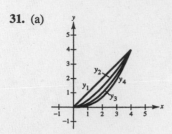

(b) y_1, y_2, y_3, y_4

(c) $y_1' = 1$, $L_1 = \int_0^4 \sqrt{2}\, dx \approx 5.657$

$y_2' = \dfrac{3}{4}x^{1/2}$, $L_2 = \int_0^4 \sqrt{1 + \dfrac{9x}{16}}\, dx \approx 5.759$

$y_3' = \dfrac{1}{2}x$, $L_3 = \int_0^4 \sqrt{1 + \dfrac{x^2}{4}}\, dx \approx 5.916$

$y_4' = \dfrac{5}{16}x^{3/2}$, $L_4 = \int_0^4 \sqrt{1 + \dfrac{25}{256}x^3}\, dx \approx 6.063$

35. $\qquad y = 20 \cosh \dfrac{x}{20}$, $\quad -20 \le x \le 20$

$y' = \sinh \dfrac{x}{20}$

$1 + (y')^2 = 1 + \sinh^2 \dfrac{x}{20} = \cosh^2 \dfrac{x}{20}$

$L = \int_{-20}^{20} \cosh \dfrac{x}{20}\, dx = 2 \int_0^{20} \cosh \dfrac{x}{20}\, dx$

$\qquad = 2(20) \sinh \dfrac{x}{20} \Big]_0^{20} = 40 \sinh(1) \approx 47.008$ m

39. $y = \dfrac{x^3}{3}$

$y' = x^2$, $\quad [0, 3]$

$S = 2\pi \int_0^3 \dfrac{x^3}{3} \sqrt{1 + x^4}\, dx$

$\quad = \dfrac{\pi}{6} \int_0^3 (1 + x^4)^{1/2}(4x^3)\, dx$

$\quad = \left[\dfrac{\pi}{9}(1 + x^4)^{3/2} \right]_0^3$

$\quad = \dfrac{\pi}{9}\big(82\sqrt{82} - 1\big) \approx 258.85$

33. $y = \dfrac{1}{3}[x^{3/2} - 3x^{1/2} + 2]$

When $x = 0$, $y = \dfrac{2}{3}$. Thus, the fleeing object has traveled $\dfrac{2}{3}$ units when it is caught.

$y' = \dfrac{1}{3}\left[\dfrac{3}{2}x^{1/2} - \dfrac{3}{2}x^{-1/2} \right] = \left(\dfrac{1}{2} \right) \dfrac{x - 1}{x^{1/2}}$

$1 + (y')^2 = 1 + \dfrac{(x - 1)^2}{4x} = \dfrac{(x + 1)^2}{4x}$

$s = \int_0^1 \dfrac{x + 1}{2x^{1/2}}\, dx = \dfrac{1}{2} \int_0^1 (x^{1/2} + x^{-1/2})\, dx$

$\quad = \dfrac{1}{2}\left[\dfrac{2}{3}x^{3/2} + 2x^{1/2} \right]_0^1 = \dfrac{4}{3} = 2\left(\dfrac{2}{3} \right)$

The pursuer has traveled twice the distance that the fleeing object has traveled when it is caught.

37. $\qquad y = \sqrt{9 - x^2}$

$y' = \dfrac{-x}{\sqrt{9 - x^2}}$

$1 + (y')^2 = \dfrac{9}{9 - x^2}$

$s = \int_0^2 \sqrt{\dfrac{9}{9 - x^2}}\, dx = \int_0^2 \dfrac{3}{\sqrt{9 - x^2}}\, dx$

$\quad = \left[3 \arcsin \dfrac{x}{3} \right]_0^2 = 3\left(\arcsin \dfrac{2}{3} - \arcsin 0 \right)$

$\quad = 3 \arcsin \dfrac{2}{3} \approx 2.1892$

41. $\qquad y = \dfrac{x^3}{6} + \dfrac{1}{2x}$

$y' = \dfrac{x^2}{2} - \dfrac{1}{2x^2}$

$1 + (y')^2 = \left(\dfrac{x^2}{2} + \dfrac{1}{2x^2} \right)^2$, $\quad [1, 2]$

$S = 2\pi \int_1^2 \left(\dfrac{x^3}{6} + \dfrac{1}{2x} \right)\left(\dfrac{x^2}{2} + \dfrac{1}{2x^2} \right) dx$

$\quad = 2\pi \int_1^2 \left(\dfrac{x^5}{12} + \dfrac{x}{3} + \dfrac{1}{4x^3} \right) dx$

$\quad = 2\pi \left[\dfrac{x^6}{72} + \dfrac{x^2}{6} - \dfrac{1}{8x^2} \right]_1^2 = \dfrac{47\pi}{16}$

43. $y = \sqrt[3]{x} + 2$

$y' = \dfrac{1}{3x^{2/3}}, \quad [1, 8]$

$S = 2\pi \displaystyle\int_1^8 x\sqrt{1 + \dfrac{1}{9x^{4/3}}}\, dx$

$= \dfrac{2\pi}{3}\displaystyle\int_1^8 x^{1/3}\sqrt{9x^{4/3} + 1}\, dx$

$= \dfrac{\pi}{18}\displaystyle\int_1^8 (9x^{4/3} + 1)^{1/2}(12x^{1/3})\, dx$

$= \left[\dfrac{\pi}{27}(9x^{4/3} + 1)^{3/2}\right]_1^8$

$= \dfrac{\pi}{27}\left(145\sqrt{145} - 10\sqrt{10}\right) \approx 199.48$

45. $y = \sin x$

$y' = \cos x, \quad [0, \pi]$

$S = 2\pi \displaystyle\int_0^\pi \sin x\sqrt{1 + \cos^2 x}\, dx$

≈ 14.4236

47. A rectifiable curve is one that has a finite arc length.

49. The precalculus formula is the surface area formula for the lateral surface of the frustum of a right circular cone. The representative element is

$$2\pi f(d_i)\sqrt{\Delta x_i^2 + \Delta y_i^2} = 2\pi f(d_i)\sqrt{1 + \left(\dfrac{\Delta y_i}{\Delta x_i}\right)^2}\,\Delta x_i.$$

51. $\quad y = \dfrac{hx}{r}$

$y' = \dfrac{h}{r}$

$1 + (y')^2 = \dfrac{r^2 + h^2}{r^2}$

$S = 2\pi \displaystyle\int_0^r x\sqrt{\dfrac{r^2 + h^2}{r^2}}\, dx$

$= \left[\dfrac{2\pi\sqrt{r^2 + h^2}}{r}\left(\dfrac{x^2}{2}\right)\right]_0^r = \pi r\sqrt{r^2 + h^2}$

53. $\quad y = \sqrt{9 - x^2}$

$y' - \dfrac{-x}{\sqrt{9 - x^2}}$

$\sqrt{1 + (y')^2} = \dfrac{3}{\sqrt{9 - x^2}}$

$S = 2\pi \displaystyle\int_0^2 \dfrac{3x}{\sqrt{9 - x^2}}\, dx$

$- -3\pi \displaystyle\int_0^2 \dfrac{-2x}{\sqrt{9 - x^2}}\, dx$

$= \left[-6\pi\sqrt{9 - x^2}\right]_0^2$

$= 6\pi\left(3 - \sqrt{5}\right) \approx 14.40$

See figure in Exercise 54.

55. $\quad y = \dfrac{1}{3}x^{1/2} - x^{3/2}$

$y' = \dfrac{1}{6}x^{-1/2} - \dfrac{3}{2}x^{1/2} = \dfrac{1}{6}(x^{-1/2} - 9x^{1/2})$

$1 + (y')^2 = 1 + \dfrac{1}{36}(x^{-1} - 18 + 81x) = \dfrac{1}{36}(x^{-1/2} + 9x^{1/2})^2$

$S = 2\pi \displaystyle\int_0^{1/3}\left(\dfrac{1}{3}x^{1/2} - x^{3/2}\right)\sqrt{\dfrac{1}{36}(x^{-1/2} + 9x^{1/2})^2}\, dx = \dfrac{2\pi}{6}\displaystyle\int_0^{1/3}\left(\dfrac{1}{3}x^{1/2} - x^{3/2}\right)(x^{-1/2} + 9x^{1/2})\, dx$

$= \dfrac{\pi}{3}\displaystyle\int_0^{1/3}\left(\dfrac{1}{3} + 2x - 9x^2\right)dx = \dfrac{\pi}{3}\left[\dfrac{1}{3}x + x^2 - 3x^3\right]_0^{1/3} = \dfrac{\pi}{27}\ \text{ft}^2 \approx 0.1164\ \text{ft}^2 \approx 16.8\ \text{in.}^2$

Amount of glass needed: $V = \dfrac{\pi}{27}\left(\dfrac{0.015}{12}\right) \approx 0.00015\ \text{ft}^3 \approx 0.25\ \text{in.}^3$

57. (a) We approximate the volume by summing six disks of thickness 3 and circumference C_i equal to the average of the given circumferences:

$$V \approx \sum_{i=1}^{6} \pi r_i^2(3) = \sum_{i=1}^{6} \pi \left(\frac{C_i}{2\pi}\right)^2 (3) = \frac{3}{4\pi} \sum_{i=1}^{6} C_i^2$$

$$= \frac{3}{4\pi}\left[\left(\frac{50 + 65.5}{2}\right)^2 + \left(\frac{65.5 + 70}{2}\right)^2 + \left(\frac{70 + 66}{2}\right)^2 + \left(\frac{66 + 58}{2}\right)^2 + \left(\frac{58 + 51}{2}\right)^2 + \left(\frac{51 + 48}{2}\right)^2\right]$$

$$= \frac{3}{4\pi}[57.75^2 + 67.75^2 + 68^2 + 62^2 + 54.5^2 + 49.5^2]$$

$$= \frac{3}{4\pi}[21813.625] = 5207.62 \text{ cubic inches}$$

(b) The lateral surface area of a frustum of a right circular cone is $\pi s(R + r)$. For the first frustum:

$$S_1 \approx \pi\left[3^2 + \left(\frac{65.5 - 50}{2\pi}\right)^2\right]^{1/2}\left[\frac{50}{2\pi} + \frac{65.5}{2\pi}\right]$$

$$= \left(\frac{50 + 65.5}{2}\right)\left[9 + \left(\frac{65.5 - 50}{2\pi}\right)^2\right]^{1/2}.$$

Adding the six frustums together:

$$S \approx \left(\frac{50 + 65.5}{2}\right)\left[9 + \left(\frac{15.5}{2\pi}\right)^2\right]^{1/2} + \left(\frac{65.5 + 70}{2}\right)\left[9 + \left(\frac{4.5}{2\pi}\right)^2\right]^{1/2} +$$

$$\left(\frac{70 + 66}{2}\right)\left[9 + \left(\frac{4}{2\pi}\right)^2\right]^{1/2} + \left(\frac{66 + 58}{2}\right)\left[9 + \left(\frac{8}{2\pi}\right)^2\right]^{1/2} +$$

$$\left(\frac{58 + 51}{2}\right)\left[9 + \left(\frac{7}{2\pi}\right)^2\right]^{1/2} + \left(\frac{51 + 48}{2}\right)\left[9 + \left(\frac{3}{2\pi}\right)^2\right]^{1/2}$$

$$\approx 224.30 + 208.96 + 208.54 + 202.06 + 174.41 + 150.37$$

$$= 1168.64$$

(c) $r = 0.00401y^3 - 0.1416y^2 + 1.232y + 7.943$

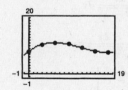

(d) $V = \displaystyle\int_0^{18} \pi r^2 \, dy \approx 5275.9 \text{ cubic inches}$

$$S = \int_0^{18} 2\pi r(y)\sqrt{1 + r'(y)^2} \, dy$$

$$\approx 1179.5 \text{ square inches}$$

59. (a) $V = \pi \displaystyle\int_1^b \frac{1}{x^2} \, dx = \left[-\frac{\pi}{x}\right]_1^b = \pi\left(1 - \frac{1}{b}\right)$

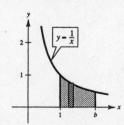

(c) $\displaystyle\lim_{b\to\infty} V = \lim_{b\to\infty} \pi\left(1 - \frac{1}{b}\right) = \pi$

(d) Since

$$\frac{\sqrt{x^4 + 1}}{x^3} > \frac{\sqrt{x^4}}{x^3} = \frac{1}{x} > 0 \text{ on } [1, b]$$

we have

$$\int_1^b \frac{\sqrt{x^4 + 1}}{x^3} \, dx > \int_1^b \frac{1}{x} \, dx = \left[\ln x\right]_1^b = \ln b$$

and $\displaystyle\lim_{b\to\infty} \ln b \to \infty$. Thus,

$$\lim_{b\to\infty} 2\pi \int_1^b \frac{\sqrt{x^4 + 1}}{x^3} \, dx = \infty.$$

(b) $S = 2\pi \displaystyle\int_1^b \frac{1}{x}\sqrt{1 + \left(-\frac{1}{x^2}\right)^2} \, dx$

$$= 2\pi \int_1^b \frac{1}{x}\sqrt{1 + \frac{1}{x^4}} \, dx$$

$$= 2\pi \int_1^b \frac{\sqrt{x^4 + 1}}{x^3} \, dx$$

61. Individual project

63. $x^{2/3} + y^{2/3} = 4$

$$y^{2/3} = 4 - x^{2/3}$$

$$y = (4 - x^{2/3})^{3/2}, \quad 0 \le x \le 8$$

$$y' = \frac{3}{2}(4 - x^{2/3})^{1/2}\left(-\frac{2}{3}x^{-1/3}\right) = \frac{-(4 - x^{2/3})^{1/2}}{x^{1/3}}$$

$$1 + (y')^2 = 1 + \frac{4 - x^{2/3}}{x^{2/3}} = \frac{4}{x^{2/3}}$$

$$S = 2\pi \int_0^8 (4 - x^{2/3})^{3/2} \sqrt{\frac{4}{x^{2/3}}}\, dx$$

$$= 4\pi \int_0^8 \frac{(4 - x^{2/3})^{3/2}}{x^{1/3}}\, dx$$

$$= \left[-\frac{12\pi}{5}(4 - x^{2/3})^{5/2} \right]_0^8 = \frac{192\pi}{5}$$

[Surface area of portion above the *x*-axis]

65. $y = kx^2, y' = 2kx$

$$1 + (y')^2 = 1 + 4k^2x^2$$

$$h = kw^2 \Rightarrow k = \frac{h}{w^2} \Rightarrow 1 + (y') = 1 + \frac{4h^2}{w^4}x^2$$

By symmetry, $C = 2\displaystyle\int_0^w \sqrt{1 + \frac{4h^2}{w^4}x^2}\, dx$.

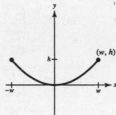

67. Let (x_0, y_0) be the point on the graph of $y^2 = x^3$ where the tangent line makes an angle of 45° with the *x*-axis.

$$y = x^{3/2}$$

$$y' = \tfrac{3}{2}x^{1/2} = 1$$

$$x_0 = \tfrac{4}{9}$$

$$L = \int_0^{4/9} \sqrt{1 + \tfrac{9}{4}x}\, dx = \tfrac{8}{27}\left(2\sqrt{2} - 1\right)$$

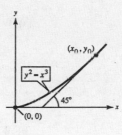

Section 7.5   Work

1. $W = Fd = (100)(10)$

$\qquad = 1000 \text{ ft} \cdot \text{lb}$

3. $W - Fd = (112)(4)$

$\qquad = 448 \text{ joules (newton-meters)}$

5. Work equals force times distance, $W = FD$.

7. Since the work equals the area under the force function, you have $(c) < (d) < (a) < (b)$.

9. $F(x) = kx$

$\qquad 5 = k(4)$

$\qquad k = \dfrac{5}{4}$

$\qquad W = \displaystyle\int_0^7 \frac{5}{4}x\, dx = \left[\frac{5}{8}x^2 \right]_0^7$

$\qquad = \dfrac{245}{8} \text{ in.} \cdot \text{lb}$

$\qquad = 30.625 \text{ in.} \cdot \text{lb} \approx 2.55 \text{ ft} \cdot \text{lb}$

11. $F(x) = kx$

$\qquad 250 = k(30) \Rightarrow k = \dfrac{25}{3}$

$\qquad W = \displaystyle\int_{20}^{50} F(x)\, dx$

$\qquad = \displaystyle\int_{20}^{50} \frac{25}{3}x\, dx = \frac{25x^2}{6} \bigg]_{20}^{50}$

$\qquad = 8750 \text{ n} \cdot \text{cm}$

$\qquad = 87.5 \text{ joules or Nm}$

13. $F(x) = kx$

$20 = k(9)$

$k = \dfrac{20}{9}$

$W = \displaystyle\int_0^{12} \dfrac{20}{9}x\,dx = \left[\dfrac{10}{9}x^2\right]_0^{12} = 160 \text{ in.} \cdot \text{lb} = \dfrac{40}{3} \text{ ft} \cdot \text{lb}$

15. $W = 18 = \displaystyle\int_0^{1/3} kx\,dx = \dfrac{kx^2}{2}\bigg]_0^{1/3} = \dfrac{k}{18} \implies k = 324$

$W = \displaystyle\int_{1/3}^{7/12} 324x\,dx = 162x^2\bigg]_{1/3}^{7/12} = 37.125 \text{ ft} \cdot \text{lbs}$

$\Big[$**Note:** $4 \text{ inches} = \tfrac{1}{3} \text{ foot}\Big]$

17. Assume that Earth has a radius of 4000 miles.

$F(x) = \dfrac{k}{x^2}$

$5 = \dfrac{k}{(4000)^2}$

$k = 80,000,000$

$F(x) = \dfrac{80,000,000}{x^2}$

(a) $W = \displaystyle\int_{4000}^{4100} \dfrac{80,000,000}{x^2}\,dx = \left[\dfrac{-80,000,000}{x}\right]_{4000}^{4100}$

$\approx 487.8 \text{ mi} \cdot \text{tons} \approx 5.15 \times 10^9 \text{ ft} \cdot \text{lb}$

(b) $W = \displaystyle\int_{4000}^{4300} \dfrac{80,000,000}{x^2}\,dx$

$\approx 1395.3 \text{ mi} \cdot \text{ton} \approx 1.47 \times 10^{10} \text{ ft} \cdot \text{ton}$

19. Assume that Earth has a radius of 4000 miles.

$F(x) = \dfrac{k}{x^2}$

$10 = \dfrac{k}{(4000)^2}$

$k = 160,000,000$

$F(x) = \dfrac{160,000,000}{x^2}$

(a) $W = \displaystyle\int_{4000}^{15,000} \dfrac{160,000,000}{x^2}\,dx = \left[-\dfrac{160,000,000}{x}\right]_{4000}^{15,000} \approx -10,666.667 + 40,000$

$= 29,333.333 \text{ mi} \cdot \text{ton}$

$\approx 2.93 \times 10^4 \text{ mi} \cdot \text{ton}$

$\approx 3.10 \times 10^{11} \text{ ft} \cdot \text{lb}$

(b) $W = \displaystyle\int_{4000}^{26,000} \dfrac{160,000,000}{x^2}\,dx = \left[-\dfrac{160,000,000}{x}\right]_{4000}^{26,000} \approx -6,153.846 + 40,000$

$= 33,846.154 \text{ mi} \cdot \text{ton}$

$\approx 3.38 \times 10^4 \text{ mi} \cdot \text{ton}$

$\approx 3.57 \times 10^{11} \text{ ft} \cdot \text{lb}$

21. Weight of each layer: $62.4(20)\,\Delta y$

Distance: $4 - y$

(a) $W = \displaystyle\int_2^4 62.4(20)(4 - y)\,dy = \left[4992y - 624y^2\right]_2^4 = 2496 \text{ ft} \cdot \text{lb}$

(b) $W = \displaystyle\int_0^4 62.4(20)(4 - y)\,dy = \left[4992y - 624y^2\right]_0^4 = 9984 \text{ ft} \cdot \text{lb}$

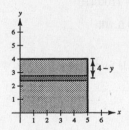

23. Volume of disk: $\pi(2)^2\,\Delta y = 4\pi\,\Delta y$

Weight of disk of water: $9800(4\pi)\,\Delta y$

Distance the disk of water is moved: $5 - y$

$W = \displaystyle\int_0^4 (5 - y)(9800)4\pi\,dy = 39,200\pi\int_0^4 (5 - y)\,dy$

$= 39,200\pi\left[5y - \dfrac{y^2}{2}\right]_0^4$

$= 39,200\pi(12) = 470,400\pi \text{ newton–meters}$

25. Volume of disk: $\pi\left(\dfrac{2}{3}y\right)^2\Delta y$

Weight of disk: $62.4\pi\left(\dfrac{2}{3}y\right)^2\Delta y$

Distance: $6 - y$

$W = \dfrac{4(62.4)\pi}{9}\displaystyle\int_0^6 (6 - y)y^2\,dy$

$= \dfrac{4}{9}(62.4)\pi\left[2y^3 - \dfrac{1}{4}y^4\right]_0^6$

$= 2995.2\pi \text{ ft} \cdot \text{lb}$

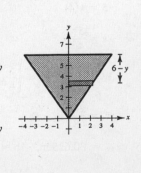

27. Volume of disk: $\pi\left(\sqrt{36 - y^2}\right)^2 \Delta y$

Weight of disk: $62.4\pi(36 - y^2)\,\Delta y$

Distance: y

$$W = 62.4\pi \int_0^6 y(36 - y^2)\,dy$$

$$= 62.4\pi \int_0^6 (36y - y^3)\,dy = 62.4\pi \left[18y^2 - \frac{1}{4}y^4\right]_0^6$$

$$= 20{,}217.6\pi \text{ ft} \cdot \text{lb}$$

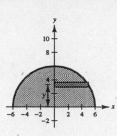

29. Volume of layer: $V = lwh = 4(2)\sqrt{(9/4) - y^2}\,\Delta y$

Weight of layer: $W = 42(8)\sqrt{(9/4) - y^2}\,\Delta y$

Distance: $\dfrac{13}{2} - y$

$$W = \int_{-1.5}^{1.5} 42(8)\sqrt{\frac{9}{4} - y^2}\left(\frac{13}{2} - y\right)dy = 336\left[\frac{13}{2}\int_{-1.5}^{1.5}\sqrt{\frac{9}{4} - y^2}\,dy - \int_{-1.5}^{1.5}\sqrt{\frac{9}{4} - y^2}\,y\,dy\right]$$

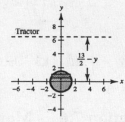

The second integral is zero since the integrand is odd and the limits of integration are symmetric to the origin. The first integral represents the area of a semicircle of radius $\frac{3}{2}$. Thus, the work is

$$W = 336\left(\frac{13}{2}\right)\pi\left(\frac{3}{2}\right)^2\left(\frac{1}{2}\right) = 2457\pi \text{ ft} \cdot \text{lb}.$$

31. Weight of section of chain: $3\,\Delta y$

Distance: $15 - y$

$$W = 3\int_0^{15}(15 - y)\,dy$$

$$= \left[-\frac{3}{2}(15 - y)^2\right]_0^{15}$$

$$= 337.5 \text{ ft} \cdot \text{lb}$$

33. The lower 5 feet of chain are raised 10 feet with a constant force.

$$W_1 = 3(5)(10) = 150 \text{ ft} \cdot \text{lb}$$

The top 10 feet of chain are raised with a variable force.

Weight per section: $3\,\Delta y$

Distance: $10 - y$

$$W_2 = 3\int_0^{10}(10 - y)\,dy$$

$$= \left[-\frac{3}{2}(10 - y)^2\right]_0^{10} = 150 \text{ ft} \cdot \text{lb}$$

$$W = W_1 + W_2 = 300 \text{ ft} \cdot \text{lb}$$

35. Weight of section of chain: $3\,\Delta y$

Distance: $15 - 2y$

$$W = 3\int_0^{7.5}(15 - 2y)\,dy = \left[-\frac{3}{4}(15 - 2y)^2\right]_0^{7.5}$$

$$= \frac{3}{4}(15)^2 = 168.75 \text{ ft} \cdot \text{lb}$$

37. Work to pull up the ball: $W_1 = 500(15) = 7500 \text{ ft} \cdot \text{lb}$

Work to wind up the top 15 feet of cable: force is variable

Weight per section: $1\,\Delta y$

Distance: $15 - x$

$$W_2 = \int_0^{15}(15 - x)\,dx = \left[-\frac{1}{2}(15 - x)^2\right]_0^{15}$$

$$= 112.5 \text{ ft} \cdot \text{lb}$$

Work to lift the lower 25 feet of cable with a constant force:

$$W_3 = (1)(25)(15) = 375 \text{ ft} \cdot \text{lb}$$

$$W = W_1 + W_2 + W_3 = 7500 + 112.5 + 375$$

$$= 7987.5 \text{ ft} \cdot \text{lb}$$

39. $p = \dfrac{k}{V}$

$1000 = \dfrac{k}{2}$

$k = 2000$

$W = \displaystyle\int_2^3 \dfrac{2000}{V}\, dV$

$= \Big[\, 2000 \ln|V| \,\Big]_2^3$

$= 2000 \ln\left(\dfrac{3}{2}\right) \approx 810.93 \text{ ft} \cdot \text{lb}$

41. $F(x) = \dfrac{k}{(2-x)^2}$

$W = \displaystyle\int_{-2}^1 \dfrac{k}{(2-x)^2}\, dx$

$= \left[\dfrac{k}{2-x}\right]_{-2}^1 = k\left(1 - \dfrac{1}{4}\right)$

$= \dfrac{3k}{4}$ (units of work)

43. $W = \displaystyle\int_0^5 1000\big[1.8 - \ln(x+1)\big]\, dx \approx 3249.44 \text{ ft} \cdot \text{lb}$

45. $W = \displaystyle\int_0^5 100x\sqrt{125 - x^3}\, dx \approx 10{,}330.3 \text{ ft} \cdot \text{lb}$

Section 7.6   Moments, Centers of Mass, and Centroids

1. $\bar{x} = \dfrac{6(-5) + 3(1) + 5(3)}{6 + 3 + 5} = -\dfrac{6}{7}$

3. $\bar{x} = \dfrac{1(7) + 1(8) + 1(12) + 1(15) + 1(18)}{1 + 1 + 1 + 1 + 1} = 12$

5. (a) $\bar{x} = \dfrac{(7+5) + (8+5) + (12+5) + (15+5) + (18+5)}{5} = 17 = 12 + 5$

(b) $\bar{x} = \dfrac{12(-6-3) + 1(-4-3) + 6(-2-3) + 3(0-3) + 11(8-3)}{12 + 1 + 6 + 3 + 11} = \dfrac{-99}{33} = -3$

7. $50x = 75(L - x) = 75(10 - x)$

$50x = 750 - 75x$

$125x = 750$

$x = 6$ feet

9. $\bar{x} = \dfrac{5(2) + 1(-3) + 3(1)}{5 + 1 + 3} = \dfrac{10}{9}$

$\bar{y} = \dfrac{5(2) + 1(1) + 3(-4)}{5 + 1 + 3} = -\dfrac{1}{9}$

$(\bar{x}, \bar{y}) = \left(\dfrac{10}{9}, -\dfrac{1}{9}\right)$

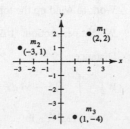

11. $\bar{x} = \dfrac{3(-2) + 4(5) + 2(7) + 1(0) + 6(-3)}{3 + 4 + 2 + 1 + 6} = \dfrac{5}{8}$

$\bar{y} = \dfrac{3(-3) + 4(5) + 2(1) + 1(0) + 6(0)}{3 + 4 + 2 + 1 + 6} = \dfrac{13}{16}$

$(\bar{x}, \bar{y}) = \left(\dfrac{5}{8}, \dfrac{13}{16}\right)$

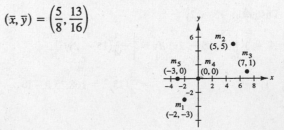

13. $m = \rho \displaystyle\int_0^4 \sqrt{x}\, dx = \left[\dfrac{2\rho}{3}x^{3/2}\right]_0^4 = \dfrac{16\rho}{3}$

$M_x = \rho \displaystyle\int_0^4 \dfrac{\sqrt{x}}{2}(\sqrt{x})\, dx = \left[\rho\dfrac{x^2}{4}\right]_0^4 = 4\rho$

$\bar{y} = \dfrac{M_x}{m} = 4\rho\left(\dfrac{3}{16\rho}\right) = \dfrac{3}{4}$

$M_y = \rho \displaystyle\int_0^4 x\sqrt{x}\, dx = \left[\rho\dfrac{2}{5}x^{5/2}\right]_0^4 = \dfrac{64\rho}{5}$

$\bar{x} = \dfrac{M_y}{m} = \dfrac{64\rho}{5}\left(\dfrac{3}{16\rho}\right) = \dfrac{12}{5}$

$(\bar{x}, \bar{y}) = \left(\dfrac{12}{5}, \dfrac{3}{4}\right)$

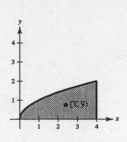

15. $m = \rho \displaystyle\int_0^1 (x^2 - x^3)\, dx = \rho\left[\dfrac{x^3}{3} - \dfrac{x^4}{4}\right]_0^1 = \dfrac{\rho}{12}$

$M_x = \rho \displaystyle\int_0^1 \dfrac{(x^2 + x^3)}{2}(x^2 - x^3)\, dx = \dfrac{\rho}{2}\displaystyle\int_0^1 (x^4 - x^6)\, dx = \dfrac{\rho}{2}\left[\dfrac{x^5}{5} - \dfrac{x^7}{7}\right]_0^1 = \dfrac{\rho}{35}$

$\bar{y} = \dfrac{M_x}{m} = \dfrac{\rho}{35}\left(\dfrac{12}{\rho}\right) = \dfrac{12}{35}$

$M_y = \rho \displaystyle\int_0^1 x(x^2 - x^3)\, dx = \rho \displaystyle\int_0^1 (x^3 - x^4)\, dx = \rho\left[\dfrac{x^4}{4} - \dfrac{x^5}{5}\right]_0^1 = \dfrac{\rho}{20}$

$\bar{x} = \dfrac{M_y}{m} = \dfrac{\rho}{20}\left(\dfrac{12}{\rho}\right) = \dfrac{3}{5}$

$(\bar{x}, \bar{y}) = \left(\dfrac{3}{5}, \dfrac{12}{35}\right)$

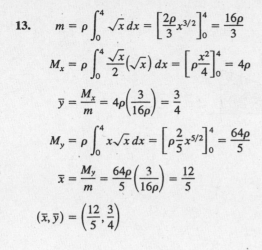

17. $m = \rho \displaystyle\int_0^3 [(-x^2 + 4x + 2) - (x + 2)]\, dx = -\rho\left[\dfrac{x^3}{3} + \dfrac{3x^2}{2}\right]_0^3 = \dfrac{9\rho}{2}$

$M_x = \rho \displaystyle\int_0^3 \left[\dfrac{(-x^2 + 4x + 2) + (x + 2)}{2}\right][(-x^2 + 4x + 2) - (x + 2)]\, dx$

$\qquad = \dfrac{\rho}{2}\displaystyle\int_0^3 (-x^2 + 5x + 4)(-x^2 + 3x)\, dx = \dfrac{\rho}{2}\displaystyle\int_0^3 (x^4 - 8x^3 + 11x^2 + 12x)\, dx$

$\qquad = \dfrac{\rho}{2}\left[\dfrac{x^5}{5} - 2x^4 + \dfrac{11x^3}{3} + 6x^2\right]_0^3 = \dfrac{99\rho}{5}$

$\bar{y} = \dfrac{M_x}{m} = \dfrac{99\rho}{5}\left(\dfrac{2}{9\rho}\right) = \dfrac{22}{5}$

$M_y = \rho \displaystyle\int_0^3 x[(-x^2 + 4x - 2) - (x + 2)]\, dx = \rho \displaystyle\int_0^3 (-x^3 + 3x^2)\, dx = \rho\left[-\dfrac{x^4}{4} + x^3\right]_0^3 = \dfrac{27\rho}{4}$

$\bar{x} = \dfrac{M_y}{m} = \dfrac{27\rho}{4}\left(\dfrac{2}{9\rho}\right) = \dfrac{3}{2}$

$(\bar{x}, \bar{y}) = \left(\dfrac{3}{2}, \dfrac{22}{5}\right)$

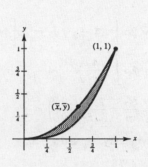

19. $m = \rho \int_0^8 x^{2/3}\,dx = \rho\left[\dfrac{3}{5}x^{5/3}\right]_0^8 = \dfrac{96\rho}{5}$

$M_x = \rho \int_0^8 \dfrac{x^{2/3}}{2}(x^{2/3})\,dx = \dfrac{\rho}{2}\left[\dfrac{3}{7}x^{7/3}\right]_0^8 = \dfrac{192\rho}{7}$

$\bar{y} = \dfrac{M_x}{m} = \dfrac{192\rho}{7}\left(\dfrac{5}{96\rho}\right) = \dfrac{10}{7}$

$M_y = \rho \int_0^8 x(x^{2/3})\,dx = \rho\left[\dfrac{3}{8}x^{8/3}\right]_0^8 = 96\rho$

$\bar{x} = \dfrac{M_y}{m} = 96\rho\left(\dfrac{5}{96\rho}\right) = 5$

$(\bar{x}, \bar{y}) = \left(5, \dfrac{10}{7}\right)$

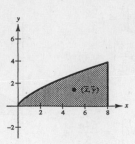

21. $m = 2\rho \int_0^2 (4 - y^2)\,dy = 2\rho\left[4y - \dfrac{y^3}{3}\right]_0^2 = \dfrac{32\rho}{3}$

$M_y = 2\rho \int_0^2 \left(\dfrac{4 - y^2}{2}\right)(4 - y^2)\,dy = \rho\left[16y - \dfrac{8}{3}y^3 + \dfrac{y^5}{5}\right]_0^2 = \dfrac{256\rho}{15}$

$\bar{x} = \dfrac{M_y}{m} = \dfrac{256\rho}{15}\left(\dfrac{3}{32\rho}\right) = \dfrac{8}{5}$

By symmetry, M_x and $\bar{y} = 0$.

$(\bar{x}, \bar{y}) = \left(\dfrac{8}{5}, 0\right)$

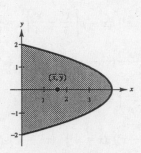

23. $m = \rho \int_0^3 [(2y - y^2) - (-y)]\,dy = \rho\left[\dfrac{3y^2}{2} - \dfrac{y^3}{3}\right]_0^3 = \dfrac{9\rho}{2}$

$M_y = \rho \int_0^3 \dfrac{[(2y - y^2) + (-y)]}{2}[(2y - y^2) - (-y)]\,dy = \dfrac{\rho}{2}\int_0^3 (y - y^2)(3y - y^2)\,dy$

$= \dfrac{\rho}{2}\int_0^3 (y^4 - 4y^3 + 3y^2)\,dy = \dfrac{\rho}{2}\left[\dfrac{y^5}{5} - y^4 + y^3\right]_0^3 = -\dfrac{27\rho}{10}$

$\bar{x} = \dfrac{M_y}{m} = -\dfrac{27\rho}{10}\left(\dfrac{2}{9\rho}\right) = -\dfrac{3}{5}$

$M_x = \rho \int_0^3 y[(2y - y^2) - (-y)]\,dy = \rho \int_0^3 (3y^2 - y^3)\,dy = \rho\left[y^3 - \dfrac{y^4}{4}\right]_0^3 = \dfrac{27\rho}{4}$

$\bar{y} = \dfrac{M_x}{m} = \dfrac{27\rho}{4}\left(\dfrac{2}{9\rho}\right) = \dfrac{3}{2}$

$(\bar{x}, \bar{y}) = \left(-\dfrac{3}{5}, \dfrac{3}{2}\right)$

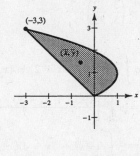

25. $A = \int_0^1 (x - x^2)\,dx = \left[\dfrac{1}{2}x^2 - \dfrac{x^3}{3}\right]_0^1 = \dfrac{1}{6}$

$M_x = \dfrac{1}{2}\int_0^1 (x^2 - x^4)\,dx = \dfrac{1}{2}\left[\dfrac{x^3}{3} - \dfrac{x^5}{5}\right]_0^1 = \dfrac{1}{2}\left(\dfrac{1}{3} - \dfrac{1}{5}\right) = \dfrac{1}{15}$

$M_y = \int_0^1 (x^2 - x^3)\,dx = \left[\dfrac{x^3}{3} - \dfrac{x^4}{4}\right]_0^1 = \left(\dfrac{1}{3} - \dfrac{1}{4}\right) = \dfrac{1}{12}$

27. $A = \int_0^3 (2x + 4)\,dx = \left[x^2 + 4x\right]_0^3 = 9 + 12 = 21$

$M_x = \dfrac{1}{2}\int_0^3 (2x + 4)^2\,dx = \int_0^3 (2x^2 + 8x + 8)\,dx$

$= \left[\dfrac{2x^3}{3} + 4x^2 + 8x\right]_0^3 = 18 + 36 + 24 = 78$

$M_y = \int_0^3 (2x^2 + 4x)\,dx = \left[\dfrac{2x^3}{3} + 2x^2\right]_0^3 = 18 + 18 = 36$

29. $m = \rho \displaystyle\int_0^5 10x\sqrt{125 - x^3}\,dx \approx 1033.0\rho$

$M_x = \rho \displaystyle\int_0^5 \left(\frac{10x\sqrt{125 - x^3}}{2}\right)\left(10x\sqrt{125 - x^3}\right)dx$

$\quad = 50\rho \displaystyle\int_0^5 x^2(125 - x^3)\,dx = \frac{3{,}124{,}375\rho}{24} \approx 130{,}208\rho$

$M_y = \rho \displaystyle\int_0^5 10x^2\sqrt{125 - x^3}\,dx = -\frac{10\rho}{3}\displaystyle\int_0^5 \sqrt{125 - x^3}\,(-3x^2)\,dx = \frac{12{,}500\sqrt{5}\rho}{9} \approx 3105.6\rho$

$\bar{x} = \dfrac{M_y}{m} \approx 3.0$

$\bar{y} = \dfrac{M_x}{m} \approx 126.0$

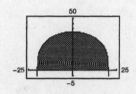

Therefore, the centroid is (3.0, 126.0).

31. $m = \rho \displaystyle\int_{-20}^{20} 5\sqrt[3]{400 - x^2}\,dx \approx 1239.76\rho$

$M_x = \rho \displaystyle\int_{-20}^{20} \frac{5\sqrt[3]{400 - x^2}}{2}\left(5\sqrt[3]{400 - x^2}\right)dx$

$\quad = \frac{25\rho}{2}\displaystyle\int_{-20}^{20}(400 - x^2)^{2/3}\,dx \approx 20064.27$

$\bar{y} = \dfrac{M_x}{m} \approx 16.18$

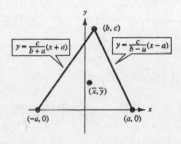

$\bar{x} = 0$ by symmetry. Therefore, the centroid is (0, 16.2).

33. $A = \dfrac{1}{2}(2a)c = ac$

$\dfrac{1}{A} = \dfrac{1}{ac}$

$\bar{x} = \left(\dfrac{1}{ac}\right)\dfrac{1}{2}\displaystyle\int_0^c \left[\left(\frac{b-a}{c}y + a\right)^2 - \left(\frac{b+a}{c}y - a\right)^2\right]dy$

$\quad = \dfrac{1}{2ac}\displaystyle\int_0^c \left[\frac{4ab}{c}y - \frac{4ab}{c^2}y^2\right]dy$

$\quad = \dfrac{1}{2ac}\left[\frac{2ab}{c}y^2 - \frac{4ab}{3c^2}y^3\right]_0^c = \dfrac{1}{2ac}\left(\frac{2}{3}abc\right) = \dfrac{b}{3}$

$\bar{y} = \dfrac{1}{ac}\displaystyle\int_0^c y\left[\left(\frac{b-a}{c}y + a\right) - \left(\frac{b+a}{c}y - a\right)\right]dy$

$\quad = \dfrac{1}{ac}\displaystyle\int_0^c y\left(-\frac{2a}{c}y + 2a\right)dy = \dfrac{2}{c}\displaystyle\int_0^c \left(y - \frac{y^2}{c}\right)dy$

$\quad = \dfrac{2}{c}\left[\frac{y^2}{2} - \frac{y^3}{3c}\right]_0^c = \dfrac{c}{3}$

$(\bar{x}, \bar{y}) = \left(\dfrac{b}{3}, \dfrac{c}{3}\right)$

From elementary geometry, $(b/3, c/3)$ is the point of intersection of the medians.

35. $A = \dfrac{c}{2}(a + b)$

$\dfrac{1}{A} = \dfrac{2}{c(a + b)}$

$\bar{x} = \dfrac{2}{c(a + b)} \displaystyle\int_0^c x\left(\dfrac{b - a}{c}x + a\right) dx = \dfrac{2}{c(a + b)} \displaystyle\int_0^c \left(\dfrac{b - a}{c}x^2 + ax\right) dx = \dfrac{2}{c(a + b)}\left[\dfrac{b - a}{c}\dfrac{x^3}{3} + \dfrac{ax^2}{2}\right]_0^c$

$= \dfrac{2}{c(a + b)}\left[\dfrac{(b - a)c^2}{3} + \dfrac{ac^2}{2}\right] = \dfrac{2}{c(a + b)}\left[\dfrac{2bc^2 - 2ac^2 + 3ac^2}{6}\right] = \dfrac{c(2b + a)}{3(a + b)} = \dfrac{(a + 2b)c}{3(a + b)}$

$\bar{y} = \dfrac{2}{c(a + b)}\dfrac{1}{2}\displaystyle\int_0^c \left(\dfrac{b - a}{c}x + a\right)^2 dx = \dfrac{1}{c(a + b)}\displaystyle\int_0^c \left[\left(\dfrac{b - a}{c}\right)^2 x^2 + \dfrac{2a(b - a)}{c}x + a^2\right] dx$

$= \dfrac{1}{c(a + b)}\left[\left(\dfrac{b - a}{c}\right)^2\dfrac{x^3}{3} + \dfrac{2a(b - a)}{c}\dfrac{x^2}{2} + a^2 x\right]_0^c = \dfrac{1}{c(a + b)}\left[\dfrac{(b - a)^2 c}{3} + ac(b - a) + a^2 c\right]$

$= \dfrac{1}{3c(a + b)}\left[(b^2 - 2ab + a^2)c + 3ac(b - a) + 3a^2 c\right]$

$= \dfrac{1}{3(a + b)}\left[b^2 - 2ab + a^2 + 3ab - 3a^2 + 3a^2\right] = \dfrac{a^2 + ab + b^2}{3(a + b)}$

Thus, $(\bar{x}, \bar{y}) = \left(\dfrac{(a + 2b)c}{3(a + b)}, \dfrac{a^2 + ab + b^2}{3(a + b)}\right)$.

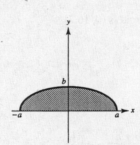

The one line passes through $(0, a/2)$ and $(c, b/2)$. It's equation is $y = \dfrac{b - a}{2c}x + \dfrac{a}{2}$. The other line

passes through $(0, -b)$ and $(c, a + b)$. It's equation is $y = \dfrac{a + 2b}{c}x - b$. $(\bar{x}, \bar{y})$ is the point of

intersection of these two lines.

37. $\bar{x} = 0$ by symmetry.

$A = \dfrac{1}{2}\pi ab$

$\dfrac{1}{A} = \dfrac{2}{\pi ab}$

$\bar{y} = \dfrac{2}{\pi ab}\dfrac{1}{2}\displaystyle\int_{-a}^a \left(\dfrac{b}{a}\sqrt{a^2 - x^2}\right)^2 dx$

$= \dfrac{1}{\pi ab}\left(\dfrac{b^2}{a^2}\right)\left[a^2 x - \dfrac{x^3}{3}\right]_{-a}^a = \dfrac{b}{\pi a^3}\left[\dfrac{4a^3}{3}\right] = \dfrac{4b}{3\pi}$

$(\bar{x}, \bar{y}) = \left(0, \dfrac{4b}{3\pi}\right)$

39. (a)

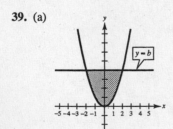

(b) $\bar{x} = 0$ by symmetry.

(c) $M_y = \displaystyle\int_{-\sqrt{b}}^{\sqrt{b}} x(b - x^2) dx = 0$ because $bx - x^3$ is odd.

(d) $\bar{y} > \dfrac{b}{2}$ since there is more area above $y = \dfrac{b}{2}$ than below.

(e) $M_x = \displaystyle\int_{-\sqrt{b}}^{\sqrt{b}} \dfrac{(b + x^2)(b - x^2)}{2} dx$

$= \displaystyle\int_{-\sqrt{b}}^{\sqrt{b}} \dfrac{b^2 - x^4}{2} dx = \dfrac{1}{2}\left[b^2 x - \dfrac{x^5}{5}\right]_{-\sqrt{b}}^{\sqrt{b}}$

$= b^2\sqrt{b} - \dfrac{b^2\sqrt{b}}{5} = \dfrac{4b^2\sqrt{b}}{5}$

$A = \displaystyle\int_{-\sqrt{b}}^{\sqrt{b}} (b - x^2) dx = \left[bx - \dfrac{x^3}{3}\right]_{-\sqrt{b}}^{\sqrt{b}}$

$= \left(b\sqrt{b} - \dfrac{b\sqrt{b}}{3}\right)2 = 4\dfrac{b\sqrt{b}}{3}$

$\bar{y} = \dfrac{M_x}{A} = \dfrac{4b^2\sqrt{b}/5}{4b\sqrt{b}/3} = \dfrac{3}{5}b$

41. (a) $\bar{x} = 0$ by symmetry.

$$A = 2\int_0^{40} f(x)\,dx = \frac{2(40)}{3(4)}[30 + 4(29) + 2(26) + 4(20) + 0] = \frac{20}{3}(278) = \frac{5560}{3}$$

$$M_x = \int_{-40}^{40} \frac{f(x)^2}{2}\,dx = \frac{40}{3(4)}[30^2 + 4(29)^2 + 2(26)^2 + 4(20)^2 + 0] = \frac{10}{3}(7216) = \frac{72,160}{3}$$

$$\bar{y} = \frac{M_x}{A} = \frac{72,160/3}{5560/3} = \frac{72,160}{5560} \approx 12.98$$

$(\bar{x}, \bar{y}) = (0, 12.98)$

(b) $y = (-1.02 \times 10^{-5})x^4 - 0.0019x^2 + 29.28$ (Use nine data points.)

(c) $\bar{y} = \dfrac{M_x}{A} \approx \dfrac{23,697.68}{1843.54} \approx 12.85$

$(\bar{x}, \bar{y}) = (0, 12.85)$

43. Centroids of the given regions: $(1, 0)$ and $(3, 0)$

Area: $A = 4 + \pi$

$$\bar{x} = \frac{4(1) + \pi(3)}{4 + \pi} = \frac{4 + 3\pi}{4 + \pi}$$

$$\bar{y} = \frac{4(0) + \pi(0)}{4 + \pi} = 0$$

$(\bar{x}, \bar{y}) = \left(\dfrac{4 + 3\pi}{4 + \pi}, 0\right) \sim (1.88, 0)$

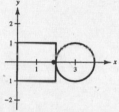

45. Centroids of the given regions: $\left(0, \dfrac{3}{2}\right)$, $(0, 5)$, and $\left(0, \dfrac{15}{2}\right)$

Area: $A = 15 + 12 + 7 = 34$

$$\bar{x} = \frac{15(0) + 12(0) + 7(0)}{34} = 0$$

$$\bar{y} = \frac{15(3/2) + 12(5) + 7(15/2)}{34} = \frac{135}{34}$$

$(\bar{x}, \bar{y}) = \left(0, \dfrac{135}{34}\right) \approx (0, 3.97)$

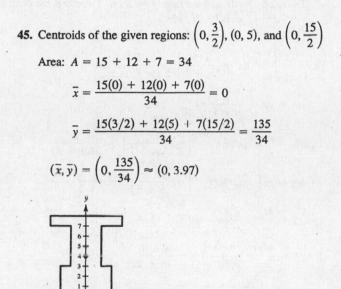

47. Centroids of the given regions: $(1, 0)$ and $(3, 0)$

Mass: $4 + 2\pi$

$$\bar{x} = \frac{4(1) + 2\pi(3)}{4 + 2\pi} = \frac{2 + 3\pi}{2 + \pi}$$

$$\bar{y} = 0$$

$(\bar{x}, \bar{y}) = \left(\dfrac{2 + 3\pi}{2 + \pi}, 0\right) \approx (2.22, 0)$

49. $r = 5$ is distance between center of circle and y-axis.
$A \approx \pi(4)^2 - 16\pi$ is area of circle. Hence,

$$V = 2\pi rA = 2\pi(5)(16\pi) = 160\pi^2 \approx 1579.14.$$

51. $A = \dfrac{1}{2}(4)(4) = 8$

$$\bar{y} = \left(\frac{1}{8}\right)\frac{1}{2}\int_0^4 (4 + x)(4 - x)\,dx = \frac{1}{16}\left[16x - \frac{x^3}{3}\right]_0^4 = \frac{8}{3}$$

$r = \bar{y} = \dfrac{8}{3}$

$$V = 2\pi rA = 2\pi\left(\frac{8}{3}\right)(8) = \frac{128\pi}{3} \approx 134.04$$

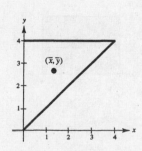

53. $m = m_1 + \cdots + m_n$

$M_y = m_1 x_1 + \cdots + m_n x_n$

$M_x = m_1 y_1 + \cdots + m_n y_n$

$\bar{x} = \dfrac{M_y}{m}, \bar{y} = \dfrac{M_x}{m}$

55. (a) Yes. $(\bar{x}, \bar{y}) = \left(\frac{5}{6}, \frac{5}{18} + 2\right) = \left(\frac{5}{6}, \frac{41}{18}\right)$

(b) Yes. $(\bar{x}, \bar{y}) = \left(\frac{5}{6} + 2, \frac{5}{18}\right) = \left(\frac{17}{6}, \frac{5}{18}\right)$

(c) Yes. $(\bar{x}, \bar{y}) = \left(\frac{5}{6}, -\frac{5}{18}\right)$

(d) No

57. The surface area of the sphere is $S = 4\pi r^2$. The arc length of C is $s = \pi r$. The distance traveled by the centroid is

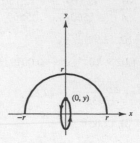

$d = \dfrac{S}{s} = \dfrac{4\pi r^2}{\pi r} = 4r.$

This distance is also the circumference of the circle of radius y.

$d = 2\pi y$

Thus, $2\pi y = 4r$ and we have $y = 2r/\pi$. Therefore, the centroid of the semicircle $y = \sqrt{r^2 - x^2}$ is $(0, 2r/\pi)$.

59. $A = \displaystyle\int_0^1 x^n\, dx = \left[\dfrac{x^{n+1}}{n+1}\right]_0^1 = \dfrac{1}{n+1}$

$m = \rho A = \dfrac{\rho}{n+1}$

$M_x = \dfrac{\rho}{2} \displaystyle\int_0^1 (x^n)^2\, dx = \left[\dfrac{\rho}{2} \cdot \dfrac{x^{2n+1}}{2n+1}\right]_0^1 = \dfrac{\rho}{2(2n+1)}$

$M_y = \rho \displaystyle\int_0^1 x(x^n)\, dx = \left[\rho \cdot \dfrac{x^{n+2}}{n+2}\right]_0^1 = \dfrac{\rho}{n+2}$

$\bar{x} = \dfrac{M_y}{m} = \dfrac{n+1}{n+2}$

$\bar{y} = \dfrac{M_x}{m} = \dfrac{n+1}{2(2n+1)} = \dfrac{n+1}{4n+2}$

Centroid: $\left(\dfrac{n+1}{n+2}, \dfrac{n+1}{4n+2}\right)$

As $n \to \infty$, $(\bar{x}, \bar{y}) \to \left(1, \frac{1}{4}\right)$. The graph approaches the x-axis and the line $x = 1$ as $n \to \infty$.

Section 7.7   Fluid Pressure and Fluid Force

1. $F = PA = [62.4(5)](3) = 936\, \text{lb}$

3. $F = 62.4(h + 2)(6) - (62.4)(h)(6)$

$= 62.4(2)(6) = 748.8\, \text{lb}$

5. $h(y) = 3 - y$

$L(y) = 4$

$F = 62.4 \displaystyle\int_0^3 (3 - y)(4)\, dy$

$= 249.6 \displaystyle\int_0^3 (3 - y)\, dy$

$= 249.6 \left[3y - \dfrac{y^2}{2}\right]_0^3$

$= 1123.2\, \text{lb}$

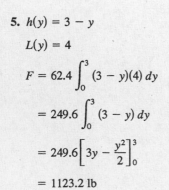

7. $h(y) = 3 - y$

$L(y) = 2\left(\dfrac{y}{3} + 1\right)$

$F = 2(62.4) \displaystyle\int_0^3 (3 - y)\left(\dfrac{y}{3} + 1\right) dy$

$= 124.8 \displaystyle\int_0^3 \left(3 - \dfrac{y^2}{3}\right) dy$

$= 124.8 \left[3y - \dfrac{y^3}{9}\right]_0^3$

$= 748.8\, \text{lb}$

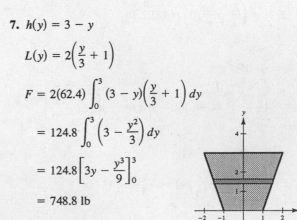

9. $h(y) = 4 - y$

$L(y) = 2\sqrt{y}$

$F = 2(62.4) \displaystyle\int_0^4 (4 - y)\sqrt{y}\, dy$

$= 124.8 \displaystyle\int_0^4 (4y^{1/2} - y^{3/2})\, dy$

$= 124.8 \left[\dfrac{8y^{3/2}}{3} - \dfrac{2y^{5/2}}{5} \right]_0^4$

$= 1064.96$ lb

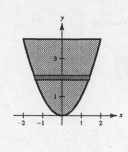

11. $h(y) = 4 - y$

$L(y) = 2$

$F = 9800 \displaystyle\int_0^2 2(4 - y)\, dy$

$= 9800 \left[8y - y^2 \right]_0^2 = 117,600$ newtons

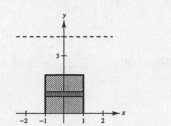

13. $h(y) = 12 - y$

$L(y) = 6 - \dfrac{2y}{3}$

$F = 9800 \displaystyle\int_0^9 (12 - y)\left(6 - \dfrac{2y}{3} \right) dy$

$= 9800 \left[72y - 7y^2 + \dfrac{2y^3}{9} \right]_0^9$

$= 2{,}381{,}400$ newtons

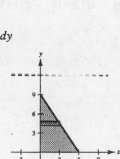

15. $h(y) = 2 - y$

$L(y) = 10$

$F = 140.7 \displaystyle\int_0^2 (2 - y)(10)\, dy$

$= 1407 \displaystyle\int_0^2 (2 - y)\, dy$

$= 1407 \left[2y - \dfrac{y^2}{2} \right]_0^2 = 2814$ lb

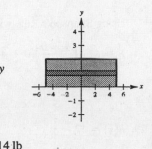

17. $h(y) = 4 - y$

$L(y) = 6$

$F = 140.7 \displaystyle\int_0^4 (4 - y)(6)\, dy$

$= 844.2 \displaystyle\int_0^4 (4 - y)\, dy$

$= 844.2 \left[4y - \dfrac{y^2}{2} \right]_0^4 = 6753.6$ lb

19. $h(y) = -y$

$L(y) = 2\left(\dfrac{1}{2} \right)\sqrt{9 - 4y^2}$

$F = 42 \displaystyle\int_{-3/2}^0 (-y)\sqrt{9 - 4y^2}\, dy$

$= \dfrac{42}{8} \displaystyle\int_{-3/2}^0 (9 - 4y^2)^{1/2}(-8y)\, dy$

$= \left[\left(\dfrac{21}{4} \right)\left(\dfrac{2}{3} \right)(9 - 4y^2)^{3/2} \right]_{-3/2}^0 = 94.5$ lb

21. $h(y) = k - y$

$L(y) = 2\sqrt{r^2 - y^2}$

$F = w \displaystyle\int_{-r}^r (k - y)\sqrt{r^2 - y^2}\,(2)\, dy$

$= w \left[2k \displaystyle\int_{-r}^r \sqrt{r^2 - y^2}\, dy + \displaystyle\int_{-r}^r \sqrt{r^2 - y^2}\,(-2y)\, dy \right]$

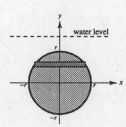

The second integral is zero since its integrand is odd and the
limits of integration are symmetric to the origin. The first integral is the area of a semicircle with radius r.

$F = w \left[(2k)\dfrac{\pi r^2}{2} + 0 \right] = wk\pi r^2$

23. $h(y) = k - y$

$L(y) = b$

$$F = w \int_{-h/2}^{h/2} (k - y)b \, dy$$

$$= wb \left[ky - \frac{y^2}{2} \right]_{-h/2}^{h/2} = wb(hk) = wkhb$$

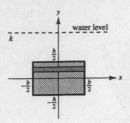

25. From Exercise 23:

$F = 64(15)(1)(1) = 960$ lb

27. $h(y) = 4 - y$

$$F = 62.4 \int_0^4 (4 - y)L(y) \, dy$$

Using Simpson's Rule with $n = 8$ we have:

$$F \approx 62.4 \left(\frac{4 - 0}{3(8)} \right) [0 + 4(3.5)(3) + 2(3)(5) + 4(2.5)(8) + 2(2)(9) + 4(1.5)(10) + 2(1)(10.25) + 4(0.5)(10.5) + 0]$$

$= 3010.8$ lb

29. $h(y) = 12 - y$

$L(y) = 2(4^{2/3} - y^{2/3})^{3/2}$

$$F = 62.4 \int_0^4 2(12 - y)(4^{2/3} - y^{2/3})^{3/2} \, dy$$

≈ 6448.73 lb

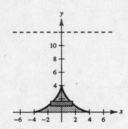

31. (a) If the fluid force is one-half of 1123.2 lb, and the height of the water is b, then

$h(y) = b - y$

$L(y) = 4$

$$F = 62.4 \int_0^b (b - y)(4) \, dy = \frac{1}{2}(1123.2)$$

$$\int_0^b (b - y) \, dy = 2.25$$

$$\left[by - \frac{y^2}{2} \right]_0^b = 2.25$$

$$b^2 - \frac{b^2}{2} = 2.25$$

$$b^2 = 4.5 \implies b \approx 2.12 \text{ ft.}$$

(b) The pressure increases with increasing depth.

33. $F = Fw = w \int_c^d h(y)L(y) \, dy$, see page 508.

Review Exercises for Chapter 7

1. $A = \int_1^5 \frac{1}{x^2}\,dx = \left[-\frac{1}{x}\right]_1^5 = \frac{4}{5}$

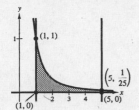

3. $A = \int_{-1}^1 \frac{1}{x^2+1}\,dx$

$ = \Big[\arctan x\Big]_{-1}^1$

$ = \frac{\pi}{4} - \left(-\frac{\pi}{4}\right) = \frac{\pi}{2}$

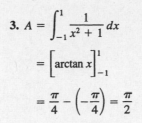

5. $A = 2\int_0^1 (x - x^3)\,dx$

$ = 2\left[\frac{1}{2}x^2 - \frac{1}{4}x^4\right]_0^1$

$ = \frac{1}{2}$

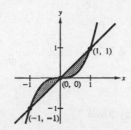

7. $A = \int_0^2 (e^2 - e^x)\,dx$

$ = \Big[xe^2 - e^x\Big]_0^2$

$ = e^2 + 1$

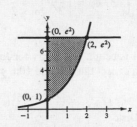

9. $A = \int_{\pi/4}^{5\pi/4} (\sin x - \cos x)\,dx$

$ = \Big[-\cos x - \sin x\Big]_{\pi/4}^{5\pi/4}$

$ = \left(\frac{1}{\sqrt{2}} + \frac{1}{\sqrt{2}}\right) - \left(-\frac{1}{\sqrt{2}} - \frac{1}{\sqrt{2}}\right)$

$ = \frac{4}{\sqrt{2}} = 2\sqrt{2}$

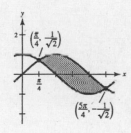

11. $A = \int_0^8 \left[(3 + 8x - x^2) - (x^2 - 8x + 3)\right]dx$

$ - \int_0^8 (16x - 2x^2)\,dx$

$ = \left[8x^2 - \frac{2}{3}x^3\right]_0^8 = \frac{512}{3} \approx 170.667$

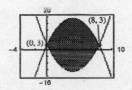

13. $y = \left(1 - \sqrt{x}\right)^2$

$A = \int_0^1 \left(1 - \sqrt{x}\right)^2 dx$

$ = \int_0^1 (1 - 2x^{1/2} + x)\,dx$

$ = \left[x - \frac{4}{3}x^{3/2} + \frac{1}{2}x^2\right]_0^1 = \frac{1}{6} \approx 0.1667$

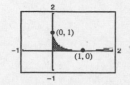

15. $x = y^2 - 2y \Rightarrow x + 1 = (y - 1)^2 \Rightarrow y = 1 \pm \sqrt{x+1}$

$A = \int_{-1}^0 \left[\left(1 + \sqrt{x+1}\right) - \left(1 - \sqrt{x+1}\right)\right]dx$

$ = \int_{-1}^0 2\sqrt{x+1}\,dx$

$A = \int_0^2 \left[0 - (y^2 - 2y)\right]dy$

$ = \int_0^2 (2y - y^2)\,dy$

$ - \left[y^2 - \frac{1}{3}y^3\right]_0^2$

$ = \frac{4}{3}$

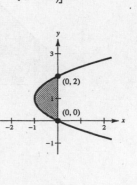

17. $A = \int_0^2 \left[1 - \left(1 - \dfrac{x}{2}\right)\right] dx + \int_2^3 \left[1 - (x - 2)\right] dx$

$= \int_0^2 \dfrac{x}{2}\, dx + \int_2^3 (3 - x)\, dx$

$y = 1 - \dfrac{x}{2} \implies x = 2 - 2y$

$y = x - 2 \implies x = y + 2,\; y = 1$

$A = \int_0^1 \left[(y + 2) - (2 - 2y)\right] dy$

$= \int_0^1 3y\, dy = \left[\dfrac{3}{2}y^2\right]_0^1 = \dfrac{3}{2}$

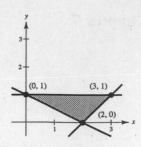

19. Job 1 is better. The salary for Job 1 is greater than the salary for Job 2 for all the years except the first and 10th years.

21. (a) Disk

$V = \pi \int_0^4 x^2\, dx = \left[\dfrac{\pi x^3}{3}\right]_0^4 = \dfrac{64\pi}{3}$

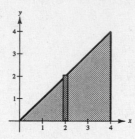

(b) Shell

$V = 2\pi \int_0^4 x^2\, dx = \left[\dfrac{2\pi}{3}x^3\right]_0^4 = \dfrac{128\pi}{3}$

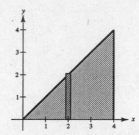

(c) Shell

$V = 2\pi \int_0^4 (4 - x)x\, dx$

$= 2\pi \int_0^4 (4x - x^2)\, dx$

$= 2\pi \left[2x^2 - \dfrac{x^3}{3}\right]_0^4 = \dfrac{64\pi}{3}$

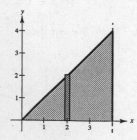

(d) Shell

$V = 2\pi \int_0^4 (6 - x)x\, dx$

$= 2\pi \int_0^4 (6x - x^2)\, dx$

$= 2\pi \left[3x^2 - \dfrac{1}{3}x^3\right]_0^4 = \dfrac{160\pi}{3}$

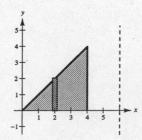

23. (a) **Shell**

$$V = 4\pi \int_0^4 x\left(\frac{3}{4}\right)\sqrt{16 - x^2}\, dx$$

$$= \left[3\pi\left(-\frac{1}{2}\right)\left(\frac{2}{3}\right)(16 - x^2)^{3/2}\right]_0^4 = 64\pi$$

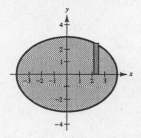

(b) **Disk**

$$V = 2\pi \int_0^4 \left[\frac{3}{4}\sqrt{16 - x^2}\right]^2 dx$$

$$= \frac{9\pi}{8}\left[16x - \frac{x^3}{3}\right]_0^4 = 48\pi$$

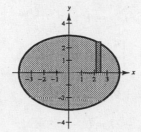

25. Shell

$$V = 2\pi \int_0^1 \frac{x}{x^4 + 1}\, dx$$

$$= \pi \int_0^1 \frac{(2x)}{(x^2)^2 + 1}\, dx$$

$$= \left[\pi \arctan(x^2)\right]_0^1$$

$$= \pi\left[\frac{\pi}{4} - 0\right] = \frac{\pi^2}{4}$$

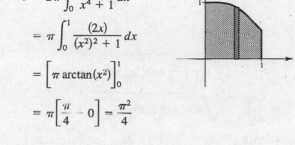

27. Shell: $V = 2\pi \int_2^6 \frac{x}{1 + \sqrt{x - 2}}\, dx$

$$u = \sqrt{x - 2}$$

$$x = u^2 + 2$$

$$dx = 2u\, du$$

$$V = 2\pi \int_2^6 \frac{x}{1 + \sqrt{x - 2}}\, dx = 4\pi \int_0^2 \frac{(u^2 + 2)u}{1 + u}\, du$$

$$= 4\pi \int_0^2 \frac{u^3 + 2u}{1 + u}\, du = 4\pi \int_0^2 \left(u^2 - u + 3 - \frac{3}{1 + u}\right) du$$

$$= 4\pi\left[\frac{1}{3}u^3 - \frac{1}{2}u^2 + 3u - 3\ln(1 + u)\right]_0^2 = \frac{4\pi}{3}(20 - 9\ln 3) \approx 42.359$$

29. Since $y \le 0$, $A = -\int_{-1}^0 x\sqrt{x + 1}\, dx$.

$$u = x + 1$$

$$x = u - 1$$

$$dx = du$$

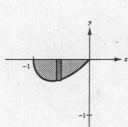

$$A = -\int_0^1 (u - 1)\sqrt{u}\, du = -\int_0^1 (u^{3/2} - u^{1/2})\, du$$

$$= -\left[\frac{2}{5}u^{5/2} - \frac{2}{3}u^{3/2}\right]_0^1 = \frac{4}{15}$$

31. From Exercise 23(a) we have: $V = 64\pi \text{ ft}^3$

$$\frac{1}{4}V = 16\pi$$

Disk: $\pi \displaystyle\int_{-3}^{y_0} \frac{16}{9}(9 - y^2)\, dy = 16\pi$

$$\frac{1}{9}\int_{-3}^{y_0} (9 - y^2)\, dy = 1$$

$$\left[9y - \frac{1}{3}y^3\right]_{-3}^{y_0} = 9$$

$$\left(9y_0 - \frac{1}{3}y_0^3\right) - (-27 + 9) = 9$$

$$y_0^3 - 27y_0 - 27 = 0$$

By Newton's Method, $y_0 \approx -1.042$ and the depth of the gasoline is $3 - 1.042 = 1.958$ ft.

33. $f(x) = \dfrac{4}{5}x^{5/4}$

$$f'(x) = x^{1/4}$$

$$1 + [f'(x)]^2 = 1 + \sqrt{x}$$

$$u = 1 + \sqrt{x}$$

$$x = (u - 1)^2$$

$$dx = 2(u - 1)\, du$$

$$s = \int_0^4 \sqrt{1 + \sqrt{x}}\, dx = 2\int_1^3 \sqrt{u}(u - 1)\, du$$

$$= 2\int_1^3 (u^{3/2} - u^{1/2})\, du$$

$$= 2\left[\frac{2}{5}u^{5/2} - \frac{2}{3}u^{3/2}\right]_1^3 = \frac{4}{15}\left[u^{3/2}(3u - 5)\right]_1^3$$

$$= \frac{8}{15}\left(1 + 6\sqrt{3}\right) \approx 6.076$$

35. $y = 300\cosh\left(\dfrac{x}{2000}\right) - 280,\ -2000 \le x \le 2000$

$$y' = \frac{3}{20}\sinh\left(\frac{x}{2000}\right)$$

$$s = \int_{-2000}^{2000} \sqrt{1 + \left[\frac{3}{20}\sinh\left(\frac{x}{2000}\right)\right]^2}\, dx$$

$$= \frac{1}{20}\int_{-2000}^{2000} \sqrt{400 + 9\sinh^2\left(\frac{x}{2000}\right)}\, dx$$

$$\approx 4018.2 \text{ ft (by Simpson's Rule or graphing utility)}$$

37. $y = \dfrac{3}{4}x$

$$y' = \frac{3}{4}$$

$$1 + (y')^2 = \frac{25}{16}$$

$$S = 2\pi \int_0^4 \left(\frac{3}{4}x\right)\sqrt{\frac{25}{16}}\, dx = \left[\left(\frac{15\pi}{8}\right)\frac{x^2}{2}\right]_0^4 = 15\pi$$

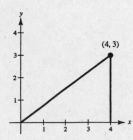

39. $F = kx$

$$4 = k(1)$$

$$F = 4x$$

$$W = \int_0^5 4x\, dx = \left[2x^2\right]_0^5$$

$$= 50 \text{ in.} \cdot \text{lb} \approx 4.167 \text{ ft} \cdot \text{lb}$$

41. Volume of disk: $\pi\left(\dfrac{1}{3}\right)^2 \Delta y$

Weight of disk: $62.4\pi\left(\dfrac{1}{3}\right)^2 \Delta y$

Distance: $175 - y$

$$W = \frac{62.4\pi}{9}\int_0^{150}(175 - y)\,dy = \frac{62.4\pi}{9}\left[175y - \frac{y^2}{2}\right]_0^{150}$$

$$= 104{,}000\pi \text{ ft} \cdot \text{lb} \approx 163.4 \text{ ft} \cdot \text{ton}$$

43. Weight of section of chain: $5\,\Delta x$

Distance moved: $10 - x$

$$W = 5\int_0^{10}(10 - x)\,dx = \left[-\frac{5}{2}(10 - x)^2\right]_0^{10} = 250 \text{ ft} \cdot \text{lb}$$

45. $W = \displaystyle\int_a^b F(x)\,dx$

$$80 = \int_0^4 ax^2\,dx = \frac{ax^3}{3}\bigg]_0^4 = \frac{64}{3}a$$

$$a = \frac{3(80)}{64} = \frac{15}{4} = 3.75$$

47. $A = \displaystyle\int_0^a \left(\sqrt{a} - \sqrt{x}\right)^2 dx = \int_0^a \left(a - 2\sqrt{a}x^{1/2} + x\right)dx = \left[ax - \frac{4}{3}\sqrt{a}x^{3/2} + \frac{1}{2}x^2\right]_0^a = \frac{a^2}{6}$

$$\frac{1}{A} = \frac{6}{a^2}$$

$$\bar{x} = \frac{6}{a^2}\int_0^a x\left(\sqrt{a} - \sqrt{x}\right)^2 dx = \frac{6}{a^2}\int_0^a \left(ax - 2\sqrt{a}x^{3/2} + x^2\right)dx = \frac{a}{5}$$

$$\bar{y} = \left(\frac{6}{a^2}\right)\frac{1}{2}\int_0^a \left(\sqrt{a} - \sqrt{x}\right)^4 dx$$

$$= \frac{3}{a^2}\int_0^a \left(a^2 - 4a^{3/2}x^{1/2} + 6ax - 4a^{1/2}x^{3/2} + x^2\right)dx$$

$$= \frac{3}{a^2}\left[a^2 x - \frac{8}{3}a^{3/2}x^{3/2} + 3ax^2 - \frac{8}{5}a^{1/2}x^{5/2} + \frac{1}{3}x^3\right]_0^a = \frac{a}{5}$$

$$(\bar{x}, \bar{y}) = \left(\frac{a}{5}, \frac{a}{5}\right)$$

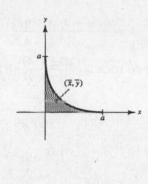

49. By symmetry, $x = 0$.

$$A = 2\int_0^1 (a^2 - x^2)\,dx = 2\left[a^2 x - \frac{x^3}{3}\right]_0^a = \frac{4a^3}{3}$$

$$\frac{1}{A} = \frac{3}{4a^3}$$

$$\bar{y} = \left(\frac{3}{4a^3}\right)\frac{1}{2}\int_{-a}^a (a^2 - x^2)^2 dx$$

$$= \frac{6}{8a^3}\int_0^a (a^4 - 2a^2 x^2 + x^4)\,dx$$

$$= \frac{6}{8a^3}\left[a^4 x - \frac{2a^2}{3}x^3 + \frac{1}{5}x^5\right]_0^a$$

$$= \frac{6}{8a^3}\left(a^5 - \frac{2}{3}a^5 + \frac{1}{5}a^5\right) = \frac{2a^2}{5}$$

$$(\bar{x}, \bar{y}) = \left(0, \frac{2a^2}{5}\right)$$

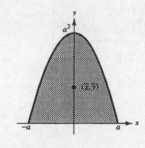

51. $\bar{y} = 0$ by symmetry.

For the trapezoid:

$$m = [(4)(6) - (1)(6)]\rho = 18\rho$$

$$M_y = \rho \int_0^6 x\left[\left(\frac{1}{6}x + 1\right) - \left(-\frac{1}{6}x - 1\right)\right] dx$$

$$= \rho \int_0^6 \left(\frac{1}{3}x^2 + 2x\right) dx = \rho\left[\frac{x^3}{9} + x^2\right]_0^6 = 60\rho$$

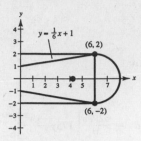

For the semicircle:

$$m = \left(\frac{1}{2}\right)(\pi)(2)^2\rho = 2\pi\rho$$

$$M_y = \rho \int_6^8 x\left[\sqrt{4 - (x - 6)^2} - \left(-\sqrt{4 - (x - 6)^2}\right)\right] dx = 2\rho \int_6^8 x\sqrt{4 - (x - 6)^2}\, dx$$

Let $u = x - 6$, then $x = u + 6$ and $dx = du$. When $x = 6$, $u = 0$. When $x = 8$, $u = 2$.

$$M_y = 2\rho \int_0^2 (u + 6)\sqrt{4 - u^2}\, du = 2\rho \int_0^2 u\sqrt{4 - u^2}\, du + 12\rho \int_0^2 \sqrt{4 - u^2}\, du$$

$$= 2\rho\left[\left(-\frac{1}{2}\right)\left(\frac{2}{3}\right)(4 - u^2)^{3/2}\right]_0^2 + 12\rho\left[\frac{\pi(2)^2}{4}\right] = \frac{16\rho}{3} + 12\pi\rho = \frac{4\rho(4 + 9\pi)}{3}$$

Thus, we have:

$$\bar{x}(18\rho + 2\pi\rho) = 60\rho + \frac{4\rho(4 + 9\pi)}{3}$$

$$\bar{x} = \frac{180\rho + 4\rho(4 + 9\pi)}{3} \cdot \frac{1}{2\rho(9 + \pi)} = \frac{2(9\pi + 49)}{3(\pi + 9)}$$

The centroid of the blade is $\left(\dfrac{2(9\pi + 49)}{3(\pi + 9)}, 0\right)$.

53. Let D = surface of liquid; ρ = weight per cubic volume.

$$F = \rho \int_c^d (D - y)[f(y) - g(y)]\, dy$$

$$= \rho\left[\int_c^d D[f(y) - g(y)]\, dy - \int_c^d y[f(y) - g(y)]\, dy\right]$$

$$= \rho\left[\int_c^d [f(y) - g(y)]\, dy\right]\left[D - \frac{\displaystyle\int_c^d y[f(y) - g(y)]\, dy}{\displaystyle\int_c^d [f(y) - g(y)]\, dy}\right]$$

$$= \rho(\text{Area})(D - \bar{y})$$

$$= \rho(\text{Area})(\text{depth of centroid})$$

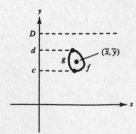

Problem Solving for Chapter 7

1. $T = \frac{1}{2}c(c^2) = \frac{1}{2}c^3$

$R = \int_0^c (cx - x^2)\, dx = \left[\frac{cx^2}{2} - \frac{x^3}{3}\right]_0^c = \frac{c^3}{2} - \frac{c^3}{3} = \frac{c^3}{6}$

$\lim_{c \to 0^+} \frac{T}{R} = \lim_{c \to 0^+} \frac{\frac{1}{2}c^3}{\frac{1}{6}c^3} = 3$

3. (a) $\frac{1}{2}V = \int_0^1 \left[\pi\left(2 + \sqrt{1 - y^2}\right)^2 - \pi\left(2 - \sqrt{1 - y^2}\right)^2\right] dy$

$= \pi \int_0^1 \left[\left(4 + 4\sqrt{1 - y^2} + (1 - y^2)\right) - \left(4 - 4\sqrt{1 - y^2} + (1 - y^2)\right)\right] dy$

$= 8\pi \int_0^1 \sqrt{1 - y^2}\, dy$ (Integral represents $1/4$ (area of circle))

$= 8\pi\left(\frac{\pi}{4}\right) = 2\pi^2 \implies V = 4\pi^2$

(b) $(x - R)^2 + y^2 = r^2 \implies x = R \pm \sqrt{r^2 - y^2}$

$\frac{1}{2}V = \int_0^r \left[\pi\left(R + \sqrt{r^2 - y^2}\right)^2 - \pi\left(R - \sqrt{r^2 - y^2}\right)^2\right] dy$

$= \pi \int_0^r 4R\sqrt{r^2 - y^2}\, dy$

$= \pi(4R)\frac{1}{4}\pi r^2 = \pi^2 r^2 R$

$V = 2\pi^2 r^2 R$

5. $V = 2(2\pi) \int_{\sqrt{r^2 - (h^2/4)}}^r x\sqrt{r^2 - x^2}\, dx$

$= -2\pi\left[\frac{2}{3}(r^2 - x^2)^{3/2}\right]_{\sqrt{r^2 - (h^2/4)}}^r$

$= \frac{-4\pi}{3}\left[-\frac{h^3}{8}\right] = \frac{\pi h^3}{6}$ which does not depend on r!

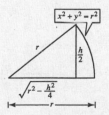

7. (a) Tangent at A: $y = x^3$, $y' = 3x^2$

$$y - 1 = 3(x - 1)$$
$$y = 3x - 2$$

To find point B:
$$x^3 = 3x - 2$$
$$x^3 - 3x + 2 = 0$$
$$(x - 1)^2(x + 2) = 0 \implies B = (-2, -8)$$

Tangent at B: $y = x^3$, $y' = 3x^2$
$$y + 8 = 12(x + 2)$$
$$y = 12x + 16$$

To find point C:
$$x^3 = 12x + 16$$
$$x^3 - 12x - 16 = 0$$
$$(x + 2)^2(x - 4) = 0 \implies C = (4, 64)$$

Area of $R = \int_{-2}^{1} (x^3 - 3x + 2)\, dx = \dfrac{27}{4}$

Area of $S = \int_{-2}^{4} (12x + 16 - x^3)\, dx = 108$

Area of $S = 16(\text{area of } R)$ $\left[\dfrac{\text{area } S}{\text{area } R} = 16\right]$

(b) Tangent at $A(a, a^3)$: $y - a^3 = 3a^2(x - a)$

$$y = 3a^2x - 2a^3$$

To find point B: $x^3 - 3a^2x + 2a^3 = 0$
$$(x - a)^2(x + 2a) = 0$$
$$\implies B = (-2a, -8a^3)$$

Tangent at B: $y + 8a^3 = 12a^2(x + 2a)$
$$y = 12a^2x + 16a^3$$

To find point C: $x^3 - 12a^2x - 16a^3 = 0$
$$(x + 2a)^2(x - 4a) = 0$$
$$\implies C = (4a, 64a^3)$$

Area of $R = \int_{-2a}^{a} [x^3 - 3a^2x + 2a^3]\, dx = \dfrac{27}{4}a^4$

Area of $S = \int_{-2a}^{4a} [12a^2x + 16a^3 - x^3]\, dx = 108a^4$

Area of $S = 16(\text{area of } R)$

9. $s(x) = \displaystyle\int_{\alpha}^{x} \sqrt{1 + f'(t)^2}\, dt$

(a) $s'(x) = \dfrac{ds}{dx} = \sqrt{1 + f'(x)^2}$

(b) $ds = \sqrt{1 + f'(x)^2}\, dx$

$$(ds)^2 = [1 + f'(x)^2](dx)^2 = \left[1 + \left(\dfrac{dy}{dx}\right)^2\right](dx)^2 = (dx)^2 + (dy)^2$$

(c) $s(x) = \displaystyle\int_{1}^{x} \sqrt{1 + \left(\dfrac{3}{2}t^{1/2}\right)^2}\, dt = \int_{1}^{x} \sqrt{1 + \dfrac{9}{4}t}\, dt$

(d) $s(2) = \displaystyle\int_{1}^{2} \sqrt{1 + \dfrac{9}{4}t}\, dt = \left[\dfrac{8}{27}\left(1 + \dfrac{9}{4}t\right)^{3/2}\right]_{1}^{2} = \dfrac{22}{27}\sqrt{22} - \dfrac{13}{27}\sqrt{13} \approx 2.0858$

This is the length of the curve $y = x^{3/2}$ from $x = 1$ to $x = 2$.

11. (a) $\bar{y} = 0$ by symmetry

$$M_y = \int_{1}^{6} x\left(\dfrac{1}{x^3} - \left(-\dfrac{1}{x^3}\right)\right) dx = \int_{1}^{6} \dfrac{2}{x^2}\, dx = \left[-2\dfrac{1}{x}\right]_{1}^{6} = \dfrac{5}{3}$$

$$m = 2\int_{1}^{6} \dfrac{1}{x^3}\, dx = \left[-\dfrac{1}{x^2}\right]_{1}^{6} = \dfrac{35}{36}$$

$$\bar{x} = \dfrac{5/3}{35/36} = \dfrac{12}{7} \qquad (\bar{x}, \bar{y}) = \left(\dfrac{12}{7}, 0\right)$$

(b) $m = 2\displaystyle\int_{1}^{b} \dfrac{1}{x^3}\, dx = \dfrac{b^2 - 1}{b^2}$

$$M_y = 2\int_{1}^{6} \dfrac{1}{x^2}\, dx = \dfrac{2(b - 1)}{b}$$

$$\bar{x} = \dfrac{2(b-1)/b}{(b^2-1)/b^2} = \dfrac{2b}{b + 1} \qquad (\bar{x}, \bar{y}) = \left(\dfrac{2b}{b + 1}, 0\right)$$

(c) $\displaystyle\lim_{b \to \infty} \bar{x} = \lim_{b \to \infty} \dfrac{2b}{b + 1} = 2 \qquad (\bar{x}, \bar{y}) = (2, 0)$

13. (a) $W = \text{area} = 2 + 4 + 6 = 12$

(b) $W = \text{area} = 3 + (1 + 1) + 2 + \dfrac{1}{2} = 7\dfrac{1}{2}$

15. Point of equilibrium: $50 - 0.5x = 0.125x$

$$x = 80, p = 10$$

$(P_0, x_0) = (10, 80)$

Consumer surplus $= \displaystyle\int_0^{80} [(50 - 0.5x) - 10]\, dx = 1600$

Producer surplus $= \displaystyle\int_0^{80} [10 - 0.125x]\, dx = 400$

17. We use Exercise 23, Section 7.7, which gives $F = wkhb$ for a rectangle plate.

Wall at shallow end

From Exercise 23: $F = 62.4(2)(4)(20) = 9984\text{ lb}$

Wall at deep end

From Exercise 23: $F = 62.4(4)(8)(20) = 39{,}936\text{ lb}$

Side wall

From Exercise 23: $F_1 = 62.4(2)(4)(40) = 19{,}968\text{ lb}$

$$F_2 = 62.4 \int_0^4 (8 - y)(10y)\, dy$$

$$= 624 \int_0^4 (8y - y^2)\, dy = 624\left[4y^2 - \frac{y^3}{3}\right]_0^4$$

$$= 26{,}624\text{ lb}$$

Total force: $F_1 + F_2 = 46{,}592\text{ lb}$

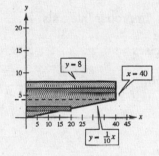

C H A P T E R 8
Integration Techniques, L'Hôpital's Rule, and Improper Integrals

Section 8.1   Basic Integration Rules 363

Section 8.2   Integration by Parts 369

Section 8.3   Trigonometric Integrals 380

Section 8.4   Trigonometric Substitution 387

Section 8.5   Partial Fractions 398

Section 8.6   Integration by Tables and Other Integration Techniques . . 404

Section 8.7   Indeterminate Forms and L'Hôpital's Rule 410

Section 8.8   Improper Integrals 417

Review Exercises 425

Problem Solving 431

CHAPTER 8
Integration Techniques, L'Hôpital's Rule, and Improper Integrals

Section 8.1   Basic Integration Rules

1. (a) $\dfrac{d}{dx}\left[2\sqrt{x^2 + 1} + C\right] = 2\left(\dfrac{1}{2}\right)(x^2 + 1)^{-1/2}(2x)$

$$= \dfrac{2x}{\sqrt{x^2 + 1}}$$

(b) $\dfrac{d}{dx}\left[\sqrt{x^2 + 1} + C\right] = \dfrac{1}{2}(x^2 + 1)^{-1/2}(2x) = \dfrac{x}{\sqrt{x^2 + 1}}$

(c) $\dfrac{d}{dx}\left[\dfrac{1}{2}\sqrt{x^2 + 1} + C\right] = \dfrac{1}{2}\left(\dfrac{1}{2}\right)(x^2 + 1)^{-1/2}(2x)$

$$= \dfrac{x}{2\sqrt{x^2 + 1}}$$

(d) $\dfrac{d}{dx}[\ln(x^2 + 1) + C] = \dfrac{2x}{x^2 + 1}$

$\displaystyle\int \dfrac{x}{\sqrt{x^2 + 1}}\, dx$ matches (b).

3. (a) $\dfrac{d}{dx}\left[\ln\sqrt{x^2 + 1} + C\right] = \dfrac{1}{2}\left(\dfrac{2x}{x^2 + 1}\right) = \dfrac{x}{x^2 + 1}$

(b) $\dfrac{d}{dx}\left[\dfrac{2x}{(x^2 + 1)^2} + C\right] = \dfrac{(x^2 + 1)^2(2) - (2x)(2)(x^2 + 1)(2x)}{(x^2 + 1)^4} = \dfrac{2(1 - 3x^2)}{(x^2 + 1)^3}$

(c) $\dfrac{d}{dx}[\arctan x + C] = \dfrac{1}{1 + x^2}$

(d) $\dfrac{d}{dx}[\ln(x^2 + 1) + C] - \dfrac{2x}{x^2 + 1}$

$\displaystyle\int \dfrac{1}{x^2 + 1}\, dx$ matches (c).

5. $\displaystyle\int (3x - 2)^4\, dx$

$u = 3x - 2, du = 3\, dx, n = 4$

Use $\displaystyle\int u^n\, du.$

7. $\displaystyle\int \dfrac{1}{\sqrt{x}\left(1 - 2\sqrt{x}\right)}\, dx$

$u = 1 - 2\sqrt{x}, du = -\dfrac{1}{\sqrt{x}}\, dx$

Use $\displaystyle\int \dfrac{du}{u}.$

9. $\displaystyle\int \dfrac{3}{\sqrt{1 - t^2}}\, dt$

$u = t, du = dt, a = 1$

Use $\displaystyle\int \dfrac{du}{\sqrt{a^2 - u^2}}.$

11. $\displaystyle\int t \sin t^2\, dt$

$u = t^2, du = 2t\, dt$

Use $\displaystyle\int \sin u\, du.$

13. $\displaystyle\int (\cos x)e^{\sin x}\, dx$

$u = \sin x, du = \cos x\, dx$

Use $\displaystyle\int e^u\, du.$

15. Let $u = x - 4, du = dx.$

$$\int 6(x - 4)^5 \, dx = 6 \int (x - 4)^5 \, dx = 6 \frac{(x - 4)^6}{6} + C$$

$$= (x - 4)^6 + C$$

17. Let $u = z - 4, du = dz.$

$$\int \frac{5}{(z - 4)^5} \, dz = 5 \int (z - 4)^{-5} \, dz = 5 \frac{(z - 4)^{-4}}{-4} + C$$

$$= \frac{-5}{4(z - 4)^4} + C$$

19. $\int \left[v + \frac{1}{(3v - 1)^3} \right] dv = \int v \, dv + \frac{1}{3} \int (3v - 1)^{-3}(3) \, dv$

$$= \frac{1}{2}v^2 - \frac{1}{6(3v - 1)^2} + C$$

21. Let $u = -t^3 + 9t + 1, du = (-3t^2 + 9) \, dt = -3(t^2 - 3) \, dt.$

$$\int \frac{t^2 - 3}{-t^3 + 9t + 1} \, dt = -\frac{1}{3} \int \frac{-3(t^2 - 3)}{-t^3 + 9t + 1} \, dt$$

$$= -\frac{1}{3} \ln|-t^3 + 9t + 1| + C$$

23. $\int \frac{x^2}{x - 1} \, dx = \int (x + 1) \, dx + \int \frac{1}{x - 1} \, dx$

$$= \frac{1}{2}x^2 + x + \ln|x - 1| + C$$

25. Let $u = 1 + e^x, du = e^x \, dx.$

$$\int \frac{e^x}{1 + e^x} \, dx = \ln(1 + e^x) + C$$

27. $\int (1 + 2x^2)^2 \, dx = \int (4x^4 + 4x^2 + 1) \, dx = \frac{4}{5}x^5 + \frac{4}{3}x^3 + x + C = \frac{x}{15}(12x^4 + 20x^2 + 15) + C$

29. Let $u = 2\pi x^2, du = 4\pi x \, dx.$

$$\int x(\cos 2\pi x^2) \, dx = \frac{1}{4\pi} \int (\cos 2\pi x^2)(4\pi x) \, dx$$

$$= \frac{1}{4\pi} \sin 2\pi x^2 + C$$

31. Let $u = \pi x, du = \pi \, dx.$

$$\int \csc(\pi x) \cot(\pi x) \, dx = \frac{1}{\pi} \int \csc(\pi x) \cot(\pi x)\pi \, dx$$

$$= -\frac{1}{\pi} \csc(\pi x) + C$$

33. Let $u = 5x, du = 5 \, dx.$

$$\int e^{5x} \, dx = \frac{1}{5} \int e^{5x}(5) \, dx = \frac{1}{5}e^{5x} + C$$

35. Let $u = 1 + e^x, du = e^x \, dx.$

$$\int \frac{2}{e^{-x} + 1} \, dx = 2 \int \left(\frac{1}{e^{-x} + 1} \right)\left(\frac{e^x}{e^x} \right) dx$$

$$= 2 \int \frac{e^x}{1 + e^x} \, dx$$

$$= 2 \ln(1 + e^x) + C$$

37. $\int \frac{\ln x^2}{x} \, dx = 2 \int (\ln x)\frac{1}{x} \, dx = 2 \frac{(\ln x)^2}{2} + C = (\ln x)^2 + C$

39. $\int \frac{1 + \sin x}{\cos x} \, dx = \int \frac{1 + \sin x}{\cos x} \cdot \frac{1 - \sin x}{1 - \sin x} \, dx$

$$= \int \frac{1 - \sin^2 x}{\cos x(1 - \sin x)} \, dx$$

$$= \int \frac{\cos^2 x}{\cos x(1 - \sin x)} \, dx$$

$$= -\int \frac{-\cos x}{1 - \sin x} \, dx$$

$$= -\ln|1 - \sin x| + C, \quad (u = 1 - \sin x)$$

Alternate Solution:

$$\int \frac{1 + \sin x}{\cos x} \, dx = \int (\sec x + \tan x) \, dx$$

$$= \ln|\sec x + \tan x| + \ln|\sec x| + C$$

$$= \ln|\sec x(\sec x + \tan x)| + C$$

41. $\dfrac{1}{\cos\theta - 1} = \dfrac{1}{\cos\theta - 1}\cdot\dfrac{\cos\theta + 1}{\cos\theta + 1} = \dfrac{\cos\theta + 1}{\cos^2\theta - 1}$

$\qquad = \dfrac{\cos\theta + 1}{-\sin^2\theta} = -\csc\theta\cdot\cot\theta - \csc^2\theta$

$\displaystyle\int\dfrac{1}{\cos\theta - 1}\,d\theta = \int(-\csc\theta\cot\theta - \csc^2\theta)\,d\theta$

$\qquad = \csc\theta + \cot\theta + C$

$\qquad = \dfrac{1}{\sin\theta} + \dfrac{\cos\theta}{\sin\theta} + C$

$\qquad = \dfrac{1 + \cos\theta}{\sin\theta} + C$

43. Let $u = 2t - 1$, $du = 2\,dt$.

$\displaystyle\int\dfrac{-1}{\sqrt{1 - (2t - 1)^2}}\,dt = -\dfrac{1}{2}\int\dfrac{2}{\sqrt{1 - (2t - 1)^2}}\,dt$

$\qquad = -\dfrac{1}{2}\arcsin(2t - 1) + C$

45. Let $u = \cos\left(\dfrac{2}{t}\right)$, $du = \dfrac{2\sin(2/t)}{t^2}\,dt$.

$\displaystyle\int\dfrac{\tan(2/t)}{t^2}\,dt = \dfrac{1}{2}\int\dfrac{1}{\cos(2/t)}\left[\dfrac{2\sin(2/t)}{t^2}\right]\,dt$

$\qquad = \dfrac{1}{2}\ln\left|\cos\left(\dfrac{2}{t}\right)\right| + C$

47. $\displaystyle\int\dfrac{3}{\sqrt{6x - x^2}}\,dx = 3\int\dfrac{1}{\sqrt{9 - (x - 3)^2}}\,dx = 3\arcsin\left(\dfrac{x - 3}{3}\right) + C$

49. $\displaystyle\int\dfrac{4}{4x^2 + 4x + 65}\,dx = \int\dfrac{1}{[x + (1/2)]^2 + 16}\,dx = \dfrac{1}{4}\arctan\left[\dfrac{x + (1/2)}{4}\right] + C = \dfrac{1}{4}\arctan\left(\dfrac{2x + 1}{8}\right) + C$

51. $\dfrac{ds}{dt} = \dfrac{t}{\sqrt{1 - t^4}}$, $\left(0, -\dfrac{1}{2}\right)$

(a)

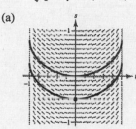

(b) $u = t^2$, $du = 2t\,dt$

$\displaystyle\int\dfrac{t}{\sqrt{1 - t^4}}\,dt = \dfrac{1}{2}\int\dfrac{2t}{\sqrt{1 - (t^2)^2}}\,dt$

$\qquad = \dfrac{1}{2}\arcsin t^2 + C$

$\left(0, -\dfrac{1}{2}\right)$: $-\dfrac{1}{2} = \dfrac{1}{2}\arcsin 0 + C \Rightarrow C = -\dfrac{1}{2}$

$s = \dfrac{1}{2}\arcsin t^2 - \dfrac{1}{2}$

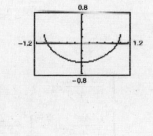

53. (a)

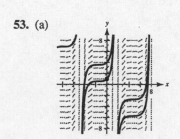

(b) $y = \displaystyle\int(\sec x + \tan x)^2\,dx$

$\qquad = \displaystyle\int(\sec^2 x + 2\sec x\tan x + \tan^2 x)\,dx$

$\qquad = \displaystyle\int(\sec^2 x + 2\sec x\tan x + (\sec^2 x - 1))\,dx$

$\qquad = \displaystyle\int(2\sec^2 x + 2\sec x\tan x - 1)\,dx$

$\qquad = 2\tan x + 2\sec x - x + C$

At $(0, 1)$: $1 = 0 + 2 - 0 + C \Rightarrow C = -1$

$y = 2\tan x + 2\sec x - x - 1$

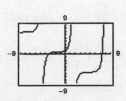

55.

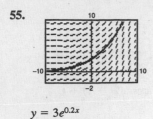

$$y = 3e^{0.2x}$$

57. $y = \int (1 + e^x)^2 \, dx$

$$= \int (e^{2x} + 2e^x + 1) \, dx$$

$$= \frac{1}{2}e^{2x} + 2e^x + x + C$$

59. $\dfrac{dy}{dx} = \dfrac{\sec^2 x}{4 + \tan^2 x}$

Let $u = \tan x$, $du = \sec^2 x \, dx$.

$$y = \int \frac{\sec^2 x}{4 + \tan^2 x} \, dx = \frac{1}{2} \arctan\left(\frac{\tan x}{2}\right) + C$$

61. Let $u = 2x$, $du = 2 \, dx$.

$$\int_0^{\pi/4} \cos 2x \, dx = \frac{1}{2}\int_0^{\pi/4} \cos 2x(2) \, dx$$

$$= \left[\frac{1}{2}\sin 2x\right]_0^{\pi/4} = \frac{1}{2}$$

63. Let $u = -x^2$, $du = -2x \, dx$.

$$\int_0^1 xe^{-x^2} \, dx = -\frac{1}{2}\int_0^1 e^{-x^2}(-2x) \, dx = \left[-\frac{1}{2}e^{-x^2}\right]_0^1$$

$$= \frac{1}{2}(1 - e^{-1}) \approx 0.316$$

65. Let $u = x^2 + 9$, $du = 2x \, dx$.

$$\int_0^4 \frac{2x}{\sqrt{x^2 + 9}} \, dx = \int_0^4 (x^2 + 9)^{-1/2}(2x) \, dx$$

$$= \left[2\sqrt{x^2 + 9}\right]_0^4 = 4$$

67. Let $u = 3x$, $du = 3 \, dx$.

$$\int_0^{2/\sqrt{3}} \frac{1}{4 + 9x^2} \, dx = \frac{1}{3}\int_0^{2/\sqrt{3}} \frac{3}{4 + (3x)^2} \, dx$$

$$= \left[\frac{1}{6}\arctan\left(\frac{3x}{2}\right)\right]_0^{2/\sqrt{3}}$$

$$= \frac{\pi}{18} \approx 0.175$$

69. $A = \displaystyle\int_0^{5/2} (-2x + 5)^{3/2} \, dx$

$$= -\frac{1}{2}\int_0^{5/2} (5 - 2x)^{3/2}(-2) \, dx$$

$$= -\frac{1}{5}(5 - 2x)^{5/2}\Big]_0^{5/2}$$

$$= 0 + \frac{1}{5}(5)^{5/2} = 5^{3/2}$$

$$= 5\sqrt{5} \approx 11.1803$$

71. $A = \displaystyle\int_0^5 \frac{3x + 2}{x^2 + 9} \, dx$

$$= \int_0^5 \frac{3x}{x^2 + 9} \, dx + \int_0^5 \frac{2}{x^2 + 9} \, dx$$

$$= \left[\frac{3}{2}\ln|x^2 + 9| + \frac{2}{3}\arctan\left(\frac{x}{3}\right)\right]_0^5$$

$$= \frac{3}{2}\ln(34) + \frac{2}{3}\arctan\left(\frac{5}{3}\right) - \frac{3}{2}\ln 9$$

$$= \frac{3}{2}\ln\left(\frac{34}{9}\right) + \frac{2}{3}\arctan\left(\frac{5}{3}\right)$$

$$\approx 2.6806$$

73. $y^2 = x^2(1 - x^2)$

$$y = \pm\sqrt{x^2(1 - x^2)}$$

$$A = 4\int_0^1 x\sqrt{1 - x^2} \, dx$$

$$= -2\int_0^1 (1 - x^2)^{1/2}(-2x) \, dx$$

$$= -\frac{4}{3}(1 - x)^{3/2}\Big]_0^1$$

$$= -\frac{4}{3}(0 - 1) = \frac{4}{3}$$

75. $\displaystyle\int \frac{1}{x^2 + 4x + 13} \, dx = \frac{1}{3}\arctan\left(\frac{x + 2}{3}\right) + C$

The antiderivatives are vertical translations of each other.

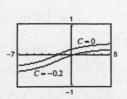

77. $\displaystyle\int \frac{1}{1 + \sin \theta}\, d\theta = \tan \theta - \sec \theta + C \quad \left(\text{or } \frac{-2}{1 + \tan(\theta/2)} \right)$

The antiderivatives are vertical translations of each other.

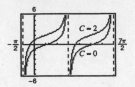

79. Power Rule: $\displaystyle\int u^n\, du = \frac{u^{n+1}}{n+1} + C, \ \ n \neq -1$

$u = x^2 + 1, \ n = 3$

81. Log Rule: $\displaystyle\int \frac{du}{u} = \ln|u| + C, \ \ u = x^2 + 1$

83. They are equivalent because

$$e^{x + C_1} = e^x \cdot e^{C_1} = Ce^x, \ C = e^{C_1}.$$

85. $\sin x + \cos x = a \sin(x + b)$

$\sin x + \cos x = a \sin x \cos b + a \cos x \sin b$

$\sin x + \cos x = (a \cos b) \sin x + (a \sin b) \cos x$

Equate coefficients of like terms to obtain the following.

$1 = a \cos b \quad \text{and} \quad 1 = a \sin b$

Thus, $a = 1/\cos b$. Now, substitute for a in $1 = a \sin b$.

$$1 = \left(\frac{1}{\cos b} \right) \sin b$$

$$1 = \tan b \implies b = \frac{\pi}{4}$$

Since $b = \dfrac{\pi}{4}$, $a = \dfrac{1}{\cos(\pi/4)} = \sqrt{2}$. Thus, $\sin x + \cos x = \sqrt{2} \sin\left(x + \dfrac{\pi}{4}\right)$.

$$\int \frac{dx}{\sin x + \cos x} = \int \frac{dx}{\sqrt{2} \sin(x + (\pi/4))} = \frac{1}{\sqrt{2}} \int \csc\left(x + \frac{\pi}{4}\right) dx = -\frac{1}{\sqrt{2}} \ln\left| \csc\left(x + \frac{\pi}{4}\right) + \cot\left(x + \frac{\pi}{4}\right) \right| + C$$

87. $f(x) = \dfrac{1}{5}(x^3 - 7x^2 + 10x)$

$\displaystyle\int_0^5 f(x)\, dx < 0$ because more area is below the x-axis than above.

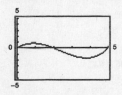

89. $\displaystyle\int_0^2 \frac{4x}{x^2 + 1}\, dx \approx 3$

Matches (a).

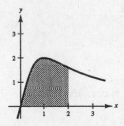

91. (a) $y = 2\pi x^2, \ \ 0 \leq x \leq 2$

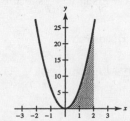

(b) $y = \sqrt{2}x, \ \ 0 \leq x \leq 2$

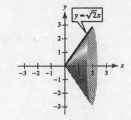

(c) $y = x, \ \ 0 \leq x \leq 2$

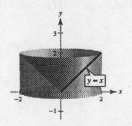

93. (a) Shell Method:

Let $u = -x^2$, $du = -2x\,dx$.

$$V = 2\pi \int_0^1 xe^{-x^2}\,dx$$

$$= -\pi \int_0^1 e^{-x^2}(-2x)\,dx$$

$$= \left[-\pi e^{-x^2} \right]_0^1$$

$$= \pi(1 - e^{-1}) \approx 1.986$$

(b) Shell Method:

$$V = 2\pi \int_0^b xe^{-x^2}\,dx$$

$$= \left[-\pi e^{-x^2} \right]_0^b$$

$$= \pi(1 - e^{-b^2}) = \frac{4}{3}$$

$$e^{-b^2} = \frac{3\pi - 4}{3\pi}$$

$$b = \sqrt{\ln\left(\frac{3\pi}{3\pi - 4}\right)} \approx 0.743$$

95.

$$y = 2\sqrt{x}$$

$$y' = \frac{1}{\sqrt{x}}$$

$$1 + (y')^2 = 1 + \frac{1}{x} = \frac{x+1}{x}$$

$$S = 2\pi \int_0^9 2\sqrt{x}\sqrt{\frac{x+1}{x}}\,dx$$

$$= 2\pi \int_0^9 2\sqrt{x+1}\,dx$$

$$= \left[4\pi\left(\frac{2}{3}\right)(x+1)^{3/2} \right]_0^9$$

$$= \frac{8\pi}{3}\left(10\sqrt{10} - 1\right)$$

$$\approx 256.545$$

97. Average value $= \dfrac{1}{b-a} \displaystyle\int_a^b f(x)\,dx$

$$= \frac{1}{3 - (-3)} \int_{-3}^3 \frac{1}{1+x^2}\,dx$$

$$= \frac{1}{6}\arctan(x)\Big]_{-3}^3$$

$$= \frac{1}{6}[\arctan(3) - \arctan(-3)]$$

$$= \frac{1}{3}\arctan(3) \approx 0.4163$$

99.

$$y = \tan(\pi x)$$

$$y' = \pi \sec^2(\pi x)$$

$$1 + (y')^2 = 1 + \pi^2 \sec^4(\pi x)$$

$$s = \int_0^{1/4} \sqrt{1 + \pi^2 \sec^4(\pi x)}\,dx$$

$$\approx 1.0320$$

101. (a) $\displaystyle\int \cos^3 x\,dx = \int (1 - \sin^2 x)\cos x\,dx$

$$= \sin x - \frac{\sin^3 x}{3} + C$$

(b) $\displaystyle\int \cos^5 x\,dx = \int (1 - \sin^2 x)^2 \cos x\,dx$

$$= \int (1 - 2\sin^2 x + \sin^4 x)\cos x\,dx$$

$$= \sin x - \frac{2}{3}\sin^3 x + \frac{\sin^5 x}{5} + C$$

(c) $\displaystyle\int \cos^7 x\,dx = \int (1 - \sin^2 x)^3 \cos x\,dx$

$$= \int (1 - 3\sin^2 x + 3\sin^4 x - \sin^6 x)\cos x\,dx$$

$$= \sin x - \sin^3 x + \frac{3}{5}\sin^5 x - \frac{1}{7}\sin^7 x + C$$

(d) $\displaystyle\int \cos^{15} x\,dx = \int (1 - \cos^2 x)^7 \cos x\,dx$

You would expand $(1 - \cos^2 x)^7$.

103. Let $f(x) = \frac{1}{2}\left(x\sqrt{x^2 + 1} + \ln\left|x + \sqrt{x^2 + 1}\right|\right) + C.$

$$f'(x) = \frac{1}{2}\left(x\frac{1}{2}(x^2 + 1)^{-1/2}(2x) + \sqrt{x^2 + 1} + \frac{1}{x + \sqrt{x^2 + 1}}\left(1 + \frac{1}{2}(x^2 + 1)^{-1/2}(2x)\right)\right)$$

$$= \frac{1}{2}\left(\frac{x^2}{\sqrt{x^2 + 1}} + \sqrt{x^2 + 1} + \frac{1}{x + \sqrt{x^2 + 1}}\left(1 + \frac{x}{\sqrt{x^2 + 1}}\right)\right)$$

$$= \frac{1}{2}\left(\frac{x^2 + (x^2 + 1)}{\sqrt{x^2 + 1}} + \frac{1}{x + \sqrt{x^2 + 1}}\left(\frac{\sqrt{x^2 + 1} + x}{\sqrt{x^2 + 1}}\right)\right)$$

$$= \frac{1}{2}\left(\frac{2x^2 + 1}{\sqrt{x^2 + 1}} + \frac{1}{\sqrt{x^2 + 1}}\right) = \frac{1}{2}\left(\frac{2(x^2 + 1)}{\sqrt{x^2 + 1}}\right) = \sqrt{x^2 + 1}$$

Thus, $\displaystyle\int \sqrt{x^2 + 1}\, dx = \frac{1}{2}\left(x\sqrt{x^2 + 1} + \ln\left|x + \sqrt{x^2 + 1}\right|\right) + C.$

Let $g(x) = \frac{1}{2}\left(x\sqrt{x^2 + 1} + \text{arcsinh}(x)\right).$

$$g'(x) = \frac{1}{2}\left(x\frac{1}{2}(x^2 + 1)^{-1/2}(2x) + \sqrt{x^2 + 1} + \frac{1}{\sqrt{x^2 + 1}}\right)$$

$$= \frac{1}{2}\left(\frac{x^2}{\sqrt{x^2 + 1}} + \sqrt{x^2 + 1} + \frac{1}{\sqrt{x^2 + 1}}\right)$$

$$= \frac{1}{2}\left(\frac{x^2 + (x^2 + 1) + 1}{\sqrt{x^2 + 1}}\right)$$

$$= \frac{1}{2}\left(\frac{2(x^2 + 1)}{\sqrt{x^2 + 1}}\right) = \sqrt{x^2 + 1}$$

Thus, $\displaystyle\int \sqrt{x^2 + 1}\, dx = \frac{1}{2}\left(x\sqrt{x^2 + 1} + \text{arcsinh}(x)\right) + C.$

Section 8.2   Integration by Parts

1. $\dfrac{d}{dx}[\sin x - x\cos x] = \cos x - (-x\sin x + \cos x) = x\sin x$

Matches (b)

3. $\dfrac{d}{dx}[x^2 e^x - 2xe^x + 2e^x] = x^2 e^x + 2xe^x - 2xe^x - 2e^x + 2e^x$

$$= x^2 e^x$$

Matches (c)

5. $\displaystyle\int xe^{2x}\, dx$

$u = x,\, dv = e^{2x}\, dx$

7. $\displaystyle\int (\ln x)^2\, dx$

$u = (\ln x)^2,\, dv = dx$

9. $\displaystyle\int x\sec^2 x\, dx$

$u = x,\, dv = \sec^2 x\, dx$

11. $dv = e^{-2x}\,dx \implies v = \int e^{-2x}\,dx = -\frac{1}{2}e^{-2x}$

$u = x \qquad \implies du = dx$

$\int xe^{-2x}\,dx = -\frac{1}{2}xe^{-2x} - \int -\frac{1}{2}e^{-2x}\,dx$

$\qquad\qquad = -\frac{1}{2}xe^{-2x} - \frac{1}{4}e^{-2x} + C$

$\qquad\qquad = \frac{-1}{4e^{2x}}(2x + 1) + C$

13. Use integration by parts three times.

(1) $dv = e^x\,dx \implies v = \int e^x\,dx = e^x$ \qquad (2) $dv = e^x\,dx \implies v = \int e^x\,dx = e^x$ \qquad (3) $dv = e^x\,dx \implies v = \int e^x\,dx = e^x$

$\quad u = x^3 \implies du = 3x^2\,dx$ \qquad\qquad $u = x^2 \implies du = 2x\,dx$ \qquad\qquad\quad $u = x \implies du = dx$

$\int x^3 e^x\,dx = x^3 e^x - 3\int x^2 e^x\,dx = x^3 e^x - 3x^2 e^x + 6\int xe^x\,dx$

$\qquad\qquad\qquad = x^3 e^x - 3x^2 e^x + 6xe^x - 6e^x + C = e^x(x^3 - 3x^2 + 6x - 6) + C$

15. $\int x^2 e^{x^3}\,dx = \frac{1}{3}\int e^{x^3}(3x^2)\,dx = \frac{1}{3}e^{x^3} + C$

17. $dv = t\,dt \qquad \implies v = \int t\,dt = \frac{t^2}{2}$

$u = \ln(t + 1) \implies du = \frac{1}{t + 1}\,dt$

$\int t\ln(t + 1)\,dt = \frac{t^2}{2}\ln(t + 1) - \frac{1}{2}\int \frac{t^2}{t + 1}\,dt$

$\qquad\qquad\qquad = \frac{t^2}{2}\ln(t + 1) - \frac{1}{2}\int\left(t - 1 + \frac{1}{t + 1}\right)dt$

$\qquad\qquad\qquad = \frac{t^2}{2}\ln(t + 1) - \frac{1}{2}\left[\frac{t^2}{2} - t + \ln(t + 1)\right] + C$

$\qquad\qquad\qquad = \frac{1}{4}[2(t^2 - 1)\ln|t + 1| - t^2 + 2t] + C$

19. Let $u = \ln x$, $du = \frac{1}{x}\,dx$.

$\int \frac{(\ln x)^2}{x}\,dx = \int (\ln x)^2\left(\frac{1}{x}\right)dx = \frac{(\ln x)^3}{3} + C$

21. $dv = \frac{1}{(2x + 1)^2}\,dx \implies v = \int (2x + 1)^{-2}\,dx$

$\qquad\qquad\qquad\qquad\qquad = -\frac{1}{2(2x + 1)}$

$u = xe^{2x} \qquad\qquad \implies du = (2xe^{2x} + e^{2x})\,dx$

$\qquad\qquad\qquad\qquad = e^{2x}(2x + 1)\,dx$

$\int \frac{xe^{2x}}{(2x + 1)^2}\,dx = -\frac{xe^{2x}}{2(2x + 1)} + \int \frac{e^{2x}}{2}\,dx$

$\qquad\qquad\qquad = \frac{-xe^{2x}}{2(2x + 1)} + \frac{e^{2x}}{4} + C = \frac{e^{2x}}{4(2x + 1)} + C$

23. Use integration by parts twice.

(1) $dv = e^x \, dx \implies v = \int e^x \, dx = e^x$ (2) $dv = e^x \, dx \implies v = \int e^x \, dx = e^x$

$\quad u = x^2 \implies du = 2x \, dx$ $\quad u = x \implies du = dx$

$$\int (x^2 - 1)e^x \, dx = \int x^2 e^x \, dx - \int e^x \, dx = x^2 e^x - 2\int xe^x \, dx - e^x$$

$$= x^2 e^x - 2\left[xe^x - \int e^x \, dx \right] - e^x = x^2 e^x - 2xe^x + e^x + C = (x-1)^2 e^x + C$$

25. $dv = \sqrt{x-1} \, dx \implies v = \int (x-1)^{1/2} \, dx = \frac{2}{3}(x-1)^{3/2}$ **27.** $dv = \cos x \, dx \implies v = \int \cos x \, dx = \sin x$

$\quad u = x \implies du = dx$ $\quad u = x \implies du = dx$

$$\int x\sqrt{x-1} \, dx = \frac{2}{3}x(x-1)^{3/2} - \frac{2}{3}\int (x-1)^{3/2} \, dx$$ $$\int x \cos x \, dx = x \sin x - \int \sin x \, dx = x \sin x + \cos x + C$$

$$= \frac{2}{3}x(x-1)^{3/2} - \frac{4}{15}(x-1)^{5/2} + C$$

$$= \frac{2(x-1)^{3/2}}{15}(3x+2) + C$$

29. Use integration by parts three times.

(1) $u = x^3, \, du = 3x^2 \, dx, \, dv = \sin x \, dx, \, v = -\cos x$ (2) $u = x^2, \, du = 2x \, dx, \, dv = \cos x \, dx, \, v = \sin x$

$$\int x^3 \sin x \, dx = -x^3 \cos x + 3\int x^2 \cos x \, dx$$ $$\int x^3 \sin x \, dx = -x^3 \cos x + 3\left[x^2 \sin x - 2\int x \sin x \, dx \right]$$

$$= -x^3 \cos x + 3x^2 \sin x - 6\int x \sin x \, dx$$

(3) $u = x, \, du = dx, \, dv = \sin x \, dx, \, v = -\cos x$

$$\int x^3 \sin x \, dx = -x^3 \cos x + 3x^2 \sin x - 6\left[-x \cos x + \int \cos x \, dx \right]$$

$$= -x^3 \cos x + 3x^2 \sin x + 6x \cos x - 6 \sin x + C$$

31. $u = t, \, du = dt, \, dv = \csc t \cot t \, dt, \, v = -\csc t$ **33.** $dv = dx \implies v = \int dx = x$

$$\int t \csc t \cot t \, dt = -t \csc t + \int \csc t \, dt$$ $\quad u = \arctan x \implies du = \dfrac{1}{1+x^2} \, dx$

$$= -t \csc t - \ln|\csc t + \cot t| + C$$ $$\int \arctan x \, dx = x \arctan x - \int \frac{x}{1+x^2} \, dx$$

$$= x \arctan x - \frac{1}{2}\ln(1+x^2) + C$$

35. Use integration by parts twice.

(1) $dv = e^{2x} \, dx \implies v = \int e^{2x} \, dx = \frac{1}{2}e^{2x}$ (2) $dv = e^{2x} \, dx \implies v = \int e^{2x} \, dx = \frac{1}{2}e^{2x}$

$\quad u = \sin x \implies du = \cos x \, dx$ $\quad u = \cos x \implies du = -\sin x \, dx$

$$\int e^{2x} \sin x \, dx = \frac{1}{2}e^{2x} \sin x - \frac{1}{2}\int e^{2x} \cos x \, dx = \frac{1}{2}e^{2x} \sin x - \frac{1}{2}\left(\frac{1}{2}e^{2x} \cos x + \frac{1}{2}\int e^{2x} \sin x \, dx \right)$$

$$\frac{5}{4}\int e^{2x} \sin x \, dx = \frac{1}{2}e^{2x} \sin x - \frac{1}{4}e^{2x} \cos x$$

$$\int e^{2x} \sin x \, dx = \frac{1}{5}e^{2x}(2 \sin x - \cos x) + C$$

37. $y' = xe^{x^2}$

$$y = \int xe^{x^2}\,dx = \frac{1}{2}e^{x^2} + C$$

39. Use integration by parts twice.

(1) $dv = \dfrac{1}{\sqrt{2+3t}}\,dt \implies v = \int (2+3t)^{-1/2}\,dt = \dfrac{2}{3}\sqrt{2+3t}$

$\quad u = t^2 \qquad\qquad \implies du = 2t\,dt$

(2) $dv = \sqrt{2+3t}\,dt \implies v = \int (2+3t)^{1/2}\,dt = \dfrac{2}{9}(2+3t)^{3/2}$

$\quad u = t \qquad\qquad \implies du = dt$

$$y = \int \frac{t^2}{\sqrt{2+3t}}\,dt = \frac{2t^2\sqrt{2+3t}}{3} - \frac{4}{3}\int t\sqrt{2+3t}\,dt$$

$$= \frac{2t^2\sqrt{2+3t}}{3} - \frac{4}{3}\left[\frac{2t}{9}(2+3t)^{3/2} - \frac{2}{9}\int (2+3t)^{3/2}\,dt\right]$$

$$= \frac{2t^2\sqrt{2+3t}}{3} - \frac{8t}{27}(2+3t)^{3/2} + \frac{16}{405}(2+3t)^{5/2} + C$$

$$= \frac{2\sqrt{2+3t}}{405}(27t^2 - 24t + 32) + C$$

41. $(\cos y)y' = 2x$

$$\int \cos y\,dy = \int 2x\,dx$$

$$\sin y = x^2 + C$$

43. (a)

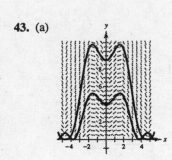

(b) $\dfrac{dy}{dx} = x\sqrt{y}\cos x, \quad (0, 4)$

$$\int \frac{dy}{\sqrt{y}} = \int x\cos x\,dx$$

$$\int y^{-1/2}\,dy = \int x\cos x\,dx \qquad (u = x, du = dx, dv = \cos x\,dx, v = \sin x)$$

$$2y^{1/2} = x\sin x - \int \sin x\,dx$$

$$= x\sin x + \cos x + C$$

$(0, 4)$: $2(4)^{1/2} = 0 + 1 + C \implies C = 3$

$$2\sqrt{y} = x\sin x + \cos x + 3$$

45. $\dfrac{dy}{dx} = \dfrac{x}{y}e^{x/8}, y(0) = 2$

47. $u = x, du = dx, dv = e^{-x/2}\, dx, v = -2e^{-x/2}$

$$\int xe^{-x/2}\, dx = -2xe^{-x/2} + \int 2e^{-x/2}\, dx = -2xe^{-x/2} - 4e^{-x/2} + C$$

Thus, $\displaystyle\int_0^4 xe^{-x/2}\, dx = \left[-2xe^{-x/2} - 4e^{-x/2}\right]_0^4$

$$= -8e^{-2} - 4e^{-2} + 4$$

$$= -12e^{-2} + 4 \approx 2.376.$$

49. See Exercise 27.

$$\int_0^{\pi/2} x \cos x\, dx = \left[x \sin x + \cos x\right]_0^{\pi/2} = \frac{\pi}{2} - 1$$

51. $u = \arccos x, du = -\dfrac{1}{\sqrt{1 - x^2}}\, dx, dv = dx, v = x$

$$\int \arccos x\, dx = x \arccos x + \int \frac{x}{\sqrt{1 - x^2}}\, dx$$

$$= x \arccos x - \sqrt{1 - x^2} + C$$

Thus, $\displaystyle\int_0^{1/2} \arccos x = \left[x \arccos x - \sqrt{1 - x^2}\right]_0^{1/2}$

$$= \frac{1}{2} \arccos\left(\frac{1}{2}\right) - \sqrt{\frac{3}{4}} + 1$$

$$= \frac{\pi}{6} - \frac{\sqrt{3}}{2} + 1 \approx 0.658.$$

53. Use integration by parts twice.

(1) $dv = e^x\, dx \implies v = \displaystyle\int e^x\, dx = e^x$

$\quad u = \sin x \implies du = \cos x\, dx$

(2) $dv = e^x\, dx \implies v = \displaystyle\int e^x\, dx = e^x$

$\quad u = \cos x \implies du = -\sin x\, dx$

$$\int e^x \sin x\, dx = e^x \sin x - \int e^x \cos x\, dx = e^x \sin x - e^x \cos x - \int e^x \sin x\, dx$$

$$2\int e^x \sin x\, dx = e^x(\sin x - \cos x)$$

$$\int e^x \sin x\, dx = \frac{e^x}{2}(\sin x - \cos x) + C$$

Thus, $\displaystyle\int_0^1 e^x \sin x\, dx = \left[\frac{e^x}{2}(\sin x - \cos x)\right]_0^1 = \frac{e}{2}(\sin 1 - \cos 1) + \frac{1}{2} = \frac{e(\sin 1 - \cos 1) + 1}{2} \approx 0.909.$

55. $dv = x^2\, dx, v = \dfrac{x^3}{3}, u = \ln x, du = \dfrac{1}{x}\, dx$

$$\int x^2 \ln x\, dx = \frac{x^3}{3} \ln x - \int \frac{x^3}{3}\left(\frac{1}{x}\right) dx$$

$$= \frac{x^3}{3} \ln x - \frac{1}{3}\int x^2\, dx$$

Hence, $\displaystyle\int_1^2 x^2 \ln x\, dx = \left[\frac{x^3}{3} \ln x - \frac{1}{9}x^3\right]_1^2$

$$= \frac{8}{3} \ln 2 - \frac{8}{9} + \frac{1}{9}$$

$$= \frac{8}{3} \ln 2 - \frac{7}{9} \approx 1.071.$$

57. $dv = x\,dx, v = \dfrac{x^2}{2}, u = \text{arcsec } x, du = \dfrac{1}{x\sqrt{x^2-1}}\,dx$

$$\int x \text{ arcsec } x\,dx = \frac{x^2}{2}\text{ arcsec } x - \int \frac{x^2/2}{x\sqrt{x^2-1}}\,dx$$

$$= \frac{x^2}{2}\text{ arcsec } x - \frac{1}{4}\int \frac{2x}{\sqrt{x^2-1}}\,dx$$

$$= \frac{x^2}{2}\text{ arcsec } x - \frac{1}{2}\sqrt{x^2-1} + C$$

Hence,

$$\int_2^4 x \text{ arcsec } x\,dx = \left[\frac{x^2}{2}\text{ arcsec } x - \frac{1}{2}\sqrt{x^2-1}\right]_2^4$$

$$= \left(8\text{ arcsec } 4 - \frac{\sqrt{15}}{2}\right) - \left(\frac{2\pi}{3} - \frac{\sqrt{3}}{2}\right)$$

$$= 8\text{ arcsec } 4 - \frac{\sqrt{15}}{2} + \frac{\sqrt{3}}{2} - \frac{2\pi}{3}$$

$$\approx 7.380.$$

59. $\displaystyle\int x^2 e^{2x}\,dx = x^2\left(\frac{1}{2}e^{2x}\right) - (2x)\left(\frac{1}{4}e^{2x}\right) + 2\left(\frac{1}{8}e^{2x}\right) + C$

$$= \frac{1}{2}x^2 e^{2x} - \frac{1}{2}xe^{2x} + \frac{1}{4}e^{2x} + C$$

$$= \frac{1}{4}e^{2x}(2x^2 - 2x + 1) + C$$

Alternate signs	u and its derivatives	v' and its antiderivatives
+	x^2	e^{2x}
−	$2x$	$\frac{1}{2}e^{2x}$
+	2	$\frac{1}{4}e^{2x}$
−	0	$\frac{1}{8}e^{2x}$

61. $\displaystyle\int x^3 \sin x\,dx = x^3(-\cos x) - 3x^2(-\sin x) + 6x\cos x - 6\sin x + C$

$$= -x^3\cos x + 3x^2\sin x + 6x\cos x - 6\sin x + C$$

$$= (3x^2 - 6)\sin x - (x^3 - 6x)\cos x + C$$

Alternate signs	u and its derivatives	v' and its antiderivatives
+	x^3	$\sin x$
−	$3x^2$	$-\cos x$
+	$6x$	$-\sin x$
−	6	$\cos x$
+	0	$\sin x$

63. $\displaystyle\int x\sec^2 x\,dx = x\tan x + \ln|\cos x| + C$

Alternate signs	u and its derivatives	v' and its antiderivatives
+	x	$\sec^2 x$
−	1	$\tan x$
+	0	$-\ln|\cos x|$

65. $u = \sqrt{x} \implies u^2 = x \implies 2u\,du = dx$

$$\int \sin \sqrt{x}\,dx = \int \sin u(2u\,du) = 2\int u \sin u\,du$$

Integration by parts: $w = u, dw = du, dv = \sin u\,du,$
$v = -\cos u$

$$2\int u \sin u\,du = 2\left(-u\cos u + \int \cos u\,du\right)$$

$$= 2(-u\cos u + \sin u) + C$$

$$= 2\left(-\sqrt{x}\cos \sqrt{x} + \sin \sqrt{x}\right) + C$$

67. Let $u = 4 - x, du = -dx, x = 4 - u.$

$$\int_0^4 x\sqrt{4 - x}\,dx = \int_4^0 (4 - u)u^{1/2}(-du)$$

$$= \int_0^4 (4u^{1/2} - u^{3/2})\,du$$

$$= \left[\frac{8}{3}u^{3/2} - \frac{2}{5}u^{5/2}\right]_0^4$$

$$= \frac{8}{3}(8) - \frac{2}{5}(32) = \frac{128}{15}$$

69. Let $w = \ln x, dw = \dfrac{1}{x}\,dx, x = e^w, dx = e^w\,dw.$

$$\int \cos(\ln x)\,dx = \int \cos w(e^w\,dw)$$

Now use integration by parts twice.

$$\int \cos w\,e^w\,dw = \cos w\,e^w + \int \sin w\,e^w\,dw \qquad [u = \cos w, dv = e^w\,dw]$$

$$= \cos w\,e^w + \left[\sin w\,e^w - \int \cos w\,e^w\,dw\right] \qquad [u = \sin w, dv = e^w\,dw]$$

$$2\int \cos w\,e^w\,dw = \cos w\,e^w + \sin w\,e^w$$

$$\int \cos w\,e^w\,dw = \frac{1}{2}e^w[\cos w + \sin w] + C$$

$$\int \cos(\ln x)\,dx = \frac{1}{2}x[\cos(\ln x) + \sin(\ln x)] + C$$

71. Integration by parts is based on the Product Rule.

73. No Substitution

75. Yes
$u = x^2, dv = e^{2x}\,dx$

77. Yes. Let $u = x$ and

$$du = \frac{1}{\sqrt{x + 1}}\,dx.$$

(Substitution also works. Let $u = \sqrt{x + 1}$.)

79. (a) $\displaystyle\int t^3 e^{-4t}\,dt = \frac{-e^{-4t}}{128}(32t^3 + 24t^2 + 12t + 3) + C$

(b)

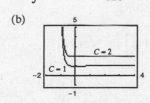

(c) The graphs are vertical translations of each other.

81. (a) $\displaystyle\int e^{-2x} \sin 3x\,dx = \frac{e^{-2x}}{13}[-2\sin 3x - 3\cos 3x] + C$

$$\int_0^{\pi/2} e^{-2x} \sin 3x\,dx = \frac{1}{13}[2e^{-\pi} + 3] \approx 0.2374$$

(b)

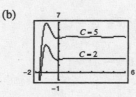

(c) The graphs are vertical translations of each other.

83. (a) $dv = \sqrt{2x - 3}\,dx \implies v = \int (2x - 3)^{1/2}\,dx = \frac{1}{3}(2x - 3)^{3/2}$

$u = 2x \qquad\qquad \implies du = 2\,dx$

$\int 2x\sqrt{2x - 3}\,dx = \frac{2}{3}x(2x - 3)^{3/2} - \frac{2}{3}\int (2x - 3)^{3/2}\,dx$

$= \frac{2}{3}x(2x - 3)^{3/2} - \frac{2}{15}(2x - 3)^{5/2} + C$

$= \frac{2}{15}(2x - 3)^{3/2}(3x + 3) + C = \frac{2}{5}(2x - 3)^{3/2}(x + 1) + C$

(b) $u = 2x - 3 \implies x = \dfrac{u + 3}{2}$ and $dx = \dfrac{1}{2}\,du$

$\int 2x\sqrt{2x - 3}\,dx = \int 2\left(\frac{u + 3}{2}\right)u^{1/2}\left(\frac{1}{2}\right)du = \frac{1}{2}\int (u^{3/2} + 3u^{1/2})\,du = \frac{1}{2}\left[\frac{2}{5}u^{5/2} + 2u^{3/2}\right] + C$

$= \frac{1}{5}u^{3/2}(u + 5) + C = \frac{1}{5}(2x - 3)^{3/2}[(2x - 3) + 5] + C = \frac{2}{5}(2x - 3)^{3/2}(x + 1) + C$

85. (a) $dv = \dfrac{x}{\sqrt{4 + x^2}}\,dx \implies v = \int (4 + x^2)^{-1/2}x\,dx = \sqrt{4 + x^2}$

$u = x^2 \qquad\qquad \implies du = 2x\,dx$

$\int \frac{x^3}{\sqrt{4 + x^2}}\,dx = x^2\sqrt{4 + x^2} - 2\int x\sqrt{4 + x^2}\,dx$

$= x^2\sqrt{4 + x^2} - \frac{2}{3}(4 + x^2)^{3/2} + C = \frac{1}{3}\sqrt{4 + x^2}(x^2 - 8) + C$

(b) $u = 4 + x^2 \implies x^2 = u - 4$ and $2x\,dx = du \implies x\,dx = \frac{1}{2}\,du$

$\int \frac{x^3}{\sqrt{4 + x^2}}\,dx = \int \frac{x^2}{\sqrt{4 + x^2}}\,x\,dx = \int \frac{u - 4}{\sqrt{u}}\frac{1}{2}\,du$

$= \frac{1}{2}\int (u^{1/2} - 4u^{-1/2})\,du = \frac{1}{2}\left(\frac{2}{3}u^{3/2} - 8u^{1/2}\right) + C$

$= \frac{1}{3}u^{1/2}(u - 12) + C = \frac{1}{3}\sqrt{4 + x^2}\,[(4 + x^2) - 12] + C = \frac{1}{3}\sqrt{4 + x^2}\,(x^2 - 8) + C$

87. $n = 0$: $\int \ln x\,dx = x(\ln x - 1) + C$

$n = 1$: $\int x \ln x\,dx = \dfrac{x^2}{4}(2 \ln x - 1) + C$

$n = 2$: $\int x^2 \ln x\,dx = \dfrac{x^3}{9}(3 \ln x - 1) + C$

$n = 3$: $\int x^3 \ln x\,dx = \dfrac{x^4}{16}(4 \ln x - 1) + C$

$n = 4$: $\int x^4 \ln x\,dx = \dfrac{x^5}{25}(5 \ln x - 1) + C$

In general, $\int x^n \ln x\,dx = \dfrac{x^{n+1}}{(n + 1)^2}[(n + 1)\ln x - 1] + C$.

89. $dv = \sin x\,dx \implies v = -\cos x$

$u = x^n \qquad\qquad \implies du = nx^{n-1}\,dx$

$\int x^n \sin x\,dx = -x^n \cos x + n\int x^{n-1} \cos x\,dx$

91. $dv = x^n \, dx \implies v = \dfrac{x^{n+1}}{n+1}$

$u = \ln x \implies du = \dfrac{1}{x} \, dx$

$\displaystyle \int x^n \ln x \, dx = \dfrac{x^{n+1}}{n+1} \ln x - \int \dfrac{x^n}{n+1} \, dx$

$\displaystyle \qquad = \dfrac{x^{n+1}}{n+1} \ln x - \dfrac{x^{n+1}}{(n+1)^2} + C$

$\displaystyle \qquad = \dfrac{x^{n+1}}{(n+1)^2}[(n+1)\ln x - 1] + C$

93. Use integration by parts twice.

(1) $dv = e^{ax} \, dx \implies v = \dfrac{1}{a} e^{ax}$

$\quad u = \sin bx \implies du = b \cos bx \, dx$

$\displaystyle \int e^{ax} \sin bx \, dx = \dfrac{e^{ax} \sin bx}{a} - \dfrac{b}{a} \int e^{ax} \cos bx \, dx$

(2) $dv = e^{ax} \, dx \implies v = \dfrac{1}{a} e^{ax}$

$\quad u = \cos bx \implies du = -b \sin bx \, dx$

$\displaystyle \qquad = \dfrac{e^{ax} \sin bx}{a} - \dfrac{b}{a}\left[\dfrac{e^{ax} \cos bx}{a} + \dfrac{b}{a} \int e^{ax} \sin bx \, dx \right] = \dfrac{e^{ax} \sin bx}{a} - \dfrac{b}{a^2} e^{ax} \cos bx - \dfrac{b^2}{a^2} \int e^{ax} \sin bx \, dx$

Therefore, $\left(1 + \dfrac{b^2}{a^2}\right) \displaystyle\int e^{ax} \sin bx \, dx = \dfrac{e^{ax}(a \sin bx - b \cos bx)}{a^2}$

$\displaystyle \qquad \int e^{ax} \sin bx \, dx = \dfrac{e^{ax}(a \sin bx - b \cos bx)}{a^2 + b^2} + C.$

95. $n = 3,$ (Use formula in Exercise 91.)

$\displaystyle \int x^3 \ln x \, dx = \dfrac{x^4}{16}[4 \ln x - 1] + C$

97. $a = 2, b = 3,$ (Use formula in Exercise 94.)

$\displaystyle \int e^{2x} \cos 3x \, dx = \dfrac{e^{2x}(2 \cos 3x + 3 \sin 3x)}{13} + C$

99. $dv = e^{-x} \, dx \implies v = -e^{-x}$

$\quad u = x \qquad \implies du = dx$

$A = \displaystyle\int_0^4 x e^{-x} \, dx = \left[-x e^{-x} \right]_0^4 + \int_0^4 e^{-x} \, dx = \dfrac{-4}{e^4} - \left[e^{-x} \right]_0^4$

$\quad = 1 - \dfrac{5}{e^4} \approx 0.908$

101. $A = \displaystyle\int_0^1 e^{-x} \sin(\pi x) \, dx$

$\quad = \left[\dfrac{e^{-x}(-\sin \pi x - \pi \cos \pi x)}{1 + \pi^2} \right]_0^1$

$\quad = \dfrac{1}{1 + \pi^2}\left(\dfrac{\pi}{e} + \pi \right)$

$\quad = \dfrac{\pi}{1 + \pi^2}\left(\dfrac{1}{e} + 1 \right)$

$\quad \approx 0.395$ (See Exercise 93.)

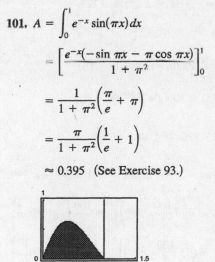

103. (a) $A = \int_1^e \ln x \, dx = \left[-x + x \ln x \right]_1^e = 1$ (See Exercise 4.)

(b) $R(x) = \ln x, r(x) = 0$

$$V = \pi \int_1^e (\ln x)^2 \, dx$$

$$= \pi \left[x(\ln x)^2 - 2x \ln x + 2x \right]_1^e$$ (Use integration by parts twice, see Exercise 7.)

$$= \pi(e - 2) \approx 2.257$$

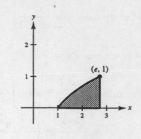

(c) $p(x) = x, h(x) = \ln x$

$$V = 2\pi \int_1^e x \ln x \, dx = 2\pi \left[\frac{x^2}{4} (-1 + 2 \ln x) \right]_1^e$$

$$= \frac{(e^2 + 1)\pi}{2} \approx 13.177$$ (See Exercise 91.)

(d) $\bar{x} = \dfrac{\int_1^e x \ln x \, dx}{1} = \dfrac{e^2 + 1}{4} \approx 2.097$

$\bar{y} = \dfrac{\frac{1}{2}\int_1^e (\ln x)^2 \, dx}{1} = \dfrac{e - 2}{2} \approx 0.359$

$(\bar{x}, \bar{y}) = \left(\dfrac{e^2 + 1}{4}, \dfrac{e - 2}{2} \right) \approx (2.097, 0.359)$

105. In Example 6, we showed that the centroid of an equivalent region was $(1, \pi/8)$. By symmetry, the centroid of this region is $(\pi/8, 1)$. You can also solve this problem directly.

$$A = \int_0^1 \left(\frac{\pi}{2} - \arcsin x \right) dx = \left[\frac{\pi}{2} x - x \arcsin x - \sqrt{1 - x^2} \right]_0^1$$ (Example 3)

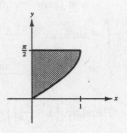

$$= \left(\frac{\pi}{2} - \frac{\pi}{2} - 0 \right) - (-1) = 1$$

$$\bar{x} = \frac{M_y}{A} = \int_0^1 x \left[\frac{\pi}{2} - \arcsin x \right] dx = \frac{\pi}{8}, \quad \bar{y} = \frac{M_x}{A} = \int_0^1 \frac{(\pi/2) + \arcsin x}{2} \left[\frac{\pi}{2} - \arcsin x \right] dx = 1$$

107. Average value $= \dfrac{1}{\pi} \int_0^{\pi} e^{-4t}(\cos 2t + 5 \sin 2t) \, dt$

$$= \frac{1}{\pi} \left[e^{-4t} \left(\frac{-4 \cos 2t + 2 \sin 2t}{20} \right) + 5e^{-4t} \left(\frac{-4 \sin 2t - 2 \cos 2t}{20} \right) \right]_0^{\pi}$$ (From Exercises 93 and 94)

$$= \frac{7}{10\pi}(1 - e^{-4\pi}) \approx 0.223$$

109. $c(t) = 100{,}000 + 4000t, r = 5\%, t_1 = 10$

$$P = \int_0^{10} (100{,}000 + 4000t)e^{-0.05t} \, dt$$

$$= 4000 \int_0^{10} (25 + t)e^{-0.05t} \, dt$$

Let $u = 25 + t, dv = e^{-0.05t} dt, du = dt, v = -\dfrac{100}{5} e^{-0.05t}$.

$$P = 4000 \left\{ \left[(25 + t) \left(-\frac{100}{5} e^{-0.05t} \right) \right]_0^{10} + \frac{100}{5} \int_0^{10} e^{-0.05t} \, dt \right\}$$

$$= 4000 \left\{ \left[(25 + t) \left(-\frac{100}{5} e^{-0.05t} \right) \right]_0^{10} - \left[\frac{10{,}000}{25} e^{-0.05t} \right]_0^{10} \right\}$$

$$\approx \$931{,}265$$

111. $\displaystyle\int_{-\pi}^{\pi} x \sin nx \, dx = \left[-\frac{x}{n} \cos nx + \frac{1}{n^2} \sin nx \right]_{-\pi}^{\pi}$

$$= -\frac{\pi}{n} \cos \pi n - \frac{\pi}{n} \cos(-\pi n)$$

$$= -\frac{2\pi}{n} \cos \pi n$$

$$= \begin{cases} -(2\pi/n), & \text{if } n \text{ is even} \\ (2\pi/n), & \text{if } n \text{ is odd} \end{cases}$$

113. Let $u = x$, $dv = \sin\left(\dfrac{n\pi}{2}x\right) dx$, $du = dx$, $v = -\dfrac{2}{n\pi}\cos\left(\dfrac{n\pi}{2}x\right)$.

$$I_1 = \int_0^1 x \sin\left(\frac{n\pi}{2}x\right) dx = \left[\frac{-2x}{n\pi}\cos\left(\frac{n\pi}{2}x\right)\right]_0^1 + \frac{2}{n\pi}\int_0^1 \cos\left(\frac{n\pi}{2}x\right) dx$$

$$= -\frac{2}{n\pi}\cos\left(\frac{n\pi}{2}\right) + \left[\left(\frac{2}{n\pi}\right)^2 \sin\left(\frac{n\pi}{2}x\right)\right]_0^1$$

$$= -\frac{2}{n\pi}\cos\left(\frac{n\pi}{2}\right) + \left(\frac{2}{n\pi}\right)^2 \sin\left(\frac{n\pi}{2}\right)$$

Let $u = (-x + 2)$, $dv = \sin\left(\dfrac{n\pi}{2}x\right) dx$, $du = -dx$, $v = -\dfrac{2}{n\pi}\cos\left(\dfrac{n\pi}{2}x\right)$.

$$I_2 = \int_1^2 (-x + 2) \sin\left(\frac{n\pi}{2}x\right) dx = \left[\frac{-2(-x+2)}{n\pi}\cos\left(\frac{n\pi}{2}x\right)\right]_1^2 - \frac{2}{n\pi}\int_1^2 \cos\left(\frac{n\pi}{2}x\right) dx$$

$$= \frac{2}{n\pi}\cos\left(\frac{n\pi}{2}\right) - \left[\left(\frac{2}{n\pi}\right)^2 \sin\left(\frac{n\pi}{2}x\right)\right]_1^2$$

$$= \frac{2}{n\pi}\cos\left(\frac{n\pi}{2}\right) + \left(\frac{2}{n\pi}\right)^2 \sin\left(\frac{n\pi}{2}\right)$$

$$h(I_1 + I_2) = b_n = h\left[\left(\frac{2}{n\pi}\right)^2 \sin\left(\frac{n\pi}{2}\right) + \left(\frac{2}{n\pi}\right)^2 \sin\left(\frac{n\pi}{2}\right)\right] = \frac{8h}{(n\pi)^2}\sin\left(\frac{n\pi}{2}\right)$$

115. **Shell Method:**

$$V = 2\pi \int_a^b x f(x)\, dx$$

$$dv = x\, dx \implies v = \frac{x^2}{2}$$

$$u = f(x) \implies du = f'(x)\, dx$$

$$V = 2\pi\left[\frac{x^2}{2}f(x) - \int \frac{x^2}{2}f'(x)\, dx\right]_a^b$$

$$= \pi\left[(b^2 f(b) - a^2 f(a)) - \int_a^b x^2 f'(x)\, dx\right]$$

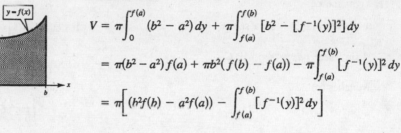

Disk Method:

$$V = \pi\int_0^{f(a)} (b^2 - a^2)\, dy + \pi\int_{f(a)}^{f(b)} [b^2 - [f^{-1}(y)]^2]\, dy$$

$$= \pi(b^2 - a^2) f(a) + \pi b^2(f(b) - f(a)) - \pi\int_{f(a)}^{f(b)} [f^{-1}(y)]^2\, dy$$

$$= \pi\left[(b^2 f(b) - a^2 f(a)) - \int_{f(a)}^{f(b)} [f^{-1}(y)]^2\, dy\right]$$

Since $x = f^{-1}(y)$, we have $f(x) = y$ and $f'(x)\, dx = dy$. When $y = f(a)$, $x = a$. When $y = f(b)$, $x = b$. Thus,

$$\int_{f(a)}^{f(b)} [f^{-1}(y)]^2\, dy = \int_a^b x^2 f'(x)\, dx$$

and the volumes are the same.

117. $f'(x) = 3x \sin(2x)$, $f(0) = 0$

(a) $f(x) = \displaystyle\int 3x \sin 2x\, dx = -\frac{3}{4}(2x \cos 2x - \sin 2x) + C$

(Parts: $u = 3x$, $dv = \sin 2x\, dx$)

$f(0) = 0 = -\dfrac{3}{4}(0) + C \implies C = 0$

$f(x) = -\dfrac{3}{4}(2x \cos 2x - \sin 2x)$

(b)

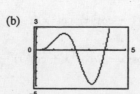

—CONTINUED—

117. —CONTINUED—

(c) Using $h = 0.05$, you obtain the points:

n	x_n	y_n
0	0	0
1	0.05	0.05
2	0.10	7.4875×10^{-4}
3	0.15	0.0037
4	0.20	0.0104
⋮	⋮	⋮
80	4.0	1.3181

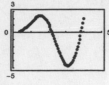

(d) Using $h = 0.1$, you obtain the points:

n	x_n	y_n
0	0	0
1	0.1	0
2	0.2	0.0060
3	0.3	0.0293
4	0.4	0.0801
⋮	⋮	⋮
40	4.0	1.0210

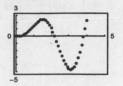

119. On $\left[0, \dfrac{\pi}{2}\right]$, $\sin x \le 1 \implies x \sin x \le x \implies \displaystyle\int_0^{\pi/2} x \sin x \, dx \le \int_0^{\pi/2} x \, dx.$

Section 8.3   Trigonometric Integrals

1. $y = \sec x$

$y' = \sec x \tan x = \sin x \sec^2 x$

$\displaystyle\int \sin x \sec^2 x \, dx = \sec x + C$

Matches (c)

3. $y = x - \tan x + \dfrac{1}{3} \tan^3 x$

$y' = 1 - \sec^2 x + \tan^2 x(\sec^2 x)$

$\quad = -\tan^2 x + \tan^2 x(1 + \tan^2 x)$

$\quad = \tan^4 x$

$\displaystyle\int \tan^4 x \, dx = x - \tan x + \dfrac{1}{3} \tan^3 x + C$

Matches (d)

5. Let $u = \cos x$, $du = -\sin x \, dx$.

$\displaystyle\int \cos^3 x \sin x \, dx = -\int \cos^3 x (-\sin x) \, dx$

$\qquad = -\dfrac{1}{4} \cos^4 x + C$

7. Let $u = \sin 2x$, $du = 2 \cos 2x \, dx$.

$\displaystyle\int \sin^5 2x \cos 2x \, dx = \dfrac{1}{2} \int \sin^5 2x (2 \cos 2x) \, dx$

$\qquad = \dfrac{1}{12} \sin^6 2x + C$

9. Let $u = \cos x$, $du = -\sin x \, dx$.

$\displaystyle\int \sin^5 x \cos^2 x \, dx = \int \sin x (1 - \cos^2 x)^2 \cos^2 x \, dx$

$\qquad = -\displaystyle\int (\cos^2 x - 2\cos^4 x + \cos^6 x)(-\sin x) \, dx = \dfrac{-1}{3} \cos^3 x + \dfrac{2}{5} \cos^5 x - \dfrac{1}{7} \cos^7 x + C$

11. $\displaystyle\int \cos^3 \theta \sqrt{\sin \theta} \, d\theta = \int \cos \theta (1 - \sin^2 \theta)(\sin \theta)^{1/2} \, d\theta$

$\qquad = \displaystyle\int [(\sin \theta)^{1/2} - (\sin \theta)^{5/2}] \cos \theta \, d\theta$

$\qquad = \dfrac{2}{3} (\sin \theta)^{3/2} - \dfrac{2}{7} (\sin \theta)^{7/2} + C$

13. $\displaystyle\int \cos^2 3x \, dx = \int \dfrac{1 + \cos 6x}{2} \, dx$

$\qquad = \dfrac{1}{2}\left(x + \dfrac{1}{6} \sin 6x\right) + C$

$\qquad = \dfrac{1}{12}(6x + \sin 6x) + C$

15. $\displaystyle\int \sin^2 \alpha \cdot \cos^2 \alpha \, d\alpha = \int \frac{1 - \cos 2\alpha}{2} \cdot \frac{1 + \cos 2\alpha}{2} \, d\alpha$

$\displaystyle\qquad = \frac{1}{4} \int (1 - \cos^2 2\alpha) \, d\alpha$

$\displaystyle\qquad = \frac{1}{4} \int \left(1 - \frac{1 + \cos 4\alpha}{2} \right) d\alpha$

$\displaystyle\qquad = \frac{1}{8} \int (1 - \cos 4\alpha) \, d\alpha$

$\displaystyle\qquad = \frac{1}{8} \left[\alpha - \frac{1}{4} \sin 4\alpha \right] + C$

$\displaystyle\qquad = \frac{1}{32} [4\alpha - \sin 4\alpha] + C$

17. Integration by parts:

$\displaystyle dv = \sin^2 x \, dx = \frac{1 - \cos 2x}{2} \implies v = \frac{x}{2} - \frac{\sin 2x}{4} = \frac{1}{4}(2x - \sin 2x)$

$u = x \implies du = dx$

$\displaystyle\int x \sin^2 x \, dx = \frac{1}{4} x (2x - \sin 2x) - \frac{1}{4} \int (2x - \sin 2x) \, dx$

$\displaystyle\qquad = \frac{1}{4} x (2x - \sin 2x) - \frac{1}{4} \left(x^2 + \frac{1}{2} \cos 2x \right) + C = \frac{1}{8}(2x^2 - 2x \sin 2x - \cos 2x) + C$

19. $\displaystyle\int_0^{\pi/2} \cos^3 x \, dx = \frac{2}{3}, \quad (n = 3)$

21. $\displaystyle\int_0^{\pi/2} \cos^7 x \, dx = \left(\frac{2}{3}\right)\left(\frac{4}{5}\right)\left(\frac{6}{7}\right) = \frac{16}{35}, \quad (n - 7)$

23. $\displaystyle\int_0^{\pi/2} \sin^6 x \, dx = \left(\frac{1}{2}\right)\left(\frac{3}{4}\right)\left(\frac{5}{6}\right)\frac{\pi}{2} = \frac{5\pi}{32}, \quad (n = 6)$

25. $\displaystyle\int \sec(3x) \, dx = \frac{1}{3} \ln|\sec 3x + \tan 3x| + C$

27. $\displaystyle\int \sec^4 5x \, dx = \int (1 + \tan^2 5x) \sec^2 5x \, dx$

$\displaystyle\qquad = \frac{1}{5} \left(\tan 5x + \frac{\tan^3 5x}{3} \right) + C$

$\displaystyle\qquad = \frac{\tan 5x}{15} (3 + \tan^2 5x) + C$

29. $\displaystyle dv = \sec^2 \pi x \, dx \implies v = \frac{1}{\pi} \tan \pi x$

$u = \sec \pi x \implies du = \pi \sec \pi x \tan \pi x \, dx$

$\displaystyle\int \sec^3 \pi x \, dx = \frac{1}{\pi} \sec \pi x \tan \pi x - \int \sec \pi x \tan^2 \pi x \, dx = \frac{1}{\pi} \sec \pi x \tan \pi x - \int \sec \pi x (\sec^2 \pi x - 1) \, dx$

$\displaystyle 2 \int \sec^3 \pi x \, dx = \frac{1}{\pi} (\sec \pi x \tan \pi x + \ln|\sec \pi x + \tan \pi x|) + C_1$

$\displaystyle\int \sec^3 \pi x \, dx = \frac{1}{2\pi} (\sec \pi x \tan \pi x + \ln|\sec \pi x + \tan \pi x|) + C$

31. $\displaystyle\int \tan^5 \frac{x}{4}\, dx = \int \left(\sec^2 \frac{x}{4} - 1 \right) \tan^3 \frac{x}{4}\, dx$

$\displaystyle = \int \tan^3 \frac{x}{4} \sec^2 \frac{x}{4}\, dx - \int \tan^3 \frac{x}{4}\, dx$

$\displaystyle = \tan^4 \frac{x}{4} - \int \left(\sec^2 \frac{x}{4} - 1 \right) \tan \frac{x}{4}\, dx$

$\displaystyle = \tan^4 \frac{x}{4} - 2 \tan^2 \frac{x}{4} - 4 \ln \left| \cos \frac{x}{4} \right| + C$

33. $u = \tan x,\ du = \sec^2 x\, dx$

$\displaystyle\int \sec^2 x \tan x\, dx = \frac{1}{2} \tan^2 x + C$

$\displaystyle\left[\text{or, } u = \sec x,\ du = \sec x \tan x\, dx, \right.$

$\displaystyle\left. \int \sec^2 x \tan x\, dx = \frac{1}{2} \sec^2 x + C. \right]$

35. $\displaystyle\int \tan^2 x \sec^2 x\, dx = \frac{\tan^3 x}{3} + C$

37. $\displaystyle\int \sec^6 4x \tan 4x\, dx = \frac{1}{4} \int \sec^5 4x (4 \sec 4x \tan 4x)\, dx$

$\displaystyle = \frac{\sec^6 4x}{24} + C$

39. Let $u = \sec x,\ du = \sec x \tan x\, dx$.

$\displaystyle\int \sec^3 x \tan x\, dx = \int \sec^2 x (\sec x \tan x)\, dx$

$\displaystyle = \frac{1}{3} \sec^3 x + C$

41. $\displaystyle\int \frac{\tan^2 x}{\sec x}\, dx = \int \frac{(\sec^2 x - 1)}{\sec x}\, dx$

$\displaystyle = \int (\sec x - \cos x)\, dx$

$\displaystyle = \ln|\sec x + \tan x| - \sin x + C$

43. $\displaystyle r = \int \sin^4(\pi\theta)\, d\theta = \frac{1}{4} \int [1 - \cos(2\pi\theta)]^2\, d\theta$

$\displaystyle = \frac{1}{4} \int [1 - 2\cos(2\pi\theta) + \cos^2(2\pi\theta)]\, d\theta$

$\displaystyle = \frac{1}{4} \int \left[1 - 2\cos(2\pi\theta) + \frac{1 + \cos(4\pi\theta)}{2} \right] d\theta$

$\displaystyle = \frac{1}{4} \left[\theta - \frac{1}{\pi} \sin(2\pi\theta) + \frac{\theta}{2} + \frac{1}{8\pi} \sin(4\pi\theta) \right] + C$

$\displaystyle = \frac{1}{32\pi} [12\pi\theta - 8\sin(2\pi\theta) + \sin(4\pi\theta)] + C$

45. $\displaystyle y = \int \tan^3 3x \sec 3x\, dx$

$\displaystyle = \int (\sec^2 3x - 1) \sec 3x \tan 3x\, dx$

$\displaystyle = \frac{1}{3} \int \sec^2 3x (3 \sec 3x \tan 3x)\, dx - \frac{1}{3} \int 3 \sec 3x \tan 3x\, dx$

$\displaystyle = \frac{1}{9} \sec^3 3x - \frac{1}{3} \sec 3x + C$

47. (a)

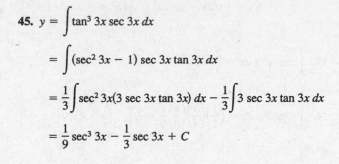

(b) $\dfrac{dy}{dx} = \sin^2 x, \quad (0, 0)$

$\displaystyle y = \int \sin^2 x\, dx$

$\displaystyle = \int \frac{1 - \cos 2x}{2}\, dx$

$\displaystyle = \frac{1}{2}x - \frac{\sin 2x}{4} + C$

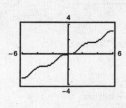

$(0, 0): \ 0 = C, \ y = \dfrac{1}{2}x - \dfrac{\sin 2x}{4}$

49. $\dfrac{dy}{dx} = \dfrac{3 \sin x}{y},\ y(0) = 2$

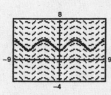

51. $\int \sin 3x \cos 2x \, dx = \dfrac{1}{2} \int (\sin 5x + \sin x) \, dx$

$$= \dfrac{-1}{2}\left(\dfrac{1}{5}\cos 5x + \cos x\right) + C$$

$$= \dfrac{-1}{10}(\cos 5x + 5\cos x) + C$$

53. $\int \sin \theta \sin 3\theta \, d\theta = \dfrac{1}{2} \int (\cos 2\theta - \cos 4\theta) \, d\theta$

$$= \dfrac{1}{2}\left(\dfrac{1}{2}\sin 2\theta - \dfrac{1}{4}\sin 4\theta\right) + C$$

$$= \dfrac{1}{8}(2\sin 2\theta - \sin 4\theta) + C$$

55. $\int \cot^3 2x \, dx = \int (\csc^2 2x - 1) \cot 2x \, dx$

$$= -\dfrac{1}{2}\int \cot 2x(-2\csc^2 2x) \, dx - \dfrac{1}{2}\int \dfrac{2\cos 2x}{\sin 2x} \, dx$$

$$= -\dfrac{1}{4}\cot^2 2x - \dfrac{1}{2}\ln|\sin 2x| + C$$

$$= \dfrac{1}{4}(\ln|\csc^2 2x| - \cot^2 2x) + C$$

57. Let $u = \cot \theta$, $du = -\csc^2 \theta \, d\theta$.

$$\int \csc^4 \theta \, d\theta = \int \csc^2 \theta(1 + \cot^2 \theta) \, d\theta$$

$$= \int \csc^2 \theta \, d\theta + \int \csc^2 \theta \cot^2 \theta \, d\theta$$

$$= -\cot \theta - \dfrac{1}{3}\cot^3 \theta + C$$

59. $\int \dfrac{\cot^2 t}{\csc t} \, dt = \int \dfrac{\csc^2 t - 1}{\csc t} \, dt$

$$= \int (\csc t - \sin t) \, dt$$

$$= \ln|\csc t - \cot t| + \cos t + C$$

61. $\int \dfrac{1}{\sec x \tan x} \, dx = \int \dfrac{\cos^2 x}{\sin x} \, dx = \int \dfrac{1 - \sin^2 x}{\sin x} \, dx$

$$= \int (\csc x - \sin x) \, dx$$

$$= \ln|\csc x - \cot x| + \cos x + C$$

63. $\int (\tan^4 t - \sec^4 t) \, dt = \int (\tan^2 t + \sec^2 t)(\tan^2 t - \sec^2 t) \, dt, \qquad (\tan^2 t - \sec^2 t = -1)$

$$= -\int (\tan^2 t + \sec^2 t) \, dt = -\int (2\sec^2 t - 1) \, dt = -2\tan t + t + C$$

65. $\displaystyle\int_{-\pi}^{\pi} \sin^2 x \, dx = 2\int_0^{\pi} \dfrac{1 - \cos 2x}{2} \, dx$

$$= \left[x - \dfrac{1}{2}\sin 2x\right]_0^{\pi} = \pi$$

67. $\displaystyle\int_0^{\pi/4} \tan^3 x \, dx = \int_0^{\pi/4} (\sec^2 x - 1) \tan x \, dx$

$$= \int_0^{\pi/4} \sec^2 x \tan x \, dx - \int_0^{\pi/4} \dfrac{\sin x}{\cos x} \, dx$$

$$= \left[\dfrac{1}{2}\tan^2 x + \ln|\cos x|\right]_0^{\pi/4}$$

$$= \dfrac{1}{2}(1 - \ln 2)$$

69. Let $u = 1 + \sin t$, $du = \cos t \, dt$.

$$\int_0^{\pi/2} \dfrac{\cos t}{1 + \sin t} \, dt = \left[\ln|1 + \sin t|\right]_0^{\pi/2} = \ln 2$$

71. Let $u = \sin x$, $du = \cos x \, dx$.

$$\int_{-\pi/2}^{\pi/2} \cos^3 x \, dx = 2\int_0^{\pi/2} (1 - \sin^2 x) \cos x \, dx$$

$$= 2\left[\sin x - \dfrac{1}{3}\sin^3 x\right]_0^{\pi/2} = \dfrac{4}{3}$$

73. $\int \cos^4 \dfrac{x}{2} \, dx = \dfrac{1}{16}[6x + 8\sin x + \sin 2x] + C$

$$= \dfrac{1}{8}\left[4\sin\dfrac{x}{2}\cos^3\dfrac{x}{2} + 6\sin\dfrac{x}{2}\cos\dfrac{x}{2} + 3x\right] + C$$

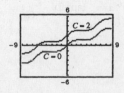

75. $\int \sec^5 \pi x \, dx = \dfrac{1}{4\pi}\left\{\sec^3 \pi x \tan \pi x + \dfrac{3}{2}[\sec \pi x \tan \pi x + \ln|\sec \pi x + \tan \pi x|]\right\} + C$

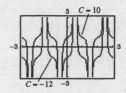

77. $\int \sec^5 \pi x \tan \pi x \, dx = \dfrac{1}{5\pi} \sec^5 \pi x + C$

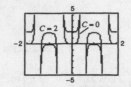

79. $\displaystyle\int_0^{\pi/4} \sin 2\theta \sin 3\theta \, d\theta = \dfrac{1}{2}\left[\sin \theta - \dfrac{1}{5}\sin 5\theta\right]_0^{\pi/4} = \dfrac{3\sqrt{2}}{10}$ **81.** $\displaystyle\int_0^{\pi/2} \sin^4 x \, dx = \dfrac{1}{4}\left[\dfrac{3x}{2} - \sin 2x + \dfrac{1}{8}\sin 4x\right]_0^{\pi/2}$

$$= \dfrac{3\pi}{16}$$

83. (a) Save one sine factor and convert the remaining sine factors to cosine. Then expand and integrate.

   (b) Save one cosine factor and convert the remaining cosine factors to sine. Then expand and integrate.

   (c) Make repeated use of the power reducing formula to convert the integrand to odd powers of the cosine.

85. (a) Let $u = \tan 3x$, $du = 3\sec^2 3x \, dx$.

$$\int \sec^4 3x \tan^3 3x \, dx = \int \sec^2 3x \tan^3 3x \sec^2 3x \, dx = \dfrac{1}{3}\int (\tan^2 3x + 1)\tan^3 3x(3\sec^2 3x) \, dx$$

$$= \dfrac{1}{3}\int (\tan^5 3x + \tan^3 3x)(3\sec^2 3x) \, dx = \dfrac{\tan^6 3x}{18} + \dfrac{\tan^4 3x}{12} + C_1$$

Or let $u = \sec 3x$, $du = 3\sec 3x \tan 3x \, dx$.

$$\int \sec^4 3x \tan^3 3x \, dx = \int \sec^3 3x \tan^2 3x \sec 3x \tan 3x \, dx$$

$$= \dfrac{1}{3}\int \sec^3 3x(\sec^2 3x - 1)(3\sec 3x \tan 3x) \, dx = \dfrac{\sec^6 3x}{18} - \dfrac{\sec^4 3x}{12} + C$$

(b)

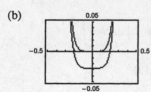

(c) $\dfrac{\sec^6 3x}{18} - \dfrac{\sec^4 3x}{12} + C = \dfrac{(1 + \tan^2 3x)^3}{18} - \dfrac{(1 + \tan^2 3x)^2}{12} + C$

$$= \dfrac{1}{18}\tan^6 3x + \dfrac{1}{6}\tan^4 3x + \dfrac{1}{6}\tan^2 3x + \dfrac{1}{18} - \dfrac{1}{12}\tan^4 3x - \dfrac{1}{6}\tan^2 3x - \dfrac{1}{12} + C$$

$$= \dfrac{\tan^6 3x}{18} + \dfrac{\tan^4 3x}{12} + \left(\dfrac{1}{18} - \dfrac{1}{12}\right) + C$$

$$= \dfrac{\tan^6 3x}{18} + \dfrac{\tan^4 3x}{12} + C_2$$

87. $A = \displaystyle\int_0^{\pi/2} (\sin x - \sin^3 x)\, dx$

$\quad = \displaystyle\int_0^{\pi/2} \sin x\, dx - \int_0^{\pi/2} \sin^3 x\, dx$

$\quad = \Big[-\cos x \Big]_0^{\pi/2} - \dfrac{2}{3} \qquad \text{(Wallis's Formula)}$

$\quad = 1 - \dfrac{2}{3} = \dfrac{1}{3}$

89. $A = \displaystyle\int_{-\pi/4}^{\pi/4} [\cos^2 x - \sin^2 x]\, dx$

$\quad = \displaystyle\int_{-\pi/4}^{\pi/4} \cos 2x\, dx$

$\quad = \dfrac{\sin 2x}{2} \Big]_{-\pi/4}^{\pi/4}$

$\quad = \dfrac{1}{2} + \dfrac{1}{2} = 1$

91. Disks

$R(x) = \tan x,\ r(x) = 0$

$V = 2\pi \displaystyle\int_0^{\pi/4} \tan^2 x\, dx$

$\quad = 2\pi \displaystyle\int_0^{\pi/4} (\sec^2 x - 1)\, dx$

$\quad = 2\pi \Big[\tan x - x \Big]_0^{\pi/4}$

$\quad = 2\pi \Big(1 - \dfrac{\pi}{4} \Big) \approx 1.348$

93. (a) $V = \pi \displaystyle\int_0^{\pi} \sin^2 x\, dx = \dfrac{\pi}{2} \int_0^{\pi} (1 - \cos 2x)\, dx = \dfrac{\pi}{2} \Big[x - \dfrac{1}{2} \sin 2x \Big]_0^{\pi} = \dfrac{\pi^2}{2}$

$\quad$ **(b)** $A = \displaystyle\int_0^{\pi} \sin x\, dx = \Big[-\cos x \Big]_0^{\pi} = 1 + 1 = 2$

$\quad\quad$ Let $u = x,\ dv = \sin x\, dx,\ du = dx,\ v = -\cos x.$

$\quad\quad \bar{x} = \dfrac{1}{A} \displaystyle\int_0^{\pi} x \sin x\, dx = \dfrac{1}{2} \Big[\Big[-x \cos x \Big]_0^{\pi} + \int_0^{\pi} \cos x\, dx \Big] = \dfrac{1}{2} \Big[-x \cos x + \sin x \Big]_0^{\pi} = \dfrac{\pi}{2}$

$\quad\quad \bar{y} = \dfrac{1}{2A} \displaystyle\int_0^{\pi} \sin^2 x\, dx = \dfrac{1}{8} \int_0^{\pi} (1 - \cos 2x)\, dx = \dfrac{1}{8} \Big[x - \dfrac{1}{2} \sin 2x \Big]_0^{\pi} = \dfrac{\pi}{8}$

$\quad\quad (\bar{x}, \bar{y}) = \Big(\dfrac{\pi}{2}, \dfrac{\pi}{8} \Big)$

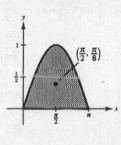

95. $dv = \sin x\, dx \implies v = -\cos x$

$\quad u = \sin^{n-1} x \implies du = (n-1) \sin^{n-2} x \cos x\, dx$

$\quad \displaystyle\int \sin^n x\, dx = -\sin^{n-1} x \cos x + (n-1) \int \sin^{n-2} x \cos^2 x\, dx = -\sin^{n-1} x \cos x + (n-1) \int \sin^{n-2} x (1 - \sin^2 x)\, dx$

$\quad\quad = -\sin^{n-1} x \cos x + (n-1) \displaystyle\int \sin^{n-2} x\, dx - (n-1) \int \sin^n x\, dx$

$\quad$ Therefore, $n \displaystyle\int \sin^n x\, dx = -\sin^{n-1} x \cos x + (n-1) \int \sin^{n-2} x\, dx$

$\quad\quad\quad \displaystyle\int \sin^n x\, dx = \dfrac{-\sin^{n-1} x \cos x}{n} + \dfrac{n-1}{n} \int \sin^{n-2} x\, dx.$

97. Let $u = \sin^{n-1} x$, $du = (n-1)\sin^{n-2} x \cos x\, dx$, $dv = \cos^m x \sin x\, dx$, $v = \dfrac{-\cos^{m+1} x}{m+1}$.

$$\int \cos^m x \sin^n x\, dx = \frac{-\sin^{n-1} x \cos^{m+1} x}{m+1} + \frac{n-1}{m+1}\int \sin^{n-2} x \cos^{m+2} x\, dx$$

$$= \frac{-\sin^{n-1} x \cos^{m+1} x}{m+1} + \frac{n-1}{m+1}\int \sin^{n-2} x \cos^m x(1 - \sin^2 x)\, dx$$

$$= \frac{-\sin^{n-1} x \cos^{m+1} x}{m+1} + \frac{n-1}{m+1}\int \sin^{n-2} x \cos^m x\, dx - \frac{n-1}{m+1}\int \sin^n x \cos^m x\, dx$$

$$\frac{m+n}{m+1}\int \cos^m x \sin^n x\, dx = \frac{-\sin^{n-1} x \cos^{m+1} x}{m+1} + \frac{n-1}{m+1}\int \sin^{n-2} x \cos^m x\, dx$$

$$\int \cos^m x \sin^n x\, dx = \frac{-\cos^{m+1} x \sin^{n-1} x}{m+n} + \frac{n-1}{m+n}\int \cos^m x \sin^{n-2} x\, dx$$

99. $\displaystyle\int \sin^5 x\, dx = -\frac{\sin^4 x \cos x}{5} + \frac{4}{5}\int \sin^3 x\, dx$

$$= -\frac{\sin^4 x \cos x}{5} + \frac{4}{5}\left[-\frac{\sin^2 x \cos x}{3} + \frac{2}{3}\int \sin x\, dx\right]$$

$$= -\frac{1}{5}\sin^4 x \cos x - \frac{4}{15}\sin^2 x \cos x - \frac{8}{15}\cos x + C$$

$$= -\frac{\cos x}{15}\left[3\sin^4 x + 4\sin^2 x + 8\right] + C$$

101. $\displaystyle\int \sec^4 \frac{2\pi x}{5}\, dx = \frac{5}{2\pi}\int \sec^4\left(\frac{2\pi x}{5}\right)\frac{2\pi}{5}\, dx$

$$= \frac{5}{2\pi}\left[\frac{1}{3}\sec^2\left(\frac{2\pi x}{5}\right)\tan\left(\frac{2\pi x}{5}\right) + \frac{2}{3}\int \sec^2\left(\frac{2\pi x}{5}\right)\frac{2\pi}{5}\, dx\right]$$

$$= \frac{5}{6\pi}\left[\sec^2\left(\frac{2\pi x}{5}\right)\tan\left(\frac{2\pi x}{5}\right) + 2\tan\left(\frac{2\pi x}{5}\right)\right] + C$$

$$= \frac{5}{6\pi}\tan\left(\frac{2\pi x}{5}\right)\left[\sec^2\left(\frac{2\pi x}{5}\right) + 2\right] + C$$

103. $f(t) = a_0 + a_1 \cos\dfrac{\pi t}{6} + b_1 \sin\dfrac{\pi t}{6}$

$$a_0 = \frac{1}{12}\int_0^{12} f(t)\, dt, \quad a_1 = \frac{1}{6}\int_0^{12} f(t)\cos\frac{\pi t}{6}\, dt, \quad b_1 = \frac{1}{6}\int_0^{12} f(t)\sin\frac{\pi t}{6}\, dt$$

(a) $a_0 \approx \dfrac{1}{12}\cdot\dfrac{(12-0)}{3(12)}[33.5 + 4(35.4) + 2(44.7) + 4(55.6) + 2(67.4) + 4(76.2) + 2(80.4) + 4(79.0) + 2(72.0)$

$$+\ 4(61.0) + 2(49.3) + 4(38.6) + 33.5]$$

≈ 57.72

$a_1 \approx -23.36$

$b_1 \approx -2.75$ (Answers will vary.)

$$H(t) \approx 57.72 - 23.36\cos\left(\frac{\pi t}{6}\right) - 2.75\sin\left(\frac{\pi t}{6}\right)$$

(b) $L(t) \approx 42.04 - 20.91\cos\left(\dfrac{\pi t}{6}\right) - 4.33\sin\left(\dfrac{\pi t}{6}\right)$

—CONTINUED—

103. —CONTINUED—

(c)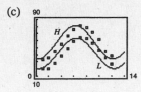

Temperature difference is greatest in the summer ($t \approx 4.9$ or end of May).

105. $\displaystyle\int_{-\pi}^{\pi} \cos(mx) \cos(nx)\, dx = \frac{1}{2}\left[\frac{\sin(m+n)x}{m+n} + \frac{\sin(m-n)x}{m-n}\right]_{-\pi}^{\pi} = 0, \quad (m \neq n)$

$\displaystyle\int_{-\pi}^{\pi} \sin(mx) \sin(nx)\, dx = \frac{1}{2}\int_{-\pi}^{\pi} [\cos(m-n)x - \cos(m+n)x]\, dx$

$\displaystyle\qquad = \frac{1}{2}\left[\frac{\sin(m-n)x}{m-n} - \frac{\sin(m+n)x}{m+n}\right]_{-\pi}^{\pi} = 0, \quad (m \neq n)$

$\displaystyle\int_{-\pi}^{\pi} \sin(mx) \cos(nx)\, dx = \frac{1}{2}\int_{-\pi}^{\pi} [\sin(m+n)x + \sin(m-n)x]\, dx$

$\displaystyle\qquad = -\frac{1}{2}\left[\frac{\cos(m+n)x}{m+n} + \frac{\cos(m-n)x}{m-n}\right]_{-\pi}^{\pi}, \quad (m \neq n)$

$\displaystyle\qquad = -\frac{1}{2}\left[\left(\frac{\cos(m+n)\pi}{m+n} + \frac{\cos(m-n)\pi}{m-n}\right) - \left(\frac{\cos(m+n)(-\pi)}{m+n} + \frac{\cos(m-n)(-\pi)}{m-n}\right)\right]$

$\displaystyle\qquad = 0, \quad \text{since } \cos(-\theta) = \cos\theta.$

$\displaystyle\int_{-\pi}^{\pi} \sin(mx) \cos(mx)\, dx = \frac{1}{m}\frac{\sin^2(mx)}{2}\bigg]_{-\pi}^{\pi} = 0$

Section 8.4   Trigonometric Substitution

1. $\displaystyle\frac{d}{dx}\left[4\ln\left|\frac{\sqrt{x^2+16}-4}{x}\right| + \sqrt{x^2+16} + C\right] = \frac{d}{dx}\left[4\ln\left|\sqrt{x^2+16}-4\right| - 4\ln|x| + \sqrt{x^2+16} + C\right]$

$\displaystyle\qquad = 4\left[\frac{x/\sqrt{x^2+16}}{\sqrt{x^2+16}-4}\right] - \frac{4}{x} + \frac{x}{\sqrt{x^2+16}}$

$\displaystyle\qquad = \frac{4x}{\sqrt{x^2+16}\left(\sqrt{x^2+16}-4\right)} - \frac{4}{x} + \frac{x}{\sqrt{x^2+16}}$

$\displaystyle\qquad = \frac{4x^2 - 4\sqrt{x^2+16}\left(\sqrt{x^2+16}-4\right) + x^2\left(\sqrt{x^2+16}-4\right)}{x\sqrt{x^2+16}\left(\sqrt{x^2+16}-4\right)}$

$\displaystyle\qquad = \frac{4x^2 - 4(x^2+16) + 16\sqrt{x^2+16} + x^2\sqrt{x^2+16} - 4x^2}{x\sqrt{x^2+16}\left(\sqrt{x^2+16}-4\right)}$

$\displaystyle\qquad = \frac{\sqrt{x^2+16}(x^2+16) - 4(x^2+16)}{x\sqrt{x^2+16}\left(\sqrt{x^2+16}-4\right)}$

$\displaystyle\qquad = \frac{(x^2+16)\left(\sqrt{x^2+16}-4\right)}{x\sqrt{x^2+16}\left(\right.}$

Indefinite integral: $\displaystyle\int \frac{\sqrt{x^2+16}}{x}\, dx,$ matches (b).

3. $\dfrac{d}{dx}\left[8\arcsin\dfrac{x}{4} - \dfrac{x\sqrt{16-x^2}}{2} + C\right] = 8\dfrac{1/4}{\sqrt{1-(x/4)^2}} - \dfrac{x(1/2)(16-x^2)^{-1/2}(-2x) + \sqrt{16-x^2}}{2}$

$$= \frac{8}{\sqrt{16-x^2}} + \frac{x^2}{2\sqrt{16-x^2}} - \frac{\sqrt{16-x^2}}{2}$$

$$= \frac{16}{2\sqrt{16-x^2}} + \frac{x^2}{2\sqrt{16-x^2}} - \frac{(16-x^2)}{2\sqrt{16-x^2}} = \frac{x^2}{\sqrt{16-x^2}}$$

Matches (a)

5. Let $x = 5\sin\theta$, $dx = 5\cos\theta\,d\theta$, $\sqrt{25-x^2} = 5\cos\theta$.

$$\int \frac{1}{(25-x^2)^{3/2}}\,dx = \int \frac{5\cos\theta}{(5\cos\theta)^3}\,d\theta$$

$$= \frac{1}{25}\int \sec^2\theta\,d\theta$$

$$= \frac{1}{25}\tan\theta + C$$

$$= \frac{x}{25\sqrt{25-x^2}} + C$$

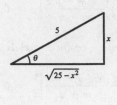

7. Same substitution as in Exercise 5

$$\int \frac{\sqrt{25-x^2}}{x}\,dx = \int \frac{25\cos^2\theta\,d\theta}{5\sin\theta} = 5\int \frac{1-\sin^2\theta}{\sin\theta}\,d\theta = 5\int (\csc\theta - \sin\theta)\,d\theta$$

$$= 5[\ln|\csc\theta - \cot\theta| + \cos\theta] + C = 5\ln\left|\frac{5-\sqrt{25-x^2}}{x}\right| + \sqrt{25-x^2} + C$$

9. Let $x = 2\sec\theta$, $dx = 2\sec\theta\tan\theta\,d\theta$, $\sqrt{x^2-4} = 2\tan\theta$.

$$\int \frac{1}{\sqrt{x^2-4}}\,dx = \int \frac{2\sec\theta\tan\theta\,d\theta}{2\tan\theta} = \int \sec\theta\,d\theta = \ln|\sec\theta + \tan\theta| + C_1$$

$$= \ln\left|\frac{x}{2} + \frac{\sqrt{x^2-4}}{2}\right| + C_1$$

$$= \ln\left|x + \sqrt{x^2-4}\right| - \ln 2 + C_1 = \ln\left|x + \sqrt{x^2-4}\right| + C$$

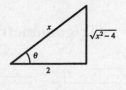

11. Same substitution as in Exercise 9

$$\int x^3\sqrt{x^2-4}\,dx = \int (8\sec^3\theta)(2\tan\theta)(2\sec\theta\tan\theta)\,d\theta = 32\int \tan^2\theta\sec^4\theta\,d\theta$$

$$= 32\int \tan^2\theta(1+\tan^2\theta)\sec^2\theta\,d\theta = 32\left(\frac{\tan^3\theta}{3} + \frac{\tan^5\theta}{5}\right) + C$$

$$= \frac{32}{15}\tan^3\theta[5 + 3\tan^2\theta] + C = \frac{32}{15}\frac{(x^2-4)^{3/2}}{8}\left[5 + 3\frac{(x^2-4)}{4}\right] + C$$

$$= \frac{1}{15}(x^2-4)^{3/2}[20 + 3(x^2-4)] + C = \frac{1}{15}(x^2-4)^{3/2}(3x^2+8) + C$$

13. Let $x = \tan\theta$, $dx = \sec^2\theta\,d\theta$, $\sqrt{1+x^2} = \sec\theta$.

$$\int x\sqrt{1+x^2}\,dx = \int \tan\theta(\sec\theta)\sec^2\theta\,d\theta = \frac{\sec^3\theta}{3} + C = \frac{1}{3}(1+x^2)^{3/2} + C$$

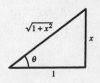

Note: This integral could have been evaluated with the Power Rule.

15. Same substitution as in Exercise 13

$$\int \frac{1}{(1 + x^2)^2}\, dx = \int \frac{1}{(\sqrt{1 + x^2})^4}\, dx = \int \frac{\sec^2 \theta\, d\theta}{\sec^4 \theta}$$

$$= \int \cos^2 \theta\, d\theta = \frac{1}{2} \int (1 + \cos 2\theta)\, d\theta$$

$$= \frac{1}{2}\left[\theta + \frac{\sin 2\theta}{2} \right]$$

$$= \frac{1}{2}[\theta + \sin \theta \cos \theta] + C$$

$$= \frac{1}{2}\left[\arctan x + \left(\frac{x}{\sqrt{1 + x^2}} \right)\left(\frac{1}{\sqrt{1 + x^2}} \right) \right] + C$$

$$= \frac{1}{2}\left[\arctan x + \frac{x}{1 + x^2} \right] + C$$

17. Let $u = 3x$, $a = 2$, and $du = 3\, dx$.

$$\int \sqrt{4 + 9x^2}\, dx = \frac{1}{3} \int \sqrt{(2)^2 + (3x)^2}\, 3\, dx$$

$$= \frac{1}{3}\left(\frac{1}{2} \right)\left(3x\sqrt{4 + 9x^2} + 4 \ln\left| 3x + \sqrt{4 + 9x^2} \right| \right) + C$$

$$= \frac{1}{2}x\sqrt{4 + 9x^2} + \frac{2}{3} \ln\left| 3x + \sqrt{4 + 9x^2} \right| + C$$

19. $\displaystyle \int \sqrt{25 - 4x^2}\, dx = \int 2\sqrt{\frac{25}{4} - x^2}\, dx, \quad a = \frac{5}{2}$

$$= 2\frac{1}{2}\left[\frac{25}{4} \arcsin\left(\frac{2x}{5} \right) + x\sqrt{\frac{25}{4} - x^2} \right] + C$$

$$= \frac{25}{4} \arcsin\left(\frac{2x}{5} \right) + \frac{x}{2}\sqrt{25 - 4x^2} + C$$

21. $\displaystyle \int \frac{x}{\sqrt{x^2 + 9}}\, dx = \frac{1}{2} \int (x^2 + 9)^{-1/2}(2x)\, dx$

$$= \sqrt{x^2 + 9} + C$$

(Power Rule)

23. $\displaystyle \int \frac{1}{\sqrt{16 - x^2}}\, dx = \arcsin\left(\frac{x}{4} \right) + C$

25. Let $x = 2 \sin \theta$, $dx = 2 \cos \theta\, d\theta$, $\sqrt{4 - x^2} = 2 \cos \theta$.

$$\int \sqrt{16 - 4x^2}\, dx = 2 \int \sqrt{4 - x^2}\, dx$$

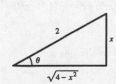

$$= 2 \int 2 \cos \theta (2 \cos \theta\, d\theta)$$

$$= 8 \int \cos^2 \theta\, d\theta$$

$$= 4 \int (1 + \cos 2\theta)\, d\theta$$

$$= 4\left[\theta + \frac{1}{2} \sin 2\theta \right] + C$$

$$= 4\theta + 4 \sin \theta \cos \theta + C$$

$$= 4 \arcsin\left(\frac{x}{2} \right) + x\sqrt{4 - x^2} + C$$

27. Let $x = 3 \sec \theta, dx = 3 \sec \theta \tan \theta \, d\theta, \sqrt{x^2 - 9} = 3 \tan \theta$.

$$\int \frac{1}{\sqrt{x^2 - 9}} \, dx = \int \frac{3 \sec \theta \tan \theta \, d\theta}{3 \tan \theta}$$

$$= \int \sec \theta \, d\theta$$

$$= \ln|\sec \theta + \tan \theta| + C_1$$

$$= \ln\left|\frac{x}{3} + \frac{\sqrt{x^2 - 9}}{3}\right| + C_1$$

$$= \ln\left|x + \sqrt{x^2 - 9}\right| + C$$

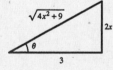

29. Let $x = \sin \theta, dx = \cos \theta \, d\theta, \sqrt{1 - x^2} = \cos \theta$.

$$\int \frac{\sqrt{1 - x^2}}{x^4} \, dx = \int \frac{\cos \theta(\cos \theta \, d\theta)}{\sin^4 \theta}$$

$$= \int \cot^2 \theta \csc^2 \theta \, d\theta$$

$$= -\frac{1}{3} \cot^3 \theta + C$$

$$= \frac{-(1 - x^2)^{3/2}}{3x^3} + C$$

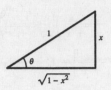

31. Same substitution as in Exercise 30

$$x = \frac{3}{2} \tan \theta, dx = \frac{3}{2} \sec^2 \theta \, d\theta$$

$$\int \frac{1}{x\sqrt{4x^2 + 9}} \, dx = \int \frac{(3/2) \sec^2 \theta \, d\theta}{(3/2) \tan \theta \, 3 \sec \theta}$$

$$= \frac{1}{3} \int \csc \theta \, d\theta$$

$$= -\frac{1}{3} \ln|\csc \theta + \cot \theta| + C$$

$$= -\frac{1}{3} \ln\left|\frac{\sqrt{4x^2 + 9} + 3}{2x}\right| + C$$

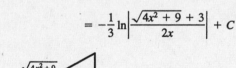

33. Let $x = \sqrt{5} \tan \theta, dx = \sqrt{5} \sec^2 \theta \, d\theta, x^2 + 5 = 5 \sec^2 \theta$.

$$\int \frac{-5x}{(x^2 + 5)^{3/2}} \, dx = \int \frac{-5\sqrt{5} \tan \theta}{(5 \sec^2 \theta)^{3/2}} \sqrt{5} \sec^2 \theta \, d\theta$$

$$= -\sqrt{5} \int \frac{\tan \theta}{\sec \theta} \, d\theta$$

$$= -\sqrt{5} \int \sin \theta \, d\theta$$

$$= \sqrt{5} \cos \theta + C$$

$$= \sqrt{5} \frac{\sqrt{5}}{\sqrt{x^2 + 5}} + C$$

$$= \frac{5}{\sqrt{x^2 + 5}} + C$$

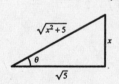

35. Let $u = 1 + e^{2x}, du = 2e^{2x} \, dx$.

$$\int e^{2x} \sqrt{1 + e^{2x}} \, dx = \frac{1}{2} \int (1 + e^{2x})^{1/2}(2e^{2x}) \, dx = \frac{1}{3}(1 + e^{2x})^{3/2} + C$$

37. Let $e^x = \sin \theta, e^x \, dx = \cos \theta \, d\theta, \sqrt{1 - e^{2x}} = \cos \theta$.

$$\int e^x \sqrt{1 - e^{2x}} \, dx = \int \cos^2 \theta \, d\theta$$

$$= \frac{1}{2} \int (1 + \cos 2\theta) \, d\theta$$

$$= \frac{1}{2}\left[\theta + \frac{\sin 2\theta}{2}\right]$$

$$= \frac{1}{2}(\theta + \sin \theta \cos \theta) + C$$

$$= \frac{1}{2}\left(\arcsin e^x + e^x \sqrt{1 - e^{2x}}\right) + C$$

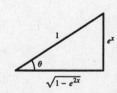

39. Let $x = \sqrt{2} \tan \theta, \, dx = \sqrt{2} \sec^2 \theta \, d\theta, \, x^2 + 2 = 2 \sec^2 \theta.$

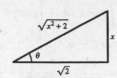

$$\int \frac{1}{4 + 4x^2 + x^4} \, dx = \int \frac{1}{(x^2 + 2)^2} \, dx = \int \frac{\sqrt{2} \sec^2 \theta \, d\theta}{4 \sec^4 \theta}$$

$$= \frac{\sqrt{2}}{4} \int \cos^2 \theta \, d\theta$$

$$= \frac{\sqrt{2}}{4} \left(\frac{1}{2} \right) \int (1 + \cos 2\theta) \, d\theta$$

$$= \frac{\sqrt{2}}{8} \left(\theta + \frac{1}{2} \sin 2\theta \right) + C$$

$$= \frac{\sqrt{2}}{8} (\theta + \sin \theta \cos \theta) + C$$

$$= \frac{\sqrt{2}}{8} \left(\arctan \frac{x}{\sqrt{2}} + \frac{x}{\sqrt{x^2 + 2}} \cdot \frac{\sqrt{2}}{\sqrt{x^2 + 2}} \right)$$

$$= \frac{1}{4} \left[\frac{x}{x^2 + 2} + \frac{1}{\sqrt{2}} \arctan \frac{x}{\sqrt{2}} \right] + C$$

41. Use integration by parts. Since $x > \frac{1}{2}$,

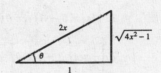

$$u = \text{arcsec } 2x \implies du = \frac{1}{x\sqrt{4x^2 - 1}} \, dx, \, dv = dx \implies v = x$$

$$\int \text{arcsec } 2x \, dx = x \, \text{arcsec } 2x - \int \frac{1}{\sqrt{4x^2 - 1}} \, dx$$

$2x = \sec \theta, \, dx = \frac{1}{2} \sec \theta \tan \theta \, d\theta, \, \sqrt{4x^2 - 1} = \tan \theta$

$$\int \text{arcsec } 2x \, dx = x \, \text{arcsec } 2x - \int \frac{(1/2) \sec \theta \tan \theta \, d\theta}{\tan \theta}$$

$$= x \, \text{arcsec } 2x - \frac{1}{2} \int \sec \theta \, d\theta$$

$$= x \, \text{arcsec } 2x - \frac{1}{2} \ln |\sec \theta + \tan \theta| + C$$

$$= x \, \text{arcsec } 2x - \frac{1}{2} \ln \left| 2x + \sqrt{4x^2 - 1} \right| + C.$$

43. $\int \dfrac{1}{\sqrt{4x - x^2}} \, dx = \int \dfrac{1}{\sqrt{4 - (x - 2)^2}} \, dx = \arcsin \left(\dfrac{x - 2}{2} \right) + C$

45. Let $x + 2 = 2 \tan \theta, \, dx = 2 \sec^2 \theta \, d\theta, \, \sqrt{(x + 2)^2 + 4} = 2 \sec \theta.$

$$\int \frac{x}{\sqrt{x^2 + 4x + 8}} \, dx = \int \frac{x}{\sqrt{(x + 2)^2 + 4}} \, dx = \int \frac{(2 \tan \theta - 2)(2 \sec^2 \theta) \, d\theta}{2 \sec \theta}$$

$$= 2 \int (\tan \theta - 1)(\sec \theta) \, d\theta$$

$$= 2[\sec \theta - \ln |\sec \theta + \tan \theta|] + C_1$$

$$= 2 \left[\frac{\sqrt{(x + 2)^2 + 4}}{2} - \ln \left| \frac{\sqrt{(x + 2)^2 + 4}}{2} + \frac{x + 2}{2} \right| \right] + C_1$$

$$= \sqrt{x^2 + 4x + 8} - 2 \left[\ln \left| \sqrt{x^2 + 4x + 8} + (x + 2) \right| - \ln 2 \right] + C_1$$

$$= \sqrt{x^2 + 4x + 8} - 2 \ln \left| \sqrt{x^2 + 4x + 8} + (x + 2) \right| + C$$

47. Let $t = \sin\theta$, $dt = \cos\theta\, d\theta$, $1 - t^2 = \cos^2\theta$.

(a) $\displaystyle\int \frac{t^2}{(1-t^2)^{3/2}}\, dt = \int \frac{\sin^2\theta\cos\theta\, d\theta}{\cos^3\theta}$

$$= \int \tan^2\theta\, d\theta = \int (\sec^2\theta - 1)\, d\theta$$

$$= \tan\theta - \theta + C$$

$$= \frac{t}{\sqrt{1-t^2}} - \arcsin t + C$$

Thus, $\displaystyle\int_0^{\sqrt{3}/2} \frac{t^2}{(1-t^2)^{3/2}}\, dt = \left[\frac{t}{\sqrt{1-t^2}} - \arcsin t\right]_0^{\sqrt{3}/2} = \frac{\sqrt{3}/2}{\sqrt{1/4}} - \arcsin\frac{\sqrt{3}}{2} = \sqrt{3} - \frac{\pi}{3} \approx 0.685.$

(b) When $t = 0$, $\theta = 0$. When $t = \sqrt{3}/2$, $\theta = \pi/3$. Thus,

$$\int_0^{\sqrt{3}/2} \frac{t^2}{(1-t^2)^{3/2}}\, dt = \Big[\tan\theta - \theta\Big]_0^{\pi/3} = \sqrt{3} - \frac{\pi}{3} \approx 0.685.$$

49. (a) Let $x = 3\tan\theta$, $dx = 3\sec^2\theta\, d\theta$, $\sqrt{x^2+9} = 3\sec\theta$.

$$\int \frac{x^3}{\sqrt{x^2+9}}\, dx = \int \frac{(27\tan^3\theta)(3\sec^2\theta\, d\theta)}{3\sec\theta}$$

$$= 27\int (\sec^2\theta - 1)\sec\theta\tan\theta\, d\theta$$

$$= 27\left[\frac{1}{3}\sec^3\theta - \sec\theta\right] + C = 9[\sec^3\theta - 3\sec\theta] + C$$

$$= 9\left[\left(\frac{\sqrt{x^2+9}}{3}\right)^3 - 3\left(\frac{\sqrt{x^2+9}}{3}\right)\right] + C = \frac{1}{3}(x^2+9)^{3/2} - 9\sqrt{x^2+9} + C$$

Thus, $\displaystyle\int_0^3 \frac{x^3}{\sqrt{x^2+9}}\, dx = \left[\frac{1}{3}(x^2+9)^{3/2} - 9\sqrt{x^2+9}\right]_0^3$

$$= \left(\frac{1}{3}(54\sqrt{2}) - 27\sqrt{2}\right) - (9 - 27)$$

$$= 18 - 9\sqrt{2} = 9(2 - \sqrt{2}) \approx 5.272.$$

(b) When $x = 0$, $\theta = 0$. When $x = 3$, $\theta = \pi/4$. Thus,

$$\int_0^3 \frac{x^3}{\sqrt{x^2+9}}\, dx = 9\left[\sec^3\theta - 3\sec\theta\right]_0^{\pi/4} = 9(2\sqrt{2} - 3\sqrt{2}) - 9(1-3) = 9(2-\sqrt{2}) \approx 5.272.$$

51. (a) Let $x = 3\sec\theta$, $dx = 3\sec\theta\tan\theta\, d\theta$, $\sqrt{x^2-9} = 3\tan\theta$.

$$\int \frac{x^2}{\sqrt{x^2-9}}\, dx = \int \frac{9\sec^2\theta}{3\tan\theta}\, 3\sec\theta\tan\theta\, d\theta$$

$$= 9\int \sec^3\theta\, d\theta$$

$$= 9\left[\frac{1}{2}\sec\theta\tan\theta + \frac{1}{2}\int \sec\theta\, d\theta\right] \quad \text{(8.3 Exercise 98 or Example 5, Section 8.2)}$$

$$= \frac{9}{2}[\sec\theta\tan\theta + \ln|\sec\theta + \tan\theta|]$$

$$= \frac{9}{2}\left[\frac{x}{3}\cdot\frac{\sqrt{x^2-9}}{3} + \ln\left|\frac{x}{3} + \frac{\sqrt{x^2-9}}{3}\right|\right]$$

—CONTINUED—

51. —CONTINUED—

Hence,

$$\int_4^6 \frac{x^2}{\sqrt{x^2-9}}\,dx = \frac{9}{2}\left[\frac{x\sqrt{x^2-9}}{9} + \ln\left|\frac{x}{3} + \frac{\sqrt{x^2-9}}{3}\right|\right]_4^6$$

$$= \frac{9}{2}\left[\left(\frac{6\sqrt{27}}{9} + \ln\left|2 + \frac{\sqrt{27}}{3}\right|\right) - \left(\frac{4\sqrt{7}}{9} + \ln\left|\frac{4}{3} + \frac{\sqrt{7}}{3}\right|\right)\right]$$

$$= 9\sqrt{3} - 2\sqrt{7} + \frac{9}{2}\left(\ln\left(\frac{6+\sqrt{27}}{3}\right) - \ln\left(\frac{4+\sqrt{7}}{3}\right)\right)$$

$$= 9\sqrt{3} - 2\sqrt{7} + \frac{9}{2}\ln\left(\frac{6+3\sqrt{3}}{4+\sqrt{7}}\right)$$

$$= 9\sqrt{3} - 2\sqrt{7} + \frac{9}{2}\ln\left(\frac{(4-\sqrt{7})(2+\sqrt{3})}{3}\right) \approx 12.644.$$

(b) When $x = 4$, $\theta = \text{arcsec}\left(\frac{4}{3}\right)$. When $x = 6$, $\theta = \text{arcsec}(2) = \frac{\pi}{3}$.

$$\int_4^6 \frac{x^2}{\sqrt{x^2-9}}\,dx = \frac{9}{2}\left[\sec\theta\tan\theta + \ln|\sec\theta + \tan\theta|\right]_{\text{arcsec}(4/3)}^{\pi/3}$$

$$= \frac{9}{2}\left[2\cdot\sqrt{3} + \ln|2 + \sqrt{3}|\right] - \frac{9}{2}\left[\frac{4}{3}\frac{\sqrt{7}}{3} + \ln\left|\frac{4}{3} + \frac{\sqrt{7}}{3}\right|\right]$$

$$= 9\sqrt{3} - 2\sqrt{7} + \frac{9}{2}\ln\left(\frac{6+3\sqrt{3}}{4+\sqrt{7}}\right) \approx 12.644$$

53. $x\dfrac{dy}{dx} = \sqrt{x^2-9}$, $x \geq 3$, $y(3) = 1$

$$y - \int \frac{\sqrt{x^2-9}}{x}\,dx$$

Let $x = 3\sec\theta$, $dx = 3\sec\theta\tan\theta\,d\theta$, $\sqrt{x^2-9} = 3\tan\theta$.

$$y = \int \frac{3\tan\theta}{3\sec\theta}3\sec\theta\tan\theta\,d\theta = 3\int\tan^2\theta\,d\theta$$

$$= 3\int(\sec^2\theta - 1)\,d\theta = 3[\tan\theta - \theta] + C$$

$$= 3\left[\frac{\sqrt{x^2-9}}{3} - \arctan\left(\frac{\sqrt{x^2-9}}{3}\right)\right] + C$$

$$= \sqrt{x^2-9} - 3\arctan\left(\frac{\sqrt{x^2-9}}{3}\right) + C$$

$y(3) = 1$: $1 = 0 - 3(0) + C \Rightarrow C = 1$

$$y = \sqrt{x^2-9} - 3\arctan\left(\frac{\sqrt{x^2-9}}{3}\right) + 1$$

55. $\displaystyle\int \frac{x^2}{\sqrt{x^2+10x+9}}\,dx = \frac{1}{2}\sqrt{x^2+10x+9}\,(x-15) + 33\ln\left|(x+5) + \sqrt{x^2+10x+9}\right| + C$

57. $\displaystyle\int \frac{x^2}{\sqrt{x^2-1}}\,dx = \frac{1}{2}\left(x\sqrt{x^2-1} + \ln\left|x + \sqrt{x^2-1}\right|\right) + C$

59. (a) $u = a\sin\theta$

(b) $u = a\tan\theta$

(c) $u = a\sec\theta$

61. (a) $u = x^2 + 9, du = 2x\,dx$

$$\int \frac{x}{x^2 + 9}\,dx = \frac{1}{2} \int \frac{du}{u} = \frac{1}{2}\ln|u| + C = \frac{1}{2}\ln(x^2 + 9) + C$$

(b) Let $x = 3\tan\theta, x^2 + 9 = 9\sec^2\theta, dx = 3\sec^2\theta\,d\theta$.

$$\int \frac{x}{x^2 + 9}\,dx = \int \frac{3\tan\theta}{9\sec^2\theta}3\sec^2\theta\,d\theta = \int \tan\theta\,d\theta$$

$$= -\ln|\cos\theta| + C_1$$

$$= -\ln\left|\frac{3}{\sqrt{x^2 + 9}}\right| + C_1$$

$$= -\ln 3 + \ln\sqrt{x^2 + 9} + C_1$$

$$= \frac{1}{2}\ln(x^2 + 9) + C_2$$

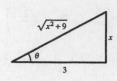

The answers are equivalent.

63. True

$$\int \frac{dx}{\sqrt{1 - x^2}} = \int \frac{\cos\theta\,d\theta}{\cos\theta} = \int d\theta$$

65. False

$$\int_0^{\sqrt{3}} \frac{dx}{\left(\sqrt{1 + x^2}\right)^3} = \int_0^{\pi/3} \frac{\sec^2\theta\,d\theta}{\sec^3\theta} = \int_0^{\pi/3} \cos\theta\,d\theta$$

67. $A = 4 \displaystyle\int_0^a \frac{b}{a}\sqrt{a^2 - x^2}\,dx$

$$= \frac{4b}{a} \int_0^a \sqrt{a^2 - x^2}\,dx$$

$$= \left[\frac{4b}{a}\left(\frac{1}{2}\right)\left(a^2 \arcsin\frac{x}{a} + x\sqrt{a^2 - x^2}\right)\right]_0^a$$

$$= \frac{2b}{a}\left(a^2\left(\frac{\pi}{2}\right)\right) = \pi ab$$

Note: See Theorem 8.2 for $\int \sqrt{a^2 - x^2}\,dx$.

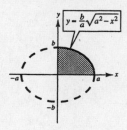

69. (a) $x^2 + (y - k)^2 = 25$

Radius of circle $= 5$

$k^2 = 5^2 + 5^2 = 50$

$k = 5\sqrt{2}$

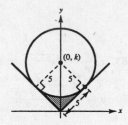

(b) Area $= $ square $- \dfrac{1}{4}$(circle)

$$= 25 - \frac{1}{4}\pi(5)^2 = 25\left(1 - \frac{\pi}{4}\right)$$

(c) Area $= r^2 - \dfrac{1}{4}\pi r^2 = r^2\left(1 - \dfrac{\pi}{4}\right)$

71. Let $x - 3 = \sin\theta, dx = \cos\theta\,d\theta, \sqrt{1 - (x - 3)^2} = \cos\theta$.

Shell Method:

$$V = 4\pi \int_2^4 x\sqrt{1 - (x - 3)^2}\,dx$$

$$= 4\pi \int_{-\pi/2}^{\pi/2} (3 + \sin\theta)\cos^2\theta\,d\theta$$

$$= 4\pi\left[\frac{3}{2}\int_{-\pi/2}^{\pi/2}(1 + \cos 2\theta)\,d\theta + \int_{-\pi/2}^{\pi/2}\cos^2\theta\sin\theta\,d\theta\right]$$

$$= 4\pi\left[\frac{3}{2}\left(\theta + \frac{1}{2}\sin 2\theta\right) - \frac{1}{3}\cos^3\theta\right]_{-\pi/2}^{\pi/2} = 6\pi^2$$

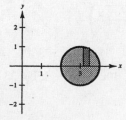

73. $y = \ln x$, $y' = \dfrac{1}{x}$, $1 + (y')^2 = 1 + \dfrac{1}{x^2} = \dfrac{x^2 + 1}{x^2}$

Let $x = \tan \theta$, $dx = \sec^2 \theta \, d\theta$, $\sqrt{x^2 + 1} = \sec \theta$.

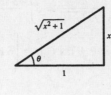

$$s = \int_1^5 \sqrt{\frac{x^2 + 1}{x^2}} \, dx = \int_1^5 \frac{\sqrt{x^2 + 1}}{x} \, dx$$

$$= \int_a^b \frac{\sec \theta}{\tan \theta} \sec^2 \theta \, d\theta = \int_a^b \frac{\sec \theta}{\tan \theta}(1 + \tan^2 \theta) \, d\theta$$

$$= \int_a^b (\csc \theta + \sec \theta \tan \theta) \, d\theta = \left[-\ln|\csc \theta + \cot \theta| + \sec \theta \right]_a^b$$

$$= \left[-\ln\left| \frac{\sqrt{x^2 + 1}}{x} + \frac{1}{x} \right| + \sqrt{x^2 + 1} \right]_1^5$$

$$= \left[-\ln\left(\frac{\sqrt{26} + 1}{5} \right) + \sqrt{26} \right] - \left[-\ln(\sqrt{2} + 1) + \sqrt{2} \right]$$

$$= \ln\left[\frac{5(\sqrt{2} + 1)}{\sqrt{26} + 1} \right] + \sqrt{26} - \sqrt{2} \approx 4.367 \quad \text{or} \quad \ln\left[\frac{\sqrt{26} - 1}{5(\sqrt{2} - 1)} \right] + \sqrt{26} - \sqrt{2}$$

75. Length of one arch of sine curve: $y = \sin x$, $y' = \cos x$

$$L_1 = \int_0^\pi \sqrt{1 + \cos^2 x} \, dx$$

Length of one arch of cosine curve: $y = \cos x$, $y' = -\sin x$

$$L_2 = \int_{-\pi/2}^{\pi/2} \sqrt{1 + \sin^2 x} \, dx$$

$$= \int_{-\pi/2}^{\pi/2} \sqrt{1 + \cos^2\left(x - \frac{\pi}{2}\right)} \, dx, \qquad u = x - \frac{\pi}{2}, \, du = dx$$

$$= \int_{-\pi}^0 \sqrt{1 + \cos^2 u} \, du$$

$$= \int_0^\pi \sqrt{1 + \cos^2 u} \, du = L_1$$

77. (a)

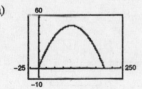

(b) $y = 0$ for $x = 200$ (range)

(c) $y = x - 0.005x^2$, $y' = 1 - 0.01x$, $1 + (y')^2 = 1 + (1 - 0.01x)^2$

Let $u = 1 - 0.01x$, $du = -0.01 \, dx$, $a = 1$.   (See Theorem 8.2.)

$$s = \int_0^{200} \sqrt{1 + (1 - 0.01x)^2} \, dx = -100 \int_0^{200} \sqrt{(1 - 0.01x)^2 + 1}(-0.01) \, dx$$

$$= -50\left[(1 - 0.01x)\sqrt{(1 - 0.01x)^2 + 1} + \ln\left| (1 - 0.01x) + \sqrt{(1 - 0.01x)^2 + 1} \right| \right]_0^{200}$$

$$= -50\left[\left(-\sqrt{2} + \ln\left| -1 + \sqrt{2} \right| \right) - \left(\sqrt{2} + \ln\left| 1 + \sqrt{2} \right| \right) \right]$$

$$= 100\sqrt{2} + 50 \ln\left(\frac{\sqrt{2} + 1}{\sqrt{2} - 1} \right) \approx 229.559$$

79. Let $x = 3 \tan \theta$, $dx = 3 \sec^2 \theta \, d\theta$, $\sqrt{x^2 + 9} = 3 \sec \theta$.

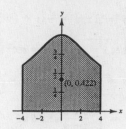

$$A = 2 \int_0^4 \frac{3}{\sqrt{x^2 + 9}} \, dx = 6 \int_0^4 \frac{dx}{\sqrt{x^2 + 9}} = 6 \int_a^b \frac{3 \sec^2 \theta \, d\theta}{3 \sec \theta}$$

$$= 6 \int_a^b \sec \theta \, d\theta = \left[6 \ln |\sec \theta + \tan \theta| \right]_a^b = \left[6 \ln \left| \frac{\sqrt{x^2 + 9} + x}{3} \right| \right]_0^4 = 6 \ln 3$$

$\bar{x} = 0$ (by symmetry)

$$\bar{y} = \frac{1}{2} \left(\frac{1}{A} \right) \int_{-4}^4 \left(\frac{3}{\sqrt{x^2 + 9}} \right)^2 dx = \frac{9}{12 \ln 3} \int_{-4}^4 \frac{1}{x^2 + 9} \, dx = \frac{3}{4 \ln 3} \left[\frac{1}{3} \arctan \frac{x}{3} \right]_{-4}^4 = \frac{2}{4 \ln 3} \arctan \frac{4}{3} \approx 0.422$$

$$(\bar{x}, \bar{y}) = \left(0, \frac{1}{2 \ln 3} \arctan \frac{4}{3} \right) \approx (0, 0.422)$$

81. $y = x^2$, $y' = 2x$, $1 + (y')^2 = 1 + 4x^2$

$$2x = \tan \theta, \, dx = \frac{1}{2} \sec^2 \theta \, d\theta, \, \sqrt{1 + 4x^2} = \sec \theta$$

(For $\int \sec^5 \theta \, d\theta$ and $\int \sec^3 \theta \, d\theta$, see Exercise 98 in Section 8.3.)

$$S = 2\pi \int_0^{\sqrt{2}} x^2 \sqrt{1 + 4x^2} \, dx = 2\pi \int_a^b \left(\frac{\tan \theta}{2} \right)^2 (\sec \theta) \left(\frac{1}{2} \sec^2 \theta \right) d\theta$$

$$= \frac{\pi}{4} \int_a^b \sec^3 \theta \tan^2 \theta \, d\theta = \frac{\pi}{4} \left[\int_a^b \sec^5 \theta \, d\theta - \int_a^b \sec^3 \theta \, d\theta \right]$$

$$= \frac{\pi}{4} \left\{ \frac{1}{4} \left[\sec^3 \theta \tan \theta + \frac{3}{2} (\sec \theta \tan \theta + \ln |\sec \theta + \tan \theta|) \right] - \frac{1}{2} (\sec \theta \tan \theta + \ln |\sec \theta + \tan \theta|) \right\}_a^b$$

$$= \frac{\pi}{4} \left[\frac{1}{4} [(1 + 4x^2)^{3/2} (2x)] - \frac{1}{8} [(1 + 4x^2)^{1/2} (2x) + \ln |\sqrt{1 + 4x^2} + 2x|] \right]_0^{\sqrt{2}}$$

$$= \frac{\pi}{4} \left[\frac{54\sqrt{2}}{4} - \frac{6\sqrt{2}}{8} - \frac{1}{8} \ln(3 + 2\sqrt{2}) \right]$$

$$= \frac{\pi}{4} \left(\frac{51\sqrt{2}}{4} - \frac{\ln(3 + 2\sqrt{2})}{8} \right) = \frac{\pi}{32} \left[102\sqrt{2} - \ln(3 + 2\sqrt{2}) \right] \approx 13.989$$

83. (a) Area of representative rectangle: $2\sqrt{1 - y^2} \, \Delta y$

Force: $2(62.4)(3 - y)\sqrt{1 - y^2} \, \Delta y$

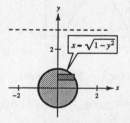

$$F = 124.8 \int_{-1}^1 (3 - y)\sqrt{1 - y^2} \, dy$$

$$= 124.8 \left[3 \int_{-1}^1 \sqrt{1 - y^2} \, dy - \int_{-1}^1 y\sqrt{1 - y^2} \, dy \right]$$

$$= 124.8 \left[\frac{3}{2} (\arcsin y + y\sqrt{1 - y^2}) + \frac{1}{2} \left(\frac{2}{3} \right)(1 - y^2)^{3/2} \right]_{-1}^1$$

$$= (62.4)3[\arcsin 1 - \arcsin(-1)] = 187.2\pi \text{ lb}$$

(b) $F = 124.8 \int_{-1}^1 (d - y)\sqrt{1 - y^2} \, dy = 124.8 \, d \int_{-1}^1 \sqrt{1 - y^2} \, dy - 124.8 \int_{-1}^1 y\sqrt{1 - y^2} \, dy$

$$= 124.8 \left(\frac{d}{2} \right) \left[\arcsin y + y\sqrt{1 - y^2} \right]_{-1}^1 - 124.8(0) = 62.4\pi d \text{ lb}$$

85. Let $u = a \sin \theta, \, du = a \cos \theta \, d\theta, \, \sqrt{a^2 - u^2} = a \cos \theta.$

$$\int \sqrt{a^2 - u^2} \, du = \int a^2 \cos^2 \theta \, d\theta = a^2 \int \frac{1 + \cos 2\theta}{2} \, d\theta$$

$$= \frac{a^2}{2}\left(\theta + \frac{1}{2}\sin 2\theta\right) + C = \frac{a^2}{2}(\theta + \sin \theta \cos \theta) + C$$

$$= \frac{a^2}{2}\left[\arcsin \frac{u}{a} + \left(\frac{u}{a}\right)\left(\frac{\sqrt{a^2 + u^2}}{a}\right)\right] + C = \frac{1}{2}\left[a^2 \arcsin \frac{u}{a} + u\sqrt{a^2 - u^2}\right] + C$$

Let $u = a \sec \theta, \, du = a \sec \theta \tan \theta \, d\theta, \, \sqrt{u^2 - a^2} = a \tan \theta.$

$$\int \sqrt{u^2 - a^2} \, du = \int a \tan \theta (a \sec \theta \tan \theta) \, d\theta = a^2 \int \tan^2 \theta \sec \theta \, d\theta$$

$$= a^2 \int (\sec^2 \theta - 1) \sec \theta \, d\theta = a^2 \int (\sec^3 \theta - \sec \theta) \, d\theta$$

$$= a^2\left[\frac{1}{2}\sec \theta \tan \theta + \frac{1}{2}\int \sec \theta \, d\theta\right] - a^2 \int \sec \theta \, d\theta = a^2\left[\frac{1}{2}\sec \theta \tan \theta - \frac{1}{2}\ln|\sec \theta + \tan \theta|\right]$$

$$= \frac{a^2}{2}\left[\frac{u}{a} \cdot \frac{\sqrt{u^2 - a^2}}{a} - \ln\left|\frac{u}{a} + \frac{\sqrt{u^2 - a^2}}{a}\right|\right] + C_1$$

$$= \frac{1}{2}\left[u\sqrt{u^2 - a^2} - a^2 \ln|u + \sqrt{u^2 - a^2}|\right] + C$$

Let $u = a \tan \theta, \, du = a \sec^2 \theta \, d\theta, \, \sqrt{u^2 + a^2} = a \sec \theta.$

$$\int \sqrt{u^2 + a^2} \, du = \int (a \sec \theta)(a \sec^2 \theta) \, d\theta$$

$$= a^2 \int \sec^3 \theta \, d\theta = a^2\left[\frac{1}{2}\sec \theta \tan \theta + \frac{1}{2}\ln|\sec \theta + \tan \theta|\right] + C_1$$

$$= \frac{a^2}{2}\left[\frac{\sqrt{u^2 + a^2}}{a} \cdot \frac{u}{a} + \ln\left|\frac{\sqrt{u^2 + a^2}}{a} + \frac{u}{a}\right|\right] + C_1 = \frac{1}{2}\left[u\sqrt{u^2 + a^2} + a^2 \ln|u + \sqrt{u^2 + a^2}|\right] + C$$

87. Large circle: $x^2 + y^2 = 25$

$$y = \sqrt{25 - x^2}, \quad \text{upper half}$$

From the right triangle, the center of the small circle is $(0, 4)$.

$$x^2 + (y - 4)^2 = 9$$

$$y = 4 + \sqrt{9 - x^2}, \quad \text{upper half}$$

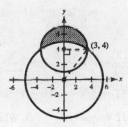

$$A = 2\int_0^3 \left[(4 + \sqrt{9 - x^2}) - \sqrt{25 - x^2}\right] dx$$

$$= 2\left[4x + \frac{1}{2}\left[9 \arcsin\left(\frac{x}{3}\right) + x\sqrt{9 - x^2}\right] - \frac{1}{2}\left[25 \arcsin\left(\frac{x}{5}\right) + x\sqrt{25 - x^2}\right]\right]_0^3$$

$$= 2\left[12 + \frac{9}{2}\arcsin(1) - \frac{25}{2}\arcsin\frac{3}{5} - 6\right]$$

$$= 12 + \frac{9\pi}{2} - 25 \arcsin\frac{3}{5} \approx 10.050$$

Section 8.5   Partial Fractions

1. $\dfrac{5}{x^2 - 10x} = \dfrac{5}{x(x - 10)} = \dfrac{A}{x} + \dfrac{B}{x - 10}$

3. $\dfrac{2x - 3}{x^3 + 10x} = \dfrac{2x - 3}{x(x^2 + 10)} = \dfrac{A}{x} + \dfrac{Bx + C}{x^2 + 10}$

5. $\dfrac{16}{x(x - 10)} = \dfrac{A}{x} + \dfrac{B}{x - 10}$

7. $\dfrac{1}{x^2 - 1} = \dfrac{1}{(x + 1)(x - 1)} = \dfrac{A}{x + 1} + \dfrac{B}{x - 1}$

$1 = A(x - 1) + B(x + 1)$

When $x = -1, 1 = -2A, A = -\frac{1}{2}.$

When $x = 1, 1 = 2B, B = \frac{1}{2}.$

$\displaystyle\int \dfrac{1}{x^2 - 1}\, dx = -\dfrac{1}{2}\int \dfrac{1}{x + 1}\, dx + \dfrac{1}{2}\int \dfrac{1}{x - 1}\, dx$

$\qquad = -\dfrac{1}{2}\ln|x + 1| + \dfrac{1}{2}\ln|x - 1| + C$

$\qquad = \dfrac{1}{2}\ln\left|\dfrac{x - 1}{x + 1}\right| + C$

9. $\dfrac{3}{x^2 + x - 2} = \dfrac{3}{(x - 1)(x + 2)} = \dfrac{A}{x - 1} + \dfrac{B}{x + 2}$

$3 = A(x + 2) + B(x - 1)$

When $x = 1, 3 = 3A, A = 1.$

When $x = -2, 3 = -3B, B = -1.$

$\displaystyle\int \dfrac{3}{x^2 + x - 2}\, dx = \int \dfrac{1}{x - 1}\, dx - \int \dfrac{1}{x + 2}\, dx$

$\qquad = \ln|x - 1| - \ln|x + 2| + C$

$\qquad = \ln\left|\dfrac{x - 1}{x + 2}\right| + C$

11. $\dfrac{5 - x}{2x^2 + x - 1} = \dfrac{5 - x}{(2x - 1)(x + 1)} = \dfrac{A}{2x - 1} + \dfrac{B}{x + 1}$

$5 - x = A(x + 1) + B(2x - 1)$

When $x = \frac{1}{2}, \frac{9}{2} = \frac{3}{2}A, A = 3.$

When $x = -1, 6 = -3B, B = -2.$

$\displaystyle\int \dfrac{5 - x}{2x^2 + x - 1}\, dx = 3\int \dfrac{1}{2x - 1}\, dx - 2\int \dfrac{1}{x + 1}\, dx$

$\qquad = \dfrac{3}{2}\ln|2x - 1| - 2\ln|x + 1| + C$

13. $\dfrac{x^2 + 12x + 12}{x(x + 2)(x - 2)} = \dfrac{A}{x} + \dfrac{B}{x + 2} + \dfrac{C}{x - 2}$

$x^2 + 12x + 12 = A(x + 2)(x - 2) + Bx(x - 2) + Cx(x + 2)$

When $x = 0, 12 = -4A, A = -3.$ When $x = -2, -8 = 8B, B = -1.$ When $x = 2, 40 = 8C, C = 5.$

$\displaystyle\int \dfrac{x^2 + 12x + 12}{x^3 - 4x}\, dx = 5\int \dfrac{1}{x - 2}\, dx - \int \dfrac{1}{x + 2}\, dx - 3\int \dfrac{1}{x}\, dx$

$\qquad = 5\ln|x - 2| - \ln|x + 2| - 3\ln|x| + C$

15. $\dfrac{2x^3 - 4x^2 - 15x + 5}{x^2 - 2x - 8} = 2x + \dfrac{x + 5}{(x - 4)(x + 2)} = 2x + \dfrac{A}{x - 4} + \dfrac{B}{x + 2}$

$x + 5 = A(x + 2) + B(x - 4)$

When $x = 4, 9 = 6A, A = \frac{3}{2}.$ When $x = -2, 3 = -6B, B = -\frac{1}{2}.$

$\displaystyle\int \dfrac{2x^3 - 4x^2 - 15x + 5}{x^2 - 2x - 8}\, dx = \int\left[2x + \dfrac{3/2}{x - 4} - \dfrac{1/2}{x + 2}\right] dx$

$\qquad = x^2 + \dfrac{3}{2}\ln|x - 4| - \dfrac{1}{2}\ln|x + 2| + C$

17. $\dfrac{4x^2 + 2x - 1}{x^2(x + 1)} = \dfrac{A}{x} + \dfrac{B}{x^2} + \dfrac{C}{x + 1}$

$4x^2 + 2x - 1 = Ax(x + 1) + B(x + 1) + Cx^2$

When $x = 0$, $B = -1$. When $x = -1$, $C = 1$. When $x = 1$, $A = 3$.

$\displaystyle \int \frac{4x^2 + 2x - 1}{x^3 + x^2}\, dx = \int \left[\frac{3}{x} - \frac{1}{x^2} + \frac{1}{x + 1} \right] dx$

$\displaystyle \qquad\qquad = 3 \ln|x| + \frac{1}{x} + \ln|x + 1| + C$

$\displaystyle \qquad\qquad = \frac{1}{x} + \ln|x^4 + x^3| + C$

19. $\dfrac{x^2 + 3x - 4}{x^3 - 4x^2 + 4x} = \dfrac{x^2 + 3x - 4}{x(x - 2)^2} = \dfrac{A}{x} + \dfrac{B}{(x - 2)} + \dfrac{C}{(x - 2)^2}$

$x^2 + 3x - 4 = A(x - 2)^2 + Bx(x - 2) + Cx$

When $x = 0$, $-4 = 4A \Rightarrow A = -1$. When $x = 2$, $6 = 2C \Rightarrow C = 3$. When $x = 1$, $0 = -1 - B + 3 \Rightarrow B = 2$.

$\displaystyle \int \frac{x^2 + 3x - 4}{x^3 - 4x^2 + 4x}\, dx = \int \frac{-1}{x}\, dx + \int \frac{2}{(x - 2)}\, dx + \int \frac{3}{(x - 2)^2}\, dx$

$\displaystyle \qquad\qquad = -\ln|x| + 2 \ln|x - 2| - \frac{3}{(x - 2)} + C$

21. $\dfrac{x^2 - 1}{x(x^2 + 1)} = \dfrac{A}{x} + \dfrac{Bx + C}{x^2 + 1}$

$x^2 - 1 = A(x^2 + 1) + (Bx + C)x$

When $x = 0$, $A = -1$. When $x = 1$, $0 = -2 + B + C$. When $x = -1$, $0 = -2 + B - C$. Solving these equations we have $A = -1$, $B = 2$, $C = 0$.

$\displaystyle \int \frac{x^2 - 1}{x^3 + x}\, dx = -\int \frac{1}{x}\, dx + \int \frac{2x}{x^2 + 1}\, dx$

$\displaystyle \qquad\qquad = -\ln|x| + \ln|x^2 + 1| + C$

$\displaystyle \qquad\qquad = \ln \left| \frac{x^2 + 1}{x} \right| + C$

23. $\dfrac{x^2}{x^4 - 2x^2 - 8} = \dfrac{A}{x - 2} + \dfrac{B}{x + 2} + \dfrac{Cx + D}{x^2 + 2}$

$x^2 = A(x + 2)(x^2 + 2) + B(x - 2)(x^2 + 2) + (Cx + D)(x + 2)(x - 2)$

When $x = 2$, $4 = 24A$. When $x = -2$, $4 = -24B$. When $x = 0$, $0 = 4A - 4B - 4D$, and when $x = 1$, $1 = 9A - 3B - 3C - 3D$. Solving these equations we have $A = \frac{1}{6}$, $B = -\frac{1}{6}$, $C = 0$, $D = \frac{1}{3}$.

$\displaystyle \int \frac{x^2}{x^4 - 2x^2 - 8}\, dx = \frac{1}{6} \left[\int \frac{1}{x - 2}\, dx - \int \frac{1}{x + 2}\, dx + 2 \int \frac{1}{x^2 + 2}\, dx \right]$

$\displaystyle \qquad\qquad = \frac{1}{6} \left[\ln \left| \frac{x - 2}{x + 2} \right| + \sqrt{2} \arctan \frac{x}{\sqrt{2}} \right] + C$

25. $\dfrac{x}{(2x - 1)(2x + 1)(4x^2 + 1)} = \dfrac{A}{2x - 1} + \dfrac{B}{2x + 1} + \dfrac{Cx + D}{4x^2 + 1}$

$$x = A(2x + 1)(4x^2 + 1) + B(2x - 1)(4x^2 + 1) + (Cx + D)(2x - 1)(2x + 1)$$

When $x = \frac{1}{2}, \frac{1}{2} = 4A$. When $x = -\frac{1}{2}, -\frac{1}{2} = -4B$. When $x = 0, 0 = A - B - D$, and when $x = 1$,
$1 = 15A + 5B + 3C + 3D$. Solving these equations we have $A = \frac{1}{8}, B = \frac{1}{8}, C = -\frac{1}{2}, D = 0$.

$$\int \frac{x}{16x^4 - 1} \, dx = \frac{1}{8}\left[\int \frac{1}{2x - 1} \, dx + \int \frac{1}{2x + 1} \, dx - 4 \int \frac{x}{4x^2 + 1} \, dx \right] = \frac{1}{16} \ln \left| \frac{4x^2 - 1}{4x^2 + 1} \right| + C$$

27. $\dfrac{x^2 + 5}{(x + 1)(x^2 - 2x + 3)} = \dfrac{A}{x + 1} + \dfrac{Bx + C}{x^2 - 2x + 3}$

$$x^2 + 5 = A(x^2 - 2x + 3) + (Bx + C)(x + 1)$$

$$= (A + B)x^2 + (-2A + B + C)x + (3A + C)$$

When $x = -1, A = 1$. By equating coefficients of like terms, we have $A + B = 1, -2A + B + C = 0$,
$3A + C = 5$. Solving these equations we have $A = 1, B = 0, C = 2$.

$$\int \frac{x^2 + 5}{x^3 - x^2 + x + 3} \, dx = \int \frac{1}{x + 1} \, dx + 2 \int \frac{1}{(x - 1)^2 + 2} \, dx$$

$$= \ln|x + 1| + \sqrt{2} \arctan\left(\frac{x - 1}{\sqrt{2}} \right) + C$$

29. $\dfrac{3}{(2x + 1)(x + 2)} = \dfrac{A}{2x + 1} + \dfrac{B}{x + 2}$

$$3 = A(x + 2) + B(2x + 1)$$

When $x = -\frac{1}{2}, A = 2$. When $x = -2, B = -1$.

$$\int_0^1 \frac{3}{2x^2 + 5x + 2} \, dx = \int_0^1 \frac{2}{2x + 1} \, dx - \int_0^1 \frac{1}{x + 2} \, dx$$

$$= \left[\ln|2x + 1| - \ln|x + 2| \right]_0^1$$

$$= \ln 2$$

31. $\dfrac{x + 1}{x(x^2 + 1)} = \dfrac{A}{x} + \dfrac{Bx + C}{x^2 + 1}$

$$x + 1 = A(x^2 + 1) + (Bx + C)x$$

When $x = 0, A = 1$. When $x = 1, 2 = 2A + B + C$.
When $x = -1, 0 = 2A + B - C$. Solving these equations we have $A = 1, B = -1, C = 1$.

$$\int_1^2 \frac{x + 1}{x(x^2 + 1)} \, dx = \int_1^2 \frac{1}{x} \, dx - \int_1^2 \frac{x}{x^2 + 1} \, dx + \int_1^2 \frac{1}{x^2 + 1} \, dx$$

$$= \left[\ln|x| - \frac{1}{2} \ln(x^2 + 1) + \arctan x \right]_1^2$$

$$= \frac{1}{2} \ln \frac{8}{5} - \frac{\pi}{4} + \arctan 2$$

$$\approx 0.557$$

33. $\displaystyle\int \frac{3x \, dx}{x^2 - 6x + 9} = 3 \ln|x - 3| - \frac{9}{x - 3} + C$

$(4, 0)$: $3 \ln|4 - 3| - \dfrac{9}{4 - 3} + C = 0 \Rightarrow C = 9$

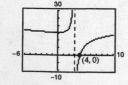

35. $\displaystyle\int \frac{x^2 + x + 2}{(x^2 + 2)^2} \, dx = \frac{\sqrt{2}}{2} \arctan \frac{x}{\sqrt{2}} - \frac{1}{2(x^2 + 2)} + C$

$(0, 1)$: $0 - \dfrac{1}{4} + C = 1 \Rightarrow C = \dfrac{5}{4}$

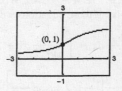

37. $\displaystyle\int \frac{2x^2 - 2x + 3}{x^3 - x^2 - x - 2} \, dx = \ln|x - 2| + \frac{1}{2} \ln|x^2 + x + 1| - \sqrt{3} \arctan\left(\frac{2x + 1}{\sqrt{3}} \right) + C$

$(3, 10)$: $0 + \dfrac{1}{2} \ln 13 - \sqrt{3} \arctan \dfrac{7}{\sqrt{3}} + C = 10 \Rightarrow C = 10 - \dfrac{1}{2} \ln 13 + \sqrt{3} \arctan \dfrac{7}{\sqrt{3}}$

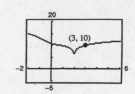

39. $\int \dfrac{1}{x^2 - 4}\, dx = \dfrac{1}{4} \ln\left|\dfrac{x - 2}{x + 2}\right| + C$

$(6, 4):\ \dfrac{1}{4} \ln\left|\dfrac{4}{8}\right| + C = 4 \Rightarrow C = 4 - \dfrac{1}{4} \ln \dfrac{1}{2} = 4 + \dfrac{1}{4} \ln 2$

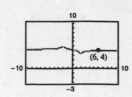

41. Let $u = \cos x\ \ du = -\sin x\, dx$.

$\dfrac{1}{u(u - 1)} = \dfrac{A}{u} + \dfrac{B}{u - 1}$

$1 = A(u - 1) + Bu$

When $u = 0, A = -1$. When $u = 1, B = 1, u = \cos x, du = -\sin x\, dx$.

$\displaystyle\int \dfrac{\sin x}{\cos x(\cos x - 1)}\, dx = -\int \dfrac{1}{u(u - 1)}\, du$

$\qquad = \displaystyle\int \dfrac{1}{u}\, du - \int \dfrac{1}{u - 1}\, du = \ln|u| - \ln|u - 1| + C = \ln\left|\dfrac{u}{u - 1}\right| + C = \ln\left|\dfrac{\cos x}{\cos x - 1}\right| + C$

43. $\displaystyle\int \dfrac{3\cos x}{\sin^2 x + \sin x - 2}\, dx = 3\int \dfrac{1}{u^2 + u - 2}\, du$

$\qquad = \ln\left|\dfrac{u - 1}{u + 2}\right| + C$

$\qquad = \ln\left|\dfrac{-1 + \sin x}{2 + \sin x}\right| + C$

(From Exercise 9 with $u = \sin x, du = \cos x\, dx$)

45. Let $u = e^x,\ du = e^x\, dx$.

$\dfrac{1}{(u - 1)(u + 4)} = \dfrac{A}{u - 1} + \dfrac{B}{u + 4}$

$1 = A(u + 4) + B(u - 1)$

When $u = 1, A = \dfrac{1}{5}$. When $u = -4, B = -\dfrac{1}{5}, u = e^x$, $du = e^x\, dx$.

$\displaystyle\int \dfrac{e^x}{(e^x - 1)(e^x + 4)}\, dx = \int \dfrac{1}{(u - 1)(u + 4)}\, du$

$\qquad = \dfrac{1}{5}\left(\displaystyle\int \dfrac{1}{u - 1}\, du - \int \dfrac{1}{u + 4}\, du\right)$

$\qquad = \dfrac{1}{5} \ln\left|\dfrac{u - 1}{u + 4}\right| + C$

$\qquad = \dfrac{1}{5} \ln\left|\dfrac{e^x - 1}{e^x + 4}\right| + C$

47. $\dfrac{1}{x(a + bx)} = \dfrac{A}{x} + \dfrac{B}{a + bx}$

$1 = A(a + bx) + Bx$

When $x = 0, 1 = aA \Rightarrow A = 1/a$.
When $x = -a/b, 1 = -(a/b)B \Rightarrow B = -b/a$.

$\displaystyle\int \dfrac{1}{x(a + bx)}\, dx = \dfrac{1}{a}\int\left(\dfrac{1}{x} - \dfrac{b}{a + bx}\right) dx$

$\qquad = \dfrac{1}{a}\Big(\ln|x| - \ln|a + bx|\Big) + C$

$\qquad = \dfrac{1}{a} \ln\left|\dfrac{x}{a + bx}\right| + C$

49. $\dfrac{x}{(a + bx)^2} = \dfrac{A}{a + bx} + \dfrac{B}{(a + bx)^2}$

$x = A(a + bx) + B$

When $x = -a/b, B = -a/b$.
When $x = 0, 0 = aA + B \Rightarrow A = 1/b$.

$\displaystyle\int \dfrac{x}{(a + bx)^2}\, dx = \int\left(\dfrac{1/b}{a + bx} + \dfrac{-a/b}{(a + bx)^2}\right) dx$

$\qquad = \dfrac{1}{b}\displaystyle\int \dfrac{1}{a + bx}\, dx - \dfrac{a}{b}\int \dfrac{1}{(a + bx)^2}\, dx$

$\qquad = \dfrac{1}{b^2} \ln|a + bx| + \dfrac{a}{b^2}\left(\dfrac{1}{a + bx}\right) + C$

$\qquad = \dfrac{1}{b^2}\left(\dfrac{a}{a + bx} + \ln|a + bx|\right) + C$

51. $\dfrac{dy}{dx} = \dfrac{6}{4 - x^2}$, $y(0) = 3$

53. Dividing x^3 by $x - 5$

55. (a) Substitution: $u = x^2 + 2x - 8$

(b) Partial fractions

(c) Trigonometric substitution (tan) or inverse tangent rule

57.
$$A = \int_0^1 \frac{12}{x^2 + 5x + 6}\, dx$$

$$\frac{12}{x^2 + 5x + 6} = \frac{12}{(x + 2)(x + 3)} = \frac{A}{x + 2} + \frac{B}{x + 3}$$

$$12 = A(x + 3) + B(x + 2)$$

Let $x = -3$: $12 = B(-1) \Rightarrow B = -12$

Let $x = -2$: $12 = A(1) \Rightarrow A = 12$

$$A = \int_0^1 \left(\frac{12}{x + 2} - \frac{12}{x + 3}\right) dx$$

$$= \left[12 \ln|x + 2| - 12 \ln|x + 3|\right]_0^1$$

$$= 12[\ln 3 - \ln 4 - \ln 2 + \ln 3]$$

$$= 12 \ln\left(\frac{9}{8}\right) \approx 1.4134$$

59. Average cost $= \dfrac{1}{80 - 75} \displaystyle\int_{75}^{80} \dfrac{124p}{(10 + p)(100 - p)}\, dp$

$$= \frac{1}{5} \int_{75}^{80} \left(\frac{-124}{(10 + p)11} + \frac{1240}{(100 - p)11}\right) dp$$

$$= \frac{1}{5}\left[\frac{-124}{11}\ln(10 + p) - \frac{1240}{11}\ln(100 - p)\right]_{75}^{80}$$

$$\approx \frac{1}{5}(24.51) = 4.9$$

Approximately $490,000

61. $V = \pi \displaystyle\int_0^3 \left(\dfrac{2x}{x^2 + 1}\right)^2 dx = 4\pi \displaystyle\int_0^3 \dfrac{x^2}{(x^2 + 1)^2}\, dx$

$$= 4\pi \int_0^3 \left(\frac{1}{x^2 + 1} - \frac{1}{(x^2 + 1)^2}\right) dx \qquad \text{(partial fractions)}$$

$$= 4\pi \left[\arctan x - \frac{1}{2}\left(\arctan x + \frac{x}{x^2 + 1}\right)\right]_0^3 \qquad \text{(trigonometric substitution)}$$

$$= 2\pi \left[\arctan x - \frac{x}{x^2 + 1}\right]_0^3 = 2\pi\left[\arctan 3 - \frac{3}{10}\right] \approx 5.963$$

$$A = \int_0^3 \frac{2x}{x^2 + 1}\, dx = \left[\ln(x^2 + 1)\right]_0^3 = \ln 10$$

$$\bar{x} = \frac{1}{A} \int_0^3 \frac{2x^2}{x^2 + 1}\, dx = \frac{1}{\ln 10} \int_0^3 \left(2 - \frac{2}{x^2 + 1}\right) dx$$

$$= \frac{1}{\ln 10}\left[2x - 2\arctan x\right]_0^3 = \frac{2}{\ln 10}[3 - \arctan 3] \approx 1.521$$

$$\bar{y} = \frac{1}{A}\left(\frac{1}{2}\right)\int_0^3 \left(\frac{2x}{x^2 + 1}\right)^2 dx = \frac{2}{\ln 10}\int_0^3 \frac{x^2}{(x^2 + 1)^2}\, dx$$

$$= \frac{2}{\ln 10}\int_0^3 \left(\frac{1}{x^2 + 1} - \frac{1}{(x^2 + 1)^2}\right) dx \qquad \text{(partial fractions)}$$

$$= \frac{2}{\ln 10}\left[\arctan x - \frac{1}{2}\left(\arctan x + \frac{x}{x^2 + 1}\right)\right]_0^3 \qquad \text{(trigonometric substitution)}$$

$$= \frac{2}{\ln 10}\left[\frac{1}{2}\arctan x - \frac{x}{2(x^2 + 1)}\right]_0^3 = \frac{1}{\ln 10}\left[\arctan x - \frac{x}{x^2 + 1}\right]_0^3 = \frac{1}{\ln 10}\left[\arctan 3 - \frac{3}{10}\right] \approx 0.412$$

(1.521, 0.412)

$(\bar{x}, \bar{y}) \approx (1.521, 0.412)$

63.

$$\frac{1}{(x+1)(n-x)} = \frac{A}{x+1} + \frac{B}{n-x}, A = B = \frac{1}{n+1}$$

$$\frac{1}{n+1}\int\left(\frac{1}{x+1} + \frac{1}{n-x}\right)dx = kt + C$$

$$\frac{1}{n+1}\ln\left|\frac{x+1}{n-x}\right| = kt + C$$

When $t = 0, x = 0, C = \frac{1}{n+1}\ln\frac{1}{n}$.

$$\frac{1}{n+1}\ln\left|\frac{x+1}{n-x}\right| = kt + \frac{1}{n+1}\ln\frac{1}{n}$$

$$\frac{1}{n+1}\left[\ln\left|\frac{x+1}{n-x}\right| - \ln\frac{1}{n}\right] = kt$$

$$\ln\frac{nx+n}{n-x} = (n+1)kt$$

$$\frac{nx+n}{n-x} = e^{(n+1)kt}$$

$$x = \frac{n[e^{(n+1)kt} - 1]}{n + e^{(n+1)kt}} \qquad \textbf{Note: } \lim_{t\to\infty} x = n$$

65.

$$\frac{x}{1+x^4} = \frac{Ax+B}{x^2 + \sqrt{2}x + 1} + \frac{Cx+D}{x^2 - \sqrt{2}x + 1}$$

$$x = (Ax+B)(x^2 - \sqrt{2}x + 1) + (Cx+D)(x^2 + \sqrt{2}x + 1)$$

$$= (A+C)x^3 + (B+D - \sqrt{2}A + \sqrt{2}C)x^2 + (A + C - \sqrt{2}D + \sqrt{2}D)x + (B+D)$$

$$0 = A + C \implies C = -A$$

$$0 = B + D - \sqrt{2}A + \sqrt{2}C \quad -2\sqrt{2}A = 0 \implies A = 0 \text{ and } C = 0$$

$$1 = A + C - \sqrt{2}B + \sqrt{2}D \quad -2\sqrt{2}B = 1 \implies B = -\frac{\sqrt{2}}{4} \text{ and } D = \frac{\sqrt{2}}{4}$$

$$0 = B + D \implies D = -B$$

Thus,

$$\int_0^1 \frac{x}{1+x^4}\,dx = \int_0^1\left[\frac{-\sqrt{2}/4}{x^2 + \sqrt{2}x + 1} + \frac{\sqrt{2}/4}{x^2 - \sqrt{2}x + 1}\right]dx$$

$$= \frac{\sqrt{2}}{4}\int_0^1\left[\frac{-1}{[x + (\sqrt{2}/2)^2 + (1/2)]} + \frac{1}{[x - (\sqrt{2}/2)^2 + (1/2)]}\right]dx$$

$$= \frac{\sqrt{2}}{4}\cdot\frac{1}{1/\sqrt{2}}\left[-\arctan\left(\frac{x + (\sqrt{2}/2)}{1/\sqrt{2}}\right) + \arctan\left(\frac{x - (\sqrt{2}/2)}{1/\sqrt{2}}\right)\right]_0^1$$

$$= \frac{1}{2}\left[-\arctan(\sqrt{2}x + 1) + \arctan(\sqrt{2}x - 1)\right]_0^1$$

$$= \frac{1}{2}\left[(-\arctan(\sqrt{2} + 1) + \arctan(\sqrt{2} - 1)) - (-\arctan 1 + \arctan(-1))\right]$$

$$= \frac{1}{2}\left[\arctan(\sqrt{2} - 1) - \arctan(\sqrt{2} + 1) + \frac{\pi}{4} + \frac{\pi}{4}\right].$$

Since $\arctan x - \arctan y = \arctan[(x-y)/(1+xy)]$, we have:

$$\int_0^1 \frac{x}{1+x^4}\,dx = \frac{1}{2}\left[\arctan\left(\frac{(\sqrt{2}-1)-(\sqrt{2}+1)}{1+(\sqrt{2}-1)(\sqrt{2}+1)}\right) + \frac{\pi}{2}\right] = \frac{1}{2}\left[\arctan\left(\frac{-2}{2}\right) + \frac{\pi}{2}\right] = \frac{1}{2}\left[-\frac{\pi}{4} + \frac{\pi}{2}\right] = \frac{\pi}{8}$$

Section 8.6   Integration by Tables and Other Integration Techniques

1. By Formula 6: $\displaystyle\int \frac{x^2}{1+x}\,dx = -\frac{x}{2}(2-x) + \ln|1+x| + C$

3. By Formula 26: $\displaystyle\int e^x \sqrt{1+e^{2x}}\,dx = \frac{1}{2}\Big[e^x\sqrt{e^{2x}+1} + \ln|e^x + \sqrt{e^{2x}+1}|\Big] + C$

$\quad u = e^x, du = e^x\,dx$

5. By Formula 44: $\displaystyle\int \frac{1}{x^2\sqrt{1-x^2}}\,dx = -\frac{\sqrt{1-x^2}}{x} + C$

7. By Formulas 50 and 48: $\displaystyle\int \sin^4(2x)\,dx = \frac{1}{2}\int \sin^4(2x)(2)\,dx$

$$= \frac{1}{2}\left[\frac{-\sin^3(2x)\cos(2x)}{4} + \frac{3}{4}\int \sin^2(2x)(2)\,dx\right]$$

$$= \frac{1}{2}\left[\frac{-\sin^3(2x)\cos(2x)}{4} + \frac{3}{8}(2x - \sin 2x \cos 2x)\right] + C$$

$$= \frac{1}{16}(6x - 3\sin 2x \cos 2x - 2\sin^3 2x \cos 2x) + C$$

9. By Formula 57: $\displaystyle\int \frac{1}{\sqrt{x}(1-\cos\sqrt{x})}\,dx = 2\int \frac{1}{1-\cos\sqrt{x}}\left(\frac{1}{2\sqrt{x}}\right)dx$

$$= -2\big(\cot\sqrt{x} + \csc\sqrt{x}\big) + C$$

$\quad u = \sqrt{x}, du = \dfrac{1}{2\sqrt{x}}\,dx$

11. By Formula 84:

$$\int \frac{1}{1+e^{2x}}\,dx = x - \frac{1}{2}\ln(1+e^{2x}) + C$$

13. By Formula 89:

$$\int x^3 \ln x\,dx = \frac{x^4}{16}\big(4\ln|x| - 1\big) + C$$

15. (a) By Formulas 83 and 82: $\displaystyle\int x^2 e^x\,dx = x^2 e^x - 2\int x e^x\,dx$

$$= x^2 e^x - 2[(x-1)e^x + C_1]$$

$$= x^2 e^x - 2x e^x + 2e^x + C$$

(b) Integration by parts: $u = x^2, du = 2x\,dx, dv = e^x\,dx, v = e^x$

$$\int x^2 e^x\,dx = x^2 e^x - \int 2x e^x\,dx$$

Parts again: $u = 2x, du = 2\,dx, dv = e^x\,dx, v = e^x$

$$\int x^2 e^x\,dx = x^2 e^x - \left[2x e^x - \int 2e^x\,dx\right] = x^2 e^x - 2x e^x + 2e^x + C$$

17. (a) By Formula: 12, $a = b = 1$, $u = x$, and

$$\int \frac{1}{x^2(x+1)} dx = \frac{-1}{1}\left(\frac{1}{x} + \frac{1}{1} \ln\left|\frac{x}{1+x}\right|\right) + C$$

$$= \frac{-1}{x} - \ln\left|\frac{x}{1+x}\right| + C$$

$$= \frac{-1}{x} + \ln\left|\frac{x+1}{x}\right| + C$$

(b) Partial fractions:

$$\frac{1}{x^2(x+1)} = \frac{A}{x} + \frac{B}{x^2} + \frac{C}{x+1}$$

$$1 = Ax(x+1) + B(x+1) + Cx^2$$

$x = 0$: $1 = B$

$x = -1$: $1 = C$

$x = 1$: $1 = 2A + 2 + 1 \Rightarrow A = -1$

$$\int \frac{1}{x^2(x+1)} dx = \int\left[\frac{-1}{x} + \frac{1}{x^2} + \frac{1}{x+1}\right] dx$$

$$= -\ln|x| - \frac{1}{x} + \ln|x+1| + C$$

$$= -\frac{1}{x} - \ln\left|\frac{x}{x+1}\right| + C$$

19. By Formula 79: $\int x \operatorname{arcsec}(x^2 + 1)\, dx = \frac{1}{2}\int \operatorname{arcsec}(x^2 + 1)(2x)\, dx$

$$= \frac{1}{2}\left[(x^2 + 1)\operatorname{arcsec}(x^2 + 1) - \ln\left((x^2 + 1) + \sqrt{x^4 + 2x^2}\right)\right] + C$$

$u = x^2 + 1$, $du = 2x\, dx$

21. By Formula 35: $\int \frac{1}{x^2\sqrt{x^2 - 4}}\, dx = \frac{\sqrt{x^2 - 4}}{4x} + C$

23. By Formula 4: $\int \frac{2x}{(1-3x)^2}\, dx = 2\int \frac{x}{(1-3x)^2}\, dx = \frac{2}{9}\left(\ln|1-3x| + \frac{1}{1-3x}\right) + C$

25. By Formula 76:

$$\int e^x \arccos e^x\, dx = e^x \arccos e^x - \sqrt{1 - e^{2x}} + C$$

$u = e^x$, $du = e^x\, dx$

27. By Formula 73:

$$\int \frac{x}{1 - \sec x^2}\, dx = \frac{1}{2}\int \frac{2x}{1 - \sec x^2}\, dx$$

$$= \frac{1}{2}(x^2 + \cot x^2 + \csc x^2) + C$$

29. By Formula 14: $\int \frac{\cos\theta}{3 + 2\sin\theta + \sin^2\theta}\, d\theta = \frac{\sqrt{2}}{2}\arctan\left(\frac{1 + \sin\theta}{\sqrt{2}}\right) + C$ $(b^2 = 4 < 12 = 4ac)$

$u = \sin\theta$, $du = \cos\theta\, d\theta$

31. By Formula 35: $\int \frac{1}{x^2\sqrt{2 + 9x^2}}\, dx = 3\int \frac{3}{(3x)^2\sqrt{\left(\sqrt{2}\right)^2 + (3x)^2}}\, dx$

$$= -\frac{3\sqrt{2 + 9x^2}}{6x} + C$$

$$= -\frac{\sqrt{2 + 9x^2}}{2x} + C$$

33. By Formula 3: $\int \frac{\ln x}{x(3 + 2\ln x)}\, dx = \frac{1}{4}\left(2\ln|x| - 3\ln|3 + 2\ln|x||\right) + C$

$u = \ln x$, $du = \frac{1}{x}\, dx$

35. By Formulas 1, 25, and 33: $\displaystyle\int \frac{x}{(x^2 - 6x + 10)^2}\,dx = \frac{1}{2}\int \frac{2x - 6 + 6}{(x^2 - 6x + 10)^2}\,dx$

$$= \frac{1}{2}\int (x^2 - 6x + 10)^{-2}(2x - 6)\,dx + 3\int \frac{1}{[(x - 3)^2 + 1]^2}\,dx$$

$$= -\frac{1}{2(x^2 - 6x + 10)} + \frac{3}{2}\left[\frac{x - 3}{x^2 - 6x + 10} + \arctan(x - 3)\right] + C$$

$$= \frac{3x - 10}{2(x^2 - 6x + 10)} + \frac{3}{2}\arctan(x - 3) + C$$

37. By Formula 31: $\displaystyle\int \frac{x}{\sqrt{x^4 - 6x^2 + 5}}\,dx = \frac{1}{2}\int \frac{2x}{\sqrt{(x^2 - 3)^2 - 4}}\,dx$

$$= \frac{1}{2}\ln\left|x^2 - 3 + \sqrt{x^4 - 6x^2 + 5}\right| + C$$

$u = x^2 - 3, \, du = 2x\,dx$

39. $\displaystyle\int \frac{x^3}{\sqrt{4 - x^2}}\,dx = \int \frac{8\sin^3 \theta(2\cos\theta\,d\theta)}{2\cos\theta}$

$$= 8\int (1 - \cos^2 \theta)\sin\theta\,d\theta$$

$$= 8\int \left[\sin\theta - \cos^2 \theta(\sin\theta)\right]d\theta$$

$$= -8\cos\theta + \frac{8\cos^3 \theta}{3} + C$$

$$= -8\frac{\sqrt{4 - x^2}}{2} + \frac{8}{3}\left(\frac{\sqrt{4 - x^2}}{2}\right)^3 + C$$

$$= \sqrt{4 - x^2}\left[-4 + \frac{1}{3}(4 - x^2)\right] + C$$

$$= \frac{-\sqrt{4 - x^2}}{3}(x^2 + 8) + C$$

$x = 2\sin\theta, \, dx = 2\cos\theta\,d\theta, \, \sqrt{4 - x^2} = 2\cos\theta$

41. By Formula 8:

$$\int \frac{e^{3x}}{(1 + e^x)^3}\,dx = \int \frac{(e^x)^2}{(1 + e^x)^3}(e^x)\,dx$$

$$= \frac{2}{1 + e^x} - \frac{1}{2(1 + e^x)^2} + \ln|1 + e^x| + C$$

$u = e^x, \, du = e^x\,dx$

43. $\displaystyle\int_0^1 xe^{x^2}\,dx$

By Formula 81:

$$\int_0^1 xe^{x^2}\,dx = \frac{1}{2}e^{x^2}\Big]_0^1 = \frac{1}{2}(e - 1)$$

45. By Formula 89:

$$\int_1^3 x^2 \ln x\,dx = \left[\frac{x^3}{9}(-1 + 3\ln|x|)\right]_1^3$$

$$= 3(-1 + 3\ln 3) + \frac{1}{9} = 9\ln 3 - \frac{26}{9}$$

47. By Formula 23, and letting $u = \sin x$:

$$\int_{-\pi/2}^{\pi/2} \frac{\cos x}{1 + \sin^2 x}\,dx = \left[\arctan(\sin x)\right]_{-\pi/2}^{\pi/2}$$

$$= \arctan(1) - \arctan(-1) = \frac{\pi}{2}$$

49. By Formulas 54 and 55:

$$\int t^3 \cos t \, dt = t^3 \sin t - 3 \int t^2 \sin t \, dt$$

$$= t^3 \sin t - 3 \left[-t^2 \cos t + 2 \int t \cos t \, dt \right]$$

$$= t^3 \sin t + 3t^2 \cos t - 6 \left[t \sin t - \int \sin t \, dt \right]$$

$$= t^3 \sin t + 3t^2 \cos t - 6t \sin t - 6 \cos t + C$$

Thus,

$$\int_0^{\pi/2} t^3 \cos t \, dt = \left[t^3 \sin t + 3t^2 \cos t - 6t \sin t - 6 \cos t \right]_0^{\pi/2}$$

$$= \left(\frac{\pi^3}{8} - 3\pi \right) + 6 = \frac{\pi^3}{8} + 6 - 3\pi$$

51. $\dfrac{u^2}{(a + bu)^2} = \dfrac{1}{b^2} - \dfrac{(2a/b)u + (a^2/b^2)}{(a + bu)^2} = \dfrac{1}{b^2} + \dfrac{A}{a + bu} + \dfrac{B}{(a + bu)^2}$

$$-\frac{2a}{b}u - \frac{a^2}{b^2} = A(a + bu) + B = (aA + B) + bAu$$

Equating the coefficients of like terms we have $aA + B = -a^2/b^2$ and $bA = -2a/b$. Solving these equations we have $A = -2a/b^2$ and $B = a^2/b^2$.

$$\int \frac{u^2}{(a + bu)^2} \, du = \frac{1}{b^2} \int du - \frac{2a}{b^2} \left(\frac{1}{b} \right) \int \frac{1}{a + bu} b \, du + \frac{a^2}{b^2} \left(\frac{1}{b} \right) \int \frac{1}{(a + bu)^2} b \, du = \frac{1}{b^2}u - \frac{2a}{b^3} \ln|a + bu| - \frac{a^2}{b^3} \left(\frac{1}{a + bu} \right) + C$$

$$= \frac{1}{b^3} \left(bu - \frac{a^2}{a + bu} - 2a \ln|a + bu| \right) + C$$

53. When we have $u^2 + a^2$:

$$u = a \tan \theta$$
$$du = a \sec^2 \theta \, d\theta$$
$$u^2 + a^2 = a^2 \sec^2 \theta$$

$$\int \frac{1}{(u^2 + a^2)^{3/2}} \, du = \int \frac{a \sec^2 \theta \, d\theta}{a^3 \sec^3 \theta}$$

$$= \frac{1}{a^2} \int \cos \theta \, d\theta$$

$$= \frac{1}{a^2} \sin \theta + C$$

$$= \frac{u}{a^2 \sqrt{u^2 + a^2}} + C$$

When we have $u^2 - a^2$:

$$u = a \sec \theta$$
$$du = a \sec \theta \tan \theta \, d\theta$$
$$u^2 - a^2 = a^2 \tan^2 \theta$$

$$\int \frac{1}{(u^2 - a^2)^{3/2}} \, du = \int \frac{a \sec \theta \tan \theta \, d\theta}{a^3 \tan^3 \theta}$$

$$= \frac{1}{a^2} \int \frac{\cos \theta}{\sin^2 \theta} \, d\theta = \frac{1}{a^2} \int \csc \theta \cot \theta \, d\theta$$

$$= -\frac{1}{a^2} \csc \theta + C$$

$$= \frac{-u}{a^2 \sqrt{u^2 - a^2}} + C$$

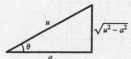

55. $\int (\arctan u)\, du = u \arctan u - \dfrac{1}{2} \int \dfrac{2u}{1 + u^2}\, du$

$\qquad\qquad\qquad = u \arctan u - \dfrac{1}{2} \ln(1 + u^2) + C$

$\qquad\qquad\qquad = u \arctan u - \ln\sqrt{1 + u^2} + C$

$\quad w = \arctan u,\ dv = du,\ dw = \dfrac{du}{1 + u^2},\ v = u$

57. $\int \dfrac{1}{x^{3/2}\sqrt{1 - x}}\, dx = \dfrac{-2\sqrt{1 - x}}{\sqrt{x}} + C$

$\quad \left(\dfrac{1}{2}, 5\right): \dfrac{-2\sqrt{1/2}}{\sqrt{1/2}} + C = 5 \Rightarrow C = 7$

$\qquad\qquad y = \dfrac{-2\sqrt{1 - x}}{\sqrt{x}} + 7$

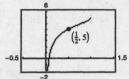

59. $\int \dfrac{1}{(x^2 - 6x + 10)^2}\, dx = \dfrac{1}{2}\left[\tan^{-1}(x - 3) + \dfrac{x - 3}{x^2 - 6x + 10} \right] + C$

$\quad (3, 0): \dfrac{1}{2}\left[0 + \dfrac{0}{10} \right] + C = 0 \Rightarrow C = 0$

$\qquad y = \dfrac{1}{2}\left[\tan^{-1}(x - 3) + \dfrac{x - 3}{x^2 - 6x + 10} \right]$

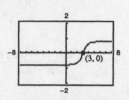

61. $\int \dfrac{1}{\sin\theta \tan\theta}\, d\theta = -\csc\theta + C$

$\quad \left(\dfrac{\pi}{4}, 2\right): -\dfrac{2}{\sqrt{2}} + C = 2 \Rightarrow C = 2 + \sqrt{2}$

$\qquad\qquad y = -\csc\theta + 2 + \sqrt{2}$

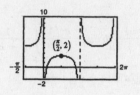

63. $\int \dfrac{1}{2 - 3\sin\theta}\, d\theta = \int \left[\dfrac{\dfrac{2\, du}{1 + u^2}}{2 - 3\left(\dfrac{2u}{1 + u^2} \right)} \right],\ u = \tan\dfrac{\theta}{2}$

$\qquad\qquad\qquad = \int \dfrac{2}{2(1 + u^2) - 6u}\, du$

$\qquad\qquad\qquad = \int \dfrac{1}{u^2 - 3u + 1}\, du$

$\qquad\qquad\qquad = \int \dfrac{1}{\left(u - \dfrac{3}{2} \right)^2 - \dfrac{5}{4}}\, du$

$\qquad\qquad\qquad = \dfrac{1}{\sqrt{5}} \ln\left| \dfrac{\left(u - \dfrac{3}{2} \right) - \dfrac{\sqrt{5}}{2}}{\left(u - \dfrac{3}{2} \right) + \dfrac{\sqrt{5}}{2}} \right| + C$

$\qquad\qquad\qquad = \dfrac{1}{\sqrt{5}} \ln\left| \dfrac{2u - 3 - \sqrt{5}}{2u - 3 + \sqrt{5}} \right| + C$

$\qquad\qquad\qquad = \dfrac{1}{\sqrt{5}} \ln\left| \dfrac{2\tan\left(\dfrac{\theta}{2} \right) - 3 - \sqrt{5}}{2\tan\left(\dfrac{\theta}{2} \right) - 3 + \sqrt{5}} \right| + C$

65. $\int_0^{\pi/2} \dfrac{1}{1 + \sin\theta + \cos\theta}\, d\theta = \int_0^1 \left[\dfrac{\dfrac{2\, du}{1 + u^2}}{1 + \dfrac{2u}{1 + u^2} + \dfrac{1 - u^2}{1 + u^2}} \right]$

$\qquad\qquad\qquad\qquad = \int_0^1 \dfrac{1}{1 + u}\, du$

$\qquad\qquad\qquad\qquad = \Big[\ln|1 + u| \Big]_0^1$

$\qquad\qquad\qquad\qquad = \ln 2$

$\qquad u = \tan\dfrac{\theta}{2}$

67. $\displaystyle\int \frac{\sin\theta}{3 - 2\cos\theta}\, d\theta = \frac{1}{2}\int \frac{2\sin\theta}{3 - 2\cos\theta}\, d\theta$

$$= \frac{1}{2}\ln|u| + C$$

$$= \frac{1}{2}\ln(3 - 2\cos\theta) + C$$

$u = 3 - 2\cos\theta, \, du = 2\sin\theta\, d\theta$

69. $\displaystyle\int \frac{\cos\sqrt{\theta}}{\sqrt{\theta}}\, d\theta = 2\int \cos\sqrt{\theta}\left(\frac{1}{2\sqrt{\theta}}\right) d\theta$

$$= 2\sin\sqrt{\theta} + C$$

$u = \sqrt{\theta}, \, du = \dfrac{1}{2\sqrt{\theta}}\, d\theta$

71. $\displaystyle A = \int_0^8 \frac{x}{\sqrt{x + 1}}\, dx$

$$= \left[\frac{-2(2 - x)}{3}\sqrt{x + 1}\right]_0^8$$

$$= 12 - \left(-\frac{4}{3}\right)$$

$$= \frac{40}{3} \approx 13.333 \text{ square units}$$

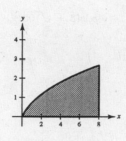

73. Arctangent Formula, Formula 23,

$$\int \frac{1}{u^2 + 1}\, du, \, u = e^x$$

75. Substitution: $u = x^2, \, du = 2x\, dx$

Then Formula 81.

77. Cannot be integrated.

79. (a) $n = 1$: $u = \ln x, \, du = \dfrac{1}{x}\, dx, \, dv = x\, dx, \, v = \dfrac{x^2}{2}$

$$\int x\ln x\, dx = \frac{x^2}{2}\ln x - \int \frac{x^2}{2}\frac{1}{x}\, dx = \frac{x^2}{2}\ln x - \frac{x^2}{4} + C$$

$n = 2$: $u = \ln x, \, du = \dfrac{1}{x}\, dx, \, dv = x^2\, dx, \, v = \dfrac{x^3}{3}$

$$\int x^2\ln x\, dx = \frac{x^3}{3}\ln x - \int \frac{x^3}{3}\frac{1}{x}\, dx = \frac{x^3}{3}\ln x - \frac{x^3}{9} + C$$

$n = 3$: $u = \ln x, \, du = \dfrac{1}{x}\, dx, \, dv = x^3\, dx, \, v = \dfrac{x^4}{4}$

$$\int x^3\ln x\, dx = \frac{x^4}{4}\ln x - \int \frac{x^4}{4}\frac{1}{x}\, dx = \frac{x^4}{4}\ln x - \frac{x^4}{16} + C$$

(b) $\displaystyle\int x^n\ln x\, dx = \frac{x^{n+1}}{n + 1}\ln x - \frac{x^{n+1}}{(n + 1)^2} + C$

81. False. You might need to convert your integral using substitution or algebra.

83. $\displaystyle W = \int_0^5 2000xe^{-x}\, dx$

$$= -2000\int_0^5 -xe^{-x}\, dx$$

$$= 2000\int_0^5 (-x)e^{-x}(-1)\, dx$$

$$= 2000\left[(-x)e^{-x} - e^{-x}\right]_0^5$$

$$= 2000\left(-\frac{6}{e^5} + 1\right)$$

$$\approx 1919.145 \text{ ft} \cdot \text{lbs}$$

85. $V = 20(2) \int_0^3 \dfrac{2}{\sqrt{1 + y^2}}\, dy$

$\qquad = \left[80 \ln\left| y + \sqrt{1 + y^2} \right| \right]_0^3$

$\qquad = 80 \ln\left(3 + \sqrt{10} \right)$

$\qquad \approx 145.5$ cubic feet

$\qquad W = 148\left(80 \ln\left(3 + \sqrt{10} \right) \right)$

$\qquad \quad = 11{,}840 \ln\left(3 + \sqrt{10} \right)$

$\qquad \quad \approx 21{,}530.4$ lb

By symmetry, $\bar{x} = 0$.

$\quad M = \rho(2) \int_0^3 \dfrac{2}{\sqrt{1 + y^2}}\, dy = \left[4\rho \ln\left| y + \sqrt{1 + y^2} \right| \right]_0^3 = 4\rho \ln\left(3 + \sqrt{10} \right)$

$\quad M_x = 2\rho \int_0^3 \dfrac{2y}{\sqrt{1 + y^2}}\, dy = \left[4\rho \sqrt{1 + y^2} \right]_0^3 = 4\rho\left(\sqrt{10} - 1 \right)$

$\quad \bar{y} = \dfrac{M_x}{M} = \dfrac{4\rho\left(\sqrt{10} - 1 \right)}{4\rho \ln\left(3 + \sqrt{10} \right)} \approx 1.19$

Centroid: $(\bar{x}, \bar{y}) \approx (0, 1.19)$

87. (a) $\displaystyle\int_0^4 \dfrac{k}{2 + 3x}\, dx = 10$

(b) $\displaystyle\int_0^4 \dfrac{15.417}{2 + 3x}\, dx$

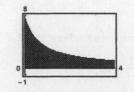

$\qquad k = \dfrac{10}{\displaystyle\int_0^4 \dfrac{1}{2 + 3x}\, dx} = \dfrac{10}{\frac{1}{3} \ln 7} \approx \dfrac{10}{0.6486}$

$\qquad \quad = 15.417 \left(= \dfrac{30}{\ln 7} \right)$

89. Let $I = \displaystyle\int_0^{\pi/2} \dfrac{dx}{1 + (\tan x)^{\sqrt{2}}}$.

For $x = \dfrac{\pi}{2} - u$, $dx = -du$, and

$\quad I = \displaystyle\int_{\pi/2}^0 \dfrac{-du}{1 + (\tan(\pi/2 - u))^{\sqrt{2}}} = \int_0^{\pi/2} \dfrac{du}{1 + (\cot u)^{\sqrt{2}}} = \int_0^{\pi/2} \dfrac{(\tan u)^{\sqrt{2}}}{(\tan u)^{\sqrt{2}} + 1}\, du.$

$\quad 2I = \displaystyle\int_0^{\pi/2} \dfrac{dx}{1 + (\tan x)^{\sqrt{2}}} + \int_0^{\pi/2} \dfrac{(\tan x)^{\sqrt{2}}}{(\tan x)^{\sqrt{2}} + 1}\, dx = \int_0^{\pi/2} dx = \dfrac{\pi}{2}$

Thus, $I = \dfrac{\pi}{4}$.

Section 8.7   Indeterminate Forms and L'Hôpital's Rule

1. $\displaystyle\lim_{x \to 0} \dfrac{\sin 5x}{\sin 2x} \approx 2.5 \left(\text{exact: } \dfrac{5}{2} \right)$

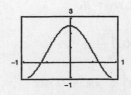

x	-0.1	-0.01	-0.001	0.001	0.01	0.1
$f(x)$	2.4132	2.4991	2.500	2.500	2.4991	2.4132

3. $\displaystyle\lim_{x \to \infty} x^5 e^{-x/100} \approx 0$

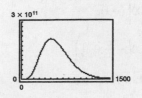

x	1	10	10^2	10^3	10^4	10^5
$f(x)$	0.9900	90,484	3.7×10^9	4.5×10^{10}	0	0

5. (a) $\lim\limits_{x \to 3} \dfrac{2(x-3)}{x^2-9} = \lim\limits_{x \to 3} \dfrac{2(x-3)}{(x+3)(x-3)} = \lim\limits_{x \to 3} \dfrac{2}{x+3} = \dfrac{1}{3}$

   (b) $\lim\limits_{x \to 3} \dfrac{2(x-3)}{x^2-9} = \lim\limits_{x \to 3} \dfrac{(d/dx)[2(x-3)]}{(d/dx)[x^2-9]} = \lim\limits_{x \to 3} \dfrac{2}{2x} = \dfrac{2}{6} = \dfrac{1}{3}$

7. (a) $\lim\limits_{x \to 3} \dfrac{\sqrt{x+1}-2}{x-3} = \lim\limits_{x \to 3} \dfrac{\sqrt{x+1}-2}{x-3} \cdot \dfrac{\sqrt{x+1}+2}{\sqrt{x+1}+2} = \lim\limits_{x \to 3} \dfrac{(x+1)-4}{(x-3)\left[\sqrt{x+1}+2\right]} = \lim\limits_{x \to 3} \dfrac{1}{\sqrt{x+1}+2} = \dfrac{1}{4}$

   (b) $\lim\limits_{x \to 3} \dfrac{\sqrt{x+1}-2}{x-3} = \lim\limits_{x \to 3} \dfrac{(d/dx)\left[\sqrt{x+1}-2\right]}{(d/dx)[x-3]} = \lim\limits_{x \to 3} \dfrac{1/(2\sqrt{x+1})}{1} = \dfrac{1}{4}$

9. (a) $\lim\limits_{x \to \infty} \dfrac{5x^2-3x+1}{3x^2-5} = \lim\limits_{x \to \infty} \dfrac{5-(3/x)+(1/x^2)}{3-(5/x^2)} = \dfrac{5}{3}$

   (b) $\lim\limits_{x \to \infty} \dfrac{5x^2-3x+1}{3x^2-5} = \lim\limits_{x \to \infty} \dfrac{(d/dx)[5x^2-3x+1]}{(d/dx)[3x^2-5]} = \lim\limits_{x \to \infty} \dfrac{10x-3}{6x} = \lim\limits_{x \to \infty} \dfrac{(d/dx)[10x-3]}{(d/dx)[6x]} = \lim\limits_{x \to \infty} \dfrac{10}{6} = \dfrac{5}{3}$

11. $\lim\limits_{x \to 2} \dfrac{x^2-x-2}{x-2} = \lim\limits_{x \to 2} \dfrac{2x-1}{1} = 3$

13. $\lim\limits_{x \to 0} \dfrac{\sqrt{4-x^2}-2}{x} = \lim\limits_{x \to 0} \dfrac{-x/\sqrt{4-x^2}}{1} = 0$

15. $\lim\limits_{x \to 0} \dfrac{e^x-(1-x)}{x} = \lim\limits_{x \to 0} \dfrac{e^x+1}{1} = 2$

17. $\lim\limits_{x \to 0^+} \dfrac{e^x-(1+x)}{x^3} = \lim\limits_{x \to 0^+} \dfrac{e^x-1}{3x^2}$

$= \lim\limits_{x \to 0^+} \dfrac{e^x}{6x} = \infty$

19. $\lim\limits_{x \to 0} \dfrac{\sin 2x}{\sin 3x} = \lim\limits_{x \to 0} \dfrac{2\cos 2x}{3\cos 3x} = \dfrac{2}{3}$

21. $\lim\limits_{x \to 0} \dfrac{\arcsin x}{x} = \lim\limits_{x \to 0} \dfrac{1/\sqrt{1-x^2}}{1} = 1$

23. $\lim\limits_{x \to \infty} \dfrac{3x^2-2x+1}{2x^2+3} = \lim\limits_{x \to \infty} \dfrac{6x-2}{4x}$

$= \lim\limits_{x \to \infty} \dfrac{6}{4} = \dfrac{3}{2}$

25. $\lim\limits_{x \to \infty} \dfrac{x^2+2x+3}{x-1} = \lim\limits_{x \to \infty} \dfrac{2x+2}{1} = \infty$

27. $\lim\limits_{x \to \infty} \dfrac{x^3}{e^{x/2}} = \lim\limits_{x \to \infty} \dfrac{3x^2}{(1/2)e^{x/2}}$

$= \lim\limits_{x \to \infty} \dfrac{6x}{(1/4)e^{x/2}} = \lim\limits_{x \to \infty} \dfrac{6}{(1/8)e^{x/2}} = 0$

29. $\lim\limits_{x \to \infty} \dfrac{x}{\sqrt{x^2+1}} = \lim\limits_{x \to \infty} \dfrac{1}{\sqrt{1+(1/x^2)}} = 1$

Note: L'Hôpital's Rule does not work on this limit. See Exercise 79.

31. $\lim\limits_{x \to \infty} \dfrac{\cos x}{x} = 0$ by Squeeze Theorem

$\left(\dfrac{\cos x}{x} \le \dfrac{1}{x}, \text{ for } x > 0\right)$

33. $\lim\limits_{x \to \infty} \dfrac{\ln x}{x^2} = \lim\limits_{x \to \infty} \dfrac{1/x}{2x} = \lim\limits_{x \to \infty} \dfrac{1}{2x^2} = 0$

35. $\lim\limits_{x \to \infty} \dfrac{e^x}{x^2} = \lim\limits_{x \to \infty} \dfrac{e^x}{2x} = \lim\limits_{x \to \infty} \dfrac{e^x}{2} = \infty$

37. (a) $\lim\limits_{x \to \infty} x \ln x$, not indeterminate

   (b) $\lim\limits_{x \to \infty} x \ln x = (\infty)(\infty) = \infty$

   (c)

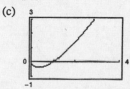

39. (a) $\lim\limits_{x\to\infty}\left(x\sin\dfrac{1}{x}\right)=(\infty)(0)$

(b) $\lim\limits_{x\to\infty}x\sin\dfrac{1}{x}=\lim\limits_{x\to\infty}\dfrac{\sin(1/x)}{1/x}$

$$=\lim\limits_{x\to\infty}\dfrac{(-1/x^2)\cos(1/x)}{-1/x^2}$$

$$=\lim\limits_{x\to\infty}\cos\left(\dfrac{1}{x}\right)=1$$

(c)

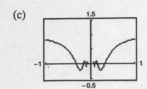

41. (a) $\lim\limits_{x\to0^+}x^{1/x}=0^\infty=0$, not indeterminate
(See Exercise 106).

(b) Let $y=x^{1/x}$.

$$\ln y=\ln x^{1/x}=\dfrac{1}{x}\ln x.$$

Since $x\to0^+$, $\dfrac{1}{x}\ln x\to(\infty)(-\infty)=-\infty$. Hence,

$$\ln y\to-\infty\implies y\to0^+.$$

Therefore, $\lim\limits_{x\to0^+}x^{1/x}=0$.

(c)

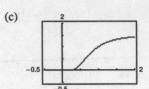

43. (a) $\lim\limits_{x\to\infty}x^{1/x}=\infty^0$

(b) Let $y=\lim\limits_{x\to\infty}x^{1/x}$.

$$\ln y=\lim\limits_{x\to\infty}\dfrac{\ln x}{x}=\lim\limits_{x\to\infty}\left(\dfrac{1/x}{1}\right)=0$$

Thus, $\ln y=0\implies y=e^0=1$. Therefore,

$$\lim\limits_{x\to\infty}x^{1/x}=1.$$

(c)

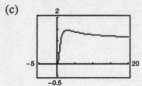

45. (a) $\lim\limits_{x\to0^+}(1+x)^{1/x}=1^\infty$

(b) Let $y=\lim\limits_{x\to0^+}(1+x)^{1/x}$.

$$\ln y=\lim\limits_{x\to0^+}\dfrac{\ln(1+x)}{x}$$

$$=\lim\limits_{x\to0^+}\left(\dfrac{1/(1+x)}{1}\right)=1$$

Thus, $\ln y=1\implies y=e^1=e$.

Therefore, $\lim\limits_{x\to0^+}(1+x)^{1/x}=e$.

(c)

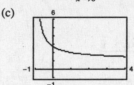

47. (a) $\lim\limits_{x\to0^+}[3(x)^{x/2}]=0^0$

(b) Let $y=\lim\limits_{x\to0^+}3(x)^{x/2}$.

$$\ln y=\lim\limits_{x\to0^+}\left[\ln 3+\dfrac{x}{2}\ln x\right]$$

$$=\lim\limits_{x\to0^+}\left[\ln 3+\dfrac{\ln x}{2/x}\right]$$

$$=\lim\limits_{x\to0^+}\ln 3+\lim\limits_{x\to0^+}\dfrac{1/x}{-2/x^2}$$

$$=\lim\limits_{x\to0^+}\ln 3-\lim\limits_{x\to0^+}\dfrac{x}{2}$$

$$=\ln 3$$

Hence, $\lim\limits_{x\to0^+}3(x)^{x/2}=3$.

(c)

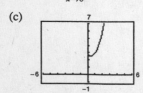

49. (a) $\lim\limits_{x\to1^+}(\ln x)^{x-1}=0^0$

(b) Let $y=\lim\limits_{x\to1^+}(\ln x)^{x-1}$

$$=\lim\limits_{x\to1^+}(x-1)\ln x=0.$$

Hence, $\lim\limits_{x\to1^+}(\ln x)^{x-1}=1$.

(c)

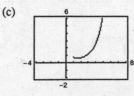

51. (a) $\lim\limits_{x \to 2^+} \left(\dfrac{8}{x^2 - 4} - \dfrac{x}{x - 2} \right) = \infty - \infty$

   (b) $\lim\limits_{x \to 2^+} \left(\dfrac{8}{x^2 - 4} - \dfrac{x}{x - 2} \right) = \lim\limits_{x \to 2^+} \dfrac{8 - x(x + 2)}{x^2 - 4}$

$$= \lim_{x \to 2^+} \frac{(2 - x)(4 + x)}{(x + 2)(x - 2)}$$

$$= \lim_{x \to 2^+} \frac{-(x + 4)}{x + 2} = \frac{-3}{2}$$

   (c)

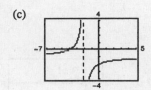

53. (a) $\lim\limits_{x \to 1^+} \left(\dfrac{3}{\ln x} - \dfrac{2}{x - 1} \right) = \infty - \infty$

   (b) $\lim\limits_{x \to 1^+} \left(\dfrac{3}{\ln x} - \dfrac{2}{x - 1} \right) = \lim\limits_{x \to 1^+} \dfrac{3x - 3 - 2 \ln x}{(x - 1)\ln x}$

$$= \lim_{x \to 1^+} \frac{3 - (2/x)}{[(x - 1)/x] + \ln x} = \infty$$

   (c)

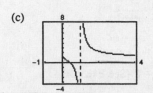

55. (a)

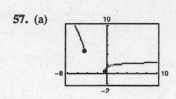

   (b) $\lim\limits_{x \to 3} \dfrac{x - 3}{\ln(2x - 5)} = \lim\limits_{x \to 3} \dfrac{1}{2/(2x - 5)}$

$$= \lim_{x \to 3} \frac{2x - 5}{2} = \frac{1}{2}$$

57. (a)

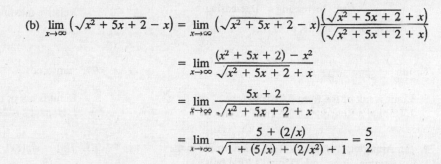

   (b) $\lim\limits_{x \to \infty} \left(\sqrt{x^2 + 5x + 2} - x \right) = \lim\limits_{x \to \infty} \left(\sqrt{x^2 + 5x + 2} - x \right)\dfrac{\left(\sqrt{x^2 + 5x + 2} + x \right)}{\left(\sqrt{x^2 + 5x + 2} + x \right)}$

$$= \lim_{x \to \infty} \frac{(x^2 + 5x + 2) - x^2}{\sqrt{x^2 + 5x + 2} + x}$$

$$= \lim_{x \to \infty} \frac{5x + 2}{\sqrt{x^2 + 5x + 2} + x}$$

$$= \lim_{x \to \infty} \frac{5 + (2/x)}{\sqrt{1 + (5/x) + (2/x^2)} + 1} = \frac{5}{2}$$

59. $\dfrac{0}{0}, \dfrac{\infty}{\infty}, 0 \cdot \infty, 1^\infty, 0^0, \infty - \infty, \infty^0$

61. (a) Let $f(x) = x^2 - 25$ and $g(x) = x - 5$.

   (b) Let $f(x) = (x - 5)^2$ and $g(x) = x^2 - 25$.

   (c) Let $f(x) = x^2 - 25$ and $g(x) = (x - 5)^3$.

63.

x	10	10^2	10^4	10^6	10^8	10^{10}
$\dfrac{(\ln x)^4}{x}$	2.811	4.498	0.720	0.036	0.001	0.000

65. $\lim\limits_{x \to \infty} \dfrac{x^2}{e^{5x}} = \lim\limits_{x \to \infty} \dfrac{2x}{5e^{5x}} = \lim\limits_{x \to \infty} \dfrac{2}{25e^{5x}} = 0$

67. $\lim\limits_{x \to \infty} \dfrac{(\ln x)^3}{x} = \lim\limits_{x \to \infty} \dfrac{3(\ln x)^2(1/x)}{1}$

$$= \lim_{x \to \infty} \frac{3(\ln x)^2}{x}$$

$$= \lim_{x \to \infty} \frac{6(\ln x)(1/x)}{1}$$

$$= \lim_{x \to \infty} \frac{6(\ln x)}{x} = \lim_{x \to \infty} \frac{6}{x} = 0$$

69. $\lim\limits_{x\to\infty} \dfrac{(\ln x)^n}{x^m} = \lim\limits_{x\to\infty} \dfrac{n(\ln x)^{n-1}/x}{mx^{m-1}}$

$\qquad = \lim\limits_{x\to\infty} \dfrac{n(\ln x)^{n-1}}{mx^m}$

$\qquad = \lim\limits_{x\to\infty} \dfrac{n(n-1)(\ln x)^{n-2}}{m^2 x^m}$

$\qquad = \cdots = \lim\limits_{x\to\infty} \dfrac{n!}{m^n x^m} = 0$

71. $y = x^{1/x},\ x > 0$

Horizontal asymptote: $y = 1$ (See Exercise 43.)

$\ln y = \dfrac{1}{x}\ln x$

$\dfrac{1}{y}\dfrac{dy}{dx} = \dfrac{1}{x}\left(\dfrac{1}{x}\right) + (\ln x)\left(-\dfrac{1}{x^2}\right)$

$\dfrac{dy}{dx} = x^{1/x}\left(\dfrac{1}{x^2}\right)(1 - \ln x) = x^{(1/x)-2}(1 - \ln x) = 0$

Critical number: $x = e$

Intervals: $(0, e)$ (e, ∞)

Sign of dy/dx: $+$ $-$

$y = f(x)$: Increasing   Decreasing

Relative maximum: $(e, e^{1/e})$

73. $y = 2xe^{-x}$

$\lim\limits_{x\to\infty} \dfrac{2x}{e^x} = \lim\limits_{x\to\infty} \dfrac{2}{e^x} = 0$

Horizontal asymptote: $y = 0$

$\dfrac{dy}{dx} = 2x(-e^{-x}) + 2e^{-x}$

$\qquad = 2e^{-x}(1 - x) = 0$

Critical number: $x = 1$

Intervals: $(-\infty, 1)$ $(1, \infty)$

Sign of dy/dx: $+$ $-$

$y = f(x)$: Increasing   Decreasing

Relative maximum: $\left(1, \dfrac{2}{e}\right)$

75. $\lim\limits_{x\to 0} \dfrac{e^{2x} - 1}{e^x} = \dfrac{0}{1} = 0$

Limit is not of the form $0/0$ or ∞/∞.
L'Hôpital's Rule does not apply.

77. $\lim\limits_{x\to\infty} x\cos\dfrac{1}{x} = \infty(1) = \infty$

Limit is not of the form $0/0$ or ∞/∞.
L'Hôpital's Rule does not apply.

79. (a) Applying L'Hôpital's Rule twice results in the original limit, so L'Hôpital's Rule fails:

$\lim\limits_{x\to\infty} \dfrac{x}{\sqrt{x^2 + 1}} = \lim\limits_{x\to\infty} \dfrac{1}{x/\sqrt{x^2 + 1}}$

$\qquad = \lim\limits_{x\to\infty} \dfrac{\sqrt{x^2 + 1}}{x}$

$\qquad = \lim\limits_{x\to\infty} \dfrac{x/\sqrt{x^2 + 1}}{1}$

$\qquad = \lim\limits_{x\to\infty} \dfrac{x}{\sqrt{x^2 + 1}}$

(b) $\lim\limits_{x\to\infty} \dfrac{x}{\sqrt{x^2 + 1}} = \lim\limits_{x\to\infty} \dfrac{x/x}{\sqrt{x^2 + 1}/x}$

$\qquad = \lim\limits_{x\to\infty} \dfrac{1}{\sqrt{1 + 1/x^2}}$

$\qquad = \dfrac{1}{\sqrt{1 + 0}} = 1$

(c)

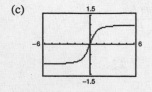

81. $f(x) = \sin(3x),\ g(x) = \sin(4x)$

$f'(x) = 3\cos(3x),\ g'(x) = 4\cos(4x)$

$y_1 = \dfrac{f(x)}{g(x)} = \dfrac{\sin 3x}{\sin 4x},\quad y_2 = \dfrac{f'(x)}{g'(x)} = \dfrac{3\cos 3x}{4\cos 4x}$

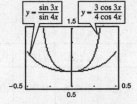

As $x \to 0$, $y_1 \to 0.75$ and $y_2 \to 0.75$

By L'Hôpital's Rule,

$\lim\limits_{x\to 0} \dfrac{\sin 3x}{\sin 4x} = \lim\limits_{x\to 0} \dfrac{3\cos 3x}{4\cos 4x} = \dfrac{3}{4}$

83. $\lim\limits_{k \to 0} \dfrac{32\left(1 - e^{-kt} + \dfrac{v_0 k e^{-kt}}{32}\right)}{k} = \lim\limits_{k \to 0} \dfrac{32(1 - e^{-kt})}{k} + \lim\limits_{k \to 0} (v_0 e^{-kt})$

$$= \lim\limits_{k \to 0} \dfrac{32(0 + te^{-kt})}{1} + \lim\limits_{k \to 0} \left(\dfrac{v_0}{e^{kt}}\right) = 32t + v_0$$

85. Let N be a fixed value for n. Then

$$\lim\limits_{x \to \infty} \dfrac{x^{N-1}}{e^x} = \lim\limits_{x \to \infty} \dfrac{(N-1)x^{N-2}}{e^x} = \lim\limits_{x \to \infty} \dfrac{(N-1)(N-2)x^{N-3}}{e^x} = \ldots = \lim\limits_{x \to \infty} \left[\dfrac{(N-1)!}{e^x}\right] = 0. \quad \text{(See Exercise 70.)}$$

87. $f(x) = x^3, g(x) = x^2 + 1, [0, 1]$

$\dfrac{f(b) - f(a)}{g(b) - g(a)} = \dfrac{f'(c)}{g'(c)}$

$\dfrac{f(1) - f(0)}{g(1) - g(0)} = \dfrac{3c^2}{2c}$

$\dfrac{1}{1} = \dfrac{3c}{2}$

$c = \dfrac{2}{3}$

89. $f(x) = \sin x, g(x) = \cos x, \left[0, \dfrac{\pi}{2}\right]$

$\dfrac{f(\pi/2) - f(0)}{g(\pi/2) - g(0)} = \dfrac{f'(c)}{g'(c)}$

$\dfrac{1}{-1} = \dfrac{\cos c}{-\sin c}$

$-1 = -\cot c$

$c = \dfrac{\pi}{4}$

91. False. L'Hôpital's Rule does not apply since

$\lim\limits_{x \to 0} (x^2 + x + 1) \neq 0.$

$\lim\limits_{x \to 0^+} \dfrac{x^2 + x + 1}{x} = \lim\limits_{x \to 0^+} \left(x + 1 + \dfrac{1}{x}\right) = 1 + \infty = \infty$

93. True

95. Area of triangle: $\dfrac{1}{2}(2x)(1 - \cos x) = x - x \cos x$

Shaded area: Area of rectangle $-$ Area under curve

$$2x(1 - \cos x) - 2\int_0^x (1 - \cos t)\, dt = 2x(1 - \cos x) - 2\left[t - \sin t\right]_0^x$$

$$= 2x(1 - \cos x) - 2(x - \sin x) = 2 \sin x - 2x \cos x$$

Ratio: $\lim\limits_{x \to 0} \dfrac{x - x \cos x}{2 \sin x - 2x \cos x} = \lim\limits_{x \to 0} \dfrac{1 + x \sin x - \cos x}{2 \cos x + 2x \sin x - 2 \cos x}$

$$= \lim\limits_{x \to 0} \dfrac{1 + x \sin x - \cos x}{2x \sin x}$$

$$= \lim\limits_{x \to 0} \dfrac{x \cos x + \sin x + \sin x}{2x \cos x + 2 \sin x}$$

$$= \lim\limits_{x \to 0} \dfrac{x \cos x + 2 \sin x}{2x \cos x + 2 \sin x} \cdot \dfrac{1/\cos x}{1/\cos x}$$

$$= \lim\limits_{x \to 0} \dfrac{x + 2 \tan x}{2x + 2 \tan x}$$

$$= \lim\limits_{x \to 0} \dfrac{1 + 2 \sec^2 x}{2 + 2 \sec^2 x} = \dfrac{3}{4}$$

97. $\lim_{x\to 0}\dfrac{4x-2\sin 2x}{2x^3}=\lim_{x\to 0}\dfrac{4-4\cos 2x}{6x^2}$

$$=\lim_{x\to 0}\dfrac{8\sin 2x}{12x}$$

$$=\lim_{x\to 0}\dfrac{16\cos 2x}{12}=\dfrac{16}{12}=\dfrac{4}{3}$$

Let $c=\dfrac{4}{3}$.

99. $\lim_{x\to 0}\dfrac{a-\cos bx}{x^2}=2$

Near $x=0$, $\cos bx\approx 1$ and $x^2\approx 0\implies a=1$.

Using L'Hôpital's Rule,

$$\lim_{x\to 0}\dfrac{1-\cos bx}{x^2}=\lim_{x\to 0}\dfrac{b\sin bx}{2x}$$

$$=\lim_{x\to 0}\dfrac{b^2\cos bx}{2}=2.$$

Hence, $b^2=4$ and $b=\pm 2$.

Answer: $a=1,\,b=\pm 2$

101. (a) $\lim_{h\to 0}\dfrac{f(x+h)-f(x-h)}{2h}=\lim_{h\to 0}\dfrac{f'(x+h)(1)-f'(x-h)(-1)}{2}$

$$=\lim_{h\to 0}\left[\dfrac{f'(x+h)+f'(x-h)}{2}\right]$$

$$=\dfrac{f'(x)+f'(x)}{2}=f'(x)$$

(b)

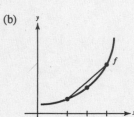

Graphically, the slope of the line joining $(x-h, f(x-h))$ and $(x+h, f(x+h))$ is approximately $f'(x)$. And, as $h\to 0$,

$$\lim_{h\to 0}\dfrac{f(x+h)-f(x-h)}{2h}=f'(x).$$

103. $g(x)=\begin{cases}e^{-1/x^2}, & x\neq 0\\ 0, & x=0\end{cases}$

$g'(0)=\lim_{x\to 0}\dfrac{g(x)-g(0)}{x-0}=\lim_{x\to 0}\dfrac{e^{-1/x^2}}{x}$

Let $y=\dfrac{e^{-1/x^2}}{x}$, then $\ln y=\ln\left(\dfrac{e^{-1/x^2}}{x}\right)=-\dfrac{1}{x^2}-\ln x=\dfrac{-1-x^2\ln x}{x^2}$. Since

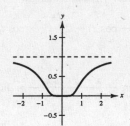

$$\lim_{x\to 0}x^2\ln x=\lim_{x\to 0}\dfrac{\ln x}{1/x^2}=\lim_{x\to 0}\dfrac{1/x}{-2/x^3}=\lim_{x\to 0}\left(-\dfrac{x^2}{2}\right)=0$$

we have $\lim_{x\to 0}\left(\dfrac{-1-x^2\ln x}{x^2}\right)=-\infty$. Thus, $\lim_{x\to 0}y=e^{-\infty}=0\implies g'(0)=0$.

Note: The graph appears to support this conclusion—the tangent line is horizontal at $(0, 0)$.

105. (a) $\lim_{x\to 0^+}(-x\ln x)$ is the form $0\cdot\infty$.

(b) $\lim_{x\to 0^+}\dfrac{-\ln x}{1/x}=\lim_{x\to 0^+}\dfrac{-1/x}{-1/x^2}=\lim_{x\to 0^+}(x)=0$

(c)

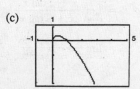

107. $\lim\limits_{x \to a} f(x)^{g(x)}$

$$y = f(x)^{g(x)}$$

$$\ln y = g(x) \ln f(x)$$

$$\lim\limits_{x \to a} g(x) \ln f(x) = (-\infty)(-\infty) = \infty$$

As $x \to a$, $\ln y \Rightarrow \infty$, and hence $y = \infty$. Thus,

$$\lim\limits_{x \to a} f(x)^{g(x)} = \infty.$$

109. (a) $\lim\limits_{x \to 0^+} x^{(\ln 2)/(1 + \ln x)}$ is of form 0^0.

Let $y = x^{(\ln 2)/(1 + \ln x)}$

$$\ln y = \frac{\ln 2}{1 + \ln x} \ln x$$

$$\lim\limits_{x \to 0^+} \ln y = \frac{\ln 2(1/x)}{1/x} = \ln 2.$$

Thus, $\lim\limits_{x \to 0^+} x^{(\ln 2)/(1 + \ln x)} = 2.$

(b) $\lim\limits_{x \to \infty} x^{(\ln 2)/(1 + \ln x)}$ is of form ∞^0.

Let $y = x^{(\ln 2)/(1 + \ln x)}$

$$\ln y = \frac{\ln 2}{1 + \ln x} \ln x$$

$$\lim\limits_{x \to \infty} \ln y = \frac{\ln 2(1/x)}{1/x} = \ln 2.$$

Thus, $\lim\limits_{x \to \infty} x^{(\ln 2)/(1 + \ln x)} = 2.$

(c) $\lim\limits_{x \to 0} (x + 1)^{(\ln 2)/(x)}$ is of form 1^∞.

Let $y = (x + 1)^{(\ln 2)/(x)}$

$$\ln y = \frac{\ln 2}{x} \ln(x + 1)$$

$$\lim\limits_{x \to 0} \ln y = \lim\limits_{x \to 0} \frac{(\ln 2)1/(x + 1)}{1} = \ln 2.$$

Thus, $\lim\limits_{x \to 0} (x + 1)^{(\ln 2)/(x)} = 2.$

111. (a) $h(x) = \dfrac{x + \sin x}{x}$

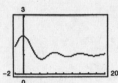

$$\lim\limits_{x \to \infty} h(x) = 1$$

(b) $h(x) = \dfrac{x + \sin x}{x} = \dfrac{x}{x} + \dfrac{\sin x}{x} = 1 + \dfrac{\sin x}{x}, x > 0$

Hence, $\lim\limits_{x \to \infty} h(x) = \lim\limits_{x \to \infty} \left[1 + \dfrac{\sin x}{x} \right] = 1 + 0 = 1.$

(c) No. $h(x)$ is not an indeterminate form.

Section 8.8   Improper Integrals

1. $\displaystyle\int_0^1 \frac{dx}{3x - 2}$ is improper because $3x - 2 = 0$ when $x = \dfrac{2}{3}$.

3. $\displaystyle\int_0^1 \frac{2x - 5}{x^2 - 5x + 6} dx = \int_0^1 \frac{2x - 5}{(x - 2)(x - 3)} dx$

is not improper because

$\dfrac{2x - 5}{(x - 2)(x - 3)}$ is continuous on $[0, 1]$.

5. Infinite discontinuity at $x = 0$.

$$\int_0^4 \frac{1}{\sqrt{x}} dx = \lim\limits_{b \to 0^+} \int_b^4 \frac{1}{\sqrt{x}} dx$$

$$= \lim\limits_{b \to 0^+} \left[2\sqrt{x} \right]_b^4$$

$$= \lim\limits_{b \to 0^+} \left(4 - 2\sqrt{b} \right) = 4$$

Converges

7. Infinite discontinuity at $x = 1$.

$$\int_0^2 \frac{1}{(x-1)^2}\, dx = \int_0^1 \frac{1}{(x-1)^2}\, dx + \int_1^2 \frac{1}{(x-1)^2}\, dx$$

$$= \lim_{b \to 1^-} \int_0^b \frac{1}{(x-1)^2}\, dx + \lim_{c \to 1^+} \int_c^2 \frac{1}{(x-1)^2}\, dx$$

$$= \lim_{b \to 1^-} \left[-\frac{1}{x-1} \right]_0^b + \lim_{c \to 1^+} \left[-\frac{1}{x-1} \right]_c^2 = (\infty - 1) + (-1 + \infty)$$

Diverges

9. Infinite limit of integration.

$$\int_0^\infty e^{-x}\, dx = \lim_{b \to \infty} \int_0^b e^{-x}\, dx$$

$$= \lim_{b \to \infty} \left[-e^{-x} \right]_0^b = 0 + 1 = 1$$

Converges

11. $\int_{-1}^1 \frac{1}{x^2}\, dx \neq -2$

because the integrand is not defined at $x = 0$.
Diverges

13. $\int_0^\infty e^{-x}\, dx \neq 0$. You need to evaluate the limit.

$$\lim_{b \to \infty} \int_0^b e^{-x}\, dx = \lim_{b \to \infty} \left[-e^{-x} \right]_0^b$$

$$= \lim_{b \to \infty} \left[-e^{-b} + 1 \right] = 1$$

15. $\int_1^\infty \frac{1}{x^2}\, dx = \lim_{b \to \infty} \int_1^b \frac{1}{x^2}\, dx$

$$= \lim_{b \to \infty} \left[-\frac{1}{x} \right]_1^b = 1$$

17. $\int_1^\infty \frac{3}{\sqrt[3]{x}}\, dx = \lim_{b \to \infty} \int_1^b 3x^{-1/3}\, dx$

$$= \lim_{b \to \infty} \left[\frac{9}{2} x^{2/3} \right]_1^b = \infty$$

Diverges

19. $\int_{-\infty}^0 xe^{-2x}\, dx = \lim_{b \to -\infty} \int_b^0 xe^{-2x}\, dx = \lim_{b \to -\infty} \frac{1}{4}\left[(-2x - 1)e^{-2x} \right]_b^0 = \lim_{b \to -\infty} \frac{1}{4}[-1 + (2b + 1)e^{-2b}] = -\infty$ (Integration by parts)

Diverges

21. $\int_0^\infty x^2 e^{-x}\, dx = \lim_{b \to \infty} \int_0^b x^2 e^{-x}\, dx = \lim_{b \to \infty} \left[-e^{-x}(x^2 + 2x + 2) \right]_0^b = \lim_{b \to \infty} \left(-\frac{b^2 + 2b + 2}{e^b} + 2 \right) = 2$

Since $\displaystyle\lim_{b \to \infty} \left(-\frac{b^2 + 2b + 2}{e^b} \right) = 0$ by L'Hôpital's Rule.

23. $\int_0^\infty e^{-x} \cos x\, dx = \lim_{b \to \infty} \frac{1}{2}\left[e^{-x}(-\cos x + \sin x) \right]_0^b$

$$= \frac{1}{2}[0 - (-1)] = \frac{1}{2}$$

25. $\int_4^\infty \frac{1}{x(\ln x)^3}\, dx = \lim_{b \to \infty} \int_4^b (\ln x)^{-3} \frac{1}{x}\, dx$

$$= \lim_{b \to \infty} \left[-\frac{1}{2}(\ln x)^{-2} \right]_4^b$$

$$= -\frac{1}{2}(\ln b)^{-2} + \frac{1}{2}(\ln 4)^{-2}$$

$$= \frac{1}{2} \frac{1}{(2 \ln 2)^2} = \frac{1}{8(\ln 2)^2}$$

27. $\displaystyle\int_{-\infty}^{\infty} \frac{2}{4+x^2}\,dx = \int_{-\infty}^{0} \frac{2}{4+x^2}\,dx + \int_{0}^{\infty} \frac{2}{4+x^2}\,dx$

$\displaystyle = \lim_{b\to-\infty}\int_{b}^{0} \frac{2}{4+x^2}\,dx + \lim_{c\to\infty}\int_{0}^{c} \frac{2}{4+x^2}\,dx$

$\displaystyle = \lim_{b\to-\infty}\left[\arctan\left(\frac{x}{2}\right)\right]_{b}^{0} + \lim_{c\to\infty}\left[\arctan\left(\frac{x}{2}\right)\right]_{0}^{c}$

$\displaystyle = \left(0 - \left(-\frac{\pi}{2}\right)\right) + \left(\frac{\pi}{2} - 0\right) = \pi$

29. $\displaystyle\int_{0}^{\infty} \frac{1}{e^x + e^{-x}}\,dx = \lim_{b\to\infty}\int_{0}^{b} \frac{e^x}{1+e^{2x}}\,dx$

$\displaystyle = \lim_{b\to\infty}\left[\arctan(e^x)\right]_{0}^{b}$

$\displaystyle = \frac{\pi}{2} - \frac{\pi}{4} = \frac{\pi}{4}$

31. $\displaystyle\int_{0}^{\infty} \cos \pi x\,dx = \lim_{b\to\infty}\left[\frac{1}{\pi}\sin \pi x\right]_{0}^{b}$

Diverges since $\sin \pi b$ does not approach a limit as $b\to\infty$.

33. $\displaystyle\int_{0}^{1} \frac{1}{x^2}\,dx = \lim_{b\to0^+}\left[\frac{-1}{x}\right]_{b}^{1} = \lim_{b\to0^+}\left[-1 + \frac{1}{b}\right] = -1 + \infty$

Diverges

35. $\displaystyle\int_{0}^{8} \frac{1}{\sqrt[3]{8-x}}\,dx = \lim_{b\to8}\int_{0}^{b} \frac{1}{\sqrt[3]{8-x}}\,dx$

$\displaystyle = \lim_{b\to8^-}\left[\frac{-3}{2}(8-x)^{2/3}\right]_{0}^{b} = 6$

37. $\displaystyle\int_{0}^{1} x\ln x\,dx = \lim_{b\to0^+}\left[\frac{x^2}{2}\ln|x| - \frac{x^2}{4}\right]_{b}^{1}$

$\displaystyle = \lim_{b\to0^+}\left[\frac{-1}{4} - \frac{b^2 \ln b}{2} + \frac{b^2}{4}\right] = \frac{-1}{4}$

since $\displaystyle\lim_{b\to0^+}(b^2 \ln b) = 0$ by L'Hôpital's Rule.

39. $\displaystyle\int_{0}^{\pi/2} \tan \theta\,d\theta = \lim_{b\to(\pi/2)^-}\left[\ln|\sec \theta|\right]_{0}^{b} = \infty$

Diverges

41. $\displaystyle\int_{2}^{4} \frac{2}{x\sqrt{x^2-4}}\,dx = \lim_{b\to2^+}\int_{b}^{4} \frac{2}{x\sqrt{x^2-4}}\,dx$

$\displaystyle = \lim_{b\to2^+}\left[\arcsec\left|\frac{x}{2}\right|\right]_{b}^{4}$

$\displaystyle = \lim_{b\to2^+}\left(\arcsec 2 - \arcsec\left(\frac{b}{2}\right)\right)$

$\displaystyle = \frac{\pi}{3} - 0 = \frac{\pi}{3}$

43. $\displaystyle\int_{2}^{4} \frac{1}{\sqrt{x^2-4}} = \lim_{b\to2^+}\left[\ln\left|x + \sqrt{x^2-4}\right|\right]_{b}^{4} = \ln\left(4 + 2\sqrt{3}\right) - \ln 2 = \ln\left(2 + \sqrt{3}\right) \approx 1.317$

45. $\displaystyle\int_{0}^{2} \frac{1}{\sqrt[3]{x-1}}\,dx = \int_{0}^{1} \frac{1}{\sqrt[3]{x-1}}\,dx + \int_{1}^{2} \frac{1}{\sqrt[3]{x-1}}\,dx$

$\displaystyle = \lim_{b\to1^-}\left[\frac{3}{2}(x-1)^{2/3}\right]_{0}^{b} + \lim_{c\to1^+}\left[\frac{3}{2}(x-1)^{2/3}\right]_{c}^{2} = \frac{-3}{2} + \frac{3}{2} = 0$

47. $\displaystyle\int_{0}^{\infty} \frac{4}{\sqrt{x}(x+6)}\,dx = \int_{0}^{1} \frac{4}{\sqrt{x}(x+6)}\,dx + \int_{1}^{\infty} \frac{4}{\sqrt{x}(x+6)}\,dx$

Let $u = \sqrt{x}$, $u^2 = x$, $2u\,du = dx$.

$\displaystyle\int \frac{4}{\sqrt{x}(x+6)}\,dx = \int \frac{4(2u\,du)}{u(u^2+6)} = 8\int \frac{du}{u^2+6} = \frac{8}{\sqrt{6}}\arctan\left(\frac{u}{\sqrt{6}}\right) + C = \frac{8}{\sqrt{6}}\arctan\left(\frac{\sqrt{x}}{\sqrt{6}}\right) + C$

Thus, $\displaystyle\int_{0}^{\infty} \frac{4}{\sqrt{x}(x+6)}\,dx = \lim_{b\to0^+}\left[\frac{8}{\sqrt{6}}\arctan\left(\frac{\sqrt{x}}{\sqrt{6}}\right)\right]_{b}^{1} + \lim_{c\to\infty}\left[\frac{8}{\sqrt{6}}\arctan\left(\frac{\sqrt{x}}{\sqrt{6}}\right)\right]_{1}^{c}$

$\displaystyle = \left(\frac{8}{\sqrt{6}}\arctan\left(\frac{1}{\sqrt{6}}\right) - \frac{8}{\sqrt{6}}0\right) + \left(\frac{8}{\sqrt{6}}\frac{\pi}{2} - \frac{8}{\sqrt{6}}\arctan\left(\frac{1}{\sqrt{6}}\right)\right)$

$\displaystyle = \frac{8\pi}{2\sqrt{6}} = \frac{2\pi\sqrt{6}}{3}.$

49. If $p = 1$, $\displaystyle\int_1^\infty \frac{1}{x}\,dx = \lim_{b \to \infty} \int_1^b \frac{1}{x}\,dx = \lim_{b \to \infty} \ln x \Big]_1^b$

$$= \lim_{b \to \infty} [\ln b] = \infty.$$

Diverges. For $p \neq 1$,

$$\int_1^\infty \frac{1}{x^p}\,dx = \lim_{b \to \infty}\left[\frac{x^{1-p}}{1-p}\right]_1^b = \lim_{b \to \infty}\left[\frac{b^{1-p}}{1-p} - \frac{1}{1-p}\right].$$

This converges to $\dfrac{1}{p-1}$ if $1 - p < 0$ or $p > 1$.

51. For $n = 1$ we have

$$\int_0^\infty xe^{-x}\,dx = \lim_{b \to \infty} \int_0^b xe^{-x}\,dx$$

$$= \lim_{b \to \infty}\left[-e^{-x}x - e^{-x}\right]_0^b \qquad \text{(Parts: } u = x, dv = e^{-x}\,dx\text{)}$$

$$= \lim_{b \to \infty}\left[-e^{-b}b - e^{-b} + 1\right]$$

$$= \lim_{b \to \infty}\left[\frac{-b}{e^b} - \frac{1}{e^b} + 1\right] = 1 \quad \text{(L'Hôpital's Rule)}.$$

Assume that $\displaystyle\int_0^\infty x^n e^{-x}\,dx$ converges. Then for $n + 1$ we have

$$\int x^{n+1} e^{-x}\,dx = -x^{n+1}e^{-x} + (n+1)\int x^n e^{-x}\,dx$$

by parts ($u = x^{n+1}$, $du = (n+1)x^n\,dx$, $dv = e^{-x}\,dx$, $v = -e^{-x}$).

Thus,

$$\int_0^\infty x^{n+1} e^{-x}\,dx = \lim_{b \to \infty}\left[-x^{n+1}e^{-x}\right]_0^b + (n+1)\int_0^\infty x^n e^{-x}\,dx = 0 + (n+1)\int_0^\infty x^n e^{-x}\,dx, \text{ which converges.}$$

53. $\displaystyle\int_0^1 \frac{1}{x^3}\,dx$ diverges.

(See Exercise 50, $p = 3 \not< 1$.)

55. $\displaystyle\int_1^\infty \frac{1}{x^3}\,dx = \frac{1}{3-1} = \frac{1}{2}$ converges.

(See Exercise 49, $p = 3$.)

57. Since $\dfrac{1}{x^2 + 5} \leq \dfrac{1}{x^2}$ on $[1, \infty)$ and $\displaystyle\int_1^\infty \frac{1}{x^2}\,dx$ converges by Exercise 49, $\displaystyle\int_1^\infty \frac{1}{x^2 + 5}\,dx$ converges.

59. Since $\dfrac{1}{\sqrt[3]{x(x-1)}} \geq \dfrac{1}{\sqrt[3]{x^2}}$ on $[2, \infty)$ and $\displaystyle\int_2^\infty \frac{1}{\sqrt[3]{x^2}}\,dx$ diverges by Exercise 49, $\displaystyle\int_2^\infty \frac{1}{\sqrt[3]{x(x-1)}}\,dx$ diverges.

61. Since $e^{-x^2} \leq e^{-x}$ on $[1, \infty)$ and $\displaystyle\int_0^\infty e^{-x}\,dx$ converges (see Exercise 9), $\displaystyle\int_0^\infty e^{-x^2}\,dx$ converges.

63. Answers will vary.

65. $\displaystyle\int_{-1}^1 \frac{1}{x^3}\,dx = \int_{-1}^0 \frac{1}{x^3}\,dx + \int_0^1 \frac{1}{x^3}\,dx$

These two integrals diverge by Exercise 50.

67. $A = \int_{-\infty}^{1} e^x \, dx$

$= \lim_{b \to -\infty} \int_{b}^{1} e^x \, dx$

$= \lim_{b \to -\infty} \left[e^x \right]_{b}^{1}$

$= \lim_{b \to -\infty} [e - e^b] = e$

69. $A = \int_{-\infty}^{\infty} \frac{1}{x^2 + 1} \, dx$

$= \lim_{b \to -\infty} \int_{b}^{0} \frac{1}{x^2 + 1} \, dx + \lim_{b \to \infty} \int_{0}^{b} \frac{1}{x^2 + 1} \, dx$

$= \lim_{b \to -\infty} \left[\arctan(x) \right]_{b}^{0} + \lim_{b \to \infty} \left[\arctan(x) \right]_{0}^{b}$

$= \lim_{b \to -\infty} [0 - \arctan(b)] + \lim_{b \to \infty} [\arctan(b) - 0]$

$= -\left(-\frac{\pi}{2}\right) + \frac{\pi}{2} = \pi$

71. (a) $A = \int_{0}^{\infty} e^{-x} \, dx$

$= \lim_{b \to \infty} \left[-e^{-x} \right]_{0}^{b} = 0 - (-1) = 1$

(b) Disk:

$V = \pi \int_{0}^{\infty} (e^{-x})^2 \, dx$

$- \lim_{b \to \infty} \pi \left[-\frac{1}{2} e^{-2x} \right]_{0}^{b} = \frac{\pi}{2}$

(c) Shell:

$V = 2\pi \int_{0}^{\infty} x e^{-x} \, dx$

$= \lim_{b \to \infty} \left\{ 2\pi \left[-e^{-x}(x + 1) \right]_{0}^{b} \right\} = 2\pi$

73. $x^{2/3} + y^{2/3} = 4$

$\frac{2}{3} x^{-1/3} + \frac{2}{3} y^{-1/3} y' = 0$

$y' = \frac{-y^{1/3}}{x^{1/3}}$

$\sqrt{1 + (y')^2} = \sqrt{1 + \frac{y^{2/3}}{x^{2/3}}} = \sqrt{\frac{x^{2/3} + y^{2/3}}{x^{2/3}}} = \sqrt{\frac{4}{x^{2/3}}} = \frac{2}{x^{1/3}}, \quad (x > 0)$

$s = 4 \int_{0}^{8} \frac{2}{x^{1/3}} \, dx = \lim_{b \to 0^+} \left[8 \cdot \frac{3}{2} x^{2/3} \right]_{b}^{8} = 48$

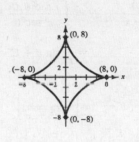

75. $(x - 2)^2 + y^2 = 1$

$2(x - 2) + 2yy' = 0$

$y' = \frac{-(x - 2)}{y}$

$\sqrt{1 + (y')^2} = \sqrt{1 + [(x - 2)^2/y^2]} = \frac{1}{y}$ (Assume $y > 0$.)

$S = 4\pi \int_{1}^{3} \frac{x}{y} \, dx = 4\pi \int_{1}^{3} \frac{x}{\sqrt{1 - (x - 2)^2}} \, dx = 4\pi \int_{1}^{3} \left[\frac{x - 2}{\sqrt{1 - (x - 2)^2}} + \frac{2}{\sqrt{1 - (x - 2)^2}} \right] dx$

$= \lim_{\substack{a \to 1^+ \\ b \to 3^-}} \left\{ 4\pi \left[-\sqrt{1 - (x - 2)^2} + 2 \arcsin(x - 2) \right]_{a}^{b} \right\} = 4\pi[0 + 2 \arcsin(1) - 2 \arcsin(-1)] = 8\pi^2$

77. (a) $F(x) = \frac{K}{x^2}, 5 = \frac{K}{(4000)^2}, K = 80,000,000$

$W = \int_{4000}^{\infty} \frac{80,000,000}{x^2} \, dx = \lim_{b \to \infty} \left[\frac{-80,000,000}{x} \right]_{4000}^{b} = 20,000 \text{ mi-ton}$

(b) $\frac{W}{2} = 10,000 = \left[\frac{-80,000,000}{x} \right]_{4000}^{b} = \frac{-80,000,000}{b} + 20,000$

$\frac{80,000,000}{b} = 10,000$

$b = 8000$

Therefore, 4000 miles *above* the earth's surface.

79. (a) $\displaystyle\int_{-\infty}^{\infty}\frac{1}{7}e^{-t/7}\,dt = \int_{0}^{\infty}\frac{1}{7}e^{-t/7}\,dt = \lim_{b\to\infty}\left[-e^{-t/7}\right]_{0}^{b} = 1$

(c) $\displaystyle\int_{0}^{\infty}t\left[\frac{1}{7}e^{-t/7}\right]dt = \lim_{b\to\infty}\left[-te^{-t/7}-7e^{-t/7}\right]_{0}^{b}$

(b) $\displaystyle\int_{0}^{4}\frac{1}{7}e^{-t/7}\,dt = \left[-e^{-t/7}\right]_{0}^{4} = -e^{-4/7}+1$

$$= 0 + 7 = 7$$

$$\approx 0.4353 = 43.53\%$$

81. (a) $\displaystyle C = 650{,}000 + \int_{0}^{5}25{,}000\,e^{-0.06t}\,dt = 650{,}000 - \left[\frac{25{,}000}{0.06}e^{-0.06t}\right]_{0}^{5} \approx \$757{,}992.41$

(b) $\displaystyle C = 650{,}000 + \int_{0}^{10}25{,}000e^{-0.06t}\,dt \approx \$837{,}995.15$

(c) $\displaystyle C = 650{,}000 + \int_{0}^{\infty}25{,}000e^{-0.06t}\,dt = 650{,}000 - \lim_{b\to\infty}\left[\frac{25{,}000}{0.06}e^{-0.06t}\right]_{0}^{b} \approx \$1{,}066{,}666.67$

83. Let $K = \dfrac{2\pi NI\,r}{k}$. Then

$$P = K\int_{c}^{\infty}\frac{1}{(r^2+x^2)^{3/2}}\,dx.$$

Let $x = r\tan\theta,\ dx = r\sec^2\theta\,d\theta,\ \sqrt{r^2+x^2} = r\sec\theta.$

$$\int\frac{1}{(r^2+x^2)^{3/2}}\,dx = \int\frac{r\sec^2\theta\,d\theta}{r^3\sec^3\theta} = \frac{1}{r^2}\int\cos\theta\,d\theta$$

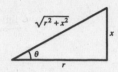

$$= \frac{1}{r^2}\sin\theta + C = \frac{1}{r^2}\frac{x}{\sqrt{r^2+x^2}} + C$$

Hence,

$$P = K\frac{1}{r^2}\lim_{b\to\infty}\left[\frac{x}{\sqrt{r^2+x^2}}\right]_{c}^{b}$$

$$= \frac{K}{r^2}\left[1 - \frac{c}{\sqrt{r^2+c^2}}\right]$$

$$= \frac{K\left(\sqrt{r^2+c^2} - c\right)}{r^2\sqrt{r^2+c^2}}$$

$$= \frac{2\pi NI\left(\sqrt{r^2+c^2} - c\right)}{kr\sqrt{r^2+c^2}}.$$

85. False. $f(x) = 1/(x+1)$ is continuous on

$[0, \infty),\ \displaystyle\lim_{x\to\infty}1/(x+1) = 0$, but

$$\int_{0}^{\infty}\frac{1}{x+1}\,dx = \lim_{b\to\infty}\left[\ln|x+1|\right]_{0}^{b} = \infty.$$

Diverges

87. True

89. (a) $\displaystyle\int_{1}^{\infty}\frac{1}{x}\,dx = \lim_{b\to\infty}\left[\ln|x|\right]_{1}^{b} = \infty$

$$\int_{1}^{\infty}\frac{1}{x^2}\,dx = \lim_{b\to\infty}\left[-\frac{1}{x}\right]_{1}^{b} = 1$$

$\displaystyle\int_{1}^{\infty}\frac{1}{x^n}\,dx$ will converge if $n > 1$

and will diverge if $n \le 1$.

(b) It would appear to converge.

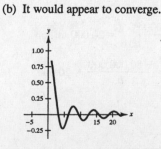

(c) Let $dv = \sin x\,dx \implies v = -\cos x$

$$u = \frac{1}{x} \implies du = -\frac{1}{x^2}\,dx.$$

$$\int_{1}^{\infty}\frac{\sin x}{x}\,dx = \lim_{b\to 0}\left[-\frac{\cos x}{x}\right]_{1}^{b} - \int_{1}^{\infty}\frac{\cos x}{x^2}\,dx$$

$$= \cos 1 - \int_{1}^{\infty}\frac{\cos x}{x^2}\,dx$$

Converges

91. $\Gamma(n) = \int_0^\infty x^{n-1} e^{-x} \, dx$

(a) $\Gamma(1) = \int_0^\infty e^{-x} \, dx = \lim_{b \to \infty} \left[-e^{-x} \right]_0^b = 1$

$\Gamma(2) = \int_0^\infty x e^{-x} \, dx = \lim_{b \to \infty} \left[-e^{-x}(x + 1) \right]_0^b = 1$

$\Gamma(3) = \int_0^\infty x^2 e^{-x} \, dx = \lim_{b \to \infty} \left[-x^2 e^{-x} - 2x e^{-x} - 2e^{-x} \right]_0^b = 2$

(b) $\Gamma(n + 1) = \int_0^\infty x^n e^{-x} \, dx = \lim_{b \to \infty} \left[-x^n e^{-x} \right]_0^b + \lim_{b \to \infty} n \int_0^b x^{n-1} e^{-x} \, dx = 0 + n\Gamma(n)$ $(u = x^n, dv = e^{-x} \, dx)$

(c) $\Gamma(n) = (n - 1)!$

93. $f(t) = 1$

$F(s) = \int_0^\infty e^{-st} \, dt - \lim_{b \to \infty} \left[-\frac{1}{s} e^{-st} \right]_0^b = \frac{1}{s}, \, s > 0$

95. $f(t) = t^2$

$F(s) = \int_0^\infty t^2 e^{-st} \, dt = \lim_{b \to \infty} \left[\frac{1}{s^3} (-s^2 t^2 - 2st - 2) e^{-st} \right]_0^b$

$= \frac{2}{s^3}, \, s > 0$

97. $f(t) = \cos at$

$F(s) = \int_0^\infty e^{-st} \cos at \, dt$

$= \lim_{b \to \infty} \left[\frac{e^{-st}}{s^2 + a^2} (-s \cos at + a \sin at) \right]_0^b$

$= 0 + \frac{s}{s^2 + a^2} = \frac{s}{s^2 + a^2}, \, s > 0$

99. $f(t) = \cosh at$

$F(s) = \int_0^\infty e^{-st} \cosh at \, dt = \int_0^\infty e^{-st} \left(\frac{e^{at} + e^{-at}}{2} \right) dt = \frac{1}{2} \int_0^\infty \left[e^{t(-s+a)} + e^{t(-s-a)} \right] dt$

$= \lim_{b \to \infty} \frac{1}{2} \left[\frac{1}{(-s + a)} e^{t(-s+a)} + \frac{1}{(-s - a)} e^{t(-s-a)} \right]_0^b = 0 - \frac{1}{2} \left[\frac{1}{(-s + a)} + \frac{1}{(-s - a)} \right]$

$= \frac{-1}{2} \left[\frac{1}{(-s + a)} + \frac{1}{(-s - a)} \right] = \frac{s}{s^2 - a^2}, \, s > |a|$

101. (a) $f(x) = \frac{1}{3\sqrt{2\pi}} e^{-(x-70)^2/18}$

$\int_{50}^{90} f(x) \, dx \approx 1.0$

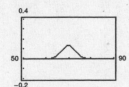

(b) $P(72 \le x < \infty) \approx 0.2525$

(c) $0.5 - P(70 \le x \le 72) \approx 0.5 - 0.2475 = 0.2525$

These are the same answers because by symmetry,

$P(70 \le x < \infty) = 0.5$

and

$0.5 = P(70 \le x < \infty)$

$= P(70 \le x \le 72) + P(72 \le x < \infty).$

103. $\displaystyle\int_0^\infty \left(\frac{1}{\sqrt{x^2+1}} - \frac{c}{x+1}\right) dx = \lim_{b\to\infty}\int_0^b \left(\frac{1}{\sqrt{x^2+1}} - \frac{c}{x+1}\right) dx$

$$= \lim_{b\to\infty}\left[\ln\left|x + \sqrt{x^2+1}\right| - c\ln|x+1|\right]_0^b$$

$$= \lim_{b\to\infty}\left[\ln\left(b + \sqrt{b^2+1}\right) - \ln(b+1)^c\right]$$

$$= \lim_{b\to\infty}\ln\left[\frac{b + \sqrt{b^2+1}}{(b+1)^c}\right]$$

This limit exists for $c = 1$, and you have

$$\lim_{b\to\infty}\ln\left[\frac{b + \sqrt{b^2+1}}{(b+1)}\right] = \ln 2.$$

105. $f(x) = \begin{cases} x\ln x, & 0 < x \le 2 \\ 0, & x = 0 \end{cases}$

$V = \pi\displaystyle\int_0^2 (x\ln x)^2\, dx$

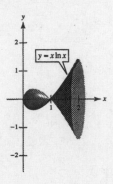

Let $u = \ln x$, $e^u = x$, $e^u\, du = dx$.

$V = \pi\displaystyle\int_{-\infty}^{\ln 2} e^{2u} u^2 (e^u\, du) = \pi\int_{-\infty}^{\ln 2} e^{3u} u^2\, du$

$$= \lim_{b\to\infty}\left[\pi\left[\frac{u^2}{3} - \frac{2u}{9} + \frac{2}{27}\right]e^{3u}\right]_b^{\ln 2}$$

$$= \pi\left[\frac{(\ln 2)^2}{3} - \frac{2\ln 2}{9} + \frac{2}{27}\right]8 \approx 2.0155$$

107. $u = \sqrt{x}$, $u^2 = x$, $2u\, du = dx$

$$\int_0^1 \frac{\sin x}{\sqrt{x}}\, dx = \int_0^1 \frac{\sin(u^2)}{u}(2u\, du) = \int_0^1 2\sin(u^2)\, du$$

Trapezoidal Rule ($n = 5$): 0.6278

109. (a)

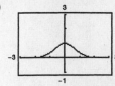

(b) Let $y = e^{-x^2}$, $0 \le x < \infty$.

$\ln y = -x^2$

$x = \sqrt{-\ln y}$ for $0 < y \le 1$

The area bounded by $y = e^{-x^2}$, $x = 0$ and $y = 0$ is

$$\int_0^\infty e^{-x^2}\, dx = \int_0^1 \sqrt{-\ln y}\, dy, \quad \left(= \frac{\sqrt{\pi}}{2}\right).$$

Review Exercises for Chapter 8

1. $\displaystyle\int x\sqrt{x^2-1}\,dx = \frac{1}{2}\int (x^2-1)^{1/2}(2x)\,dx$

$$= \frac{1}{2}\frac{(x^2-1)^{3/2}}{3/2}+C$$

$$= \frac{1}{3}(x^2-1)^{3/2}+C$$

3. $\displaystyle\int \frac{x}{x^2-1}\,dx = \frac{1}{2}\int \frac{2x}{x^2-1}\,dx$

$$= \frac{1}{2}\ln|x^2-1|+C$$

5. Let $u = \ln(2x)$, $du = \dfrac{1}{x}dx$.

$$\int_1^e \frac{\ln(2x)}{x}\,dx = \int_{\ln 2}^{1+\ln 2} u\,du$$

$$= \frac{u^2}{2}\Big]_{\ln 2}^{1+\ln 2}$$

$$= \frac{1}{2}\left[1 + 2\ln 2 + (\ln 2)^2 - (\ln 2)^2\right]$$

$$= \frac{1}{2} + \ln 2 \approx 1.1931$$

7. $\displaystyle\int \frac{16}{\sqrt{16-x^2}}\,dx = 16\arcsin\left(\frac{x}{4}\right)+C$

9. $\displaystyle\int e^{2x}\sin 3x\,dx = -\frac{1}{3}e^{2x}\cos 3x + \frac{2}{3}\int e^{2x}\cos 3x\,dx$

$$= -\frac{1}{3}e^{2x}\cos 3x + \frac{2}{3}\left(\frac{1}{3}e^{2x}\sin 3x - \frac{2}{3}\int e^{2x}\sin 3x\,dx\right)$$

$$\frac{13}{9}\int e^{2x}\sin 3x\,dx = -\frac{1}{3}e^{2x}\cos 3x + \frac{2}{9}e^{2x}\sin 3x$$

$$\int e^{2x}\sin 3x\,dx = \frac{e^{2x}}{13}(2\sin 3x - 3\cos 3x)+C$$

(1) $dv = \sin 3x\,dx \implies v = -\dfrac{1}{3}\cos 3x$

$\quad u = e^{2x} \qquad \implies du = 2e^{2x}\,dx$

(2) $dv = \cos 3x\,dx \implies v = \dfrac{1}{3}\sin 3x$

$\quad u = e^{2x} \qquad \implies du = 2e^{2x}\,dx$

11. $u = x$, $du = dx$, $dv = (x-5)^{1/2}\,dx$, $v = \dfrac{2}{3}(x-5)^{3/2}$

$$\int x\sqrt{x-5}\,dx = \frac{2}{3}x(x-5)^{3/2} - \int \frac{2}{3}(x-5)^{3/2}\,dx$$

$$= \frac{2}{3}x(x-5)^{3/2} - \frac{4}{15}(x-5)^{5/2}+C$$

$$= (x-5)^{3/2}\left[\frac{2}{3}x - \frac{4}{15}(x-5)\right]+C$$

$$= (x-5)^{3/2}\left[\frac{6}{15}x + \frac{4}{3}\right]+C$$

$$= \frac{2}{15}(x-5)^{3/2}[3x+10]+C$$

13. $\displaystyle\int x^2\sin 2x\,dx = -\frac{1}{2}x^2\cos 2x + \int x\cos 2x\,dx$

$$= -\frac{1}{2}x^2\cos 2x + \frac{1}{2}x\sin 2x - \frac{1}{2}\int \sin 2x\,dx$$

$$= -\frac{1}{2}x^2\cos 2x + \frac{x}{2}\sin 2x + \frac{1}{4}\cos 2x + C$$

(1) $dv = \sin 2x\,dx \implies v = -\dfrac{1}{2}\cos 2x$

$\quad u = x^2 \qquad \implies du = 2x\,dx$

(2) $dv = \cos 2x\,dx \implies v = \dfrac{1}{2}\sin 2x$

$\quad u = x \qquad \implies du = dx$

15. $\displaystyle\int x \arcsin 2x \, dx = \frac{x^2}{2} \arcsin 2x - \int \frac{x^2}{\sqrt{1-4x^2}} \, dx$

$\displaystyle = \frac{x^2}{2} \arcsin 2x - \frac{1}{8} \int \frac{2(2x)^2}{\sqrt{1-(2x)^2}} \, dx$

$\displaystyle = \frac{x^2}{2} \arcsin 2x - \frac{1}{8}\left(\frac{1}{2}\right)\left[-(2x)\sqrt{1-4x^2} + \arcsin 2x\right] + C \quad \text{(by Formula 43 of Integration Tables)}$

$\displaystyle = \frac{1}{16}\left[(8x^2-1)\arcsin 2x + 2x\sqrt{1-4x^2}\right] + C$

$dv = x \, dx \quad \Longrightarrow \quad v = \dfrac{x^2}{2}$

$u = \arcsin 2x \Longrightarrow du = \dfrac{2}{\sqrt{1-4x^2}} \, dx$

17. $\displaystyle\int \cos^3(\pi x - 1) \, dx = \int [1 - \sin^2(\pi x - 1)]\cos(\pi x - 1) \, dx$

$\displaystyle = \frac{1}{\pi}\left[\sin(\pi x - 1) - \frac{1}{3}\sin^3(\pi x - 1)\right] + C$

$\displaystyle = \frac{1}{3\pi}\sin(\pi x - 1)[3 - \sin^2(\pi x - 1)] + C$

$\displaystyle = \frac{1}{3\pi}\sin(\pi x - 1)[3 - (1 - \cos^2(\pi x - 1))] + C$

$\displaystyle = \frac{1}{3\pi}\sin(\pi x - 1)[2 + \cos^2(\pi x - 1)] + C$

19. $\displaystyle\int \sec^4\!\left(\frac{x}{2}\right) dx = \int \left[\tan^2\!\left(\frac{x}{2}\right) + 1\right]\sec^2\!\left(\frac{x}{2}\right) dx$

$\displaystyle = \int \tan^2\!\left(\frac{x}{2}\right)\sec^2\!\left(\frac{x}{2}\right) dx + \int \sec^2\!\left(\frac{x}{2}\right) dx$

$\displaystyle = \frac{2}{3}\tan^3\!\left(\frac{x}{2}\right) + 2\tan\!\left(\frac{x}{2}\right) + C = \frac{2}{3}\left[\tan^3\!\left(\frac{x}{2}\right) + 3\tan\!\left(\frac{x}{2}\right)\right] + C$

21. $\displaystyle\int \frac{1}{1-\sin\theta} \, d\theta = \int \frac{1}{1-\sin\theta} \cdot \frac{1+\sin\theta}{1+\sin\theta} \, d\theta = \int \frac{1+\sin\theta}{\cos^2\theta} \, d\theta = \int (\sec^2\theta + \sec\theta\tan\theta) \, d\theta = \tan\theta + \sec\theta + C$

23. $\displaystyle A = \int_{\pi/4}^{3\pi/4} \sin^4 x \, dx.$ Using the Table of Integrals,

$\displaystyle\int \sin^4 x \, dx = -\frac{\sin^3 x \cos x}{4} + \frac{3}{4}\int \sin^2 x \, dx$

$\displaystyle = \frac{-\sin^3 x \cos x}{4} + \frac{3}{4}\left[\frac{1}{2}(x - \sin x \cos x)\right] + C$

$\displaystyle\int_{\pi/4}^{3\pi/4} \sin^4 x \, dx = \left[\frac{-\sin^3 x \cos x}{4} + \frac{3}{8}x - \frac{3}{8}\sin x \cos x\right]_{\pi/4}^{3\pi/4}$

$\displaystyle = \left(\frac{1}{16} + \frac{9\pi}{32} + \frac{3}{16}\right) - \left(\frac{-1}{16} + \frac{3\pi}{32} - \frac{3}{16}\right)$

$\displaystyle = \frac{3\pi}{16} + \frac{1}{2} \approx 1.0890$

25. $\displaystyle\int \frac{-12}{x^2\sqrt{4-x^2}} \, dx = \int \frac{-24\cos\theta \, d\theta}{(4\sin^2\theta)(2\cos\theta)}$

$\displaystyle = -3\int \csc^2\theta \, d\theta$

$\displaystyle = 3\cot\theta + C$

$\displaystyle = \frac{3\sqrt{4-x^2}}{x} + C$

$x = 2\sin\theta, \, dx = 2\cos\theta \, d\theta, \, \sqrt{4-x^2} = 2\cos\theta$

27.　$x = 2 \tan \theta$

$dx = 2 \sec^2 \theta \, d\theta$

$4 + x^2 = 4 \sec^2 \theta$

$$\int \frac{x^3}{\sqrt{4 + x^2}} \, dx = \int \frac{8 \tan^3 \theta}{2 \sec \theta} 2 \sec^2 \theta \, d\theta$$

$$= 8 \int \tan^3 \theta \sec \theta \, d\theta$$

$$= 8 \int (\sec^2 \theta - 1) \tan \theta \sec \theta \, d\theta$$

$$= 8 \left[\frac{\sec^3 \theta}{3} - \sec \theta \right] + C$$

$$= 8 \left[\frac{(x^2 + 4)^{3/2}}{24} - \frac{\sqrt{x^2 + 4}}{2} \right] + C$$

$$= \sqrt{x^2 + 4} \left[\frac{1}{3}(x^2 + 4) - 4 \right] + C$$

$$= \frac{1}{3} x^2 \sqrt{x^2 + 4} - \frac{8}{3} \sqrt{x^2 + 4} + C$$

$$= \frac{1}{3} (x^2 + 4)^{1/2} (x^2 - 8) + C$$

29. $\int_{-2}^{0} \sqrt{4 - x^2} \, dx = \frac{1}{2} \left[4 \arcsin\left(\frac{x}{2}\right) + x\sqrt{4 - x^2} \right]_{-2}^{0}$

$$= \frac{1}{2} [0 - 4 \arcsin(-1)]$$

$$= \frac{1}{2} \left[-4\left(\frac{-\pi}{2}\right) \right]$$

$$= \pi$$

Note: The integral represents the area of a quarter circle of radius 2: $A = \frac{1}{4}(\pi 2^2) = \pi$.

31. (a) Let $x = 2 \tan \theta, dx = 2 \sec^2 \theta \, d\theta$.

$$\int \frac{x^3}{\sqrt{4 + x^2}} \, dx = \int \frac{8 \tan^3 \theta}{2 \sec \theta} 2 \sec^2 \theta \, d\theta$$

$$= 8 \int \tan^3 \theta \sec \theta \, d\theta$$

$$= 8 \int \frac{\sin^3 \theta}{\cos^4 \theta} \, d\theta$$

$$= 8 \int (1 - \cos^2 \theta) \cos^{-4} \theta \sin \theta \, d\theta$$

$$= 8 \int (\cos^{-4} \theta - \cos^{-2} \theta) \sin \theta \, d\theta$$

$$= 8 \left[\frac{\cos^{-3} \theta}{3} - \frac{\cos^{-1} \theta}{-1} \right] + C$$

$$= \frac{8}{3} \sec \theta (\sec^2 \theta - 3) + C$$

$$= \frac{8}{3} \frac{\sqrt{4 + x^2}}{2} \left(\frac{4 + x^2}{4} - 3 \right) + C$$

$$= \frac{1}{3} \sqrt{4 + x^2} \, (x^2 - 8) + C$$

(b) $\int \frac{x^3}{\sqrt{4 + x^2}} \, dx = \int \frac{x^2}{\sqrt{4 + x^2}} x \, dx$

$$= \int \frac{(u^2 - 4)u \, du}{u}$$

$$= \int (u^2 - 4) \, du$$

$$= \frac{1}{3} u^3 - 4u + C$$

$$= \frac{u}{3} (u^2 - 12) + C$$

$$= \frac{\sqrt{4 + x^2}}{3} (x^2 - 8) + C$$

$u^2 = 4 + x^2, 2u \, du = 2x \, dx$

(c) $\int \frac{x^3}{\sqrt{4 + x^2}} \, dx = x^2 \sqrt{4 + x^2} - \int 2x \sqrt{4 + x^2} \, dx$

$$= x^2 \sqrt{4 + x^2} - \frac{2}{3}(4 + x^2)^{3/2} + C = \frac{\sqrt{4 + x^2}}{3} (x^2 - 8) + C$$

$dv = \frac{x}{\sqrt{4 + x^2}} \, dx \implies v = \sqrt{4 + x^2}$

$u = x^2 \qquad\qquad \implies du = 2x \, dx$

33. $\dfrac{x - 28}{x^2 - x - 6} = \dfrac{A}{x - 3} + \dfrac{B}{x + 2}$

$\qquad x - 28 = A(x + 2) + B(x - 3)$

$\qquad x = -2 \implies -30 = B(-5) \implies B = 6$

$\qquad x = 3 \quad \implies -25 = A(5) \quad \implies A = -5$

$\qquad \displaystyle\int \dfrac{x - 28}{x^2 - x - 6}\, dx = \int \left(\dfrac{-5}{x - 3} + \dfrac{6}{x + 2} \right) dx = -5\ln|x - 3| + 6\ln|x + 2| + C$

35. $\dfrac{x^2 + 2x}{(x - 1)(x^2 + 1)} = \dfrac{A}{x - 1} + \dfrac{Bx + C}{x^2 + 1}$

$\qquad x^2 + 2x = A(x^2 + 1) + (Bx + C)(x - 1)$

$\text{Let } x = 1:\ 3 = 2A \implies A = \dfrac{3}{2} \quad \text{Let } x = 0:\ 0 = A - C \implies C = \dfrac{3}{2} \quad \text{Let } x = 2:\ 8 = 5A + 2B + C \implies B = -\dfrac{1}{2}$

$\qquad \displaystyle\int \dfrac{x^2 + 2x}{x^3 - x^2 + x - 1}\, dx = \dfrac{3}{2} \int \dfrac{1}{x - 1}\, dx - \dfrac{1}{2} \int \dfrac{x - 3}{x^2 + 1}\, dx$

$\qquad\qquad = \dfrac{3}{2} \int \dfrac{1}{x - 1}\, dx - \dfrac{1}{4} \int \dfrac{2x}{x^2 + 1}\, dx + \dfrac{3}{2} \int \dfrac{1}{x^2 + 1}\, dx$

$\qquad\qquad = \dfrac{3}{2} \ln|x - 1| - \dfrac{1}{4} \ln|x^2 + 1| + \dfrac{3}{2} \arctan x + C$

$\qquad\qquad = \dfrac{1}{4}[6 \ln|x - 1| - \ln(x^2 + 1) + 6 \arctan x] + C$

37. $\dfrac{x^2}{x^2 + 2x - 15} = 1 + \dfrac{15 - 2x}{x^2 + 2x - 15}$

$\qquad \dfrac{15 - 2x}{(x - 3)(x + 5)} = \dfrac{A}{x - 3} + \dfrac{B}{x + 5}$

$\qquad 15 - 2x = A(x + 5) + B(x - 3)$

$\text{Let } x = 3:\quad 9 = 8A \implies A = \dfrac{9}{8}$

$\text{Let } x = -5:\ 25 = -8B \implies B = -\dfrac{25}{8}$

$\qquad \displaystyle\int \dfrac{x^2}{x^2 + 2x - 15}\, dx = \int dx + \dfrac{9}{8} \int \dfrac{1}{x - 3}\, dx - \dfrac{25}{8} \int \dfrac{1}{x + 5}\, dx = x + \dfrac{9}{8} \ln|x - 3| - \dfrac{25}{8} \ln|x + 5| + C$

39. $\displaystyle\int \dfrac{x}{(2 + 3x)^2}\, dx = \dfrac{1}{9}\left[\dfrac{2}{2 + 3x} + \ln|2 + 3x| \right] + C$

(Formula 4)

41. Let $u = x^2$, $du = 2x\, dx$.

$\displaystyle\int_0^{\sqrt{\pi/2}} \dfrac{x}{1 + \sin x^2}\, dx = \dfrac{1}{2} \int_0^{\pi/4} \dfrac{1}{1 + \sin u}\, du$

$\qquad\qquad = \dfrac{1}{2}\left[\tan u - \sec u \right]_0^{\pi/4}$

$\qquad\qquad = \dfrac{1}{2}\left[(1 - \sqrt{2}) - (0 - 1) \right]$

$\qquad\qquad = 1 - \dfrac{\sqrt{2}}{2}$

43. $\displaystyle\int \frac{x}{x^2 + 4x + 8}\, dx = \frac{1}{2}\left[\ln|x^2 + 4x + 8| - 4\int \frac{1}{x^2 + 4x + 8}\, dx\right]$ (Formula 15)

$\displaystyle = \frac{1}{2}\left[\ln|x^2 + 4x + 8|\right] - 2\left[\frac{2}{\sqrt{32 - 16}}\arctan\left(\frac{2x + 4}{\sqrt{32 - 16}}\right)\right] + C$ (Formula 14)

$\displaystyle = \frac{1}{2}\ln|x^2 + 4x + 8| - \arctan\left(1 + \frac{x}{2}\right) + C$

45. $\displaystyle\int \frac{1}{\sin \pi x \cos \pi x}\, dx = \frac{1}{\pi}\int \frac{1}{\sin \pi x \cos \pi x}(\pi)\, dx \quad (u = \pi x)$

$\displaystyle = \frac{1}{\pi}\ln|\tan \pi x| + C$ (Formula 58)

47. $dv = dx \quad \Rightarrow \quad v = x$

$\displaystyle u = (\ln x)^n \Rightarrow du = n(\ln x)^{n-1}\frac{1}{x}dx$

$\displaystyle \int (\ln x)^n\, dx = x(\ln x)^n - n\int (\ln x)^{n-1}\, dx$

49. $\displaystyle\int \theta \sin \theta \cos \theta\, d\theta = \frac{1}{2}\int \theta \sin 2\theta\, d\theta$

$\displaystyle = -\frac{1}{4}\theta \cos 2\theta + \frac{1}{4}\int \cos 2\theta\, d\theta = -\frac{1}{4}\theta \cos 2\theta + \frac{1}{8}\sin 2\theta + C = \frac{1}{8}(\sin 2\theta - 2\theta \cos 2\theta) + C$

$\displaystyle dv = \sin 2\theta\, d\theta \Rightarrow v = -\frac{1}{2}\cos 2\theta$

$u = \theta \qquad\quad \Rightarrow du = d\theta$

51. $\displaystyle\int \frac{x^{1/4}}{1 + x^{1/2}}\, dx = 4\int \frac{u(u^3)}{1 + u^2}\, du$

$\displaystyle = 4\int \left(u^2 - 1 + \frac{1}{u^2 + 1}\right) du$

$\displaystyle = 4\left(\frac{1}{3}u^3 - u + \arctan u\right) + C$

$\displaystyle = \frac{4}{3}\left[x^{3/4} - 3x^{1/4} + 3\arctan(x^{1/4})\right] + C$

$u = \sqrt[4]{x}, x = u^4, dx = 4u^3\, du$

53. $\displaystyle\int \sqrt{1 + \cos x}\, dx = \int \frac{\sqrt{1 + \cos x}}{1}\cdot\frac{\sqrt{1 - \cos x}}{\sqrt{1 - \cos x}}\, dx$

$\displaystyle = \int \frac{\sin x}{\sqrt{1 - \cos x}}\, dx$

$\displaystyle = \int (1 - \cos x)^{-1/2}(\sin x)\, dx$

$\displaystyle = 2\sqrt{1 - \cos x} + C$

$u = 1 - \cos x, du = \sin x\, dx$

55. $\displaystyle\int \cos x \ln(\sin x)\, dx = \sin x \ln(\sin x) - \int \cos x\, dx$

$\displaystyle = \sin x \ln(\sin x) - \sin x + C$

$dv = \cos x\, dx \Rightarrow v = \sin x$

$\displaystyle u = \ln(\sin x) \Rightarrow du = \frac{\cos x}{\sin x}dx$

57. $\displaystyle y = \int \frac{9}{x^2 - 9}\, dx = \frac{3}{2}\ln\left|\frac{x - 3}{x + 3}\right| + C$

(by Formula 24 of Integration Tables)

59. $\displaystyle y = \int \ln(x^2 + x)\, dx = x\ln|x^2 + x| - \int \frac{2x^2 + x}{x^2 + x}\, dx$

$\displaystyle = x\ln|x^2 + x| - \int \frac{2x + 1}{x + 1}\, dx$

$\displaystyle = x\ln|x^2 + x| - \int 2\, dx + \int \frac{1}{x + 1}\, dx$

$\displaystyle = x\ln|x^2 + x| - 2x + \ln|x + 1| + C$

$dv = dx \qquad\quad \Rightarrow v = x$

$\displaystyle u = \ln(x^2 + x) \Rightarrow du = \frac{2x + 1}{x^2 + x}dx$

61. $\displaystyle\int_2^{\sqrt{5}} x(x^2 - 4)^{3/2}\, dx = \left[\frac{1}{5}(x^2 - 4)^{5/2}\right]_2^{\sqrt{5}} = \frac{1}{5}$

63. $\int_1^4 \frac{\ln x}{x} dx = \left[\frac{1}{2}(\ln x)^2 \right]_1^4 = \frac{1}{2}(\ln 4)^2 = 2(\ln 2)^2 \approx 0.961$

65. $\int_0^\pi x \sin x \, dx = \left[-x \cos x + \sin x \right]_0^\pi = \pi$

67. $A = \int_0^4 x\sqrt{4-x} \, dx = \int_2^0 (4-u^2)u(-2u) \, du$

$\qquad = \int_2^0 2(u^4 - 4u^2) \, du$

$\qquad = \left[2\left(\frac{u^5}{5} - \frac{4u^3}{3} \right) \right]_2^0 = \frac{128}{15}$

$\qquad u = \sqrt{4-x}, \, x = 4 - u^2, \, dx = -2u \, du$

69. By symmetry, $\bar{x} = 0, A = \frac{1}{2}\pi$.

$\qquad \bar{y} = \frac{2}{\pi}\left(\frac{1}{2} \right)\int_{-1}^1 \left(\sqrt{1-x^2} \right)^2 dx = \frac{1}{\pi}\left[x - \frac{1}{3}x^3 \right]_{-1}^1 = \frac{4}{3\pi}$

$\qquad (\bar{x}, \bar{y}) = \left(0, \frac{4}{3\pi} \right)$

71. $s = \int_0^\pi \sqrt{1 + \cos^2 x} \, dx \approx 3.82$

73. $\lim_{x \to 1} \left[\frac{(\ln x)^2}{x-1} \right] = \lim_{x \to 1} \left[\frac{2(1/x)\ln x}{1} \right] = 0$

75. $\lim_{x \to \infty} \frac{e^{2x}}{x^2} = \lim_{x \to \infty} \frac{2e^{2x}}{2x} = \lim_{x \to \infty} \frac{4e^{2x}}{2} = \infty$

77. $y = \lim_{x \to \infty} (\ln x)^{2/x}$

$\qquad \ln y = \lim_{x \to \infty} \frac{2 \ln(\ln x)}{x} = \lim_{x \to \infty} \left[\frac{2/(x \ln x)}{1} \right] = 0$

$\qquad$ Since $\ln y = 0, y = 1$.

79. $\lim_{n \to \infty} 1000\left(1 + \frac{0.09}{n} \right)^n = 1000 \lim_{n \to \infty} \left(1 + \frac{0.09}{n} \right)^n$

Let $y = \lim_{n \to \infty} \left(1 + \frac{0.09}{n} \right)^n$.

$\ln y = \lim_{n \to \infty} n \ln\left(1 + \frac{0.09}{n} \right) = \lim_{n \to \infty} \frac{\ln\left(1 + \frac{0.09}{n} \right)}{\frac{1}{n}} = \lim_{n \to \infty} \left(\frac{\frac{-0.09/n^2}{1 + (0.09/n)}}{-\frac{1}{n^2}} \right) = \lim_{n \to \infty} \frac{0.09}{1 + \left(\frac{0.09}{n} \right)} = 0.09$

Thus, $\ln y = 0.09 \implies y = e^{0.09}$ and $\lim_{n \to \infty} 1000\left(1 + \frac{0.09}{n} \right)^n = 1000e^{0.09} \approx 1094.17$.

81. $\int_0^{16} \frac{1}{\sqrt[4]{x}} dx = \lim_{b \to 0^+} \left[\frac{4}{3}x^{3/4} \right]_b^{16} = \frac{32}{3}$

Converges

83. $\int_1^\infty x^2 \ln x \, dx = \lim_{b \to \infty} \left[\frac{x^3}{9}(-1 + 3\ln x) \right]_1^b = \infty$

Diverges

85. Let $u = \ln x, du = \frac{1}{x} dx, dv = x^{-2} dx, v = -x^{-1}$.

$\qquad \int \frac{\ln x}{x^2} dx = \frac{-\ln x}{x} + \int \frac{1}{x^2} dx = \frac{-\ln x}{x} - \frac{1}{x} + C$

$\qquad \int_1^\infty \frac{\ln x}{x^2} dx = \lim_{b \to \infty} \left[\frac{-\ln x}{x} - \frac{1}{x} \right]_1^b$

$\qquad = \lim_{b \to \infty} \left(\frac{-\ln b}{b} - \frac{1}{b} \right) - (-1)$

$\qquad = 0 + 1 = 1$

87. $\int_0^{t_0} 500{,}000 e^{-0.05t} \, dt = \left[\frac{500{,}000}{-0.05} e^{-0.05t} \right]_0^{t_0}$

$\qquad = \frac{-500{,}000}{0.05}(e^{-0.05t_0} - 1)$

$\qquad = 10{,}000{,}000(1 - e^{-0.05t_0})$

(a) $t_0 = 20$: $\$6{,}321{,}205.59$

(b) $t_0 \to \infty$: $\$10{,}000{,}000$

89. (a) $P(13 \leq x < \infty) = \frac{1}{0.95\sqrt{2\pi}} \int_{13}^\infty e^{-(x-12.9)^2/2(0.95)^2} \, dx \approx 0.4581$

(b) $P(15 \leq x < \infty) = \frac{1}{0.95\sqrt{2\pi}} \int_{15}^\infty e^{-(x-12.9)^2/2(0.95)^2} \, dx \approx 0.0135$

Problem Solving for Chapter 8

1. (a) $\int_{-1}^{1} (1-x^2)\, dx = \left[x - \frac{x^3}{3}\right]_{-1}^{1} = 2\left(1 - \frac{1}{3}\right) = \frac{4}{3}$

$\int_{-1}^{1} (1-x^2)^2\, dx = \int_{-1}^{1} (1 - 2x^2 + x^4)\, dx = \left[x - \frac{2x^3}{3} + \frac{x^5}{5}\right]_{-1}^{1} = 2\left(1 - \frac{2}{3} + \frac{1}{5}\right) = \frac{16}{15}$

(b) Let $x = \sin u,\, dx = \cos u\, du,\, 1 - x^2 = 1 - \sin^2 u = \cos^2 u$.

$$\int_{-1}^{1} (1-x^2)^n\, dx = \int_{-\pi/2}^{\pi/2} (\cos^2 u)^n \cos u\, du$$

$$= \int_{-\pi/2}^{\pi/2} \cos^{2n+1} u\, du$$

$$= 2\left[\frac{2}{3} \cdot \frac{4}{5} \cdot \frac{6}{7} \cdots \frac{(2n)}{(2n+1)}\right] \qquad \text{(Wallis's Formula)}$$

$$= 2\left[\frac{2^2 \cdot 4^2 \cdot 6^2 \cdots (2n)^2}{2 \cdot 3 \cdot 4 \cdot 5 \cdots (2n)(2n+1)}\right] = \frac{2(2^{2n})(n!)^2}{(2n+1)!} = \frac{2^{2n+1}(n!)^2}{(2n+1)!}$$

3. $\qquad \lim_{x\to\infty} \left(\frac{x+c}{x-c}\right)^x = 9$

$\lim_{x\to\infty} x \ln\left(\frac{x+c}{x-c}\right) = \ln 9$

$\lim_{x\to\infty} \frac{\ln(x+c) - \ln(x-c)}{1/x} = \ln 9$

$\lim_{x\to\infty} \frac{\dfrac{1}{x+c} - \dfrac{1}{x-c}}{-\dfrac{1}{x^2}} = \ln 9$

$\lim_{x\to\infty} \frac{-2c}{(x+c)(x-c)}(-x^2) = \ln 9$

$\lim_{x\to\infty} \left(\frac{2cx^2}{x^2 - c^2}\right) = \ln 9$

$2c = \ln 9$

$2c = 2 \ln 3$

$c = \ln 3$

5. $\sin\theta = \dfrac{PB}{OP} = PB,\ \cos\theta = OB$

$AQ = \widehat{AP} = \theta$

$BR = OR + OB = OR + \cos\theta$

The triangles $\triangle AQR$ and $\triangle BPR$ are similar:

$$\frac{AR}{AQ} = \frac{BR}{BP} \implies \frac{OR+1}{\theta} = \frac{OR + \cos\theta}{\sin\theta}$$

$$\sin\theta\,(OR) + \sin\theta = (OR)\theta + \theta\cos\theta$$

$$OR = \frac{\theta\cos\theta - \sin\theta}{\sin\theta - \theta}$$

$\lim_{\theta\to 0^+} OR = \lim_{\theta\to 0^+} \frac{\theta\cos\theta - \sin\theta}{\sin\theta - \theta}$

$= \lim_{\theta\to 0^+} \frac{-\theta\sin\theta + \cos\theta - \cos\theta}{\cos\theta - 1}$

$= \lim_{\theta\to 0^+} \frac{-\theta\sin\theta}{\cos\theta - 1}$

$= \lim_{\theta\to 0^+} \frac{-\sin\theta - \theta\cos\theta}{-\sin\theta}$

$= \lim_{\theta\to 0^+} \frac{\cos\theta + \cos\theta - \theta\sin\theta}{\cos\theta}$

$= 2$

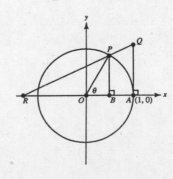

7. (a)

Area ≈ 0.2986

(b) Let $x = 3 \tan \theta, dx = 3 \sec^2 \theta \, d\theta, x^2 + 9 = 9 \sec^2 \theta.$

$$\int \frac{x^2}{(x^2 + 9)^{3/2}} \, dx = \int \frac{9 \tan^2 \theta}{(9 \sec^2 \theta)^{3/2}} (3 \sec^2 \theta \, d\theta)$$

$$= \int \frac{\tan^2 \theta}{\sec \theta} \, d\theta$$

$$= \int \frac{\sin^2 \theta}{\cos \theta} \, d\theta$$

$$= \int \frac{1 - \cos^2 \theta}{\cos \theta} \, d\theta$$

$$= \ln|\sec \theta + \tan \theta| - \sin \theta + C$$

$$\text{Area} = \int_0^4 \frac{x^2}{(x^2 + 9)^{3/2}} \, dx = \left[\ln|\sec \theta + \tan \theta| - \sin \theta \right]_0^{\tan^{-1}(4/3)}$$

$$= \left[\ln\left(\frac{\sqrt{x^2 + 9}}{3} + \frac{x}{3} \right) - \frac{x}{\sqrt{x^2 + 9}} \right]_0^4$$

$$= \ln\left(\frac{5}{3} + \frac{4}{3} \right) - \frac{4}{5} = \ln 3 - \frac{4}{5}$$

(c) $x = 3 \sinh u, dx = 3 \cosh u \, du, x^2 + 9 = 9 \sinh^2 u + 9 = 9 \cosh^2 u$

$$A = \int_0^4 \frac{x^2}{(x^2 + 9)^{3/2}} \, dx = \int_0^{\sinh^{-1}(4/3)} \frac{9 \sinh^2 u}{(9 \cosh^2 u)^{3/2}} (3 \cosh u \, du) = \int_0^{\sinh^{-1}(4/3)} \tanh^2 u \, du$$

$$= \int_0^{\sinh^{-1}(4/3)} (1 - \operatorname{sech}^2 u) \, du = \left[u - \tanh u \right]_0^{\sinh^{-1}(4/3)}$$

$$= \sinh^{-1}\left(\frac{4}{3} \right) - \tanh\left(\sinh^{-1}\left(\frac{4}{3} \right) \right) = \ln\left(\frac{4}{3} + \sqrt{\frac{16}{9} + 1} \right) - \tanh\left[\ln\left(\frac{4}{3} + \sqrt{\frac{16}{9} + 1} \right) \right]$$

$$= \ln\left(\frac{4}{3} + \frac{5}{3} \right) - \tanh\left(\ln\left(\frac{4}{3} + \frac{5}{3} \right) \right) = \ln 3 - \tanh(\ln 3)$$

$$= \ln 3 - \frac{3 - (1/3)}{3 + (1/3)} = \ln 3 - \frac{4}{5}$$

9. $y = \ln(1 - x^2), y' = \dfrac{-2x}{1 - x^2}$

$1 + (y')^2 = 1 + \dfrac{4x^2}{(1 - x^2)^2} = \dfrac{1 - 2x^2 + x^4 + 4x^2}{(1 - x^2)^2} = \left(\dfrac{1 + x^2}{1 - x^2}\right)^2$

$\begin{aligned}
\text{Arc length} &= \int_0^{1/2} \sqrt{1 + (y')^2}\, dx \\[2mm]
&= \int_0^{1/2} \left(\dfrac{1 + x^2}{1 - x^2}\right) dx \\[2mm]
&= \int_0^{1/2} \left(-1 + \dfrac{2}{1 - x^2}\right) dx \\[2mm]
&= \int_0^{1/2} \left(-1 + \dfrac{1}{x + 1} + \dfrac{1}{1 - x}\right) dx \\[2mm]
&= \Big[-x + \ln(1 + x) - \ln(1 - x) \Big]_0^{1/2} \\[2mm]
&= \left(-\dfrac{1}{2} + \ln\dfrac{3}{2} - \ln\dfrac{1}{2}\right) \\[2mm]
&= -\dfrac{1}{2} + \ln 3 - \ln 2 + \ln 2 \\[2mm]
&= \ln 3 - \dfrac{1}{2} \approx 0.5986
\end{aligned}$

11. Consider $\displaystyle\int \dfrac{1}{\ln x}\, dx$.

Let $u = \ln x,\ du = \dfrac{1}{x}\, dx,\ x = e^u$. Then $\displaystyle\int \dfrac{1}{\ln x}\, dx = \int \dfrac{1}{u} e^u\, du = \int \dfrac{e^u}{u}\, du$.

If $\displaystyle\int \dfrac{1}{\ln x}\, dx$ were elementary, then $\displaystyle\int \dfrac{e^u}{u}\, du$ would be too, which is false.

Hence, $\displaystyle\int \dfrac{1}{\ln x}\, dx$ is not elementary.

13. $\begin{aligned}
x^4 + 1 &= (x^2 + ax + b)(x^2 + cx + d) \\
&= x^4 + (a + c)x^3 + (ac + b + d)x^2 + (ad + bc)x + bd
\end{aligned}$

$a = -c, b = d = 1, a = \sqrt{2}$

$x^4 + 1 = \left(x^2 + \sqrt{2}x + 1\right)\left(x^2 - \sqrt{2}x + 1\right)$

$\begin{aligned}
\int_0^1 \dfrac{1}{x^4 + 1}\, dx &= \int_0^1 \dfrac{Ax + B}{x^2 + \sqrt{2}x + 1}\, dx + \int_0^1 \dfrac{Cx + D}{x^2 - \sqrt{2}x + 1}\, dx \\[2mm]
&= \int_0^1 \dfrac{\dfrac{1}{2} + \dfrac{\sqrt{2}}{4}x}{x^2 + \sqrt{2}x + 1}\, dx - \int_0^1 \dfrac{-\dfrac{1}{2} + \dfrac{\sqrt{2}}{4}x}{x^2 + \sqrt{2}x + 1}\, dx \\[2mm]
&= \dfrac{\sqrt{2}}{4}\Big[\arctan(\sqrt{2}x + 1) + \arctan(\sqrt{2}x - 1) \Big]_0^1 + \dfrac{\sqrt{2}}{8}\Big[\ln(x^2 + \sqrt{2}x + 1) - \ln(x^2 - \sqrt{2}x + 1) \Big]_0^1 \\[2mm]
&= \dfrac{\sqrt{2}}{4}\Big[\arctan(\sqrt{2} + 1) + \arctan(\sqrt{2} - 1) \Big] + \dfrac{\sqrt{2}}{8}\Big[\ln(2 + \sqrt{2}) - \ln(2 - \sqrt{2}) \Big] - \dfrac{\sqrt{2}}{4}\Big[\dfrac{\pi}{4} - \dfrac{\pi}{4} \Big] - \dfrac{\sqrt{2}}{8}[0] \\[2mm]
&\approx 0.5554 + 0.3116 \\[2mm]
&\approx 0.8670
\end{aligned}$

15. Using a graphing utility,

(a) $\lim\limits_{x \to 0^+} \left(\cot x + \dfrac{1}{x} \right) = \infty$

(b) $\lim\limits_{x \to 0^+} \left(\cot x - \dfrac{1}{x} \right) = 0$

(c) $\lim\limits_{x \to 0^+} \left(\cot x + \dfrac{1}{x} \right)\left(\cot x - \dfrac{1}{x} \right) \approx -\dfrac{2}{3}$.

Analytically,

(a) $\lim\limits_{x \to 0^+} \left(\cot x + \dfrac{1}{x} \right) = \infty + \infty = \infty$

(b) $\lim\limits_{x \to 0^+} \left(\cot x - \dfrac{1}{x} \right) = \lim\limits_{x \to 0^+} \dfrac{x \cot x - 1}{x} = \lim\limits_{x \to 0^+} \dfrac{x \cos x - \sin x}{x \sin x}$

$$= \lim\limits_{x \to 0^+} \dfrac{\cos x - x \sin x - \cos x}{\sin x + x \cos x} = \lim\limits_{x \to 0^+} \dfrac{-x \sin x}{\sin x + x \cos x}$$

$$= \lim\limits_{x \to 0^+} \dfrac{-\sin x - x \cos x}{\cos x + \cos x - x \sin x} = 0.$$

(c) $\left(\cot x + \dfrac{1}{x} \right)\left(\cot x - \dfrac{1}{x} \right) = \cot^2 x - \dfrac{1}{x^2}$

$$= \dfrac{x^2 \cot^2 x - 1}{x^2}$$

$$\lim\limits_{x \to 0^+} \dfrac{x^2 \cot^2 x - 1}{x^2} = \lim\limits_{x \to 0^+} \dfrac{2x \cot^2 x - 2x^2 \cot x \csc^2 x}{2x}$$

$$= \lim\limits_{x \to 0^+} \dfrac{\cot^2 x - x \cot x \csc^2 x}{1}$$

$$= \lim\limits_{x \to 0^+} \dfrac{\cos^2 x \sin x - x \cos x}{\sin^3 x}$$

$$= \lim\limits_{x \to 0^+} \dfrac{(1 - \sin^2 x)\sin x - x \cos x}{\sin^3 x}$$

$$= \lim\limits_{x \to 0^+} \dfrac{\sin x - x \cos x}{\sin^3 x} - 1$$

Now, $\lim\limits_{x \to 0^+} \dfrac{\sin x - x \cos x}{\sin^3 x} = \lim\limits_{x \to 0^+} \dfrac{\cos x - \cos x + x \sin x}{3 \sin^2 x \cos x}$

$$= \lim\limits_{x \to 0^+} \dfrac{x}{3 \sin x \cdot \cos x}$$

$$= \lim\limits_{x \to 0^+} \left(\dfrac{x}{\sin x} \right)\dfrac{1}{3 \cos x} = \dfrac{1}{3}.$$

Thus, $\lim\limits_{x \to 0^+} \left(\cot x + \dfrac{1}{x} \right)\left(\cot x - \dfrac{1}{x} \right) = \dfrac{1}{3} - 1 = -\dfrac{2}{3}.$

The form $0 \cdot \infty$ is indeterminant.

17. $\dfrac{x^3 - 3x^2 + 1}{x^4 - 13x^2 + 12x} = \dfrac{P_1}{x} + \dfrac{P_2}{x - 1} + \dfrac{P_3}{x + 4} + \dfrac{P_4}{x - 3} \Rightarrow c_1 = 0,\ c_2 = 1,\ c_3 = -4,\ c_4 = 3$

$N(x) = x^3 - 3x^2 + 1$

$D'(x) = 4x^3 - 26x + 12$

$P_1 = \dfrac{N(0)}{D'(0)} = \dfrac{1}{12}$

$P_2 = \dfrac{N(1)}{D'(1)} = \dfrac{-1}{-10} = \dfrac{1}{10}$

$P_3 = \dfrac{N(-4)}{D'(-4)} = \dfrac{-111}{-140} = \dfrac{111}{140}$

$P_4 = \dfrac{N(3)}{D'(3)} = \dfrac{1}{42}$

Thus, $\dfrac{x^3 - 3x^2 + 1}{x^4 - 13x^2 + 12x} = \dfrac{1/12}{x} + \dfrac{1/10}{x - 1} + \dfrac{111/140}{x + 4} + \dfrac{1/42}{x - 3}.$

19. By parts,

$$\int_a^b f(x)g''(x)\,dx = \left[f(x)g'(x) \right]_a^b - \int_a^b f'(x)g'(x)\,dx\, [u = f(x),\, dv = g''(x)\,dx]$$

$$= -\int_a^b f'(x)g'(x)\,dx$$

$$= \left[-f'(x)g(x) \right]_a^b + \int_a^b g(x)f''(x)\,dx\, [u = f'(x),\, dv = g'(x)\,dx]$$

$$= \int_a^b f''(x)g(x)\,dx.$$

21.

$$\int_2^\infty \left[\frac{1}{x^5} + \frac{1}{x^{10}} + \frac{1}{x^{15}} \right] dx < \int_2^\infty \frac{1}{x^5 - 1}\,dx < \int_2^\infty \left[\frac{1}{x^5} + \frac{1}{x^{10}} + \frac{2}{x^{15}} \right] dx$$

$$\lim_{b \to \infty} \left[-\frac{1}{4x^4} - \frac{1}{9x^9} - \frac{1}{14x^{14}} \right]_2^b < \int_2^\infty \frac{1}{x^5 - 1}\,dx < \lim_{b \to \infty} \left[-\frac{1}{4x^4} - \frac{1}{9x^9} - \frac{1}{7x^{14}} \right]_2^b$$

$$0.015846 < \int_2^\infty \frac{1}{x^5 - 1}\,dx < 0.015851$$

C H A P T E R 9
Infinite Series

Section 9.1   Sequences **437**

Section 9.2   Series and Convergence **444**

Section 9.3   The Integral Test and p-Series **451**

Section 9.4   Comparisons of Series **457**

Section 9.5   Alternating Series **460**

Section 9.6   The Ratio and Root Tests **465**

Section 9.7   Taylor Polynomials and Approximations **472**

Section 9.8   Power Series **477**

Section 9.9   Representation of Functions by Power Series **483**

Section 9.10   Taylor and Maclaurin Series **487**

Review Exercises **494**

Problem Solving **499**

CHAPTER 9
Infinite Series

Section 9.1   Sequences

1. $a_n = 2^n$

$a_1 = 2^1 = 2$

$a_2 = 2^2 = 4$

$a_3 = 2^3 = 8$

$a_4 = 2^4 = 16$

$a_5 = 2^5 = 32$

3. $a_n = \left(-\dfrac{1}{2}\right)^n$

$a_1 = \left(-\dfrac{1}{2}\right)^1 = -\dfrac{1}{2}$

$a_2 = \left(-\dfrac{1}{2}\right)^2 = \dfrac{1}{4}$

$a_3 = \left(-\dfrac{1}{2}\right)^3 = -\dfrac{1}{8}$

$a_4 = \left(-\dfrac{1}{2}\right)^4 = \dfrac{1}{16}$

$a_5 = \left(-\dfrac{1}{2}\right)^5 = -\dfrac{1}{32}$

5. $a_n = \sin\dfrac{n\pi}{2}$

$a_1 = \sin\dfrac{\pi}{2} = 1$

$a_2 = \sin \pi = 0$

$a_3 = \sin\dfrac{3\pi}{2} = -1$

$a_4 = \sin 2\pi = 0$

$a_5 = \sin\dfrac{5\pi}{2} = 1$

7. $a_n = \dfrac{(-1)^{n(n+1)/2}}{n^2}$

$a_1 = \dfrac{(-1)^1}{1^2} = -1$

$a_2 = \dfrac{(-1)^3}{2^2} = -\dfrac{1}{4}$

$a_3 = \dfrac{(-1)^6}{3^2} = \dfrac{1}{9}$

$a_4 = \dfrac{(-1)^{10}}{4^2} = \dfrac{1}{16}$

$a_5 = \dfrac{(-1)^{15}}{5^2} = -\dfrac{1}{25}$

9. $a_n = 5 - \dfrac{1}{n} + \dfrac{1}{n^2}$

$a_1 = 5 - 1 + 1 = 5$

$a_2 = 5 - \dfrac{1}{2} + \dfrac{1}{4} = \dfrac{19}{4}$

$a_3 = 5 - \dfrac{1}{3} + \dfrac{1}{9} = \dfrac{43}{9}$

$a_4 = 5 - \dfrac{1}{4} + \dfrac{1}{16} = \dfrac{77}{16}$

$a_5 = 5 - \dfrac{1}{5} + \dfrac{1}{25} = \dfrac{121}{25}$

11. $a_1 = 3,\, a_{k+1} = 2(a_k - 1)$

$a_2 = 2(a_1 - 1)$

$\quad = 2(3 - 1) = 4$

$a_3 = 2(a_2 - 1)$

$\quad = 2(4 - 1) = 6$

$a_4 = 2(a_3 - 1)$

$\quad = 2(6 - 1) = 10$

$a_5 = 2(a_4 - 1)$

$\quad = 2(10 - 1) = 18$

13. $a_1 = 32,\, a_{k+1} = \dfrac{1}{2}a_k$

$a_2 = \dfrac{1}{2}a_1 = \dfrac{1}{2}(32) = 16$

$a_3 = \dfrac{1}{2}a_2 = \dfrac{1}{2}(16) = 8$

$a_4 = \dfrac{1}{2}a_3 = \dfrac{1}{2}(8) = 4$

$a_5 = \dfrac{1}{2}a_4 = \dfrac{1}{2}(4) = 2$

15. $a_n = \dfrac{8}{n+1},\, a_1 = 4,\, a_2 = \dfrac{8}{3}$,

decreases to 0; matches (f).

17. $a_n = 4(0.5)^{n-1},\, a_1 = 4,\, a_2 = 2$,

decreases to 0; matches (e).

19. $a_n(-1)^n,\, a_1 = -1,\, a_2 = 1$,

$a_3 = -1$, etc.; matches (d).

21.

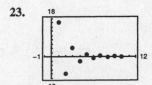

$a_n = \dfrac{2}{3}n,\, n = 1,\ldots,10$

23.

$a_n = 16(-0.5)^{n-1},\, n = 1,\ldots,10$

25. $a_n = 3n - 1$

$a_5 = 3(5) - 1 = 14$

$a_6 = 3(6) - 1 = 17$

Add 3 to preceding term.

27. $a_{n+1} = 2a_n,\, a_1 = 5$

$a_5 = 2(40) = 80$

$a_6 = 2(80) = 160$

29. $a_n = \dfrac{3}{(-2)^{n-1}}$

$a_5 = \dfrac{3}{(-2)^4} = \dfrac{3}{16}$

$a_6 = \dfrac{3}{(-2)^5} = -\dfrac{3}{32}$

Multiply the preceding term by $-\frac{1}{2}$.

31. $\dfrac{10!}{8!} = \dfrac{8!(9)(10)}{8!}$

$= (9)(10) = 90$

33. $\dfrac{(n+1)!}{n!} = \dfrac{n!(n+1)}{n!} = n+1$

35. $\dfrac{(2n-1)!}{(2n+1)!} = \dfrac{(2n-1)!}{(2n-1)!(2n)(2n+1)}$

$= \dfrac{1}{2n(2n+1)}$

37. $\displaystyle\lim_{n\to\infty} \dfrac{5n^2}{n^2+2} = 5$

39. $\displaystyle\lim_{n\to\infty} \dfrac{2n}{\sqrt{n^2+1}} = \lim_{n\to\infty} \dfrac{2}{\sqrt{1+(1/n^2)}} = \dfrac{2}{1} = 2$

41. $\displaystyle\lim_{n\to\infty} \sin\left(\dfrac{1}{n}\right) = 0$

43.

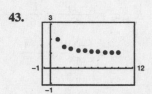

The graph seems to indicate that the sequence converges to 1. Analytically,

$\displaystyle\lim_{n\to\infty} a_n = \lim_{n\to\infty} \dfrac{n+1}{n} = \lim_{x\to\infty} \dfrac{x+1}{x} = \lim_{x\to\infty} 1 = 1.$

45.

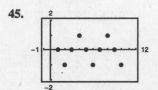

The graph seems to indicate that the sequence diverges. Analytically, the sequence is

$\{a_n\} = \{0, -1, 0, 1, 0, -1, \ldots\}.$

Hence, $\displaystyle\lim_{n\to\infty} a_n$ does not exist.

47. $\displaystyle\lim_{n\to\infty} (-1)^n\left(\dfrac{n}{n+1}\right)$

does not exist (oscillates between -1 and 1), diverges.

49. $\displaystyle\lim_{n\to\infty} \dfrac{3n^2 - n + 4}{2n^2 + 1} = \dfrac{3}{2}$, converges

51. $a_n = \dfrac{1 \cdot 3 \cdot 5 \cdots (2n-1)}{(2n)^n}$

$= \dfrac{1}{2n} \cdot \dfrac{3}{2n} \cdot \dfrac{5}{2n} \cdots \dfrac{2n-1}{2n} < \dfrac{1}{2n}$

Thus, $\displaystyle\lim_{n\to\infty} a_n = 0$, converges.

53. $\displaystyle\lim_{n\to\infty} \dfrac{1+(-1)^n}{n} = 0$, converges

55. $\displaystyle\lim_{n\to\infty} \dfrac{\ln(n^3)}{2n} = \lim_{n\to\infty} \dfrac{3}{2} \dfrac{\ln(n)}{n}$

$= \displaystyle\lim_{n\to\infty} \dfrac{3}{2}\left(\dfrac{1}{n}\right) = 0$, converges

(L'Hôpital's Rule)

57. $\displaystyle\lim_{n\to\infty} \left(\dfrac{3}{4}\right)^n = 0$, converges

59. $\displaystyle\lim_{n\to\infty} \dfrac{(n+1)!}{n!} = \lim_{n\to\infty} (n+1)$

$= \infty$, diverges

61. $\displaystyle\lim_{n\to\infty} \left(\dfrac{n-1}{n} - \dfrac{n}{n-1}\right) = \lim_{n\to\infty} \dfrac{(n-1)^2 - n^2}{n(n-1)}$

$= \displaystyle\lim_{n\to\infty} \dfrac{1 - 2n}{n^2 - n} = 0$, converges

63. $\lim\limits_{n\to\infty} \dfrac{n^p}{e^n} = 0$, converges

$(p > 0, n \geq 2)$

65. $a_n = \left(1 + \dfrac{k}{n}\right)^n$

$\lim\limits_{n\to\infty} \left(1 + \dfrac{k}{n}\right)^n = \lim\limits_{u\to 0} [(1 + u)^{1/u}]^k = e^k$

where $u = k/n$, converges

67. $\lim\limits_{n\to\infty} \dfrac{\sin n}{n} = \lim\limits_{n\to\infty} (\sin n)\dfrac{1}{n} = 0$,

converges (because $(\sin n)$ is bounded)

69. $a_n = 3n - 2$

71. $a_n - n^2 - 2$

73. $a_n = \dfrac{n+1}{n+2}$

75. $a_n = 1 + \dfrac{1}{n} = \dfrac{n+1}{n}$

77. $a_n = \dfrac{n}{(n+1)(n+2)}$

79. $a_n = \dfrac{(-1)^{n-1}}{1 \cdot 3 \cdot 5 \cdots (2n-1)}$

$= \dfrac{(-1)^{n-1} 2^n n!}{(2n)!}$

81. $a_n = (2n)!, n = 1, 2, 3, \ldots$

83. $a_n = 4 - \dfrac{1}{n} < 4 - \dfrac{1}{n+1} = a_{n+1}$,

monotonic; $|a_n| < 4$, bounded

85. $\dfrac{n}{2^{n+2}} \overset{?}{\geq} \dfrac{n+1}{2^{(n+1)+2}}$

$2^{n+3}n \overset{?}{\geq} 2^{n+2}(n+1)$

$2n \overset{?}{\geq} n+1$

$n \geq 1$

Hence, $n \geq 1$

$2n \geq n+1$

$2^{n+3}n \geq 2^{n+2}(n+1)$

$\dfrac{n}{2^{n+2}} \geq \dfrac{n+1}{2^{(n+1)+2}}$

$a_n \geq a_{n+1}.$

True; monotonic; $|a_n| \leq \frac{1}{8}$, bounded

87. $a_n = (-1)^n\left(\dfrac{1}{n}\right)$

$a_1 = -1$

$a_2 = \dfrac{1}{2}$

$a_3 = -\dfrac{1}{3}$

Not monotonic; $|a_n| \leq 1$, bounded

89. $a_n = \left(\dfrac{2}{3}\right)^n > \left(\dfrac{2}{3}\right)^{n+1} = a_{n+1}$

Monotonic; $|a_n| \leq \frac{2}{3}$, bounded

91. $a_n = \sin\left(\dfrac{n\pi}{6}\right)$

$a_1 = 0.500$

$a_2 = 0.8660$

$a_3 = 1.000$

$a_4 = 0.8660$

Not monotonic; $|a_n| \leq 1$, bounded

93. $a_n = \dfrac{\cos n}{n}$

$a_1 = 0.5403$

$a_2 = -0.2081$

$a_3 = -0.3230$

$a_4 = -0.1634$

Not monotonic; $|a_n| \leq 1$, bounded

95. (a) $a_n = 5 + \dfrac{1}{n}$

$\left|5 + \dfrac{1}{n}\right| \leq 6 \Rightarrow \{a_n\}$, bounded

$a_n = 5 + \dfrac{1}{n} > 5 + \dfrac{1}{n+1}$

$= a_{n+1} \Rightarrow \{a_n\}$, monotonic

Therefore, $\{a_n\}$ converges.

(b)

$\lim\limits_{n\to\infty} \left(5 + \dfrac{1}{n}\right) = 5$

97. (a) $a_n = \frac{1}{3}\left(1 - \frac{1}{3^n}\right)$

$\left|\frac{1}{3}\left(1 - \frac{1}{3^n}\right)\right| < \frac{1}{3} \implies \{a_n\}$, bounded

$a_n = \frac{1}{3}\left(1 - \frac{1}{3^n}\right) < \frac{1}{3}\left(1 - \frac{1}{3^{n+1}}\right)$

$= a_{n+1} \implies \{a_n\}$, monotonic

Therefore, $\{a_n\}$ converges.

(b)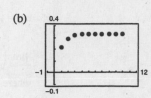

$\lim_{n \to \infty}\left[\frac{1}{3}\left(1 - \frac{1}{3^n}\right)\right] = \frac{1}{3}$

99. $\{a_n\}$ has a limit because it is a bounded, monotonic sequence. The limit is less than or equal to 4, and greater than or equal to 2.

$2 \le \lim_{n \to \infty} a_n \le 4$

101. $A_n = P\left(1 + \frac{r}{12}\right)^n$

(a) No, the sequence $\{A_n\}$ diverges, $\lim_{n \to \infty} A_n = \infty$.
The amount will grow arbitrarily large over time.

(b) $P = 9000, r = 0.055, A_n = 9000\left(1 + \frac{0.055}{12}\right)^n$

$A_1 = 9041.25$	$A_6 = 9250.35$
$A_2 = 9082.69$	$A_7 = 9292.75$
$A_3 = 9124.32$	$A_8 = 9335.34$
$A_4 = 9166.14$	$A_9 = 9378.13$
$A_5 = 9208.15$	$A_{10} = 9421.11$

103. (a) A sequence is a function whose domain is the set of positive integers.

(b) A sequence converges if it has a limit. See the definition.

(c) A sequence is monotonic if its terms are nondecreasing, or nonincreasing.

(d) A sequence is bounded if it is bounded below ($a_n \ge N$ for some N) and bounded above ($a_n \le M$ for some M).

105. $a_n = 10 - \frac{1}{n}$

107. $a_n = \frac{3n}{4n + 1}$

109. (a) $A_n = (0.8)^n (2.5)$ billion

(b) $A_1 = \$2$ billion

$A_2 = \$1.6$ billion

$A_3 = \$1.28$ billion

$A_4 = \$1.024$ billion

(c) $\lim_{n \to \infty} (0.8)^n (2.5) = 0$

111. (a) $a_n = -6.60n^2 + 151.7n + 387$

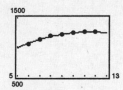

(b) In 2008, $n = 18$ and $a_{18} \approx 979$ endangered species.

113. $a_n = \frac{10^n}{n!}$

(a) $a_9 = a_{10} = \frac{10^9}{9!}$

$= \frac{1,000,000,000}{362,880}$

$= \frac{1,562,500}{567}$

(b) Decreasing

(c) Factorials increase more rapidly than exponentials.

115. $\{a_n\} = \left\{\sqrt[n]{n}\right\} = \{n^{1/n}\}$

$a_1 = 1^{1/1} = 1$

$a_2 = \sqrt{2} \approx 1.4142$

$a_3 = \sqrt[3]{3} \approx 1.4422$

$a_4 = \sqrt[4]{4} \approx 1.4142$

$a_5 = \sqrt[5]{5} \approx 1.3797$

$a_6 = \sqrt[6]{6} \approx 1.3480$

Let $y = \lim\limits_{n\to\infty} n^{1/n}$.

$\ln y = \lim\limits_{n\to\infty}\left(\dfrac{1}{n}\ln n\right) = \lim\limits_{n\to\infty}\dfrac{\ln n}{n} - \lim\limits_{n\to\infty}\dfrac{1/n}{n} = 0$

Since $\ln y = 0$, we have $y = e^0 = 1$. Therefore, $\lim\limits_{n\to\infty}\sqrt[n]{n} = 1$.

117. True

119. True

121. $a_{n+2} = a_n + a_{n+1}$

(a)
$a_1 = 1$	$a_7 = 8 + 5 = 13$
$a_2 = 1$	$a_8 = 13 + 8 = 21$
$a_3 = 1 + 1 = 2$	$a_9 = 21 + 13 = 34$
$a_4 = 2 + 1 = 3$	$a_{10} = 34 + 21 = 55$
$a_5 = 3 + 2 = 5$	$a_{11} = 55 + 34 = 89$
$a_6 = 5 + 3 = 8$	$a_{12} = 89 + 55 = 144$

(b) $b_n = \dfrac{a_{n+1}}{a_n}, n \geq 1$

$b_1 = \dfrac{1}{1} = 1$	$b_6 = \dfrac{13}{8}$
$b_2 = \dfrac{2}{1} = 2$	$b_7 = \dfrac{21}{13}$
$b_3 = \dfrac{3}{2}$	$b_8 = \dfrac{34}{21}$
$b_4 = \dfrac{5}{3}$	$b_9 = \dfrac{55}{34}$
$b_5 = \dfrac{8}{5}$	$b_{10} = \dfrac{89}{55}$

(c) $1 + \dfrac{1}{b_{n-1}} = 1 + \dfrac{1}{a_n/a_{n-1}}$

$= 1 + \dfrac{a_{n-1}}{a_n}$

$= \dfrac{a_n + a_{n-1}}{a_n} = \dfrac{a_{n+1}}{a_n} = b_n$

(d) If $\lim\limits_{n\to\infty} b_n = \rho$, then $\lim\limits_{n\to\infty}\left(1 + \dfrac{1}{b_{n-1}}\right) = \rho$.

Since $\lim\limits_{n\to\infty} b_n = \lim\limits_{n\to\infty} b_{n-1}$, we have

$1 + (1/\rho) = \rho.$

$\rho + 1 = \rho^2$

$0 = \rho^2 - \rho - 1$

$\rho = \dfrac{1 \pm \sqrt{1 + 4}}{2} = \dfrac{1 \pm \sqrt{5}}{2}$

Since a_n, and thus b_n, is positive, $\rho = \dfrac{1 + \sqrt{5}}{2} \approx 1.6180$.

123. (a) $a_1 = \sqrt{2} \approx 1.4142$

$a_2 = \sqrt{2 + \sqrt{2}} \approx 1.8478$

$a_3 = \sqrt{2 + \sqrt{2 + \sqrt{2}}} \approx 1.9616$

$a_4 = \sqrt{2 + \sqrt{2 + \sqrt{2 + \sqrt{2}}}} \approx 1.9904$

$a_5 = \sqrt{2 + \sqrt{2 + \sqrt{2 + \sqrt{2 + \sqrt{2}}}}} \approx 1.9976$

(b) $a_n = \sqrt{2 + a_{n-1}}, \quad n \geq 2, a_1 = \sqrt{2}$

—CONTINUED—

123. **—CONTINUED—**

(c) We first use mathematical induction to show that $a_n \leq 2$; clearly $a_1 \leq 2$. So assume $a_k \leq 2$. Then

$$a_k + 2 \leq 4$$

$$\sqrt{a_k + 2} \leq 2$$

$$a_{k+1} \leq 2.$$

Now we show that $\{a_n\}$ is an increasing sequence. Since $a_n \geq 0$ and $a_n \leq 2$,

$$(a_n - 2)(a_n + 1) \leq 0$$

$$a_n^2 - a_n - 2 \leq 0$$

$$a_n^2 \leq a_n + 2$$

$$a_n \leq \sqrt{a_n + 2}$$

$$a_n \leq a_{n+1}.$$

Since $\{a_n\}$ is a bounding increasing sequence, it converges to some number L, by Theorem 9.5.

$$\lim_{n \to \infty} a_n = L \implies \sqrt{2 + L} = L \implies 2 + L = L^2 \implies L^2 - L - 2 = 0$$

$$\implies (L - 2)(L + 1) = 0 \implies L = 2 \quad (L \neq -1)$$

125. (a) We use mathematical induction to show that

$$a_n \leq \frac{1 + \sqrt{1 + 4k}}{2}.$$

[Note that if $k = 2$, then $a_n \leq 3$, and if $k = 6$, then $a_n \leq 3$.] Clearly,

$$a_1 = \sqrt{k} \leq \frac{\sqrt{1 + 4k}}{2} \leq \frac{1 + \sqrt{1 + 4k}}{2}.$$

Before proceeding to the induction step, note that

$$2 + 2\sqrt{1 + 4k} + 4k = 2 + 2\sqrt{1 + 4k} + 4k$$

$$\frac{1 + \sqrt{1 + 4k}}{2} + k = \frac{1 + 2\sqrt{1 + 4k} + 1 + 4k}{4}$$

$$\frac{1 + \sqrt{1 + 4k}}{2} + k = \left[\frac{1 + \sqrt{1 + 4k}}{2}\right]^2$$

$$\sqrt{\frac{1 + \sqrt{1 + 4k}}{2} + k} = \frac{1 + \sqrt{1 + 4k}}{2}.$$

So assume $a_n \leq \dfrac{1 + \sqrt{1 + 4k}}{2}$. Then

$$a_n + k \leq \frac{1 + \sqrt{1 + 4k}}{2} + k$$

$$\sqrt{a_n + k} \leq \sqrt{\frac{1 + \sqrt{1 + 4k}}{2} + k}$$

$$a_{n+1} \leq \frac{1 + \sqrt{1 + 4k}}{2}.$$

$\{a_n\}$ is increasing because

$$\left(a_n - \frac{1 + \sqrt{1 + 4k}}{2}\right)\left(a_n - \frac{1 - \sqrt{1 + 4k}}{2}\right) \leq 0$$

$$a_n^2 - a_n - k \leq 0$$

$$a_n^2 \leq a_n + k$$

$$a_n \leq \sqrt{a_n + k}$$

$$a_n \leq a_{n+1}.$$

(b) Since $\{a_n\}$ is bounded and increasing, it has a limit L.

(c) $\displaystyle\lim_{n \to \infty} a_n = L$ implies that

$$L = \sqrt{k + L} \implies L^2 = k + L$$

$$\implies L^2 - L - k = 0$$

$$\implies L = \frac{1 \pm \sqrt{1 + 4k}}{2}.$$

Since $L > 0$, $L = \dfrac{1 + \sqrt{1 + 4k}}{2}$.

127. (a) $f(x) = \sin x, a_n = n \sin \frac{1}{n}$

$f'(x) = \cos x, \ f'(0) = 1$

$\lim_{n \to \infty} a_n = \lim_{n \to \infty} n \sin \frac{1}{n} = \lim_{n \to \infty} \frac{\sin(1/n)}{(1/n)} = 1 = f'(0)$

(b) $f'(0) = \lim_{h \to 0^+} \dfrac{f(0 + h) - f(0)}{h}$

$= \lim_{h \to 0^+} \dfrac{f(h)}{h}$

$= \lim_{n \to \infty} \dfrac{f(1/n)}{(1/n)}$

$= \lim_{n \to \infty} nf\left(\dfrac{1}{n}\right)$

$= \lim_{n \to \infty} a_n$

129. (a)

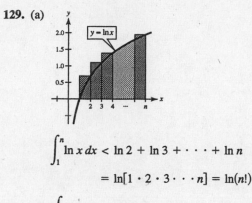

$\displaystyle\int_1^n \ln x \, dx < \ln 2 + \ln 3 + \cdots + \ln n$

$= \ln[1 \cdot 2 \cdot 3 \cdots n] = \ln(n!)$

(b)

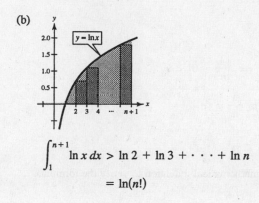

$\displaystyle\int_1^{n+1} \ln x \, dx > \ln 2 + \ln 3 + \cdots + \ln n$

$= \ln(n!)$

(c) $\displaystyle\int \ln x \, dx = x \ln x - x + C$

$\displaystyle\int_1^n \ln x \, dx = n \ln n - n + 1 = \ln n^n - n + 1$

From part (a): $\ln n^n - n + 1 < \ln(n!)$

$e^{\ln n^n - n + 1} < n!$

$\dfrac{n^n}{e^{n-1}} < n!$

$\displaystyle\int_1^{n+1} \ln x \, dx = (n + 1)\ln(n + 1) - (n + 1) + 1 = \ln(n + 1)^{n+1} - n$

From part (b): $\ln(n + 1)^{n+1} - n > \ln(n!)$

$e^{\ln(n+1)^{n+1} - n} > n!$

$\dfrac{(n + 1)^{n+1}}{e^n} > n!$

(d) $\dfrac{n^n}{e^{n-1}} < n! < \dfrac{(n + 1)^{n+1}}{e^n}$

$\dfrac{n}{e^{1-(1/n)}} < \sqrt[n]{n!} < \dfrac{(n + 1)^{(n+1)/n}}{e}$

$\dfrac{1}{e^{1-(1/n)}} < \dfrac{\sqrt[n]{n!}}{n} < \dfrac{(n + 1)^{1+(1/n)}}{ne}$

$\lim_{n \to \infty} \dfrac{1}{e^{1-(1/n)}} = \dfrac{1}{e}$

$\lim_{n \to \infty} \dfrac{(n + 1)^{1+(1/n)}}{ne} = \lim_{n \to \infty} \dfrac{(n + 1)}{n} \dfrac{(n + 1)^{1/n}}{e}$

$= (1)\dfrac{1}{e} = \dfrac{1}{e}$

By the Squeeze Theorem, $\lim_{n \to \infty} \dfrac{\sqrt[n]{n!}}{n} = \dfrac{1}{e}$.

(e) $n = 20$: $\dfrac{\sqrt[20]{20!}}{20} \approx 0.4152$

$n = 50$: $\dfrac{\sqrt[50]{50!}}{50} \approx 0.3897$

$n = 100$: $\dfrac{\sqrt[100]{100!}}{100} \approx 0.3799$

$\dfrac{1}{e} \approx 0.3679$

131. For a given $\epsilon > 0$, we must find $M > 0$ such that

$$|a_n - L| = \left|\frac{1}{n^3}\right| < \varepsilon$$

whenever $n > M$. That is,

$$n^3 > \frac{1}{\varepsilon} \text{ or } n > \left(\frac{1}{\varepsilon}\right)^{1/3}.$$

So, let $\varepsilon > 0$ be given. Let M be an integer satisfying $M > (1/\varepsilon)^{1/3}$. For $n > M$, we have

$$n > \left(\frac{1}{\varepsilon}\right)^{1/3}$$

$$n^3 > \frac{1}{\varepsilon}$$

$$\varepsilon > \frac{1}{n^3} \implies \left|\frac{1}{n^3} - 0\right| < \varepsilon.$$

Thus, $\lim_{n \to \infty} \dfrac{1}{n^3} = 0$.

135. $T_n = n! + 2^n$

We use mathematical induction to verify the formula.

$T_0 = 1 + 1 = 2$

$T_1 = 1 + 2 = 3$

$T_2 = 2 + 4 = 6$

Assume $T_k = k! + 2^k$. Then

$$T_{k+1} = (k + 1 + 4)T_k - 4(k + 1)T_{k-1} + (4(k + 1) - 8)T_{k-2}$$

$$= (k + 5)[k! + 2^k] - 4(k + 1)((k - 1)! + 2^{k-1}) + (4k - 4)((k - 2)! + 2^{k-2})$$

$$= [(k + 5)(k)(k - 1) - 4(k + 1)(k - 1) + 4(k - 1)](k - 2)! + [(k + 5)4 - 8(k + 1) + 4(k - 1)]2^{k-2}$$

$$= [k^2 + 5k - 4k - 4 + 4](k - 1)! + 8 \cdot 2^{k-2}$$

$$= (k + 1)! + 2^{k+1}.$$

By mathematical induction, the formula is valid for all n.

Section 9.2   Series and Convergence

1. $S_1 = 1$

$S_2 = 1 + \frac{1}{4} = 1.2500$

$S_3 = 1 + \frac{1}{4} + \frac{1}{9} \approx 1.3611$

$S_4 = 1 + \frac{1}{4} + \frac{1}{9} + \frac{1}{16} \approx 1.4236$

$S_5 = 1 + \frac{1}{4} + \frac{1}{9} + \frac{1}{16} + \frac{1}{25} \approx 1.4636$

5. $S_1 = 3$

$S_2 = 3 + \frac{3}{2} = 4.5$

$S_3 = 3 + \frac{3}{2} + \frac{3}{4} = 5.250$

$S_4 = 3 + \frac{3}{2} + \frac{3}{4} + \frac{3}{8} = 5.625$

$S_5 = 3 + \frac{3}{2} + \frac{3}{4} + \frac{3}{8} + \frac{3}{16} = 5.8125$

133. If $\{a_n\}$ is bounded, monotonic and nonincreasing, then $a_1 \geq a_2 \geq a_3 \geq \cdots \geq a_n \geq \cdots$. Then $-a_1 \leq -a_2 \leq -a_3 \leq \cdots \leq -a_n \leq \cdots$ is a bounded, monotonic, nondecreasing sequence which converges by the first half of the theorem. Since $\{-a_n\}$ converges, then so does $\{a_n\}$.

3. $S_1 = 3$

$S_2 = 3 - \frac{9}{2} = -1.5$

$S_3 = 3 - \frac{9}{2} + \frac{27}{4} = 5.25$

$S_4 = 3 - \frac{9}{2} + \frac{27}{4} - \frac{81}{8} = -4.875$

$S_5 = 3 - \frac{9}{2} + \frac{27}{4} - \frac{81}{8} + \frac{243}{16} = 10.3125$

7. $\displaystyle\sum_{n=0}^{\infty} 3\left(\frac{3}{2}\right)^n$

Geometric series

$$r = \frac{3}{2} > 1$$

Diverges by Theorem 9.6

9. $\displaystyle\sum_{n=0}^{\infty} 1000(1.055)^n$

Geometric series

$r = 1.055 > 1$

Diverges by Theorem 9.6

11. $\displaystyle\sum_{n=1}^{\infty} \frac{n}{n+1}$

$\displaystyle\lim_{n\to\infty} \frac{n}{n+1} = 1 \neq 0$

Diverges by Theorem 9.9

13. $\displaystyle\sum_{n=1}^{\infty} \frac{n^2}{n^2+1}$

$\displaystyle\lim_{n\to\infty} \frac{n^2}{n^2+1} = 1 \neq 0$

Diverges by Theorem 9.9

15. $\displaystyle\sum_{n=1}^{\infty} \frac{2^n+1}{2^{n+1}}$

$\displaystyle\lim_{n\to\infty} \frac{2^n+1}{2^{n+1}} = \lim_{n\to\infty} \frac{1+2^{-n}}{2} = \frac{1}{2} \neq 0$

Diverges by Theorem 9.9

17. $\displaystyle\sum_{n=0}^{\infty} \frac{9}{4}\left(\frac{1}{4}\right)^n = \frac{9}{4}\left[1 + \frac{1}{4} + \frac{1}{16} + \cdots\right]$

$S_0 = \frac{9}{4}, S_1 = \frac{9}{4}\cdot\frac{5}{4} = \frac{45}{16}, S_2 = \frac{9}{4}\cdot\frac{21}{16} \approx 2.95, \ldots$

Matches graph (c). Analytically, the series is geometric:

$$\sum_{n=0}^{\infty} \left(\frac{9}{4}\right)\left(\frac{1}{4}\right)^n = \frac{9/4}{1-1/4} = \frac{9/4}{3/4} = 3$$

19. $\displaystyle\sum_{n=0}^{\infty} \frac{15}{4}\left(-\frac{1}{4}\right)^n = \frac{15}{4}\left[1 - \frac{1}{4} + \frac{1}{16} - \cdots\right]$

$S_0 = \frac{15}{4}, S_1 = \frac{45}{16}, S_2 \approx 3.05, \ldots$

Matches graph (a). Analytically, the series is geometric:

$$\sum_{n=0}^{\infty} \frac{15}{4}\left(-\frac{1}{4}\right)^n = \frac{15/4}{1-(-1/4)} = \frac{15/4}{5/4} = 3$$

21. $\displaystyle\sum_{n=0}^{\infty} \frac{17}{3}\left(-\frac{1}{2}\right)^n = \frac{17}{3}\left[1 - \frac{1}{2} + \frac{1}{4} - \cdots\right]$

$S_0 = \frac{17}{3}, S_1 = \frac{17}{6}, \ldots$

Matches (f). The series is geometric:

$$\sum_{n=0}^{\infty} \frac{17}{3}\left(-\frac{1}{2}\right)^n = \frac{17}{3}\frac{1}{1-(-1/2)} = \frac{17}{3}\frac{2}{3} = \frac{34}{9}$$

23. $\displaystyle\sum_{n=1}^{\infty} \frac{1}{n(n+1)} = \sum_{n=1}^{\infty}\left(\frac{1}{n} - \frac{1}{n+1}\right) = \left(1 - \frac{1}{2}\right) + \left(\frac{1}{2} - \frac{1}{3}\right) + \left(\frac{1}{3} - \frac{1}{4}\right) + \left(\frac{1}{4} - \frac{1}{5}\right) + \cdots, \quad S_n = 1 - \frac{1}{n+1}$

$\displaystyle\sum_{n=1}^{\infty} \frac{1}{n(n+1)} = \lim_{n\to\infty} S_n = \lim_{n\to\infty}\left(1 - \frac{1}{n+1}\right) = 1$

25. $\displaystyle\sum_{n=0}^{\infty} 2\left(\frac{3}{4}\right)^n$

Geometric series with $r = \frac{3}{4} < 1$

Converges by Theorem 9.6

27. $\displaystyle\sum_{n=0}^{\infty} (0.9)^n$

Geometric series with $r = 0.9 < 1$

Converges by Theorem 9.6

29. (a) $\displaystyle\sum_{n=1}^{\infty} \frac{6}{n(n+3)} = 2\sum_{n=1}^{\infty}\left(\frac{1}{n} - \frac{1}{n+3}\right)$

$$= 2\left[\left(1 - \frac{1}{4}\right) + \left(\frac{1}{2} - \frac{1}{5}\right) + \left(\frac{1}{3} - \frac{1}{6}\right) + \left(\frac{1}{4} - \frac{1}{7}\right) + \cdots\right]$$

$$\left(S_n = 2\left[1 + \frac{1}{2} + \frac{1}{3} - \left(\frac{1}{n+1} + \frac{1}{n+2} + \frac{1}{n+3}\right)\right]\right)$$

$$= 2\left[1 + \frac{1}{2} + \frac{1}{3}\right] = \frac{11}{3} \approx 3.667$$

(b)

n	5	10	20	50	100
S_n	2.7976	3.1643	3.3936	3.5513	3.6078

(c)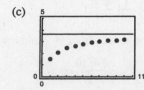

(d) The terms of the series decrease in magnitude slowly. Thus, the sequence of partial sums approaches the sum slowly.

31. (a) $\displaystyle\sum_{n=1}^{\infty} 2(0.9)^{n-1} = \sum_{n=0}^{\infty} 2(0.9)^n = \frac{2}{1-0.9} = 20$

(b)

n	5	10	20	50	100
S_n	8.1902	13.0264	17.5685	19.8969	19.9995

(c)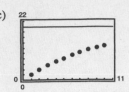

(d) The terms of the series decrease in magnitude slowly. Thus, the sequence of partial sums approaches the sum slowly.

33. (a) $\displaystyle\sum_{n=1}^{\infty} 10(0.25)^{n-1} = \frac{10}{1-0.25} = \frac{40}{3} \approx 13.3333$

(b)

n	5	10	20	50	100
S_n	13.3203	13.3333	13.3333	13.3333	13.3333

(c)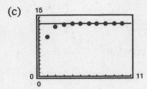

(d) The terms of the series decrease in magnitude rapidly. Thus, the sequence of partial sums approaches the sum rapidly.

35. $\displaystyle\sum_{n=2}^{\infty} \frac{1}{n^2 - 1} = \sum_{n=2}^{\infty} \left(\frac{1/2}{n-1} - \frac{1/2}{n+1}\right) = \frac{1}{2}\sum_{n=2}^{\infty} \left(\frac{1}{n-1} - \frac{1}{n+1}\right)$

$$= \frac{1}{2}\left[\left(1 - \frac{1}{3}\right) + \left(\frac{1}{2} - \frac{1}{4}\right) + \left(\frac{1}{3} - \frac{1}{5}\right) + \left(\frac{1}{4} - \frac{1}{6}\right) + \cdots\right]$$

$$= \frac{1}{2}\left(1 + \frac{1}{2}\right) = \frac{3}{4}$$

37. $\displaystyle\sum_{n=1}^{\infty} \frac{8}{(n+1)(n+2)} = 8\sum_{n=1}^{\infty} \left(\frac{1}{n+1} - \frac{1}{n+2}\right) = 8\left[\left(\frac{1}{2} - \frac{1}{3}\right) + \left(\frac{1}{3} - \frac{1}{4}\right) + \left(\frac{1}{4} - \frac{1}{5}\right) + \cdots\right] = 8\left(\frac{1}{2}\right) = 4$

39. $\displaystyle\sum_{n=0}^{\infty} \left(\frac{1}{2}\right)^n = \frac{1}{1-(1/2)} = 2$

41. $\displaystyle\sum_{n=0}^{\infty} \left(-\frac{1}{2}\right)^n = \frac{1}{1-(-1/2)} = \frac{2}{3}$

43. $\displaystyle\sum_{n=0}^{\infty} \left(\frac{1}{10}\right)^n = \frac{1}{1-(1/10)} = \frac{10}{9}$

45. $\displaystyle\sum_{n=0}^{\infty} 3\left(-\frac{1}{3}\right)^n = \frac{3}{1-(-1/3)} = \frac{9}{4}$

47. $\displaystyle\sum_{n=0}^{\infty} \left(\frac{1}{2^n} - \frac{1}{3^n}\right) = \sum_{n=0}^{\infty} \left(\frac{1}{2}\right)^n - \sum_{n=0}^{\infty} \left(\frac{1}{3}\right)^n$

$$= \frac{1}{1-(1/2)} - \frac{1}{1-(1/3)}$$

$$= 2 - \frac{3}{2} = \frac{1}{2}$$

49. Note that $\sin(1) \approx 0.8415 < 1$. The series $\displaystyle\sum_{n=1}^{\infty} [\sin(1)]^n$ is geometric with $r = \sin(1) < 1$. Thus,

$$\sum_{n=1}^{\infty} [\sin(1)]^n = \sin(1)\sum_{n=0}^{\infty} [\sin(1)]^n = \frac{\sin(1)}{1-\sin(1)} \approx 5.3080.$$

51. (a) $0.\overline{4} = \displaystyle\sum_{n=0}^{\infty} \frac{4}{10}\left(\frac{1}{10}\right)^n$

(b) Geometric series with $a = \frac{4}{10}$ and $r = \frac{1}{10}$

$$S = \frac{a}{1-r} = \frac{4/10}{1-(1/10)} = \frac{4}{9}$$

53. (a) $0.\overline{81} = \displaystyle\sum_{n=0}^{\infty} \frac{81}{100}\left(\frac{1}{100}\right)^n$

(b) Geometric series with $a = \frac{81}{100}$ and $r = \frac{1}{100}$

$$S = \frac{a}{1-r} = \frac{81/100}{1-(1/100)} = \frac{81}{99} = \frac{9}{11}$$

55. (a) $0.0\overline{75} = \displaystyle\sum_{n=0}^{\infty} \frac{3}{40}\left(\frac{1}{100}\right)^n$

(b) Geometric series with $a = \frac{3}{40}$ and $r = \frac{1}{100}$

$$S = \frac{a}{1-r} = \frac{3/40}{99/100} = \frac{5}{66}$$

57. $\displaystyle\sum_{n=1}^{\infty} \frac{n+10}{10n+1}$

$\displaystyle\lim_{n\to\infty} \frac{n+10}{10n+1} = \frac{1}{10} \neq 0$

Diverges by Theorem 9.9

59. $\displaystyle\sum_{n=1}^{\infty}\left(\frac{1}{n} - \frac{1}{n+2}\right) = \left(1 - \frac{1}{3}\right) + \left(\frac{1}{2} - \frac{1}{4}\right) + \left(\frac{1}{3} - \frac{1}{5}\right) + \left(\frac{1}{4} - \frac{1}{6}\right) + \cdots = 1 + \frac{1}{2} = \frac{3}{2}$, converges

61. $\displaystyle\sum_{n=1}^{\infty} \frac{3n-1}{2n+1}$

$\displaystyle\lim_{n\to\infty} \frac{3n-1}{2n+1} = \frac{3}{2} \neq 0$

Diverges by Theorem 9.9

63. $\displaystyle\sum_{n=0}^{\infty} \frac{4}{2^n} = 4 \sum_{n=0}^{\infty}\left(\frac{1}{2}\right)^n$

Geometric series with $r = \frac{1}{2}$

Converges by Theorem 9.6

65. $\displaystyle\sum_{n=0}^{\infty} (1.075)^n$

Geometric series with $r = 1.075$

Diverges by Theorem 9.6

67. Since $n > \ln(n)$, the terms $a_n = \dfrac{n}{\ln(n)}$

do not approach 0 as $n \to \infty$. Hence, the series

$\displaystyle\sum_{n=2}^{\infty} \frac{n}{\ln(n)}$ diverges.

69. For $k \neq 0$,

$$\lim_{n\to\infty}\left(1 + \frac{k}{n}\right)^n = \lim_{n\to\infty}\left[\left(1 + \frac{k}{n}\right)^{n/k}\right]^k$$

$$= e^k \neq 0.$$

For $k = 0$, $\displaystyle\lim_{n\to\infty} (1 + 0)^n = 1 \neq 0$.

Hence, $\displaystyle\sum_{n=1}^{\infty}\left[1 + \frac{k}{n}\right]^n$ diverges.

71. $\displaystyle\lim_{n\to\infty} \arctan n = \frac{\pi}{2} \neq 0$

Hence, $\displaystyle\sum_{n=1}^{\infty} \arctan n$ diverges.

73. See definition on page 606.

75. The series given by

$$\sum_{n=0}^{\infty} ar^n = a + ar + ar^2 + \cdots + ar^n + \cdots, \quad a \neq 0$$

is a geometric series with ratio r. When $0 < |r| < 1$, the series converges to $a/(1-r)$. The series diverges if $|r| \geq 1$.

77. $a_n = \dfrac{n+1}{n}$

$\displaystyle\lim_{n\to\infty} a_n = \lim_{n\to\infty} \frac{n+1}{n} = 1$ ($\{a_n\}$ converges)

$\displaystyle\sum_{n=1}^{\infty} a_n$ diverges because $\displaystyle\lim_{n\to\infty} a_n \neq 0$.

79. $\displaystyle\sum_{n=1}^{\infty} \frac{x^n}{2^n} = \sum_{n=1}^{\infty}\left(\frac{x}{2}\right)^n = \frac{x}{2}\sum_{n=0}^{\infty}\left(\frac{x}{2}\right)^n$

Geometric series: converges for $\left|\dfrac{x}{2}\right| < 1$ or $|x| < 2$

$f(x) = \dfrac{x}{2}\displaystyle\sum_{n=0}^{\infty}\left(\frac{x}{2}\right)^n$

$\quad = \dfrac{x}{2}\dfrac{1}{1-(x/2)} = \dfrac{x}{2}\dfrac{2}{2-x} = \dfrac{x}{2-x}, \quad |x| < 2$

81. $\displaystyle\sum_{n=1}^{\infty} (x-1)^n = (x-1)\sum_{n=0}^{\infty}(x-1)^n$

Geometric series: converges for $|x - 1| < 1 \Rightarrow 0 < x < 2$

$f(x) = (x-1)\displaystyle\sum_{n=0}^{\infty}(x-1)^n$

$\quad = (x-1)\dfrac{1}{1-(x-1)} = \dfrac{x-1}{2-x}, \quad 0 < x < 2$

83. $\sum_{n=0}^{\infty} (-1)^n x^n = \sum_{n=0}^{\infty} (-x)^n$

Geometric series: converges for

$|-x| < 1 \Rightarrow |x| < 1 \Rightarrow -1 < x < 1$

$f(x) = \sum_{n=0}^{\infty} (-x)^n = \dfrac{1}{1+x}, \quad -1 < x < 1$

85. $\sum_{n=0}^{\infty} \left(\dfrac{1}{x}\right)^n$

Geometric series: converges if $\left|\dfrac{1}{x}\right| < 1$

$\Rightarrow |x| > 1 \Rightarrow x < -1 \text{ or } x > 1$

$f(x) = \sum_{n=0}^{\infty} \left(\dfrac{1}{x}\right)^n = \dfrac{1}{1-(1/x)} = \dfrac{x}{x-1}, \quad x > 1 \text{ or } x < -1$

87. (a) Yes, the new series will still diverge. (b) Yes, the new series will converge.

89. (a) x is the common ratio.

(b) $1 + x + x^2 + \cdots = \sum_{n=0}^{\infty} x^n = \dfrac{1}{1-x}, \quad |x| < 1$

(c) $y_1 = \dfrac{1}{1-x}$

$y_2 = S_3 = 1 + x + x^2$

$y_3 = S_5 = 1 + x + x^2 + x^3 + x^4$

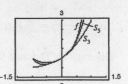

91. $f(x) = 3\left[\dfrac{1 - 0.5^x}{1 - 0.5}\right]$

Horizontal asymptote: $y = 6$

$\sum_{n=0}^{\infty} 3\left(\dfrac{1}{2}\right)^n$

$S = \dfrac{3}{1 - (1/2)} = 6$

The horizontal asymptote is the sum of the series.
$f(n)$ is the nth partial sum.

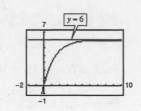

93. $\dfrac{1}{n(n+1)} < 0.0001$

$10{,}000 < n^2 + n$

$0 < n^2 + n - 10{,}000$

$n = \dfrac{-1 \pm \sqrt{1^2 - 4(1)(-10{,}000)}}{2}$

Choosing the positive value for n we have $n \approx 99.5012$.
The first *term* that is less than 0.0001 is $n = 100$.

$\left(\dfrac{1}{8}\right)^n < 0.0001$

$10{,}000 < 8^n$

This inequality is true when $n = 5$. This series converges at a faster rate.

95. $\sum_{i=0}^{n-1} 8000(0.9)^i = \dfrac{8000[1 - (0.9)^{(n-1)+1}]}{1 - 0.9}$

$= 80{,}000(1 - 0.9^n), \quad n > 0$

97. $\sum_{i=0}^{n-1} 100(0.75)^i = \dfrac{100[1 - 0.75^{(n-1)+1}]}{1 - 0.75}$

$= 400(1 - 0.75^n) \text{ million dollars}$

Sum = 400 million dollars

99. $D_1 = 16$

$D_2 = \underbrace{0.81(16)}_{\text{up}} + \underbrace{0.81(16)}_{\text{down}} = 32(0.81)$

$D_3 = 16(0.81)^2 + 16(0.81)^2 = 32(0.81)^2$

$\vdots$

$D = 16 + 32(0.81) + 32(0.81)^2 + \cdots$

$= -16 + \sum_{n=0}^{\infty} 32(0.81)^n = -16 + \dfrac{32}{1 - 0.81}$

$\approx 152.42 \text{ feet}$

101. $P(n) = \dfrac{1}{2}\left(\dfrac{1}{2}\right)^n$

$P(2) = \dfrac{1}{2}\left(\dfrac{1}{2}\right)^2 = \dfrac{1}{8}$

$\sum_{n=0}^{\infty} \dfrac{1}{2}\left(\dfrac{1}{2}\right)^n = \dfrac{1/2}{1 - (1/2)} = 1$

103. (a) $\displaystyle\sum_{n=1}^{\infty}\left(\frac{1}{2}\right)^n = \sum_{n=0}^{\infty}\frac{1}{2}\left(\frac{1}{2}\right)^n$

$$= \frac{1}{2}\,\frac{1}{(1-(1/2))} = 1$$

(b) No, the series is not geometric.

(c) $\displaystyle\sum_{n=1}^{\infty} n\left(\frac{1}{2}\right)^n = 2$

105. (a) $64 + 32 + 16 + 8 + 4 + 2 = 126$ in.^{2}

(b) $\displaystyle\sum_{n=0}^{\infty} 64\left(\frac{1}{2}\right)^n = \frac{64}{1-(1/2)} = 128$ in.^{2}

Note: This is one-half of the area of the original square!

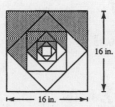

16 in.

16 in.

107. $\displaystyle\sum_{n=1}^{20} 50{,}000\left(\frac{1}{1.06}\right)^n = \frac{50{,}000}{1.06}\sum_{i=0}^{19}\left(\frac{1}{1.06}\right)^i$

$$= \frac{50{,}000}{1.06}\left[\frac{1-1.06^{-20}}{1-1.06^{-1}}\right], \quad [n=20, r=1.06^{-1}]$$

$$\approx 573{,}496.06$$

109. $w = \displaystyle\sum_{i=0}^{n-1} 0.01(2)^i = \frac{0.01(1-2^n)}{1-2} = 0.01(2^n - 1)$

(a) When $n = 29$: $w = \$5{,}368{,}709.11$

(b) When $n = 30$: $w = \$10{,}737{,}418.23$

(c) When $n = 31$: $w = \$21{,}474{,}836.47$

111. $P = 50, r = 0.03, t = 20$

(a) $A = 50\left(\dfrac{12}{0.03}\right)\left[\left(1+\dfrac{0.03}{12}\right)^{12(20)} - 1\right] \approx \$16{,}415.10$

(b) $A = \dfrac{50(e^{0.03(20)} - 1)}{e^{0.03/12} - 1} \approx \$16{,}421.83$

113. $P = 100, r = 0.04, t = 40$

(a) $A = 100\left(\dfrac{12}{0.04}\right)\left[\left(1+\dfrac{0.04}{12}\right)^{12(40)} - 1\right] \approx \$118{,}196.13$

(b) $A = \dfrac{100(e^{0.04(40)} - 1)}{e^{0.04/12} - 1} \approx \$118{,}393.43$

115. (a) $a_n = 3484.1363(1.0502)^n = 3484.1363e^{0.04897n}$

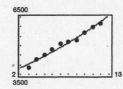

(b) Adding the ten values $a_3, a_4, \ldots, a_{12}$ you obtain

$$\sum_{n=3}^{12} a_n = 50{,}809, \text{ or } \$50{,}809{,}000{,}000.$$

(c) $\displaystyle\sum_{n=3}^{12} 3484.1363e^{0.04897n} \approx 50{,}803, \text{ or } \$50{,}803{,}000{,}000$

(Answers will vary.)

117. False. $\displaystyle\lim_{n\to\infty}\frac{1}{n} = 0$, but $\displaystyle\sum_{n=1}^{\infty}\frac{1}{n}$ diverges.

119. False; $\displaystyle\sum_{n=1}^{\infty} ar^n = \left(\frac{a}{1-r}\right) - a$

The formula requires that the geometric series begins with $n = 0$.

121. True

$$0.74999\ldots = 0.74 + \frac{9}{10^3} + \frac{9}{10^4} + \cdots$$

$$= 0.74 + \frac{9}{10^3} \sum_{n=0}^{\infty} \left(\frac{1}{10}\right)^n$$

$$= 0.74 + \frac{9}{10^3} \cdot \frac{1}{1 - (1/10)}$$

$$= 0.74 + \frac{9}{10^3} \cdot \frac{10}{9}$$

$$= 0.74 + \frac{1}{100} = 0.75$$

123. By letting $S_0 = 0$, we have

$$a_n = \sum_{k=1}^{n} a_k - \sum_{k=1}^{n-1} a_k = S_n - S_{n-1}. \text{ Thus,}$$

$$\sum_{n=1}^{\infty} a_n = \sum_{n=1}^{\infty} (S_n - S_{n-1})$$

$$= \sum_{n=1}^{\infty} (S_n - S_{n-1} + c - c)$$

$$= \sum_{n=1}^{\infty} [(c - S_{n-1}) - (c - S_n)].$$

125. Let $\sum a_n = \sum_{n=0}^{\infty} 1$ and $\sum b_n = \sum_{n=0}^{\infty} (-1)$.

Both are divergent series.

$$\sum (a_n + b_n) = \sum_{n=0}^{\infty} [1 + (-1)] = \sum_{n=0}^{\infty} [1 - 1] = 0$$

127. Suppose, on the contrary, that $\Sigma\, ca_n$ converges. Because $c \neq 0$,

$$\sum \left(\frac{1}{c}\right) ca_n = \sum a_n$$

converges. This is a contradiction since $\Sigma\, ca_n$ diverged. Hence, $\Sigma\, ca_n$ diverges.

129. (a) $\dfrac{1}{a_{n+1}a_{n+2}} - \dfrac{1}{a_{n+2}a_{n+3}} = \dfrac{a_{n+3} - a_{n+1}}{a_{n+1}a_{n+2}a_{n+3}}$

$$= \frac{a_{n+2}}{a_{n+1}a_{n+2}a_{n+3}}$$

$$= \frac{1}{a_{n+1}a_{n+3}}$$

(b) $S_n = \sum_{k=0}^{n} \dfrac{1}{a_{k+1}a_{k+3}}$

$$= \sum_{k=0}^{n} \left[\frac{1}{a_{k+1}a_{k+2}} - \frac{1}{a_{k+2}a_{k+3}} \right]$$

$$= \left[\frac{1}{a_1 a_2} - \frac{1}{a_2 a_3} \right] + \left[\frac{1}{a_2 a_3} - \frac{1}{a_3 a_4} \right] + \cdots + \left[\frac{1}{a_{n+1}a_{n+2}} - \frac{1}{a_{n+2}a_{n+3}} \right]$$

$$= \frac{1}{a_1 a_2} - \frac{1}{a_{n+2}a_{n+3}}$$

$$= 1 - \frac{1}{a_{n+2}a_{n+3}}$$

$$\sum_{n=0}^{\infty} \frac{1}{a_{n+1}a_{n+3}} = \lim_{n \to \infty} S_n = \lim_{n \to \infty} \left[1 - \frac{1}{a_{n+2}a_{n+3}} \right] = 1$$

131. $\dfrac{1}{r} + \dfrac{1}{r^2} + \dfrac{1}{r^3} + \cdots = \displaystyle\sum_{n=0}^{\infty} \frac{1}{r}\left(\frac{1}{r}\right)^n$

$$= \frac{1/r}{1 - (1/r)}$$

$$= \frac{1}{r-1} \quad \left(\text{since } \left|\frac{1}{r}\right| < 1\right)$$

This is a geometric series which converges if

$$\left|\frac{1}{r}\right| < 1 \iff |r| > 1.$$

133. Let H represent the half-life of the drug. If a patient receives n equal doses of P units each of this drug, administered at equal time interval of length t, the total amount of the drug in the patient's system at the time the last dose is administered is given by $T_n = P + Pe^{kt} + Pe^{2kt} + \cdots + Pe^{(n-1)kt}$ where $k = -(\ln 2)/H$. One time interval *after* the last dose is administered is given by $T_{n+1} = Pe^{kt} + Pe^{2kt} + Pe^{3kt} + \cdots + Pe^{nkt}$. Two time intervals *after* the last dose is administered is given by $T_{n+2} = Pe^{2kt} + Pe^{3kt} + Pe^{4kt} + \cdots + Pe^{(n+1)kt}$ and so on. Since $k < 0$, $T_{n+s} \to 0$ as $s \to \infty$, where s is an integer.

135. $f(1) = 0, f(2) = 1, f(3) = 2, f(4) = 4, \ldots$

In general: $f(n) = \begin{cases} n^2/4, & n \text{ even} \\ (n^2 - 1)/4, & n \text{ odd.} \end{cases}$

(See below for a proof of this.)

$x + y$ and $x - y$ are either both odd or both even. If both even, then

$$f(x + y) - f(x - y) = \frac{(x + y)^2}{4} - \frac{(x - y)^2}{4} = xy.$$

If both odd,

$$f(x + y) - f(x - y) = \frac{(x + y)^2 - 1}{4} - \frac{(x - y)^2 - 1}{4}$$

$$= xy.$$

Proof by induction that the formula for $f(n)$ is correct. It is true for $n = 1$. Assume that the formula is valid for k. If k is even, then $f(k) = k^2/4$ and

$$f(k + 1) = f(k) + \frac{k}{2} = \frac{k^2}{4} + \frac{k}{2}$$

$$= \frac{k^2 + 2k}{4} = \frac{(k + 1)^2 - 1}{4}.$$

The argument is similar if k is odd.

Section 9.3   The Integral Test and *p*-Series

1. $\displaystyle\sum_{n=1}^{\infty} \frac{1}{n + 1}$

Let $f(x) = \dfrac{1}{x + 1}$.

f is positive, continuous, and decreasing for $x \geq 1$.

$$\int_1^{\infty} \frac{1}{x + 1}\, dx = \left[\ln(x + 1) \right]_1^{\infty} = \infty$$

Diverges by Theorem 9.10

3. $\displaystyle\sum_{n=1}^{\infty} e^{-n}$

Let $f(x) = e^{-x}$.

f is positive, continuous, and decreasing for $x \geq 1$.

$$\int_1^{\infty} e^{-x}\, dx = \left[-e^{-x} \right]_1^{\infty} = \frac{1}{e}$$

Converges by Theorem 9.10

5. $\displaystyle\sum_{n=1}^{\infty} \frac{1}{n^2 + 1}$

Let $f(x) = \dfrac{1}{x^2 + 1}$.

f is positive, continuous, and decreasing for $x \geq 1$.

$$\int_1^{\infty} \frac{1}{x^2 + 1}\, dx = \left[\arctan x \right]_1^{\infty} = \frac{\pi}{4}$$

Converges by Theorem 9.10

7. $\displaystyle\sum_{n=1}^{\infty} \frac{\ln(n + 1)}{n + 1}$

Let $f(x) = \dfrac{\ln(x + 1)}{x + 1}$.

f is positive, continuous, and decreasing for $x \geq 2$ since

$$f'(x) = \frac{1 - \ln(x + 1)}{(x + 1)^2} < 0 \text{ for } x \geq 2.$$

$$\int_1^{\infty} \frac{\ln(x + 1)}{x + 1}\, dx = \left[\frac{[\ln(x + 1)]^2}{2} \right]_1^{\infty} = \infty$$

Diverges by Theorem 9.10

9. $\displaystyle\sum_{n=1}^{\infty} \frac{1}{\sqrt{n}\left(\sqrt{n}+1\right)}$

Let $f(x) = \dfrac{1}{\sqrt{x}\left(\sqrt{x}+1\right)}$, $f'(x) = -\dfrac{1+2\sqrt{x}}{2x^{3/2}\left(\sqrt{x}+1\right)^2} < 0.$

f is positive, continuous, and decreasing for $x \geq 1.$

$\displaystyle\int_1^{\infty} \frac{1}{\sqrt{x}\left(\sqrt{x}+1\right)}\,dx = \left[2\ln\left(\sqrt{x}+1\right)\right]_1^{\infty} = \infty,$ diverges

Hence, the series diverges by Theorem 9.10.

11. $\displaystyle\sum_{n=1}^{\infty} \frac{1}{\sqrt{n+1}}$

Let $f(x) = \dfrac{1}{\sqrt{x+1}}$, $f'(x) = \dfrac{-1}{2(x+1)^{3/2}} < 0.$

f is positive, continuous, and decreasing for $x \geq 1.$

$\displaystyle\int_1^{\infty} \frac{1}{\sqrt{x+1}}\,dx = \left[2\sqrt{x+1}\right]_1^{\infty} = \infty,$ diverges

Hence, the series diverges by Theorem 9.10.

13. $\displaystyle\sum_{n=1}^{\infty} \frac{\ln n}{n^2}$

Let $f(x) = \dfrac{\ln x}{x^2}$, $f'(x) = \dfrac{1 - 2\ln x}{x^3}.$

f is positive, continuous, and decreasing for $x > e^{1/2} \approx 1.6.$

$\displaystyle\int_1^{\infty} \frac{\ln x}{x^2}\,dx = \left[\frac{-(\ln x + 1)}{x}\right]_1^{\infty} = 1,$ converges

Hence, the series converges by Theorem 9.10.

15. $\displaystyle\sum_{n=1}^{\infty} \frac{\arctan n}{n^2+1}$

Let $f(x) = \dfrac{\arctan x}{x^2+1}$, $f'(x) = \dfrac{1 - 2x\arctan x}{(x^2+1)^2} < 0$ for $x \geq 1.$

f is positive, continuous, and decreasing for $x \geq 1.$

$\displaystyle\int_1^{\infty} \frac{\arctan x}{x^2+1}\,dx = \left[\frac{(\arctan x)^2}{2}\right]_1^{\infty} = \frac{3\pi^2}{32},$ converges

Hence, the series converges by Theorem 9.10.

17. $\displaystyle\sum_{n=1}^{\infty} \frac{2n}{n^2+1}$

Let $f(x) = \dfrac{2x}{x^2+1}$, $f'(x) = 2\dfrac{1 - x^2}{(x^2+1)^2} < 0$ for $x > 1.$

f is positive, continuous, and decreasing for $x > 1.$

$\displaystyle\int_1^{\infty} \frac{2x}{x^2+1}\,dx = \left[\ln(x^2+1)\right]_1^{\infty} = \infty,$ diverges

Hence, the series diverges by Theorem 9.10.

19. $\displaystyle\sum_{n=1}^{\infty} \frac{n^{k-1}}{n^k+c}$

Let $f(x) = \dfrac{x^{k-1}}{x^k+c}.$

f is positive, continuous, and decreasing for
$x > \sqrt[k]{c(k-1)}$ since

$f'(x) = \dfrac{x^{k-2}[c(k-1) - x^k]}{(x^k+c)^2} < 0$

for $x > \sqrt[k]{c(k-1)}.$

$\displaystyle\int_1^{\infty} \frac{x^{k-1}}{x^k+c}\,dx = \left[\frac{1}{k}\ln(x^k+c)\right]_1^{\infty} = \infty$

Diverges by Theorem 9.10

21. Let $f(x) = \dfrac{(-1)^x}{x}$, $f(n) = a_n.$

The function f is not positive for $x \geq 1.$

23. Let $f(x) = \dfrac{2 + \sin x}{x}$, $f(n) = a_n.$

The function f is not decreasing for $x \geq 1.$

25. $\displaystyle\sum_{n=1}^{\infty} \frac{1}{n^3}$

Let $f(x) = \dfrac{1}{x^3}.$

f is positive, continuous, and decreasing for $x \geq 1.$

$\displaystyle\int_1^{\infty} \frac{1}{x^3}\,dx = \left[-\frac{1}{2x^2}\right]_1^{\infty} = \frac{1}{2}$

Converges by Theorem 9.10

27. $\displaystyle\sum_{n=1}^{\infty} \frac{1}{\sqrt{n}}$

Let $f(x) = \dfrac{1}{\sqrt{x}}$, $f'(x) = \dfrac{-1}{2x^{3/2}} < 0$ for $x \geq 1.$

f is positive, continuous, and decreasing for $x \geq 1.$

$\displaystyle\int_1^{\infty} \frac{1}{\sqrt{x}}\,dx = \left[2\sqrt{x}\right]_1^{\infty} = \infty,$ diverges

Hence, the series diverges by Theorem 9.10.

29. $\sum_{n=1}^{\infty} \dfrac{1}{\sqrt[5]{n}} = \sum_{n=1}^{\infty} \dfrac{1}{n^{1/5}}$

Divergent p-series
with $p = \frac{1}{5} < 1$

31. $\sum_{n=1}^{\infty} \dfrac{1}{n^{1/2}}$

Divergent p-series
with $p = \frac{1}{2} < 1$

33. $\sum_{n=1}^{\infty} \dfrac{1}{n^{3/2}}$

Convergent p-series
with $p = \frac{3}{2} > 1$

35. $\sum_{n=1}^{\infty} \dfrac{1}{n^{1.04}}$

Convergent p-series
with $p = 1.04 > 1$

37. $\sum_{n=1}^{\infty} \dfrac{2}{n^{3/4}}$

$S_1 = 2$

$S_2 \approx 3.1892$

$S_3 \approx 4.0666$

$S_4 \approx 4.7740$

Matches (c), diverges

39. $\sum_{n=1}^{\infty} \dfrac{2}{\sqrt{n^{\pi}}} = \sum_{n=1}^{\infty} \dfrac{2}{n^{\pi/2}}$

$S_1 = 2$

$S_2 \approx 2.6732$

$S_3 \approx 3.0293$

$S_4 \approx 3.2560$

Matches (b), converges

Note: The partial sums for 39
and 41 are very similar because
$\pi/2 \approx 3/2$.

41. $\sum_{n=1}^{\infty} \dfrac{2}{n\sqrt{n}} = \sum_{n=1}^{\infty} \dfrac{2}{n^{3/2}}$

$S_1 = 2$

$S_2 \approx 2.7071$

$S_3 \approx 3.0920$

$S_4 \approx 3.3420$

Matches (d), converges

Note: The partial sums for 39
and 41 are very similar because
$\pi/2 \approx 3/2$.

43. (a)

n	5	10	20	50	100
S_n	3.7488	3.75	3.75	3.75	3.75

The partial sums approach the sum 3.75 very rapidly.

(b)

n	5	10	20	50	100
S_n	1.4636	1.5498	1.5962	1.6251	1.635

The partial sums approach the sum $\pi^2/6 \approx 1.6449$
slower than the series in part (a).

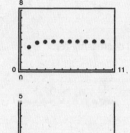

45. Let f be positive, continuous, and decreasing for $x \geq 1$
and $a_n = f(n)$. Then,

$$\sum_{n=1}^{\infty} a_n \text{ and } \int_{1}^{\infty} f(x)\,dx$$

either both converge or both diverge (Theorem 9.10).
See Example 1, page 618.

47. Your friend is not correct. The series

$$\sum_{n=10,000}^{\infty} \frac{1}{n} = \frac{1}{10,000} + \frac{1}{10,001} + \cdots$$

is the harmonic series, starting with the 10,000th term,
and hence diverges.

49. The area under the rectangles is greater than the area
under the graph of $y = 1/\sqrt{x}, x \geq 1$.

$$\sum_{n=1}^{\infty} \frac{1}{\sqrt{n}} > \int_{1}^{\infty} \frac{1}{\sqrt{x}}\,dx$$

Since $\int_{1}^{\infty} \dfrac{1}{\sqrt{x}}\,dx = \left[2\sqrt{x}\right]_{1}^{\infty} = \infty$ diverges, the series

$$\sum_{n=1}^{\infty} \frac{1}{\sqrt{n}} \text{ diverges.}$$

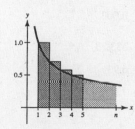

51. $\sum_{n=2}^{\infty} \dfrac{1}{n(\ln n)^p}$

If $p = 1$, then the series diverges by the Integral Test.
If $p \neq 1$,

$$\int_{2}^{\infty} \frac{1}{x(\ln x)^p}\,dx = \int_{2}^{\infty} (\ln x)^{-p} \frac{1}{x}\,dx = \left[\frac{(\ln x)^{-p+1}}{-p+1}\right]_{2}^{\infty}.$$

Converges for $-p + 1 < 0$ or $p > 1$

53. $\sum_{n=1}^{\infty} \dfrac{n}{(1 + n^2)^p}$

If $p = 1$, $\sum_{n=1}^{\infty} \dfrac{n}{1 + n^2}$ diverges (see Example 1). Let

$$f(x) = \frac{x}{(1 + x^2)^p}, \quad p \neq 1$$

$$f'(x) = \frac{1 - (2p - 1)x^2}{(1 + x^2)^{p+1}}.$$

For a fixed $p > 0$, $p \neq 1$, $f'(x)$ is eventually negative. f is positive, continuous, and eventually decreasing.

$$\int_1^{\infty} \frac{x}{(1 + x^2)^p} \, dx = \left[\frac{1}{(x^2 + 1)^{p-1}(2 - 2p)} \right]_1^{\infty}$$

For $p > 1$, this integral converges. For $0 < p < 1$, it diverges.

55. $\sum_{n=2}^{\infty} \dfrac{1}{n(\ln n)^p}$ converges for $p > 1$.

Hence, $\sum_{n=2}^{\infty} \dfrac{1}{n \ln n}$, diverges.

57. $\sum_{n=2}^{\infty} \dfrac{1}{n(\ln n)^p}$ converges for $p > 1$.

Hence, $\sum_{n=2}^{\infty} \dfrac{1}{n(\ln n)^2}$, converges.

59. Since f is positive, continuous, and decreasing for $x \geq 1$ and $a_n = f(n)$, we have,

$$R_N = S - S_N = \sum_{n=1}^{\infty} a_n - \sum_{n=1}^{N} a_n = \sum_{n=N+1}^{\infty} a_n > 0.$$

Also,

$$R_N = S - S_N = \sum_{n=N+1}^{\infty} a_n \leq a_{N+1} + \int_{N+1}^{\infty} f(x) \, dx$$

$$\leq \int_N^{\infty} f(x) \, dx.$$

Thus, $0 \leq R_N \leq \displaystyle\int_N^{\infty} f(x) \, dx.$

61. $S_6 = 1 + \dfrac{1}{2^4} + \dfrac{1}{3^4} + \dfrac{1}{4^4} + \dfrac{1}{5^4} + \dfrac{1}{6^4} \approx 1.0811$

$$R_6 \leq \int_6^{\infty} \frac{1}{x^4} \, dx = \left[-\frac{1}{3x^3} \right]_6^{\infty} \approx 0.0015$$

$$1.0811 \leq \sum_{n=1}^{\infty} \frac{1}{n^4} \leq 1.0811 + 0.0015 = 1.0826$$

63. $S_{10} = \dfrac{1}{2} + \dfrac{1}{5} + \dfrac{1}{10} + \dfrac{1}{17} + \dfrac{1}{26} + \dfrac{1}{37} + \dfrac{1}{50} + \dfrac{1}{65} + \dfrac{1}{82} + \dfrac{1}{101} \approx 0.9818$

$$R_{10} \leq \int_{10}^{\infty} \frac{1}{x^2 + 1} \, dx = \Big[\arctan x \Big]_{10}^{\infty} = \frac{\pi}{2} - \arctan 10 \approx 0.0997$$

$$0.9818 \leq \sum_{n=1}^{\infty} \frac{1}{n^2 + 1} \leq 0.9818 + 0.0997 = 1.0815$$

65. $S_4 = \dfrac{1}{e} + \dfrac{2}{e^4} + \dfrac{3}{e^9} + \dfrac{4}{e^{16}} \approx 0.4049$

$$R_4 \leq \int_4^{\infty} xe^{-x^2} \, dx = \left[-\frac{1}{2} e^{-x^2} \right]_4^{\infty} = \frac{e^{-16}}{2} \approx 5.6 \times 10^{-8}$$

$$0.4049 \leq \sum_{n=1}^{\infty} ne^{-n^2} \leq 0.4049 + 5.6 \times 10^{-8}$$

67. $0 \leq R_N \leq \displaystyle\int_N^{\infty} \frac{1}{x^4} \, dx = \left[-\frac{1}{3x^3} \right]_N^{\infty} = \frac{1}{3N^3} < 0.001$

$$\frac{1}{N^3} < 0.003$$

$$N^3 > 333.33$$

$$N > 6.93$$

$$N \geq 7$$

69. $R_N \leq \displaystyle\int_N^\infty e^{-5x}\,dx = \left[-\dfrac{1}{5}e^{-5x}\right]_N^\infty = \dfrac{e^{-5N}}{5} < 0.001$

$\dfrac{1}{e^{5N}} < 0.005$

$e^{5N} > 200$

$5N > \ln 200$

$N > \dfrac{\ln 200}{5}$

$N > 1.0597$

$N \geq 2$

71. $R_N \leq \displaystyle\int_N^\infty \dfrac{1}{x^2+1}\,dx = \Big[\arctan x\Big]_N^\infty$

$\qquad = \dfrac{\pi}{2} - \arctan N < 0.001$

$-\arctan N < -1.5698$

$\arctan N > 1.5698$

$N > \tan 1.5698$

$N \geq 1004$

73. (a) $\displaystyle\sum_{n=2}^\infty \dfrac{1}{n^{1.1}}$. This is a convergent p-series with $p = 1.1 > 1$. $\displaystyle\sum_{n=2}^\infty \dfrac{1}{n\ln n}$ is a divergent series. Use the Integral Test.

$f(x) = \dfrac{1}{x\ln x}$ is positive, continuous, and decreasing for $x \geq 2$.

$\displaystyle\int_2^\infty \dfrac{1}{x\ln x}\,dx = \Big[\ln|\ln x|\Big]_2^\infty = \infty$

(b) $\displaystyle\sum_{n=2}^6 \dfrac{1}{n^{1.1}} = \dfrac{1}{2^{1.1}} + \dfrac{1}{3^{1.1}} + \dfrac{1}{4^{1.1}} + \dfrac{1}{5^{1.1}} + \dfrac{1}{6^{1.1}} \approx 0.4665 + 0.2987 + 0.2176 + 0.1703 + 0.1393$

$\displaystyle\sum_{n=2}^6 \dfrac{1}{n\ln n} = \dfrac{1}{2\ln 2} + \dfrac{1}{3\ln 3} + \dfrac{1}{4\ln 4} + \dfrac{1}{5\ln 5} + \dfrac{1}{6\ln 6} \approx 0.7213 + 0.3034 + 0.1803 + 0.1243 + 0.0930$

For $n \geq 4$, the terms of the convergent series seem to be larger than those of the divergent series!

(c) $\dfrac{1}{n^{1.1}} < \dfrac{1}{n\ln n}$

$n\ln n < n^{1.1}$

$\ln n < n^{0.1}$

This inequality holds when $n \geq 3.5 \times 10^{15}$. Or, $n > e^{40}$. Then $\ln e^{40} = 40 < (e^{40})^{0.1} = e^4 \approx 55$.

75. (a) Let $f(x) = 1/x$. f is positive, continuous, and decreasing on $[1, \infty)$.

$S_n - 1 \leq \displaystyle\int_1^n \dfrac{1}{x}\,dx$

$S_n - 1 \leq \ln n$

Hence, $S_n \leq 1 + \ln n$. Similarly,

$S_n \geq \displaystyle\int_1^{n+1} \dfrac{1}{x}\,dx = \ln(n+1)$.

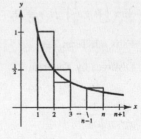

Thus, $\ln(n+1) \leq S_n \leq 1 + \ln n$.

(b) Since $\ln(n+1) \leq S_n \leq 1 + \ln n$, we have $\ln(n+1) - \ln n \leq S_n - \ln n \leq 1$. Also, since $\ln x$ is an increasing function, $\ln(n+1) - \ln n > 0$ for $n \geq 1$. Thus, $0 \leq S_n - \ln n \leq 1$ and the sequence $\{a_n\}$ is bounded.

(c) $a_n - a_{n+1} = [S_n - \ln n] - [S_{n+1} - \ln(n+1)] = \displaystyle\int_n^{n+1} \dfrac{1}{x}\,dx - \dfrac{1}{n+1} \geq 0$

Thus, $a_n \geq a_{n+1}$ and the sequence is decreasing.

(d) Since the sequence is bounded and monotonic, it converges to a limit, γ.

(e) $a_{100} = S_{100} - \ln 100 \approx 0.5822$ (Actually $\gamma \approx 0.577216$.)

77. $\displaystyle\sum_{n=2}^{\infty} x^{\ln n}$

(a) $x = 1$: $\displaystyle\sum_{n=2}^{\infty} 1^{\ln n} = \sum_{n=2}^{\infty} 1$, diverges

(b) $x = \dfrac{1}{e}$: $\displaystyle\sum_{n=2}^{\infty} \left(\dfrac{1}{e}\right)^{\ln n} = \sum_{n=2}^{\infty} e^{-\ln n} = \sum_{n=2}^{\infty} \dfrac{1}{n}$, diverges

(c) Let x be given, $x > 0$. Put $x = e^{-p} \iff \ln x = -p$.

$$\sum_{n=2}^{\infty} x^{\ln n} = \sum_{n=2}^{\infty} e^{-p \ln n} = \sum_{n=2}^{\infty} n^{-p} = \sum_{n=2}^{\infty} \dfrac{1}{n^p}$$

This series converges for $p > 1 \implies x < \dfrac{1}{e}$.

79. $\displaystyle\sum_{n=1}^{\infty} \dfrac{1}{2n-1}$

Let $f(x) = \dfrac{1}{2x-1}$.

f is positive, continuous, and decreasing for $x \geq 1$.

$$\int_1^{\infty} \dfrac{1}{2x-1}\, dx = \left[\ln\sqrt{2x-1}\right]_1^{\infty} = \infty$$

Diverges by Theorem 9.10

81. $\displaystyle\sum_{n=1}^{\infty} \dfrac{1}{n\sqrt[4]{n}} = \sum_{n=1}^{\infty} \dfrac{1}{n^{5/4}}$

p-series with $p = \dfrac{5}{4}$

Converges by Theorem 9.11

83. $\displaystyle\sum_{n=0}^{\infty} \left(\dfrac{2}{3}\right)^n$

Geometric series with $r = \dfrac{2}{3}$

Converges by Theorem 9.6

85. $\displaystyle\sum_{n=1}^{\infty} \dfrac{n}{\sqrt{n^2+1}}$

$$\lim_{n\to\infty} \dfrac{n}{\sqrt{n^2+1}} = \lim_{n\to\infty} \dfrac{1}{\sqrt{1+(1/n^2)}} = 1 \neq 0$$

Diverges by Theorem 9.9

87. $\displaystyle\sum_{n=1}^{\infty} \left(1 + \dfrac{1}{n}\right)^n$

$$\lim_{n\to\infty} \left(1 + \dfrac{1}{n}\right)^n = e \neq 0$$

Fails nth-Term Test

Diverges by Theorem 9.9

89. $\displaystyle\sum_{n=2}^{\infty} \dfrac{1}{n(\ln n)^3}$

Let $f(x) = \dfrac{1}{x(\ln x)^3}$.

f is positive, continuous and decreasing for $x \geq 2$.

$$\int_2^{\infty} \dfrac{1}{x(\ln x)^3}\, dx = \int_2^{\infty} (\ln x)^{-3}\dfrac{1}{x}\, dx = \left[\dfrac{(\ln x)^{-2}}{-2}\right]_2^{\infty} = \left[-\dfrac{1}{2(\ln x)^2}\right]_2^{\infty} = \dfrac{1}{2(\ln 2)^2}$$

Converges by Theorem 9.10. See Exercise 51.

Section 9.4   Comparisons of Series

1. (a) $\sum_{n=1}^{\infty} \frac{6}{n^{3/2}} = \frac{6}{1} + \frac{6}{2^{3/2}} + \cdots; \ S_1 = 6$

$\sum_{n=1}^{\infty} \frac{6}{n^{3/2} + 3} = \frac{6}{4} + \frac{6}{2^{3/2} + 3} + \cdots; \ S_1 = \frac{3}{2}$

$\sum_{n=1}^{\infty} \frac{6}{n\sqrt{n^2 + 0.5}} = \frac{6}{1\sqrt{1.5}} + \frac{6}{2\sqrt{4.5}} + \cdots; \ S_1 = \frac{6}{\sqrt{1.5}} \approx 4.9$

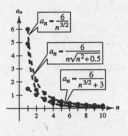

(b) The first series is a p-series. It converges $\left(p = \frac{3}{2} > 1 \right)$.

(c) The magnitude of the terms of the other two series are less than the corresponding terms at the convergent p-series. Hence, the other two series converge.

(d) The smaller the magnitude of the terms, the smaller the magnitude of the terms of the sequence of partial sums.

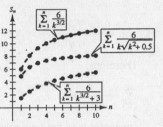

3. $0 < \dfrac{1}{n^2 + 1} < \dfrac{1}{n^2}$

Therefore,

$\sum_{n=1}^{\infty} \dfrac{1}{n^2 + 1}$

converges by comparison with the convergent p-series

$\sum_{n=1}^{\infty} \dfrac{1}{n^2}.$

5. $\dfrac{1}{n-1} > \dfrac{1}{n} > 0$ for $n \geq 2$

Therefore,

$\sum_{n=2}^{\infty} \dfrac{1}{n-1}$

diverges by comparison with the divergent p-series

$\sum_{n=2}^{\infty} \dfrac{1}{n}.$

7. $0 < \dfrac{1}{3^n + 1} < \dfrac{1}{3^n}$

Therefore,

$\sum_{n=0}^{\infty} \dfrac{1}{3^n + 1}$

converges by comparison with the convergent geometric series

$\sum_{n=0}^{\infty} \left(\dfrac{1}{3} \right)^n.$

9. For $n \geq 3$, $\dfrac{\ln n}{n+1} > \dfrac{1}{n+1} > 0.$

Therefore,

$\sum_{n=1}^{\infty} \dfrac{\ln n}{n+1}$

diverges by comparison with the divergent series

$\sum_{n=1}^{\infty} \dfrac{1}{n+1}.$

Note: $\sum_{n=1}^{\infty} \dfrac{1}{n+1}$ diverges by the Integral Test.

11. For $n > 3$, $\dfrac{1}{n^2} > \dfrac{1}{n!} > 0.$

Therefore,

$\sum_{n=0}^{\infty} \dfrac{1}{n!}$

converges by comparison with the convergent p-series

$\sum_{n=1}^{\infty} \dfrac{1}{n^2}.$

13. $0 < \dfrac{1}{e^{n^2}} \leq \dfrac{1}{e^n}$

Therefore,

$\sum_{n=0}^{\infty} \dfrac{1}{e^{n^2}}$

converges by comparison with the convergent geometric series

$\sum_{n=0}^{\infty} \left(\dfrac{1}{e} \right)^n.$

15. $\lim_{n \to \infty} \dfrac{n/(n^2 + 1)}{1/n} = \lim_{n \to \infty} \dfrac{n^2}{n^2 + 1} = 1$

Therefore,

$\sum_{n=1}^{\infty} \dfrac{n}{n^2 + 1}$

diverges by a limit comparison with the divergent p-series

$\sum_{n=1}^{\infty} \dfrac{1}{n}.$

17. $\lim_{n \to \infty} \dfrac{1/\sqrt{n^2 + 1}}{1/n} = \lim_{n \to \infty} \dfrac{n}{\sqrt{n^2 + 1}} = 1$

Therefore,

$\sum_{n=0}^{\infty} \dfrac{1}{\sqrt{n^2 + 1}}$

diverges by a limit comparison with the divergent p-series

$\sum_{n=1}^{\infty} \dfrac{1}{n}.$

19. $\lim\limits_{n\to\infty} \dfrac{\dfrac{2n^2 - 1}{3n^5 + 2n + 1}}{1/n^3} = \lim\limits_{n\to\infty} \dfrac{2n^5 - n^3}{3n^5 + 2n + 1} = \dfrac{2}{3}$

Therefore,

$\sum\limits_{n=1}^{\infty} \dfrac{2n^2 - 1}{3n^5 + 2n + 1}$

converges by a limit comparison with the convergent p-series

$\sum\limits_{n=1}^{\infty} \dfrac{1}{n^3}.$

21. $\lim\limits_{n\to\infty} \dfrac{\dfrac{n + 3}{n(n + 2)}}{1/n} = \lim\limits_{n\to\infty} \dfrac{n^2 + 3n}{n^2 + 2n} = 1$

Therefore,

$\sum\limits_{n=1}^{\infty} \dfrac{n + 3}{n(n + 2)}$

diverges by a limit comparison with the divergent p-series

$\sum\limits_{n=1}^{\infty} \dfrac{1}{n}.$

23. $\lim\limits_{n\to\infty} \dfrac{1/\left(n\sqrt{n^2 + 1}\right)}{1/n^2} = \lim\limits_{n\to\infty} \dfrac{n^2}{n\sqrt{n^2 + 1}} = 1$

Therefore,

$\sum\limits_{n=1}^{\infty} \dfrac{1}{n\sqrt{n^2 + 1}}$

converges by a limit comparison with the convergent p-series

$\sum\limits_{n=1}^{\infty} \dfrac{1}{n^2}.$

25. $\lim\limits_{n\to\infty} \dfrac{(n^{k-1})/(n^k + 1)}{1/n} = \lim\limits_{n\to\infty} \dfrac{n^k}{n^k + 1} = 1$

Therefore,

$\sum\limits_{n=1}^{\infty} \dfrac{n^{k-1}}{n^k + 1}$

diverges by a limit comparison with the divergent p-series

$\sum\limits_{n=1}^{\infty} \dfrac{1}{n}.$

27. $\lim\limits_{n\to\infty} \dfrac{\sin(1/n)}{1/n} = \lim\limits_{n\to\infty} \dfrac{(-1/n^2)\cos(1/n)}{-1/n^2}$

$\qquad = \lim\limits_{n\to\infty} \cos\left(\dfrac{1}{n}\right) = 1$

Therefore,

$\sum\limits_{n=1}^{\infty} \sin\left(\dfrac{1}{n}\right)$

diverges by a limit comparison with the divergent p-series

$\sum\limits_{n=1}^{\infty} \dfrac{1}{n}.$

29. $\sum\limits_{n=1}^{\infty} \dfrac{\sqrt{n}}{n} = \sum\limits_{n=1}^{\infty} \dfrac{1}{\sqrt{n}}$

Diverges

p-series with $p = \frac{1}{2}$

31. $\sum\limits_{n=1}^{\infty} \dfrac{1}{3^n + 2}$

Converges

Direct comparison with $\sum\limits_{n=1}^{\infty} \left(\dfrac{1}{3}\right)^n$

33. $\sum\limits_{n=1}^{\infty} \dfrac{n}{2n + 3}$

Diverges; nth-Term Test

$\lim\limits_{n\to\infty} \dfrac{n}{2n + 3} = \dfrac{1}{2} \neq 0$

35. $\sum\limits_{n=1}^{\infty} \dfrac{n}{(n^2 + 1)^2}$

Converges; Integral Test

37. $\lim\limits_{n\to\infty} \dfrac{a_n}{1/n} = \lim\limits_{n\to\infty} na_n.$ By given conditions $\lim\limits_{n\to\infty} na_n$ is finite and nonzero. Therefore,

$\sum\limits_{n=1}^{\infty} a_n$

diverges by a limit comparison with the p-series

$\sum\limits_{n=1}^{\infty} \dfrac{1}{n}.$

39. $\dfrac{1}{2} + \dfrac{2}{5} + \dfrac{3}{10} + \dfrac{4}{17} + \dfrac{5}{26} + \cdots = \sum\limits_{n=1}^{\infty} \dfrac{n}{n^2 + 1},$

which diverges since the degree of the numerator is only one less than the degree of the denominator.

41. $\displaystyle\sum_{n=1}^{\infty}\frac{1}{n^3+1}$

converges since the degree of the numerator is three less than the degree of the denominator.

43. $\displaystyle\lim_{n\to\infty}n\left(\frac{n^3}{5n^4+3}\right)=\lim_{n\to\infty}\frac{n^4}{5n^4+3}=\frac{1}{5}\neq 0$

Therefore, $\displaystyle\sum_{n=1}^{\infty}\frac{n^3}{5n^4+3}$ diverges.

45. $\displaystyle\frac{1}{200}+\frac{1}{400}+\frac{1}{600}+\cdots=\sum_{n=1}^{\infty}\frac{1}{200n}$

diverges, (harmonic)

47. $\displaystyle\frac{1}{201}+\frac{1}{204}+\frac{1}{209}+\frac{1}{216}=\sum_{n=1}^{\infty}\frac{1}{200+n^2}$

converges

49. Some series diverge or converge very slowly. You cannot decide convergence or divergence of a series by comparing the first few terms.

51. See Theorem 9.13, page 626. One example is

$\displaystyle\sum_{n=2}^{\infty}\frac{1}{\sqrt{n-1}}$ diverges because $\displaystyle\lim_{n\to\infty}\frac{1/\sqrt{n-1}}{1/\sqrt{n}}=1$ and

$\displaystyle\sum_{n=2}^{\infty}\frac{1}{\sqrt{n}}$ diverges (p-series).

53.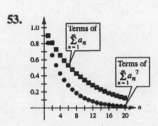

For $0<a_n<1,\ 0<a_n^{\ 2}<a_n<1$. Hence, the lower terms are those of $\sum a_n^{\ 2}$.

55. False. Let $a_n=\dfrac{1}{n^3}$ and $b_n=\dfrac{1}{n^2}$. $0<a_n\le b_n$ and both $\displaystyle\sum_{n=1}^{\infty}\frac{1}{n^3}$ and $\displaystyle\sum_{n=1}^{\infty}\frac{1}{n^2}$ converge.

57. True

59. True

61. Since $\displaystyle\sum_{n=1}^{\infty}b_n$ converges, $\displaystyle\lim_{n\to\infty}b_n=0$. There exists N such that $b_n<1$ for $n>N$. Thus, $a_nb_n<a_n$ for $n>N$ and

$\displaystyle\sum_{n=1}^{\infty}a_nb_n$ converges by comparison to the convergent

series $\displaystyle\sum_{i=1}^{\infty}a_n$.

63. $\displaystyle\sum\frac{1}{n^2}$ and $\displaystyle\sum\frac{1}{n^3}$ both converge, and hence so does

$$\sum\left(\frac{1}{n^2}\right)\left(\frac{1}{n^3}\right)=\sum\frac{1}{n^5}.$$

65. Suppose $\displaystyle\lim_{n\to\infty}\frac{a_n}{b_n}=0$ and $\sum b_n$ converges.

From the definition of limit of a sequence, there exists $M>0$ such that

$$\left|\frac{a_n}{b_n}-0\right|<1$$

whenever $n>M$. Hence, $a_n<b_n$ for $n>M$. From the Comparison Test, $\sum a_n$ converges.

67. (a) Let $\sum a_n = \sum \dfrac{1}{(n+1)^3}$, and $\sum b_n = \sum \dfrac{1}{n^2}$, converges.

$$\lim_{n\to\infty} \frac{a_n}{b_n} = \lim_{n\to\infty} \frac{1/[(n+1)^3]}{1/(n^2)} = \lim_{n\to\infty} \frac{n^2}{(n+1)^3} = 0$$

By Exercise 65, $\displaystyle\sum_{n=1}^{\infty} \frac{1}{(n+1)^3}$ converges.

(b) Let $\sum a_n = \sum \dfrac{1}{\sqrt{n}\pi^n}$, and $\sum b_n = \sum \dfrac{1}{\pi^n}$, converges.

$$\lim_{n\to\infty} \frac{a_n}{b_n} = \lim_{n\to\infty} \frac{1/(\sqrt{n}\pi^n)}{1/(\pi^n)} = \lim_{n\to\infty} \frac{1}{\sqrt{n}} = 0$$

By Exercise 65, $\displaystyle\sum_{n=1}^{\infty} \frac{1}{\sqrt{n}\pi^n}$ converges.

69. Since $\lim\limits_{n\to\infty} a_n = 0$, the terms of $\sum \sin(a_n)$ are positive for sufficiently large n. Since

$$\lim_{n\to\infty} \frac{\sin(a_n)}{a_n} = 1 \text{ and } \sum a_n$$

converges, so does $\sum \sin(a_n)$.

71. The series diverges. For $n > 1$,

$$n < 2^n$$
$$n^{1/n} < 2$$
$$\frac{1}{n^{1/n}} > \frac{1}{2}$$
$$\frac{1}{n^{(n+1)/n}} > \frac{1}{2n}$$

Since $\sum \dfrac{1}{2n}$ diverges, so does $\sum \dfrac{1}{n^{(n+1)/n}}$.

Section 9.5   Alternating Series

1. $\displaystyle\sum_{n=1}^{\infty} \frac{6}{n^2}$

$S_1 = 6$

$S_2 = 7.5$

$S_3 \approx 8.1667$

Matches (d).

3. $\displaystyle\sum_{n=1}^{\infty} \frac{3}{n!}$

$S_1 = 3$

$S_2 = 4.5$

$S_3 = 5.0$

Matches (a).

5. $\displaystyle\sum_{n=1}^{\infty} \frac{10}{n2^n}$

$S_1 = 5$

$S_2 = 6.25$

Matches (e).

7. $\displaystyle\sum_{n=1}^{\infty} \frac{(-1)^{n-1}}{2n-1} = \frac{\pi}{4} \approx 0.7854$

(a)

n	1	2	3	4	5	6	7	8	9	10
S_n	1	0.6667	0.8667	0.7238	0.8349	0.7440	0.8209	0.7543	0.8131	0.7605

(b)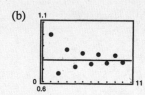

(c) The points alternate sides of the horizontal line that represents the sum of the series. The distance between successive points and the line decreases.

(d) The distance in part (c) is always less than the magnitude of the next term of the series.

9. $\displaystyle\sum_{n=1}^{\infty} \frac{(-1)^{n-1}}{n^2} = \frac{\pi^2}{12} \approx 0.8225$

(a)

n	1	2	3	4	5	6	7	8	9	10
S_n	1	0.75	0.8611	0.7986	0.8386	0.8108	0.8312	0.8156	0.8280	0.8180

(b)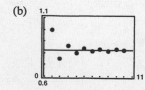

(c) The points alternate sides of the horizontal line that represents the sum of the series. The distance between successive points and the line decreases.

(d) The distance in part (c) is always less than the magnitude of the next term in the series.

11. $\displaystyle\sum_{n=1}^{\infty} \frac{(-1)^{n+1}}{n}$

$$a_{n+1} = \frac{1}{n+1} < \frac{1}{n} = a_n$$

$$\lim_{n\to\infty} \frac{1}{n} = 0$$

Converges by Theorem 9.14

13. $\displaystyle\sum_{n=1}^{\infty} \frac{(-1)^{n+1}}{2n-1}$

$$a_{n+1} = \frac{1}{2(n+1)-1} < \frac{1}{2n-1} = a_n$$

$$\lim_{n\to\infty} \frac{1}{2n-1} = 0$$

Converges by Theorem 9.14

15. $\displaystyle\sum_{n=1}^{\infty} \frac{(-1)^n n^2}{n^2+1}$

$$\lim_{n\to\infty} \frac{n^2}{n^2+1} = 1. \text{ Thus, } \lim_{n\to\infty} a_n \neq 0.$$

Diverges by the *n*th-Term Test

17. $\displaystyle\sum_{n=1}^{\infty} \frac{(-1)^n}{\sqrt{n}}$

$$a_{n+1} = \frac{1}{\sqrt{n+1}} < \frac{1}{\sqrt{n}} = a_n$$

$$\lim_{n\to\infty} \frac{1}{\sqrt{n}} = 0$$

Converges by Theorem 9.14

19. $\displaystyle\sum_{n=1}^{\infty} \frac{(-1)^{n+1}(n+1)}{\ln(n+1)}$

$$\lim_{n\to\infty} \frac{n+1}{\ln(n+1)} = \lim_{n\to\infty} \frac{1}{1/(n+1)} = \lim_{n\to\infty} (n+1) = \infty$$

Diverges by the *n*th-Term Test

21. $\displaystyle\sum_{n=1}^{\infty} \sin\left[\frac{(2n-1)\pi}{2}\right] = \sum_{n=1}^{\infty} (-1)^{n+1}$

Diverges by the *n*th-Term Test

23. $\displaystyle\sum_{n=1}^{\infty} \cos n\pi = \sum_{n=1}^{\infty} (-1)^n$

Diverges by the *n*th-Term Test

25. $\displaystyle\sum_{n=0}^{\infty} \frac{(-1)^n}{n!}$

$$a_{n+1} = \frac{1}{(n+1)!} < \frac{1}{n!} = a_n$$

$$\lim_{n\to\infty} \frac{1}{n!} = 0$$

Converges by Theorem 9.14

27. $\displaystyle\sum_{n=1}^{\infty} \frac{(-1)^{n+1}\sqrt{n}}{n+2}$

$$a_{n+1} = \frac{\sqrt{n+1}}{(n+1)+2}$$

$$< \frac{\sqrt{n}}{n+2} \text{ for } n \geq 2$$

$$\lim_{n\to\infty} \frac{\sqrt{n}}{n+2} = 0$$

Converges by Theorem 9.14

29. $\displaystyle\sum_{n=1}^{\infty} \frac{(-1)^{n+1}n!}{1\cdot3\cdot5\cdots(2n-1)}$

$$a_{n+1} = \frac{(n+1)!}{1\cdot3\cdot5\cdots(2n-1)(2n+1)} = \frac{n!}{1\cdot3\cdot5\cdots(2n-1)} \cdot \frac{n+1}{2n+1} = a_n\left(\frac{n+1}{2n+1}\right) < a_n$$

$$\lim_{n\to\infty} a_n = \lim_{n\to\infty} \frac{n!}{1\cdot3\cdot5\cdots(2n-1)}$$

$$= \lim_{n\to\infty} \frac{1\cdot2\cdot3\cdots n}{1\cdot3\cdot5\cdots(2n-1)}$$

$$= \lim_{n\to\infty} 2\left[\frac{3}{3}\cdot\frac{4}{5}\cdot\frac{5}{7}\cdots\frac{n}{2n-3}\right]\cdot\frac{1}{2n-1} = 0$$

Converges by Theorem 9.14

31. $\displaystyle\sum_{n=1}^{\infty} \frac{(-1)^{n+1}(2)}{e^n - e^{-n}} = \sum_{n=1}^{\infty} \frac{(-1)^{n+1}(2e^n)}{e^{2n} - 1}$

Let $f(x) = \dfrac{2e^x}{e^{2x} - 1}$. Then

$$f'(x) = \frac{-2e^x(e^{2x} + 1)}{(e^{2x} - 1)^2} < 0.$$

Thus, $f(x)$ is decreasing. Therefore, $a_{n+1} < a_n$, and

$$\lim_{n \to \infty} \frac{2e^n}{e^{2n} - 1} = \lim_{n \to \infty} \frac{2e^n}{2e^{2n}} = \lim_{n \to \infty} \frac{1}{e^n} = 0.$$

The series converges by Theorem 9.14.

33. $S_6 = \displaystyle\sum_{n=1}^{6} \frac{3(-1)^{n+1}}{n^2} = 2.4325$

$2.4325 - 0.0612 \le S \le 2.4325 + 0.0612$

$$|R_6| = |S - S_6| \le a_7 = \frac{3}{49} \approx 0.0612$$

$2.3713 \le S \le 2.4937$

35. $S_6 = \displaystyle\sum_{n=0}^{5} \frac{2(-1)^n}{n!} \approx 0.7333$

$0.7333 - 0.002778 \le S \le 0.7333 + 0.002778$

$$|R_6| = |S - S_6| \le a_7 = \frac{2}{6!} = 0.002778$$

$0.7305 \le S \le 0.7361$

37. $\displaystyle\sum_{n=0}^{\infty} \frac{(-1)^n}{n!}$

(a) By Theorem 9.15,

$$|R_N| \le a_{N+1} = \frac{1}{(N + 1)!} < 0.001.$$

This inequality is valid when $N = 6$.

(b) We may approximate the series by

$$\sum_{n=0}^{6} \frac{(-1)^n}{n!} = 1 - 1 + \frac{1}{2} - \frac{1}{6} + \frac{1}{24} - \frac{1}{120} + \frac{1}{720}$$

$$\approx 0.368.$$

(7 terms. Note that the sum begins with $n = 0$.)

39. $\displaystyle\sum_{n=0}^{\infty} \frac{(-1)^n}{(2n + 1)!}$

(a) By Theorem 9.15,

$$|R_N| \le a_{N+1} = \frac{1}{[2(N + 1) + 1]!} < 0.001.$$

This inequality is valid when $N = 2$.

(b) We may approximate the series by

$$\sum_{n=0}^{2} \frac{(-1)^n}{(2n + 1)!} = 1 - \frac{1}{6} + \frac{1}{120} \approx 0.842.$$

(3 terms. Note that the sum begins with $n = 0$.)

41. $\displaystyle\sum_{n=1}^{\infty} \frac{(-1)^{n+1}}{n}$

(a) By Theorem 9.15,

$$|R_N| \le a_{N+1} = \frac{1}{N + 1} < 0.001.$$

This inequality is valid when $N = 1000$.

(b) We may approximate the series by

$$\sum_{n=1}^{1000} \frac{(-1)^{n+1}}{n} = 1 - \frac{1}{2} + \frac{1}{3} - \frac{1}{4} + \cdots - \frac{1}{1000}$$

$$\approx 0.693.$$

(1000 terms)

43. $\displaystyle\sum_{n=1}^{\infty} \frac{(-1)^{n+1}}{n^3}$

By Theorem 9.15,

$$|R_N| \le a_{N+1} = \frac{1}{(N + 1)^3} < 0.001$$

$$\Rightarrow (N + 1)^3 > 1000 \Rightarrow N + 1 > 10.$$

Use 10 terms.

45. $\displaystyle\sum_{n=1}^{\infty} \frac{(-1)^{n+1}}{2n^3 - 1}$

By Theorem 9.15,

$$|R_N| \le a_{N+1} = \frac{1}{2(N + 1)^3 - 1} < 0.001.$$

This inequality is valid when $N = 7$. Use 7 terms.

47. $\displaystyle\sum_{n=1}^{\infty} \frac{(-1)^{n+1}}{(n+1)^2}$

$\displaystyle\sum_{n=1}^{\infty} \frac{1}{(n+1)^2}$ converges by comparison to the *p*-series

$\displaystyle\sum_{n=1}^{\infty} \frac{1}{n^2}.$

Therefore, the given series converges absolutely.

51. $\displaystyle\sum_{n=1}^{\infty} \frac{(-1)^{n+1} n^2}{(n+1)^2}$

$\displaystyle\lim_{n\to\infty} \frac{n^2}{(n+1)^2} = 1$

Therefore, the series diverges by the *n*th-Term Test.

55. $\displaystyle\sum_{n=2}^{\infty} \frac{(-1)^n n}{n^3 - 1}$

$\displaystyle\sum_{n=2}^{\infty} \frac{n}{n^3 - 1}$ converges by a limit comparison to

the convergent *p*-series $\displaystyle\sum_{n=2}^{\infty} \frac{1}{n^2}$. Therefore, the given series converges absolutely.

59. $\displaystyle\sum_{n=0}^{\infty} \frac{\cos n\pi}{n+1} = \sum_{n=0}^{\infty} \frac{(-1)^n}{n+1}$

The given series converges by the Alternating Series Test, but

$\displaystyle\sum_{n=0}^{\infty} \frac{|\cos n\pi|}{n+1} = \sum_{n=0}^{\infty} \frac{1}{n+1}$

diverges by a limit comparison to the divergent harmonic series,

$\displaystyle\sum_{n=1}^{\infty} \frac{1}{n}.$

$\displaystyle\lim_{n\to\infty} \frac{|\cos n\pi|/(n+1)}{1/n} = 1$, therefore the series converges conditionally.

49. $\displaystyle\sum_{n=1}^{\infty} \frac{(-1)^{n+1}}{\sqrt{n}}$

The given series converges by the Alternating Series Test, but does not converge absolutely since

$\displaystyle\sum_{n=1}^{\infty} \frac{1}{\sqrt{n}}$

is a divergent *p*-series. Therefore, the series converges conditionally.

53. $\displaystyle\sum_{n=2}^{\infty} \frac{(-1)^n}{\ln n}$

The given series converges by the Alternating Series Test but does not converge absolutely since the series

$\displaystyle\sum_{n=2}^{\infty} \frac{1}{\ln n}$ diverges by comparison to the harmonic series

$\displaystyle\sum_{n=1}^{\infty} \frac{1}{n}.$ Therefore, the series converges conditionally.

57. $\displaystyle\sum_{n=0}^{\infty} \frac{(-1)^n}{(2n+1)!}$

$\displaystyle\sum_{n=0}^{\infty} \frac{1}{(2n+1)!}$

is convergent by comparison to the convergent geometric series

$\displaystyle\sum_{n=0}^{\infty} \left(\frac{1}{2}\right)^n$

since

$\displaystyle\frac{1}{(2n+1)!} < \frac{1}{2^n}$ for $n > 0$.

Therefore, the given series converges absolutely.

61. $\displaystyle\sum_{n=1}^{\infty} \frac{\cos n\pi}{n^2} = \sum_{n=1}^{\infty} \frac{(-1)^n}{n^2}$

$\displaystyle\sum_{n=1}^{\infty} \frac{1}{n^2}$ is a convergent *p*-series.

Therefore, the given series converges absolutely.

63. An alternating series is a series whose terms alternate in sign. See Theorem 9.14.

65. $\Sigma\, a_n$ is absolutely convergent if $\Sigma|a_n|$ converges. $\Sigma\, a_n$ is conditionally convergent if $\Sigma|a_n|$ diverges, but $\Sigma\, a_n$ converges.

67. False. Let $a_n = \dfrac{(-1)^n}{n}$.

Then $\sum a_n$ converges and $\sum (-a_n)$ converges.

But, $\sum|a_n| = \sum \dfrac{1}{n}$ diverges.

69. True. $S_{100} = -1 + \dfrac{1}{2} - \dfrac{1}{3} + \cdots + \dfrac{1}{100}$

Since the next term $-\dfrac{1}{101}$ is negative, S_{100} is an overestimate of the sum.

71. $\displaystyle\sum_{n=1}^{\infty} (-1)^n \dfrac{1}{n^p}$

If $p = 0$, then $\displaystyle\sum_{n=1}^{\infty} (-1)^n$ diverges.

If $p < 0$, then $\displaystyle\sum_{n=1}^{\infty} (-1)^n n^{-p}$ diverges.

If $p > 0$, then $\displaystyle\lim_{n\to\infty} \dfrac{1}{n^p} = 0$ and

$$a_{n+1} = \dfrac{1}{(n+1)^p} < \dfrac{1}{n^p} = a_n.$$

Therefore, the series converges for $p > 0$.

73. Since

$$\sum_{n=1}^{\infty} |a_n|$$

converges we have $\displaystyle\lim_{n\to\infty} |a_n| = 0$. Thus, there must exist an $N > 0$ such that $|a_N| < 1$ for all $n > N$ and it follows that $a_n^2 \le |a_n|$ for all $n > N$. Hence, by the Comparison Test,

$$\sum_{n=1}^{\infty} a_n^2$$

converges. Let $a_n = 1/n$ to see that the converse is false.

75. $\displaystyle\sum_{n=1}^{\infty} \dfrac{1}{n^2}$ converges, hence so does $\displaystyle\sum_{n=1}^{\infty} \dfrac{1}{n^4}$.

77. (a) No, the series does not satisfy $a_{n+1} \le a_n$ for all n. For example, $\frac{1}{9} < \frac{1}{8}$.

(b) Yes, the series converges.

$$S_{2n} = \dfrac{1}{2} - \dfrac{1}{3} + \cdots + \dfrac{1}{2^n} - \dfrac{1}{3^n}$$

$$= \left(\dfrac{1}{2} + \cdots + \dfrac{1}{2^n} \right) - \left(\dfrac{1}{3} + \cdots + \dfrac{1}{3^n} \right)$$

$$= \left(1 + \dfrac{1}{2} + \cdots + \dfrac{1}{2^n} \right) - \left(1 + \dfrac{1}{3} + \cdots + \dfrac{1}{3^n} \right)$$

As $n \to \infty$, $S_{2n} \to \dfrac{1}{1 - (1/2)} - \dfrac{1}{1 - (1/3)} = 2 - \dfrac{3}{2} = \dfrac{1}{2}$.

79. $\displaystyle\sum_{n=1}^{\infty} \dfrac{10}{n^{3/2}} = 10 \sum_{n=1}^{\infty} \dfrac{1}{n^{3/2}}$,

convergent p-series

81. Diverges by nth-Term Test

$$\lim_{n\to\infty} a_n = \infty$$

83. Convergent geometric series

$$\left(r = \dfrac{7}{8} < 1 \right)$$

85. Convergent geometric series $\left(r = 1/\sqrt{e} \right)$ or Integral Test

87. Converges (absolutely) by Alternating Series Test

89. The first term of the series is zero, not one. You cannot regroup series terms arbitrarily.

91. $s = 1 - \dfrac{1}{2} + \dfrac{1}{3} - \dfrac{1}{4} + \cdots$

$S = 1 + \dfrac{1}{3} - \dfrac{1}{2} + \dfrac{1}{5} + \dfrac{1}{7} - \dfrac{1}{4} + \dfrac{1}{9} + \dfrac{1}{11} - \dfrac{1}{6} + \cdots$

(i) $s_{4n} = 1 - \dfrac{1}{2} + \dfrac{1}{3} - \dfrac{1}{4} + \dfrac{1}{5} - \dfrac{1}{6} + \cdots + \dfrac{1}{4n-1} - \dfrac{1}{4n}$

$\dfrac{1}{2} s_{2n} = \dfrac{1}{2} - \dfrac{1}{4} + \dfrac{1}{6} - \dfrac{1}{8} + \dfrac{1}{10} - \cdots + \dfrac{1}{4n-2} - \dfrac{1}{4n}$

Adding: $s_{4n} + \dfrac{1}{2} s_{2n} = 1 + \dfrac{1}{3} - \dfrac{1}{2} + \dfrac{1}{5} + \dfrac{1}{7} - \dfrac{1}{4} + \cdots + \dfrac{1}{4n-3} + \dfrac{1}{4n-1} - \dfrac{1}{2n} = S_{3n}$

(ii) $\lim\limits_{n\to\infty} s_n = s$ (In fact, $s = \ln 2$.)

$s \neq 0$ since $s > \dfrac{1}{2}$.

$S = \lim\limits_{n\to\infty} S_{3n} = s_{4n} + \dfrac{1}{2} s_{2n} = s + \dfrac{1}{2} s = \dfrac{3}{2} s$

Thus, $S \neq s$.

Section 9.6   The Ratio and Root Tests

1. $\dfrac{(n+1)!}{(n-2)!} = \dfrac{(n+1)(n)(n-1)(n-2)!}{(n-2)!}$

$\quad = (n+1)(n)(n-1)$

3. Use the Principle of Mathematical Induction. When $k = 1$, the formula is valid since $1 = \dfrac{(2(1))!}{2^1 \cdot 1!}$. Assume that

$1 \cdot 3 \cdot 5 \cdots (2n-1) = \dfrac{(2n)!}{2^n n!}$

and show that

$1 \cdot 3 \cdot 5 \cdots (2n-1)(2n+1) = \dfrac{(2n+2)!}{2^{n+1}(n+1)!}.$

To do this, note that:

$1 \cdot 3 \cdot 5 \cdots (2n-1)(2n+1) = [1 \cdot 3 \cdot 5 \cdots (2n-1)](2n+1)$

$= \dfrac{(2n)!}{2^n n!} \cdot (2n+1)$ (Induction hypothesis)

$= \dfrac{(2n)!(2n+1)}{2^n n!} \cdot \dfrac{(2n+2)}{2(n+1)}$

$= \dfrac{(2n)!(2n+1)(2n+2)}{2^{n+1} n!(n+1)}$

$= \dfrac{(2n+2)!}{2^{n+1}(n+1)!}$

The formula is valid for all $n \geq 1$.

5. $\displaystyle\sum_{n=1}^{\infty} n\left(\dfrac{3}{4}\right)^n = 1\left(\dfrac{3}{4}\right) + 2\left(\dfrac{9}{16}\right) + \cdots$

$S_1 = \dfrac{3}{4},\ S_2 \approx 1.875$

Matches (d).

7. $\displaystyle\sum_{n=1}^{\infty} \dfrac{(-3)^{n+1}}{n!} = 9 - \dfrac{3^3}{2} + \cdots$

$S_1 = 9$

Matches (f).

9. $\displaystyle\sum_{n=1}^{\infty} \left(\dfrac{4n}{5n-3}\right)^n = \dfrac{4}{2} + \left(\dfrac{8}{7}\right)^2 + \cdots$

$S_1 = 2,\ S_2 = 3.31$

Matches (a).

11. (a) Ratio Test: $\lim\limits_{n\to\infty}\left|\dfrac{a_{n+1}}{a_n}\right| = \lim\limits_{n\to\infty}\dfrac{(n+1)^2(5/8)^{n+1}}{n^2(5/8)^n} = \lim\limits_{n\to\infty}\left(\dfrac{n+1}{n}\right)^2\dfrac{5}{8} = \dfrac{5}{8} < 1.$ Converges

(b)

n	5	10	15	20	25
S_n	9.2104	16.7598	18.8016	19.1878	19.2491

(c)

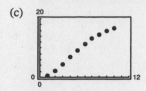

(d) The sum is approximately 19.26.

(e) The more rapidly the terms of the series approach 0, the more rapidly the sequence of the partial sums approaches the sum of the series.

13. $\displaystyle\sum_{n=0}^{\infty}\dfrac{n!}{3^n}$

$\lim\limits_{n\to\infty}\left|\dfrac{a_{n+1}}{a_n}\right| = \lim\limits_{n\to\infty}\left|\dfrac{(n+1)!}{3^{n+1}}\cdot\dfrac{3^n}{n!}\right|$

$\qquad = \lim\limits_{n\to\infty}\dfrac{n+1}{3} = \infty$

Therefore, by the Ratio Test, the series diverges.

15. $\displaystyle\sum_{n=1}^{\infty}n\left(\dfrac{3}{4}\right)^n$

$\lim\limits_{n\to\infty}\left|\dfrac{a_{n+1}}{a_n}\right| = \lim\limits_{n\to\infty}\left|\dfrac{(n+1)(3/4)^{n+1}}{n(3/4)^n}\right|$

$\qquad = \lim\limits_{n\to\infty}\left|\dfrac{3(n+1)}{4n}\right| = \dfrac{3}{4}$

Therefore, by the Ratio Test, the series converges.

17. $\displaystyle\sum_{n=1}^{\infty}\dfrac{n}{2^n}$

$\lim\limits_{n\to\infty}\left|\dfrac{a_{n+1}}{a_n}\right| = \lim\limits_{n\to\infty}\left|\dfrac{n+1}{2^{n+1}}\cdot\dfrac{2^n}{n}\right|$

$\qquad = \lim\limits_{n\to\infty}\dfrac{n+1}{2n} = \dfrac{1}{2}$

Therefore, by the Ratio Test, the series converges.

19. $\displaystyle\sum_{n=1}^{\infty}\dfrac{2^n}{n^2}$

$\lim\limits_{n\to\infty}\left|\dfrac{a_{n+1}}{a_n}\right| = \lim\limits_{n\to\infty}\left|\dfrac{2^{n+1}}{(n+1)^2}\cdot\dfrac{n^2}{2^n}\right|$

$\qquad = \lim\limits_{n\to\infty}\dfrac{2n^2}{(n+1)^2} = 2$

Therefore, by the Ratio Test, the series diverges.

21. $\displaystyle\sum_{n=0}^{\infty}\dfrac{(-1)^n 2^n}{n!}$

$\lim\limits_{n\to\infty}\left|\dfrac{a_{n+1}}{a_n}\right| = \lim\limits_{n\to\infty}\left|\dfrac{2^{n+1}}{(n+1)!}\cdot\dfrac{n!}{2^n}\right|$

$\qquad = \lim\limits_{n\to\infty}\dfrac{2}{n+1} = 0$

Therefore, by the Ratio Test, the series converges.

23. $\displaystyle\sum_{n=1}^{\infty}\dfrac{n!}{n3^n}$

$\lim\limits_{n\to\infty}\left|\dfrac{a_{n+1}}{a_n}\right| = \lim\limits_{n\to\infty}\left|\dfrac{(n+1)!}{(n+1)3^{n+1}}\cdot\dfrac{n3^n}{n!}\right|$

$\qquad = \lim\limits_{n\to\infty}\dfrac{n}{3} = \infty$

Therefore, by the Ratio Test, the series diverges.

25. $\displaystyle\sum_{n=0}^{\infty}\dfrac{4^n}{n!}$

$\lim\limits_{n\to\infty}\left|\dfrac{a_{n+1}}{a_n}\right| = \lim\limits_{n\to\infty}\left|\dfrac{4^{n+1}}{(n+1)!}\cdot\dfrac{n!}{4^n}\right|$

$\qquad = \lim\limits_{n\to\infty}\dfrac{4}{n+1} = 0$

Therefore, by the Ratio Test, the series converges.

27. $\displaystyle\sum_{n=0}^{\infty} \frac{3^n}{(n+1)^n}$

$$\lim_{n\to\infty}\left|\frac{a_{n+1}}{a_n}\right| = \lim_{n\to\infty}\left|\frac{3^{n+1}}{(n+2)^{n+1}}\cdot\frac{(n+1)^n}{3^n}\right| = \lim_{n\to\infty}\frac{3(n+1)^n}{(n+2)^{n+1}} = \lim_{n\to\infty}\frac{3}{n+2}\left(\frac{n+1}{n+2}\right)^n = (0)\left(\frac{1}{e}\right) = 0$$

To find $\displaystyle\lim_{n\to\infty}\left(\frac{n+1}{n+2}\right)^n$, let $y = \displaystyle\lim_{n\to\infty}\left(\frac{n+1}{n+2}\right)^n$. Then,

$$\ln y = \lim_{n\to\infty} n\ln\left(\frac{n+1}{n+2}\right) = \lim_{n\to\infty}\frac{\ln[(n+1)/(n+2)]}{1/n} = \frac{0}{0}$$

$$\ln y = \lim_{n\to\infty}\frac{[(1)/(n+1)] - [(1)/(n+2)]}{-(1/n^2)} = -1 \text{ by L'Hôpital's Rule}$$

$$y = e^{-1} = \frac{1}{e}.$$

Therefore, by the Ratio Test, the series converges.

29. $\displaystyle\sum_{n=0}^{\infty} \frac{4^n}{3^n + 1}$

$$\lim_{n\to\infty}\left|\frac{a_{n+1}}{a_n}\right| = \lim_{n\to\infty}\left|\frac{4^{n+1}}{3^{n+1}+1}\cdot\frac{3^n+1}{4^n}\right|$$

$$= \lim_{n\to\infty}\frac{4(3^n+1)}{3^{n+1}+1}$$

$$= \lim_{n\to\infty}\frac{4(1+1/3^n)}{3+1/3^n} = \frac{4}{3}$$

Therefore, by the Ratio Test, the series diverges.

31. $\displaystyle\sum_{n=0}^{\infty} \frac{(-1)^{n+1}n!}{1\cdot 3\cdot 5\cdots(2n+1)}$

$$\lim_{n\to\infty}\left|\frac{a_{n+1}}{a_n}\right| = \lim_{n\to\infty}\left|\frac{(n+1)!}{1\cdot 3\cdot 5\cdots(2n+1)(2n+3)}\cdot\frac{1\cdot 3\cdot 5\cdots(2n+1)}{n!}\right| = \lim_{n\to\infty}\frac{n+1}{2n+3} = \frac{1}{2}$$

Therefore, by the Ratio Test, the series converges.

Note: The first few terms of this series are $-1 + \dfrac{1}{1\cdot 3} - \dfrac{2!}{1\cdot 3\cdot 5} + \dfrac{3!}{1\cdot 3\cdot 5\cdot 7} - \cdots$.

33. $\displaystyle\sum_{n=1}^{\infty} \frac{1}{n^{3/2}}$

$$\lim_{n\to\infty}\left|\frac{a_{n+1}}{a_n}\right| = \lim_{n\to\infty}\left|\frac{1}{(n+1)^{3/2}}\cdot\frac{n^{3/2}}{1}\right|$$

$$= \lim_{n\to\infty}\left(\frac{n}{n+1}\right)^{3/2} = 1$$

Ratio Test is inconclusive.

35. $\displaystyle\sum_{n=1}^{\infty} \frac{1}{n^4}$

$$\lim_{n\to\infty}\left|\frac{a_{n+1}}{a_n}\right| = \lim_{n\to\infty}\left|\frac{1}{(n+1)^4}\cdot\frac{n^4}{1}\right| = \lim_{n\to\infty}\left(\frac{n}{n+1}\right)^4 = 1$$

37. $\displaystyle\sum_{n=1}^{\infty} \left(\frac{n}{2n+1}\right)^n$

$$\lim_{n\to\infty}\sqrt[n]{|a_n|} = \lim_{n\to\infty}\sqrt[n]{\left(\frac{n}{2n+1}\right)^n}$$

$$= \lim_{n\to\infty}\frac{n}{2n+1} = \frac{1}{2}$$

Therefore, by the Root Test, the series converges.

39. $\displaystyle\sum_{n=2}^{\infty} \left(\frac{2n+1}{n-1}\right)^n$

$$\lim_{n\to\infty}\sqrt[n]{|a_n|} = \lim_{n\to\infty}\sqrt[n]{\left(\frac{2n+1}{n-1}\right)^n}$$

$$= \lim_{n\to\infty}\left(\frac{2n+1}{n-1}\right) = 2$$

Since $2 > 1$, the series diverges.

41. $\sum\limits_{n=2}^{\infty} \dfrac{(-1)^n}{(\ln n)^n}$

$$\lim_{n\to\infty} \sqrt[n]{|a_n|} = \lim_{n\to\infty} \sqrt[n]{\left|\dfrac{(-1)^n}{(\ln n)^n}\right|}$$

$$= \lim_{n\to\infty} \dfrac{1}{|\ln n|} = 0$$

Therefore, by the Root Test, the series converges.

43. $\sum\limits_{n=1}^{\infty} \left(2\sqrt[n]{n} + 1\right)^n$

$$\lim_{n\to\infty} \sqrt[n]{|a_n|} = \lim_{n\to\infty} \sqrt[n]{\left(2\sqrt[n]{n} + 1\right)^n} = \lim_{n\to\infty} \left(2\sqrt[n]{n} + 1\right)$$

To find $\lim\limits_{n\to\infty} \sqrt[n]{n}$, let $y = \lim\limits_{n\to\infty} \sqrt[n]{n}$. Then

$$\ln y = \lim_{n\to\infty} \left(\ln \sqrt[n]{n}\right) = \lim_{n\to\infty} \dfrac{1}{n} \ln n = \lim_{n\to\infty} \dfrac{\ln n}{n} = \lim_{n\to\infty} \dfrac{1/n}{1} = 0.$$

Thus, $\ln y = 0$, so $y = e^0 = 1$ and $\lim\limits_{n\to\infty} \left(2\sqrt[n]{n} + 1\right) = 2(1) + 1 = 3$. Therefore, by the Root Test, the series diverges.

45. $\sum\limits_{n=1}^{\infty} \dfrac{n}{4^n}$

$$\lim_{n\to\infty} \sqrt[n]{|a_n|} = \lim_{n\to\infty} \sqrt[n]{\dfrac{n}{4^n}}$$

$$= \lim_{n\to\infty} \dfrac{n^{1/n}}{4} = \dfrac{1}{4}$$

Since $\frac{1}{4} < 1$, the series converges.

Note: You can use L'Hôpital's Rule to show $\lim\limits_{n\to\infty} n^{1/n} = 1$.

Let $y = n^{1/n} \implies \ln y = \dfrac{\ln n}{n}$

$$\lim_{n\to\infty} \dfrac{\ln n}{n} = \lim_{n\to\infty} \dfrac{1/n}{1} = 0.$$

Hence, $\ln y \to 0 \implies y = n^{1/n} \to 1$.

47. $\sum\limits_{n=1}^{\infty} \left(\dfrac{1}{n} - \dfrac{1}{n^2}\right)^n$

$$\lim_{n\to\infty} \sqrt[n]{|a_n|} = \lim_{n\to\infty} \sqrt[n]{\left(\dfrac{1}{n} - \dfrac{1}{n^2}\right)^n}$$

$$= \lim_{n\to\infty} \left(\dfrac{1}{n} - \dfrac{1}{n^2}\right) = 0 - 0 = 0 < 1$$

Hence, the series converges.

49. $\sum\limits_{n=2}^{\infty} \dfrac{n}{(\ln n)^n}$

$$\lim_{n\to\infty} \sqrt[n]{|a_n|} = \lim_{n\to\infty} \sqrt[n]{\dfrac{n}{(\ln n)^n}}$$

$$= \lim_{n\to\infty} \dfrac{n^{1/n}}{\ln n} = 0$$

Since $0 < 1$, the series converges by the Root Test.

51. $\sum\limits_{n=1}^{\infty} \dfrac{(-1)^{n+1} 5}{n}$

$$a_{n+1} = \dfrac{5}{n+1} < \dfrac{5}{n} = a_n$$

$$\lim_{n\to\infty} \dfrac{5}{n} = 0$$

Therefore, by the Alternating Series Test, the series converges (conditional convergence).

53. $\sum\limits_{n=1}^{\infty} \dfrac{3}{n\sqrt{n}} = 3 \sum\limits_{n=1}^{\infty} \dfrac{1}{n^{3/2}}$

This is convergent p-series.

55. $\sum\limits_{n=1}^{\infty} \dfrac{2n}{n+1}$

$$\lim_{n\to\infty} \dfrac{2n}{n+1} = 2 \neq 0$$

This diverges by the nth-Term Test for Divergence.

57. $\displaystyle\sum_{n=1}^{\infty} \frac{(-1)^n 3^{n-2}}{2^n} = \sum_{n=1}^{\infty} \frac{(-1)^n 3^n 3^{-2}}{2^n} = \sum_{n=1}^{\infty} \frac{1}{9}\left(-\frac{3}{2}\right)^n$

Since $|r| = \frac{3}{2} > 1$, this is a divergent geometric series.

59. $\displaystyle\sum_{n=1}^{\infty} \frac{10n + 3}{n2^n}$

$\displaystyle\lim_{n\to\infty} \frac{(10n + 3)/n2^n}{1/2^n} = \lim_{n\to\infty} \frac{10n + 3}{n} = 10$

Therefore, the series converges by a Limit Comparison Test with the geometric series

$$\sum_{n=0}^{\infty} \left(\frac{1}{2}\right)^n.$$

61. $\displaystyle\sum_{n=1}^{\infty} \frac{\cos n}{2^n}$

$\displaystyle\left|\frac{\cos n}{2^n}\right| \le \frac{1}{2^n}$

Therefore, the series

$$\sum_{n=1}^{\infty} \left|\frac{\cos n}{2^n}\right|$$

converges by comparison with the geometric series

$$\sum_{n=0}^{\infty} \left(\frac{1}{2}\right)^n.$$

63. $\displaystyle\sum_{n=1}^{\infty} \frac{n7^n}{n!}$

$\displaystyle\lim_{n\to\infty} \left|\frac{a_{n+1}}{a_n}\right| = \lim_{n\to\infty} \left|\frac{(n + 1)7^{n+1}}{(n + 1)!} \cdot \frac{n!}{n7^n}\right| = \lim_{n\to\infty} \frac{7}{n} = 0$

Therefore, by the Ratio Test, the series converges.

65. $\displaystyle\sum_{n=1}^{\infty} \frac{(-1)^n 3^{n-1}}{n!}$

$\displaystyle\lim_{n\to\infty} \left|\frac{a_{n+1}}{a_n}\right| = \lim_{n\to\infty} \left|\frac{3^n}{(n + 1)!} \cdot \frac{n!}{3^{n-1}}\right| = \lim_{n\to\infty} \frac{3}{n + 1} = 0$

Therefore, by the Ratio Test, the series converges.

(Absolutely)

67. $\displaystyle\sum_{n=1}^{\infty} \frac{(-3)^n}{3 \cdot 5 \cdot 7 \cdots (2n + 1)}$

$\displaystyle\lim_{n\to\infty} \left|\frac{a_{n+1}}{a_n}\right| = \lim_{n\to\infty} \left|\frac{(-3)^{n+1}}{3 \cdot 5 \cdot 7 \cdots (2n + 1)(2n + 3)} \cdot \frac{3 \cdot 5 \cdot 7 \cdots (2n + 1)}{(-3)^n}\right| = \lim_{n\to\infty} \frac{3}{2n + 3} = 0$

Therefore, by the Ratio Test, the series converges.

69. (a) and (c)

$\displaystyle\sum_{n=1}^{\infty} \frac{n5^n}{n!} = \sum_{n=0}^{\infty} \frac{(n + 1)5^{n+1}}{(n + 1)!}$

$\displaystyle = 5 + \frac{(2)(5)^2}{2!} + \frac{(3)(5)^3}{3!} + \frac{(4)(5)^4}{4!} + \cdots$

71. (a) and (b) are the same.

73. Replace n with $n + 1$.

$\displaystyle\sum_{n=1}^{\infty} \frac{n}{4^n} = \sum_{n=0}^{\infty} \frac{n + 1}{4^{n+1}}$

75. Since

$\displaystyle\frac{3^{10}}{2^{10} 10!} \approx 1.59 \times 10^{-5}$, use 9 terms.

$\displaystyle\sum_{k=1}^{9} \frac{(-3)^k}{2^k k!} \approx -0.7769$

77. $\lim\limits_{n\to\infty}\left|\dfrac{a_{n+1}}{a_n}\right| = \lim\limits_{n\to\infty}\left|\dfrac{(4n-1)/(3n+2)a_n}{a_n}\right|$

$\qquad\qquad = \lim\limits_{n\to\infty}\dfrac{4n-1}{3n+2} = \dfrac{4}{3} > 1$

The series diverges by the Ratio Test.

81. $\lim\limits_{n\to\infty}\left|\dfrac{a_{n+1}}{a_n}\right| = \lim\limits_{n\to\infty}\left|\dfrac{(1+(1)/(n))a_n}{a_n}\right|$

$\qquad\qquad = \lim\limits_{n\to\infty}\left(1+\dfrac{1}{n}\right) = 1$

The Ratio Test is inconclusive.

But, $\lim\limits_{n\to\infty} a_n \neq 0$, so the series diverges.

85. $\displaystyle\sum_{n=3}^{\infty}\dfrac{1}{(\ln n)^n}$

$\lim\limits_{n\to\infty}\sqrt[n]{|a_n|} = \lim\limits_{n\to\infty}\sqrt[n]{\dfrac{1}{(\ln n)^n}} = \lim\limits_{n\to\infty}\dfrac{1}{\ln n} = 0$

Therefore, by the Root Test, the series converges.

89. $\lim\limits_{n\to\infty}\left|\dfrac{a_{n+1}}{a_n}\right| = \lim\limits_{n\to\infty}\left|\dfrac{(x+1)^{n+1}/(n+1)}{x^n/n}\right|$

$\qquad\qquad = \lim\limits_{n\to\infty}\left|\dfrac{n}{n+1}(x+1)\right| = |x+1|$

For the series to converge,

$\qquad |x+1| < 1 \implies -1 < x+1 < 1$

$\qquad\qquad\qquad \implies -2 < x < 0.$

For $x=0$, $\displaystyle\sum_{n=1}^{\infty}\dfrac{(-1)^n}{n}$, converges.

For $x=-2$, $\displaystyle\sum_{n=1}^{\infty}\dfrac{(-1)^n(-1)^n}{n} = \sum_{n=1}^{\infty}\dfrac{1}{n}$, diverges.

Answer: $-2 < x \leq 0$

93. See Theorem 9.17, page 597.

97. The series converges absolutely. See Theorem 9.17.

79. $\lim\limits_{n\to\infty}\left|\dfrac{a_{n+1}}{a_n}\right| = \lim\limits_{n\to\infty}\left|\dfrac{(\sin n+1)/(\sqrt{n})a_n}{a_n}\right|$

$\qquad\qquad = \lim\limits_{n\to\infty}\dfrac{\sin n+1}{\sqrt{n}} = 0 < 1$

The series converges by the Ratio Test.

83. $\lim\limits_{n\to\infty}\left|\dfrac{a_{n+1}}{a_n}\right| = \lim\limits_{n\to\infty}\left|\dfrac{\dfrac{1\cdot 2\cdots n(n+1)}{1\cdot 3\cdots(2n-1)(2n+1)}}{\dfrac{1\cdot 2\cdots n}{1\cdot 3\cdots(2n-1)}}\right|$

$\qquad\qquad = \lim\limits_{n\to\infty}\dfrac{n+1}{2n+1} = \dfrac{1}{2} < 1$

The series converges by the Ratio Test.

87. $\lim\limits_{n\to\infty}\left|\dfrac{a_{n+1}}{a_n}\right| = \lim\limits_{n\to\infty}\left|\dfrac{2(x/3)^{n+1}}{2(x/3)^n}\right|$

$\qquad\qquad = \lim\limits_{n\to\infty}\left|\dfrac{x}{3}\right| = \left|\dfrac{x}{3}\right|$

For the series to converge: $\left|\dfrac{x}{3}\right| < 1 \implies -3 < x < 3.$

For $x=3$, the series diverges.

For $x=-3$, the series diverges.

Answer: $-3 < x < 3$

91. $\lim\limits_{n\to\infty}\left|\dfrac{a_{n+1}}{a_n}\right| = \lim\limits_{n\to\infty}\dfrac{(n+1)!\left|\dfrac{x}{2}\right|^{n+1}}{n!\left|\dfrac{x}{2}\right|^n}$

$\qquad\qquad = \lim\limits_{n\to\infty}(n+1)\left|\dfrac{x}{2}\right| = \infty$

The series converges only at $x=0$.

95. No. Let $a_n = \dfrac{1}{n+10,000}$.

The series $\displaystyle\sum_{n=1}^{\infty}\dfrac{1}{n+10,000}$ diverges.

99. First, let

$$\lim_{n \to \infty} \sqrt[n]{|a_n|} = r < 1$$

and choose R such that $0 \le r < R < 1$. There must exist some $N > 0$ such that $\sqrt[n]{|a_n|} < R$ for all $n > N$. Thus, for $n > N$ $|a_n| < R^n$ and since the geometric series

$$\sum_{n=0}^{\infty} R^n$$

converges, we can apply the Comparison Test to conclude that

$$\sum_{n=1}^{\infty} |a_n|$$

converges which in turn implies that $\displaystyle\sum_{n=1}^{\infty} a_n$ converges.

Second, let

$$\lim_{n \to \infty} \sqrt[n]{|a_n|} = r > R > 1.$$

Then there must exist some $M > 0$ such that $\sqrt[n]{|a_n|} > R$ for infinitely many $n > M$. Thus, for infinitely many $n > M$, we have $|a_n| > R^n > 1$ which implies that $\lim_{n \to \infty} a_n \ne 0$ which in turn implies that

$$\sum_{n=1}^{\infty} a_n \text{ diverges.}$$

101. Ratio Test:

$$\lim_{n \to \infty} \left| \frac{a_{n+1}}{a_n} \right| = \lim_{n \to \infty} \frac{n(\ln n)^p}{(n+1)(\ln(n+1))^p} = 1, \text{ inconclusive.}$$

Root Test:

$$\lim_{n \to \infty} \sqrt[n]{|a_n|} = \lim_{n \to \infty} \sqrt[n]{\frac{1}{n(\ln n)^p}} = \lim_{n \to \infty} \frac{1}{n^{1/n}(\ln n)^{p/n}}.$$

$\lim_{n \to \infty} n^{1/n} = 1.$ Furthermore, let $y = (\ln n)^{p/n} \longrightarrow$

$$\ln y = \frac{p}{n} \ln (\ln n). \lim_{n \to \infty} \ln y = \lim_{n \to \infty} \frac{p \ln (\ln n)}{n}$$

$$= \lim_{n \to \infty} \frac{p}{\ln(n)(1/n)} = 0 \implies \lim_{n \to \infty} (\ln n)^{p/n} = 1.$$

Thus, $\lim_{n \to \infty} \dfrac{1}{n^{1/n}(\ln n)^{p/n}} = 1$, inconclusive.

103. For $n = 1, 2, 3, \ldots, -|a_n| \le a_n \le |a_n| \implies -\sum_{n=1}^{k} |a_n| \le \sum_{n=1}^{k} a_n \le \sum_{n=1}^{k} |a_n|.$

Taking limits as $k \to \infty$, $-\sum_{n=1}^{\infty} |a_n| \le \sum_{n=1}^{\infty} a_n \le \sum_{n=1}^{\infty} |a_n| \implies \left| \sum_{n=1}^{\infty} a_n \right| \le \sum_{n=1}^{\infty} |a_n|.$

105. Using the Ratio Test,

$$\lim_{n \to \infty} \left| \frac{a_{n+1}}{a_n} \right| = \lim_{n \to \infty} \left[\frac{n!}{(n+1)^n} \left(\frac{19}{7} \right)^n \bigg/ \frac{(n-1)!}{n^{n-1}} \left(\frac{19}{7} \right)^{n-1} \right]$$

$$= \lim_{n \to \infty} \left[\frac{n \cdot n^{n-1}}{(n+1)^n} \left(\frac{19}{7} \right) \right]$$

$$= \lim_{n \to \infty} \left[\frac{1}{(1 + (1/n))^n} \left(\frac{19}{7} \right) \right]$$

$$= \frac{19}{7} \cdot \frac{1}{e} < 1$$

Hence, the series converges.

Section 9.7   Taylor Polynomials and Approximations

1. $y = -\frac{1}{2}x^2 + 1$

Parabola

Matches (d)

3. $y = e^{-1/2}[(x + 1) + 1]$

Linear

Matches (a)

5. $f(x) = \dfrac{4}{\sqrt{x}} = 4x^{-1/2}$ $f(1) = 4$

$f'(x) = -2x^{-3/2}$ $f'(1) = -2$

$P_1(x) = f(1) + f'(1)(x - 1)$

$\quad\quad = 4 + (-2)(x - 1)$

$P_1(x) = -2x + 6$

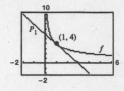

P_1 is called the first degree Taylor polynomial for f at c.

7. $f(x) = \sec x$ $f\left(\dfrac{\pi}{4}\right) = \sqrt{2}$

$f'(x) = \sec x \tan x$ $f'\left(\dfrac{\pi}{4}\right) = \sqrt{2}$

$P_1(x) = f\left(\dfrac{\pi}{4}\right) + f'\left(\dfrac{\pi}{4}\right)\left(x - \dfrac{\pi}{4}\right)$

$P_1(x) = \sqrt{2} + \sqrt{2}\left(x - \dfrac{\pi}{4}\right)$

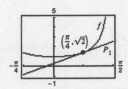

P_1 is called the first degree Taylor polynomial for f at c.

9. $f(x) = \dfrac{4}{\sqrt{x}} = 4x^{-1/2}$ $f(1) = 4$

$f'(x) = -2x^{-3/2}$ $f'(1) = -2$

$f''(x) = 3x^{-5/2}$ $f''(1) = 3$

$P_2 = f(1) + f'(1)(x - 1) + \dfrac{f''(1)}{2}(x - 1)^2$

$\quad = 4 - 2(x - 1) + \dfrac{3}{2}(x - 1)^2$

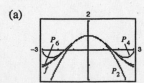

x	0	0.8	0.9	1.0	1.1	1.2	2
$f(x)$	Error	4.4721	4.2164	4.0	3.8139	3.6515	2.8284
$P_2(x)$	7.5	4.46	4.215	4.0	3.815	3.66	3.5

11. $f(x) = \cos x$

$P_2(x) = 1 - \frac{1}{2}x^2$

$P_4(x) = 1 - \frac{1}{2}x^2 + \frac{1}{24}x^4$

$P_6(x) = 1 - \frac{1}{2}x^2 + \frac{1}{24}x^4 - \frac{1}{720}x^6$

(a)

(b) $f'(x) = -\sin x$ $P_2'(x) = -x$

$f''(x) = -\cos x$ $P_2''(x) = -1$

$f''(0) = P_2''(0) = -1$

$f'''(x) = \sin x$ $P_4'''(x) = x$

$f^{(4)}(x) = \cos x$ $P_4^{(4)}(x) = 1$

$f^{(4)}(0) = 1 = P_4^{(4)}(0)$

$f^{(5)}(x) = -\sin x$ $P_6^{(5)}(x) = -x$

$f^{(6)}(x) = -\cos x$ $P^{(6)}(x) = -1$

$f^{(6)}(0) = -1 = P_6^{(6)}(0)$

(c) In general, $f^{(n)}(0) = P_n^{(n)}(0)$ for all n.

13. $f(x) = e^{-x}$ $f(0) = 1$

     $f'(x) = -e^{-x}$ $f'(0) = -1$

     $f''(x) = e^{-x}$ $f''(0) = 1$

     $f'''(x) = -e^{-x}$ $f'''(0) = -1$

$$P_3(x) = f(0) + f'(0)x + \frac{f''(0)}{2!}x^2 + \frac{f'''(0)}{3!}x^3$$

$$= 1 - x + \frac{x^2}{2} - \frac{x^3}{6}$$

15. $f(x) = e^{2x}$ $f(0) = 1$

     $f'(x) = 2e^{2x}$ $f'(0) = 2$

     $f''(x) = 4e^{2x}$ $f''(0) = 4$

     $f'''(x) = 8e^{2x}$ $f'''(0) = 8$

     $f^{(4)}(x) = 16^{2x}$ $f^{(4)}(0) = 16$

$$P_4(x) = 1 + 2x + \frac{4}{2!}x^2 + \frac{8}{3!}x^3 + \frac{16}{4!}x^4$$

$$= 1 + 2x + 2x^2 + \frac{4}{3}x^3 + \frac{2}{3}x^4$$

17. $f(x) = \sin x$ $f(0) = 0$

     $f'(x) = \cos x$ $f'(0) = 1$

     $f''(x) = -\sin x$ $f''(0) = 0$

     $f'''(x) = -\cos x$ $f'''(0) = -1$

     $f^{(4)}(x) = \sin x$ $f^{(4)}(0) = 0$

     $f^{(5)}(x) = \cos x$ $f^{(5)}(0) = 1$

$$P_5(x) = 0 + (1)x + \frac{0}{2!}x^2 + \frac{-1}{3!}x^3 + \frac{0}{4!}x^4 + \frac{1}{5!}x^5$$

$$= x - \frac{1}{6}x^3 + \frac{1}{120}x^5$$

19. $f(x) = xe^x$ $f(0) = 0$

     $f'(x) = xe^x + e^x$ $f'(0) = 1$

     $f''(x) = xe^x + 2e^x$ $f''(0) = 2$

     $f'''(x) = xe^x + 3e^x$ $f'''(0) = 3$

     $f^{(4)}(x) = xe^x + 4e^x$ $f^{(4)}(0) = 4$

$$P_4(x) = 0 + x + \frac{2}{2!}x^2 + \frac{3}{3!}x^3 + \frac{4}{4!}x^4$$

$$= x + x^2 + \frac{1}{2}x^3 + \frac{1}{6}x^4$$

21. $f(x) = \dfrac{1}{x + 1}$ $f(0) = 1$

     $f'(x) = -\dfrac{1}{(x + 1)^2}$ $f'(0) = -1$

     $f''(x) = \dfrac{2}{(x + 1)^3}$ $f''(0) = 2$

     $f'''(x) = \dfrac{-6}{(x + 1)^4}$ $f'''(0) = -6$

     $f^{(4)}(x) = \dfrac{24}{(x + 1)^5}$ $f^{(4)}(0) = 24$

$$P_4(x) = 1 - x + \frac{2}{2!}x^2 + \frac{-6}{3!}x^3 + \frac{24}{4!}x^4$$

$$= 1 - x + x^2 - x^3 + x^4$$

23. $f(x) = \sec x$ $f(0) = 1$

     $f'(x) = \sec x \tan x$ $f'(0) = 0$

     $f''(x) = \sec^3 x + \sec x \tan^2 x$ $f''(0) = 1$

$$P_2(x) = 1 + 0x + \frac{1}{2!}x^2 = 1 + \frac{1}{2}x^2$$

25. $f(x) = \dfrac{1}{x}$ $f(1) = 1$

     $f'(x) = -\dfrac{1}{x^2}$ $f'(1) = -1$

     $f''(x) = \dfrac{2}{x^3}$ $f''(1) = 2$

     $f'''(x) = -\dfrac{6}{x^4}$ $f'''(1) = -6$

     $f^{(4)}(x) = \dfrac{24}{x^5}$ $f^{(4)}(1) = 24$

$$P_4(x) = 1 - (x - 1) + \frac{2}{2!}(x - 1)^2 + \frac{-6}{3!}(x - 1)^3$$

$$+ \frac{24}{4!}(x - 1)^4$$

$$= 1 - (x - 1) + (x - 1)^2 - (x - 1)^3 + (x - 1)^4$$

27. $f(x) = \sqrt{x}$ $f(1) = 1$

     $f'(x) = \dfrac{1}{2\sqrt{x}}$ $f'(1) = \dfrac{1}{2}$

     $f''(x) = -\dfrac{1}{4x\sqrt{x}}$ $f''(1) = -\dfrac{1}{4}$

     $f'''(x) = \dfrac{3}{8x^2\sqrt{x}}$ $f'''(1) = \dfrac{3}{8}$

     $f^{(4)}(x) = -\dfrac{15}{16x^3\sqrt{x}}$ $f^{(4)}(1) = -\dfrac{15}{16}$

$$P_4(x) = 1 + \frac{1}{2}(x - 1) - \frac{1}{8}(x - 1)^2$$

$$+ \frac{1}{16}(x - 1)^3 - \frac{5}{128}(x - 1)^4$$

29. $f(x) = \ln x$ $f(1) = 0$

$f'(x) = \dfrac{1}{x}$ $f'(1) = 1$

$f''(x) = -\dfrac{1}{x^2}$ $f''(1) = -1$

$f'''(x) = \dfrac{2}{x^3}$ $f'''(1) = 2$

$f^{(4)}(x) = -\dfrac{6}{x^4}$ $f^{(4)}(1) = -6$

$P_4(x) = 0 + (x - 1) - \dfrac{1}{2}(x - 1)^2$

$\qquad + \dfrac{1}{3}(x - 1)^3 - \dfrac{1}{4}(x - 1)^4$

31. $f(x) = \tan x$

$f'(x) = \sec^2 x$

$f''(x) = 2 \sec^2 x \tan x$

$f'''(x) = 4 \sec^2 x \tan^2 x + 2 \sec^4 x$

$f^{(4)}(x) = 8 \sec^2 x \tan^3 x + 16 \sec^4 x \tan x$

$f^{(5)}(x) = 16 \sec^2 x \tan^4 x + 88 \sec^4 x \tan^2 x + 16 \sec^6 x$

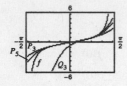

(a) $n = 3, c = 0$

$$P_3(x) = 0 + x + \frac{0}{2!}x^2 + \frac{2}{3!}x^3 = x + \frac{1}{3}x^3$$

(b) $n = 3, c = \dfrac{\pi}{4}$

$$Q_3(x) = 1 + 2\left(x - \frac{\pi}{4}\right) + \frac{4}{2!}\left(x - \frac{\pi}{4}\right)^2 + \frac{16}{3!}\left(x - \frac{\pi}{4}\right)^3$$

$$= 1 + 2\left(x - \frac{\pi}{4}\right) + 2\left(x - \frac{\pi}{4}\right)^2 + \frac{8}{3}\left(x - \frac{\pi}{4}\right)^3$$

33. $f(x) = \sin x$

$P_1(x) = x$

$P_3(x) = x - \dfrac{1}{6}x^3$

$P_5(x) = x - \dfrac{1}{6}x^3 + \dfrac{1}{120}x^5$

(a)

x	0.00	0.25	0.50	0.75	1.00
$\sin x$	0.0000	0.2474	0.4794	0.6816	0.8415
$P_1(x)$	0.0000	0.2500	0.5000	0.7500	1.0000
$P_3(x)$	0.0000	0.2474	0.4792	0.6797	0.8333
$P_5(x)$	0.0000	0.2474	0.4794	0.6817	0.8417

(b)

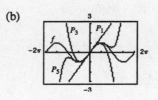

(c) As the distance increases, the accuracy decreases.

35. $f(x) = \arcsin x$

(a) $P_3(x) = x + \dfrac{x^3}{6}$

(b)

x	-0.75	-0.50	-0.25	0	0.25	0.50	0.75
$f(x)$	-0.848	-0.524	-0.253	0	0.253	0.524	0.848
$P_3(x)$	-0.820	-0.521	-0.253	0	0.253	0.521	0.820

(c)

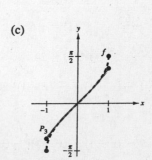

37. $f(x) = \cos x$

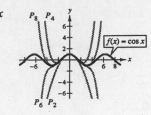

39. $f(x) = \ln(x^2 + 1)$

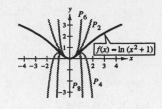

41. $f(x) = e^{-x} \approx 1 - x + \dfrac{x^2}{2} - \dfrac{x^3}{6}$

$f\left(\dfrac{1}{2}\right) \approx 0.6042$

43. $f(x) = \ln x \approx (x - 1) - \frac{1}{2}(x - 1)^2 + \frac{1}{3}(x - 1)^3 - \frac{1}{4}(x - 1)^4$

$f(1.2) \approx 0.1823$

45. $f(x) = \cos x$, $f^{(5)}(x) = -\sin x \Rightarrow$ Max on $[0, 0.3]$ is 1.

$R_4(x) \le \dfrac{1}{5!}(0.3)^5 = 2.025 \times 10^{-5}$

Note: you could use $R_5(x)$: $f^{(6)}(x) = -\cos x$, max on $[0, 0.3]$ is 1.

$R_5(x) \le \dfrac{1}{6!}(0.3)^6 = 1.0125 \times 10^{-6}$

Exact error: $0.000001 = 1.0 \times 10^{-6}$

47. $f(x) = \arcsin x$; $f^{(4)}(x) = \dfrac{x(6x^2 + 9)}{(1 - x^2)^{7/2}} \Rightarrow$ Max on $[0, 0.4]$ is $f^{(4)}(0.4) \approx 7.3340$.

$R_3(x) \le \dfrac{7.3340}{4!}(0.4)^4 \approx 0.00782 = 7.82 \times 10^{-3}$. The exact error is 8.5×10^{-4}. [Note: You could use R_4.]

49. $g(x) = \sin x$

$\left|g^{(n+1)}(x)\right| \le 1$ for all x.

$R_n(x) \le \dfrac{1}{(n + 1)!}(0.3)^{n+1} < 0.001$

By trial and error, $n = 3$.

51. $f(x) = e^x$

$f^{(n+1)}(x) = e^x$

Max on $[0, 0.6]$ is $e^{0.6} \approx 1.8221$.

$R_n \le \dfrac{1.8221}{(n + 1)!}(0.6)^{n+1} < 0.001$

By trial and error, $n = 5$.

53. $f(x) = \ln(x + 1)$

$f^{(n+1)}(x) = \dfrac{(-1)^n n!}{(x + 1)^{n+1}} \Rightarrow$ Max on $[0, 0.5]$ is $n!$.

$R_n \le \dfrac{n!}{(n + 1)!}(0.5)^{n+1} = \dfrac{(0.5)^{n+1}}{n + 1} < 0.0001$

By trial and error, $n = 9$. (See Example 9.) Using 9 terms, $\ln(1.5) \approx 0.4055$.

55. $f(x) = e^{-\pi x}$, $f(1.3)$

$f'(x) = (-\pi)e^{-\pi x}$

$f^{(n+1)}(x) = (-\pi)^{n+1}e^{-\pi x} \le \left|(-\pi)^{n+1}\right|$ on $[0, 1.3]$

$|R_n| \le \dfrac{(\pi)^{n+1}}{(n + 1)!}(1.3)^{n+1} < 0.0001$

By trial and error, $n = 16$.

57. $f(x) = e^x \approx 1 + x + \dfrac{x^2}{2} + \dfrac{x^3}{6}$, $x < 0$

$$R_3(x) = \frac{e^z}{4!}x^4 < 0.001$$

$$e^z x^4 < 0.024$$

$$|xe^{z/4}| < 0.3936$$

$$|x| < \frac{0.3936}{e^{z/4}} < 0.3936, \; z < 0$$

$$-0.3936 < x < 0$$

59. $f(x) = \cos x \approx 1 - \dfrac{x^2}{2!} + \dfrac{x^4}{4!}$, fifth degree polynomial

$$|f^{(n+1)}(x)| \le 1 \text{ for all } x \text{ and all } n.$$

$$|R_5(x)| \le \frac{1}{6!}|x|^6 < 0.001$$

$$|x|^6 < 0.72$$

$$|x| < 0.9467$$

$$-0.9467 < x < 0.9467$$

Note: Use a graphing utility to graph $y = \cos x - (1 - x^2/2 + x^4/24)$ in the viewing window $[-0.9467, 0.9467] \times [-0.001, 0.001]$ to verify the answer.

61. The graph of the approximating polynomial P and the elementary function f both pass through the point $(c, f(c))$ and the slopes of P and f agree at $(c, f(c))$. Depending on the degree of P, the nth derivatives of P and f agree at $(c, f(c))$.

63. See definition on page 650.

65. The accuracy increases as the degree increases (for values within the interval of convergence).

67. (a) $f(x) = e^x$

$$P_4(x) = 1 + x + \frac{1}{2}x^2 + \frac{1}{6}x^3 + \frac{1}{24}x^4$$

$$g(x) = xe^x$$

$$Q_5(x) = x + x^2 + \frac{1}{2}x^3 + \frac{1}{6}x^4 + \frac{1}{24}x^5$$

$$Q_5(x) = x P_4(x)$$

(b) $f(x) = \sin x$

$$P_5(x) = x - \frac{x^3}{3!} + \frac{x^5}{5!}$$

$$g(x) = x \sin x$$

$$Q_6(x) = x P_5(x) = x^2 - \frac{x^4}{3!} + \frac{x^6}{5!}$$

(c) $g(x) = \dfrac{\sin x}{x} = \dfrac{1}{x} P_5(x) = 1 - \dfrac{x^2}{3!} + \dfrac{x^4}{5!}$

69. (a) $Q_2(x) = -1 + \dfrac{\pi^2(x+2)^2}{32}$

(b) $R_2(x) = -1 + \dfrac{\pi^2(x-6)^2}{32}$

(c) No. The polynomial will be linear. Translations are possible at $x = -2 + 8n$.

71. Let f be an even function and P_n be the nth Maclaurin polynomial for f. Since f is even, f' is odd, f'' is even, f''' is odd, etc. All of the odd derivatives of f are odd and thus, all of the odd powers of x will have coefficients of zero. P_n will only have terms with even powers of x.

73. As you move away from $x = c$, the Taylor Polynomial becomes less and less accurate.

Section 9.8   Power Series

1. Centered at 0

3. Centered at 2

5. $\sum\limits_{n=0}^{\infty} (-1)^n \dfrac{x^n}{n+1}$

$L = \lim\limits_{n\to\infty} \left| \dfrac{u_{n+1}}{u_n} \right| = \lim\limits_{n\to\infty} \left| \dfrac{(-1)^{n+1}x^{n+1}}{n+2} \cdot \dfrac{n+1}{(-1)^n x^n} \right|$

$\quad = \lim\limits_{n\to\infty} \left| \dfrac{n+1}{n+2} \right| |x| = |x|$

$|x| < 1 \Rightarrow R = 1$

7. $\sum\limits_{n=1}^{\infty} \dfrac{(2x)^n}{n^2}$

$L = \lim\limits_{n\to\infty} \left| \dfrac{u_{n+1}}{u_n} \right| = \lim\limits_{n\to\infty} \left| \dfrac{(2x)^{n+1}}{(n+1)^2} \cdot \dfrac{n^2}{(2x)^n} \right|$

$\quad = \lim\limits_{n\to\infty} \left| \dfrac{2n^2 x}{(n+1)^2} \right| = 2|x|$

$2|x| < 1 \Rightarrow R = \dfrac{1}{2}$

9. $\sum\limits_{n=0}^{\infty} \dfrac{(2x)^{2n}}{(2n)!}$

$L = \lim\limits_{n\to\infty} \left| \dfrac{u_{n+1}}{u_n} \right| = \lim\limits_{n\to\infty} \left| \dfrac{(2x)^{2n+2}/(2n+2)!}{(2x)^{2n}/(2n)!} \right|$

$\quad = \lim\limits_{n\to\infty} \left| \dfrac{(2x)^2}{(2n+2)(2n+1)} \right| = 0$

Thus, the series converges for all x. $R = \infty$.

11. $\sum\limits_{n=0}^{\infty} \left(\dfrac{x}{2} \right)^n$

Since the series is geometric, it converges only if $|x/2| < 1$ or $-2 < x < 2$.

13. $\sum\limits_{n=1}^{\infty} \dfrac{(-1)^n x^n}{n}$

$\lim\limits_{n\to\infty} \left| \dfrac{u_{n+1}}{u_n} \right| = \lim\limits_{n\to\infty} \left| \dfrac{(-1)^{n+1}x^{n+1}}{n+1} \cdot \dfrac{n}{(-1)^n x^n} \right|$

$\quad = \lim\limits_{n\to\infty} \left| \dfrac{nx}{n+1} \right| = |x|$

Interval: $-1 < x < 1$

When $x = 1$, the alternating series $\sum\limits_{n=1}^{\infty} \dfrac{(-1)^n}{n}$ converges.

When $x = -1$, the p-series $\sum\limits_{n=1}^{\infty} \dfrac{1}{n}$ diverges.

Therefore, the interval of convergence is $-1 < x \le 1$.

15. $\sum\limits_{n=0}^{\infty} \dfrac{x^n}{n!}$

$\lim\limits_{n\to\infty} \left| \dfrac{u_{n+1}}{u_n} \right| = \lim\limits_{n\to\infty} \left| \dfrac{x^{n+1}}{(n+1)!} \cdot \dfrac{n!}{x^n} \right|$

$\quad = \lim\limits_{n\to\infty} \left| \dfrac{x}{n+1} \right| = 0$

The series converges for all x. Therefore, the interval of convergence is $-\infty < x < \infty$.

17. $\sum\limits_{n=0}^{\infty} (2n)! \left(\dfrac{x}{2} \right)^n$

$\lim\limits_{n\to\infty} \left| \dfrac{u_{n+1}}{u_n} \right| = \lim\limits_{n\to\infty} \left| \dfrac{(2n+2)!x^{n+1}}{2^{n+1}} \cdot \dfrac{2^n}{(2n)!x^n} \right| = \lim\limits_{n\to\infty} \left| \dfrac{(2n+2)(2n+1)x}{2} \right| = \infty$

Therefore, the series converges only for $x = 0$.

19. $\sum\limits_{n=1}^{\infty} \dfrac{(-1)^{n+1}x^n}{4^n}$

Since the series is geometric, it converges only if $|x/4| < 1$ or $-4 < x < 4$.

21. $\displaystyle\sum_{n=1}^{\infty} \frac{(-1)^{n+1}(x-5)^n}{n5^n}$

$$\lim_{n\to\infty} \left|\frac{u_{n+1}}{u_n}\right| = \lim_{n\to\infty} \left|\frac{(-1)^{n+2}(x-5)^{n+1}}{(n+1)5^{n+1}} \cdot \frac{n5^n}{(-1)^{n+1}(x-5)^n}\right| = \lim_{n\to\infty} \left|\frac{n(x-5)}{5(n+1)}\right| = \frac{1}{5}|x-5|$$

$R = 5$

Center: $x = 5$

Interval: $-5 < x - 5 < 5$ or $0 < x < 10$

When $x = 0$, the p-series $\displaystyle\sum_{n=1}^{\infty} \frac{-1}{n}$ diverges. When $x = 10$, the alternating series $\displaystyle\sum_{n=1}^{\infty} \frac{(-1)^{n+1}}{n}$ converges.

Therefore, the interval of convergence is $0 < x \le 10$.

23. $\displaystyle\sum_{n=0}^{\infty} \frac{(-1)^{n+1}(x-1)^{n+1}}{n+1}$

$$\lim_{n\to\infty} \left|\frac{u_{n+1}}{u_n}\right| = \lim_{n\to\infty} \left|\frac{(-1)^{n+2}(x-1)^{n+2}}{n+2} \cdot \frac{n+1}{(-1)^{n+1}(x-1)^{n+1}}\right| = \lim_{n\to\infty} \left|\frac{(n+1)(x-1)}{n+2}\right| = |x-1|$$

$R = 1$

Center: $x = 1$

Interval: $-1 < x - 1 < 1$ or $0 < x < 2$

When $x = 0$, the series $\displaystyle\sum_{n=0}^{\infty} \frac{1}{n+1}$ diverges by the integral test.

When $x = 2$, the alternating series $\displaystyle\sum_{n=0}^{\infty} \frac{(-1)^{n+1}}{n+1}$ converges.

Therefore, the interval of convergence is $0 < x \le 2$.

25. $\displaystyle\sum_{n=1}^{\infty} \left(\frac{x-3}{3}\right)^{n-1}$ is geometric. It converges if

$$\left|\frac{x-3}{3}\right| < 1 \implies |x-3| < 3 \implies 0 < x < 6.$$

Interval convergence: $0 < x < 6$

27. $\displaystyle\sum_{n=1}^{\infty} \frac{n}{n+1}(-2x)^{n-1}$

$$\lim_{n\to\infty} \left|\frac{u_{n+1}}{u_n}\right| = \lim_{n\to\infty} \left|\frac{(n+1)(-2x)^n}{n+2} \cdot \frac{n+1}{n(-2x)^{n-1}}\right|$$

$$= \lim_{n\to\infty} \left|\frac{(-2x)(n+1)^2}{n(n+2)}\right| = 2|x|$$

$R = \dfrac{1}{2}$

Interval: $-\dfrac{1}{2} < x < \dfrac{1}{2}$

When $x = -\dfrac{1}{2}$, the series $\displaystyle\sum_{n=1}^{\infty} \frac{n}{n+1}$ diverges by the nth Term Test.

When $x = \dfrac{1}{2}$, the alternating series $\displaystyle\sum_{n=1}^{\infty} \frac{(-1)^{n-1}n}{n+1}$ diverges.

Therefore, the interval of convergence is $-\dfrac{1}{2} < x < \dfrac{1}{2}$.

29. $\displaystyle\sum_{n=0}^{\infty} \frac{x^{2n+1}}{(2n+1)!}$

$$\lim_{n\to\infty} \left|\frac{u_{n+1}}{u_n}\right| = \lim_{n\to\infty} \left|\frac{x^{2n+3}}{(2n+3)!} \cdot \frac{(2n+1)!}{x^{2n+1}}\right|$$

$$= \lim_{n\to\infty} \left|\frac{x^2}{(2n+2)(2n+3)}\right| = 0$$

Therefore, the interval of convergence is $-\infty < x < \infty$.

31. $\displaystyle\sum_{n=1}^{\infty} \frac{2 \cdot 3 \cdot 4 \cdots (n+1)x^n}{n!} = \sum_{n=1}^{\infty} (n+1)x^n$

$$\lim_{n\to\infty} \left| \frac{a_{n+1}}{a_n} \right| = \lim_{n\to\infty} \left| \frac{(n+2)x^{n+1}}{(n+1)x^n} \right| = \lim_{n\to\infty} \left| \frac{n+2}{n+1}x \right| = |x|$$

Converges if $|x| < 1 \Rightarrow -1 < x < 1$.

At $x = \pm 1$, diverges.

Interval of convergence: $-1 < x < 1$

33. $\displaystyle\sum_{n=1}^{\infty} \frac{(-1)^{n+1}3 \cdot 7 \cdot 11 \cdots (4n-1)(x-3)^n}{4^n}$

$$\lim_{n\to\infty} \left| \frac{u_{n+1}}{u_n} \right| = \lim_{n\to\infty} \left| \frac{(-1)^{n+2} \cdot 3 \cdot 7 \cdot 11 \cdots (4n-1)(4n+3)(x-3)^{n+1}}{4^{n+1}} \cdot \frac{4^n}{(-1)^{n+1} \cdot 3 \cdot 7 \cdot 11 \cdots (4n-1)(x-3)^n} \right|$$

$$= \lim_{n\to\infty} \left| \frac{(4n+3)(x-3)}{4} \right| = \infty$$

$R = 0$

Center: $x = 3$

Therefore, the series converges only for $x = 3$.

35. $\displaystyle\sum_{n=1}^{\infty} \frac{(x-c)^{n-1}}{c^{n-1}}$

$$\lim_{n\to\infty} \left| \frac{u_{n+1}}{u_n} \right| = \lim_{n\to\infty} \left| \frac{(x-c)^n}{c^n} \cdot \frac{c^{n-1}}{(x-c)^{n-1}} \right| = \frac{1}{c}|x-c|$$

$R = c$

Center: $x = c$

Interval: $-c < x - c < c$ or $0 < x < 2c$

When $x = 0$, the series $\displaystyle\sum_{n=1}^{\infty} (-1)^{n-1}$ diverges.

When $x = 2c$, the series $\displaystyle\sum_{n=1}^{\infty} 1$ diverges.

Therefore, the interval of convergence is $0 < x < 2c$.

37. $\displaystyle\sum_{n=0}^{\infty} \left(\frac{x}{k}\right)^n$

Since the series is geometric, it converges only if $|r/k| < 1$ or $-k < r < k$.

39. $\displaystyle\sum_{n=1}^{\infty} \frac{k(k+1) \cdots (k+n-1)x^n}{n!}$

$$\lim_{n\to\infty} \left| \frac{u_{n+1}}{u_n} \right| = \lim_{n\to\infty} \left| \frac{k(k+1) \cdots (k+n-1)(k+n)x^{n+1}}{(n+1)!} \cdot \frac{n!}{k(k+1) \cdots (k+n-1)x^n} \right| = \lim_{n\to\infty} \left| \frac{(k+n)x}{n+1} \right| = |x|$$

$R = 1$

When $x = \pm 1$, the series diverges and the interval of convergence is $-1 < x < 1$.

$$\left[\frac{k(k+1) \cdots (k+n-1)}{1 \cdot 2 \cdots n} \geq 1 \right]$$

41. $\displaystyle\sum_{n=0}^{\infty} \frac{x^n}{n!} = 1 + \frac{x}{1} + \frac{x^2}{2} + \cdots = \sum_{n=1}^{\infty} \frac{x^{n-1}}{(n-1)!}$

43. $\displaystyle\sum_{n=0}^{\infty} \frac{x^{2n+1}}{(2n+1)!} = \sum_{n=1}^{\infty} \frac{x^{2n-1}}{(2n-1)!}$

Replace n with $n-1$.

45. (a) $f(x) = \sum_{n=0}^{\infty} \left(\frac{x}{2}\right)^n, -2 < x < 2$ (Geometric)

(b) $f'(x) = \sum_{n=1}^{\infty} \left(\frac{n}{2}\right)\left(\frac{x}{2}\right)^{n-1}, -2 < x < 2$

(c) $f''(x) = \sum_{n=2}^{\infty} \left(\frac{n}{2}\right)\left(\frac{n-1}{2}\right)\left(\frac{x}{2}\right)^{n-2}, -2 < x < 2$

(d) $\int f(x)\, dx = \sum_{n=0}^{\infty} \frac{2}{n+1}\left(\frac{x}{2}\right)^{n+1}, -2 \le x < 2$

47. (a) $f(x) = \sum_{n=0}^{\infty} \frac{(-1)^{n+1}(x-1)^{n+1}}{n+1}, 0 < x \le 2$

(b) $f'(x) = \sum_{n=0}^{\infty} (-1)^{n+1}(x-1)^n, 0 < x < 2$

(c) $f''(x) = \sum_{n=1}^{\infty} (-1)^{n+1}n(x-1)^{n-1}, 0 < x < 2$

(d) $\int f(x)\, dx = \sum_{n=1}^{\infty} \frac{(-1)^{n+1}(x-1)^{n+2}}{(n+1)(n+2)}, 0 \le x \le 2$

49. $g(1) = \sum_{n=0}^{\infty} \left(\frac{1}{3}\right)^n = 1 + \frac{1}{3} + \frac{1}{9} + \cdots$

$S_1 = 1, S_2 = 1.33$. Matches (c).

51. $g(3.1) = \sum_{n=0}^{\infty} \left(\frac{3.1}{3}\right)^n$ diverges. Matches (b).

53. $g\left(\frac{1}{8}\right) = \sum_{n=0}^{\infty} \left[2\left(\frac{1}{8}\right)\right]^n = \sum_{n=0}^{\infty} \left(\frac{1}{4}\right)^n = 1 + \frac{1}{4} + \frac{1}{16} + \cdots$, converges

$S_1 = 1, S_2 = 1.25, S_3 = 1.3125$ Matches (b).

55. $g\left(\frac{9}{16}\right) = \sum_{n=0}^{\infty} \left[2\left(\frac{9}{16}\right)\right]^n = \sum_{n=0}^{\infty} \left(\frac{9}{8}\right)^n$, diverges

$S_1 = 1, S_2 = \frac{17}{8}$ Matches (d).

57. A series of the form

$$\sum_{n=0}^{\infty} a_n(x-c)^n$$

is called a power series centered at c.

59. A single point, an interval, or the entire real line.

61. Answers will vary.

$\sum_{n=1}^{\infty} \frac{x^n}{n}$ converges for $-1 \le x < 1$. At $x = -1$, the convergence is conditional because $\sum \frac{1}{n}$ diverges.

$\sum_{n=1}^{\infty} \frac{x^n}{n^2}$ converges for $-1 \le x \le 1$. At $x = \pm 1$, the convergence is absolute.

63. (a) $f(x) = \sum_{n=0}^{\infty} \frac{(-1)^n x^{2n+1}}{(2n+1)!}, -\infty < x < \infty$

(See Exercise 29.)

$g(x) = \sum_{n=0}^{\infty} \frac{(-1)^n x^{2n}}{(2n)!}, -\infty < x < \infty$

(b) $f'(x) = \sum_{n=0}^{\infty} \frac{(-1)^n x^{2n}}{(2n)!} = g(x)$

(c) $g'(x) = \sum_{n=1}^{\infty} \frac{(-1)^n x^{2n-1}}{(2n-1)!} = \sum_{n=0}^{\infty} \frac{(-1)^{n+1} x^{2n+1}}{(2n+1)!}$

$= -\sum_{n=0}^{\infty} \frac{(-1)^n x^{2n+1}}{(2n+1)!} = -f(x)$

(d) $f(x) = \sin x$ and $g(x) = \cos x$

65. $y = \sum_{n=0}^{\infty} \frac{(-1)^n x^{2n+1}}{(2n+1)!} = \sum_{n=1}^{\infty} \frac{(-1)^{n-1} x^{2n-1}}{(2n-1)!}$

$y' = \sum_{n=0}^{\infty} \frac{(-1)^n (2n+1) x^{2n}}{(2n+1)!} = \sum_{n=0}^{\infty} \frac{(-1)^n x^{2n}}{(2n)!}$

$y'' = \sum_{n=1}^{\infty} \frac{(-1)^n (2n) x^{2n-1}}{(2n)!} = \sum_{n=1}^{\infty} \frac{(-1)^n x^{2n-1}}{(2n-1)!}$

$y'' + y = \sum_{n=1}^{\infty} \frac{(-1)^n x^{2n-1}}{(2n-1)!} + \sum_{n=1}^{\infty} \frac{(-1)^{n-1} x^{2n-1}}{(2n-1)!} = 0$

67. $y = \sum_{n=0}^{\infty} \frac{x^{2n+1}}{(2n+1)!} = \sum_{n=1}^{\infty} \frac{x^{2n-1}}{(2n-1)!}$

$y' = \sum_{n=0}^{\infty} \frac{(2n+1) x^{2n}}{(2n+1)!} = \sum_{n=0}^{\infty} \frac{x^{2n}}{(2n)!}$

$y'' = \sum_{n=1}^{\infty} \frac{(2n) x^{2n-1}}{(2n)!} = \sum_{n=1}^{\infty} \frac{x^{2n-1}}{(2n-1)!} = y$

$y'' - y = 0$

69. $y = \displaystyle\sum_{n=0}^{\infty} \frac{x^{2n}}{2^n n!}$ $y' = \displaystyle\sum_{n=1}^{\infty} \frac{2nx^{2n-1}}{2^n n!}$ $y'' = \displaystyle\sum_{n=1}^{\infty} \frac{2n(2n-1)x^{2n-2}}{2^n n!}$

$y'' - xy' - y = \displaystyle\sum_{n=1}^{\infty} \frac{2n(2n-1)x^{2n-2}}{2^n n!} - \sum_{n=1}^{\infty} \frac{2nx^{2n}}{2^n n!} - \sum_{n=0}^{\infty} \frac{x^{2n}}{2^n n!}$

$\displaystyle = \sum_{n=1}^{\infty} \frac{2n(2n-1)x^{2n-2}}{2^n n!} - \sum_{n=0}^{\infty} \frac{(2n+1)x^{2n}}{2^n n!}$

$\displaystyle = \sum_{n=0}^{\infty} \left[\frac{(2n+2)(2n+1)x^{2n}}{2^{n+1}(n+1)!} - \frac{(2n+1)x^{2n}}{2^n n!} \cdot \frac{2(n+1)}{2(n+1)} \right]$

$\displaystyle = \sum_{n=0}^{\infty} \frac{2(n+1)x^{2n}\left[(2n+1)-(2n+1)\right]}{2^{n+1}(n+1)!} = 0$

71. $J_0(x) = \displaystyle\sum_{k=0}^{\infty} \frac{(-1)^k x^{2k}}{2^{2k}(k!)^2}$

(a) $\displaystyle\lim_{k\to\infty} \left| \frac{u_{k+1}}{u_k} \right| = \lim_{k\to\infty} \left| \frac{(-1)^{k+1} x^{2k+2}}{2^{2k+2}[(k+1)!]^2} \cdot \frac{2^{2k}(k!)^2}{(-1)^k x^{2k}} \right| = \lim_{k\to\infty} \left| \frac{(-1)x^2}{2^2(k+1)^2} \right| = 0$

Therefore, the interval of convergence is $-\infty < x < \infty$.

(b) $J_0 = \displaystyle\sum_{k=0}^{\infty} (-1)^k \frac{x^{2k}}{4^k(k!)^2}$

$J_0' = \displaystyle\sum_{k=1}^{\infty} (-1)^k \frac{2kx^{2k-1}}{4^k(k!)^2} = \sum_{k=0}^{\infty} (-1)^{k+1} \frac{(2k+2)x^{2k+1}}{4^{k+1}[(k+1)!]^2}$

$J_0'' = \displaystyle\sum_{k=1}^{\infty} (-1)^k \frac{2k(2k-1)x^{2k-2}}{4^k(k!)^2} = \sum_{k=0}^{\infty} (-1)^{k+1} \frac{(2k+2)(2k+1)x^{2k}}{4^{k+1}[(k+1)!]^2}$

$x^2 J_0'' + x J_0' + x^2 J_0 = \displaystyle\sum_{k=0}^{\infty} (-1)^{k+1} \frac{2(2k+1)x^{2k+2}}{4^{k+1}(k+1)!k!} + \sum_{k=0}^{\infty} (-1)^{k+1} \frac{2x^{2k+2}}{4^{k+1}(k+1)!k!} + \sum_{k=0}^{\infty} (-1)^k \frac{x^{2k+2}}{4^k(k!)^2}$

$\displaystyle = \sum_{k=0}^{\infty} \frac{(-1)^k x^{2k+2}}{4^k(k!)^2} \left[(-1)\frac{2(2k+1)}{4(k+1)} + (-1)\frac{2}{4(k+1)} + 1 \right]$

$\displaystyle = \sum_{k=0}^{\infty} \frac{(-1)^k x^{2k+2}}{4^k(k!)^2} \left[\frac{-4k-2}{4k+4} - \frac{2}{4k+4} + \frac{4k+4}{4k+4} \right] = 0$

(c) $P_6(x) = 1 - \dfrac{x^2}{4} + \dfrac{x^4}{64} - \dfrac{x^6}{2304}$

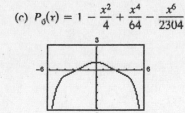

(d) $\displaystyle\int_0^1 J_0\,dx = \int_0^1 \sum_{k=0}^{\infty} \frac{(-1)^k x^{2k}}{4^k(k!)^2}\,dx$

$\displaystyle = \left[\sum_{k=0}^{\infty} \frac{(-1)^k x^{2k+1}}{4^k(k!)^2(2k+1)} \right]_0^1$

$\displaystyle = \sum_{k=0}^{\infty} \frac{(-1)^k}{4^k(k!)^2(2k+1)}$

$\displaystyle = 1 - \frac{1}{12} + \frac{1}{320} \approx 0.92$

(integral is approximately 0.9197304101)

73. $f(x) = \displaystyle\sum_{n=0}^{\infty} (-1)^n \frac{x^{2n}}{(2n)!} = \cos x$

75. $f(x) = \displaystyle\sum_{n=0}^{\infty} (-1)^n x^n = \sum_{n=0}^{\infty} (-x)^n$ Geometric

$\displaystyle = \frac{1}{1-(-x)} = \frac{1}{1+x}$ for $-1 < x < 1$

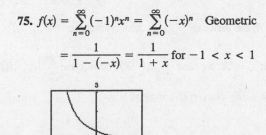

77. $\displaystyle\sum_{n=0}^{\infty}\left(\frac{x}{2}\right)^n$

(a) $\displaystyle\sum_{n=0}^{\infty}\left(\frac{3/4}{2}\right)^n = \sum_{n=0}^{\infty}\left(\frac{3}{8}\right)^n = \frac{1}{1-(3/8)} = \frac{8}{5} = 1.6$

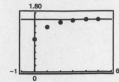

(c) The alternating series converges more rapidly. The partial sums of the series of positive terms approach the sum from below. The partial sums of the alternating series alternate sides of the horizontal line representing the sum.

(b) $\displaystyle\sum_{n=0}^{\infty}\left(\frac{-3/4}{2}\right)^n = \sum_{n=0}^{\infty}\left(-\frac{3}{8}\right)^n$

$$= \frac{1}{1-(-3/8)} = \frac{8}{11} \approx 0.7272$$

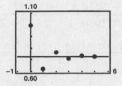

(d) $\displaystyle\sum_{n=0}^{N}\left(\frac{3}{2}\right)^n > M$

M	10	100	1000	10,000
N	4	9	15	21

79. False;

$$\sum_{n=0}^{\infty}\frac{(-1)^n x^n}{n2^n}$$

converges for $x = 2$ but diverges for $x = -2$.

81. True; the radius of convergence is $R = 1$ for both series.

83. $\displaystyle\lim_{n\to\infty}\left|\frac{a_{n+1}}{a_n}\right| = \lim_{n\to\infty}\left|\frac{(n+1+p)!}{(n+1)!(n+1+q)!}x^{n+1}\middle/\frac{(n+p)!}{n!(n+q)!}x^n\right| = \lim_{n\to\infty}\left|\frac{(n+1+p)x}{(n+1)(n+1+q)}\right| = 0$

Thus, the series converges for all x: $R = \infty$.

85. (a) $\displaystyle f(x) = \sum_{n=0}^{\infty}c_n x^n,\ c_{n+3} = c_n$

$= c_0 + c_1 x + c_2 x^2 + c_0 x^3 + c_1 x^4 + c_2 x^5 + c_0 x^6 + \cdots$

$S_{3n} = c_0(1 + x^3 + \cdots + x^{3n}) + c_1 x(1 + x^3 + \cdots + x^{3n}) + c_2 x^2(1 + x^3 + \cdots + x^{3n})$

$\displaystyle\lim_{n\to\infty} S_{3n} = c_0 \sum_{n=0}^{\infty}x^{3n} + c_1 x \sum_{n=0}^{\infty}x^{3n} + c_2 x^2 \sum_{n=0}^{\infty}x^{3n}$

Each series is geometric, $R = 1$, and the interval of convergence is $(-1, 1)$.

(b) For $|x| < 1, f(x) = c_0\dfrac{1}{1-x^3} + c_1 x\dfrac{1}{1-x^3} + c_2 x^2\dfrac{1}{1-x^3} = \dfrac{c_0 + c_1 x + c_2 x^2}{1-x^3}$.

87. At $x_0 + R$, the series is $\displaystyle\sum_{n=0}^{\infty}c_n(x_0 - x_0 - R)^n = \sum_{n=0}^{\infty}c_n(-R)^n$ which converges.

At $x_0 - R$, the series is $\displaystyle\sum_{n=0}^{\infty}c_n R^n$, which diverges.

Hence, at $x_0 + R$, we have $\displaystyle\sum_{n=0}^{\infty}c_n(-R)^n$ converges, $\displaystyle\sum_{n=0}^{\infty}\left|c_n(-R)^n\right| = \sum_{n=0}^{\infty}c_n R^n$ diverges.

$\Rightarrow$ The series converges conditionally.

Section 9.9 Representation of Functions by Power Series

1. (a) $\dfrac{1}{2-x} = \dfrac{1/2}{1-(x/2)} = \dfrac{a}{1-r}$

$= \displaystyle\sum_{n=0}^{\infty} \dfrac{1}{2}\left(\dfrac{x}{2}\right)^n = \sum_{n=0}^{\infty} \dfrac{x^n}{2^{n+1}}$

This series converges on $(-2, 2)$.

(b)

$$\require{enclose}\begin{array}{r}\frac{1}{2}+\frac{x}{4}+\frac{x^2}{8}+\frac{x^3}{16}+\cdots\\[2pt]2-x\enclose{longdiv}{1}\end{array}$$

$$\begin{array}{r}1-\dfrac{x}{2}\\\hline\dfrac{x}{2}\\[6pt]\dfrac{x}{2}-\dfrac{x^2}{4}\\\hline\dfrac{x^2}{4}\\[6pt]\dfrac{x^2}{4}-\dfrac{x^3}{8}\\\hline\dfrac{x^3}{8}\\[6pt]\dfrac{x^3}{8}-\dfrac{x^4}{16}\\\hline\vdots\end{array}$$

3. (a) $\dfrac{1}{2+x} = \dfrac{1/2}{1-(-x/2)} = \dfrac{a}{1-r}$

$= \displaystyle\sum_{n=0}^{\infty} \dfrac{1}{2}\left(-\dfrac{x}{2}\right)^n = \sum_{n=0}^{\infty} \dfrac{(-1)^n x^n}{2^{n+1}}$

This series converges on $(-2, 2)$.

(b)

$$\begin{array}{r}\frac{1}{2}-\frac{x}{4}+\frac{x^2}{8}-\frac{x^3}{16}+\cdots\\[2pt]2+x\enclose{longdiv}{1}\end{array}$$

$$\begin{array}{r}1+\dfrac{x}{2}\\\hline-\dfrac{x}{2}\\[6pt]-\dfrac{x}{2}-\dfrac{x^2}{4}\\\hline\dfrac{x^2}{4}\\[6pt]\dfrac{x^2}{4}+\dfrac{x^3}{8}\\\hline-\dfrac{x^3}{8}\\[6pt]-\dfrac{x^3}{8}-\dfrac{x^4}{16}\\\hline\vdots\end{array}$$

5. Writing $f(x)$ in the form $a/(1-r)$, we have

$$\frac{1}{2-x} = \frac{1}{-3-(x-5)} = \frac{-1/3}{1+(1/3)(x-5)}$$

which implies that $a = -1/3$ and $r = (-1/3)(x-5)$. Therefore, the power series for $f(x)$ is given by

$$\frac{1}{2-x} = \sum_{n=0}^{\infty} ar^n = \sum_{n=0}^{\infty} -\frac{1}{3}\left[-\frac{1}{3}(x-5)\right]^n$$

$$= \sum_{n=0}^{\infty} \frac{(x-5)^n}{(-3)^{n+1}}, \; |x-5| < 3 \text{ or } 2 < x < 8.$$

7. Writing $f(x)$ in the form $a/(1-r)$, we have

$$\frac{3}{2x-1} = \frac{-3}{1-2x} = \frac{a}{1-r}$$

which implies that $a = -3$ and $r = 2x$. Therefore, the power series for $f(x)$ is given by

$$\frac{3}{2x-1} = \sum_{n=0}^{\infty} ar^n = \sum_{n=0}^{\infty} (-3)(2x)^n$$

$$= -3\sum_{n=0}^{\infty} (2x)^n, \; |2x| < 1 \text{ or } -\frac{1}{2} < x < \frac{1}{2}.$$

9. Writing $f(x)$ in the form $a/(1-r)$, we have

$$\frac{1}{2x-5} = \frac{-1}{11-2(x+3)}$$

$$= \frac{-1/11}{1-(2/11)(x+3)} = \frac{a}{1-r}$$

which implies that $a = -1/11$ and $r = (2/11)(x+3)$. Therefore, the power series for $f(x)$ is given by

$$\frac{1}{2x-5} = \sum_{n=0}^{\infty} ar^n = \sum_{n=0}^{\infty} \left(-\frac{1}{11}\right)\left[\frac{2}{11}(x+3)\right]^n$$

$$= -\sum_{n=0}^{\infty} \frac{2^n(x+3)^n}{11^{n+1}},$$

$$|x+3| < \frac{11}{2} \text{ or } -\frac{17}{2} < x < \frac{5}{2}.$$

11. Writing $f(x)$ in the form $a/(1-r)$, we have

$$\frac{3}{x+2} = \frac{3}{2+x} = \frac{3/2}{1+(1/2)x} = \frac{a}{1-r}$$

which implies that $a = 3/2$ and $r = (-1/2)x$. Therefore, the power series for $f(x)$ is given by

$$\frac{3}{x+2} = \sum_{n=0}^{\infty} ar^n = \sum_{n=0}^{\infty} \frac{3}{2}\left(-\frac{1}{2}x\right)^n$$

$$= 3\sum_{n=0}^{\infty} \frac{(-1)^n x^n}{2^{n+1}} = \frac{3}{2}\sum_{n=0}^{\infty}\left(-\frac{x}{2}\right)^n,$$

$$|x| < 2 \text{ or } -2 < x < 2.$$

13. $\dfrac{3x}{x^2 + x - 2} = \dfrac{2}{x + 2} + \dfrac{1}{x - 1} = \dfrac{2}{2 + x} + \dfrac{1}{-1 + x} = \dfrac{1}{1 + (1/2)x} + \dfrac{-1}{1 - x}$

Writing $f(x)$ as a sum of two geometric series, we have

$$\frac{3x}{x^2 + x - 2} = \sum_{n=0}^{\infty} \left(-\frac{1}{2}x\right)^n + \sum_{n=0}^{\infty} (-1)(x)^n = \sum_{n=0}^{\infty} \left[\frac{1}{(-2)^n} - 1\right]x^n.$$

The interval of convergence is $-1 < x < 1$ since

$$\lim_{n \to \infty} \left|\frac{u_{n+1}}{u_n}\right| = \lim_{n \to \infty} \left|\frac{(1 - (-2)^{n+1})x^{n+1}}{(-2)^{n+1}} \cdot \frac{(-2)^n}{(1 - (-2)^n)x^n}\right| = \lim_{n \to \infty} \left|\frac{(1 - (-2)^{n+1})x}{-2 - (-2)^{n+1}}\right| = |x|.$$

15. $\dfrac{2}{1 - x^2} = \dfrac{1}{1 - x} + \dfrac{1}{1 + x}$

Writing $f(x)$ as a sum of two geometric series, we have

$$\frac{2}{1 - x^2} = \sum_{n=0}^{\infty} x^n + \sum_{n=0}^{\infty} (-x)^n = \sum_{n=0}^{\infty} (1 + (-1)^n)x^n = \sum_{n=0}^{\infty} 2x^{2n}.$$

The interval of convergence is $|x^2| < 1$ or $-1 < x < 1$ since $\lim_{n \to \infty} \left|\dfrac{u_{n+1}}{u_n}\right| = \lim_{n \to \infty} \left|\dfrac{2x^{2n+2}}{2x^{2n}}\right| = |x^2|.$

17. $\dfrac{1}{1 + x} = \displaystyle\sum_{n=0}^{\infty} (-1)^n x^n$

$\dfrac{1}{1 - x} = \displaystyle\sum_{n=0}^{\infty} (-1)^n (-x)^n = \sum_{n=0}^{\infty} (-1)^{2n} x^n = \sum_{n=0}^{\infty} x^n$

$h(x) = \dfrac{-2}{x^2 - 1} = \dfrac{1}{1 + x} + \dfrac{1}{1 - x} = \displaystyle\sum_{n=0}^{\infty} (-1)^n x^n + \sum_{n=0}^{\infty} x^n = \sum_{n=0}^{\infty} [(-1)^n + 1]x^n$

$= 2 + 0x + 2x^2 + 0x^3 + 2x^4 + 0x^5 + 2x^6 + \cdots = \displaystyle\sum_{n=0}^{\infty} 2x^{2n}, \; -1 < x < 1$ (See Exercise 15.)

19. By taking the first derivative, we have $\dfrac{d}{dx}\left[\dfrac{1}{x + 1}\right] = \dfrac{-1}{(x + 1)^2}$. Therefore,

$$\frac{-1}{(x + 1)^2} = \frac{d}{dx}\left[\sum_{n=0}^{\infty} (-1)^n x^n\right] = \sum_{n=1}^{\infty} (-1)^n n x^{n-1}$$

$$= \sum_{n=0}^{\infty} (-1)^{n+1}(n + 1)x^n, \; -1 < x < 1.$$

21. By integrating, we have $\displaystyle\int \dfrac{1}{x + 1} \, dx = \ln(x + 1)$. Therefore,

$$\ln(x + 1) = \int \left[\sum_{n=0}^{\infty} (-1)^n x^n\right] dx = C + \sum_{n=0}^{\infty} \frac{(-1)^n x^{n+1}}{n + 1}, \; -1 < x \le 1.$$

To solve for C, let $x = 0$ and conclude that $C = 0$. Therefore,

$$\ln(x + 1) = \sum_{n=0}^{\infty} \frac{(-1)^n x^{n+1}}{n + 1}, \; -1 < x \le 1.$$

23. $\dfrac{1}{x^2 + 1} = \displaystyle\sum_{n=0}^{\infty} (-1)^n (x^2)^n = \sum_{n=0}^{\infty} (-1)^n x^{2n}, \; -1 < x < 1$

25. Since, $\dfrac{1}{x + 1} = \displaystyle\sum_{n=0}^{\infty} (-1)^n x^n$, we have $\dfrac{1}{4x^2 + 1} = \displaystyle\sum_{n=0}^{\infty} (-1)^n (4x^2)^n = \sum_{n=0}^{\infty} (-1)^n 4^n x^{2n} = \sum_{n=0}^{\infty} (-1)^n (2x)^{2n}, \; -\dfrac{1}{2} < x < \dfrac{1}{2}.$

27. $x - \dfrac{x^2}{2} \le \ln(x+1) \le x - \dfrac{x^2}{2} + \dfrac{x^3}{3}$

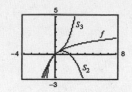

x	0.0	0.2	0.4	0.6	0.8	1.0
$x - \dfrac{x^2}{2}$	0.000	0.180	0.320	0.420	0.480	0.500
$\ln(x+1)$	0.000	0.182	0.336	0.470	0.588	0.693
$x - \dfrac{x^2}{2} + \dfrac{x^3}{3}$	0.000	0.183	0.341	0.492	0.651	0.833

29. $\displaystyle\sum_{n=1}^{\infty} \frac{(-1)^{n+1}(x-1)^n}{n} = \frac{(x-1)}{1} - \frac{(x-1)^2}{2} + \frac{(x-1)^3}{3} - \cdots$

(a)

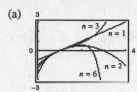

(b) From Example 4,

$$\sum_{n=1}^{\infty} \frac{(-1)^{n+1}(x-1)^n}{n} = \sum_{n=0}^{\infty} \frac{(-1)^n(x-1)^{n+1}}{n+1}$$

$$= \ln x, \quad 0 < x \le 2, \quad R = 1.$$

(c) $x = 0.5$:

$$\sum_{n=1}^{\infty} \frac{(-1)^{n+1}(-1/2)^n}{n} = \sum_{n=1}^{\infty} \frac{-(1/2)^n}{n} \approx -0.693147$$

(d) This is an approximation of $\ln\left(\frac{1}{2}\right)$. The error is approximately 0. $\left[\text{The error is less than the first omitted term, } 1/(51 \cdot 2^{51}) \approx 8.7 \times 10^{-18}.\right]$

31. $g(x) = x$ line

Matches (c)

33. $g(x) = x - \dfrac{x^3}{3} + \dfrac{x^5}{5}$

Matches (a)

In Exercises 35–37, $\arctan x = \displaystyle\sum_{n=0}^{\infty} (-1)^n \frac{x^{2n+1}}{2n+1}$.

35. $\arctan \dfrac{1}{4} = \displaystyle\sum_{n=0}^{\infty} (-1)^n \frac{(1/4)^{2n+1}}{2n+1} = \sum_{n=0}^{\infty} \frac{(-1)^n}{(2n+1)4^{2n+1}} = \frac{1}{4} - \frac{1}{192} + \frac{1}{5120} + \cdots$

Since $\frac{1}{5120} < 0.001$, we can approximate the series by its first two terms: $\arctan \frac{1}{4} \approx \frac{1}{4} - \frac{1}{192} \approx 0.245$.

37. $\dfrac{\arctan x^2}{x} = \dfrac{1}{x} \displaystyle\sum_{n=0}^{\infty} (-1)^n \frac{(x^2)^{2n+1}}{2n+1} = \sum_{n=0}^{\infty} (-1)^2 \frac{x^{4n+1}}{2n+1}$

$\displaystyle\int \frac{\arctan x^2}{x}\, dx = \sum_{n=0}^{\infty} (-1)^n \frac{x^{4n+2}}{(4n+2)(2n+1)} + C$ (Note: $C = 0$)

$\displaystyle\int_0^{1/2} \frac{\arctan x^2}{x}\, dx = \sum_{n=0}^{\infty} (-1)^n \frac{1}{(4n+2)(2n+1)2^{4n+2}} = \frac{1}{8} - \frac{1}{1152} + \cdots$

Since $\dfrac{1}{1152} < 0.001$, we can approximate the series by its first term: $\displaystyle\int_0^{1/2} \frac{\arctan x^2}{x}\, dx \approx 0.125$.

In Exercises 39–43, use $\dfrac{1}{1-x} = \displaystyle\sum_{n=0}^{\infty} x^n$, $|x| < 1$.

39. $\dfrac{1}{(1-x)^2} = \dfrac{d}{dx}\left[\dfrac{1}{1-x}\right] = \dfrac{d}{dx}\left[\displaystyle\sum_{n=0}^{\infty} x^n\right] = \displaystyle\sum_{n=1}^{\infty} nx^{n-1}$, $|x| < 1$

41. $\dfrac{1+x}{(1-x)^2} = \dfrac{1}{(1-x)^2} + \dfrac{x}{(1-x)^2}$

$\qquad = \displaystyle\sum_{n=1}^{\infty} n(x^{n-1} + x^n)$, $|x| < 1$

$\qquad = \displaystyle\sum_{n=0}^{\infty} (2n+1)x^n$, $|x| < 1$

43. $P(n) = \left(\dfrac{1}{2}\right)^n$

$E(n) = \displaystyle\sum_{n=1}^{\infty} nP(n) = \sum_{n=1}^{\infty} n\left(\dfrac{1}{2}\right)^n = \dfrac{1}{2}\sum_{n=1}^{\infty} n\left(\dfrac{1}{2}\right)^{n-1}$

$\qquad = \dfrac{1}{2}\dfrac{1}{[1-(1/2)]^2} = 2$

Since the probability of obtaining a head on a single toss is $\frac{1}{2}$, it is expected that, on average, a head will be obtained in two tosses.

45. Replace x with $(-x)$.

47. Replace x with $(-x)$ and multiply the series by 5.

49. Let $\arctan x + \arctan y = \theta$. Then,

$\qquad \tan(\arctan x + \arctan y) = \tan\theta$

$\qquad \dfrac{\tan(\arctan x) + \tan(\arctan y)}{1 - \tan(\arctan x)\tan(\arctan y)} = \tan\theta$

$\qquad\qquad \dfrac{x+y}{1-xy} = \tan\theta$

$\arctan\left(\dfrac{x+y}{1-xy}\right) = \theta$. Therefore, $\arctan x + \arctan y = \arctan\left(\dfrac{x+y}{1-xy}\right)$ for $xy \neq 1$.

51. (a) $2\arctan\dfrac{1}{2} = \arctan\dfrac{1}{2} + \arctan\dfrac{1}{2} = \arctan\left[\dfrac{\frac{1}{2}+\frac{1}{2}}{1-(1/2)^2}\right] = \arctan\dfrac{4}{3}$

$\qquad 2\arctan\dfrac{1}{2} - \arctan\dfrac{1}{7} = \arctan\dfrac{4}{3} + \arctan\left(-\dfrac{1}{7}\right) = \arctan\left[\dfrac{(4/3)-(1/7)}{1+(4/3)(1/7)}\right] = \arctan\dfrac{25}{25} = \arctan 1 = \dfrac{\pi}{4}$

(b) $\pi = 8\arctan\dfrac{1}{2} - 4\arctan\dfrac{1}{7} \approx 8\left[\dfrac{1}{2} - \dfrac{(0.5)^3}{3} + \dfrac{(0.5)^5}{5} - \dfrac{(0.5)^7}{7}\right] - 4\left[\dfrac{1}{7} - \dfrac{(1/7)^3}{3} + \dfrac{(1/7)^5}{5} - \dfrac{(1/7)^7}{7}\right] \approx 3.14$

53. From Exercise 21, we have

$\ln(x+1) = \displaystyle\sum_{n=0}^{\infty} \dfrac{(-1)^n x^{n+1}}{n+1} = \sum_{n=1}^{\infty} \dfrac{(-1)^{n-1}x^n}{n}$

$\qquad = \displaystyle\sum_{n=1}^{\infty} \dfrac{(-1)^{n+1}x^n}{n}$.

Thus, $\displaystyle\sum_{n=1}^{\infty} (-1)^{n+1}\dfrac{1}{2^n n} = \sum_{n=1}^{\infty} \dfrac{(-1)^{n+1}(1/2)^n}{n}$

$\qquad\qquad = \ln\left(\dfrac{1}{2}+1\right) = \ln\dfrac{3}{2} \approx 0.4055$.

55. From Exercise 53, we have

$\displaystyle\sum_{n=1}^{\infty} (-1)^{n+1}\dfrac{2^n}{5^n n} = \sum_{n=1}^{\infty} \dfrac{(-1)^{n+1}(2/5)^n}{n}$

$\qquad = \ln\left(\dfrac{2}{5}+1\right) = \ln\dfrac{7}{5} \approx 0.3365$.

57. From Exercise 56, we have

$\displaystyle\sum_{n=0}^{\infty} (-1)^n\dfrac{1}{2^{2n+1}(2n+1)} = \sum_{n=0}^{\infty} (-1)^n \dfrac{(1/2)^{2n+1}}{2n+1}$

$\qquad = \arctan\dfrac{1}{2} \approx 0.4636$.

59. $f(x) = \arctan x$ is an odd function (symmetric to the origin).

61. The series in Exercise 56 converges to its sum at a slower rate because its terms approach 0 at a much slower rate.

63. Because the first series is the derivative of the second series, the second series converges for $|x + 1| < 4$ (and perhaps at the endpoints, $x = 3$ and $x = -5$.)

65. $\displaystyle\sum_{n=0}^{\infty} \frac{(-1)^n}{3^n(2n+1)}$

From Example 5 we have $\arctan x = \displaystyle\sum_{n=0}^{\infty} (-1)^n \frac{x^{2n+1}}{2n+1}$.

$$\sum_{n=0}^{\infty} \frac{(-1)^n}{3^n(2n+1)} = \sum_{n=0}^{\infty} \frac{(-1)^n}{(\sqrt{3})^{2n}(2n+1)} \frac{\sqrt{3}}{\sqrt{3}}$$

$$= \sqrt{3} \sum_{n=0}^{\infty} \frac{(-1)^n(1/\sqrt{3})^{2n+1}}{2n+1}$$

$$= \sqrt{3} \arctan\left(\frac{1}{\sqrt{3}}\right)$$

$$= \sqrt{3} \left(\frac{\pi}{6}\right) \approx 0.9068997$$

Section 9.10   Taylor and Maclaurin Series

1. For $c = 0$, we have:

$$f(x) = e^{2x}$$

$$f^{(n)}(x) = 2^n e^{2x} \Longrightarrow f^{(n)}(0) = 2^n$$

$$e^{2x} = 1 + 2x + \frac{4x^2}{2!} + \frac{8x^3}{3!} + \frac{16x^4}{4!} + \cdots = \sum_{n=0}^{\infty} \frac{(2x)^n}{n!}.$$

3. For $c = \pi/4$, we have:

$$f(x) = \cos(x) \qquad f\left(\frac{\pi}{4}\right) = \frac{\sqrt{2}}{2}$$

$$f'(x) = -\sin(x) \qquad f'\left(\frac{\pi}{4}\right) = -\frac{\sqrt{2}}{2}$$

$$f''(x) = -\cos(x) \qquad f''\left(\frac{\pi}{4}\right) = -\frac{\sqrt{2}}{2}$$

$$f'''(x) = \sin(x) \qquad f'''\left(\frac{\pi}{4}\right) = \frac{\sqrt{2}}{2}$$

$$f^{(4)}(x) = \cos(x) \qquad f^{(4)}\left(\frac{\pi}{4}\right) = \frac{\sqrt{2}}{2}$$

and so on. Therefore, we have:

$$\cos x = \sum_{n=0}^{\infty} \frac{f^{(n)}(\pi/4)[x - (\pi/4)]^n}{n!}$$

$$= \frac{\sqrt{2}}{2}\left[1 - \left(x - \frac{\pi}{4}\right) - \frac{[x - (\pi/4)]^2}{2!} + \frac{[x - (\pi/4)]^3}{3!} + \frac{[x - (\pi/4)]^4}{4!} - \cdots\right]$$

$$= \frac{\sqrt{2}}{2} \sum_{n=0}^{\infty} \frac{(-1)^{n(n+1)/2}[x - (\pi/4)]^n}{n!}.$$

[Note: $(-1)^{n(n+1)/2} = 1, -1, -1, 1, 1, -1, -1, 1, \ldots$]

5. For $c = 1$, we have,

$$f(x) = \ln x \qquad f(1) = 0$$

$$f'(x) = \frac{1}{x} \qquad f'(1) = 1$$

$$f''(x) = -\frac{1}{x^2} \qquad f''(1) = -1$$

$$f'''(x) = \frac{2}{x^3} \qquad f'''(1) = 2$$

$$f^{(4)}(x) = -\frac{6}{x^4} \qquad f^{(4)}(1) = -6$$

$$f^{(5)}(x) = \frac{24}{x^5} \qquad f^{(5)}(1) = 24$$

and so on. Therefore, we have:

$$\ln x = \sum_{n=0}^{\infty} \frac{f^{(n)}(1)(x-1)^n}{n!}$$

$$= 0 + (x-1) - \frac{(x-1)^2}{2!} + \frac{2(x-1)^3}{3!} - \frac{6(x-1)^4}{4!} + \frac{24(x-1)^5}{5!} - \cdots$$

$$= (x-1) - \frac{(x-1)^2}{2} + \frac{(x-1)^3}{3} - \frac{(x-1)^4}{4} + \frac{(x-1)^5}{5} - \cdots$$

$$= \sum_{n=0}^{\infty} (-1)^n \frac{(x-1)^{n+1}}{n+1}.$$

7. For $c = 0$, we have:

$$f(x) = \sin 2x \qquad\qquad f(0) = 0$$

$$f'(x) = 2\cos 2x \qquad\qquad f'(0) = 2$$

$$f''(x) = -4\sin 2x \qquad\qquad f''(0) = 0$$

$$f'''(x) = -8\cos 2x \qquad\qquad f'''(0) = -8$$

$$f^{(4)}(x) = 16\sin 2x \qquad\qquad f^{(4)}(0) = 0$$

$$f^{(5)}(x) = 32\cos 2x \qquad\qquad f^{(5)}(0) = 32$$

$$f^{(6)}(x) = -64\sin 2x \qquad\qquad f^{(6)}(0) = 0$$

$$f^{(7)}(x) = -128\cos 2x \qquad\qquad f^{(7)}(0) = -128$$

and so on. Therefore, we have:

$$\sin 2x = \sum_{n=0}^{\infty} \frac{f^{(n)}(0)x^n}{n!} = 0 + 2x + \frac{0x^2}{2!} - \frac{8x^3}{3!} + \frac{0x^4}{4!} + \frac{32x^5}{5!} + \frac{0x^6}{6!} - \frac{128x^7}{7!} + \cdots$$

$$= 2x - \frac{8x^3}{3!} + \frac{32x^5}{5!} - \frac{128x^7}{7!} + \cdots = \sum_{n=0}^{\infty} \frac{(-1)^n(2x)^{2n+1}}{(2n+1)!}.$$

9. For $c = 0$, we have:

$$f(x) = \sec(x) \qquad\qquad\qquad\qquad\qquad f(0) = 1$$

$$f'(x) = \sec(x)\tan(x) \qquad\qquad\qquad\qquad f'(0) = 0$$

$$f''(x) = \sec^3(x) + \sec(x)\tan^2(x) \qquad\qquad f''(0) = 1$$

$$f'''(x) = 5\sec^3(x)\tan(x) + \sec(x)\tan^3(x) \qquad f'''(0) = 0$$

$$f^{(4)}(x) = 5\sec^5(x) + 18\sec^3(x)\tan^2(x) + \sec(x)\tan^4(x) \qquad f^{(4)}(0) = 5$$

$$\sec(x) = \sum_{n=0}^{\infty} \frac{f^{(n)}(0)x^n}{n!} = 1 + \frac{x^2}{2!} + \frac{5x^4}{4!} + \cdots.$$

11. The Maclaurin series for $f(x) = \cos x$ is $\displaystyle\sum_{n=0}^{\infty} \frac{(-1)^n x^{2n}}{(2n)!}$.

Because $f^{(n+1)}(x) = \pm\sin x$ or $\pm\cos x$, we have $|f^{(n+1)}(z)| \le 1$ for all z. Hence by Taylor's Theorem,

$$0 \le |R_n(x)| = \left|\frac{f^{(n+1)}(z)}{(n+1)!}x^{n+1}\right| \le \frac{|x|^{n+1}}{(n+1)!}.$$

Since $\displaystyle\lim_{n\to\infty} \frac{|x|^{n+1}}{(n+1)!} = 0$, it follows that $R_n(x) \to 0$ as $n \to \infty$. Hence, the Maclaurin series for $\cos x$ converges to $\cos x$ for all x.

13. The Maclaurin series for $f(x) = \sinh x$ is $\displaystyle\sum_{n=0}^{\infty} \frac{x^{2n+1}}{(2n+1)!}$.

$f^{(n+1)}(x) = \sinh x$ (or $\cosh x$). For fixed x,

$$0 \le |R_n(x)| = \left|\frac{f^{(n+1)}(z)}{(n+1)!}x^{n+1}\right| = \left|\frac{\sinh(z)}{(n+1)!}x^{n+1}\right| \to 0 \text{ as } n \to \infty.$$

(The argument is the same if $f^{(n+1)}(x) = \cosh x$). Hence, the Maclaurin series for $\sinh x$ converges to $\sinh x$ for all x.

15. Since $(1+x)^{-k} = 1 - kx + \dfrac{k(k+1)x^2}{2!} - \dfrac{k(k+1)(k+2)x^3}{3!} + \cdots$, we have

$$(1+x)^{-2} = 1 - 2x + \frac{2(3)x^2}{2!} - \frac{2(3)(4)x^3}{3!} + \frac{2(3)(4)(5)x^4}{4!} - \cdots = 1 - 2x + 3x^2 - 4x^3 + 5x^4 - \cdots$$

$$= \sum_{n=0}^{\infty} (-1)^n (n+1) x^n.$$

17. $\dfrac{1}{\sqrt{4+x^2}} = \left(\dfrac{1}{2}\right)\left[1 + \left(\dfrac{x}{2}\right)^2\right]^{-1/2}$ and since $(1+x)^{-1/2} = 1 + \displaystyle\sum_{n=1}^{\infty} \frac{(-1)^n 1 \cdot 3 \cdot 5 \cdots (2n-1)x^n}{2^n n!}$, we have

$$\frac{1}{\sqrt{4+x^2}} = \frac{1}{2}\left[1 + \sum_{n=1}^{\infty} \frac{(-1)^n 1 \cdot 3 \cdot 5 \cdots (2n-1)(x/2)^{2n}}{2^n n!}\right] = \frac{1}{2} + \sum_{n=1}^{\infty} \frac{(-1)^n 1 \cdot 3 \cdot 5 \cdots (2n-1)x^{2n}}{2^{3n+1}n!}.$$

19. Since $(1+x)^{1/2} = 1 + \dfrac{x}{2} + \displaystyle\sum_{n=2}^{\infty} \frac{(-1)^{n+1} 1 \cdot 3 \cdot 5 \cdots (2n-3)x^n}{2^n n!}$

we have $(1+x^2)^{1/2} = 1 + \dfrac{x^2}{2} + \displaystyle\sum_{n=2}^{\infty} \frac{(-1)^{n+1} 1 \cdot 3 \cdot 5 \cdots (2n-3)x^{2n}}{2^n n!}$.

21. $e^x = \displaystyle\sum_{n=0}^{\infty} \frac{x^n}{n!} = 1 + x + \frac{x^2}{2!} + \frac{x^3}{3!} + \frac{x^4}{4!} + \frac{x^5}{5!} + \cdots$

$e^{x^2/2} = \displaystyle\sum_{n=0}^{\infty} \frac{(x^2/2)^n}{n!} = \sum_{n=0}^{\infty} \frac{x^{2n}}{2^n n!} = 1 + \frac{x^2}{2} + \frac{x^4}{2^2 2!} + \frac{x^6}{2^3 3!} + \frac{x^8}{2^4 4!} + \cdots$

23. $\sin x = \displaystyle\sum_{n=0}^{\infty} \frac{(-1)^n x^{2n+1}}{(2n+1)!}$

$\sin 3x = \displaystyle\sum_{n=0}^{\infty} \frac{(-1)^n (3x)^{2n+1}}{(2n+1)!}$

25. $\cos x = \displaystyle\sum_{n=0}^{\infty} \frac{(-1)^n x^{2n}}{(2n)!} = 1 - \frac{x^2}{2!} + \frac{x^4}{4!} - \cdots$

$\cos x^{3/2} = \displaystyle\sum_{n=0}^{\infty} \frac{(-1)^n (x^{3/2})^{2n}}{(2n)!}$

$= \displaystyle\sum_{n=0}^{\infty} \frac{(-1)^n x^{3n}}{(2n)!}$

$= 1 - \dfrac{x^3}{2!} + \dfrac{x^6}{4!} - \cdots$

27.
$$e^x = 1 + x + \frac{x^2}{2!} + \frac{x^3}{3!} + \frac{x^4}{4!} + \frac{x^5}{5!} + \cdots$$

$$e^{-x} = 1 - x + \frac{x^2}{2!} - \frac{x^3}{3!} + \frac{x^4}{4!} - \frac{x^5}{5!} + \cdots$$

$$e^x - e^{-x} = 2x + \frac{2x^3}{3!} + \frac{2x^5}{5!} + \frac{2x^7}{7!} + \cdots$$

$$\sinh(x) = \frac{1}{2}(e^x - e^{-x})$$

$$= x + \frac{x^3}{3!} + \frac{x^5}{5!} + \frac{x^7}{7!} + \cdots = \sum_{n=0}^{\infty} \frac{x^{2n+1}}{(2n+1)!}$$

29. $\cos^2(x) = \frac{1}{2}[1 + \cos(2x)]$

$$= \frac{1}{2}\left[1 + 1 - \frac{(2x)^2}{2!} + \frac{(2x)^4}{4!} - \frac{(2x)^6}{6!} - \cdots\right] = \frac{1}{2}\left[1 + \sum_{n=0}^{\infty} \frac{(-1)^n(2x)^{2n}}{(2n)!}\right]$$

31. $x \sin x = x\left(x - \frac{x^3}{3!} + \frac{x^5}{5!} - \cdots\right)$

$$= x^2 - \frac{x^4}{3!} + \frac{x^6}{5!} - \cdots$$

$$= \sum_{n=0}^{\infty} \frac{(-1)^n x^{2n+2}}{(2n+1)!}$$

33. $\dfrac{\sin x}{x} = \dfrac{x - (x^3/3!) + (x^5/5!) - \cdots}{x}$

$$= 1 - \frac{x^2}{2!} + \frac{x^4}{4!} - \cdots = \sum_{n=0}^{\infty} \frac{(-1)^n x^{2n}}{(2n+1)!}, \, x \neq 0$$

35.
$$e^{ix} = 1 + ix + \frac{(ix)^2}{2!} + \frac{(ix)^3}{3!} + \frac{(ix)^4}{4!} + \cdots = 1 + ix - \frac{x^2}{2!} - \frac{ix^3}{3!} + \frac{x^4}{4!} + \frac{ix^5}{5!} - \frac{x^6}{6!} - \cdots$$

$$e^{-ix} = 1 - ix + \frac{(-ix)^2}{2!} + \frac{(-ix)^3}{3!} + \frac{(-ix)^4}{4!} + \cdots = 1 - ix - \frac{x^2}{2!} + \frac{ix^3}{3!} + \frac{x^4}{4!} - \frac{ix^5}{5!} - \frac{x^6}{6!} + \cdots$$

$$e^{ix} - e^{-ix} = 2ix - \frac{2ix^3}{3!} + \frac{2ix^5}{5!} - \frac{2ix^7}{7!} + \cdots$$

$$\frac{e^{ix} - e^{-ix}}{2i} = x - \frac{x^3}{3!} + \frac{x^5}{5!} - \frac{x^7}{7!} + \cdots = \sum_{n=0}^{\infty} \frac{(-1)^n x^{2n+1}}{(2n+1)!} = \sin(x)$$

37. $f(x) = e^x \sin x$

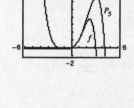

$$= \left(1 + x + \frac{x^2}{2} + \frac{x^3}{6} + \frac{x^4}{24} + \cdots\right)\left(x - \frac{x^3}{6} + \frac{x^5}{120} - \cdots\right)$$

$$= x + x^2 + \left(\frac{x^3}{2} - \frac{x^3}{6}\right) + \left(\frac{x^4}{6} - \frac{x^4}{6}\right) + \left(\frac{x^5}{120} - \frac{x^5}{12} + \frac{x^5}{24}\right) + \cdots$$

$$= x + x^2 + \frac{x^3}{3} - \frac{x^5}{30} + \cdots$$

39. $h(x) = \cos x \ln(1 + x)$

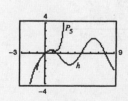

$$= \left(1 - \frac{x^2}{2} + \frac{x^4}{24} - \cdots\right)\left(x - \frac{x^2}{2} + \frac{x^3}{3} - \frac{x^4}{4} + \frac{x^5}{5} - \cdots\right)$$

$$= x - \frac{x^2}{2} + \left(\frac{x^3}{3} - \frac{x^3}{2}\right) + \left(\frac{x^4}{4} - \frac{x^4}{4}\right) + \left(\frac{x^5}{5} - \frac{x^5}{6} + \frac{x^5}{24}\right) + \cdots$$

$$= x - \frac{x^2}{2} - \frac{x^3}{6} + \frac{3x^5}{40} + \cdots$$

41. $g(x) = \dfrac{\sin x}{1 + x}$. Divide the series for $\sin x$ by $(1 + x)$.

$$g(x) = x - x^2 + \frac{5x^3}{6} - \frac{5x^4}{6} + \cdots$$

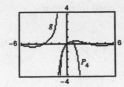

$$
\begin{array}{r}
x - x^2 + \dfrac{5x^3}{6} - \dfrac{5x^4}{6} + \\[2mm]
1 + x \overline{)\; x + 0x^2 - \dfrac{x^3}{6} + 0x^4 + \dfrac{x^5}{120} + \cdots} \\[2mm]
\underline{x + x^2} \\[2mm]
-x^2 - \dfrac{x^3}{6} \\[2mm]
\underline{-x^2 - \dfrac{x^3}{6}} \\[2mm]
\dfrac{5x^3}{6} + 0x^4 \\[2mm]
\underline{\dfrac{5x^3}{6} + \dfrac{5x^4}{6}} \\[2mm]
-\dfrac{5x^4}{6} + \dfrac{x^5}{120} \\[2mm]
\underline{-\dfrac{5x^4}{6} - \dfrac{5x^5}{6}} \\[2mm]
\vdots
\end{array}
$$

43. $y = x^2 - \dfrac{x^4}{3!} = x\left(x - \dfrac{x^3}{3!}\right)$

     $f(x) = x \sin x$

     Matches (c)

45. $y = x + x^2 + \dfrac{x^3}{2!} = x\left(1 + x + \dfrac{x^2}{2!}\right)$

     $f(x) = xe^x$

     Matches (a)

47. $\displaystyle\int_0^x (e^{-t^2} - 1)\, dt = \int_0^x \left[\left(\sum_{n=0}^{\infty} \frac{(-1)^n t^{2n}}{n!}\right) - 1\right] dt$

$$= \int_0^x \left[\sum_{n=0}^{\infty} \frac{(-1)^{n+1} t^{2n+2}}{(n+1)!}\right] dt = \left[\sum_{n=0}^{\infty} \frac{(-1)^{n+1} t^{2n+3}}{(2n+3)(n+1)!}\right]_0^x = \sum_{n=0}^{\infty} \frac{(-1)^{n+1} x^{2n+3}}{(2n+3)(n+1)!}$$

49. Since $\ln x = \displaystyle\sum_{n=0}^{\infty} \frac{(-1)^n (x-1)^{n+1}}{n+1} = (x-1) - \frac{(x-1)^2}{2} + \frac{(x-1)^3}{3} - \frac{(x-1)^4}{4} + \cdots, \quad (0 < x \le 2)$

we have $\ln 2 = 1 - \dfrac{1}{2} + \dfrac{1}{3} - \dfrac{1}{4} + \cdots = \displaystyle\sum_{n=1}^{\infty} (-1)^{n+1} \frac{1}{n} \approx 0.6931.$ (10,001 terms)

51. Since $e^x = \displaystyle\sum_{n=0}^{\infty} \frac{x^n}{n!} = 1 + x + \frac{x^2}{2!} + \frac{x^3}{3!} + \cdots,$

we have $e^2 = 1 + 2 + \dfrac{2^2}{2!} + \dfrac{2^3}{3!} + \cdots = \displaystyle\sum_{n=0}^{\infty} \frac{2^n}{n!} \approx 7.3891.$ (12 terms)

53. Since

$$\cos x = \sum_{n=0}^{\infty} \frac{(-1)^n x^{2n}}{(2n)!} = 1 - \frac{x^2}{2!} + \frac{x^4}{4!} - \frac{x^6}{6!} + \frac{x^8}{8!} - \cdots$$

$$1 - \cos x = \frac{x^2}{2!} - \frac{x^4}{4!} + \frac{x^6}{6!} - \frac{x^8}{8!} + \cdots = \sum_{n=0}^{\infty} \frac{(-1)^n x^{2n+2}}{(2n+2)!}$$

$$\frac{1 - \cos x}{x} = \frac{x}{2!} - \frac{x^3}{4!} + \frac{x^5}{6!} - \frac{x^7}{8!} + \cdots = \sum_{n=0}^{\infty} \frac{(-1)^n x^{2n+1}}{(2n+2)!}$$

we have $\displaystyle\lim_{x \to 0} \frac{1 - \cos x}{x} = \lim_{x \to 0} \sum_{n=0}^{\infty} \frac{(-1)^n x^{2n+1}}{(2n+2)!} = 0.$

55. $\displaystyle\int_0^1 \frac{\sin x}{x}\,dx = \int_0^1 \left[\sum_{n=0}^{\infty} \frac{(-1)^n x^{2n}}{(2n+1)!}\right] dx = \left[\sum_{n=0}^{\infty} \frac{(-1)^n x^{2n+1}}{(2n+1)(2n+1)!}\right]_0^1 = \sum_{n=0}^{\infty} \frac{(-1)^n}{(2n+1)(2n+1)!}$

Since $1/(7 \cdot 7!) < 0.0001$, we need three terms:

$$\int_0^1 \frac{\sin x}{x}\,dx = 1 - \frac{1}{3 \cdot 3!} + \frac{1}{5 \cdot 5!} - \cdots \approx 0.9461. \quad \text{(using three nonzero terms)}$$

Note: We are using $\displaystyle\lim_{x \to 0^+} \frac{\sin x}{x} = 1$.

57. $\displaystyle\int_{0.1}^{0.3} \sqrt{1 + x^3}\,dx = \int_{0.1}^{0.3} \left(1 + \frac{x^3}{2} - \frac{x^6}{8} + \frac{x^9}{16} - \frac{5x^{12}}{128} + \cdots\right) dx = \left[x + \frac{x^4}{8} - \frac{x^7}{56} + \frac{x^{10}}{160} - \frac{5x^{13}}{1664} + \cdots\right]_{0.1}^{0.3}$

Since $\frac{1}{56}(0.3^7 - 0.1^7) < 0.0001$, we need two terms.

$$\int_{0.1}^{0.3} \sqrt{1 + x^3}\,dx = \left[(0.3 - 0.1) + \frac{1}{8}(0.3^4 - 0.1^4)\right] \approx 0.201.$$

59. $\displaystyle\int_0^{\pi/2} \sqrt{x}\cos x\,dx = \int_0^{\pi/2} \left[\sum_{n=0}^{\infty} \frac{(-1)^n x^{(4n+1)/2}}{(2n)!}\right] dx = \left[\sum_{n=0}^{\infty} \frac{(-1)^n x^{(4n+3)/2}}{\left(\frac{4n+3}{2}\right)(2n)!}\right]_0^{\pi/2} = \left[\sum_{n=0}^{\infty} \frac{(-1)^n 2 x^{(4n+3)/2}}{(4n+3)(2n)!}\right]_0^{\pi/2}$

Since $2(\pi/2)^{23/2}/(23 \cdot 10!) < 0.0001$, we need five terms.

$$\int_0^1 \sqrt{x}\cos x\,dx = 2\left[\frac{(\pi/2)^{3/2}}{3} - \frac{(\pi/2)^{7/2}}{14} + \frac{(\pi/2)^{11/2}}{264} - \frac{(\pi/2)^{15/2}}{10,800} + \frac{(\pi/2)^{19/2}}{766,080}\right] \approx 0.7040.$$

61. From Exercise 21, we have

$$\frac{1}{\sqrt{2\pi}}\int_0^1 e^{-x^2/2}\,dx = \frac{1}{\sqrt{2\pi}}\int_0^1 \sum_{n=0}^{\infty} \frac{(-1)^n x^{2n}}{2^n n!}\,dx = \frac{1}{\sqrt{2\pi}}\left[\sum_{n=0}^{\infty} \frac{(-1)^n x^{2n+1}}{2^n n!(2n+1)}\right]_0^1 = \frac{1}{\sqrt{2\pi}}\sum_{n=0}^{\infty} \frac{(-1)^n}{2^n n!(2n+1)}$$

$$\approx \frac{1}{\sqrt{2\pi}}\left[1 - \frac{1}{2 \cdot 1 \cdot 3} + \frac{1}{2^2 \cdot 2! \cdot 5} - \frac{1}{2^3 \cdot 3! \cdot 7}\right] \approx 0.3414.$$

63. $f(x) = x\cos 2x = \displaystyle\sum_{n=0}^{\infty} \frac{(-1)^n 4^n x^{2n+1}}{(2n)!}$

$P_5(x) = x - 2x^3 + \dfrac{2x^5}{3}$

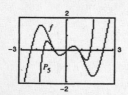

The polynomial is a reasonable approximation on the interval $\left[-\frac{3}{4}, \frac{3}{4}\right]$.

65. $f(x) = \sqrt{x}\ln x,\ c = 1$

$P_5(x) = (x - 1) - \dfrac{(x-1)^3}{24} + \dfrac{(x-1)^4}{24} - \dfrac{71(x-1)^5}{1920}$

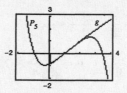

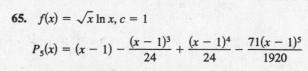

The polynomial is a reasonable approximation on the interval $\left[\frac{1}{4}, 2\right]$.

67. See Guidelines, page 680.

69. (a) Replace x with $(-x)$.

    (b) Replace x with $3x$.

(c) Multiply series by x.

(d) Replace x with $2x$, then replace x with $-2x$, and add the two together.

71. $y = \left(\tan\theta - \dfrac{g}{kv_0\cos\theta}\right)x - \dfrac{g}{k^2}\ln\left(1 - \dfrac{kx}{v_0\cos\theta}\right)$

$= (\tan\theta)x - \dfrac{gx}{kv_0\cos\theta} - \dfrac{g}{k^2}\left[-\dfrac{kx}{v_0\cos\theta} - \dfrac{1}{2}\left(\dfrac{kx}{v_0\cos\theta}\right)^2 - \dfrac{1}{3}\left(\dfrac{kx}{v_0\cos\theta}\right)^3 - \dfrac{1}{4}\left(\dfrac{kx}{v_0\cos\theta}\right)^4 - \cdots\right]$

$= (\tan\theta)x - \dfrac{gx}{kv_0\cos\theta} + \dfrac{gx}{kv_0\cos\theta} + \dfrac{gx^2}{2v_0^2\cos^2\theta} + \dfrac{gkx^3}{3v_0^3\cos^3\theta} + \dfrac{gk^2x^4}{4v_0^4\cos^4\theta} + \cdots\right]$

$= (\tan\theta)x + \dfrac{gx^2}{2v_0^2\cos^2\theta} + \dfrac{kgx^3}{3v_0^3\cos^3\theta} + \dfrac{k^2gx^4}{4v_0^4\cos^4\theta} + \cdots$

73. $f(x) = \begin{cases} e^{-1/x^2}, & x \neq 0 \\ 0, & x = 0 \end{cases}$

(a)

(b) $f'(0) = \lim\limits_{x\to 0}\dfrac{f(x) - f(0)}{x - 0} = \lim\limits_{x\to 0}\dfrac{e^{-1/x^2} - 0}{x}$

Let $y = \lim\limits_{x\to 0}\dfrac{e^{-1/x^2}}{x}$. Then

$\ln y = \lim\limits_{x\to 0}\ln\left(\dfrac{e^{-1/x^2}}{x}\right) = \lim\limits_{x\to 0^+}\left[-\dfrac{1}{x^2} - \ln x\right] = \lim\limits_{x\to 0^+}\left[\dfrac{-1 - x^2\ln x}{x^2}\right] = -\infty.$

Thus, $y = e^{-\infty} = 0$ and we have $f'(0) = 0$.

(c) $\sum\limits_{n=0}^{\infty}\dfrac{f^{(n)}(0)}{n!}x^n = f(0) + \dfrac{f'(0)x}{1!} + \dfrac{f''(0)x^2}{2!} + \cdots = 0 \neq f(x)$ This series converges to f at $x = 0$ only.

75. By the Ratio Test: $\lim\limits_{n\to\infty}\left|\dfrac{x^{n+1}}{(n+1)!}\cdot\dfrac{n!}{x^n}\right| = \lim\limits_{n\to\infty}\dfrac{|x|}{n+1} = 0$ which shows that $\sum\limits_{n=0}^{\infty}\dfrac{x^n}{n!}$ converges for all x.

77. $\binom{5}{3} = \dfrac{5\cdot 4\cdot 3}{3!} = \dfrac{60}{6} = 10$

79. $\binom{0.5}{4} = \dfrac{(0.5)(-0.5)(-1.5)(-2.5)}{4!}$

$= -0.0390625 = -\dfrac{5}{128}$

81. $(1 + x)^k = \sum\limits_{n=0}^{\infty}\binom{k}{n}x^n$

Example: $(1 + x)^2 = \sum\limits_{n=0}^{\infty}\binom{2}{n}x^n = 1 + 2x + x^2$

83. $g(x) = \dfrac{x}{1 - x - x^2} = a_0 + a_1x + a_2x^2 + \cdots$

$x = (1 - x - x^2)(a_0 + a_1x + a_2x^2 + \cdots)$

$x = a_0 + (a_1 - a_0)x + (a_2 - a_1 - a_0)x^2 + (a_3 - a_2 - a_1)x^3 + \cdots$

Equating coefficients,

$a_0 = 0$

$a_1 - a_0 = 1 \implies a_1 = 1$

$a_2 - a_1 - a_0 = 0 \implies a_2 = 1$

$a_3 - a_2 - a_1 = 0 \implies a_3 = 2$

$a_4 = a_3 + a_2 = 3$, etc.

In general, $a_n = a_{n-1} + a_{n-2}$. The coefficients are the Fibonacci numbers.

Review Exercises for Chapter 9

1. $a_n = \dfrac{1}{n!}$

3. $a_n = 4 + \dfrac{2}{n}$: $6, 5, 4.67, \ldots$

Matches (a)

5. $a_n = 10(0.3)^{n-1}$: $10, 3, \ldots$

Matches (d)

7. $a_n = \dfrac{5n + 2}{n}$

The sequence seems to converge to 5.

$$\lim_{n\to\infty} a_n = \lim_{n\to\infty} \frac{5n + 2}{n}$$

$$= \lim_{n\to\infty} \left(5 + \frac{2}{n}\right) = 5 \cdot$$

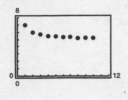

9. $\lim\limits_{n\to\infty} \dfrac{n + 1}{n^2} = 0$

Converges

11. $\lim\limits_{n\to\infty} \dfrac{n^3}{n^2 + 1} = \infty$

13. $\lim\limits_{n\to\infty} \left(\sqrt{n + 1} - \sqrt{n}\right) = \lim\limits_{n\to\infty} \left(\sqrt{n + 1} - \sqrt{n}\right)\dfrac{\sqrt{n + 1} + \sqrt{n}}{\sqrt{n + 1} + \sqrt{n}} = \lim\limits_{n\to\infty} \dfrac{1}{\sqrt{n + 1} + \sqrt{n}} = 0$ Converges

15. $\lim\limits_{n\to\infty} \dfrac{\sin\sqrt{n}}{\sqrt{n}} = 0$

Converges

17. $A_n = 5000\left(1 + \dfrac{0.05}{4}\right)^n = 5000(1.0125)^n$

$n = 1, 2, 3$

(a) $A_1 = 5062.50$ $A_5 \approx 5320.41$

$A_2 \approx 5125.78$ $A_6 \approx 5386.92$

$A_3 \approx 5189.85$ $A_7 \approx 5454.25$

$A_4 \approx 5254.73$ $A_8 \approx 5522.43$

(b) $A_{40} \approx 8218.10$

19. (a)

k	5	10	15	20	25
S_k	13.2	113.3	873.8	6448.5	50,500.3

The series diverges $\left(\text{geometric } r = \tfrac{3}{2} > 1\right)$.

(b)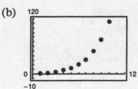

21. (a)

k	5	10	15	20	25
S_k	0.4597	0.4597	0.4597	0.4597	0.4597

The series converges by the Alternating Series Test.

(b)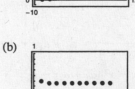

23. Converges. Geometric series, $r = 0.82$, $|r| < 1$.

25. Diverges. nth-Term Test. $\lim\limits_{n\to\infty} a_n \neq 0$.

27. $\sum\limits_{n=0}^{\infty} \left(\dfrac{2}{3}\right)^n$ Geometric series with $a = 1$ and $r = \dfrac{2}{3}$.

$$S = \frac{a}{1 - r} = \frac{1}{1 - (2/3)} = \frac{1}{1/3} = 3$$

29. $\sum\limits_{n=0}^{\infty} \left(\dfrac{1}{2^n} - \dfrac{1}{3^n}\right) = \sum\limits_{n=0}^{\infty} \left(\dfrac{1}{2}\right)^n - \sum\limits_{n=0}^{\infty} \left(\dfrac{1}{3}\right)^n$

$$= \frac{1}{1 - (1/2)} - \frac{1}{1 - (1/3)} = 2 - \frac{3}{2} = \frac{1}{2}$$

31. (a) $0.\overline{09} = 0.09 + 0.0009 + 0.000009 + \cdots = 0.09(1 + 0.01 + 0.0001 + \cdots) = \sum_{n=0}^{\infty} (0.09)(0.01)^n$

(b) $0.\overline{09} = \dfrac{0.09}{1 - 0.01} = \dfrac{1}{11}$

33. $D_1 = 8$

$D_2 = 0.7(8) + 0.7(8) = 16(0.7)$

$\vdots$

$D = 8 + 16(0.7) + 16(0.7)^2 + \cdots + 16(0.7)^n + \cdots$

$= -8 + \sum_{n=0}^{\infty} 16(0.7)^n = -8 + \dfrac{16}{1 - 0.7} = 45\frac{1}{3}$ meters

35. See Exercise 110 in Section 9.2.

$A = \dfrac{P(e^{rt} - 1)}{e^{r/12} - 1}$

$= \dfrac{200(e^{(0.06)(2)} - 1)}{e^{0.06/12} - 1}$

$\approx \$5087.14$

37. $\displaystyle\int_1^{\infty} x^{-4} \ln(x)\, dx = \lim_{b \to \infty} \left[-\dfrac{\ln x}{3x^3} - \dfrac{1}{9x^3} \right]_1^b = 0 + \dfrac{1}{9} = \dfrac{1}{9}$

By the Integral Test, the series converges.

39. $\displaystyle\sum_{n=1}^{\infty} \left(\dfrac{1}{n^2} - \dfrac{1}{n} \right) = \sum_{n=1}^{\infty} \dfrac{1}{n^2} - \sum_{n=1}^{\infty} \dfrac{1}{n}$

Since the second series is a divergent p-series while the first series is a convergent p-series, the difference diverges.

41. $\displaystyle\sum_{n=1}^{\infty} \dfrac{1}{\sqrt{n^3 + 2n}}$

$\displaystyle\lim_{n \to \infty} \dfrac{1/\sqrt{n^3 + 2n}}{1/(n^{3/2})} = \lim_{n \to \infty} \dfrac{n^{3/2}}{\sqrt{n^3 + 2n}} = 1$

By a limit comparison test with the convergent p-series $\displaystyle\sum_{n=1}^{\infty} \dfrac{1}{n^{3/2}}$, the series converges.

43. $\displaystyle\sum_{n=1}^{\infty} \dfrac{1 \cdot 3 \cdot 5 \cdots (2n - 1)}{2 \cdot 4 \cdot 6 \cdots (2n)}$

$a_n = \dfrac{1 \cdot 3 \cdot 5 \cdots (2n - 1)}{2 \cdot 4 \cdot 6 \cdots (2n)}$

$= \left(\dfrac{3}{2} \cdot \dfrac{5}{4} \cdots \dfrac{2n - 1}{2n - 2} \right)\dfrac{1}{2n} > \dfrac{1}{2n}$

Since $\displaystyle\sum_{n=1}^{\infty} \dfrac{1}{2n} = \dfrac{1}{2} \sum_{n=1}^{\infty} \dfrac{1}{n}$ diverges (harmonic series), so does the original series.

45. Converges by the Alternating Series Test. (Conditional convergence)

47. Diverges by the nth-Term Test.

49. $\displaystyle\sum_{n=1}^{\infty} \dfrac{n}{e^{n^2}}$

$\displaystyle\lim_{n \to \infty} \left| \dfrac{a_{n+1}}{a_n} \right| = \lim_{n \to \infty} \left| \dfrac{n + 1}{e^{(n+1)^2}} \cdot \dfrac{e^{n^2}}{n} \right|$

$= \lim_{n \to \infty} \left| \dfrac{e^{n^2}(n + 1)}{e^{n^2 + 2n + 1}n} \right|$

$= \lim_{n \to \infty} \left(\dfrac{1}{e^{2n+1}} \right)\left(\dfrac{n + 1}{n} \right)$

$= (0)(1) = 0 < 1$

By the Ratio Test, the series converges.

51. $\displaystyle\sum_{n=1}^{\infty} \dfrac{2^n}{n^3}$

$\displaystyle\lim_{n \to \infty} \left| \dfrac{a_{n+1}}{a_n} \right| = \lim_{n \to \infty} \left| \dfrac{2^{n+1}}{(n + 1)^3} \cdot \dfrac{n^3}{2^n} \right| = \lim_{n \to \infty} \dfrac{2n^3}{(n + 1)^3} = 2$

Therefore, by the Ratio Test, the series diverges.

53. (a) Ratio Test: $\displaystyle\lim_{n \to \infty} \left| \dfrac{a_{n+1}}{a_n} \right| = \lim_{n \to \infty} \dfrac{(n + 1)(3/5)^{n+1}}{n(3/5)^n} = \lim_{n \to \infty} \left(\dfrac{n + 1}{n} \right)\left(\dfrac{3}{5} \right) = \dfrac{3}{5} < 1$, Converges

(b)

x	5	10	15	20	25
S_n	2.8752	3.6366	3.7377	3.7488	3.7499

(c)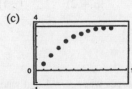

(d) The sum is approximately 3.75.

55. (a) $\int_N^\infty \frac{1}{x^2}\,dx = \left[-\frac{1}{x}\right]_N^\infty = \frac{1}{N}$

N	5	10	20	30	40
$\sum\limits_{n=1}^N \dfrac{1}{n^2}$	1.4636	1.5498	1.5962	1.6122	1.6202
$\int_N^\infty \dfrac{1}{x^2}\,dx$	0.2000	0.1000	0.0500	0.0333	0.0250

(b) $\int_N^\infty \frac{1}{x^5}\,dx = \left[-\frac{1}{4x^4}\right]_N^\infty = \frac{1}{4N^4}$

N	5	10	20	30	40
$\sum\limits_{n=1}^N \dfrac{1}{n^5}$	1.0367	1.0369	1.0369	1.0369	1.0369
$\int_N^\infty \dfrac{1}{x^5}\,dx$	0.0004	0.0000	0.0000	0.0000	0.0000

The series in part (b) converges more rapidly. The integral values represent the remainders of the partial sums.

57. $f(x) = e^{-x/2}$ $\qquad f(0) = 1 \qquad P_3(x) = f(0) + f'(0)x + f''(0)\dfrac{x^2}{2!} + f'''(0)\dfrac{x^3}{3!}$

$f'(x) = -\dfrac{1}{2}e^{-x/2} \qquad f'(0) = -\dfrac{1}{2} \qquad\qquad = 1 - \dfrac{1}{2}x + \dfrac{1}{4}\dfrac{x^2}{2!} - \dfrac{1}{8}\dfrac{x^3}{3!}$

$f''(x) = \dfrac{1}{4}e^{-x/2} \qquad f''(0) = \dfrac{1}{4} \qquad\qquad = 1 - \dfrac{1}{2}x + \dfrac{1}{8}x^2 - \dfrac{1}{48}x^3$

$f'''(x) = -\dfrac{1}{8}e^{-x/2} \qquad f'''(0) = -\dfrac{1}{8}$

59. Since $\dfrac{(95\pi)^9}{180^9 \cdot 9!} < 0.001$, use four terms.

$$\sin 95^\circ = \sin\left(\frac{95\pi}{180}\right) \approx \frac{95\pi}{180} - \frac{(95\pi)^3}{180^3 3!} + \frac{(95\pi)^5}{180^5 5!} - \frac{(95\pi)^7}{180^7 7!} \approx 0.99594$$

61. $\ln(1.75) \approx (0.75) - \dfrac{(0.75)^2}{2} + \dfrac{(0.75)^3}{3} - \dfrac{(0.75)^4}{4} + \dfrac{(0.75)^5}{5} - \dfrac{(0.75)^6}{6} + \cdots - \dfrac{(0.75)^{14}}{14} \approx 0.559062$

63. $f(x) = \cos x,\ c = 0$ $\qquad$ **(a)** $R_n(x) \le \dfrac{(0.5)^{n+1}}{(n+1)!} < 0.001$ $\qquad$ **(c)** $R_n(x) \le \dfrac{(0.5)^{n+1}}{(n+1)!} < 0.0001$

$R_n(x) = \dfrac{f^{(n+1)}(z)}{(n+1)!}x^{n+1}$ $\qquad\qquad$ This inequality is true for $n = 4$. $\qquad$ This inequality is true for $n = 5$.

$|f^{(n+1)}(z)| \le 1 \implies R_n(x) \le \dfrac{x^{n+1}}{(n+1)!}$ $\qquad$ **(b)** $R_n(x) \le \dfrac{(1)^{n+1}}{(n+1)!} < 0.001$ $\qquad$ **(d)** $R_n(x) \le \dfrac{2^{n+1}}{(n+1)!} < 0.0001$

$\qquad\qquad\qquad\qquad\qquad\qquad\qquad$ This inequality is true for $n = 6$. $\qquad$ This inequality is true for $n = 10$.

65. $\sum\limits_{n=0}^\infty \left(\dfrac{x}{10}\right)^n$

Geometric series which converges only if $|x/10| < 1$ or $-10 < x < 10$.

67. $\sum\limits_{n=0}^\infty \dfrac{(-1)^n(x-2)^n}{(n+1)^2}$ $\qquad\qquad\qquad$ **69.** $\sum\limits_{n=0}^\infty n!(x-2)^n$

$\lim\limits_{n\to\infty}\left|\dfrac{u_{n+1}}{u_n}\right| = \lim\limits_{n\to\infty}\left|\dfrac{(-1)^{n+1}(x-2)^{n+1}}{(n+2)^2}\cdot\dfrac{(n+1)^2}{(-1)^n(x-2)^n}\right|$ $\qquad$ $\lim\limits_{n\to\infty}\left|\dfrac{u_{n+1}}{u_n}\right| = \lim\limits_{n\to\infty}\left|\dfrac{(n+1)!(x-2)^{n+1}}{n!(x-2)^n}\right| = \infty$

$\qquad\qquad\qquad = |x-2|$ $\qquad\qquad\qquad\qquad\qquad\qquad$ which implies that the series converges only at the center

$R = 1$ $\qquad\qquad\qquad\qquad\qquad\qquad\qquad\qquad\qquad\qquad$ $x = 2$.

Center: 2

Since the series converges when $x = 1$ and when $x = 3$,
the interval of convergence is $1 \le x \le 3$.

71.
$$y = \sum_{n=0}^{\infty} (-1)^n \frac{x^{2n}}{4^n(n!)^2}$$

$$y' = \sum_{n=1}^{\infty} \frac{(-1)^n(2n)x^{2n-1}}{4^n(n!)^2} = \sum_{n=0}^{\infty} \frac{(-1)^{n+1}(2n+2)x^{2n+1}}{4^{n+1}[(n+1)!]^2}$$

$$y'' = \sum_{n=0}^{\infty} \frac{(-1)^{n+1}(2n+2)(2n+1)x^{2n}}{4^{n+1}[(n+1)!]^2}$$

$$x^2y'' + xy' + x^2y = \sum_{n=0}^{\infty} \frac{(-1)^{n+1}(2n+2)(2n+1)x^{2n+2}}{4^{n+1}[(n+1)!]^2} + \sum_{n=0}^{\infty} \frac{(-1)^{n+1}(2n+2)x^{2n+2}}{4^{n+1}[(n+1)!]^2} + \sum_{n=0}^{\infty} (-1)^n \frac{x^{2n+2}}{4^n(n!)^2}$$

$$= \sum_{n=0}^{\infty} \left[(-1)^{n+1} \frac{(2n+2)(2n+1)}{4^{n+1}[(n+1)!]^2} + \frac{(-1)^{n+1}(2n+2)}{4^{n+1}[(n+1)!]^2} + \frac{(-1)^n}{4^n(n!)^2} \right] x^{2n+2}$$

$$= \sum_{n=0}^{\infty} \left[\frac{(-1)^{n+1}(2n+2)(2n+1+1)}{4^{n+1}[(n+1)!]^2} + (-1)^n \frac{1}{4^n(n!)^2} \right] x^{2n+2}$$

$$= \sum_{n=0}^{\infty} \left[\frac{(-1)^{n+1}4(n+1)^2}{4^{n+1}[(n+1)!]^2} + (-1)^n \frac{1}{4^n(n!)^2} \right] x^{2n+2}$$

$$= \sum_{n=0}^{\infty} \left[\frac{(-1)^{n+1}1}{4^n(n!)^2} + (-1)^n \frac{1}{4^n(n!)^2} \right] x^{2n+2} = 0$$

73. $\dfrac{2}{3-x} = \dfrac{2/3}{1-(x/3)} = \dfrac{a}{1-r}$

$$\sum_{n=0}^{\infty} \frac{2}{3}\left(\frac{x}{3}\right)^n = \sum_{n=0}^{\infty} \frac{2x^n}{3^{n+1}}$$

75. $g(x) = \dfrac{2}{3-x}$. Power series $\displaystyle\sum_{n=0}^{\infty} \frac{2}{3}\left(\frac{x}{3}\right)^n$

Derivative: $\displaystyle\sum_{n=1}^{\infty} \frac{2}{3}n\left(\frac{x}{3}\right)^{n-1}\left(\frac{1}{3}\right) = \sum_{n=1}^{\infty} \frac{2}{9}n\left(\frac{x}{3}\right)^{n-1}$

$$= \sum_{n=0}^{\infty} \frac{2}{9}(n+1)\left(\frac{x}{3}\right)^n$$

77. $1 + \dfrac{2}{3}x + \dfrac{4}{9}x^2 + \dfrac{8}{27}x^3 + \cdots = \displaystyle\sum_{n=0}^{\infty}\left(\frac{2x}{3}\right)^n = \dfrac{1}{1-(2x/3)} = \dfrac{3}{3-2x}, \quad -\dfrac{3}{2} < x < \dfrac{3}{2}$

79. $f(x) = \sin x$

$f'(x) = \cos x$

$f''(x) = -\sin x$

$f'''(x) = -\cos x, \cdots$

$$\sin(x) = \sum_{n=0}^{\infty} \frac{f^{(n)}(x)[x - (3\pi/4)]^n}{n!}$$

$$= \frac{\sqrt{2}}{2} - \frac{\sqrt{2}}{2}\left(x - \frac{3\pi}{4}\right) - \frac{\sqrt{2}}{2\cdot 2!}\left(x - \frac{3\pi}{4}\right)^2 + \cdots = \frac{\sqrt{2}}{2}\sum_{n=0}^{\infty} \frac{(-1)^{n(n+1)/2}[x - (3\pi/4)]^n}{n!}$$

81. $3^x = (e^{\ln(3)})^x = e^{x\ln(3)}$ and since $e^x = \displaystyle\sum_{n=0}^{\infty} \frac{x^n}{n!}$, we have $3^x = \displaystyle\sum_{n=0}^{\infty} \frac{(x\ln 3)^n}{n!} = 1 + x\ln 3 + \frac{x^2[\ln 3]^2}{2!} + \frac{x^3[\ln 3]^3}{3!} + \frac{x^4[\ln 3]^4}{4!} + \cdots$

83. $f(x) = \dfrac{1}{x}$

$f'(x) = -\dfrac{1}{x^2}$

$f''(x) = \dfrac{2}{x^3}$

$f'''(x) = -\dfrac{6}{x^4}, \cdots$

$$\frac{1}{x} = \sum_{n=0}^{\infty} \frac{f^{(n)}(-1)(x+1)^n}{n!} = \sum_{n=0}^{\infty} \frac{-n!(x+1)^n}{n!} = -\sum_{n=0}^{\infty}(x+1)^n, \quad -2 < x < 0$$

85. $(1 + x)^k = 1 + kx + \dfrac{k(k-1)x^2}{2!} + \dfrac{k(k-1)(k-2)x^3}{3!} + \cdots$

$(1 + x)^{1/5} = 1 + \dfrac{x}{5} + \dfrac{(1/5)(-4/5)x^2}{2!} + \dfrac{1/5(-4/5)(-9/5)x^3}{3!} + \cdots$

$\qquad = 1 + \dfrac{1}{5}x - \dfrac{1 \cdot 4x^2}{5^2 2!} + \dfrac{1 \cdot 4 \cdot 9x^3}{5^3 3!} - \cdots$

$\qquad = 1 + \dfrac{x}{5} + \displaystyle\sum_{n=2}^{\infty} \dfrac{(-1)^{n+1} 4 \cdot 9 \cdot 14 \cdots (5n-6)x^n}{5^n n!}$

$\qquad = 1 + \dfrac{x}{5} - \dfrac{2}{25}x^2 + \dfrac{6}{125}x^3 - \cdots$

87. $\ln x = \displaystyle\sum_{n=1}^{\infty} (-1)^{n+1} \dfrac{(x-1)^n}{n}, \qquad 0 < x \le 2$

$\ln\left(\dfrac{5}{4}\right) = \displaystyle\sum_{n=1}^{\infty} (-1)^{n+1} \left(\dfrac{(5/4)-1}{n}\right)^n$

$\qquad = \displaystyle\sum_{n=1}^{\infty} (-1)^{n+1} \dfrac{1}{4^n n} \approx 0.2231$

89. $e^x = \displaystyle\sum_{n=0}^{\infty} \dfrac{x^n}{n!}, \quad -\infty < x < \infty$

$e^{1/2} = \displaystyle\sum_{n=0}^{\infty} \dfrac{(1/2)^n}{n!} = \sum_{n=0}^{\infty} \dfrac{1}{2^n n!} \approx 1.6487$

91. $\cos x = \displaystyle\sum_{n=0}^{\infty} (-1)^n \dfrac{x^{2n}}{(2n)!}, \quad -\infty < x < \infty$

$\cos\left(\dfrac{2}{3}\right) = \displaystyle\sum_{n=0}^{\infty} (-1)^n \dfrac{2^{2n}}{3^{2n}(2n)!} \approx 0.7859$

93. The series for Exercise 41 converges very slowly because the terms approach 0 at a slow rate.

95. (a) $f(x) = e^{2x} \qquad f(0) = 1$

$\quad f'(x) = 2e^{2x} \qquad f'(0) = 2$

$\quad f''(x) = 4e^{2x} \qquad f''(0) = 4$

$\quad f'''(x) = 8e^{2x} \qquad f'''(0) = 8$

$\quad P(x) = 1 + 2x + \dfrac{4x^2}{2!} + \dfrac{8x^3}{3!} = 1 + 2x + 2x^2 + \dfrac{4}{3}x^3$

(b) $\quad e^x = \displaystyle\sum_{n=0}^{\infty} \dfrac{x^n}{n!}, \; e^{2x} = \sum_{n=0}^{\infty} \dfrac{(2x)^n}{n!}$

$\quad P(x) = 1 + 2x + 2x^2 + \dfrac{4}{3}x^3$

(c) $e^x \cdot e^x = \left(1 + x + \dfrac{x^2}{2!} + \cdots\right)\left(1 + x + \dfrac{x^2}{2!} + \cdots\right)$

$\quad P = 1 + 2x + 2x^2 + \dfrac{4}{3}x^3$

97. $\sin t = \displaystyle\sum_{n=0}^{\infty} \dfrac{(-1)^n t^{2n+1}}{(2n+1)!}$

$\dfrac{\sin t}{t} = \displaystyle\sum_{n=0}^{\infty} \dfrac{(-1)^n t^{2n}}{(2n+1)!}$

$\displaystyle\int_0^x \dfrac{\sin t}{t} \, dt = \left[\sum_{n=0}^{\infty} \dfrac{(-1)^n t^{2n+1}}{(2n+1)(2n+1)!}\right]_0^x$

$\qquad = \displaystyle\sum_{n=0}^{\infty} \dfrac{(-1)^n x^{2n+1}}{(2n+1)(2n+1)!}$

99. $\dfrac{1}{1+t} = \displaystyle\sum_{n=0}^{\infty} (-1)^n t^n$

$\ln(1+t) = \displaystyle\int \dfrac{1}{1+t} \, dt = \sum_{n=0}^{\infty} \dfrac{(-1)^n t^{n+1}}{n+1}$

$\dfrac{\ln(t+1)}{t} = \displaystyle\sum_{n=0}^{\infty} \dfrac{(-1)^n t^n}{n+1}$

$\displaystyle\int_0^x \dfrac{\ln(t+1)}{t} \, dt = \left[\sum_{n=0}^{\infty} \dfrac{(-1)^n t^{n+1}}{(n+1)^2}\right]_0^x = \sum_{n=0}^{\infty} \dfrac{(-1)^n x^{n+1}}{(n+1)^2}$

101. $\arctan x = x - \dfrac{x^3}{3} + \dfrac{x^5}{5} - \dfrac{x^7}{7} + \dfrac{x^9}{9} - \cdots$

$\dfrac{\arctan x}{\sqrt{x}} = \sqrt{x} - \dfrac{x^{5/2}}{3} + \dfrac{x^{9/2}}{5} - \dfrac{x^{13/2}}{7} + \dfrac{x^{17/2}}{9} - \cdots$

$\displaystyle\lim_{x \to 0^+} \dfrac{\arctan x}{\sqrt{x}} = 0$

By L'Hôpital's Rule, $\displaystyle\lim_{x \to 0^+} \dfrac{\arctan x}{\sqrt{x}} = \lim_{x \to 0^+} \dfrac{\left(\dfrac{1}{1+x^2}\right)}{\left(\dfrac{1}{2\sqrt{x}}\right)} = \lim_{x \to 0^+} \dfrac{2\sqrt{x}}{1+x^2} = 0.$

Problem Solving for Chapter 9

1. (a) $1\left(\frac{1}{3}\right) + 2\left(\frac{1}{9}\right) + 4\left(\frac{1}{27}\right) + \cdots = \sum_{n=0}^{\infty} \frac{1}{3}\left(\frac{2}{3}\right)^n$

$$= \frac{1/3}{1-(2/3)} = 1$$

(b) $0, \frac{1}{3}, \frac{2}{3}, 1,$ etc.

(c) $\lim_{n\to\infty} C_n = 1 - \sum_{n=0}^{\infty} \frac{1}{3}\left(\frac{2}{3}\right)^n = 1 - 1 = 0$

3. If there are n rows, then $a_n = \dfrac{n(n+1)}{2}$.

For one circle,

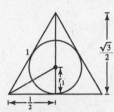

$$a_1 = 1 \text{ and } r_1 = \frac{1}{3}\left(\frac{\sqrt{3}}{2}\right) = \frac{\sqrt{3}}{6} = \frac{1}{2\sqrt{3}}.$$

For three circles,

$a_2 = 3$ and $1 = 2\sqrt{3}r_2 + 2r_2$

$$r_2 = \frac{1}{2+2\sqrt{3}}.$$

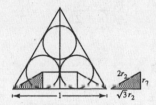

For six circles,

$a_3 = 6$ and $1 = 2\sqrt{3}r_3 + 4r_3$

$$r_3 = \frac{1}{2\sqrt{3}+4}.$$

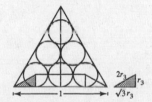

Continuing this pattern, $r_n = \dfrac{1}{2\sqrt{3}+2(n-1)}.$

Total Area $= (\pi r_n^2)a_n = \pi\left(\dfrac{1}{2\sqrt{3}+2(n-1)}\right)^2 \dfrac{n(n+1)}{2}$

$$A_n = \frac{\pi}{2} \frac{n(n+1)}{\left[2\sqrt{3}+2(n-1)\right]^2}$$

$\lim_{n\to\infty} A_n = \dfrac{\pi}{2} \cdot \dfrac{1}{4} = \dfrac{\pi}{8}$

5. (a) $\sum a_n x^n = 1 + 2x + 3x^2 + x^3 + 2x^4 + 3x^5 + \cdots$

$= (1 + x^3 + x^6 + \cdots) + 2(x + x^4 + x^7 + \cdots) + 3(x^2 + x^5 + x^8 + \cdots)$

$= (1 + x^3 + x^6 + \cdots)[1 + 2x + 3x^2]$

$= (1 + 2x + 3x^2)\dfrac{1}{1-x^3}$

$R = 1$ because each series in the second line has $R = 1$.

—CONTINUED—

5. —CONTINUED—

(b) $\sum a_n x^n = (a_0 + a_1 x + \cdots + a_{p-1} x^{p-1}) + (a_0 x^p + a_1 x^{p+1} + \cdots) + \cdots$

$\qquad = a_0(1 + x^p + \cdots) + a_1 x(1 + x^p + \cdots) + \cdots + a_{p-1} x^{p-1}(1 + x^p + \cdots)$

$\qquad = (a_0 + a_1 x + \cdots + a_{p-1} x^{p-1})(1 + x^p + \cdots)$

$\qquad = (a_0 + a_1 x + \cdots + a_{p-1} x^{p-1})\dfrac{1}{1 - x^p}$

$R = 1$

(Assume all $a_n > 0$.)

7. (a) $\qquad e^x = 1 + x + \dfrac{x^2}{2!} + \cdots = \displaystyle\sum_{n=0}^{\infty} \dfrac{x^n}{n!}$

$xe^x = \displaystyle\sum_{n=0}^{\infty} \dfrac{x^{n+1}}{n!}$

$\displaystyle\int xe^x \, dx = xe^x - e^x + C = \sum_{n=0}^{\infty} \dfrac{x^{n+2}}{(n+2)n!}$

Letting $x = 0$, you have $C = 1$. Letting $x = 1$,

$e - e + 1 = \displaystyle\sum_{n=0}^{\infty} \dfrac{1}{(n+2)n!} = \dfrac{1}{2} + \sum_{n=1}^{\infty} \dfrac{1}{(n+2)n!}.$

Thus, $\displaystyle\sum_{n=1}^{\infty} \dfrac{1}{(n+2)n!} = \dfrac{1}{2}.$

(b) Differentiating,

$xe^x + e^x = \displaystyle\sum_{n=0}^{\infty} \dfrac{(n+1)x^n}{n!}.$

Letting $x = 1$,

$2e = \displaystyle\sum_{n=0}^{\infty} \dfrac{n+1}{n!} \approx 5.4366.$

9. Let $a_1 = \displaystyle\int_0^{\pi} \dfrac{\sin x}{x} \, dx, a_2 = -\int_{\pi}^{2\pi} \dfrac{\sin x}{x} \, dx, a_3 = \int_{2\pi}^{3\pi} \dfrac{\sin x}{x} \, dx,$ etc.

Then,

$\displaystyle\int_0^{\infty} \dfrac{\sin x}{x} \, dx = a_1 - a_2 + a_3 - a_4 + \cdots.$

Since $\displaystyle\lim_{n \to \infty} a_n = 0$ and $a_{n+1} < a_n$, this series converges.

11. (a) $a_1 = 3.0$

$a_2 \approx 1.73205$

$a_3 \approx 2.17533$

$a_4 \approx 2.27493$

$a_5 \approx 2.29672$

$a_6 \approx 2.30146$

$\displaystyle\lim_{n \to \infty} a_n = \dfrac{1 + \sqrt{13}}{2}$ [See part (b) for proof.]

(b) Use mathematical induction to show the sequence is increasing. Clearly, $a_2 = \sqrt{a + a_1} = \sqrt{a\sqrt{a}} > \sqrt{a} = a_1$.

Now assume $a_n > a_{n-1}$. Then

$a_n + a > a_{n-1} + a$

$\overline{\sqrt{a_n + a} > \sqrt{a_{n-1} + a}}$

$a_{n+1} > a_n.$

Use mathematical induction to show that the sequence is bounded above by a. Clearly, $a_1 = \sqrt{a} < a$.

—CONTINUED—

11. —CONTINUED—

Now assume $a_n < a$. Then $a > a_n$ and $a - 1 > 1$ implies

$$a(a - 1) > a_n(1)$$

$$a^2 - a > a_n$$

$$a^2 > a_n + a$$

$$a > \sqrt{a_n + a} = a_{n+1}.$$

Hence, the sequence converges to some number L. To find L, assume $a_{n+1} \approx a_n \approx L$:

$$L = \sqrt{a + L} \implies L^2 = a + L \implies L^2 - L - a = 0$$

$$L = \frac{1 \pm \sqrt{1 + 4a}}{2}.$$

Hence, $L = \dfrac{1 + \sqrt{1 + 4a}}{2}$.

13. (a) $\displaystyle\sum_{n=1}^{\infty} \frac{1}{2^{n+(-1)^n}} = \frac{1}{2^{1-1}} + \frac{1}{2^{2+1}} + \frac{1}{2^{3-1}} + \frac{1}{2^{4+1}} + \frac{1}{2^{5-1}} + \cdots$

$$S_1 = \frac{1}{2^0} = 1$$

$$S_1 = 1 + \frac{1}{8} = \frac{9}{8}$$

$$S_3 = \frac{9}{8} + \frac{1}{4} = \frac{11}{8}$$

$$S_4 = \frac{11}{8} + \frac{1}{32} = \frac{45}{32}$$

$$S_5 = \frac{45}{32} + \frac{1}{16} = \frac{47}{32}$$

(b) $\dfrac{a_{n+1}}{a_n} = \dfrac{2^{n+(-1)^n}}{2^{(n+1)+(-1)^{n+1}}} = \dfrac{2^{(-1)^n}}{2^{1+(-1)^{n+1}}}$

This sequence is $\frac{1}{8}, 2, \frac{1}{8}, 2, \ldots$ which diverges.

(c) $\sqrt[n]{\dfrac{1}{2^{n+(-1)^n}}} = \left(\dfrac{1}{2^n \cdot 2^{(-1)^n}}\right)^{1/n}$

$$= \frac{1}{2 \cdot \sqrt[n]{2^{(-1)^n}}} \to \frac{1}{2} < 1 \text{ converges}$$

because $\left\{2^{(-1)^n}\right\} = \frac{1}{2}, 2, \frac{1}{2}, 2, \ldots$ and $\sqrt[n]{1/2} \to 1$ and $\sqrt[n]{2} \to 1$.

15. $S_6 = 130 + 70 + 40 = 240$

$S_7 = 240 + 130 + 70 = 440$

$S_8 = 440 + 240 + 130 = 810$

$S_9 = 810 + 440 + 240 = 1490$

$S_{10} = 1490 + 810 + 440 = 2740$

17. (a) $\displaystyle\sum_{n=1}^{\infty} \frac{1}{2n} = \frac{1}{2} \sum_{n=1}^{\infty} \frac{1}{n}$ diverges (harmonic)

(b) Let $f(x) = \sin x$. By the Mean Value Theorem,

$$|f(x) - f(y)| = f'(c)|x - y| = \cos(c)|x - y| \le |x - y|,$$

where c is between x and y. Thus,

$$0 \le \left|\sin\left(\frac{1}{2n}\right) - \sin\left(\frac{1}{2n + 1}\right)\right|$$

$$\le \left|\frac{1}{2n} - \frac{1}{2n + 1}\right|$$

$$= \frac{1}{2n(2n + 1)}$$

Because $\displaystyle\sum_{n=1}^{\infty} \frac{1}{2n(2n + 1)}$ converges, the Comparison Theorem tells us that $\displaystyle\sum_{n=1}^{\infty} \left[\sin\left(\frac{1}{2n}\right) - \sin\left(\frac{1}{2n + 1}\right)\right]$ converges.

CHAPTER 10
Conics, Parametric Equations, and Polar Coordinates

Section 10.1   Conics and Calculus 503

Section 10.2   Plane Curves and Parametric Equations 515

Section 10.3   Parametric Equations and Calculus 519

Section 10.4   Polar Coordinates and Polar Graphs 526

Section 10.5   Area and Arc Length in Polar Coordinates 534

Section 10.6   Polar Equations of Conics and Kepler's Laws 540

Review Exercises 544

Problem Solving 553

CHAPTER 10
Conics, Parametric Equations, and Polar Coordinates

Section 10.1   Conics and Calculus

1. $y^2 = 4x$

   Vertex: $(0, 0)$

   $p = 1 > 0$

   Opens to the right
   Matches graph (h).

3. $(x + 3)^2 = -2(y - 2)$

   Vertex: $(-3, 2)$

   $p = -\frac{1}{2} < 0$

   Opens downward
   Matches graph (e).

5. $\dfrac{x^2}{9} + \dfrac{y^2}{4} = 1$

   Center: $(0, 0)$
   Ellipse
   Matches (f)

7. $\dfrac{y^2}{16} - \dfrac{x^2}{1} = 1$

   Hyperbola
   Center: $(0, 0)$

   Vertical transverse axis
   Matches (c)

9. $y^2 = -6x = 4\left(-\frac{3}{2}\right)x$

   Vertex: $(0, 0)$

   Focus: $\left(-\frac{3}{2}, 0\right)$

   Directrix: $x = \frac{3}{2}$

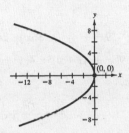

11. $(x + 3) + (y - 2)^2 = 0$

$$(y - 2)^2 = 4\left(-\tfrac{1}{4}\right)(x + 3)$$

   Vertex: $(-3, 2)$

   Focus: $(\,3.25, 2)$

   Directrix: $x = -2.75$

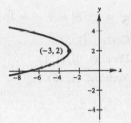

13. $y^2 - 4y - 4x = 0$

$$y^2 - 4y + 4 = 4x + 4$$

$$(y - 2)^2 = 4(1)(x + 1)$$

   Vertex: $(-1, 2)$

   Focus: $(0, 2)$

   Directrix: $x = -2$

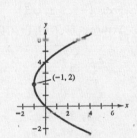

15. $x^2 + 4x + 4y - 4 = 0$

$$x^2 + 4x + 4 = -4y + 4 + 4$$

$$(x + 2)^2 = 4(-1)(y - 2)$$

   Vertex: $(-2, 2)$

   Focus: $(-2, 1)$

   Directrix: $y = 3$

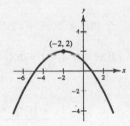

17. $y^2 + x + y = 0$

$$y^2 + y + \tfrac{1}{4} = -x + \tfrac{1}{4}$$

$$\left(y + \tfrac{1}{2}\right)^2 = 4\left(-\tfrac{1}{4}\right)\left(x - \tfrac{1}{4}\right)$$

   Vertex: $\left(\frac{1}{4}, -\frac{1}{2}\right)$

   Focus: $\left(0, -\frac{1}{2}\right)$

   Directrix: $x = \frac{1}{2}$

   $y_1 = -\frac{1}{2} \pm \sqrt{\frac{1}{4} - x}$

   $y_2 = -\frac{1}{2} - \sqrt{\frac{1}{4} - x}$

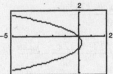

19. $y^2 - 4x - 4 = 0$

$$y^2 = 4x + 4$$

$$= 4(1)(x + 1)$$

   Vertex: $(-1, 0)$

   Focus: $(0, 0)$

   Directrix: $x = -2$

   $y_1 = 2\sqrt{x + 1}$

   $y_2 = -2\sqrt{x + 1}$

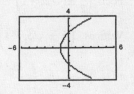

21. $$(y - 2)^2 = 4(-2)(x - 3)$$

$$y^2 - 4y + 8x - 20 = 0$$

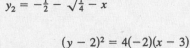

23.
$$(x - h)^2 = 4p(y - k)$$
$$x^2 = 4(6)(y - 4)$$
$$x^2 - 24y + 96 = 0$$

25.
$$y = 4 - x^2$$
$$x^2 + y - 4 = 0$$

27. Since the axis of the parabola is vertical, the form of the equation is $y = ax^2 + bx + c$. Now, substituting the values of the given coordinates into this equation, we obtain

$$3 = c, \ 4 = 9a + 3b + c, \ 11 = 16a + 4b + c.$$

Solving this system, we have $a = \frac{5}{3}$, $b = -\frac{14}{3}$, $c = 3$.

Therefore,

$$y = \tfrac{5}{3}x^2 - \tfrac{14}{3}x + 3 \text{ or } 5x^2 - 14x - 3y + 9 = 0.$$

29. $x^2 + 4y^2 = 4$

$$\frac{x^2}{4} + \frac{y^2}{1} = 1$$

$a^2 = 4$, $b^2 = 1$, $c^2 = 3$

Center: $(0, 0)$

Foci: $\left(\pm\sqrt{3}, 0\right)$

Vertices: $(\pm 2, 0)$

$$e = \frac{\sqrt{3}}{2}$$

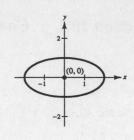

31. $\dfrac{(x - 1)^2}{9} + \dfrac{(y - 5)^2}{25} = 1$

$a^2 = 25$, $b^2 = 9$, $c^2 = 16$

Center: $(1, 5)$

Foci: $(1, 9), (1, 1)$

Vertices: $(1, 10), (1, 0)$

$$e = \frac{4}{5}$$

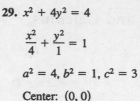

33.
$$9x^2 + 4y^2 + 36x - 24y + 36 = 0$$
$$9(x^2 + 4x + 4) + 4(y^2 - 6y + 9) = -36 + 36 + 36$$
$$= 36$$
$$\frac{(x + 2)^2}{4} + \frac{(y - 3)^2}{9} = 1$$

$a^2 = 9$, $b^2 = 4$, $c^2 = 5$

Center: $(-2, 3)$

Foci: $\left(-2, 3 \pm \sqrt{5}\right)$

Vertices: $(-2, 6), (-2, 0)$

$$e = \frac{\sqrt{5}}{3}$$

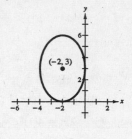

35.
$$12x^2 + 20y^2 - 12x + 40y - 37 = 0$$
$$12\left(x^2 - x + \frac{1}{4}\right) + 20(y^2 + 2y + 1) = 37 + 3 + 20$$
$$= 60$$
$$\frac{[x - (1/2)]^2}{5} + \frac{(y + 1)^2}{3} = 1$$

$a^2 = 5$, $b^2 = 3$, $c^2 = 2$

Center: $\left(\dfrac{1}{2}, -1\right)$

Foci: $\left(\dfrac{1}{2} \pm \sqrt{2}, -1\right)$

Vertices: $\left(\dfrac{1}{2} \pm \sqrt{5}, -1\right)$

Solve for y:

$$20(y^2 + 2y + 1) = -12x^2 + 12x + 37 + 20$$
$$(y + 1)^2 = \frac{57 + 12x - 12x^2}{20}$$
$$y = -1 \pm \sqrt{\frac{57 + 12x - 12x^2}{20}}$$

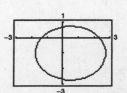

(Graph each of these separately.)

37. $x^2 + 2y^2 - 3x + 4y + 0.25 = 0$

$$\left(x^2 - 3x + \frac{9}{4}\right) + 2(y^2 + 2y + 1) = -\frac{1}{4} + \frac{9}{4} + 2 = 4$$

$$\frac{[x - (3/2)]^2}{4} + \frac{(y + 1)^2}{2} = 1$$

$a^2 = 4,\ b^2 = 2,\ c^2 = 2$

Center: $\left(\dfrac{3}{2}, -1\right)$ Solve for y:

Foci: $\left(\dfrac{3}{2} + \sqrt{2}, -1\right)$ $2(y^2 + 2y + 1) = -x^2 + 3x - \dfrac{1}{4} + 2$

Vertices: $\left(-\dfrac{1}{2}, -1\right), \left(\dfrac{7}{2}, -1\right)$ $(y + 1)^2 = \dfrac{1}{2}\left(\dfrac{7}{4} + 3x - x^2\right)$

$$y = -1 \pm \sqrt{\frac{7 + 12x - 4x^2}{8}}$$

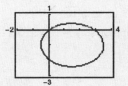

(Graph each of these separately.)

39. Center: $(0, 0)$

Focus: $(2, 0)$

Vertex: $(3, 0)$

Horizontal major axis

$a = 3,\ c = 2 \implies b = \sqrt{5}$

$$\frac{x^2}{9} + \frac{y^2}{5} = 1$$

41. Vertices: $(3, 1), (3, 9)$

Minor axis length: 6

Vertical major axis

Center: $(3, 5)$

$a = 4,\ b = 3$

$$\frac{(x - 3)^2}{9} + \frac{(y - 5)^2}{16} = 1$$

43. Center: $(0, 0)$

Horizontal major axis

Points on ellipse: $(3, 1), (4, 0)$

Since the major axis is horizontal,

$$\left(\frac{x^2}{a^2}\right) + \left(\frac{y^2}{b^2}\right) = 1.$$

Substituting the values of the coordinates of the given points into this equation, we have

$$\left(\frac{9}{a^2}\right) + \left(\frac{1}{b^2}\right) = 1, \text{ and } \frac{16}{a^2} = 1.$$

The solution to this system is $a^2 = 16,\ b^2 = 16/7$.

Therefore,

$$\frac{x^2}{16} + \frac{y^2}{16/7} = 1, \frac{x^2}{16} + \frac{7y^2}{16} = 1.$$

45. $\dfrac{y^2}{1} - \dfrac{x^2}{4} = 1$

$a = 1,\ b = 2,\ c = \sqrt{5}$

Center: $(0, 0)$

Vertices: $(0, \pm 1)$

Foci: $\left(0, \pm\sqrt{5}\right)$

Asymptotes: $y = \pm\dfrac{1}{2}x$

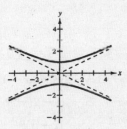

47. $\dfrac{(x - 1)^2}{4} - \dfrac{(y + 2)^2}{1} = 1$

$a = 2,\ b = 1,\ c = \sqrt{5}$

Center: $(1, -2)$

Vertices: $(-1, -2), (3, -2)$

Foci: $\left(1 \pm \sqrt{5}, -2\right)$

Asymptotes: $y = -2 \pm \dfrac{1}{2}(x - 1)$

49. $9x^2 - y^2 - 36x - 6y + 18 = 0$

$9(x^2 - 4x + 4) - (y^2 + 6y + 9) = -18 + 36 - 9$

$$\frac{(x-2)^2}{1} - \frac{(y+3)^2}{9} = 1$$

$a = 1, b = 3, c = \sqrt{10}$

Center: $(2, -3)$

Vertices: $(1, -3), (3, -3)$

Foci: $\left(2 \pm \sqrt{10}, -3\right)$

Asymptotes: $y = -3 \pm 3(x - 2)$

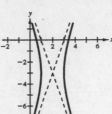

51. $x^2 - 9y^2 + 2x - 54y - 80 = 0$

$(x^2 + 2x + 1) - 9(y^2 + 6y + 9) = 80 + 1 - 81 = 0$

$(x + 1)^2 - 9(y + 3)^2 = 0$

$$y + 3 = \pm\frac{1}{3}(x + 1)$$

Degenerate hyperbola is two lines intersecting at $(-1, -3)$.

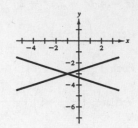

53. $9y^2 - x^2 + 2x + 54y + 62 = 0$

$9(y^2 + 6y + 9) - (x^2 - 2x + 1) = -62 - 1 + 81 = 18$

$$\frac{(y+3)^2}{2} - \frac{(x-1)^2}{18} = 1$$

$a = \sqrt{2}, b = 3\sqrt{2}, c = 2\sqrt{5}$

Center: $(1, -3)$

Vertices: $\left(1, -3 \pm \sqrt{2}\right)$

Foci: $\left(1, -3 \pm 2\sqrt{5}\right)$

Solve for y:

$$9(y^2 + 6y + 9) = x^2 - 2x - 62 + 81$$

$$(y + 3)^2 = \frac{x^2 - 2x + 19}{9}$$

$$y = -3 \pm \frac{1}{3}\sqrt{x^2 - 2x + 19}$$

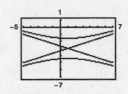

(Graph each curve separately.)

55. $3x^2 - 2y^2 - 6x - 12y - 27 = 0$

$3(x^2 - 2x + 1) - 2(y^2 + 6y + 9) = 27 + 3 - 18 = 12$

$$\frac{(x-1)^2}{4} - \frac{(y+3)^2}{6} = 1$$

$a = 2, b = \sqrt{6}, c = \sqrt{10}$

Center: $(1, -3)$

Vertices: $(-1, -3), (3, -3)$

Foci: $\left(1 \pm \sqrt{10}, -3\right)$

Solve for y:

$$2(y^2 + 6y + 9) = 3x^2 - 6x - 27 + 18$$

$$(y + 3)^2 = \frac{3x^2 - 6x - 9}{2}$$

$$y = -3 \pm \sqrt{\frac{3(x^2 - 2x - 3)}{2}}$$

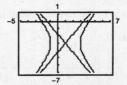

(Graph each curve separately.)

57. Vertices: $(\pm 1, 0)$

Asymptotes: $y = \pm 3x$

Horizontal transverse axis

Center: $(0, 0)$

$$a = 1, \pm\frac{b}{a} = \pm\frac{b}{1} = \pm 3 \implies b = 3$$

Therefore, $\dfrac{x^2}{1} - \dfrac{y^2}{9} = 1$.

59. Vertices: $(2, \pm 3)$

Point on graph: $(0, 5)$

Vertical transverse axis

Center: $(2, 0)$

$a = 3$

Therefore, the equation is of the form $\dfrac{y^2}{9} - \dfrac{(x-2)^2}{b^2} = 1$.

Substituting the coordinates of the point $(0, 5)$, we have

$$\frac{25}{9} - \frac{4}{b^2} = 1 \quad \text{or} \quad b^2 = \frac{9}{4}.$$

Therefore, the equation is $\dfrac{y^2}{9} - \dfrac{(x-2)^2}{9/4} = 1$.

61. Center: $(0, 0)$

Vertex: $(0, 2)$

Focus: $(0, 4)$

Vertical transverse axis

$a = 2$, $c = 4$, $b^2 = c^2 - a^2 = 12$

Therefore, $\dfrac{y^2}{4} - \dfrac{x^2}{12} = 1$.

63. Vertices: $(0, 2)$, $(6, 2)$

Asymptotes: $y = \dfrac{2}{3}x$, $y = 4 - \dfrac{2}{3}x$

Horizontal transverse axis

Center: $(3, 2)$

$a = 3$

Slopes of asymptotes: $\pm \dfrac{b}{a} = \pm \dfrac{2}{3}$

Thus, $b = 2$.

Therefore,

$$\frac{(x-3)^2}{9} - \frac{(y-2)^2}{4} = 1.$$

65. (a) $\dfrac{x^2}{9} - y^2 - 1$, $\dfrac{2x}{9} - 2yy' - 0$, $\dfrac{x}{9y} - y'$

At $x = 6$: $y = \pm\sqrt{3}$, $y' = \dfrac{\pm 6}{9\sqrt{3}} = \dfrac{\pm 2\sqrt{3}}{9}$

At $\left(6, \sqrt{3}\right)$: $y - \sqrt{3} = \dfrac{2\sqrt{3}}{9}(x - 6)$

or $2x - 3\sqrt{3}y - 3 = 0$

At $\left(6, -\sqrt{3}\right)$: $y + \sqrt{3} = \dfrac{-2\sqrt{3}}{9}(x - 6)$

or $2x + 3\sqrt{3}y - 3 = 0$

(b) From part (a) we know that the slopes of the normal lines must be $\mp 9/\left(2\sqrt{3}\right)$.

At $\left(6, \sqrt{3}\right)$: $y - \sqrt{3} = -\dfrac{9}{2\sqrt{3}}(x - 6)$

or $9x + 2\sqrt{3}y - 60 = 0$

At $\left(6, -\sqrt{3}\right)$: $y + \sqrt{3} = \dfrac{9}{2\sqrt{3}}(x - 6)$

or $9x - 2\sqrt{3}y - 60 = 0$

67. $x^2 + 4y^2 - 6x + 16y + 21 = 0$

$(x^2 - 6x + 9) + 4(y^2 + 4y + 4) = -21 + 9 + 16$

$(x - 3)^2 + 4(y + 2)^2 = 4$

Ellipse

69. $y^2 - 4y - 4x = 0$

$y^2 - 4y + 4 = 4x + 4$

$(y - 2)^2 = 4(x + 1)$

Parabola

71. $4x^2 + 4y^2 - 16y + 15 = 0$

$4x^2 + 4(y^2 - 4y + 4) = -15 + 16$

$4x^2 + 4(y - 2)^2 = 1$

Circle   (Ellipse)

73. $9x^2 + 9y^2 - 36x + 6y + 34 = 0$

$9(x^2 - 4x + 4) + 9\left(y^2 + \tfrac{2}{3}y + \tfrac{1}{9}\right) = -34 + 36 + 1$

$9(x - 2)^2 + 9\left(y + \tfrac{1}{3}\right)^2 = 3$

Circle   (Ellipse)

75. $3(x - 1)^2 = 6 + 2(y + 1)^2$

$3(x - 1)^2 - 2(y + 1)^2 = 6$

$\dfrac{(x - 1)^2}{2} - \dfrac{(y + 1)^2}{3} = 1$

Hyperbola

77. (a) A parabola is the set of all points (x, y) that are equidistant from a fixed line (directrix) and a fixed point (focus) not on the line.

(b) $(x - h)^2 = 4p(y - k)$ or $(y - k)^2 = 4p(x - h)$

(c) See Theorem 10.2.

79. (a) A hyperbola is the set of all points (x, y) for which the absolute value of the difference between the distances from two distance fixed points (foci) is constant.

(b) $\dfrac{(x - h)^2}{a^2} - \dfrac{(y - k)^2}{b^2} = 1$ or $\dfrac{(y - k)^2}{a^2} - \dfrac{(x - h)^2}{b^2} = 1$

(c) $y = k \pm \dfrac{b}{a}(x - h)$ or $y = k \pm \dfrac{a}{b}(x - h)$

83. $y = ax^2$

$y' = 2ax$

The equation of the tangent line is

$$y - ax_0^2 = 2ax_0(x - x_0) \text{ or } y = 2ax_0x - ax_0^2.$$

Let $y = 0$.

Then: $-ax_0^2 = 2ax_0x - 2ax_0^2$

$ax_0^2 = 2ax_0x$

$x = \dfrac{x_0}{2}$

81. Assume that the vertex is at the origin.

$$x^2 = 4py$$

$$(3)^2 = 4p(1)$$

$$\frac{9}{4} = p$$

The pipe is located

$\frac{9}{4}$ meters from the vertex.

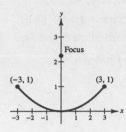

Therefore, $\left(\dfrac{x_0}{2}, 0\right)$ is the x-intercept.

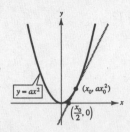

85. (a) Consider the parabola $x^2 = 4py$. Let m_0 be the slope of the one tangent line at (x_1, y_1) and therefore, $-1/m_0$ is the slope of the second at (x_2, y_2). Differentiating,

$2x = 4py'$ or $y' = \dfrac{x}{2p}$, and we have:

$$m_0 = \frac{1}{2p}x_1 \text{ or } x_1 = 2pm_0$$

$$\frac{-1}{m_0} = \frac{1}{2p}x_2 \text{ or } x_2 = \frac{-2p}{m_0}.$$

Substituting these values of x into the equation $x^2 = 4py$, we have the coordinates of the points of tangency $(2pm_0, pm_0^2)$ and $(-2p/m_0, p/m_0^2)$ and the equations of the tangent lines are

$$(y - pm_0^2) = m_0(x - 2pm_0) \quad \text{and}$$

$$\left(y - \frac{p}{m_0^2}\right) = \frac{-1}{m_0}\left(x + \frac{2p}{m_0}\right).$$

The point of intersection of these lines is

$\left(\dfrac{p(m_0^2 - 1)}{m_0}, -p\right)$ and is on the directrix, $y = -p$.

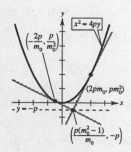

(b) $x^2 - 4x - 4y + 8 = 0$

$$(x - 2)^2 = 4(y - 1)$$

Vertex: $(2, 1)$

$$2x - 4 - 4\frac{dy}{dx} = 0$$

$$\frac{dy}{dx} = \frac{1}{2}x - 1$$

At $(-2, 5)$, $\dfrac{dy}{dx} = -2$. At $\left(3, \dfrac{5}{4}\right)$, $\dfrac{dy}{dx} = \dfrac{1}{2}$.

Tangent line at $(-2, 5)$:

$$y - 5 = -2(x + 2) \implies 2x + y - 1 = 0.$$

Tangent line at $\left(3, \dfrac{5}{4}\right)$:

$$y - \frac{5}{4} = \frac{1}{2}(x - 3) \implies 2x - 4y - 1 = 0.$$

Since $m_1m_2 = (-2)\left(\dfrac{1}{2}\right) = -1$, the lines are perpendicular.

Point of intersection: $-2x + 1 = \dfrac{1}{2}x - \dfrac{1}{4}$

$$-\frac{5}{2}x = -\frac{5}{4}$$

$$x = \frac{1}{2}$$

$$y = 0$$

Directrix: $y = 0$ and the point of intersection $\left(\dfrac{1}{2}, 0\right)$ lies

on this line.

87. $y = x - x^2$

$$\frac{dy}{dx} = 1 - 2x$$

At the point of tangency (x_1, y_1) on the mountain, $m = 1 - 2x_1$. Also, $m = \dfrac{y_1 - 1}{x_1 + 1}$.

$$\frac{y_1 - 1}{x_1 + 1} = 1 - 2x_1$$

$$(x_1 - x_1{}^2) - 1 = (1 - 2x_1)(x_1 + 1)$$

$$-x_1{}^2 + x_1 - 1 = -2x_1{}^2 - x_1 + 1$$

$$x_1{}^2 + 2x_1 - 2 = 0$$

$$x_1 = \frac{-2 \pm \sqrt{2^2 - 4(1)(-2)}}{2(1)} = \frac{-2 \pm 2\sqrt{3}}{2} = -1 \pm \sqrt{3}$$

Choosing the positive value for x_1, we have $x_1 = -1 + \sqrt{3}$.

$$m = 1 - 2\left(-1 + \sqrt{3}\right) = 3 - 2\sqrt{3}$$

$$m = \frac{0 - 1}{x_0 + 1} = -\frac{1}{x_0 + 1}$$

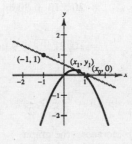

Thus, $-\dfrac{1}{x_0 + 1} = 3 - 2\sqrt{3}$

$$\frac{-1}{3 - 2\sqrt{3}} = x_0 + 1$$

$$\frac{3 + 2\sqrt{3}}{3} \quad 1 = x_0$$

$$\frac{2\sqrt{3}}{3} = x_0.$$

The closest the receiver can be to the hill is $\left(2\sqrt{3}/3\right) - 1 \approx 0.155$.

---

89.   Parabola   Circle

Vertex: $(0, 4)$ Center: $(0, k)$

$\quad x^2 = 4p(y - 4)$ Radius: 8

$\quad 4^2 = 4p(0 - 4)$ $x^2 + (y - k)^2 = 64$

$\quad\quad\quad p = -1$ $4^2 + (0 - k)^2 = 64$

$\quad x^2 = -4(y - 4)$ 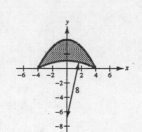 $k^2 = 48$

$\quad\quad y = 4 - \dfrac{x^2}{4}$ $k = -4\sqrt{3}$ (Center is on the negative y-axis.)

                                             $x^2 + \left(y + 4\sqrt{3}\right)^2 = 64$

                                                  $y = -4\sqrt{3} \pm \sqrt{64 - x^2}$

Since the y-value is positive when $x = 0$, we have $y = -4\sqrt{3} + \sqrt{64 - x^2}$.

$$A = 2\int_0^4 \left[\left(4 - \frac{x^2}{4}\right) - \left(-4\sqrt{3} + \sqrt{64 - x^2}\right)\right] dx$$

$$= 2\left[4x - \frac{x^3}{12} + 4\sqrt{3}x - \frac{1}{2}\left(x\sqrt{64 - x^2} + 64\arcsin\frac{x}{8}\right)\right]_0^4$$

$$= 2\left[16 - \frac{64}{12} + 16\sqrt{3} - 2\sqrt{48} - 32\arcsin\frac{1}{2}\right]$$

$$= \frac{16\left(4 + 3\sqrt{3} - 2\pi\right)}{3} \approx 15.536 \text{ square feet}$$

91. (a) Assume that $y = ax^2$.

$$20 = a(60)^2 \Rightarrow a = \frac{2}{360} = \frac{1}{180} \Rightarrow y = \frac{1}{180}x^2$$

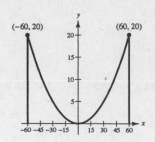

(b) $f(x) = \frac{1}{180}x^2, f'(x) = \frac{1}{90}x$

$$S = 2\int_0^{60} \sqrt{1 + \left(\frac{1}{90}x\right)^2}\, dx = \frac{2}{90}\int_0^{60} \sqrt{90^2 + x^2}\, dx$$

$$= \frac{2}{90}\frac{1}{2}\left[x\sqrt{90^2 + x^2} + 90^2 \ln\left| x + \sqrt{90^2 + x^2}\right| \right]_0^{60} \quad \text{(Formula 26)}$$

$$= \frac{1}{90}[60\sqrt{11{,}700} + 90^2 \ln(60 + \sqrt{11{,}700}) - 90^2 \ln 90]$$

$$= \frac{1}{90}[1800\sqrt{13} + 90^2 \ln(60 + 30\sqrt{13}) - 90^2 \ln 90]$$

$$= 20\sqrt{13} + 90 \ln\left(\frac{60 + 30\sqrt{13}}{90}\right)$$

$$= 10\left[2\sqrt{13} + 9 \ln\left(\frac{2 + \sqrt{13}}{3}\right)\right] \approx 128.4 \text{ m}$$

93. $x^2 = 4py, p = \frac{1}{4}, \frac{1}{2}, 1, \frac{3}{2}, 2$

As p increases, the graph becomes wider.

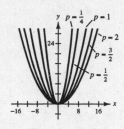

95. (a) At the vertices we notice that the string is horizontal and has a length of $2a$.

(b) The thumbtacks are located at the foci and the length of string is the constant sum of the distances from the foci.

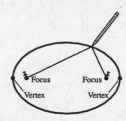

97.

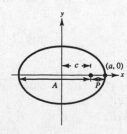

99. $e = \frac{c}{a}$

$A + P = 2a$

$$a = \frac{A + P}{2}$$

$$c = a - P = \frac{A + P}{2} - P = \frac{A - P}{2}$$

$$e = \frac{c}{a} = \frac{(A - P)/2}{(A + P)/2} = \frac{A - P}{A + P}$$

101. $e = \dfrac{A - P}{A + P} = \dfrac{35.29 - 0.59}{35.29 + 0.59} = 0.9671$

103. $\dfrac{x^2}{10^2} + \dfrac{y^2}{5^2} = 1$

$\dfrac{2x}{10^2} + \dfrac{2yy'}{5^2} = 0$

$y' = \dfrac{-5^2 x}{10^2 y} = \dfrac{-x}{4y}$

At $(-8, 3)$: $y' = \dfrac{8}{12} = \dfrac{2}{3}$

The equation of the tangent line is $y - 3 = \frac{2}{3}(x + 8)$. It will cross the y-axis when $x = 0$ and $y = \frac{2}{3}(8) + 3 = \frac{25}{3}$.

105. $16x^2 + 9y^2 + 96x + 36y + 36 = 0$

$32x + 18yy' + 96 + 36y' = 0$

$y'(18y + 36) = -(32x + 96)$

$y' = \dfrac{-(32x + 96)}{18y + 36}$

$y' = 0$ when $x = -3$. y' is undefined when $y = -2$.

At $x = -3, y = 2$ or -6.

Endpoints of major axis: $(-3, 2), (-3, -6)$

At $y = -2, x = 0$ or -6.

Endpoints of minor axis: $(0, -2), (-6, -2)$

Note: Equation of ellipse is $\dfrac{(x + 3)^2}{9} + \dfrac{(y + 2)^2}{16} = 1$

107. (a) $A = 4\displaystyle\int_0^2 \frac{1}{2}\sqrt{4 - x^2}\, dx = \left[x\sqrt{4 - x^2} + 4\arcsin\!\left(\frac{x}{2}\right) \right]_0^2 = 2\pi$ $\left[\text{or, } A = \pi ab = \pi(2)(1) = 2\pi\right]$

(b) **Disk:** $V = 2\pi\displaystyle\int_0^2 \frac{1}{4}(4 - x^2)\, dx = \frac{1}{2}\pi\left[4x - \frac{1}{3}x^3 \right]_0^2 = \frac{8\pi}{3}$

$y = \dfrac{1}{2}\sqrt{4 - x^2}$

$y' = \dfrac{-x}{2\sqrt{4 - x^2}}$

$\sqrt{1 + (y')^2} = \sqrt{1 + \dfrac{x^2}{16 - 4x^2}} = \sqrt{\dfrac{16 - 3x^2}{4y}}$

$S = 2(2\pi)\displaystyle\int_0^2 y\left(\dfrac{\sqrt{16 - 3x^2}}{4y} \right) dx = \pi\int_0^2 \sqrt{16 - 3x^2}\, dx$

$= \dfrac{\pi}{2\sqrt{3}}\left[\sqrt{3}\,x\sqrt{16 - 3x^2} + 16\arcsin\!\left(\dfrac{\sqrt{3}\,x}{4}\right) \right]_0^2$

$= \dfrac{2\pi}{9}\left(9 + 4\sqrt{3}\,\pi\right) \approx 21.48$

(c) **Shell:** $V = 2\pi\displaystyle\int_0^2 x\sqrt{4 - x^2}\, dx = -\pi\int_0^2 -2x(4 - x^2)^{1/2}\, dx = -\dfrac{2\pi}{3}\left[(4 - x^2)^{3/2} \right]_0^2 = \dfrac{16\pi}{3}$

$x = 2\sqrt{1 - y^2}$

$x' = \dfrac{-2y}{\sqrt{1 - y^2}}$

$\sqrt{1 + (x')^2} = \sqrt{1 + \dfrac{4y^2}{1 - y^2}} = \dfrac{\sqrt{1 + 3y^2}}{\sqrt{1 - y^2}}$

$S = 2(2\pi)\displaystyle\int_0^1 2\sqrt{1 - y^2}\,\dfrac{\sqrt{1 + 3y^2}}{\sqrt{1 - y^2}}\, dy = 8\pi\int_0^1 \sqrt{1 + 3y^2}\, dy$

$= \dfrac{8\pi}{2\sqrt{3}}\left[\sqrt{3}\,y\sqrt{1 + 3y^2} + \ln\left| \sqrt{3}\,y + \sqrt{1 + 3y^2} \right| \right]_0^1$

$= \dfrac{4\pi}{3}\left| 6 + \sqrt{3}\,\ln\!\left(2 + \sqrt{3}\right) \right| \approx 34.69$

109. From Example 5,

$$C = 4a \int_0^{\pi/2} \sqrt{1 - e^2 \sin^2 \theta} \, d\theta$$

For $\dfrac{x^2}{25} + \dfrac{y^2}{49} = 1$, we have

$$a = 7, b = 5, c = \sqrt{49 - 25} = 2\sqrt{6}, e = \frac{c}{a} = \frac{2\sqrt{6}}{7}.$$

$$C = 4(7) \int_0^{\pi/2} \sqrt{1 - \frac{24}{49} \sin^2 \theta} \, d\theta$$

$$\approx 28(1.3558) \approx 37.9614$$

111. Area circle $= \pi r^2 = 100\pi$

Area ellipse $= \pi ab = \pi a(10)$

$$2(100\pi) = 10\pi a \implies a = 20$$

Hence, the length of the major axis is $2a = 40$.

113. The transverse axis is horizontal since $(2, 2)$ and $(10, 2)$ are the foci (see definition of hyperbola).

Center: $(6, 2)$

$$c = 4, 2a = 6, b^2 = c^2 - a^2 = 7$$

Therefore, the equation is

$$\frac{(x - 6)^2}{9} - \frac{(y - 2)^2}{7} = 1.$$

115. $2a = 10 \implies a = 5$

$$c = 6 \implies b = \sqrt{11}$$

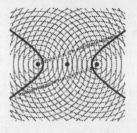

117. Time for sound of bullet hitting target to reach (x, y): $\dfrac{2c}{v_m} + \dfrac{\sqrt{(x - c)^2 + y^2}}{v_s}$

Time for sound of rifle to reach (x, y): $\dfrac{\sqrt{(x + c)^2 + y^2}}{v_s}$

Since the times are the same, we have: $\dfrac{2c}{v_m} + \dfrac{\sqrt{(x - c)^2 + y^2}}{v_s} = \dfrac{\sqrt{(x + c)^2 + y^2}}{v_s}$

$$\frac{4c^2}{v_m^2} + \frac{4c}{v_m v_s} \sqrt{(x - c)^2 + y^2} + \frac{(x - c)^2 + y^2}{v_s^2} = \frac{(x + c)^2 + y^2}{v_s^2}$$

$$\sqrt{(x - c)^2 + y^2} = \frac{v_m^2 x - v_s^2 c}{v_s v_m}$$

$$\left(1 - \frac{v_m^2}{v_s^2}\right)x^2 + y^2 = \left(\frac{v_s^2}{v_m^2} - 1\right)c^2$$

$$\frac{x^2}{c^2 v_s^2 / v_m^2} - \frac{y^2}{c^2(v_m^2 - v_s^2)/v_m^2} = 1$$

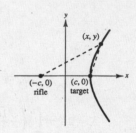

119. The point (x, y) lies on the line between $(0, 10)$ and $(10, 0)$. Thus, $y = 10 - x$.

The point also lies on the hyperbola $(x^2/36) - (y^2/64) = 1$.

Using substitution, we have:

$$\frac{x^2}{36} - \frac{(10 - x)^2}{64} = 1$$

$$16x^2 - 9(10 - x)^2 = 576$$

$$7x^2 + 180x - 1476 = 0$$

$$x = \frac{-180 \pm \sqrt{180^2 - 4(7)(-1476)}}{2(7)}$$

$$= \frac{-180 \pm 192\sqrt{2}}{14} = \frac{-90 \pm 96\sqrt{2}}{7}$$

Choosing the positive value for x we have:

$$x = \frac{-90 + 96\sqrt{2}}{7} \approx 6.538 \text{ and}$$

$$y = \frac{160 - 96\sqrt{2}}{7} \approx 3.462$$

121. $\frac{x^2}{a^2} + \frac{2y^2}{b^2} = 1 \implies \frac{2y^2}{b^2} = 1 - \frac{x^2}{a^2}$. Let $c^2 = a^2 - b^2$.

$$\frac{x^2}{a^2 - b^2} - \frac{2y^2}{b^2} = 1 \implies \frac{2y^2}{b^2} = \frac{x^2}{a^2 - b^2} - 1$$

$$1 - \frac{x^2}{a^2} = \frac{x^2}{a^2 - b^2} - 1 \implies 2 = x^2\left(\frac{1}{a^2} + \frac{1}{a^2 - b^2}\right)$$

$$x^2 = \frac{2a^2(a^2 - b^2)}{2a^2 - b^2} \implies x = \pm\frac{\sqrt{2}a\sqrt{a^2 - b^2}}{\sqrt{2a^2 - b^2}} = \pm\frac{\sqrt{2}ac}{\sqrt{2a^2 - b^2}}$$

$$\frac{2y^2}{b^2} = 1 - \frac{1}{a^2}\left(\frac{2a^2c^2}{2a^2 - b^2}\right) \implies \frac{2y^2}{b^2} = \frac{b^2}{2a^2 - b^2}$$

$$y^2 = \frac{b^4}{2(2a^2 - b^2)} \implies y = \pm\frac{b^2}{\sqrt{2}\sqrt{2a^2 - b^2}}$$

There are four points of intersection: $\left(\dfrac{\sqrt{2}ac}{\sqrt{2a^2 - b^2}}, \pm\dfrac{b^2}{\sqrt{2}\sqrt{2a^2 - b^2}}\right)$, $\left(-\dfrac{\sqrt{2}ac}{\sqrt{2a^2 - b^2}}, \pm\dfrac{b^2}{\sqrt{2}\sqrt{2a^2 - b^2}}\right)$

$$\frac{x^2}{a^2} + \frac{2y^2}{b^2} = 1 \implies \frac{2x}{a^2} + \frac{4yy'}{b^2} = 0 \implies y'_e = -\frac{b^2x}{2a^2y}$$

$$\frac{x^2}{a^2 - b^2} - \frac{2y^2}{b^2} = 1 \implies \frac{2x}{c^2} - \frac{4yy'}{b^2} = 0 \implies y'_h = \frac{b^2x}{2c^2y}$$

At $\left(\dfrac{\sqrt{2}ac}{\sqrt{2a^2 - b^2}}, \dfrac{b^2}{\sqrt{2}\sqrt{2a^2 - b^2}}\right)$, the slopes of the tangent lines are:

$$y'_e = \frac{-b^2\left(\dfrac{\sqrt{2}ac}{\sqrt{2a^2 - b^2}}\right)}{2a^2\left(\dfrac{b^2}{\sqrt{2}\sqrt{2a^2 - b^2}}\right)} = -\frac{c}{a} \quad \text{and} \quad y'_h = \frac{b^2\left(\dfrac{\sqrt{2}ac}{\sqrt{2a^2 - b^2}}\right)}{2c^2\left(\dfrac{b^2}{\sqrt{2}\sqrt{2a^2 - b^2}}\right)} = \frac{a}{c}.$$

Since the slopes are negative reciprocals, the tangent lines are perpendicular. Similarly, the curves are perpendicular at the other three points of intersection.

123. False. See the definition of a parabola.

125. True

127. True

129. Let $\dfrac{x^2}{a^2} + \dfrac{y^2}{b^2} = 1$ be the equation of the ellipse with $a > b > 0$. Let $(\pm c, 0)$ be the foci,

$c^2 = a^2 - b^2$. Let (u, v) be a point on the tangent line at $P(x, y)$, as indicated in the figure.

$$x^2b^2 + y^2a^2 = a^2b^2$$

$$2xb^2 + 2yy'a^2 = 0$$

$$y' = -\frac{b^2x}{a^2y} \quad \text{Slope at } P(x, y)$$

Now, $\dfrac{y - v}{x - u} = -\dfrac{b^2x}{a^2y}$

$$y^2a^2 - a^2vy = -b^2x^2 + b^2xu$$

$$y^2a^2 + x^2b^2 = a^2vy + b^2ux$$

$$a^2b^2 = a^2vy + b^2ux$$

—CONTINUED—

129. —CONTINUED—

Since there is a right angle at (u, v),

$$\frac{v}{u} = \frac{a^2 y}{b^2 x}$$

$$vb^2 x = a^2 uy.$$

We have two equations:

$$a^2 vy + b^2 ux = a^2 b^2$$

$$a^2 uy - b^2 vx = 0.$$

Multiplying the first by v and the second by u, and adding,

$$a^2 v^2 y + a^2 u^2 y = a^2 b^2 v$$

$$y[u^2 + v^2] = b^2 v$$

$$yd^2 = b^2 v$$

$$v = \frac{yd^2}{b^2}.$$

Similarly, $u = \dfrac{xd^2}{a^2}$.

From the figure, $u = d \cos \theta$ and $v = d \sin \theta$. Thus, $\cos \theta = \dfrac{xd}{a^2}$ and $\sin \theta = \dfrac{yd}{b^2}$.

$$\cos^2 \theta + \sin^2 \theta = \frac{x^2 d^2}{a^4} + \frac{y^2 d^2}{b^4} = 1$$

$$x^2 b^4 d^2 + y^2 a^4 d^2 = a^4 b^4$$

$$d^2 = \frac{a^4 b^4}{x^2 b^4 + y^2 a^4}$$

Let $r_1 = PF_1$ and $r_2 = PF_2$, $r_1 + r_2 = 2a$.

$$r_1 r_2 = \frac{1}{2}[(r_1 + r_2)^2 - r_1^2 - r_2^2]$$

$$= \frac{1}{2}[4a^2 - (x + c)^2 - y^2 - (x - c)^2 - y^2]$$

$$= 2a^2 - x^2 - y^2 - c^2$$

$$= a^2 + b^2 - x^2 - y^2$$

Finally, $d^2 r_1 r_2 = \dfrac{a^4 b^4}{x^2 b^4 + y^2 a^4} \cdot [a^2 + b^2 - x^2 - y^2]$

$$= \frac{a^4 b^4}{b^2(b^2 x^2) + a^2(a^2 y^2)} \cdot [a^2 + b^2 - x^2 - y^2]$$

$$= \frac{a^4 b^4}{b^2(a^2 b^2 - a^2 y^2) + a^2(a^2 b^2 - b^2 x^2)} \cdot [a^2 + b^2 - x^2 - y^2]$$

$$= \frac{a^4 b^4}{a^2 b^2 [a^2 + b^2 - x^2 - y^2]} \cdot [a^2 + b^2 - x^2 - y^2]$$

$$= a^2 b^2, \quad \text{a constant!}$$

Section 10.2 Plane Curves and Parametric Equations

1. $x = \sqrt{t},\ y = 1 - t$

(a)

t	0	1	2	3	4
x	0	1	$\sqrt{2}$	$\sqrt{3}$	2
y	1	0	-1	-2	-3

(b)

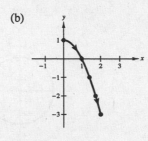

(c)

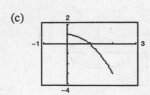

(d) $x^2 = t$

$y = 1 - x^2,\ x \geq 0$

3. $x = 3t - 1$

$y = 2t + 1$

$y = 2\left(\dfrac{x+1}{3}\right) + 1$

$2x - 3y + 5 = 0$

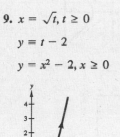

5. $x = t + 1$

$y = t^2$

$y = (x - 1)^2$

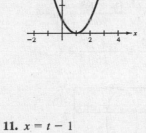

7. $x = t^3$

$y = \frac{1}{2}t^2$

$x = t^3$ implies $t = x^{1/3}$

$y = \frac{1}{2}x^{2/3}$

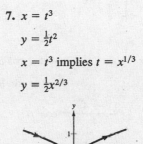

9. $x = \sqrt{t},\ t \geq 0$

$y = t - 2$

$y = x^2 - 2,\ x \geq 0$

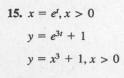

11. $x = t - 1$

$y = \dfrac{t}{t - 1}$

$y = \dfrac{x + 1}{x}$

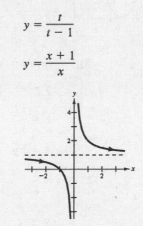

13. $x = 2t$

$y = |t - 2|$

$y = \left|\dfrac{x}{2} - 2\right| = \dfrac{|x - 4|}{2}$

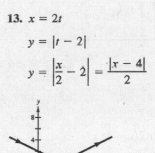

15. $x = e^t,\ x > 0$

$y = e^{3t} + 1$

$y = x^3 + 1,\ x > 0$

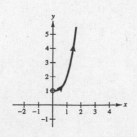

17. $x = \sec \theta$

$y = \cos \theta$

$0 \le \theta < \dfrac{\pi}{2}, \dfrac{\pi}{2} < \theta \le \pi$

$xy = 1$

$y = \dfrac{1}{x}$

$|x| \ge 1, |y| \le 1$

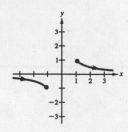

19. $x = 3 \cos \theta, \ y = 3 \sin \theta$

Squaring both equations and adding, we have

$x^2 + y^2 = 9.$

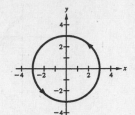

21. $x = 4 \sin 2\theta$

$y = 2 \cos 2\theta$

$\dfrac{x^2}{16} = \sin^2 2\theta$

$\dfrac{y^2}{4} = \cos^2 2\theta$

$\dfrac{x^2}{16} + \dfrac{y^2}{4} = 1$

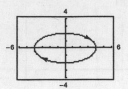

23.

$x = 4 + 2 \cos \theta$

$y = -1 + \sin \theta$

$\dfrac{(x - 4)^2}{4} = \cos^2 \theta$

$\dfrac{(y + 1)^2}{1} = \sin^2 \theta$

$\dfrac{(x - 4)^2}{4} + \dfrac{(y + 1)^2}{1} = 1$

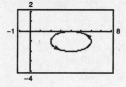

25.

$x = 4 + 2 \cos \theta$

$y = -1 + 4 \sin \theta$

$\dfrac{(x - 4)^2}{4} = \cos^2 \theta$

$\dfrac{(y + 1)^2}{16} = \sin^2 \theta$

$\dfrac{(x - 4)^2}{4} + \dfrac{(y + 1)^2}{16} = 1$

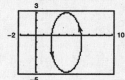

27.

$x = 4 \sec \theta$

$y = 3 \tan \theta$

$\dfrac{x^2}{16} = \sec^2 \theta$

$\dfrac{y^2}{9} = \tan^2 \theta$

$\dfrac{x^2}{16} - \dfrac{y^2}{9} = 1$

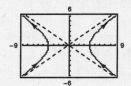

29. $x = t^3$

$y = 3 \ln t$

$y = 3 \ln \sqrt[3]{x} = \ln x$

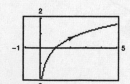

31. $x = e^{-t}$

$y = e^{3t}$

$e^t = \dfrac{1}{x}$

$e^t = \sqrt[3]{y}$

$\sqrt[3]{y} = \dfrac{1}{x}$

$y = \dfrac{1}{x^3}$

$x > 0$

$y > 0$

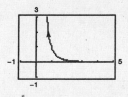

33. By eliminating the parameters in (a) – (d), we get $y = 2x + 1$. They differ from each other in orientation and in restricted domains. These curves are all smooth except for (b).

(a) $x = t$, $y = 2t + 1$

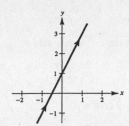

(b) $x = \cos \theta$ $y = 2 \cos \theta + 1$

     $-1 \leq x \leq 1$ $-1 \leq y \leq 3$

$$\frac{dx}{d\theta} = \frac{dy}{d\theta} = 0 \text{ when } \theta = 0, \pm\pi, \pm 2\pi, \ldots$$

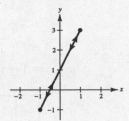

(c) $x = e^{-t}$ $y = 2e^{-t} + 1$

     $x > 0$ $y > 1$

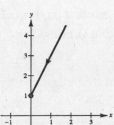

(d) $x = e^{t}$ $y = 2e^{t} + 1$

     $x > 0$ $y > 1$

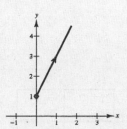

35. The curves are identical on $0 < \theta < \pi$. They are both smooth. Represent $y = 2(1 - x^2)$

37. (a)

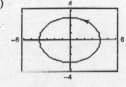

(b) The orientation of the second curve is reversed.

(c) The orientation will be reversed.

(d) Many answers possible. For example, $x = 1 + t$, $y = 1 + 2t$, and $x = 1 - t$, $x = 1 - 2t$.

39.
$$x = x_1 + t(x_2 - x_1)$$
$$y = y_1 + t(y_2 - y_1)$$
$$\frac{x - x_1}{x_2 - x_1} = t$$
$$y = y_1 + \left(\frac{x - x_1}{x_2 - x_1}\right)(y_2 - y_1)$$
$$y - y_1 = \frac{y_2 - y_1}{x_2 - x_1}(x - x_1)$$
$$y - y_1 = m(x - x_1)$$

41.
$$x = h + a \cos \theta$$
$$y = k + b \sin \theta$$
$$\frac{x - h}{a} = \cos \theta$$
$$\frac{y - k}{b} = \sin \theta$$
$$\frac{(x - h)^2}{a^2} + \frac{(y - k)^2}{b^2} = 1$$

43. From Exercise 39 we have

$$x = 5t$$
$$y = -2t.$$

Solution not unique

45. From Exercise 40 we have

$$x = 2 + 4 \cos \theta$$
$$y = 1 + 4 \sin \theta.$$

Solution not unique

47. From Exercise 41 we have

$$a = 5, c = 4 \implies b = 3$$
$$x = 5 \cos \theta$$
$$y = 3 \sin \theta.$$

Center: $(0, 0)$
Solution not unique

49. From Exercise 42 we have

$a = 4, c = 5 \implies b = 3$

$x = 4 \sec \theta$

$y = 3 \tan \theta.$

Center: $(0, 0)$
Solution not unique

51. $y = 3x - 2$

Example

$x = t, \qquad y = 3t - 2$

$x = t - 3, \quad y = 3t - 11$

53. $y = x^3$

Example

$x = t, \qquad y = t^3$

$x = \sqrt[3]{t}, \qquad y = t$

$x = \tan t, \qquad y = \tan^3 t$

55. $x = 2(\theta - \sin \theta)$

$y = 2(1 - \cos \theta)$

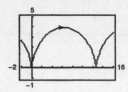

Not smooth at $\theta = 2n\pi$

57. $x = \theta - \frac{3}{2} \sin \theta$

$y = 1 - \frac{3}{2} \cos \theta$

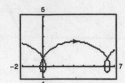

59. $x = 3 \cos^3 \theta$

$y = 3 \sin^3 \theta$

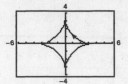

Not smooth at $(x, y) = (\pm 3, 0)$
and $(0, \pm 3)$, or $\theta = \frac{1}{2}n\pi$.

61. $x = 2 \cot \theta$

$y = 2 \sin^2 \theta$

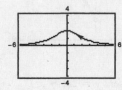

Smooth everywhere

63. See definition on page 709.

65. A plane curve C, represented by $x = f(t), y = g(t)$, is smooth if f' and g' are continuous and not simultaneously 0. See page 714.

67. When the circle has rolled θ radians, we know that the center is at $(a\theta, a)$.

$$\sin \theta = \sin(180° - \theta) = \frac{|AC|}{b} = \frac{|BD|}{b} \quad \text{or} \quad |BD| = b \sin \theta$$

$$\cos \theta = -\cos(180° - \theta) = \frac{|AP|}{-b} \quad \text{or} \quad |AP| = -b \cos \theta$$

Therefore, $x = a\theta - b \sin \theta$ and $y = a - b \cos \theta$.

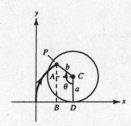

69. False

$x = t^2 \implies x \geq 0$

$y = t^2 \implies y \geq 0$

The graph of the parametric equations is only a portion of the line $y = x$.

71. (a) $100 \text{ mi/hr} = \dfrac{(100)(5280)}{3600} = \dfrac{440}{3} \text{ ft/sec}$

$$x = (v_0 \cos \theta)t = \left(\dfrac{440}{3} \cos \theta\right)t$$

$$y = h + (v_0 \sin \theta)t - 16t^2$$

$$= 3 + \left(\dfrac{440}{3} \sin \theta\right)t - 16t^2$$

(b)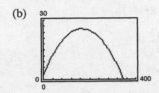

It is not a home run—when $x = 400$, $y < 10$.

(c)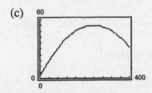

Yes, it's a home run when $x = 400$, $y > 10$.

(d) We need to find the angle θ (and time t) such that

$$x = \left(\dfrac{440}{3} \cos \theta\right)t = 400$$

$$y = 3 + \left(\dfrac{440}{3} \sin \theta\right)t - 16t^2 = 10.$$

From the first equation $t = 1200/440 \cos \theta$. Substituting into the second equation,

$$10 = 3 + \left(\dfrac{440}{3} \sin \theta\right)\left(\dfrac{1200}{440 \cos \theta}\right) - 16\left(\dfrac{1200}{440 \cos \theta}\right)^2$$

$$7 = 400 \tan \theta - 16\left(\dfrac{120}{44}\right)^2 \sec^2 \theta$$

$$= 400 \tan \theta - 16\left(\dfrac{120}{44}\right)^2 (\tan^2 \theta + 1).$$

We now solve the quadratic for $\tan \theta$:

$$16\left(\dfrac{120}{44}\right)^2 \tan^2 \theta - 400 \tan \theta + 7 + 16\left(\dfrac{120}{44}\right)^2 = 0$$

$$\tan \theta \approx 0.35185 \implies \theta \approx 19.4°$$

Section 10.3   Parametric Equations and Calculus

1. $\dfrac{dy}{dx} = \dfrac{dy/dt}{dx/dt} = \dfrac{-4}{2t} = \dfrac{-2}{t}$

3. $\dfrac{dy}{dx} = \dfrac{dy/d\theta}{dx/d\theta} = \dfrac{-2 \cos \theta \sin \theta}{2 \sin \theta \cos \theta} = -1$

$$\left[\text{Note: } x + y = 1 \implies y = 1 - x \text{ and } \dfrac{dy}{d\theta} = -1\right]$$

5. $x = 2t$, $y = 3t - 1$

$$\dfrac{dy}{dx} = \dfrac{dy/dt}{dx/dt} = \dfrac{3}{2}$$

$$\dfrac{d^2y}{dx^2} = 0 \quad \text{Line}$$

7. $x = t + 1$, $y = t^2 + 3t$

$$\dfrac{dy}{dx} = \dfrac{2t + 3}{1} = 1 \text{ when } t = -1.$$

$$\dfrac{d^2y}{dx^2} = 2 \quad \text{concave upwards}$$

9. $x = 2 \cos \theta$, $y = 2 \sin \theta$

$$\dfrac{dy}{dx} = \dfrac{2 \cos \theta}{-2 \sin \theta} = -\cot \theta = -1 \text{ when } \theta = \dfrac{\pi}{4}.$$

$$\dfrac{d^2y}{dx^2} = \dfrac{\csc^2 \theta}{-2 \sin \theta} = \dfrac{-\csc^3 \theta}{2} = -\sqrt{2} \text{ when } \theta = \dfrac{\pi}{4}.$$

concave downward

11. $x = 2 + \sec \theta$, $y = 1 + 2 \tan \theta$

$$\dfrac{dy}{dx} = \dfrac{2 \sec^2 \theta}{\sec \theta \tan \theta}$$

$$= \dfrac{2 \sec \theta}{\tan \theta} = 2 \csc \theta = 4 \text{ when } \theta = \dfrac{\pi}{6}.$$

$$\dfrac{d^2y}{dx^2} = \dfrac{d\left[\dfrac{dy}{dx}\right]}{\dfrac{dx}{d\theta}} = \dfrac{-2 \csc \theta \cot \theta}{\sec \theta \tan \theta}$$

$$= -2 \cot^3 \theta = -6\sqrt{3} \text{ when } \theta = \dfrac{\pi}{6}.$$

concave downward

13. $x = \cos^3 \theta$, $y = \sin^3 \theta$

$$\frac{dy}{dx} = \frac{3 \sin^2 \theta \cos \theta}{-3 \cos^2 \theta \sin \theta}$$

$$= -\tan \theta = -1 \text{ when } \theta = \frac{\pi}{4}.$$

$$\frac{d^2y}{dx^2} = \frac{-\sec^2 \theta}{-3 \cos^2 \theta \sin \theta} = \frac{1}{3 \cos^4 \theta \sin \theta}$$

$$= \frac{\sec^4 \theta \csc \theta}{3} = \frac{4\sqrt{2}}{3} \text{ when } \theta = \frac{\pi}{4}.$$

concave upward

15. $x = 2 \cot \theta$, $y = 2 \sin^2 \theta$

$$\frac{dy}{dx} = \frac{4 \sin \theta \cos \theta}{-2 \csc^2 \theta} = -2 \sin^3 \theta \cos \theta$$

At $\left(-\frac{2}{\sqrt{3}}, \frac{3}{2}\right)$, $\theta = \frac{2\pi}{3}$, and $\frac{dy}{dx} = \frac{3\sqrt{3}}{8}$.

Tangent line:
$$y - \frac{3}{2} = \frac{3\sqrt{3}}{8}\left(x + \frac{2}{\sqrt{3}}\right)$$

$$3\sqrt{3}x - 8y + 18 = 0$$

At $(0, 2)$, $\theta = \frac{\pi}{2}$, and $\frac{dy}{dx} = 0$.

Tangent line: $y - 2 = 0$

At $\left(2\sqrt{3}, \frac{1}{2}\right)$, $\theta = \frac{\pi}{6}$, and $\frac{dy}{dx} = -\frac{\sqrt{3}}{8}$.

Tangent line:
$$y - \frac{1}{2} = -\frac{\sqrt{3}}{8}(x - 2\sqrt{3})$$

$$\sqrt{3}x + 8y - 10 = 0$$

17. $x = 2t$, $y = t^2 - 1$, $t = 2$

(a)

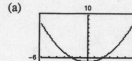

(b) At $t = 2$, $(x, y) = (4, 3)$, and

$$\frac{dx}{dt} = 2, \frac{dy}{dt} = 4, \frac{dy}{dx} = 2.$$

(c) $\frac{dy}{dx} = 2$. At $(4, 3)$, $y - 3 = 2(x - 4)$

$$y = 2x - 5.$$

(d)

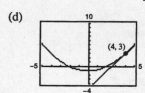

19. $x = t^2 - t + 2$, $y = t^3 - 3t$, $t = -1$

(a)

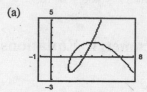

(b) At $t = -1$, $(x, y) = (4, 2)$, and

$$\frac{dx}{dt} = -3, \frac{dy}{dt} = 0, \frac{dy}{dx} = 0.$$

(c) $\frac{dy}{dx} = 0$. At $(4, 2)$, $y - 2 = 0(x - 4)$

$$y = 2$$

(d)

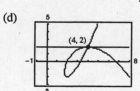

21. $x = 2 \sin 2t$, $y = 3 \sin t$ crosses itself at the origin, $(x, y) = (0, 0)$.

At this point, $t = 0$ or $t = \pi$.

$$\frac{dy}{dx} = \frac{3 \cos t}{4 \cos 2t}$$

At $t = 0$: $\frac{dy}{dx} = \frac{3}{4}$ and $y = \frac{3}{4}x$.   Tangent Line

At $t = \pi$, $\frac{dy}{dx} = -\frac{3}{4}$ and $y = \frac{-3}{4}x$.   Tangent Line

23. $x = t^2 - t$, $y = t^3 - 3t - 1$ crosses itself at the point $(x, y) = (2, 1)$.

At this point, $t = -1$ or $t = 2$.

$$\frac{dy}{dx} = \frac{3t^2 - 3}{2t - 1}$$

At $t = -1$, $\frac{dy}{dx} = 0$ and $y = 1$.   Tangent Line

At $t = 2$, $\frac{dy}{dt} = \frac{9}{3} = 3$ and $y - 1 = 3(x - 2)$ or $y = 3x - 5$.   Tangent Line

25. $x = \cos \theta + \theta \sin \theta$, $y = \sin \theta - \theta \cos \theta$

Horizontal tangents: $\frac{dy}{d\theta} = \theta \sin \theta = 0$ when $\theta = \pm \pi, \pm 2\pi, \pm 3\pi, \ldots$

   Points: $(-1, [2n - 1]\pi)$, $(1, 2n\pi)$ where n is an integer.   Points shown: $(1, 0)$, $(-1, \pi)$, $(1, -2\pi)$

Vertical tangents: $\frac{dx}{d\theta} = \theta \cos \theta = 0$ when $\theta = \pm \frac{\pi}{2}, \pm \frac{3\pi}{2}, \pm \frac{5\pi}{2}, \ldots$

   Note: $\theta = 0$ corresponds to the cusp at $(x, y) = (1, 0)$.

$$\frac{dy}{dx} = \frac{\theta \sin \theta}{\theta \cos \theta} = \tan \theta = 0 \text{ at } \theta = 0$$

   Points: $\left(\frac{(-1)^{n+1}(2n - 1)\pi}{2}, (-1)^{n+1} \right)$ Points shown: $\left(\frac{\pi}{2}, 1 \right)$, $\left(-\frac{3\pi}{2}, -1 \right)$, $\left(\frac{5\pi}{2}, 1 \right)$

27. $x = 1 - t$, $y = t^2$

Horizontal tangents: $\frac{dy}{dt} = 2t = 0$ when $t = 0$

   Point: $(1, 0)$

Vertical tangents: $\frac{dx}{dt} = -1 \neq 0$; none

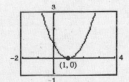

29. $x = 1 - t$, $y = t^3 - 3t$

Horizontal tangents: $\frac{dy}{dt} = 3t^2 - 3 = 0$ when $t = \pm 1$.

   Points: $(0, -2)$, $(2, 2)$

Vertical tangents: $\frac{dx}{dt} = -1 \neq 0$; none

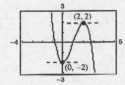

31. $x = 3 \cos \theta$, $y = 3 \sin \theta$

Horizontal tangents: $\frac{dy}{d\theta} = 3 \cos \theta = 0$ when $\theta = \frac{\pi}{2}, \frac{3\pi}{2}$.

   Points: $(0, 3)$, $(0, -3)$

Vertical tangents: $\frac{dx}{d\theta} = -3 \sin \theta = 0$ when $\theta = 0, \pi$.

   Points: $(3, 0)$, $(-3, 0)$

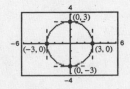

33. $x = 4 + 2 \cos \theta$, $y = -1 + \sin \theta$

Horizontal tangents: $\frac{dy}{d\theta} = \cos \theta = 0$ when $\theta = \frac{\pi}{2}, \frac{3\pi}{2}$.

   Points: $(4, 0)$, $(4, -2)$

Vertical tangents: $\frac{dx}{d\theta} = -2 \sin \theta = 0$ when $\theta = 0, \pi$.

   Points: $(6, -1)$, $(2, -1)$

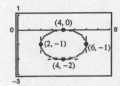

35. $x = \sec\theta$, $y = \tan\theta$

Horizontal tangents: $\dfrac{dy}{d\theta} = \sec^2\theta \neq 0$; none

Vertical tangents: $\dfrac{dx}{d\theta} = \sec\theta\tan\theta = 0$ when $x = 0, \pi$.

Points: $(1, 0)$, $(-1, 0)$

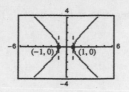

37. $x = t^2$, $y = t^3 - t$

$\dfrac{dy}{dx} = \dfrac{3t^2 - 1}{2t}$

$\dfrac{d^2y}{dx^2} = \left[\dfrac{\dfrac{2t(6t) - (3t^2 - 1)2}{4t^2}}{2t} \right]$

$= \dfrac{6t^2 + 2}{8t^3} = \dfrac{1 + 3t^2}{4t^3}$

Concave upward for $t > 0$

Concave downward for $t < 0$

39. $x = 2t + \ln t$, $y = 2t - \ln t$, $t > 0$

$\dfrac{dy}{dx} = \dfrac{2 - (1/t)}{2 + (1/t)} = \dfrac{2t - 1}{2t + 1}$

$\dfrac{d^2y}{dx^2} = \left[\dfrac{(2t + 1)2 - (2t - 1)2}{(2t + 1)^2} \right] \bigg/ \left(2 + \dfrac{1}{t} \right)$

$= \dfrac{4}{(2t + 1)^2} \cdot \dfrac{t}{2t + 1}$

$= \dfrac{4t}{(2t + 1)^3}$

Because $t > 0$, $\dfrac{d^2y}{dx^2} > 0$

Concave upward for $t > 0$

41. $x = \sin t$, $y = \cos t$, $0 < t < \pi$

$\dfrac{dy}{dx} = -\dfrac{\sin t}{\cos t} = -\tan t$

$\dfrac{d^2y}{dx^2} = -\dfrac{\sec^2 t}{\cos t} = -\dfrac{1}{\cos^3 t}$

Concave upward on $\pi/2 < t < \pi$

Concave downward on $0 < t < \pi/2$

43. $x = 2t - t^2$, $y = 2t^{3/2}$, $1 \leq t \leq 2$

$\dfrac{dx}{dt} = 2 - 2t$, $\dfrac{dy}{dt} = 3t^{1/2}$

$s = \int_a^b \sqrt{\left(\dfrac{dx}{dt}\right)^2 + \left(\dfrac{dy}{dt}\right)^2}\, dt$

$= \int_1^2 \sqrt{(2 - 2t)^2 + (3t^{1/2})^2}\, dt$

$= \int_1^2 \sqrt{4 - 8t + 4t^2 + 9t}\, dt$

$= \int_1^2 \sqrt{4t^2 + t + 4}\, dt$

45. $x = e^t + 2$, $y = 2t + 1$, $-2 \leq t \leq 2$

$\dfrac{dx}{dt} = e^t$, $\dfrac{dy}{dt} = 2$

$s = \int_a^b \sqrt{\left(\dfrac{dx}{dt}\right)^2 + \left(\dfrac{dy}{dt}\right)^2}\, dt$

$= \int_{-2}^2 \sqrt{e^{2t} + 4}\, dt$

47. $x = t^2$, $y = 2t$, $0 \leq t \leq 2$

$\dfrac{dx}{dt} = 2t$, $\dfrac{dy}{dt} = 2$, $\left(\dfrac{dx}{dt}\right)^2 + \left(\dfrac{dy}{dt}\right)^2 = 4t^2 + 4 = 4(t^2 + 1)$

$s = 2\int_0^2 \sqrt{t^2 + 1}\, dt$

$= \left[t\sqrt{t^2 + 1} + \ln|t + \sqrt{t^2 + 1}| \right]_0^2$

$= 2\sqrt{5} + \ln(2 + \sqrt{5}) \approx 5.916$

49. $x = e^{-t} \cos t$, $y = e^{-t} \sin t$, $0 \le t \le \dfrac{\pi}{2}$

$\dfrac{dx}{dt} = -e^{-t}(\sin t + \cos t)$, $\dfrac{dy}{dt} = e^{-t}(\cos t - \sin t)$

$s = \displaystyle\int_0^{\pi/2} \sqrt{\left(\dfrac{dx}{dt}\right)^2 + \left(\dfrac{dy}{dt}\right)^2}\, dt$

$= \displaystyle\int_0^{\pi/2} \sqrt{2e^{-2t}}\, dt = -\sqrt{2}\int_0^{\pi/2} e^{-t}(-1)\, dt$

$= \left[-\sqrt{2}e^{-t}\right]_0^{\pi/2} = \sqrt{2}(1 - e^{-\pi/2}) \approx 1.12$

51. $x = \sqrt{t}$, $y = 3t - 1$, $\dfrac{dx}{dt} = \dfrac{1}{2\sqrt{t}}$, $\dfrac{dy}{dt} = 3$

$s = \displaystyle\int_0^1 \sqrt{\dfrac{1}{4t} + 9}\, dt = \dfrac{1}{2}\int_0^1 \dfrac{\sqrt{1 + 36t}}{\sqrt{t}}\, dt$

$= \dfrac{1}{6}\displaystyle\int_0^6 \sqrt{1 + u^2}\, du$

$= \dfrac{1}{12}\left[\ln\left(\sqrt{1 + u^2} + u\right) + u\sqrt{1 + u^2}\right]_0^6$

$= \dfrac{1}{12}\left[\ln\left(\sqrt{37} + 6\right) + 6\sqrt{37}\right] \approx 3.249$

$u = 6\sqrt{t}$, $du = \dfrac{3}{\sqrt{t}}\, dt$

53. $x = a\cos^3 \theta$, $y = a\sin^3 \theta$, $\dfrac{dx}{d\theta} = -3a\cos^2 \theta \sin \theta$,

$\dfrac{dy}{d\theta} = 3a\sin^2 \theta \cos \theta$

$s = 4\displaystyle\int_0^{\pi/2} \sqrt{9a^2 \cos^4 \theta \sin^2 \theta + 9a^2 \sin^4 \theta \cos^2 \theta}\, d\theta$

$= 12a\displaystyle\int_0^{\pi/2} \sin \theta \cos \theta \sqrt{\cos^2 \theta + \sin^2 \theta}\, d\theta$

$= 6a\displaystyle\int_0^{\pi/2} \sin 2\theta\, d\theta = \left[-3a\cos 2\theta\right]_0^{\pi/2} = 6a$

55. $x = a(\theta - \sin \theta)$, $y = a(1 - \cos \theta)$,

$\dfrac{dx}{d\theta} = a(1 - \cos \theta)$, $\dfrac{dy}{d\theta} = a\sin \theta$

$s = 2\displaystyle\int_0^{\pi} \sqrt{a^2(1 - \cos \theta)^2 + a^2 \sin^2 \theta}\, d\theta$

$= 2\sqrt{2}a\displaystyle\int_0^{\pi} \sqrt{1 - \cos \theta}\, d\theta$

$= 2\sqrt{2}a\displaystyle\int_0^{\pi} \dfrac{\sin \theta}{\sqrt{1 + \cos \theta}}\, d\theta$

$= \left[-4\sqrt{2}a\sqrt{1 + \cos \theta}\right]_0^{\pi} = 8a$

57. $x = (90\cos 30°)t$, $y = (90\sin 30°)t - 16t^2$

(a)

(b) Range: 219.2 ft, $\left(t = \dfrac{45}{16}\right)$

(c) $\dfrac{dx}{dt} = 90\cos 30°$, $\dfrac{dy}{dt} = 90\sin 30° - 32t$

$y = 0$ for $t = \dfrac{45}{16}$.

$s = \displaystyle\int_0^{45/16} \sqrt{(90\cos 30°)^2 + (90\sin 30° - 32t)^2}\, dt$

≈ 230.8 ft

59. $x = \dfrac{4t}{1 + t^3}$, $y = \dfrac{4t^2}{1 + t^3}$

(a) $x^3 + y^3 = 4xy$

(b) $\dfrac{dy}{dt} = \dfrac{(1 + t^3)(8t) - 4t^2(3t^2)}{(1 + t^3)^2}$

$= \dfrac{4t(2 - t^3)}{(1 + t^3)^2} = 0$ when $t = 0$ or $t = \sqrt[3]{2}$.

Points: $(0, 0)$, $\left(\dfrac{4\sqrt[3]{2}}{3}, \dfrac{4\sqrt[3]{4}}{3}\right) \approx (1.6799, 2.1165)$

(c) $s = 2\displaystyle\int_0^1 \sqrt{\left[\dfrac{4(1 - 2t^3)}{(1 + t^3)^2}\right]^2 + \left[\dfrac{4t(2 - t^3)}{(1 + t^3)^2}\right]^2}\, dt = 2\int_0^1 \sqrt{\dfrac{16}{(1 + t^3)^4}[t^8 + 4t^6 - 4t^5 - 4t^3 + 4t^2 + 1]}\, dt$

$= 8\displaystyle\int_0^1 \dfrac{\sqrt{t^8 + 4t^6 - 4t^5 - 4t^3 + 4t^2 + 1}}{(1 + t^3)^2}\, dt \approx 6.557$

61. (a) $x = t - \sin t$ $x = 2t - \sin(2t)$

$y = 1 - \cos t$ $y = 1 - \cos(2t)$

$0 \le t \le 2\pi$ $0 \le t \le \pi$

(b) The average speed of the particle on the second path is twice the average speed of a particle on the first path.

(c) $x = \frac{1}{2}t - \sin\left(\frac{1}{2}t\right)$

$y = 1 - \cos\left(\frac{1}{2}t\right)$

The time required for the particle to traverse the same path is $t = 4\pi$.

63. $x = 4t, \dfrac{dx}{dt} = 4$

$y = t + 1, \dfrac{dy}{dt} = 1$

$S = 2\pi \displaystyle\int_0^2 (t + 1)\sqrt{4^2 + 1^2}\, dt$

$= 2\pi \displaystyle\int_0^2 \sqrt{17}(t + 1)\, dt = 8\pi\sqrt{17}$

≈ 103.6249

65. $x = \cos^2 \theta, \dfrac{dx}{d\theta} = -2\cos\theta\sin\theta$

$y = \cos\theta, \dfrac{dy}{d\theta} = -\sin\theta$

$S = 2\pi \displaystyle\int_0^{\pi/2} \cos\theta\sqrt{4\cos^2\theta\sin^2\theta + \sin^2\theta}\, d\theta$

≈ 5.3304

67. $x = t,\ y = 2t,\ \dfrac{dx}{dt} = 1,\ \dfrac{dy}{dt} = 2$

(a) $S = 2\pi\displaystyle\int_0^4 2t\sqrt{1 + 4}\, dt = 4\sqrt{5}\pi\displaystyle\int_0^4 t\, dt$

$= \left[2\sqrt{5}\pi t^2\right]_0^4 = 32\pi\sqrt{5}$

(b) $S = 2\pi\displaystyle\int_0^4 t\sqrt{1 + 4}\, dt = 2\sqrt{5}\pi\displaystyle\int_0^4 t\, dt$

$= \left[\sqrt{5}\pi t^2\right]_0^4 = 16\pi\sqrt{5}$

69. $x = 4\cos\theta,\ y = 4\sin\theta,\ \dfrac{dx}{d\theta} = -4\sin\theta,\ \dfrac{dy}{d\theta} = 4\cos\theta$

$S = 2\pi\displaystyle\int_0^{\pi/2} 4\cos\theta\sqrt{(-4\sin\theta)^2 + (4\cos\theta)^2}\, d\theta$

$= 32\pi\displaystyle\int_0^{\pi/2} \cos\theta\, d\theta = \left[32\pi\sin\theta\right]_0^{\pi/2} = 32\pi$

71. $x = a\cos^3\theta,\ y = a\sin^3\theta,\ \dfrac{dx}{d\theta} = -3a\cos^2\theta\sin\theta,\ \dfrac{dy}{d\theta} = 3a\sin^2\theta\cos\theta$

$S = 4\pi\displaystyle\int_0^{\pi/2} a\sin^3\theta\sqrt{9a^2\cos^4\theta\sin^2\theta + 9a^2\sin^4\theta\cos^2\theta}\, d\theta = 12a^2\pi\displaystyle\int_0^{\pi/2} \sin^4\theta\cos\theta\, d\theta = \dfrac{12\pi a^2}{5}\left[\sin^5\theta\right]_0^{\pi/2} = \dfrac{12}{5}\pi a^2$

73. $\dfrac{dy}{dx} = \dfrac{dy/dt}{dx/dt}$

See Theorem 10.7.

75. One possible answer is the graph given by

$x = t,\ y = -t.$

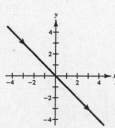

77. $s = \displaystyle\int_a^b \sqrt{\left(\dfrac{dx}{dt}\right)^2 + \left(\dfrac{dy}{dt}\right)^2}\, dt$

See Theorem 10.8.

79. Let y be a continuous function of x on $a \le x \le b$. Suppose that $x = f(t)$, $y = g(t)$, and $f(t_1) = a$, $f(t_2) = b$. Then using integration by substitution, $dx = f'(t)\,dt$ and

$$\int_a^b y\,dx = \int_{t_1}^{t_2} g(t)f'(t)\,dt.$$

81. $x = 2\sin^2\theta$

$y = 2\sin^2\theta\tan\theta$

$\dfrac{dx}{d\theta} = 4\sin\theta\cos\theta$

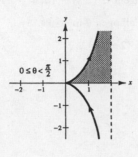

$0 \le \theta < \dfrac{\pi}{2}$

$$A = \int_0^{\pi/2} 2\sin^2\theta\tan\theta(4\sin\theta\cos\theta)\,d\theta = 8\int_0^{\pi/2}\sin^4\theta\,d\theta$$

$$= 8\left[\frac{-\sin^3\theta\cos\theta}{4} - \frac{3}{8}\sin\theta\cos\theta + \frac{3}{8}\theta\right]_0^{\pi/2} = \frac{3\pi}{2}$$

83. πab is area of ellipse (d).   **85.** $6\pi a^2$ is area of cardioid (f).   **87.** $\frac{8}{3}ab$ is area of hourglass (a).

89. $x = \sqrt{t},\ y = 4 - t,\ 0 < t < 4$

$$A = \int_0^2 y\,dx = \int_0^4 (4-t)\frac{1}{2\sqrt{t}}\,dt = \frac{1}{2}\int_0^4 (4t^{-1/2} - t^{1/2})\,dt = \left[\frac{1}{2}\left(8\sqrt{t} - \frac{2}{3}t\sqrt{t}\right)\right]_0^4 = \frac{16}{3}$$

$$\bar{x} = \frac{1}{A}\int_0^2 yx\,dx = \frac{3}{16}\int_0^4 (4-t)\sqrt{t}\left(\frac{1}{2\sqrt{t}}\right)dt = \frac{3}{32}\int_0^4 (4-t)\,dt = \left[\frac{3}{32}\left(4t - \frac{t^2}{2}\right)\right]_0^4 = \frac{3}{4}$$

$$\bar{y} = \frac{1}{A}\int_0^2 \frac{y^2}{2}\,dx = \frac{3}{32}\int_0^4 (4-t)^2\frac{1}{2\sqrt{t}}\,dt = \frac{3}{64}\int_0^4 [16t^{-1/2} - 8t^{1/2} + t^{3/2}]\,dt = \frac{3}{64}\left[32\sqrt{t} - \frac{16}{3}t\sqrt{t} + \frac{2}{5}t^2\sqrt{t}\right]_0^4 = \frac{8}{5}$$

$$(\bar{x}, \bar{y}) = \left(\frac{3}{4}, \frac{8}{5}\right)$$

91. $x = 3\cos\theta,\ y = 3\sin\theta,\ \dfrac{dx}{d\theta} = -3\sin\theta$

$$V = 2\pi\int_{\pi/2}^0 (3\sin\theta)^2(-3\sin\theta)\,d\theta$$

$$= -54\pi\int_{\pi/2}^0 \sin^3\theta\,d\theta$$

$$= -54\pi\int_{\pi/2}^0 (1 - \cos^2\theta)\sin\theta\,d\theta$$

$$= -54\pi\left[-\cos\theta + \frac{\cos^3\theta}{3}\right]_{\pi/2}^0 = 36\pi \text{ (Sphere)}$$

93. $x = a(\theta - \sin\theta),\ y = a(1 - \cos\theta)$

(a) $\dfrac{dy}{d\theta} = a\sin\theta,\ \dfrac{dx}{d\theta} = a(1 - \cos\theta)$

$\dfrac{dy}{dx} = \dfrac{a\sin\theta}{a(1-\cos\theta)} = \dfrac{\sin\theta}{1-\cos\theta}$

$\dfrac{d^2y}{dx^2} = \left[\dfrac{(1-\cos\theta)\cos\theta - \sin\theta(\sin\theta)}{(1-\cos\theta)^2}\right]\Big/[a(1-\cos\theta)]$

$= \dfrac{\cos\theta - 1}{a(1-\cos\theta)^3} = \dfrac{-1}{a(\cos\theta - 1)^2}$

(b) At $\theta = \dfrac{\pi}{6},\ x = a\left(\dfrac{\pi}{6} - \dfrac{1}{2}\right),\ y = a\left(1 - \dfrac{\sqrt{3}}{2}\right),$

$\dfrac{dy}{dx} = \dfrac{1/2}{1 - \sqrt{3}/2} = 2 + \sqrt{3}.$

Tangent line:

$$y - a\left(1 - \frac{\sqrt{3}}{2}\right) = \left(2 + \sqrt{3}\right)\left(x - a\left(\frac{\pi}{6} - \frac{1}{2}\right)\right)$$

—CONTINUED—

93. —CONTINUED—

(c) $\dfrac{dy}{dx} = \dfrac{\sin\theta}{1 - \cos\theta} = 0 \implies \sin\theta = 0, 1 - \cos\theta \neq 0$

Points of horizontal tangency: $(x, y) = (a(2n + 1)\pi, 2a)$

(d) Concave downward on all open θ-intervals:

$\ldots, (-2\pi, 0), (0, 2\pi), (2\pi, 4\pi), \ldots$

(e) $s = \displaystyle\int_0^{2\pi} \sqrt{a^2 \sin^2\theta + a^2(1 - \cos\theta)^2}\, d\theta$

$= a\displaystyle\int_0^{2\pi} \sqrt{2 - 2\cos\theta}\, d\theta$

$= a\displaystyle\int_0^{2\pi} \sqrt{4\sin^2\theta/2}\, d\theta$

$= 2a\displaystyle\int_0^{2\pi} \sin\dfrac{\theta}{2}\, d\theta$

$= -4a\cos\left(\dfrac{\theta}{2}\right)\Big]_0^{2\pi} = 8a$

95. $x = t + u = r\cos\theta + r\theta\sin\theta$

$= r(\cos\theta + \theta\sin\theta)$

$y = v - w = r\sin\theta - r\theta\cos\theta$

$= r(\sin\theta - \theta\cos\theta)$

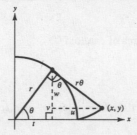

97. (a)

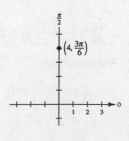

(b) $x = \dfrac{1 - t^2}{1 + t^2}$, $y = \dfrac{2t}{1 + t^2}$, $-20 \leq t \leq 20$

The graph (for $-\infty < t < \infty$) is the circle $x^2 + y^2 = 1$, except the point $(-1, 0)$.

Verify: $x^2 + y^2 = \left(\dfrac{1 - t^2}{1 + t^2}\right)^2 + \left(\dfrac{2t}{1 + t^2}\right)^2 = \dfrac{1 - 2t^2 + t^4 + 4t^2}{(1 + t^2)^2} = \dfrac{(1 + t^2)^2}{(1 + t^2)^2} = 1$

(c) As t increases from -20 to 0, the speed increases, and as t increases from 0 to 20, the speed decreases.

99. False. $\dfrac{d^2y}{dx^2} = \dfrac{\dfrac{d}{dt}\left[\dfrac{g'(t)}{f'(t)}\right]}{f'(t)} = \dfrac{f'(t)g''(t) - g'(t)f''(t)}{[f'(t)]^3}$

Section 10.4 Polar Coordinates and Polar Graphs

1. $\left(4, \dfrac{\pi}{2}\right)$

$x = 4\cos\left(\dfrac{\pi}{2}\right) = 0$

$y = 4\sin\left(\dfrac{\pi}{2}\right) = 4$

$(x, y) = (0, 4)$

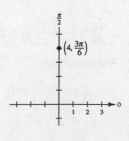

3. $\left(-4, -\dfrac{\pi}{3}\right)$

$x = -4\cos\left(-\dfrac{\pi}{3}\right) = -2$

$y = -4\sin\left(-\dfrac{\pi}{3}\right) = 2\sqrt{3}$

$(x, y) = \left(-2, 2\sqrt{3}\right)$

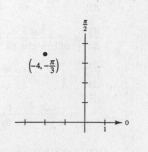

5. $\left(\sqrt{2}, 2.36\right)$

$x = \sqrt{2}\cos(2.36) \approx -1.004$

$y = \sqrt{2}\sin(2.36) \approx 0.996$

$(x, y) = (-1.004, 0.996)$

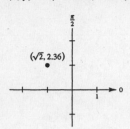

7. $(r, \theta) = \left(5, \dfrac{3\pi}{4}\right)$

$(x, y) = (-3.5355, 3.5355)$

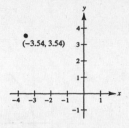

9. $(r, \theta) = (-3.5, 2.5)$

$(x, y) = (2.804, -2.095)$

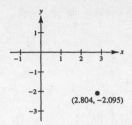

11. $(x, y) = (1, 1)$

$r = \pm\sqrt{2}$

$\tan\theta = 1$

$\theta = \dfrac{\pi}{4}, \dfrac{5\pi}{4}, \left(\sqrt{2}, \dfrac{\pi}{4}\right), \left(-\sqrt{2}, \dfrac{5\pi}{4}\right)$

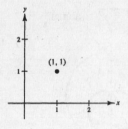

13. $(x, y) = (-3, 4)$

$r = \pm\sqrt{9 + 16} = \pm 5$

$\tan\theta = -\dfrac{4}{3}$

$\theta \approx 2.214, 5.356, (5, 2.214),$

$(-5, 5.356)$

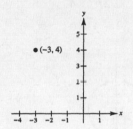

15. $(x, y) = \left(\sqrt{3}, -1\right)$

$r = \sqrt{3 + 1} = 2$

$\tan\theta = -\sqrt{3}/3$

$(r, \theta) = (2, 11\pi/6)$

$= (-2, 5\pi/6)$

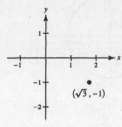

17. $(x, y) = (3, -2)$

$(r, \theta) = (3.606, -0.588)$

19. $(x, y) = \left(\dfrac{5}{7}, \dfrac{4}{3}\right)$

$(r, \theta) = (2.833, 0.490)$

21. (a) $(x, y) = (4, 3.5)$

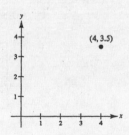

(b) $(r, \theta) = (4, 3.5)$

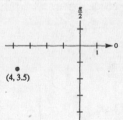

23. $r = 2\sin\theta$ circle

Matches (c)

25. $r = 3(1 + \cos\theta)$

Cardioid

Matches (a)

27. $x^2 + y^2 = a^2$

$r = a$

29. $\quad y = 4$

$r\sin\theta = 4$

$r = 4\csc\theta$

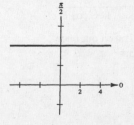

31.
$$3x - y + 2 = 0$$
$$3r \cos \theta - r \sin \theta + 2 = 0$$
$$r(3 \cos \theta - \sin \theta) = -2$$
$$r = \frac{-2}{3 \cos \theta - \sin \theta}$$

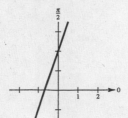

33.
$$y^2 = 9x$$
$$r^2 \sin^2 \theta = 9r \cos \theta$$
$$r = \frac{9 \cos \theta}{\sin^2 \theta}$$
$$r = 9 \csc^2 \theta \cos \theta$$

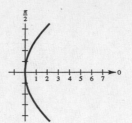

35.
$$r = 3$$
$$r^2 = 9$$
$$x^2 + y^2 = 9$$

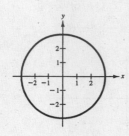

37.
$$r = \sin \theta$$
$$r^2 = r \sin \theta$$
$$x^2 + y^2 = y$$
$$x^2 + \left(y - \frac{1}{2}\right)^2 = \frac{1}{4}$$
$$x^2 + y^2 - y = 0$$

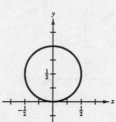

39.
$$r = \theta$$
$$\tan r = \tan \theta$$
$$\tan \sqrt{x^2 + y^2} = \frac{y}{x}$$
$$\sqrt{x^2 + y^2} = \arctan \frac{y}{x}$$

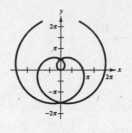

41.
$$r = 3 \sec \theta$$
$$r \cos \theta = 3$$
$$x = 3$$
$$x - 3 = 0$$

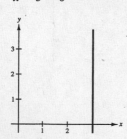

43. $r = 3 - 4 \cos \theta$
$$0 \le \theta < 2\pi$$

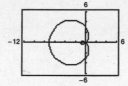

45. $r = 2 + \sin \theta$
$$0 \le \theta < 2\pi$$

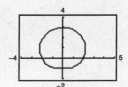

47. $r = \dfrac{2}{1 + \cos \theta}$

Traced out once on $-\pi < \theta < \pi$

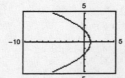

49. $r = 2 \cos\left(\dfrac{3\theta}{2}\right)$
$$0 \le \theta < 4\pi$$

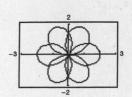

51. $r^2 = 4 \sin 2\theta$

$r_1 = 2\sqrt{\sin 2\theta}$

$r_2 = -2\sqrt{\sin 2\theta}$

$0 \le \theta < \dfrac{\pi}{2}$

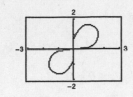

53.
$$r = 2(h \cos \theta + k \sin \theta)$$
$$r^2 = 2r(h \cos \theta + k \sin \theta)$$
$$r^2 = 2[h(r \cos \theta) + k(r \sin \theta)]$$
$$x^2 + y^2 = 2(hx + ky)$$
$$x^2 + y^2 - 2hx - 2ky = 0$$
$$(x^2 - 2hx + h^2) + (y^2 - 2ky + k^2) = 0 + h^2 + k^2 \qquad \text{Radius: } \sqrt{h^2 + k^2}$$
$$(x - h)^2 + (y - k)^2 = h^2 + k^2 \qquad\qquad \text{Center: } (h, k)$$

55. $\left(4, \dfrac{2\pi}{3}\right), \left(2, \dfrac{\pi}{6}\right)$

$d = \sqrt{4^2 + 2^2 - 2(4)(2) \cos\left(\dfrac{2\pi}{3} - \dfrac{\pi}{6}\right)}$

$ = \sqrt{20 - 16 \cos \dfrac{\pi}{2}} = 2\sqrt{5} \approx 4.5$

57. $(2, 0.5), (7, 1.2)$

$d = \sqrt{2^2 + 7^2 - 2(2)(7) \cos(0.5 - 1.2)}$

$ = \sqrt{53 - 28 \cos(-0.7)} \approx 5.6$

59. $r = 2 + 3 \sin \theta$

$\dfrac{dy}{dx} = \dfrac{3 \cos \theta \sin \theta + \cos \theta (2 + 3 \sin \theta)}{3 \cos \theta \cos \theta - \sin \theta (2 + 3 \sin \theta)}$

$\phantom{\dfrac{dy}{dx}} = \dfrac{2 \cos \theta (3 \sin \theta + 1)}{3 \cos 2\theta - 2 \sin \theta} = \dfrac{2 \cos \theta (3 \sin \theta + 1)}{6 \cos^2 \theta - 2 \sin \theta - 3}$

At $\left(5, \dfrac{\pi}{2}\right)$, $\dfrac{dy}{dx} = 0$.

At $(2, \pi)$, $\dfrac{dy}{dx} = -\dfrac{2}{3}$.

At $\left(-1, \dfrac{3\pi}{2}\right)$, $\dfrac{dy}{dx} = 0$.

61. (a), (b) $r = 3(1 - \cos \theta)$

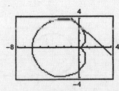

$(r, \theta) = \left(3, \dfrac{\pi}{2}\right) \implies (x, y) = (0, 3)$

Tangent line: $y - 3 = -1(x - 0)$

$\phantom{\text{Tangent line: }} y = -x + 3$

(c) At $\theta = \dfrac{\pi}{2}, \dfrac{dy}{dx} = -1.0$.

63. (a), (b) $r = 3 \sin \theta$

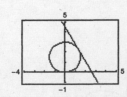

(c) At $\theta = \dfrac{\pi}{3}, \dfrac{dy}{dx} = -\sqrt{3} \approx -1.732$.

$(r, \theta) = \left(\dfrac{3\sqrt{3}}{2}, \dfrac{\pi}{3}\right) \implies (x, y) = \left(\dfrac{3\sqrt{3}}{4}, \dfrac{9}{4}\right)$

Tangent line: $y - \dfrac{9}{4} = -\sqrt{3}\left(x - \dfrac{3\sqrt{3}}{4}\right)$

$\phantom{\text{Tangent line: }} y = -\sqrt{3}x + \dfrac{9}{2}$

65. $r = 1 - \sin\theta$

$\dfrac{dy}{d\theta} = (1 - \sin\theta)\cos\theta - \cos\theta\sin\theta$

$\qquad = \cos\theta(1 - 2\sin\theta) = 0$

$\cos\theta = 0$ or $\sin\theta = \dfrac{1}{2} \implies \theta = \dfrac{\pi}{2}, \dfrac{3\pi}{2}, \dfrac{\pi}{6}, \dfrac{5\pi}{6}$

Horizontal tangents: $\left(2, \dfrac{3\pi}{2}\right), \left(\dfrac{1}{2}, \dfrac{\pi}{6}\right), \left(\dfrac{1}{2}, \dfrac{5\pi}{6}\right)$

$\dfrac{dx}{d\theta} = (-1 + \sin\theta)\sin\theta - \cos\theta\cos\theta$

$\qquad = -\sin\theta + \sin^2\theta + \sin^2\theta - 1$

$\qquad = 2\sin^2\theta - \sin\theta - 1$

$\qquad = (2\sin\theta + 1)(\sin\theta - 1) = 0$

$\sin\theta = 1$ or $\sin\theta = -\dfrac{1}{2} \implies \theta = \dfrac{\pi}{2}, \dfrac{7\pi}{6}, \dfrac{11\pi}{6}$

Vertical tangents: $\left(\dfrac{3}{2}, \dfrac{7\pi}{6}\right), \left(\dfrac{3}{2}, \dfrac{11\pi}{6}\right)$

67. $r = 2\csc\theta + 3$

$\dfrac{dy}{d\theta} = (2\csc\theta + 3)\cos\theta + (-2\csc\theta\cot\theta)\sin\theta$

$\qquad = 3\cos\theta = 0$

$\theta = \dfrac{\pi}{2}, \dfrac{3\pi}{2}$

Horizontal: $\left(5, \dfrac{\pi}{2}\right), \left(1, \dfrac{3\pi}{2}\right)$

69. $r = 4\sin\theta\cos^2\theta$

Horizontal tangents:

$(r, \theta) = (0, 0), (1.4142, 0.7854),$

$(1.4142, 2.3562)$

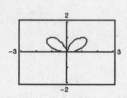

71. $r = 2\csc\theta + 5$

Horizontal tangents:

$(r, \theta) = \left(7, \dfrac{\pi}{2}\right), \left(3, \dfrac{3\pi}{2}\right)$

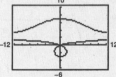
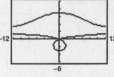

73.
$$r = 3\sin\theta$$
$$r^2 = 3r\sin\theta$$
$$x^2 + y^2 = 3y$$
$$x^2 + \left(y - \dfrac{3}{2}\right)^2 = \dfrac{9}{4}$$

Circle $r = \dfrac{3}{2}$

Center: $\left(0, \dfrac{3}{2}\right)$

Tangent at the pole: $\theta = 0$

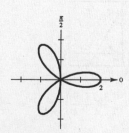

75. $r = 2(1 - \sin\theta)$

Cardioid

Symmetric to y-axis, $\theta = \dfrac{\pi}{2}$

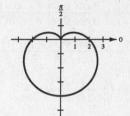

77. $r = 2\cos(3\theta)$

Rose curve with three petals

Symmetric to the polar axis

Relative extrema: $(2, 0), \left(-2, \dfrac{\pi}{3}\right), \left(2, \dfrac{2\pi}{3}\right)$

θ	0	$\dfrac{\pi}{6}$	$\dfrac{\pi}{4}$	$\dfrac{\pi}{3}$	$\dfrac{\pi}{2}$	$\dfrac{2\pi}{3}$	$\dfrac{5\pi}{6}$	π
r	2	0	$-\sqrt{2}$	-2	0	2	0	-2

Tangents at the pole: $\theta = \dfrac{\pi}{6}, \dfrac{\pi}{2}, \dfrac{5\pi}{6}$

79. $r = 3 \sin 2\theta$

Rose curve with four petals

Symmetric to the polar axis, $\theta = \dfrac{\pi}{2}$, and pole

Relative extrema: $\left(\pm 3, \dfrac{\pi}{4}\right), \left(\pm 3, \dfrac{5\pi}{4}\right)$

Tangents at the pole: $\theta = 0, \dfrac{\pi}{2}$

($\theta = \pi, \ 3\pi/2$ give the same tangents.)

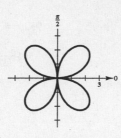

81. $r = 5$

Circle radius: 5

$x^2 + y^2 = 25$

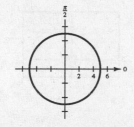

83. $r = 4(1 + \cos \theta)$

Cardioid

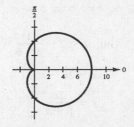

85. $r = 3 - 2 \cos \theta$

Limaçon

Symmetric to polar axis

θ	0	$\dfrac{\pi}{3}$	$\dfrac{\pi}{2}$	$\dfrac{2\pi}{3}$	π
r	1	2	3	4	5

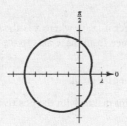

87. $r = 3 \csc \theta$

$r \sin \theta = 3$

$y = 3$

Horizontal line

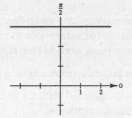

89. $r = 2\theta$

Spiral of Archimedes

Symmetric to $\theta = \dfrac{\pi}{2}$

θ	0	$\dfrac{\pi}{4}$	$\dfrac{\pi}{2}$	$\dfrac{3\pi}{4}$	π	$\dfrac{5\pi}{4}$	$\dfrac{3\pi}{2}$
r	0	$\dfrac{\pi}{2}$	π	$\dfrac{3\pi}{2}$	2π	$\dfrac{5\pi}{2}$	3π

Tangent at the pole: $\theta = 0$

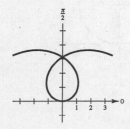

91. $r^2 = 4 \cos(2\theta)$

$r = 2\sqrt{\cos 2\theta}, \quad 0 \le \theta \le 2\pi$

Lemniscate

Symmetric to the polar axis, $\theta = \dfrac{\pi}{2}$, and pole

Relative extrema: $(\pm 2, 0)$

θ	0	$\dfrac{\pi}{6}$	$\dfrac{\pi}{4}$
r	± 2	$\pm\sqrt{2}$	0

Tangents at the pole: $\theta = \dfrac{\pi}{4}, \dfrac{3\pi}{4}$

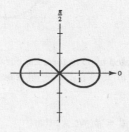

93. Since

$$r = 2 - \sec \theta = 2 - \frac{1}{\cos \theta},$$

the graph has polar axis symmetry and the tangents at the pole are

$$\theta = \frac{\pi}{3}, \frac{-\pi}{3}.$$

Furthermore,

$$r \Rightarrow -\infty \text{ as } \theta \Rightarrow \frac{\pi}{2}^-$$

$$r \Rightarrow \infty \text{ as } \theta \Rightarrow -\frac{\pi}{2}^+.$$

Also, $r = 2 - \dfrac{1}{\cos \theta} = 2 - \dfrac{r}{r \cos \theta} = 2 - \dfrac{r}{x}$

$$rx = 2x - r$$

$$r = \frac{2x}{1 + x}.$$

Thus, $r \Rightarrow \pm\infty$ as $x \Rightarrow -1$.

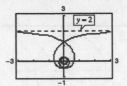

95. $r = \dfrac{2}{\theta}$

Hyperbolic spiral

$r \Rightarrow \infty$ as $\theta \Rightarrow 0$

$$r = \frac{2}{\theta} \Rightarrow \theta = \frac{2}{r} = \frac{2 \sin \theta}{r \sin \theta} = \frac{2 \sin \theta}{y}$$

$$y = \frac{2 \sin \theta}{\theta}$$

$$\lim_{\theta \to 0} \frac{2 \sin \theta}{\theta} = \lim_{\theta \to 0} \frac{2 \cos \theta}{1} = 2$$

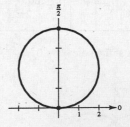

97. The rectangular coordinate system consists of all points of the form (x, y) where x is the directed distance from the y-axis to the point, and y is the directed distance from the x-axis to the point. Every point has a unique representation.

The polar coordinate system uses (r, θ) to designate the location of a point.

r is the directed distance to the origin and θ is the angle the point makes with the positive x-axis, measured clockwise.

Points do not have a unique polar representation.

99. $r = a$, circle

$\theta = b$, line

101. $r = 4 \sin \theta$

(a) $0 \le \theta \le \dfrac{\pi}{2}$

(b) $\dfrac{\pi}{2} \le \theta \le \pi$

(c) $-\dfrac{\pi}{2} \le \theta \le \dfrac{\pi}{2}$

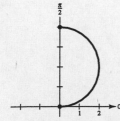   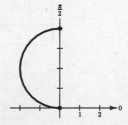   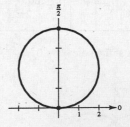

103. Let the curve $r = f(\theta)$ be rotated by ϕ to form the curve $r = g(\theta)$.

If (r_1, θ_1) is a point on $r = f(\theta)$, then $(r_1, \theta_1 + \phi)$ is on $r = g(\theta)$. That is,

$$g(\theta_1 + \phi) = r_1 = f(\theta_1).$$

Letting $\theta = \theta_1 + \phi$, or $\theta_1 = \theta - \phi$, we see that

$$g(\theta) = g(\theta_1 + \phi) = f(\theta_1) = f(\theta - \phi).$$

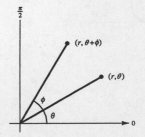

105. $r = 2 - \sin \theta$

(a) $r = 2 - \sin\left(\theta - \dfrac{\pi}{4}\right) = 2 - \dfrac{\sqrt{2}}{2}(\sin \theta - \cos \theta)$

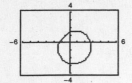

(b) $r = 2 - \sin\left(\theta - \dfrac{\pi}{2}\right) = 2 - (-\cos \theta) = 2 + \cos \theta$

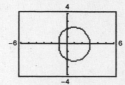

(c) $r = 2 - \sin(\theta - \pi) = 2 - (-\sin \theta) = 2 + \sin \theta$

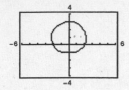

(d) $r = 2 - \sin\left(\theta - \dfrac{3\pi}{2}\right) = 2 - \cos \theta$

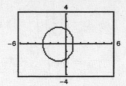

107. (a) $r = 1 - \sin \theta$

(b) $r = 1 - \sin\left(\theta - \dfrac{\pi}{4}\right)$

Rotate the graph of

$r = 1 - \sin \theta$

through the angle $\pi/4$.

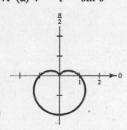

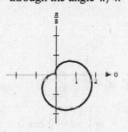

109. $\tan \psi = \dfrac{r}{dr/d\theta} = \dfrac{2(1 - \cos \theta)}{2 \sin \theta}$

At $\theta = \pi$, $\tan \psi$ is undefined $\Rightarrow \psi = \dfrac{\pi}{2}$.

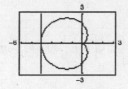

111. $r = 2 \cos 3\theta$

$\tan \psi = \dfrac{r}{dr/d\theta} = \dfrac{2 \cos 3\theta}{-6 \sin 3\theta} = -\dfrac{1}{3} \cot 3\theta$

At $\theta = \dfrac{\pi}{4}$, $\tan \psi = -\dfrac{1}{3} \cot\left(\dfrac{3\pi}{4}\right) = \dfrac{1}{3}$.

$\psi = \arctan\left(\dfrac{1}{3}\right) \approx 18.4°$

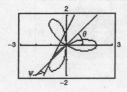

113. $r = \dfrac{6}{1 - \cos \theta} = 6(1 - \cos \theta)^{-1}$

$\Rightarrow \dfrac{dr}{d\theta} = \dfrac{6 \sin \theta}{(1 - \cos \theta)^2}$

$\tan \psi = \dfrac{r}{\dfrac{dr}{d\theta}} = \dfrac{\dfrac{6}{(1 - \cos \theta)}}{\dfrac{-6 \sin \theta}{(1 - \cos \theta)^2}} = \dfrac{1 - \cos \theta}{-\sin \theta}$

At $\theta = \dfrac{2\pi}{3}$, $\tan \psi = \dfrac{1 - \left(-\dfrac{1}{2}\right)}{-\dfrac{\sqrt{3}}{2}} = -\sqrt{3}$.

$\psi = \dfrac{\pi}{3}, (60°)$

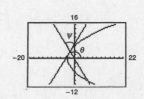

115. True

117. True

Section 10.5   Area and Arc Length in Polar Coordinates

1. $A = \dfrac{1}{2}\displaystyle\int_{\alpha}^{\beta} [f(\theta)]^2\, d\theta$

$= \dfrac{1}{2}\displaystyle\int_{\pi/2}^{\pi} (2\sin\theta)^2\, d\theta$

$= 2\displaystyle\int_{\pi/2}^{\pi} \sin^2\theta\, d\theta$

3. $A = \dfrac{1}{2}\displaystyle\int_{\alpha}^{\beta} [f(\theta)]^2\, d\theta$

$= \dfrac{1}{2}\displaystyle\int_{\pi/2}^{3\pi/2} (1 - \sin\theta)^2\, d\theta$

5. (a) $r = 8\sin\theta$

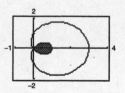

$A = \pi(4)^2 = 16\pi$

(b) $A = 2\left(\dfrac{1}{2}\right)\displaystyle\int_{0}^{\pi/2} [8\sin\theta]^2\, d\theta$

$= 64\displaystyle\int_{0}^{\pi/2} \sin^2\theta\, d\theta$

$= 32\displaystyle\int_{0}^{\pi/2} (1 - \cos 2\theta)\, d\theta$

$= 32\left[\theta - \dfrac{\sin 2\theta}{2}\right]_{0}^{\pi/2} = 16\pi$

7. $A = 2\left[\dfrac{1}{2}\displaystyle\int_{0}^{\pi/6} (2\cos 3\theta)^2\, d\theta\right] = 2\left[\theta + \dfrac{1}{6}\sin 6\theta\right]_{0}^{\pi/6} = \dfrac{\pi}{3}$

9. $A = 2\left[\dfrac{1}{2}\displaystyle\int_{0}^{\pi/4} (\cos 2\theta)^2\, d\theta\right]$

$= \dfrac{1}{2}\left[\theta + \dfrac{1}{4}\sin 4\theta\right]_{0}^{\pi/4} = \dfrac{\pi}{8}$

11. $A = 2\left[\dfrac{1}{2}\displaystyle\int_{-\pi/2}^{\pi/2} (1 - \sin\theta)^2\, d\theta\right]$

$= \left[\dfrac{3}{2}\theta + 2\cos\theta - \dfrac{1}{4}\sin 2\theta\right]_{-\pi/2}^{\pi/2} = \dfrac{3\pi}{2}$

13. $A = 2\dfrac{1}{2}\left[\displaystyle\int_{2\pi/3}^{\pi} (1 + 2\cos\theta)^2\, d\theta\right]$

$= \left[3\theta + 4\sin\theta + \sin 2\theta\right]_{2\pi/3}^{\pi} = \dfrac{2\pi - 3\sqrt{3}}{2}$

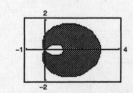

15. The area inside the outer loop is

$2\left[\dfrac{1}{2}\displaystyle\int_{0}^{2\pi/3} (1 + 2\cos\theta)^2\, d\theta\right] = \left[3\theta + 4\sin\theta + \sin 2\theta\right]_{0}^{2\pi/3} = \dfrac{4\pi + 3\sqrt{3}}{2}.$

From the result of Exercise 13, the area between the loops is

$A = \left(\dfrac{4\pi + 3\sqrt{3}}{2}\right) - \left(\dfrac{2\pi - 3\sqrt{3}}{2}\right) = \pi + 3\sqrt{3}.$

17. $r = 1 + \cos\theta$

$r = 1 - \cos\theta$

Solving simultaneously,

$$1 + \cos\theta = 1 - \cos\theta$$

$$2\cos\theta = 0$$

$$\theta = \frac{\pi}{2}, \frac{3\pi}{2}.$$

Replacing r by $-r$ and θ by $\theta + \pi$ in the first equation and solving, $-1 + \cos\theta = 1 - \cos\theta$, $\cos\theta = 1$, $\theta = 0$. Both curves pass through the pole, $(0, \pi)$, and $(0, 0)$, respectively.

Points of intersection: $\left(1, \frac{\pi}{2}\right), \left(1, \frac{3\pi}{2}\right), (0, 0)$

19. $r = 1 + \cos\theta$

$r = 1 - \sin\theta$

Solving simultaneously,

$$1 + \cos\theta = 1 - \sin\theta$$

$$\cos\theta = -\sin\theta$$

$$\tan\theta = -1$$

$$\theta = \frac{3\pi}{4}, \frac{7\pi}{4}.$$

Replacing r by $-r$ and θ by $\theta + \pi$ in the first equation and solving, $-1 + \cos\theta = 1 - \sin\theta$, $\sin\theta + \cos\theta = 2$, which has no solution. Both curves pass through the pole, $(0, \pi)$, and $(0, \pi/2)$, respectively.

Points of intersection:

$$\left(\frac{2 - \sqrt{2}}{2}, \frac{3\pi}{4}\right), \left(\frac{2 + \sqrt{2}}{2}, \frac{7\pi}{4}\right), (0, 0)$$

21. $r = 4 - 5\sin\theta$

$r = 3\sin\theta$

Solving simultaneously,

$$4 - 5\sin\theta - 3\sin\theta$$

$$\sin\theta - \frac{1}{2}$$

$$\theta = \frac{\pi}{6}, \frac{5\pi}{6}.$$

Both curves pass through the pole, $(0, \arcsin 4/5)$, and $(0, 0)$, respectively.

Points of intersection: $\left(\frac{3}{2}, \frac{\pi}{6}\right), \left(\frac{3}{2}, \frac{5\pi}{6}\right), (0, 0)$

23. $r = \frac{\theta}{2}$

$r - 2$

Solving simultaneously, we have

$\theta/2 = 2, \theta = 4$.

Points of intersection:

$(2, 4), (-2, -4)$

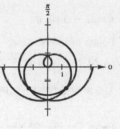

25. $r = 4\sin 2\theta$

$r = 2$

$r = 4\sin 2\theta$ is the equation of a rose curve with four petals and is symmetric to the polar axis, $\theta = \pi/2$, and the pole. Also, $r = 2$ is the equation of a circle of radius 2 centered at the pole. Solving simultaneously,

$4\sin 2\theta = 2$

$$2\theta = \frac{\pi}{6}, \frac{5\pi}{6}$$

$$\theta = \frac{\pi}{12}, \frac{5\pi}{12}.$$

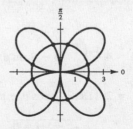

Therefore, the points of intersection for one petal are $(2, \pi/12)$ and $(2, 5\pi/12)$. By symmetry, the other points of intersection are $(2, 7\pi/12)$, $(2, 11\pi/12)$, $(2, 13\pi/12)$, $(2, 17\pi/12)$, $(2, 19\pi/12)$, and $(2, 23\pi/12)$.

27. $r = 2 + 3\cos\theta$

$$r = \frac{\sec\theta}{2}$$

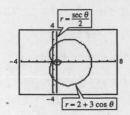

The graph of $r = 2 + 3\cos\theta$ is a limaçon with an inner loop ($b > a$) and is symmetric to the polar axis. The graph of $r = (\sec\theta)/2$ is the vertical line $x = 1/2$. Therefore, there are four points of intersection. Solving simultaneously,

$$2 + 3\cos\theta = \frac{\sec\theta}{2}$$

$$6\cos^2\theta + 4\cos\theta - 1 = 0$$

$$\cos\theta = \frac{-2 \pm \sqrt{10}}{6}$$

$$\theta = \arccos\left(\frac{-2 + \sqrt{10}}{6}\right) \approx 1.376$$

$$\theta = \arccos\left(\frac{-2 - \sqrt{10}}{6}\right) \approx 2.6068.$$

Points of intersection: $(-0.581, \pm 2.607)$, $(2.581, \pm 1.376)$

29. $r = \cos\theta$

$r = 2 - 3\sin\theta$

Points of intersection:

$(0, 0)$, $(0.935, 0.363)$, $(0.535, -1.006)$

The graphs reach the pole at different times (θ values).

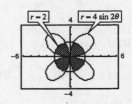

31. From Exercise 25, the points of intersection for one petal are $(2, \pi/12)$ and $(2, 5\pi/12)$. The area within one petal is

$$A = \frac{1}{2}\int_0^{\pi/12} (4\sin 2\theta)^2\, d\theta + \frac{1}{2}\int_{\pi/12}^{5\pi/12} (2)^2\, d\theta + \frac{1}{2}\int_{5\pi/12}^{\pi/2} (4\sin 2\theta)^2\, d\theta$$

$$= 16\int_0^{\pi/12} \sin^2(2\theta)\, d\theta + 2\int_{\pi/12}^{5\pi/12} d\theta \quad \text{(by symmetry of the petal)}$$

$$= 8\left[\theta - \frac{1}{4}\sin 4\theta\right]_0^{\pi/12} + \left[2\theta\right]_{\pi/12}^{5\pi/12} = \frac{4\pi}{3} - \sqrt{3}.$$

Total area $= 4\left(\dfrac{4\pi}{3} - \sqrt{3}\right) = \dfrac{16\pi}{3} - 4\sqrt{3} = \dfrac{4}{3}\left(4\pi - 3\sqrt{3}\right)$

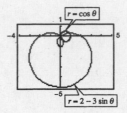

33. $A = 4\left[\dfrac{1}{2}\displaystyle\int_0^{\pi/2} (3 - 2\sin\theta)^2\, d\theta\right]$

$= 2\left[11\theta + 12\cos\theta - \sin(2\theta)\right]_0^{\pi/2} = 11\pi - 24$

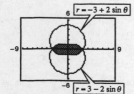

35. $A = 2\left[\dfrac{1}{2}\displaystyle\int_0^{\pi/6} (4\sin\theta)^2\, d\theta + \dfrac{1}{2}\int_{\pi/6}^{\pi/2} (2)^2\, d\theta\right]$

$= 16\left[\dfrac{1}{2}\theta - \dfrac{1}{4}\sin(2\theta)\right]_0^{\pi/6} + \left[4\theta\right]_{\pi/6}^{\pi/2}$

$= \dfrac{8\pi}{3} - 2\sqrt{3} = \dfrac{2}{3}\left(4\pi - 3\sqrt{3}\right)$

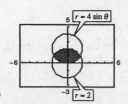

37. $A = 2\left[\dfrac{1}{2}\displaystyle\int_0^\pi [a(1 + \cos \theta)]^2\, d\theta\right] - \dfrac{a^2\pi}{4}$

$\qquad = a^2\left[\dfrac{3}{2}\theta + 2\sin\theta + \dfrac{\sin 2\theta}{4}\right]_0^\pi - \dfrac{a^2\pi}{4}$

$\qquad = \dfrac{3a^2\pi}{2} - \dfrac{a^2\pi}{4} = \dfrac{5a^2\pi}{4}$

39. $A = \dfrac{\pi a^2}{8} + \dfrac{1}{2}\displaystyle\int_{\pi/2}^\pi [a(1 + \cos\theta)]^2\, d\theta$

$\qquad = \dfrac{\pi a^2}{8} + \dfrac{a^2}{2}\displaystyle\int_{\pi/2}^\pi \left(\dfrac{3}{2} + 2\cos\theta + \dfrac{\cos 2\theta}{2}\right) d\theta$

$\qquad = \dfrac{\pi a^2}{8} + \dfrac{a^2}{2}\left[\dfrac{3}{2}\theta + 2\sin\theta + \dfrac{\sin 2\theta}{4}\right]_{\pi/2}^\pi$

$\qquad = \dfrac{\pi a^2}{8} + \dfrac{a^2}{2}\left[\dfrac{3\pi}{2} - \dfrac{3\pi}{4} - 2\right] = \dfrac{a^2}{2}[\pi - 2]$

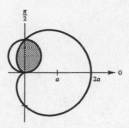

41. (a) $r = a\cos^2\theta$

$\qquad r^3 = ar^2\cos^2\theta$

$\qquad (x^2 + y^2)^{3/2} = ax^2$

(b)

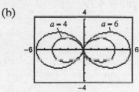

(c) $A = 4\left(\dfrac{1}{2}\right)\displaystyle\int_0^{\pi/2}\left[(6\cos^2\theta)^2 - (4\cos^2\theta)^2\right] d\theta$

$\qquad = 40\displaystyle\int_0^{\pi/2}\cos^4\theta\, d\theta$

$\qquad = 10\displaystyle\int_0^{\pi/2}(1 + \cos 2\theta)^2\, d\theta$

$\qquad = 10\displaystyle\int_0^{\pi/2}\left(1 + 2\cos 2\theta + \dfrac{1 - \cos 4\theta}{2}\right) d\theta$

$\qquad = 10\left[\dfrac{3}{2}\theta + \sin 2\theta + \dfrac{1}{8}\sin 4\theta\right]_0^{\pi/2} = \dfrac{15\pi}{2}$

43. $r = a\cos(n\theta)$

For $n = 1$:

$r = a\cos\theta$

$A = \pi\left(\dfrac{a}{2}\right)^2 = \dfrac{\pi a^2}{4}$

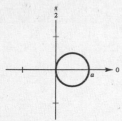

For $n = 2$:

$r = a\cos 2\theta$

$A = 8\left(\dfrac{1}{2}\right)\displaystyle\int_0^{\pi/4}(a\cos 2\theta)^2\, d\theta = \dfrac{\pi a^2}{2}$

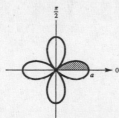

For $n = 3$:

$r = a\cos 3\theta$

$A = 6\left(\dfrac{1}{2}\right)\displaystyle\int_0^{\pi/6}(a\cos 3\theta)^2\, d\theta = \dfrac{\pi a^2}{4}$

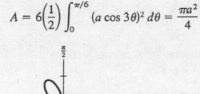

For $n = 4$:

$r = a\cos 4\theta$

$A = 16\left(\dfrac{1}{2}\right)\displaystyle\int_0^{\pi/8}(a\cos 4\theta)^2\, d\theta = \dfrac{\pi a^2}{2}$

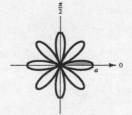

In general, the area of the region enclosed by $r = a\cos(n\theta)$ for $n = 1, 2, 3, \ldots$ is $(\pi a^2)/4$ if n is odd and is $(\pi a^2)/2$ if n is even.

45. $r = a$

$r' = 0$

$$s = \int_0^{2\pi} \sqrt{a^2 + 0^2}\, d\theta = \Big[a\theta\Big]_0^{2\pi} = 2\pi a$$

(circumference of circle of radius a)

47. $r = 1 + \sin\theta$

$r' = \cos\theta$

$$s = 2\int_{\pi/2}^{3\pi/2} \sqrt{(1 + \sin\theta)^2 + (\cos\theta)^2}\, d\theta$$

$$= 2\sqrt{2}\int_{\pi/2}^{3\pi/2} \sqrt{1 + \sin\theta}\, d\theta$$

$$= 2\sqrt{2}\int_{\pi/2}^{3\pi/2} \frac{-\cos\theta}{\sqrt{1 - \sin\theta}}\, d\theta$$

$$= \Big[4\sqrt{2}\sqrt{1 - \sin\theta}\Big]_{\pi/2}^{3\pi/2}$$

$$= 4\sqrt{2}\left(\sqrt{2} - 0\right) = 8$$

49. $r = 2\theta, \; 0 \le \theta \le \dfrac{\pi}{2}$

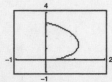

Length ≈ 4.16

51. $r = \dfrac{1}{\theta}, \; \pi \le \theta \le 2\pi$

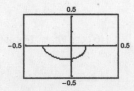

Length ≈ 0.71

53. $r = \sin(3\cos\theta), \; 0 \le \theta \le \pi$

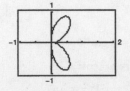

Length ≈ 4.39

55. $r = 6\cos\theta$

$r' = -6\sin\theta$

$$S = 2\pi\int_0^{\pi/2} 6\cos\theta\sin\theta\sqrt{36\cos^2\theta + 36\sin^2\theta}\, d\theta$$

$$= 72\pi\int_0^{\pi/2} \sin\theta\cos\theta\, d\theta$$

$$= \Big[36\pi\sin^2\theta\Big]_0^{\pi/2}$$

$$= 36\pi$$

57. $r = e^{a\theta}$

$r' = ae^{a\theta}$

$$S = 2\pi\int_0^{\pi/2} e^{a\theta}\cos\theta\sqrt{(e^{a\theta})^2 + (ae^{a\theta})^2}\, d\theta$$

$$= 2\pi\sqrt{1 + a^2}\int_0^{\pi/2} e^{2a\theta}\cos\theta\, d\theta$$

$$= 2\pi\sqrt{1 + a^2}\left[\frac{e^{2a\theta}}{4a^2 + 1}(2a\cos\theta + \sin\theta)\right]_0^{\pi/2}$$

$$= \frac{2\pi\sqrt{1 + a^2}}{4a^2 + 1}(e^{\pi a} - 2a)$$

59. $r = 4\cos 2\theta$

$r' = -8\sin 2\theta$

$$S = 2\pi\int_0^{\pi/4} 4\cos 2\theta\sin\theta\sqrt{16\cos^2 2\theta + 64\sin^2\theta\, 2\theta}\, d\theta$$

$$= 32\pi\int_0^{\pi/4} \cos 2\theta\sin\theta\sqrt{\cos^2 2\theta + 4\sin^2 2\theta}\, d\theta \approx 21.87$$

61. $\quad$ Area $= \dfrac{1}{2}\int_\alpha^\beta [f(\theta)]^2\, d\theta = \dfrac{1}{2}\int_\alpha^\beta r^2\, d\theta$

$\quad$ Arc length $= \displaystyle\int_\alpha^\beta \sqrt{f(\theta)^2 + f'(\theta)^2}\, d\theta = \int_\alpha^\beta \sqrt{r^2 + \left(\dfrac{dr}{d\theta}\right)^2}\, d\theta$

63. (a) is correct: $s \approx 33.124$.

65. Revolve $r = 2$ about the line $r = 5 \sec \theta$.

$f(\theta) = 2, f'(\theta) = 0$

$$S = 2\pi \int_0^{2\pi} (5 - 2 \cos \theta)\sqrt{2^2 + 0^2}\, d\theta$$

$$= 4\pi \int_0^{2\pi} (5 - 2 \cos \theta)\, d\theta$$

$$= 4\pi \left[5\theta - 2 \sin \theta \right]_0^{2\pi}$$

$$= 40\pi^2$$

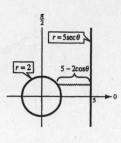

67. $r = 8 \cos \theta, 0 \le \theta \le \pi$

(a) $A = \dfrac{1}{2} \displaystyle\int_0^\pi r^2\, d\theta = \dfrac{1}{2} \displaystyle\int_0^\pi 64 \cos^2 \theta\, d\theta = 32 \displaystyle\int_0^\pi \dfrac{1 + \cos 2\theta}{2}\, d\theta = 16 \left[\theta + \dfrac{\sin 2\theta}{2} \right]_0^\pi = 16\pi$

(Area circle $= \pi r^2 = \pi 4^2 = 16\pi$)

(b)

θ	0.2	0.4	0.6	0.8	1.0	1.2	1.4
A	6.32	12.14	17.06	20.80	23.27	24.60	25.08

(c), (d) For $\frac{1}{4}$ of area $(4\pi \approx 12.57)$: 0.42

For $\frac{1}{2}$ of area $(8\pi \approx 25.13)$: 1.57 $(\pi/2)$

For $\frac{3}{4}$ of area $(12\pi \approx 37.70)$: 2.73

(e) No, it does not depend on the radius.

69. $\qquad\qquad r = a \sin \theta + b \cos \theta$

$$r^2 = ar \sin \theta + br \cos \theta$$

$$x^2 + y^2 = ay + bx$$

$x^2 + y^2 - bx - ay = 0$ represents a circle.

71. (a) $r = \theta, \theta \ge 0$

As a increases, the spiral opens more rapidly. If $\theta < 0$, the spiral is reflected about the y-axis.

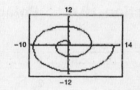

(b) $r = a\theta, \theta \ge 0$, crosses the polar axis for $\theta = n\pi$, n and integer. To see this

$$r = a\theta \implies r \sin \theta = y = a\theta \sin \theta = 0$$

for $\theta = n\pi$. The points are $(r, \theta) = (an\pi, n\pi)$, $n = 1, 2, 3, \ldots$.

(c) $f(\theta) = \theta, f'(\theta) = 1$

$$s = \int_0^{2\pi} \sqrt{\theta^2 + 1}\, d\theta = \frac{1}{2}\left[\ln\left(\sqrt{x^2 + 1} + x\right) + x\sqrt{x^2 + 1}\right]_0^{2\pi}$$

$$= \frac{1}{2} \ln\left(\sqrt{4\pi^2 + 1} + 2\pi\right) + \pi\sqrt{4\pi^2 + 1} \approx 21.2563$$

(d) $A = \dfrac{1}{2} \displaystyle\int_\alpha^\beta r^2\, dr$

$$= \frac{1}{2} \int_0^{2\pi} \theta^2\, d\theta$$

$$= \frac{\theta^3}{6}\bigg]_0^{2\pi} = \frac{4}{3}\pi^3$$

73. The smaller circle has equation $r = a \cos \theta$.

The area of the shaded lune is:

$$A = 2\left(\frac{1}{2}\right)\int_0^{\pi/4} [(a \cos \theta)^2 - 1] \, d\theta$$

$$= \int_0^{\pi/4} \left[\frac{a^2}{2}(1 + \cos 2\theta) - 1\right] d\theta$$

$$= \left[\frac{a^2}{2}\left(\theta + \frac{\sin 2\theta}{2}\right) - \theta\right]_0^{\pi/4}$$

$$= \frac{a^2}{2}\left(\frac{\pi}{4} + \frac{1}{2}\right) - \frac{\pi}{4}$$

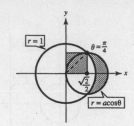

This equals the area of the square, $\left(\dfrac{\sqrt{2}}{2}\right)^2 = \dfrac{1}{2}$.

$$\frac{a^2}{2}\left(\frac{\pi}{4} + \frac{1}{2}\right) - \frac{\pi}{4} = \frac{1}{2}$$

$$\pi a^2 + 2a^2 - 2\pi - 4 = 0$$

$$a^2 = \frac{4 + 2\pi}{2 + \pi} = 2$$

$$a = \sqrt{2}$$

Smaller circle: $r = \sqrt{2} \cos \theta$

75. False. $f(\theta) = 1$ and $g(\theta) = -1$ have the same graphs.

77. In parametric form,

$$s = \int_a^b \sqrt{\left(\frac{dx}{dt}\right)^2 + \left(\frac{dy}{dt}\right)^2} \, dt.$$

Using θ instead of t, we have $x = r \cos \theta = f(\theta) \cos \theta$ and $y = r \sin \theta = f(\theta) \sin \theta$. Thus,

$$\frac{dx}{d\theta} = f'(\theta) \cos \theta - f(\theta) \sin \theta \text{ and } \frac{dy}{d\theta} = f'(\theta) \sin \theta + f(\theta) \cos \theta.$$

It follows that

$$\left(\frac{dx}{d\theta}\right)^2 + \left(\frac{dy}{d\theta}\right)^2 = [f(\theta)]^2 + [f'(\theta)]^2.$$

Therefore, $s = \displaystyle\int_\alpha^\beta \sqrt{[f(\theta)]^2 + [f'(\theta)]^2} \, d\theta.$

Section 10.6   Polar Equations of Conics and Kepler's Laws

1. $r = \dfrac{2e}{1 + e \cos \theta}$

(a) $e = 1, r = \dfrac{2}{1 + \cos \theta}$, parabola

(b) $e = 0.5, r = \dfrac{1}{1 + 0.5 \cos \theta} = \dfrac{2}{2 + \cos \theta}$, ellipse

(c) $e = 1.5, r = \dfrac{3}{1 + 1.5 \cos \theta} = \dfrac{6}{2 + 3 \cos \theta}$, hyperbola

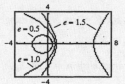

3. $r = \dfrac{2e}{1 - e \sin \theta}$

(a) $e = 1, r = \dfrac{2}{1 - \sin \theta}$, parabola

(b) $e = 0.5, r = \dfrac{1}{1 - 0.5 \sin \theta} = \dfrac{2}{2 - \sin \theta}$, ellipse

(c) $e = 1.5, r = \dfrac{3}{1 - 1.5 \sin \theta} = \dfrac{6}{2 - 3 \sin \theta}$, hyperbola

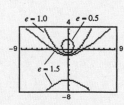

5. $r = \dfrac{4}{1 + e \sin \theta}$

(a)

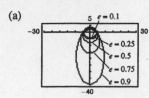

The conic is an ellipse. As $e \to 1^-$, the ellipse becomes more elliptical, and as $e \to 0^+$, it becomes more circular.

(c)

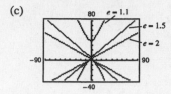

The conic is a hyperbola. As $e \to 1^+$, the hyperbolas opens more slowly, and as $e \to \infty$, they open more rapidly.

(b)

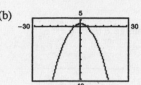

The conic is a parabola.

7. Parabola; Matches (c)

9. Hyperbola; Matches (a)

11. Ellipse; Matches (b)

13. $r = \dfrac{-1}{1 - \sin \theta}$

Parabola because $e = 1$; $d = -1$

Distance from pole to directrix: $|d| = 1$

Directrix: $y = 1$

Vertex: $(r, \theta) = \left(-\dfrac{1}{2}, \dfrac{3\pi}{2}\right)$

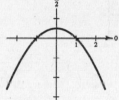

15. $r = \dfrac{6}{2 + \cos \theta} = \dfrac{3}{1 + (1/2) \cos \theta}$

Ellipse because $e = 1/2$; $d = 6$

Directrix: $x = 6$

Distance from pole to directrix: $|d| = 6$

Vertices: $(r, \theta) = (2, 0), (6, \pi)$

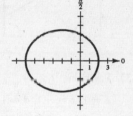

17. $r(2 + \sin \theta) = 4$

$r = \dfrac{4}{2 + \sin \theta} = \dfrac{2}{1 + (1/2) \sin \theta}$

Ellipse because $e = 1/2$; $d = 4$

Directrix: $y = 4$

Distance from pole to directrix: $|d| = 4$

Vertices: $(r, \theta) = (4/3, \pi/2), (4, 3\pi/2)$

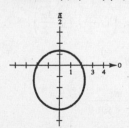

19. $r = \dfrac{5}{-1 + 2 \cos \theta} = \dfrac{-5}{1 - 2 \cos \theta}$

Hyperbola because $e - 2 > 1$; $d = -5/2$

Directrix: $x = 5/2$

Distance from pole to directrix: $|d| = 5/2$

Vertices: $(r, \theta) = (5, 0), (-5/3, \pi)$

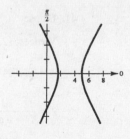

21. $r = \dfrac{3}{2 + 6 \sin \theta} = \dfrac{3/2}{1 + 3 \sin \theta}$

Hyperbola because $e = 3 > 0$; $d = 1/2$

Directrix: $y = 1/2$

Distance from pole to directrix: $|d| = 1/2$

Vertices: $(r, \theta) = (3/8, \pi/2), (-3/4, 3\pi/2)$

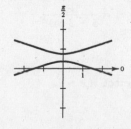

23. $r = 3/(-4 + 2 \sin \theta)$

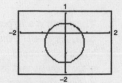

Ellipse

25. $r = -1/(1 - \cos \theta)$

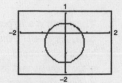

Parabola

27. $r = \dfrac{-1}{1 - \sin\left(\theta - \dfrac{\pi}{4}\right)}$

Rotate the graph of

$r = \dfrac{-1}{1 - \sin \theta}$

counterclockwise through the angle $\dfrac{\pi}{4}$.

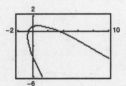

29. $r = \dfrac{6}{2 + \cos\left(\theta + \dfrac{\pi}{6}\right)}$

Rotate the graph of

$r = \dfrac{6}{2 + \cos \theta}$

clockwise through the angle $\dfrac{\pi}{6}$.

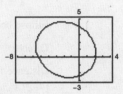

31. Change θ to $\theta + \dfrac{\pi}{4}$:

$r = \dfrac{5}{5 + 3 \cos\left(\theta + \dfrac{\pi}{4}\right)}$

33. Parabola

$e = 1, x = -1, d = 1$

$r = \dfrac{ed}{1 - e \cos \theta} = \dfrac{1}{1 - \cos \theta}$

35. Ellipse

$e = \dfrac{1}{2}, y = 1, d = 1$

$r = \dfrac{ed}{1 + e \sin \theta}$

$= \dfrac{1/2}{1 + (1/2) \sin \theta}$

$= \dfrac{1}{2 + \sin \theta}$

37. Hyperbola

$e = 2, x = 1, d = 1$

$r = \dfrac{ed}{1 + e \cos \theta} = \dfrac{2}{1 + 2 \cos \theta}$

39. Parabola

Vertex: $\left(1, -\dfrac{\pi}{2}\right)$

$e = 1, d = 2, r = \dfrac{2}{1 - \sin \theta}$

41. Ellipse

Vertices: $(2, 0), (8, \pi)$

$e = \dfrac{3}{5}, d = \dfrac{16}{3}$

$r = \dfrac{ed}{1 + e \cos \theta}$

$= \dfrac{16/5}{1 + (3/5) \cos \theta}$

$= \dfrac{16}{5 + 3 \cos \theta}$

43. Hyperbola

Vertices: $\left(1, \dfrac{3\pi}{2}\right), \left(9, \dfrac{3\pi}{2}\right)$

$e = \dfrac{5}{4}, d = \dfrac{9}{5}$

$r = \dfrac{ed}{1 - e \sin \theta}$

$= \dfrac{9/4}{1 - (5/4) \sin \theta}$

$= \dfrac{9}{4 - 5 \sin \theta}$

45. Ellipse if $0 < e < 1$, parabola if $e = 1$, hyperbola if $e > 1$.

47. (a) Hyperbola ($e = 2 > 1$)

(b) Ellipse $\left(e = \dfrac{1}{10} < 1 \right)$

(c) Parabola ($e = 1$)

(d) Rotated hyperbola ($e = 3$)

49.
$$\frac{x^2}{a^2} + \frac{y^2}{b^2} = 1$$
$$x^2 b^2 + y^2 a^2 = a^2 b^2$$
$$b^2 r^2 \cos^2 \theta + a^2 r^2 \sin^2 \theta = a^2 b^2$$
$$r^2 [b^2 \cos^2 \theta + a^2(1 - \cos^2 \theta)] = a^2 b^2$$
$$r^2 [a^2 + \cos^2 \theta (b^2 - a^2)] = a^2 b^2$$
$$r^2 = \frac{a^2 b^2}{a^2 + (b^2 - a^2)\cos^2 \theta} = \frac{a^2 b^2}{a^2 - c^2 \cos^2 \theta}$$
$$= \frac{b^2}{1 - (c/a)^2 \cos^2 \theta} = \frac{b^2}{1 - e^2 \cos^2 \theta}$$

51. $a = 5, c = 4, e = \dfrac{4}{5}, b = 3$

$$r^2 = \frac{9}{1 - (16/25)\cos^2 \theta}$$

53. $a = 3, b = 4, c = 5, e = \dfrac{5}{3}$

$$r^2 = \frac{-16}{1 - (25/9)\cos^2 \theta}$$

55. $A = 2 \left[\dfrac{1}{2} \displaystyle\int_0^\pi \left(\dfrac{3}{2 - \cos \theta} \right)^2 d\theta \right]$

$$= 9 \int_0^\pi \frac{1}{(2 - \cos \theta)^2} \, d\theta \approx 10.88$$

57. Vertices: $(123{,}000 + 4000, 0) = (127{,}000, 0)$

$$(119 + 4000, \pi) = (4119, \pi)$$

$$a = \frac{127{,}000 + 4119}{2} = 65{,}559.5$$

$$c = 65{,}559.5 - 4119 = 61{,}440.5$$

$$e = \frac{c}{a} = \frac{122{,}881}{131{,}119} \approx 0.93717$$

$$r = \frac{ed}{1 - e\cos\theta}$$

$$\theta = 0: \; r = \frac{ed}{1 - e}, \quad \theta = \pi: \; r = \frac{ed}{1 + e}$$

$$2a = 2(65{,}559.5) = \frac{ed}{1 - e} + \frac{ed}{1 + e}$$

$$131{,}119 = d\left(\frac{e}{1 - e} + \frac{e}{1 + e} \right) = d\left(\frac{2e}{1 - e^2} \right)$$

$$d = \frac{131{,}119(1 - e^2)}{2e} \approx 8514.1397$$

$$r = \frac{7979.21}{1 - 0.93717\cos\theta} = \frac{1{,}046{,}226{,}000}{131{,}119 - 122{,}881\cos\theta}$$

When $\theta = 60° = \dfrac{\pi}{3}, r \approx 15{,}015$.

Distance between earth and the satellite is $r - 4000 \approx 11{,}015$ miles.

59. $a = 1.496 \times 10^8, e = 0.0167$

$$r = \frac{(1 - e^2)a}{1 - e\cos\theta} = \frac{149{,}558{,}278.1}{1 - 0.0167\cos\theta}$$

Perihelion distance: $a(1 - e) \approx 147{,}101{,}680$ km

Aphelion distance: $a(1 + e) \approx 152{,}098{,}320$ km

61. $a = 5.906 \times 10^9, e = 0.2488$

$$r = \frac{(1 - e^2)a}{1 - e\cos\theta} = \frac{5{,}540{,}410{,}095}{1 - 0.2488\cos\theta}$$

Perihelion distance: $a(1 - e) \approx 4{,}436{,}587{,}200$ km

Aphelion distance: $a(1 + e) \approx 7{,}375{,}412{,}800$ km

63. $r = \dfrac{5.540 \times 10^9}{1 - 0.2488 \cos \theta}$

(a) $A = \dfrac{1}{2} \displaystyle\int_0^{\pi/9} r^2 \, d\theta \approx 9.368 \times 10^{18} \text{ km}^2$

$$248 \left[\frac{\dfrac{1}{2}\displaystyle\int_0^{\pi/9} r^2 \, d\theta}{\dfrac{1}{2}\displaystyle\int_0^{2\pi} r^2 \, d\theta} \right] \approx 21.89 \text{ yrs}$$

(b) $\dfrac{1}{2} \displaystyle\int_\pi^{\alpha} r^2 \, d\theta = 9.368 \times 10^8$

By trial and error, $\alpha = \pi + 0.9018$

$0.9018 > \pi/9 \approx 0.3491$ because the rays in part (a) are longer than those in part (b).

(c) For part (a)

$s = \displaystyle\int_0^{\pi/9} \sqrt{r^2 + (dr/d\theta)^2} \, d\theta \approx 2.563 \times 10^9 \text{ km}$

Average per year $= \dfrac{2.563 \times 10^9}{21.89} \approx 1.17 \times 10^8 \text{ km/yr}$

For part (b)

$s = \displaystyle\int_\pi^{\pi+0.9018} \sqrt{r^2 + (dr/d\theta)^2} \, d\theta \approx 4.133 \times 10^9$

Average per year $= \dfrac{4.133 \times 10^9}{21.89} \approx 1.89 \times 10^8 \text{ km/yr}$

65. $r_1 = a + c, \; r_0 = a - c, \; r_1 - r_0 = 2c, \; r_1 + r_0 = 2a$

$e = \dfrac{c}{a} = \dfrac{r_1 - r_0}{r_1 + r_0}$

$\dfrac{1 + e}{1 - e} = \dfrac{1 + \dfrac{c}{a}}{1 - \dfrac{c}{a}} = \dfrac{a + c}{a - c} = \dfrac{r_1}{r_0}$

67. $r_1 = \dfrac{ed}{1 + \sin \theta}$ and $r_2 = \dfrac{ed}{1 - \sin \theta}$

Points of intersection: $(ed, 0), (ed, \pi)$

$r_1: \dfrac{dy}{dx} = \dfrac{\left(\dfrac{ed}{1 + \sin \theta}\right)(\cos \theta) + \left(\dfrac{-ed \cos \theta}{(1 + \sin \theta)^2}\right)(\sin \theta)}{\left(\dfrac{-ed}{1 + \sin \theta}\right)(\sin \theta) + \left(\dfrac{-ed \cos \theta}{(1 + \sin \theta)^2}\right)(\cos \theta)}$

At $(ed, 0), \dfrac{dy}{dx} = -1$. At $(ed, \pi), \dfrac{dy}{dx} = 1$.

$r_2: \dfrac{dy}{dx} = \dfrac{\left(\dfrac{ed}{1 - \sin \theta}\right)(\cos \theta) + \left(\dfrac{ed \cos \theta}{(1 - \sin \theta)^2}\right)(\sin \theta)}{\left(\dfrac{-ed}{1 - \sin \theta}\right)(\sin \theta) + \left(\dfrac{ed \cos \theta}{(1 - \sin \theta)^2}\right)(\cos \theta)}$

At $(ed, 0), \dfrac{dy}{dx} = 1$. At $(ed, \pi), \dfrac{dy}{dx} = -1$.

Therefore, at $(ed, 0)$ we have $m_1 m_2 = (-1)(1) = -1$, and at (ed, π) we have $m_1 m_2 = 1(-1) = -1$. The curves intersect at right angles.

Review Exercises for Chapter 10

1. $4x^2 + y^2 = 4$

Ellipse

Vertex: $(1, 0)$.

Matches (e)

3. $y^2 = -4x$

Parabola opening to left.

Matches (b)

5. $x^2 + 4y^2 = 4$

Ellipse

Vertex: $(0, 1)$

Matches (a)

7. $16x^2 + 16y^2 - 16x + 24y - 3 = 0$

$$\left(x^2 - x + \frac{1}{4}\right) + \left(y^2 + \frac{3}{2}y + \frac{9}{16}\right) = \frac{3}{16} + \frac{1}{4} + \frac{9}{16}$$

$$\left(x - \frac{1}{2}\right)^2 + \left(y + \frac{3}{4}\right)^2 = 1$$

Circle

Center: $\left(\frac{1}{2}, \frac{3}{4}\right)$

Radius: 1

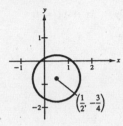

9. $3x^2 - 2y^2 + 24x + 12y + 24 = 0$

$$3(x^2 + 8x + 16) - 2(y^2 - 6y + 9) = -24 + 48 - 18$$

$$\frac{(x + 4)^2}{2} - \frac{(y - 3)^2}{3} = 1$$

Hyperbola

Center: $(-4, 3)$

Vertices: $(-4 \pm \sqrt{2}, 3)$

Asymptotes:

$$y = 3 \pm \sqrt{\frac{3}{2}}(x + 4)$$

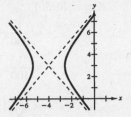

11. $3x^2 + 2y^2 - 12x + 12y + 29 = 0$

$$3(x^2 - 4x + 4) + 2(y^2 + 6y + 9) = -29 + 12 + 18$$

$$\frac{(x - 2)^2}{1/3} + \frac{(y + 3)^2}{1/2} = 1$$

Ellipse

Center: $(2, -3)$

Vertices: $\left(2, -3 \pm \frac{\sqrt{2}}{2}\right)$

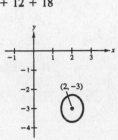

13. Vertex: $(0, 2)$

Directrix: $x = -3$

Parabola opens to the right

$p = 3$

$(y - 2)^2 = 4(3)(x - 0)$

$y^2 - 4y - 12x + 4 = 0$

15. Vertices: $(-3, 0), (7, 0)$

Foci: $(0, 0), (4, 0)$

Horizontal major axis

Center: $(2, 0)$

$a = 5, c = 2, b = \sqrt{21}$

$$\frac{(x - 2)^2}{25} + \frac{y^2}{21} = 1$$

17. Vertices: $(\pm 4, 0)$

Foci: $(\pm 6, 0)$

Center: $(0, 0)$

Horizontal transverse axis

$a = 4, c = 6, b = \sqrt{36 - 16} = 2\sqrt{5}$

$$\frac{x^2}{16} - \frac{y^2}{20} = 1$$

19. $\dfrac{x^2}{9} + \dfrac{y^2}{4} = 1, a = 3, b = 2, c = \sqrt{5}, e = \dfrac{\sqrt{5}}{3}$

By Example 5 of Section 10.1,

$$C = 12 \int_0^{\pi/2} \sqrt{1 - \left(\frac{5}{9}\right) \sin^2 \theta} \, d\theta \approx 15.87.$$

21. $y = x - 2$ has a slope of 1. The perpendicular slope is -1.

$y = x^2 - 2x + 2$

$\dfrac{dy}{dx} = 2x - 2 = -1$ when $x = \dfrac{1}{2}$ and $y = \dfrac{5}{4}$.

Perpendicular line: $\quad y - \dfrac{5}{4} = -1\left(x - \dfrac{1}{2}\right)$

$$4x + 4y - 7 = 0$$

23. $y = \dfrac{1}{200}x^2$

(a) $x^2 = 200y$

$x^2 = 4(50)y$

Focus: $(0, 50)$

(b) $\qquad y = \dfrac{1}{200}x^2$

$y' = \dfrac{1}{100}x$

$\sqrt{1 + (y')^2} = \sqrt{1 + \dfrac{x^2}{10,000}}$

$S = 2\pi \displaystyle\int_0^{100} x\sqrt{1 + \dfrac{x^2}{10,000}}\, dx \approx 38,294.49$

25. $x = 1 + 4t,\ y = 2 - 3t$

$t = \dfrac{x-1}{4} \Rightarrow y = 2 - 3\!\left(\dfrac{x-1}{4}\right)$

$y = -\dfrac{3}{4}x + \dfrac{11}{4}$

$4y + 3x - 11 = 0$

Line

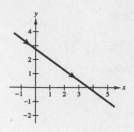

27. $x = 6\cos\theta,\ y = 6\sin\theta$

$\left(\dfrac{x}{6}\right)^2 + \left(\dfrac{y}{6}\right)^2 = 1$

$x^2 + y^2 = 36$

Circle

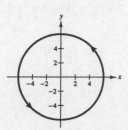

29. $x = 2 + \sec\theta,\ y = 3 + \tan\theta$

$(x - 2)^2 = \sec^2\theta = 1 + \tan^2\theta = 1 + (y - 3)^2$

$(x - 2)^2 - (y - 3)^2 = 1$

Hyperbola

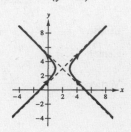

31. $x = 3 + (3 - (-2))t = 3 + 5t$

$y = 2 + (2 - 6)t = 2 - 4t$

(other answers possible)

33. $\dfrac{(x+3)^2}{16} + \dfrac{(y-4)^2}{9} = 1$

Let $\dfrac{(x+3)^2}{16} = \cos^2\theta$ and $\dfrac{(y-4)^2}{9} = \sin^2\theta$.

Then $x = -3 + 4\cos\theta$ and $y = 4 + 3\sin\theta$.

35. $x = \cos 3\theta + 5\cos\theta$

$y = \sin 3\theta + 5\sin\theta$

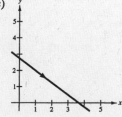

37. $x = 1 + 4t$

$y = 2 - 3t$

(a) $\dfrac{dy}{dx} = -\dfrac{3}{4}$

No horizontal tangents

(b) $t = \dfrac{x-1}{4}$

$y = 2 - \dfrac{3}{4}(x - 1) = \dfrac{-3x + 11}{4}$

(c)

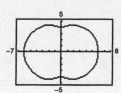

39. $x = \dfrac{1}{t}$

$y = 2t + 3$

(a) $\dfrac{dy}{dx} = \dfrac{2}{-1/t^2} = -2t^2$

No horizontal tangents, $(t \neq 0)$

(b) $t = \dfrac{1}{x}$

$y = \dfrac{2}{x} + 3$

(c)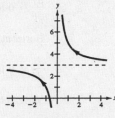

41. $x = \dfrac{1}{2t + 1}$

$y = \dfrac{1}{t^2 - 2t}$

(a) $\dfrac{dy}{dx} = \dfrac{\dfrac{-(2t - 2)}{(t^2 - 2t)^2}}{\dfrac{-2}{(2t + 1)^2}} = \dfrac{(t - 1)(2t + 1)^2}{t^2(t - 2)^2} = 0$ when $t = 1$.

Point of horizontal tangency: $\left(\dfrac{1}{3}, -1\right)$

(b) $2t + 1 = \dfrac{1}{x} \implies t = \dfrac{1}{2}\left(\dfrac{1}{x} - 1\right)$

$y = \dfrac{1}{\dfrac{1}{2}\left(\dfrac{1 - x}{x}\right)\left[\dfrac{1}{2}\left(\dfrac{1 - x}{x}\right) - 2\right]}$

$= \dfrac{4x^2}{(1 - x)^2 - 4x(1 - x)} = \dfrac{4x^2}{(5x - 1)(x - 1)}, \quad (x \neq 0)$

(c)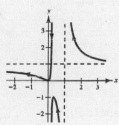

43. $x = 3 + 2\cos\theta$

$y = 2 + 5\sin\theta$

(a) $\dfrac{dy}{dx} = \dfrac{5\cos\theta}{-2\sin\theta} = -2.5\cot\theta = 0$ when $\theta = \dfrac{\pi}{2}, \dfrac{3\pi}{2}$.

Points of horizontal tangency: $(3, 7), (3, -3)$

(b) $\dfrac{(x - 3)^2}{4} + \dfrac{(y - 2)^2}{25} = 1$

(c)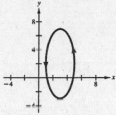

45. $x = \cos^3\theta$

$y = 4\sin^3\theta$

(a) $\dfrac{dy}{dx} = \dfrac{12\sin^2\theta\cos\theta}{3\cos^2\theta(-\sin\theta)}$

$= \dfrac{-4\sin\theta}{\cos\theta} = -4\tan\theta = 0$

when $\theta = 0, \pi$.

But, $\dfrac{dy}{dt} = \dfrac{dx}{dt} = 0$ at $\theta = 0, \pi$. Hence no points of horizontal tangency.

(b) $x^{2/3} + \left(\dfrac{y}{4}\right)^{2/3} = 1$

(c)

![graph for exercise 45]

47. $x = 4 - t, \quad y = t^2$

$$\frac{dx}{dt} = -1, \quad \frac{dy}{dt} = 2t$$

$$\frac{dy}{dt} = 0 \text{ for } t = 0.$$

Horizontal tangent at $t = 0$: $(x, y) = (4, 0)$

No vertical tangents

49. $x = 2 + 2 \sin \theta, y = 1 + \cos \theta$

$$\frac{dx}{d\theta} = 2 \cos \theta, \frac{dy}{d\theta} = -\sin \theta$$

$$\frac{dy}{d\theta} = 0 \text{ for } \theta = 0, \pi, 2\pi, \ldots$$

Horizontal tangents: $(x, y) = (2, 2), (2, 0)$

$$\frac{dx}{d\theta} = 0 \text{ for } \theta = \frac{\pi}{2}, \frac{3\pi}{2}, \ldots$$

Vertical tangents: $(x, y) = (4, 1), (0, 1)$

51. $x = \cot \theta$

$y = \sin 2\theta = 2 \sin \theta \cos \theta$

(a), (c)

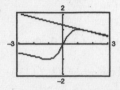

(b) At $\theta = \frac{\pi}{6}, \frac{dx}{d\theta} = -4, \frac{dy}{d\theta} = 1$, and $\frac{dy}{dx} = -\frac{1}{4}$.

53. $x = r(\cos \theta + \theta \sin \theta)$

$\quad y = r(\sin \theta - \theta \cos \theta)$

$$\frac{dx}{d\theta} = r\theta \cos \theta$$

$$\frac{dy}{d\theta} = r\theta \sin \theta$$

$$s = r \int_0^{\pi} \sqrt{\theta^2 \cos^2 \theta + \theta^2 \sin^2 \theta} \, d\theta$$

$$= r \int_0^{\pi} \theta \, d\theta = \frac{r}{2} \left[\theta^2 \right]_0^{\pi} = \frac{1}{2} \pi^2 r$$

55. $x = t, y = 3t, 0 \le t \le 2$

$$\frac{dx}{dt} = 1, \frac{dy}{dt} = -3, \sqrt{\left(\frac{dx}{dt}\right)^2 + \left(\frac{dy}{dt}\right)^2} = \sqrt{1 + 9} = \sqrt{10}$$

(a) $S = 2\pi \int_0^2 3t\sqrt{10} \, dt = 6\sqrt{10} \, \pi \left[\frac{t^2}{2}\right]_0^2 = 12\sqrt{10} \, \pi$

(b) $S = 2\pi \int_0^2 \sqrt{10} \, dt = 2\pi\sqrt{10} \left[t \right]_0^2 = 4\pi\sqrt{10}$

57. $x = 3 \sin \theta, y = 2 \cos \theta$

$$A = \int_a^b y \, dx = \int_{-\pi/2}^{\pi/2} 2 \cos \theta (3 \cos \theta) \, d\theta$$

$$= 6 \int_{-\pi/2}^{\pi/2} \frac{1 + \cos 2\theta}{2} \, d\theta$$

$$= 3 \left[\theta + \frac{\sin 2\theta}{2} \right]_{-\pi/2}^{\pi/2}$$

$$= 3 \left[\frac{\pi}{2} + \frac{\pi}{2} \right] = 3\pi$$

59. $(r, \theta) = \left(3, \frac{\pi}{2}\right)$

$$(x, y) = \left(3 \cos \frac{\pi}{2}, 3 \sin \frac{\pi}{2}\right)$$

$$= (0, 3)$$

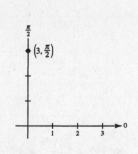

61. $(r, \theta) = \left(\sqrt{3}, 1.56\right)$

$$(x, y) = \left(\sqrt{3} \cos(1.56), \sqrt{3} \sin(1.56)\right)$$

$$\approx (0.0187, 1.7319)$$

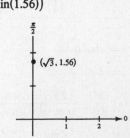

63. $(x, y) = (4, -4)$

$$r = \sqrt{4^2 + (-4)^2} = 4\sqrt{2}$$

$$\theta = \frac{7\pi}{4}$$

$$(r, \theta) = \left(4\sqrt{2}, \frac{7\pi}{4}\right), \left(-4\sqrt{2}, \frac{3\pi}{4}\right)$$

65.
$$r = 3\cos\theta$$
$$r^2 = 3r\cos\theta$$
$$x^2 + y^2 = 3x$$
$$x^2 + y^2 - 3x = 0$$

67.
$$r = -2(1 + \cos\theta)$$
$$r^2 = -2r(1 + \cos\theta)$$
$$x^2 + y^2 = -2\left(\pm\sqrt{x^2 + y^2}\right) - 2x$$
$$(x^2 + y^2 + 2x)^2 = 4(x^2 + y^2)$$

69.
$$r^2 = \cos 2\theta = \cos^2\theta - \sin^2\theta$$
$$r^4 = r^2\cos^2\theta - r^2\sin^2\theta$$
$$(x^2 + y^2)^2 = x^2 - y^2$$

71.
$$r = 4\cos 2\theta\sec\theta$$
$$= 4(2\cos^2\theta - 1)\left(\frac{1}{\cos\theta}\right)$$
$$r\cos\theta = 8\cos^2\theta - 4$$
$$x = 8\left(\frac{x^2}{x^2 + y^2}\right) - 4$$
$$x^3 + xy^2 = 4x^2 - 4y^2$$
$$y^2 = x^2\left(\frac{4 - x}{4 + x}\right)$$

73.
$$(x^2 + y^2)^2 = ax^2y$$
$$r^4 = a(r^2\cos^2\theta)(r\sin\theta)$$
$$r = a\cos^2\theta\sin\theta$$

75.
$$x^2 + y^2 = a^2\left(\arctan\frac{y}{x}\right)^2$$
$$r^2 = a^2\theta^2$$

77. $r = 4$

Circle of radius 4

Centered at the pole

Symmetric to polar axis,

$\theta = \pi/2$, and pole

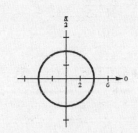

79. $r = -\sec\theta = \dfrac{-1}{\cos\theta}$

$r\cos\theta = -1, \ x = -1$

Vertical line

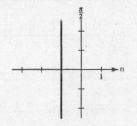

81. $r = -2(1 + \cos\theta)$

Cardioid

Symmetric to polar axis

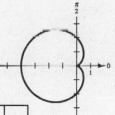

θ	0	$\dfrac{\pi}{3}$	$\dfrac{\pi}{2}$	$\dfrac{2\pi}{3}$	π
r	-4	-3	-2	-1	0

83. $r = 4 - 3\cos\theta$

Limaçon

Symmetric to polar axis

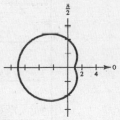

θ	0	$\dfrac{\pi}{3}$	$\dfrac{\pi}{2}$	$\dfrac{2\pi}{3}$	π
r	1	$\dfrac{5}{2}$	4	$\dfrac{11}{2}$	7

85. $r = -3\cos 2\theta$

Rose curve with four petals

Symmetric to polar axis, $\theta = \dfrac{\pi}{2}$, and pole

Relative extrema: $(-3, 0), \left(3, \dfrac{\pi}{2}\right), (-3, \pi), \left(3, \dfrac{3\pi}{2}\right)$

Tangents at the pole: $\theta = \dfrac{\pi}{4}, \dfrac{3\pi}{4}$

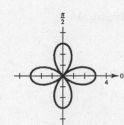

87. $r^2 = 4 \sin^2 2\theta$

$r = \pm 2 \sin(2\theta)$

Rose curve with four petals

Symmetric to the polar axis, $\theta = \dfrac{\pi}{2}$, and pole

Relative extrema: $\left(\pm 2, \dfrac{\pi}{4} \right), \left(\pm 2, \dfrac{3\pi}{4} \right)$

Tangents at the pole: $\theta = 0, \dfrac{\pi}{2}$

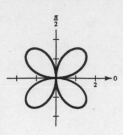

89. $r = \dfrac{3}{\cos \theta - (\pi/4)}$

Graph of $r = 3 \sec \theta$ rotated
through an angle of $\pi/4$

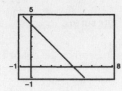

91. $r = 4 \cos 2\theta \sec \theta$

Strophoid

Symmetric to the polar axis

$r \Rightarrow \infty$ as $\theta \Rightarrow \dfrac{\pi^-}{2}$

$r \Rightarrow \infty$ as $\theta \Rightarrow \dfrac{-\pi^+}{2}$

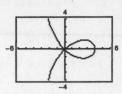

93. $r = 1 - 2 \cos \theta$

(a) The graph has polar symmetry and the tangents at the pole are $\theta = \dfrac{\pi}{3}, -\dfrac{\pi}{3}$.

(b) $\dfrac{dy}{dx} = \dfrac{2 \sin^2 \theta + (1 - 2 \cos \theta) \cos \theta}{2 \sin \theta \cos \theta - (1 - 2 \cos \theta) \sin \theta}$

Horizontal tangents: $-4 \cos^2 \theta + \cos \theta + 2 = 0$, $\cos \theta = \dfrac{-1 \pm \sqrt{1 + 32}}{-8} = \dfrac{1 \pm \sqrt{33}}{8}$

When $\cos \theta = \dfrac{1 \pm \sqrt{33}}{8}$, $r = 1 - 2 \left(\dfrac{1 + \sqrt{33}}{8} \right) = \dfrac{3 \mp \sqrt{33}}{4}$,

$$\left[\dfrac{3 - \sqrt{33}}{4}, \arccos\left(\dfrac{1 + \sqrt{33}}{8} \right) \right] \approx (-0.686, 0.568)$$

$$\left[\dfrac{3 - \sqrt{33}}{4}, -\arccos\left(\dfrac{1 + \sqrt{33}}{8} \right) \right] \approx (-0.686, -0.568)$$

$$\left[\dfrac{3 + \sqrt{33}}{4}, \arccos\left(\dfrac{1 - \sqrt{33}}{8} \right) \right] \approx (2.186, 2.206)$$

$$\left[\dfrac{3 + \sqrt{33}}{4}, -\arccos\left(\dfrac{1 - \sqrt{33}}{8} \right) \right] \approx (2.186, -2.206).$$

Vertical tangents: $\sin \theta (4 \cos \theta - 1) = 0$, $\sin \theta = 0$, $\cos \theta = \dfrac{1}{4}$, $\theta = 0, \pi$, $\theta = \pm \arccos\left(\dfrac{1}{4} \right)$, $(-1, 0)$, $(3, \pi)$

$$\left(\dfrac{1}{2}, \pm \arccos \dfrac{1}{4} \right) \approx (0.5, \pm 1.318)$$

(c)

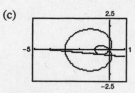

95. Circle: $r = 3 \sin \theta$

$$\frac{dy}{dx} = \frac{3 \cos \theta \sin \theta + 3 \sin \theta \cos \theta}{3 \cos \theta \cos \theta - 3 \sin \theta \sin \theta} = \frac{\sin 2\theta}{\cos^2 \theta - \sin^2 \theta} = \tan 2\theta \text{ at } \theta = \frac{\pi}{6}, \frac{dy}{dx} = \sqrt{3}$$

Limaçon: $r = 4 - 5 \sin \theta$

$$\frac{dy}{dx} = \frac{-5 \cos \theta \sin \theta + (4 - 5 \sin \theta) \cos \theta}{-5 \cos \theta \cos \theta - (4 - 5 \sin \theta) \sin \theta} \text{ at } \theta = \frac{\pi}{6}, \frac{dy}{dx} = \frac{\sqrt{3}}{9}$$

Let α be the angle between the curves:

$$\tan \alpha = \frac{\sqrt{3} - (\sqrt{3}/9)}{1 + (1/3)} = \frac{2\sqrt{3}}{3}.$$

Therefore, $\alpha = \arctan\left(\dfrac{2\sqrt{3}}{3}\right) \approx 49.1°.$

97. $r = 1 + \cos \theta, r = 1 - \cos \theta$

The points $(1, \pi/2)$ and $(1, 3\pi/2)$ are the two points of intersection (other than the pole). The slope of the graph of $r = 1 + \cos \theta$ is

$$m_1 = \frac{dy}{dx} = \frac{r' \sin \theta + r \cos \theta}{r' \cos \theta - r \sin \theta} = \frac{-\sin^2 \theta + \cos \theta(1 + \cos \theta)}{-\sin \theta \cos \theta - \sin \theta(1 + \cos \theta)}.$$

At $(1, \pi/2)$, $m_1 = -1/-1 = 1$ and at $(1, 3\pi/2)$, $m_1 = -1/1 = -1$. The slope of the graph of $r = 1 - \cos \theta$ is

$$m_2 = \frac{dy}{dx} = \frac{\sin^2 \theta + \cos \theta(1 - \cos \theta)}{\sin \theta \cos \theta - \sin \theta(1 - \cos \theta)}.$$

At $(1, \pi/2)$, $m_2 = 1/-1 = -1$ and at $(1, 3\pi/2)$, $m_2 = 1/1 = 1$. In both cases, $m_1 = -1/m_2$ and we conclude that the graphs are orthogonal at $(1, \pi/2)$ and $(1, 3\pi/2)$.

99. $r = 2 + \cos \theta$

$$A = 2\left[\frac{1}{2} \int_0^\pi (2 + \cos \theta)^2 \, d\theta\right] \approx 14.14, \left(\frac{9\pi}{2}\right)$$

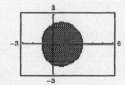

101. $r^2 = 4 \sin 2\theta$

$$A = 2\left[\frac{1}{2} \int_0^{\pi/2} 4 \sin 2\theta \, d\theta\right] = 4$$

103. $r = \sin \theta \cos^2 \theta$

$$A = 2\left[\frac{1}{2} \int_0^{\pi/2} (\sin \theta \cos^2 \theta)^2 \, d\theta\right]$$

$$\approx 0.10, \left(\frac{\pi}{32}\right)$$

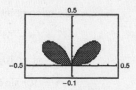

105. $r = 3, r^2 = 18 \sin 2\theta$

$$9 = r^2 = 18 \sin 2\theta$$

$$\sin 2\theta = \frac{1}{2}$$

$$\theta = \frac{\pi}{12}$$

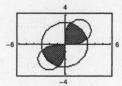

$$A = 2\left[\frac{1}{2} \int_0^{\pi/12} 18 \sin 2\theta \, d\theta + \frac{1}{2} \int_{\pi/12}^{5\pi/12} 9 \, d\theta + \frac{1}{2} \int_{5\pi/12}^{\pi/2} 18 \sin 2\theta \, d\theta\right]$$

$$\approx 1.2058 + 9.4248 + 1.2058 \approx 11.84$$

107. $r = a(1 - \cos \theta), 0 \le \theta \le \pi$

$$\frac{dr}{d\theta} = a \sin \theta$$

$$s = \int_0^\pi \sqrt{a^2(1 - \cos \theta)^2 + a^2 \sin^2 \theta} \, d\theta$$

$$= \sqrt{2} a \int_0^\pi \sqrt{1 - \cos \theta} \, d\theta$$

$$= \sqrt{2} a \int_0^\pi \frac{\sin \theta}{\sqrt{1 + \cos \theta}} \, d\theta$$

$$= -2\sqrt{2} a \left[(1 + \cos \theta)^{1/2} \right]_0^\pi = 4a$$

109. $f(\theta) = 1 + 4 \cos \theta$

$f'(\theta) = -4 \sin \theta$

$\sqrt{f(\theta)^2 + f'(\theta)^2} = \sqrt{(1 + 4 \cos \theta)^2 + (-4 \sin \theta)^2} = \sqrt{17 + 8 \cos \theta}$

$$S = 2\pi \int_0^{\pi/2} (1 + 4 \cos \theta) \sin \theta \sqrt{17 + 8 \cos \theta} \, d\theta$$

$$= \frac{34\pi\sqrt{17}}{5} \approx 88.08$$

111. $r = \dfrac{2}{1 - \sin \theta}, e = 1$

Parabola

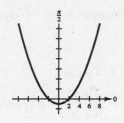

113. $r = \dfrac{6}{3 + 2 \cos \theta} = \dfrac{2}{1 + (2/3) \cos \theta}, e = \dfrac{2}{3}$

Ellipse

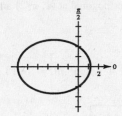

115. $r = \dfrac{4}{2 - 3 \sin \theta} = \dfrac{2}{1 - (3/2)\sin \theta}, e = \dfrac{3}{2}$

Hyperbola

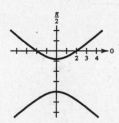

117. Circle

Center: $\left(5, \dfrac{\pi}{2}\right) = (0, 5)$ in rectangular coordinates

Solution point: $(0, 0)$

$x^2 + (y - 5)^5 = 25$

$x^2 + y^2 - 10y = 0$

$r^2 - 10r \sin \theta = 0$

$r = 10 \sin \theta$

119. Parabola

Vertex: $(2, \pi)$

Focus: $(0, 0)$

$e = 1, d = 4$

$r = \dfrac{4}{1 - \cos \theta}$

121. Ellipse

Vertices: $(5, 0), (1, \pi)$

Focus: $(0, 0)$

$a = 3, c = 2, e = \dfrac{2}{3}, d = \dfrac{5}{2}$

$r = \dfrac{\left(\dfrac{2}{3}\right)\left(\dfrac{5}{2}\right)}{1 - \left(\dfrac{2}{3}\right) \cos \theta} = \dfrac{5}{3 - 2 \cos \theta}$

Problem Solving for Chapter 10

1. (a)

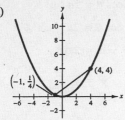

(b) $x^2 = 4y$

$2x = 4y'$

$y' = \dfrac{1}{2}x$

$y - 4 = 2(x - 4) \implies y = 2x - 4$ Tangent line at $(4, 4)$

$y - \dfrac{1}{4} = -\dfrac{1}{2}(x + 1) \implies y = -\dfrac{1}{2}x - \dfrac{1}{4}$ Tangent line at $\left(-1, \dfrac{1}{4}\right)$

Tangent lines have slopes of 2 and $-1/2 \implies$ perpendicular.

(c) Intersection:

$2x - 4 = -\dfrac{1}{2}x - \dfrac{1}{4}$

$8x - 16 = -2x - 1$

$10x = 15$

$x = \dfrac{3}{2} \implies \left(\dfrac{3}{2}, -1\right)$

Point of intersection, $(3/2, -1)$, is on directrix $y = -1$.

3. Consider $x^2 = 4py$ with focus $F = (0, p)$.

Let $P(a, b)$ be point on parabola.

$2x = 4py' \implies y' = \dfrac{x}{2p}$

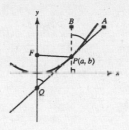

$y - b = \dfrac{a}{2p}(x - a)$ Tangent line at P

For $x = 0$, $y = b + \dfrac{a}{2p}(-a) = b - \dfrac{a^2}{2p} = b - \dfrac{4pb}{2p} = -b$.

Thus, $Q = (0, -b)$.

$\triangle FQP$ is isosceles because

$|FQ| = p + b$

$|FP| = \sqrt{(a - 0)^2 + (b - p)^2} = \sqrt{a^2 + b^2 - 2bp + p^2}$

$\qquad = \sqrt{4pb + b^2 - 2bp + p^2}$

$\qquad = \sqrt{(b + p)^2}$

$\qquad = b + p.$

Thus, $\angle FQP = \angle BPA = \angle FPQ$.

5. (a) In $\triangle OCB$, $\cos\theta = \dfrac{2a}{OB} \implies OB = 2a \cdot \sec\theta$.

In $\triangle OAC$, $\cos\theta = \dfrac{OA}{2a} \implies OA = 2a \cdot \cos\theta$.

$r = OP = AB = OB - OA = 2a(\sec\theta - \cos\theta)$

$\qquad = 2a\left(\dfrac{1}{\cos\theta} - \cos\theta\right)$

$\qquad = 2a \cdot \dfrac{\sin^2\theta}{\cos\theta}$

$\qquad = 2a \cdot \tan\theta\sin\theta$

—CONTINUED—

5. —CONTINUED—

(b) $x = r \cos \theta = (2a \tan \theta \sin \theta)\cos \theta = 2a \sin^2 \theta$

$y = r \sin \theta = (2a \tan \theta \sin \theta)\sin \theta = 2a \tan \theta \cdot \sin^2 \theta, \ -\dfrac{\pi}{2} < \theta < \dfrac{\pi}{2}$

Let $t = \tan \theta, \ -\infty < t < \infty$.

Then $\sin^2 \theta = \dfrac{t^2}{1 + t^2}$ and $x = 2a\dfrac{t^2}{1 + t^2}, \ y = 2a\dfrac{t^3}{1 + t^2}$.

(triangle diagram with hypotenuse $\sqrt{1+t^2}$, vertical side t, horizontal side 1, angle θ)

(c) $\qquad r = 2a \tan \theta \sin \theta$

$r \cos \theta = 2a \sin^2 \theta$

$r^3 \cos \theta = 2a \, r^2 \sin^2 \theta$

$(x^2 + y^2)x = 2ay^2$

$y^2 = \dfrac{x^3}{(2a - x)}$

7. (a) $y^2 = \dfrac{t^2(1 - t^2)^2}{(1 + t^2)^2}, x^2 = \dfrac{(1 - t^2)^2}{(1 + t^2)^2}$

$\dfrac{1 - x}{1 + x} = \dfrac{1 - \left(\dfrac{1 - t^2}{1 + t^2}\right)}{1 + \left(\dfrac{1 - t^2}{1 + t^2}\right)} = \dfrac{2t^2}{2} = t^2$

Thus, $y^2 = x^2\left(\dfrac{1 - x}{1 + x}\right)$.

(b) $\qquad r^2 \sin^2 \theta = r^2 \cos^2 \theta\left(\dfrac{1 - r \cos \theta}{1 + r \cos \theta}\right)$

$\sin^2 \theta(1 + r \cos \theta) = \cos^2 \theta(1 - r \cos \theta)$

$r \cos \theta \sin^2 \theta + \sin^2 \theta = \cos^2 \theta - r \cos^3 \theta$

$r \cos \theta(\sin^2 \theta + \cos^2 \theta) = \cos^2 \theta - \sin^2 \theta$

$r \cos \theta = \cos 2\theta$

$r = \cos 2\theta \cdot \sec \theta$

(c)

(graph showing loop curve through origin)

(d) $r(\theta) = 0$ for $\theta = \dfrac{\pi}{4}, \dfrac{3\pi}{4}$.

Thus, $y = x$ and $y = -x$ are tangent lines to curve at the origin.

(e) $y'(t) = \dfrac{(1 + t^2)(1 - 3t^2) - (t - t^3)(2t)}{(1 + t^2)^2} = \dfrac{1 - 4t^2 - t^4}{(1 + t^2)^2} = 0$

$t^4 + 4t^2 - 1 = 0 \Rightarrow t^2 = -2 \pm \sqrt{5} \Rightarrow x = \dfrac{1 - (-2 \pm \sqrt{5})}{1 + (-2 \pm \sqrt{5})} = \dfrac{3 \mp \sqrt{5}}{-1 \pm \sqrt{5}}$

$= \dfrac{3 - \sqrt{5}}{-1 + \sqrt{5}} = \dfrac{\sqrt{5} - 1}{2}$

$\left(\dfrac{\sqrt{5} - 1}{2}, \pm\dfrac{\sqrt{5} - 1}{2}\sqrt{-2 + \sqrt{5}}\right)$

9. (a)

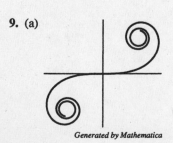

Generated by Mathematica

(b) $(-x, -y) = \left(-\displaystyle\int_0^t \cos \dfrac{\pi u^2}{2}\, du, \ -\displaystyle\int_0^t \sin \dfrac{\pi u^2}{2}\, du\right)$ is on

the curve whenever (x, y) is on the curve.

(c) $x'(t) = \cos \dfrac{\pi t^2}{2}, y'(t) = \sin \dfrac{\pi t^2}{2}, x'(t)^2 + y'(t)^2 = 1$

Thus, $s = \displaystyle\int_0^a dt = a$.

On $[-\pi, \pi], s = 2\pi$.

11. $r = \dfrac{ab}{a \sin \theta + b \cos \theta}, 0 \leq \theta \leq \dfrac{\pi}{2}$

$r(a \sin \theta + b \cos \theta) = ab$

$ay + bx = ab$

$\dfrac{y}{b} + \dfrac{x}{a} = 1$

Line segment

Area $= \dfrac{1}{2}ab$

13. Let (r, θ) be on the graph.

$\sqrt{r^2 + 1 + 2r \cos \theta}\sqrt{r^2 + 1 - 2r \cos \theta} = 1$

$(r^2 + 1)^2 - 4r^2 \cos^2 \theta = 1$

$r^4 + 2r^2 + 1 - 4r^2 \cos^2 \theta = 1$

$r^2(r^2 - 4 \cos^2 \theta + 2) = 0$

$r^2 = 4 \cos^2 \theta - 2$

$r^2 = 2(2 \cos^2 \theta - 1)$

$r^2 = 2 \cos 2\theta$

15. (a) The first plane makes an angle of 70° with the positive x-axis, and is 150 miles from P:

$x_1 = \cos 70°(150 - 375t)$

$y_1 = \sin 70°(150 - 375t)$

Similarly for the second plane,

$x_2 = \cos 135°(190 - 450t)$

$\quad = \cos 45°(-190 + 450t)$

$y_2 = \sin 135°(190 - 450t)$

$\quad = \sin 45°(190 - 450t).$

(b) $d = \sqrt{(x_2 - x_1)^2 + (y_2 - y_1)^2}$

$\quad = [[\cos 45(-190 + 450t) - \cos 70(150 - 375t)]^2 + [\sin 45(190 - 450t) - \sin 70(150 - 375t)]^2]^{1/2}$

(c)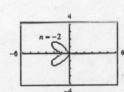

The minimum distance is 7.59 miles when $t = 0.4145$.

17.

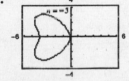 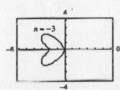 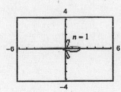 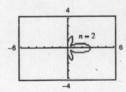

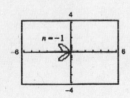 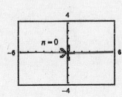 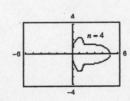 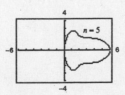

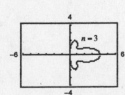 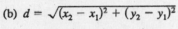

$n = 1, 2, 3, 4, 5$ produce "bells"; $n = -1, -2, -3, -4, -5$ produce "hearts".

CHAPTER 11
Vectors and the Geometry of Space

Section 11.1   Vectors in the Plane 557

Section 11.2   Space Coordinates and Vectors in Space 562

Section 11.3   The Dot Product of Two Vectors 567

Section 11.4   The Cross Product of Two Vectors in Space 572

Section 11.5   Lines and Planes in Space 575

Section 11.6   Surfaces in Space 582

Section 11.7   Cylindrical and Spherical Coordinates 585

Review Exercises 590

Problem Solving 594

CHAPTER 11
Vectors and the Geometry of Space

Section 11.1   Vectors in the Plane

1. (a) $\mathbf{v} = \langle 5 - 1, 3 - 1 \rangle = \langle 4, 2 \rangle$

(b)

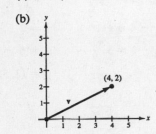

3. (a) $\mathbf{v} = \langle -4 - 3, -2 - (-2) \rangle = \langle -7, 0 \rangle$

(b)

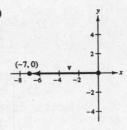

5. $\mathbf{u} = \langle 5 - 3, 6 - 2 \rangle = \langle 2, 4 \rangle$

$\mathbf{v} = \langle 1 - (-1), 8 - 4 \rangle = \langle 2, 4 \rangle$

$\mathbf{u} = \mathbf{v}$

7. $\mathbf{u} = \langle 6 - 0, -2 - 3 \rangle = \langle 6, -5 \rangle$

$\mathbf{v} = \langle 9 - 3, 5 - 10 \rangle = \langle 6, -5 \rangle$

$\mathbf{u} = \mathbf{v}$

9. (b) $\mathbf{v} = \langle 5 - 1, 5 - 2 \rangle = \langle 4, 3 \rangle$

(a) and (c).

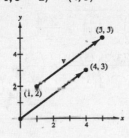

11. (b) $\mathbf{v} = \langle 6 - 10, -1 - 2 \rangle = \langle -4, -3 \rangle$

(a) and (c).

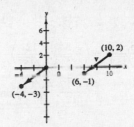

13. (b) $\mathbf{v} = \langle 6 - 6, 6 - 2 \rangle = \langle 0, 4 \rangle$

(a) and (c).

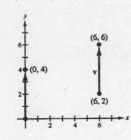

15. (b) $\mathbf{v} = \langle \frac{1}{2} - \frac{3}{2}, 3 - \frac{4}{3} \rangle = \langle -1, \frac{5}{3} \rangle$

(a) and (c).

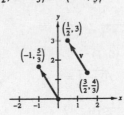

17. (a) $2\mathbf{v} = \langle 4, 6 \rangle$

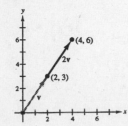

(b) $-3\mathbf{v} = \langle -6, -9 \rangle$

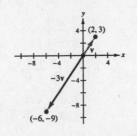

(c) $\frac{7}{2}\mathbf{v} = \langle 7, \frac{21}{2} \rangle$

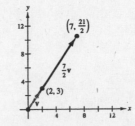

(d) $\frac{2}{3}\mathbf{v} = \langle \frac{4}{3}, 2 \rangle$

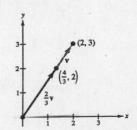

557

19.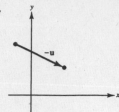

21.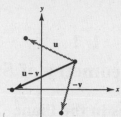

23. (a) $\frac{2}{3}\mathbf{u} = \frac{2}{3}\langle 4, 9 \rangle = \langle \frac{8}{3}, 6 \rangle$ (b) $\mathbf{v} - \mathbf{u} = \langle 2, -5 \rangle - \langle 4, 9 \rangle = \langle -2, -14 \rangle$ (c) $2\mathbf{u} + 5\mathbf{v} = 2\langle 4, 9 \rangle + 5\langle 2, -5 \rangle = \langle 18, -7 \rangle$

25. $\mathbf{v} = \frac{3}{2}(2\mathbf{i} - \mathbf{j}) = 3\mathbf{i} - \frac{3}{2}\mathbf{j}$

$= \langle 3, -\frac{3}{2} \rangle$

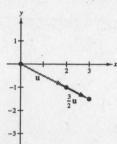

27. $\mathbf{v} = (2\mathbf{i} - \mathbf{j}) + 2(\mathbf{i} + 2\mathbf{j})$

$= 4\mathbf{i} + 3\mathbf{j} = \langle 4, 3 \rangle$

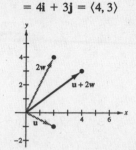

29. $u_1 - 4 = -1$ $u_1 = 3$

$u_2 - 2 = 3$ $u_2 = 5$

$Q = (3, 5)$

31. $\|\mathbf{v}\| = \sqrt{16 + 9} = 5$

33. $\|\mathbf{v}\| = \sqrt{36 + 25} = \sqrt{61}$

35. $\|\mathbf{v}\| = \sqrt{0 + 16} = 4$

37. $\|\mathbf{u}\| = \sqrt{3^2 + 12^2} = \sqrt{153}$

$\mathbf{v} = \dfrac{\mathbf{u}}{\|\mathbf{u}\|} = \dfrac{\langle 3, 12 \rangle}{\sqrt{153}} = \left\langle \dfrac{3}{\sqrt{153}}, \dfrac{12}{\sqrt{153}} \right\rangle$

$= \left\langle \dfrac{\sqrt{17}}{17}, \dfrac{4\sqrt{17}}{17} \right\rangle$ unit vector

39. $\|\mathbf{u}\| = \sqrt{\left(\dfrac{3}{2}\right)^2 + \left(\dfrac{5}{2}\right)^2} = \dfrac{\sqrt{34}}{2}$

$\mathbf{v} = \dfrac{\mathbf{u}}{\|\mathbf{u}\|} = \dfrac{\langle (3/2), (5/2) \rangle}{\sqrt{34}/2} = \left\langle \dfrac{3}{\sqrt{34}}, \dfrac{5}{\sqrt{34}} \right\rangle$

$= \left\langle \dfrac{3\sqrt{34}}{34}, \dfrac{5\sqrt{34}}{34} \right\rangle$ unit vector

41. $\mathbf{u} = \langle 1, -1 \rangle, \mathbf{v} = \langle -1, 2 \rangle$

(a) $\|\mathbf{u}\| = \sqrt{1 + 1} = \sqrt{2}$

(b) $\|\mathbf{v}\| = \sqrt{1 + 4} = \sqrt{5}$

(c) $\mathbf{u} + \mathbf{v} = \langle 0, 1 \rangle$

$\|\mathbf{u} + \mathbf{v}\| = \sqrt{0 + 1} = 1$

(d) $\dfrac{\mathbf{u}}{\|\mathbf{u}\|} = \dfrac{1}{\sqrt{2}}\langle 1, -1 \rangle$

$\left\| \dfrac{\mathbf{u}}{\|\mathbf{u}\|} \right\| = 1$

(e) $\dfrac{\mathbf{v}}{\|\mathbf{v}\|} = \dfrac{1}{\sqrt{5}}\langle -1, 2 \rangle$

$\left\| \dfrac{\mathbf{v}}{\|\mathbf{v}\|} \right\| = 1$

(f) $\dfrac{\mathbf{u} + \mathbf{v}}{\|\mathbf{u} + \mathbf{v}\|} = \langle 0, 1 \rangle$

$\left\| \dfrac{\mathbf{u} + \mathbf{v}}{\|\mathbf{u} + \mathbf{v}\|} \right\| = 1$

43. $\mathbf{u} = \left\langle 1, \dfrac{1}{2} \right\rangle, \mathbf{v} = \langle 2, 3 \rangle$

(a) $\|\mathbf{u}\| = \sqrt{1 + \dfrac{1}{4}} = \dfrac{\sqrt{5}}{2}$

(b) $\|\mathbf{v}\| = \sqrt{4 + 9} = \sqrt{13}$

(c) $\mathbf{u} + \mathbf{v} = \left\langle 3, \dfrac{7}{2} \right\rangle$

$\|\mathbf{u} + \mathbf{v}\| = \sqrt{9 + \dfrac{49}{4}} = \dfrac{\sqrt{85}}{2}$

(d) $\dfrac{\mathbf{u}}{\|\mathbf{u}\|} = \dfrac{2}{\sqrt{5}}\left\langle 1, \dfrac{1}{2} \right\rangle$ (e) $\dfrac{\mathbf{v}}{\|\mathbf{v}\|} = \dfrac{1}{\sqrt{13}}\langle 2, 3 \rangle$

$\left\| \dfrac{\mathbf{u}}{\|\mathbf{u}\|} \right\| = 1$ $\left\| \dfrac{\mathbf{v}}{\|\mathbf{v}\|} \right\| = 1$

(f) $\dfrac{\mathbf{u} + \mathbf{v}}{\|\mathbf{u} + \mathbf{v}\|} = \dfrac{2}{\sqrt{85}}\left\langle 3, \dfrac{7}{2} \right\rangle$

$\left\| \dfrac{\mathbf{u} + \mathbf{v}}{\|\mathbf{u} + \mathbf{v}\|} \right\| = 1$

45. $\mathbf{u} = \langle 2, 1 \rangle$

$\|\mathbf{u}\| = \sqrt{5} \approx 2.236$

$\mathbf{v} = \langle 5, 4 \rangle$

$\|\mathbf{v}\| = \sqrt{41} \approx 6.403$

$\mathbf{u} + \mathbf{v} = \langle 7, 5 \rangle$

$\|\mathbf{u} + \mathbf{v}\| = \sqrt{74} \approx 8.602$

$\|\mathbf{u} + \mathbf{v}\| \leq \|\mathbf{u}\| + \|\mathbf{v}\|$

$\sqrt{74} \leq \sqrt{5} + \sqrt{41}$

47. $\dfrac{\mathbf{u}}{\|\mathbf{u}\|} = \dfrac{1}{\sqrt{2}} \langle 1, 1 \rangle$

$4\left(\dfrac{\mathbf{u}}{\|\mathbf{u}\|} \right) = 2\sqrt{2}\, \langle 1, 1 \rangle$

$\mathbf{v} = \langle 2\sqrt{2}, 2\sqrt{2} \rangle$

49. $\dfrac{\mathbf{u}}{\|\mathbf{u}\|} = \dfrac{1}{2\sqrt{3}} \langle \sqrt{3}, 3 \rangle$

$2\left(\dfrac{\mathbf{u}}{\|\mathbf{u}\|} \right) = \dfrac{1}{\sqrt{3}} \langle \sqrt{3}, 3 \rangle$

$\mathbf{v} = \langle 1, \sqrt{3} \rangle$

51. $\mathbf{v} = 3[(\cos 0°)\mathbf{i} + (\sin 0°)\mathbf{j}]$

$= 3\mathbf{i} = \langle 3, 0 \rangle$

53. $\mathbf{v} = 2[(\cos 150°)\mathbf{i} + (\sin 150°)\mathbf{j}]$

$= -\sqrt{3}\mathbf{i} + \mathbf{j} = \langle -\sqrt{3}, 1 \rangle$

55. $\mathbf{u} = \mathbf{i}$

$\mathbf{v} = \dfrac{3\sqrt{2}}{2}\mathbf{i} + \dfrac{3\sqrt{2}}{2}\mathbf{j}$

$\mathbf{u} + \mathbf{v} = \left(\dfrac{2 + 3\sqrt{2}}{2} \right)\mathbf{i} + \dfrac{3\sqrt{2}}{2}\mathbf{j}$

57. $\mathbf{u} = 2(\cos 4)\mathbf{i} + 2(\sin 4)\mathbf{j}$

$\mathbf{v} = (\cos 2)\mathbf{i} + (\sin 2)\mathbf{j}$

$\mathbf{u} + \mathbf{v} = (2\cos 4 + \cos 2)\mathbf{i} + (2\sin 4 + \sin 2)\mathbf{j}$

59. A scalar is a real number. A vector is represented by a directed line segment. A vector has both length and direction.

61. (a) Vector. The velocity has both magnitude and direction.

    (b) Scalar. The price is a number.

For Exercises 63–67, $a\mathbf{u} + b\mathbf{w} = a(\mathbf{i} + 2\mathbf{j}) + b(\mathbf{i} - \mathbf{j}) = (a + b)\mathbf{i} + (2a - b)\mathbf{j}$.

63. $\mathbf{v} = 2\mathbf{i} + \mathbf{j}$. Therefore, $a + b = 2, 2a - b = 1$. Solving simultaneously, we have $a = 1$, $b = 1$.

65. $\mathbf{v} = 3\mathbf{i}$. Therefore, $a + b = 3, 2a - b = 0$. Solving simultaneously, we have $a = 1$, $b = 2$.

67. $\mathbf{v} = \mathbf{i} + \mathbf{j}$. Therefore, $a + b = 1, 2a - b = 1$. Solving simultaneously, we have $a = \frac{2}{3}$, $b = \frac{1}{3}$.

69. $f(x) = x^2, f'(x) = 2x, f'(3) = 6$

    (a) $m = 6$. Let $\mathbf{w} = \langle 1, 6 \rangle, \|\mathbf{w}\| = \sqrt{37}$.

$\pm \dfrac{\mathbf{w}}{\|\mathbf{w}\|} = \pm \dfrac{1}{\sqrt{37}} \langle 1, 6 \rangle$ unit tangent vectors

    (b) $m = -\dfrac{1}{6}$. Let $\mathbf{w} = \langle -6, 1 \rangle, \|\mathbf{w}\| = \sqrt{37}$.

$\pm \dfrac{\mathbf{w}}{\|\mathbf{w}\|} = \pm \dfrac{1}{\sqrt{37}} \langle -6, 1 \rangle$ unit normal vectors

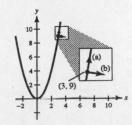

71. $f(x) = x^3, f'(x) = 3x^2 = 3$ at $x = 1$.

    (a) $m = 3$. Let $\mathbf{w} = \langle 1, 3 \rangle$, then

$\dfrac{\mathbf{w}}{\|\mathbf{w}\|} = \pm \dfrac{1}{\sqrt{10}} \langle 1, 3 \rangle.$

    (b) $m = -\dfrac{1}{3}$. Let $\mathbf{w} = \langle 3, -1 \rangle$, then

$\dfrac{\mathbf{w}}{\|\mathbf{w}\|} = \pm \dfrac{1}{\sqrt{10}} \langle 3, -1 \rangle.$

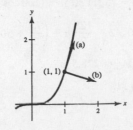

73. $f(x) = \sqrt{25 - x^2}$

$f'(x) = \dfrac{-x}{\sqrt{25 - x^2}} = \dfrac{-3}{4}$ at $x = 3$.

(a) $m = -\dfrac{3}{4}$. Let $\mathbf{w} = \langle -4, 3 \rangle$, then

$\dfrac{\mathbf{w}}{\|\mathbf{w}\|} = \pm\dfrac{1}{5}\langle -4, 3 \rangle$.

(b) $m = \dfrac{4}{3}$. Let $\mathbf{w} = \langle 3, 4 \rangle$, then

$\dfrac{\mathbf{w}}{\|\mathbf{w}\|} = \pm\dfrac{1}{5}\langle 3, 4 \rangle$.

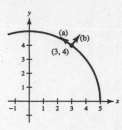

75. $\mathbf{u} = \dfrac{\sqrt{2}}{2}\mathbf{i} + \dfrac{\sqrt{2}}{2}\mathbf{j}$

$\mathbf{u} + \mathbf{v} = \sqrt{2}\,\mathbf{j}$

$\mathbf{v} = (\mathbf{u} + \mathbf{v}) - \mathbf{u} = -\dfrac{\sqrt{2}}{2}\mathbf{i} + \dfrac{\sqrt{2}}{2}\mathbf{j}$

77. Programs will vary.

79. $\|\mathbf{F}_1\| = 2,\ \theta_{\mathbf{F}_1} = 33°$

$\|\mathbf{F}_2\| = 3,\ \theta_{\mathbf{F}_2} = -125°$

$\|\mathbf{F}_3\| = 2.5,\ \theta_{\mathbf{F}_3} = 110°$

$\|\mathbf{R}\| = \|\mathbf{F}_1 + \mathbf{F}_2 + \mathbf{F}_3\| \approx 1.33$

$\theta_{\mathbf{R}} = \theta_{\mathbf{F}_1 + \mathbf{F}_2 + \mathbf{F}_3} \approx 132.5°$

81. (a) $180(\cos 30°\mathbf{i} + \sin 30°\mathbf{j}) + 275\mathbf{i} \approx 430.88\mathbf{i} + 90\mathbf{j}$

Direction: $\alpha \approx \arctan\left(\dfrac{90}{430.88}\right) \approx 0.206\ (\approx 11.8°)$

Magnitude: $\sqrt{430.88^2 + 90^2} \approx 440.18$ newtons

(b) $M = \sqrt{(275 + 180 \cos \theta)^2 + (180 \sin \theta)^2}$

$\alpha = \arctan\left[\dfrac{180 \sin \theta}{275 + 180 \cos \theta}\right]$

(c)

θ	0°	30°	60°	90°	120°	150°	180°
M	455	440.2	396.9	328.7	241.9	149.3	95
α	0°	11.8°	23.1°	33.2°	40.1°	37.1°	0

(d)

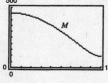

(e) M decreases because the forces change from acting in the same direction to acting in the opposite direction as θ increases from 0° to 180°.

83. $\mathbf{F}_1 + \mathbf{F}_2 + \mathbf{F}_3 = (75 \cos 30°\mathbf{i} + 75 \sin 30°\mathbf{j}) + (100 \cos 45°\mathbf{i} + 100 \sin 45°\mathbf{j}) + (125 \cos 120°\mathbf{i} + 125 \sin 120°\mathbf{j})$

$= \left(\dfrac{75}{2}\sqrt{3} + 50\sqrt{2} - \dfrac{125}{2}\right)\mathbf{i} + \left(\dfrac{75}{2} + 50\sqrt{2} + \dfrac{125}{2}\sqrt{3}\right)\mathbf{j}$

$\|\mathbf{R}\| = \|\mathbf{F}_1 + \mathbf{F}_2 + \mathbf{F}_3\| \approx 228.5$ lb

$\theta_{\mathbf{R}} = \theta_{\mathbf{F}_1 + \mathbf{F}_2 + \mathbf{F}_3} \approx 71.3°$

85. (a) The forces act along the same direction. $\theta = 0°$.

(b) The forces cancel out each other. $\theta = 180°$.

(c) No, the magnitude of the resultant can not be greater than the sum.

87. $(-4, -1), (6, 5), (10, 3)$

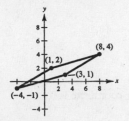   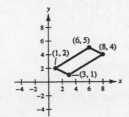   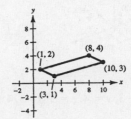

89. $\mathbf{u} = \overrightarrow{CB} = \|\mathbf{u}\|(\cos 30°\,\mathbf{i} + \sin 30°\,\mathbf{j})$

$\mathbf{v} = \overrightarrow{CA} = \|\mathbf{v}\|(\cos 130°\,\mathbf{i} + \sin 130°\,\mathbf{j})$

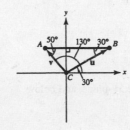

Vertical components: $\|\mathbf{u}\| \sin 30° + \|\mathbf{v}\| \sin 130° = 2000$

Horizontal components: $\|\mathbf{u}\| \cos 30° + \|\mathbf{v}\| \cos 130° = 0$

Solving this system, you obtain

$\|\mathbf{u}\| \approx 1305.5$ pounds and $\|\mathbf{v}\| \approx 1758.8$ pounds.

91. Horizontal component $= \|\mathbf{v}\| \cos \theta = 1200 \cos 6° \approx 1193.43$ ft/sec

Vertical component $= \|\mathbf{v}\| \sin \theta = 1200 \sin 6° \approx 125.43$ ft/sec

93. $\mathbf{u} = 900[\cos 148°\,\mathbf{i} + \sin 148°\,\mathbf{j}]$

$\mathbf{v} = 100[\cos 45°\,\mathbf{i} + \sin 45°\,\mathbf{j}]$

$\mathbf{u} + \mathbf{v} = [900 \cos 148° + 100 \cos 45°]\mathbf{i} + [900 \sin 148° + 100 \sin 45°]\mathbf{j}$

$\approx -692.53\,\mathbf{i} + 547.64\,\mathbf{j}$

$\theta \approx \arctan\left(\dfrac{547.64}{-692.53}\right) \approx -38.34°; \quad 38.34°$ North of West

$\|\mathbf{u} + \mathbf{v}\| \approx \sqrt{(-692.53)^2 + (547.64)^2} \approx 882.9$ km/hr

95. True   **97.** True   **99.** False

$$\|a\mathbf{i} + b\mathbf{j}\| = \sqrt{2}\,|a|$$

101. $\|\mathbf{u}\| = \sqrt{\cos^2 \theta + \sin^2 \theta} = 1$,

$\|\mathbf{v}\| = \sqrt{\sin^2 \theta + \cos^2 \theta} = 1$

103. Let $\mathbf{u}$ and $\mathbf{v}$ be the vectors that determine the parallelogram, as indicated in the figure. The two diagonals are $\mathbf{u} + \mathbf{v}$ and $\mathbf{v} - \mathbf{u}$. Therefore, $\mathbf{r} = x(\mathbf{u} + \mathbf{v})$, $\mathbf{s} = y(\mathbf{v} - \mathbf{u})$. But,

$\mathbf{u} = \mathbf{r} - \mathbf{s}$

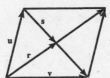

$= x(\mathbf{u} + \mathbf{v}) - y(\mathbf{v} - \mathbf{u}) = (x + y)\mathbf{u} + (x - y)\mathbf{v}$.

Therefore, $x + y = 1$ and $x - y = 0$. Solving we have $x = y = \frac{1}{2}$.

105. The set is a circle of radius 5, centered at the origin.

$\|\mathbf{u}\| = \|\langle x, y \rangle\| = \sqrt{x^2 + y^2} = 5 \implies x^2 + y^2 = 25$

Section 11.2   Space Coordinates and Vectors in Space

1.

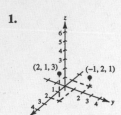

3.

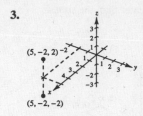

5. $A(2, 3, 4)$

$B(-1, -2, 2)$

7. $x = -3, y = 4, z = 5$: $(-3, 4, 5)$

9. $y = z = 0, x = 10$: $(10, 0, 0)$

11. The z-coordinate is 0.

13. The point is 6 units above the xy-plane.

15. The point is on the plane parallel to the yz-plane that passes through $x = 4$.

17. The point is to the left of the xz-plane.

19. The point is on or between the planes $y = 3$ and $y = -3$.

21. The point (x, y, z) is 3 units below the xy-plane, and below either quadrant I or III.

23. The point could be above the xy-plane and thus above quadrants II or IV, or below the xy-plane, and thus below quadrants I or III.

25. $d = \sqrt{(5 - 0)^2 + (2 - 0)^2 + (6 - 0)^2}$
$= \sqrt{25 + 4 + 36} = \sqrt{65}$

27. $d = \sqrt{(6 - 1)^2 + (-2 - (-2))^2 + (-2 - 4)^2}$
$= \sqrt{25 + 0 + 36} = \sqrt{61}$

29. $A(0, 0, 0), B(2, 2, 1), C(2, -4, 4)$

$|AB| = \sqrt{4 + 4 + 1} = 3$

$|AC| = \sqrt{4 + 16 + 16} = 6$

$|BC| = \sqrt{0 + 36 + 9} = 3\sqrt{5}$

$|BC|^2 = |AB|^2 + |AC|^2$

Right triangle

31. $A(1, -3, -2), B(5, -1, 2), C(-1, 1, 2)$

$|AB| = \sqrt{16 + 4 + 16} = 6$

$|AC| = \sqrt{4 + 16 + 16} = 6$

$|BC| = \sqrt{36 + 4 + 0} = 2\sqrt{10}$

Since $|AB| = |AC|$, the triangle is isosceles.

33. The z-coordinate is changed by 5 units:

$(0, 0, 5), (2, 2, 6), (2, -4, 9)$

35. $\left(\dfrac{5 + (-2)}{2}, \dfrac{-9 + 3}{2}, \dfrac{7 + 3}{2} \right) = \left(\dfrac{3}{2}, -3, 5 \right)$

37. Center: $(0, 2, 5)$

Radius: 2

$(x - 0)^2 + (y - 2)^2 + (z - 5)^2 = 4$

$x^2 + y^2 + z^2 - 4y - 10z + 25 = 0$

39. Center: $\dfrac{(2, 0, 0) + (0, 6, 0)}{2} = (1, 3, 0)$

Radius: $\sqrt{10}$

$(x - 1)^2 + (y - 3)^2 + (z - 0)^2 = 10$

$x^2 + y^2 + z^2 - 2x - 6y = 0$

41. $x^2 + y^2 + z^2 - 2x + 6y + 8z + 1 = 0$

$(x^2 - 2x + 1) + (y^2 + 6y + 9) + (z^2 + 8z + 16) = -1 + 1 + 9 + 16$

$(x - 1)^2 + (y + 3)^2 + (z + 4)^2 = 25$

Center: $(1, -3, -4)$

Radius: 5

43. $9x^2 + 9y^2 + 9z^2 - 6x + 18y + 1 = 0$

$$x^2 + y^2 + z^2 - \frac{2}{3}x + 2y + \frac{1}{9} = 0$$

$$\left(x^2 - \frac{2}{3}x + \frac{1}{9}\right) + (y^2 + 2y + 1) + z^2 = -\frac{1}{9} + \frac{1}{9} + 1$$

$$\left(x - \frac{1}{3}\right)^2 + (y + 1)^2 + (z - 0)^2 = 1$$

Center: $\left(\frac{1}{3}, -1, 0\right)$

Radius: 1

45. $x^2 + y^2 + z^2 \leq 36$

Solid ball of radius 6 centered at origin.

47.
$$x^2 + y^2 + z^2 < 4x - 6y + 8z - 13$$

$$(x^2 - 4x + 4) + (y^2 + 6y + 9) + (z^2 - 8z + 16) < 4 + 9 + 16 - 13$$

$$(x - 2)^2 + (y + 3)^2 + (z - 4)^2 < 16$$

Interior of sphere of radius 4 centered at $(2, -3, 4)$.

49. (a) $\mathbf{v} = (2 - 4)\mathbf{i} + (4 - 2)\mathbf{j} + (3 - 1)\mathbf{k}$

$\quad = -2\mathbf{i} + 2\mathbf{j} + 2\mathbf{k} = \langle -2, 2, 2 \rangle$

(b)

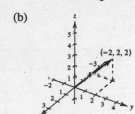

51. (a) $\mathbf{v} = (0 - 3)\mathbf{i} + (3 - 3)\mathbf{j} + (3 - 0)\mathbf{k}$

$\quad = -3\mathbf{i} + 3\mathbf{k} = \langle -3, 0, 3 \rangle$

(b)

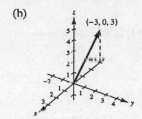

53. $\langle 4 - 3, 1 - 2, 6 - 0 \rangle = \langle 1, -1, 6 \rangle$

$\|\langle 1, -1, 6 \rangle\| = \sqrt{1 + 1 + 36} = \sqrt{38}$

Unit vector: $\dfrac{\langle 1, -1, 6 \rangle}{\sqrt{38}} = \left\langle \dfrac{1}{\sqrt{38}}, \dfrac{-1}{\sqrt{38}}, \dfrac{6}{\sqrt{38}} \right\rangle$

55. $\langle -5 - (-4), 3 - 3, 0 - 1 \rangle = \langle -1, 0, -1 \rangle$

$\|\langle -1, 0, -1 \rangle\| = \sqrt{1 + 1} = \sqrt{2}$

Unit vector: $\left\langle \dfrac{-1}{\sqrt{2}}, 0, \dfrac{-1}{\sqrt{2}} \right\rangle$

57. (b) $\mathbf{v} = (3 + 1)\mathbf{i} + (3 - 2)\mathbf{j} + (4 - 3)\mathbf{k}$

$\quad = 4\mathbf{i} + \mathbf{j} + \mathbf{k} = \langle 4, 1, 1 \rangle$

(a) and (c).

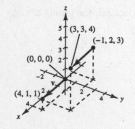

59. $(q_1, q_2, q_3) - (0, 6, 2) = (3, -5, 6)$

$Q = (3, 1, 8)$

61. (a) $2\mathbf{v} = \langle 2, 4, 4 \rangle$ (b) $-\mathbf{v} = \langle -1, -2, -2 \rangle$

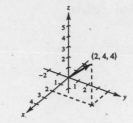

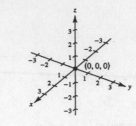

(c) $\frac{3}{2}\mathbf{v} = \langle \frac{3}{2}, 3, 3 \rangle$ (d) $0\mathbf{v} = \langle 0, 0, 0 \rangle$

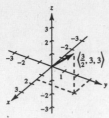

63. $\mathbf{z} = \mathbf{u} - \mathbf{v} = \langle 1, 2, 3 \rangle - \langle 2, 2, -1 \rangle = \langle -1, 0, 4 \rangle$

65. $\mathbf{z} = 2\mathbf{u} + 4\mathbf{v} - \mathbf{w} = \langle 2, 4, 6 \rangle + \langle 8, 8, -4 \rangle - \langle 4, 0, -4 \rangle = \langle 6, 12, 6 \rangle$

67. $2\mathbf{z} - 3\mathbf{u} = 2\langle z_1, z_2, z_3 \rangle - 3\langle 1, 2, 3 \rangle = \langle 4, 0, -4 \rangle$

$\quad 2z_1 - 3 = 4 \implies z_1 = \frac{7}{2}$

$\quad 2z_2 - 6 = 0 \implies z_2 = 3$

$\quad 2z_3 - 9 = -4 \implies z_3 = \frac{5}{2}$

$\quad \mathbf{z} = \langle \frac{7}{2}, 3, \frac{5}{2} \rangle$

69. (a) and (b) are parallel since $\langle -6, -4, 10 \rangle = -2\langle 3, 2, -5 \rangle$ and $\langle 2, \frac{4}{3}, -\frac{10}{3} \rangle = \frac{2}{3}\langle 3, 2, -5 \rangle$.

71. $\mathbf{z} = -3\mathbf{i} + 4\mathbf{j} + 2\mathbf{k}$

(a) is parallel since $-6\mathbf{i} + 8\mathbf{j} + 4\mathbf{k} = 2\mathbf{z}$.

73. $P(0, -2, -5), Q(3, 4, 4), R(2, 2, 1)$

$\quad \overrightarrow{PQ} = \langle 3, 6, 9 \rangle$

$\quad \overrightarrow{PR} = \langle 2, 4, 6 \rangle$

$\quad \langle 3, 6, 9 \rangle = \frac{3}{2}\langle 2, 4, 6 \rangle$

Therefore, $\overrightarrow{PQ}$ and $\overrightarrow{PR}$ are parallel. The points are collinear.

75. $P(1, 2, 4), Q(2, 5, 0), R(0, 1, 5)$

$\quad \overrightarrow{PQ} = \langle 1, 3, -4 \rangle$

$\quad \overrightarrow{PR} = \langle -1, -1, 1 \rangle$

Since $\overrightarrow{PQ}$ and $\overrightarrow{PR}$ are not parallel, the points are not collinear.

77. $A(2, 9, 1), B(3, 11, 4), C(0, 10, 2), D(1, 12, 5)$

$\quad \overrightarrow{AB} = \langle 1, 2, 3 \rangle$

$\quad \overrightarrow{CD} = \langle 1, 2, 3 \rangle$

$\quad \overrightarrow{AC} = \langle -2, 1, 1 \rangle$

$\quad \overrightarrow{BD} = \langle -2, 1, 1 \rangle$

Since $\overrightarrow{AB} = \overrightarrow{CD}$ and $\overrightarrow{AC} = \overrightarrow{BD}$, the given points form the vertices of a parallelogram.

79. $\|\mathbf{v}\| = 0$

81. $\mathbf{v} = \langle 1, -2, -3 \rangle$

$\|\mathbf{v}\| = \sqrt{1 + 4 + 9} = \sqrt{14}$

83. $\mathbf{v} = \langle 0, 3, -5 \rangle$

$\|\mathbf{v}\| = \sqrt{0 + 9 + 25} = \sqrt{34}$

85. $\mathbf{u} = \langle 2, -1, 2 \rangle$

$\|\mathbf{u}\| = \sqrt{4 + 1 + 4} = 3$

(a) $\dfrac{\mathbf{u}}{\|\mathbf{u}\|} = \dfrac{1}{3}\langle 2, -1, 2 \rangle$

(b) $-\dfrac{\mathbf{u}}{\|\mathbf{u}\|} = -\dfrac{1}{3}\langle 2, -1, 2 \rangle$

87. $\mathbf{u} = \langle 3, 2, -5 \rangle$

$\|\mathbf{u}\| = \sqrt{9 + 4 + 25} = \sqrt{38}$

(a) $\dfrac{\mathbf{u}}{\|\mathbf{u}\|} = \dfrac{1}{\sqrt{38}}\langle 3, 2, -5 \rangle$

(b) $-\dfrac{\mathbf{u}}{\|\mathbf{u}\|} = -\dfrac{1}{\sqrt{38}}\langle 3, 2, -5 \rangle$

89. Programs will vary.

91. $c\mathbf{v} = \langle 2c, 2c, -c \rangle$

$\|c\mathbf{v}\| = \sqrt{4c^2 + 4c^2 + c^2} = 5$

$9c^2 = 25$

$c = \pm\dfrac{5}{3}$

93. $\mathbf{v} = 10\dfrac{\mathbf{u}}{\|\mathbf{u}\|} = 10\left\langle 0, \dfrac{1}{\sqrt{2}}, \dfrac{1}{\sqrt{2}} \right\rangle$

$= \left\langle 0, \dfrac{10}{\sqrt{2}}, \dfrac{10}{\sqrt{2}} \right\rangle$

95. $\mathbf{v} = \dfrac{3}{2}\dfrac{\mathbf{u}}{\|\mathbf{u}\|} = \dfrac{3}{2}\left\langle \dfrac{2}{3}, \dfrac{-2}{3}, \dfrac{1}{3} \right\rangle = \left\langle 1, -1, \dfrac{1}{2} \right\rangle$

97. $\mathbf{v} = 2[\cos(\pm 30°)\mathbf{j} + \sin(\pm 30°)\mathbf{k}]$

$= \sqrt{3}\,\mathbf{j} \pm \mathbf{k} = \langle 0, \sqrt{3}, \pm 1 \rangle$

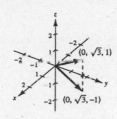

99. $\mathbf{v} = \langle -3, -6, 3 \rangle$

$\dfrac{2}{3}\mathbf{v} = \langle -2, -4, 2 \rangle$

$(4, 3, 0) + (-2, -4, 2) = (2, -1, 2)$

101. (a)

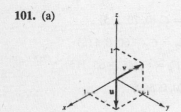

(c) $a\mathbf{i} + (a + b)\mathbf{j} + b\mathbf{k} = \mathbf{i} + 2\mathbf{i} + \mathbf{k}$

$a = 1, b = 1$

$\mathbf{w} = \mathbf{u} + \mathbf{v}$

(b) $\mathbf{w} = a\mathbf{u} + b\mathbf{v} = a\mathbf{i} + (a + b)\mathbf{j} + b\mathbf{k} = \mathbf{0}$

$a = 0, a + b = 0, b = 0$

Thus, a and b are both zero.

(d) $a\mathbf{i} + (a + b)\mathbf{j} + b\mathbf{k} = \mathbf{i} + 2\mathbf{j} + 3\mathbf{k}$

$a = 1, a + b = 2, b = 3$

Not possible

103. x_0 is directed distance to yz-plane.

y_0 is directed distance to xz-plane.

z_0 is directed distance to xy-plane.

105. $(x - x_0)^2 + (y - y_0)^2 + (z - z_0)^2 = r^2$

107.

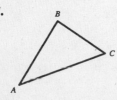

$\overrightarrow{AB} + \overrightarrow{BC} = \overrightarrow{AC}$

Hence, $\overrightarrow{AB} + \overrightarrow{BC} + \overrightarrow{CA} = \overrightarrow{AC} + \overrightarrow{CA} = \mathbf{0}$

109. (a) The height of the right triangle is $h = \sqrt{L^2 - 18^2}$.
The vector $\overrightarrow{PQ}$ is given by

$$\overrightarrow{PQ} = \langle 0, -18, h \rangle.$$

The tension vector **T** in each wire is

$$\mathbf{T} = c\langle 0, -18, h \rangle \text{ where } ch = \frac{24}{3} = 8.$$

Hence, $\mathbf{T} = \dfrac{8}{h}\langle 0, -18, h \rangle$ and

$$T = \|\mathbf{T}\| = \frac{8}{h}\sqrt{18^2 + h^2} = \frac{8}{\sqrt{L^2 - 18^2}}\sqrt{18^2 + (L^2 - 18^2)} = \frac{8L}{\sqrt{L^2 - 18^2}}.$$

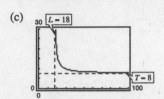

(b)

L	20	25	30	35	40	45	50
T	18.4	11.5	10	9.3	9.0	8.7	8.6

(c)

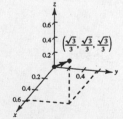

$x = 18$ is a vertical asymptote and $y = 8$ is a horizontal asymptote.

(d) $\displaystyle \lim_{L \to 18^+} \frac{8L}{\sqrt{L^2 - 18^2}} = \infty$

$$\lim_{L \to \infty} \frac{8L}{\sqrt{L^2 - 18^2}} = \lim_{L \to \infty} \frac{8}{\sqrt{1 - (18/L)^2}} = 8$$

(e) From the table, $T = 10$ implies $L = 30$ inches.

111. Let α be the angle between **v** and the coordinate axes.

$$\mathbf{v} = (\cos \alpha)\mathbf{i} + (\cos \alpha)\mathbf{j} + (\cos \alpha)\mathbf{k}$$

$$\|\mathbf{v}\| = \sqrt{3}\cos\alpha = 1$$

$$\cos\alpha = \frac{1}{\sqrt{3}} = \frac{\sqrt{3}}{3}$$

$$\mathbf{v} = \frac{\sqrt{3}}{3}(\mathbf{i} + \mathbf{j} + \mathbf{k}) = \frac{\sqrt{3}}{3}\langle 1, 1, 1 \rangle$$

113. $\overrightarrow{AB} = \langle 0, 70, 115 \rangle$, $\mathbf{F}_1 = C_1\langle 0, 70, 115 \rangle$

$\overrightarrow{AC} = \langle -60, 0, 115 \rangle$, $\mathbf{F}_2 = C_2\langle -60, 0, 115 \rangle$

$\overrightarrow{AD} = \langle 45, -65, 115 \rangle$, $\mathbf{F}_3 = C_3\langle 45, -65, 115 \rangle$

$\mathbf{F} = \mathbf{F}_1 + \mathbf{F}_2 + \mathbf{F}_3 = \langle 0, 0, 500 \rangle$

Thus:

$$-60C_2 + 45C_3 = 0$$
$$70C_1 - 65C_3 = 0$$
$$115(C_1 + C_2 + C_3) = 500$$

Solving this system yields $C_1 = \dfrac{104}{69}$, $C_2 = \dfrac{28}{23}$, and $C_3 = \dfrac{112}{69}$.

Thus:

$$\|\mathbf{F}_1\| \approx 202.919N$$
$$\|\mathbf{F}_2\| \approx 157.909N$$
$$\|\mathbf{F}_3\| \approx 226.521N$$

115. $d(AP) = 2d(BP)$

$$\sqrt{x^2 + (y + 1)^2 + (z - 1)^2} = 2\sqrt{(x - 1)^2 + (y - 2)^2 + z^2}$$

$$x^2 + y^2 + z^2 + 2y - 2z + 2 = 4(x^2 + y^2 + z^2 - 2x - 4y + 5)$$

$$0 = 3x^2 + 3y^2 + 3z^2 - 8x - 18y + 2z + 18$$

$$-6 + \frac{16}{9} + 9 + \frac{1}{9} = \left(x^2 - \frac{8}{3}x + \frac{16}{9}\right) + (y^2 - 6y + 9) + \left(z^2 + \frac{2}{3}z + \frac{1}{9}\right)$$

$$\frac{44}{9} = \left(x - \frac{4}{3}\right)^2 + (y - 3)^2 + \left(z + \frac{1}{3}\right)^2$$

Sphere; center: $\left(\frac{4}{3}, 3, -\frac{1}{3}\right)$, radius: $\frac{2\sqrt{11}}{3}$

Section 11.3   The Dot Product of Two Vectors

1. $\mathbf{u} = \langle 3, 4 \rangle$, $\mathbf{v} = \langle 2, -3 \rangle$

(a) $\mathbf{u} \cdot \mathbf{v} = 3(2) + 4(-3) = -6$

(b) $\mathbf{u} \cdot \mathbf{u} = 3(3) + 4(4) = 25$

(c) $\|\mathbf{u}\|^2 = 25$

(d) $(\mathbf{u} \cdot \mathbf{v})\mathbf{v} = -6\langle 2, -3 \rangle = \langle -12, 18 \rangle$

(e) $\mathbf{u} \cdot (2\mathbf{v}) = 2(\mathbf{u} \cdot \mathbf{v}) = 2(-6) = -12$

3. (a) $\mathbf{u} \cdot \mathbf{v} = \langle 5, -1 \rangle \cdot \langle -3, 2 \rangle = 5(-3) + (-1)(2) = -17$

(b) $\mathbf{u} \cdot \mathbf{u} = \langle 5, -1 \rangle \cdot \langle 5, -1 \rangle = 5(5) + (-1)(-1) = 26$

(c) $\|\mathbf{u}\|^2 = \mathbf{u} \cdot \mathbf{u} = 26$

(d) $(\mathbf{u} \cdot \mathbf{v})\mathbf{v} = (-17)\langle -3, 2 \rangle = \langle 51, -34 \rangle$

(e) $\mathbf{u} \cdot (2\mathbf{v}) = 2(\mathbf{u} \cdot \mathbf{v}) = 2(-17) = -34$

5. $\mathbf{u} = \langle 2, -3, 4 \rangle$, $\mathbf{v} = \langle 0, 6, 5 \rangle$

(a) $\mathbf{u} \cdot \mathbf{v} = 2(0) + (-3)(6) + (4)(5) = 2$

(b) $\mathbf{u} \cdot \mathbf{u} = 2(2) + (-3)(-3) + 4(4) = 29$

(c) $\|\mathbf{u}\|^2 = 29$

(d) $(\mathbf{u} \cdot \mathbf{v})\mathbf{v} = 2\langle 0, 6, 5 \rangle = \langle 0, 12, 10 \rangle$

(e) $\mathbf{u} \cdot (2\mathbf{v}) = 2(\mathbf{u} \cdot \mathbf{v}) = 2(2) = 4$

7. $\mathbf{u} = 2\mathbf{i} - \mathbf{j} + \mathbf{k}$, $\mathbf{v} = \mathbf{i} - \mathbf{k}$

(a) $\mathbf{u} \cdot \mathbf{v} = 2(1) + (-1)(0) + 1(-1) = 1$

(b) $\mathbf{u} \cdot \mathbf{u} = 2(2) + (-1)(-1) + (1)(1) = 6$

(c) $\|\mathbf{u}\|^2 = 6$

(d) $(\mathbf{u} \cdot \mathbf{v})\mathbf{v} = \mathbf{v} = \mathbf{i} - \mathbf{k}$

(e) $\mathbf{u} \cdot (2\mathbf{v}) = 2(\mathbf{u} \cdot \mathbf{v}) = 2$

9. $\dfrac{\mathbf{u} \cdot \mathbf{v}}{\|\mathbf{u}\| \, \|\mathbf{v}\|} = \cos \theta$

$\mathbf{u} \cdot \mathbf{v} = (8)(5) \cos \dfrac{\pi}{3} = 20$

11. $\mathbf{u} = \langle 1, 1 \rangle$, $\mathbf{v} = \langle 2, -2 \rangle$

$\cos \theta = \dfrac{\mathbf{u} \cdot \mathbf{v}}{\|\mathbf{u}\| \, \|\mathbf{v}\|} = \dfrac{0}{\sqrt{2}\sqrt{8}} = 0$

$\theta = \dfrac{\pi}{2}$

13. $\mathbf{u} = 3\mathbf{i} + \mathbf{j}$, $\mathbf{v} = -2\mathbf{i} + 4\mathbf{j}$

$\cos \theta = \dfrac{\mathbf{u} \cdot \mathbf{v}}{\|\mathbf{u}\| \, \|\mathbf{v}\|} = \dfrac{-2}{\sqrt{10}\sqrt{20}} = \dfrac{-1}{5\sqrt{2}}$

$\theta = \arccos\left(-\dfrac{1}{5\sqrt{2}}\right) \approx 98.1°$

15. $\mathbf{u} = \langle 1, 1, 1 \rangle$, $\mathbf{v} = \langle 2, 1, -1 \rangle$

$\cos \theta = \dfrac{\mathbf{u} \cdot \mathbf{v}}{\|\mathbf{u}\| \, \|\mathbf{v}\|} = \dfrac{2}{\sqrt{3}\sqrt{6}} = \dfrac{\sqrt{2}}{3}$

$\theta = \arccos \dfrac{\sqrt{2}}{3} \approx 61.9°$

17. $\mathbf{u} = 3\mathbf{i} + 4\mathbf{j}, \mathbf{v} = -2\mathbf{j} + 3\mathbf{k}$

$$\cos\theta = \frac{\mathbf{u} \cdot \mathbf{v}}{\|\mathbf{u}\|\,\|\mathbf{v}\|} = \frac{-8}{5\sqrt{13}} = \frac{-8\sqrt{13}}{65}$$

$$\theta = \arccos\left(-\frac{8\sqrt{13}}{65}\right) \approx 116.3°$$

19. $\mathbf{u} = \langle 4, 0 \rangle, \mathbf{v} = \langle 1, 1 \rangle$

$\mathbf{u} \neq c\mathbf{v} \implies$ not parallel

$\mathbf{u} \cdot \mathbf{v} = 4 \neq 0 \implies$ not orthogonal

Neither

21. $\mathbf{u} = \langle 4, 3 \rangle, \mathbf{v} = \left\langle \frac{1}{2}, -\frac{2}{3} \right\rangle$

$\mathbf{u} \neq c\mathbf{v} \implies$ not parallel

$\mathbf{u} \cdot \mathbf{v} = 0 \implies$ orthogonal

23. $\mathbf{u} = \mathbf{j} + 6\mathbf{k}, \mathbf{v} = \mathbf{i} - 2\mathbf{j} - \mathbf{k}$

$\mathbf{u} \neq c\mathbf{v} \implies$ not parallel

$\mathbf{u} \cdot \mathbf{v} = -8 \neq 0 \implies$ not orthogonal

Neither

25. $\mathbf{u} = \langle 2, -3, 1 \rangle, \mathbf{v} = \langle -1, -1, -1 \rangle$

$\mathbf{u} \neq c\mathbf{v} \implies$ not parallel

$\mathbf{u} \cdot \mathbf{v} = 0 \implies$ orthogonal

27. The vector $\langle 1, 2, 0 \rangle$ joining $(1, 2, 0)$ and $(0, 0, 0)$ is perpendicular to the vector $\langle -2, 1, 0 \rangle$ joining $(-2, 1, 0)$ and $(0, 0, 0)$:

$$\langle 1, 2, 0 \rangle \cdot \langle -2, 1, 0 \rangle = 0$$

The triangle is a right triangle.

29. The vectors forming the sides of the triangle are:

$\mathbf{u} = \langle 0 - 2, 1 + 3, 2 - 4 \rangle = \langle -2, 4, -2 \rangle$

$\mathbf{v} = \langle -1 - 2, 2 + 3, 0 - 4 \rangle = \langle -3, 5, -4 \rangle$

$\mathbf{w} = \langle -1 - 0, 2 - 1, 0 - 2 \rangle = \langle -1, 1, -2 \rangle$

$\mathbf{u} \cdot \mathbf{v} = 6 + 9 + 8 = 23 > 0$

$\mathbf{u} \cdot \mathbf{w} = 2 + 4 + 4 = 10 > 0$

$\mathbf{v} \cdot \mathbf{w} = 4 + 5 + 8 = 17 > 0$

The triangle has three acute angles.

31. $\mathbf{u} = \mathbf{i} + 2\mathbf{j} + 2\mathbf{k}, \|\mathbf{u}\| = 3$

$$\cos\alpha = \frac{1}{3}$$

$$\cos\beta = \frac{2}{3}$$

$$\cos\gamma = \frac{2}{3}$$

$$\cos^2\alpha + \cos^2\beta + \cos^2\gamma = \frac{1}{9} + \frac{4}{9} + \frac{4}{9} = 1$$

33. $\mathbf{u} = \langle 0, 6, -4 \rangle, \|\mathbf{u}\| = \sqrt{52} = 2\sqrt{13}$

$$\cos\alpha = 0$$

$$\cos\beta = \frac{3}{\sqrt{13}}$$

$$\cos\gamma = -\frac{2}{\sqrt{13}}$$

$$\cos^2\alpha + \cos^2\beta + \cos^2\gamma = 0 + \frac{9}{13} + \frac{4}{13} = 1$$

35. $\mathbf{u} = \langle 3, 2, -2 \rangle \quad \|\mathbf{u}\| = \sqrt{17}$

$$\cos\alpha = \frac{3}{\sqrt{17}} \implies \alpha \approx 0.7560 \text{ or } 43.3°$$

$$\cos\beta = \frac{2}{\sqrt{17}} \implies \beta \approx 1.0644 \text{ or } 61.0°$$

$$\cos\gamma = \frac{-2}{\sqrt{17}} \implies \gamma \approx 2.0772 \text{ or } 119.0°$$

37. $\mathbf{u} = \langle -1, 5, 2 \rangle \quad \|\mathbf{u}\| = \sqrt{30}$

$$\cos\alpha = \frac{-1}{\sqrt{30}} \implies \alpha \approx 1.7544 \text{ or } 100.5°$$

$$\cos\beta = \frac{5}{\sqrt{30}} \implies \beta \approx 0.4205 \text{ or } 24.1°$$

$$\cos\gamma = \frac{2}{\sqrt{30}} \implies \gamma \approx 1.1970 \text{ or } 68.6°$$

39. F_1: $C_1 = \dfrac{50}{\|F_1\|} \approx 4.3193$

    F_2: $C_2 = \dfrac{80}{\|F_2\|} \approx 5.4183$

    $F = F_1 + F_2$

        $\approx 4.3193\langle 10, 5, 3\rangle + 5.4183\langle 12, 7, -5\rangle$

        $= \langle 108.2126, 59.5246, -14.1336\rangle$

    $\|F\| \approx 124.310\text{ lb}$

    $\cos\alpha \approx \dfrac{108.2126}{\|F\|} \implies \alpha \approx 29.48°$

    $\cos\beta \approx \dfrac{59.5246}{\|F\|} \implies \beta \approx 61.39°$

    $\cos\gamma \approx \dfrac{-14.1336}{\|F\|} \implies \gamma \approx 96.53°$

41. $\overrightarrow{OA} = \langle 0, 10, 10\rangle$

    $\cos\alpha = \dfrac{0}{\sqrt{0^2 + 10^2 + 10^2}} = 0 \implies \alpha = 90°$

    $\cos\beta = \cos\gamma = \dfrac{10}{\sqrt{0^2 + 10^2 + 10^2}}$

        $= \dfrac{1}{\sqrt{2}} \implies \beta = \gamma = 45°$

43. $w_2 = u - w_1 = \langle 6, 7\rangle - \langle 2, 8\rangle = \langle 4, -1\rangle$

45. $w_2 = u - w_1 = \langle 0, 3, 3\rangle - \langle -2, 2, 2\rangle = \langle 2, 1, 1\rangle$

47. $u = \langle 2, 3\rangle$, $v = \langle 5, 1\rangle$

    (a) $w_1 = \left(\dfrac{u \cdot v}{\|v\|^2}\right)v = \dfrac{13}{26}\langle 5, 1\rangle = \left\langle \dfrac{5}{2}, \dfrac{1}{2}\right\rangle$

    (b) $w_2 = u - w_1 = \left\langle -\dfrac{1}{2}, \dfrac{5}{2}\right\rangle$

49. $u = \langle 2, 1, 2\rangle$, $v = \langle 0, 3, 4\rangle$

    (a) $w_1 = \left(\dfrac{u \cdot v}{\|v\|^2}\right)v$

        $= \dfrac{11}{25}\langle 0, 3, 4\rangle = \left\langle 0, \dfrac{33}{25}, \dfrac{44}{25}\right\rangle$

    (b) $w_2 = u - w_1 = \left\langle 2, -\dfrac{8}{25}, \dfrac{6}{25}\right\rangle$

51. $u \cdot v = \langle u_1, u_2, u_3\rangle \cdot \langle v_1, v_2, v_3\rangle = u_1 v_1 + u_2 v_2 + u_3 v_3$

53. (a) Orthogonal, $\theta = \dfrac{\pi}{2}$

    (b) Acute, $0 < \theta < \dfrac{\pi}{2}$

    (c) Obtuse, $\dfrac{\pi}{2} < \theta < \pi$

55. See page 784. Direction cosines of $v = \langle v_1, v_2, v_3\rangle$ are

    $\cos\alpha = \dfrac{v_1}{\|v\|}, \cos\beta = \dfrac{v_2}{\|v\|}, \cos\gamma = \dfrac{v_3}{\|v\|}.$

    α, β, and γ are the direction angles. See Figure 11.26.

57. (a) $\left(\dfrac{u \cdot v}{\|v\|^2}\right)v = u \implies u = cv \implies u$ and v are parallel.

    (b) $\left(\dfrac{u \cdot v}{\|v\|^2}\right)v = 0 \implies u \cdot v = 0 \implies u$ and v are orthogonal.

59. $u = \langle 3240, 1450, 2235\rangle$

    $v = \langle 1.35, 2.65, 1.85\rangle$

    $u \cdot v = 3240(1.35) + 1450(2.65) + 2235(1.85)$

        $= \$12{,}351.25$

    This represents the total amount that the restaurant earned on its three products.

61. Programs will vary.

63. Programs will vary.

65. Because u appears to be perpendicular to v, the projection of u onto v is 0. Analytically,

    $\text{proj}_v u = \dfrac{u \cdot v}{\|v\|^2}v = \dfrac{\langle 2, -3\rangle \cdot \langle 6, 4\rangle}{\|\langle 6, 4\rangle\|^2}\langle 6, 4\rangle = 0\langle 6, 4\rangle = 0.$

67. $\mathbf{u} = \frac{1}{2}\mathbf{i} - \frac{2}{3}\mathbf{j}$. Want $\mathbf{u} \cdot \mathbf{v} = 0$.

$\mathbf{v} = 8\mathbf{i} + 6\mathbf{j}$ and $-\mathbf{v} = -8\mathbf{i} - 6\mathbf{j}$ are orthogonal to $\mathbf{u}$.

69. $\mathbf{u} = \langle 3, 1, -2 \rangle$. Want $\mathbf{u} \cdot \mathbf{v} = 0$.

$\mathbf{v} = \langle 0, 2, 1 \rangle$ and $-\mathbf{v} = \langle 0, -2, -1 \rangle$ are orthogonal to $\mathbf{u}$.

71. (a) Gravitational Force $\mathbf{F} = -48,000\,\mathbf{j}$

$\mathbf{v} = \cos 10°\mathbf{i} + \sin 10°\mathbf{j}$

$\mathbf{w}_1 = \dfrac{\mathbf{F} \cdot \mathbf{v}}{\|\mathbf{v}\|^2}\mathbf{v} = (\mathbf{F} \cdot \mathbf{v})\mathbf{v} = (-48,000)(\sin 10°)\mathbf{v}$

$\approx -8335.1(\cos 10°\,\mathbf{i} + \sin 10°\mathbf{j})$

$\|\mathbf{w}_1\| \approx 8335.1$ lb

(b) $\mathbf{w}_2 = \mathbf{F} \cdot \mathbf{w}_1 = -48,000\mathbf{j} + 8335.1(\cos 10°\mathbf{i} + \sin 10°\mathbf{j})$

$= 8208.5\mathbf{i} - 46,552.6\mathbf{j}$

$\|\mathbf{w}_2\| \approx 47,270.8$ lb

73. $\mathbf{F} = 85\left(\dfrac{1}{2}\mathbf{i} + \dfrac{\sqrt{3}}{2}\mathbf{j}\right)$

$\mathbf{v} = 10\mathbf{i}$

$W = \mathbf{F} \cdot \mathbf{v} = 425$ ft · lb

75. False.

Let $\mathbf{u} = \langle 2, 4 \rangle$, $\mathbf{v} = \langle 1, 7 \rangle$ and $\mathbf{w} = \langle 5, 5 \rangle$. Then $\mathbf{u} \cdot \mathbf{v} = 2 + 28 = 30$ and $\mathbf{u} \cdot \mathbf{w} = 10 + 20 = 30$.

77. Let s = length of a side.

$\mathbf{v} = \langle s, s, s \rangle$

$\|\mathbf{v}\| = s\sqrt{3}$

$\cos \alpha = \cos \beta = \cos \gamma = \dfrac{s}{s\sqrt{3}} = \dfrac{1}{\sqrt{3}}$

$\alpha = \beta = \gamma = \arccos\left(\dfrac{1}{\sqrt{3}}\right) \approx 54.7°$

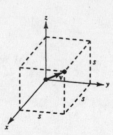

79. (a) The graphs $y_1 = x^2$ and $y_2 = x^{1/3}$ intersect at $(0, 0)$ and $(1, 1)$.

$y'_1 = 2x$ and $y'_2 = \dfrac{1}{3x^{2/3}}$.

At $(0, 0)$, $\pm\langle 1, 0 \rangle$ is tangent to y_1 and $\pm\langle 0, 1 \rangle$ is tangent to y_2.

At $(1, 1)$, $y'_1 = 2$ and $y'_2 = \dfrac{1}{3}$.

$\pm\dfrac{1}{\sqrt{5}}\langle 1, 2 \rangle$ is tangent to y_1, $\pm\dfrac{1}{\sqrt{10}}\langle 3, 1 \rangle$ is tangent to y_2.

(b) At $(0, 0)$, the vectors are perpendicular $(90°)$.

At $(1, 1)$, $\cos \theta = \dfrac{\dfrac{1}{\sqrt{5}}\langle 1, 2 \rangle \cdot \dfrac{1}{\sqrt{10}}\langle 3, 1 \rangle}{(1)(1)} = \dfrac{5}{\sqrt{50}} = \dfrac{1}{\sqrt{2}}$

$\theta = 45°$.

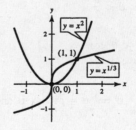

81. (a) The graphs of $y_1 = 1 - x^2$ and $y^2 = x^2 - 1$ intersect at $(1, 0)$ and $(-1, 0)$.

$y'_1 = -2x$ and $y'_2 = 2x$.

At $(1, 0)$, $y'_1 = -2$ and $y'_2 = 2$.

$\pm\dfrac{1}{\sqrt{5}}\langle 1, -2\rangle$ is tangent to y_1, $\pm\dfrac{1}{\sqrt{5}}\langle 1, 2\rangle$ is tangent to y_2.

At $(-1, 0)$, $y'_1 = 2$ and $y'_2 = -2$.

$\pm\dfrac{1}{\sqrt{5}}\langle 1, 2\rangle$ is tangent to y_1, $\pm\dfrac{1}{\sqrt{5}}\langle 1, -2\rangle$ is tangent to y_2.

(b) At $(1, 0)$, $\cos\theta = \dfrac{1}{\sqrt{5}}\langle 1, -2\rangle \cdot \dfrac{-1}{\sqrt{5}}\langle 1, -2\rangle = \dfrac{3}{5}$.

$\theta \approx 0.9273$ or $53.13°$

By symmetry, the angle is the same at $(-1, 0)$.

83. In a rhombus, $\|\mathbf{u}\| = \|\mathbf{v}\|$. The diagonals are $\mathbf{u} + \mathbf{v}$ and $\mathbf{u} - \mathbf{v}$.

$(\mathbf{u} + \mathbf{v}) \cdot (\mathbf{u} - \mathbf{v}) = (\mathbf{u} + \mathbf{v}) \cdot \mathbf{u} - (\mathbf{u} + \mathbf{v}) \cdot \mathbf{v}$

$= \mathbf{u} \cdot \mathbf{u} + \mathbf{v} \cdot \mathbf{u} - \mathbf{u} \cdot \mathbf{v} - \mathbf{v} \cdot \mathbf{v}$

$= \|\mathbf{u}\|^2 - \|\mathbf{v}\|^2 = 0$

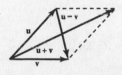

Therefore, the diagonals are orthogonal.

85. (a)

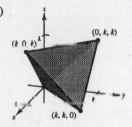

(b) Length of each edge:

$\sqrt{k^2 + k^2 + 0^2} = k\sqrt{2}$

(c) $\cos\theta = \dfrac{k^2}{\left(k\sqrt{2}\right)\left(k\sqrt{2}\right)} = \dfrac{1}{2}$

$\theta = \arccos\left(\dfrac{1}{2}\right) = 60°$

(d) $\overrightarrow{r_1} = \langle k, k, 0\rangle - \left\langle \dfrac{k}{2}, \dfrac{k}{2}, \dfrac{k}{2}\right\rangle = \left\langle \dfrac{k}{2}, \dfrac{k}{2}, -\dfrac{k}{2}\right\rangle$

$\overrightarrow{r_2} = \langle 0, 0, 0\rangle - \left\langle \dfrac{k}{2}, \dfrac{k}{2}, \dfrac{k}{2}\right\rangle = \left\langle -\dfrac{k}{2}, -\dfrac{k}{2}, -\dfrac{k}{2}\right\rangle$

$\cos\theta = \dfrac{-\dfrac{k^2}{4}}{\left(\dfrac{k}{2}\right)^2 \cdot 3} = -\dfrac{1}{3}$

$\theta = 109.5°$

87. $\|\mathbf{u} - \mathbf{v}\|^2 = (\mathbf{u} - \mathbf{v}) \cdot (\mathbf{u} - \mathbf{v})$

$= (\mathbf{u} - \mathbf{v}) \cdot \mathbf{u} - (\mathbf{u} - \mathbf{v}) \cdot \mathbf{v}$

$= \mathbf{u} \cdot \mathbf{u} - \mathbf{v} \cdot \mathbf{u} - \mathbf{u} \cdot \mathbf{v} + \mathbf{v} \cdot \mathbf{v}$

$= \|\mathbf{u}\|^2 - \mathbf{u} \cdot \mathbf{v} - \mathbf{u} \cdot \mathbf{v} + \|\mathbf{v}\|^2$

$= \|\mathbf{u}\|^2 + \|\mathbf{v}\|^2 - 2\mathbf{u} \cdot \mathbf{v}$

89. $\|\mathbf{u} + \mathbf{v}\|^2 = (\mathbf{u} + \mathbf{v}) \cdot (\mathbf{u} + \mathbf{v})$

$= (\mathbf{u} + \mathbf{v}) \cdot \mathbf{u} + (\mathbf{u} + \mathbf{v}) \cdot \mathbf{v}$

$= \mathbf{u} \cdot \mathbf{u} + \mathbf{v} \cdot \mathbf{u} + \mathbf{u} \cdot \mathbf{v} + \mathbf{v} \cdot \mathbf{v}$

$= \|\mathbf{u}\|^2 + 2\mathbf{u} \cdot \mathbf{v} + \|\mathbf{v}\|^2$

$\le \|\mathbf{u}\|^2 + 2\|\mathbf{u}\|\,\|\mathbf{v}\| + \|\mathbf{v}\|^2$

$\le (\|\mathbf{u}\| + \|\mathbf{v}\|)^2$

Therefore, $\|\mathbf{u} + \mathbf{v}\| \le \|\mathbf{u}\| + \|\mathbf{v}\|$.

Section 11.4   The Cross Product of Two Vectors in Space

1. $\mathbf{j} \times \mathbf{i} = \begin{vmatrix} \mathbf{i} & \mathbf{j} & \mathbf{k} \\ 0 & 1 & 0 \\ 1 & 0 & 0 \end{vmatrix} = -\mathbf{k}$

3. $\mathbf{j} \times \mathbf{k} = \begin{vmatrix} \mathbf{i} & \mathbf{j} & \mathbf{k} \\ 0 & 1 & 0 \\ 0 & 0 & 1 \end{vmatrix} = \mathbf{i}$

5. $\mathbf{i} \times \mathbf{k} = \begin{vmatrix} \mathbf{i} & \mathbf{j} & \mathbf{k} \\ 1 & 0 & 0 \\ 0 & 0 & 1 \end{vmatrix} = -\mathbf{j}$

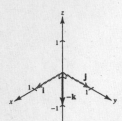

7. (a) $\mathbf{u} \times \mathbf{v} = \begin{vmatrix} \mathbf{i} & \mathbf{j} & \mathbf{k} \\ -2 & 3 & 4 \\ 3 & 7 & 2 \end{vmatrix} = \langle -22, 16, -23 \rangle$

   (b) $\mathbf{v} \times \mathbf{u} = -(\mathbf{u} \times \mathbf{v}) = \langle 22, -16, 23 \rangle$

   (c) $\mathbf{v} \times \mathbf{v} = \begin{vmatrix} \mathbf{i} & \mathbf{j} & \mathbf{k} \\ 3 & 7 & 2 \\ 3 & 7 & 2 \end{vmatrix} = \mathbf{0}$

9. (a) $\mathbf{u} \times \mathbf{v} = \begin{vmatrix} \mathbf{i} & \mathbf{j} & \mathbf{k} \\ 7 & 3 & 2 \\ 1 & -1 & 5 \end{vmatrix} = \langle 17, -33, -10 \rangle$

   (b) $\mathbf{v} \times \mathbf{u} = -(\mathbf{u} \times \mathbf{v}) = \langle -17, 33, 10 \rangle$

   (c) $\mathbf{v} \times \mathbf{v} = \mathbf{0}$

11. $\mathbf{u} = \langle 2, -3, 1 \rangle, \mathbf{v} = \langle 1, -2, 1 \rangle$

$\mathbf{u} \times \mathbf{v} = \begin{vmatrix} \mathbf{i} & \mathbf{j} & \mathbf{k} \\ 2 & -3 & 1 \\ 1 & -2 & 1 \end{vmatrix} = -\mathbf{i} - \mathbf{j} - \mathbf{k} = \langle -1, -1, -1 \rangle$

$\mathbf{u} \cdot (\mathbf{u} \times \mathbf{v}) = 2(-1) + (-3)(-1) + (1)(-1) = 0 \implies \mathbf{u} \perp \mathbf{u} \times \mathbf{v}$

$\mathbf{v} \cdot (\mathbf{u} \times \mathbf{v}) = 1(-1) + (-2)(-1) + (1)(-1) = 0 \implies \mathbf{v} \perp \mathbf{u} \times \mathbf{v}$

13. $\mathbf{u} = \langle 12, -3, 0 \rangle, \mathbf{v} = \langle -2, 5, 0 \rangle$

$\mathbf{u} \times \mathbf{v} = \begin{vmatrix} \mathbf{i} & \mathbf{j} & \mathbf{k} \\ 12 & -3 & 0 \\ -2 & 5 & 0 \end{vmatrix} = 54\mathbf{k} = \langle 0, 0, 54 \rangle$

$\mathbf{u} \cdot (\mathbf{u} \times \mathbf{v}) = 12(0) + (-3)(0) + 0(54)$

$\qquad = 0 \implies \mathbf{u} \perp \mathbf{u} \times \mathbf{v}$

$\mathbf{v} \cdot (\mathbf{u} \times \mathbf{v}) = -2(0) + 5(0) + 0(54)$

$\qquad = 0 \implies \mathbf{v} \perp \mathbf{u} \times \mathbf{v}$

15. $\mathbf{u} = \mathbf{i} + \mathbf{j} + \mathbf{k}, \mathbf{v} = 2\mathbf{i} + \mathbf{j} - \mathbf{k}$

$\mathbf{u} \times \mathbf{v} = \begin{vmatrix} \mathbf{i} & \mathbf{j} & \mathbf{k} \\ 1 & 1 & 1 \\ 2 & 1 & -1 \end{vmatrix} = -2\mathbf{i} + 3\mathbf{j} - \mathbf{k} = \langle -2, 3, -1 \rangle$

$\mathbf{u} \cdot (\mathbf{u} \times \mathbf{v}) = 1(-2) + 1(3) + 1(-1)$

$\qquad = 0 \implies \mathbf{u} \perp \mathbf{u} \times \mathbf{v}$

$\mathbf{v} \cdot (\mathbf{u} \times \mathbf{v}) = 2(-2) + 1(3) + (-1)(-1)$

$\qquad = 0 \implies \mathbf{v} \perp \mathbf{u} \times \mathbf{v}$

$(-\mathbf{v}) \times \mathbf{u} = -(\mathbf{v} \times \mathbf{u}) = \mathbf{u} \times \mathbf{v}$

17.

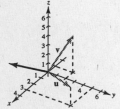

19.

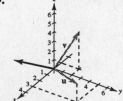

21. $\mathbf{u} = \langle 4, -3.5, 7 \rangle$

$\mathbf{v} = \langle -1, 8, 4 \rangle$

$\mathbf{u} \times \mathbf{v} = \left\langle -70, -23, \dfrac{57}{2} \right\rangle$

$\dfrac{\mathbf{u} \times \mathbf{v}}{\|\mathbf{u} \times \mathbf{v}\|} = \left\langle \dfrac{-140}{\sqrt{24{,}965}}, \dfrac{-46}{\sqrt{24{,}965}}, \dfrac{57}{\sqrt{24{,}965}} \right\rangle$

23. $\mathbf{u} = -3\mathbf{i} + 2\mathbf{j} - 5\mathbf{k}$

$\mathbf{v} = \dfrac{1}{2}\mathbf{i} - \dfrac{3}{4}\mathbf{j} + \dfrac{1}{10}\mathbf{k}$

$\mathbf{u} \times \mathbf{v} = \left\langle -\dfrac{71}{20}, -\dfrac{11}{5}, \dfrac{5}{4} \right\rangle$

$\dfrac{\mathbf{u} \times \mathbf{v}}{\|\mathbf{u} \times \mathbf{v}\|} = \dfrac{20}{\sqrt{7602}} \left\langle -\dfrac{71}{20}, -\dfrac{11}{5}, \dfrac{5}{4} \right\rangle$

$= \left\langle -\dfrac{71}{\sqrt{7602}}, -\dfrac{44}{\sqrt{7602}}, \dfrac{25}{\sqrt{7602}} \right\rangle$

25. Programs will vary.

27. $\mathbf{u} = \mathbf{j}$

$\mathbf{v} = \mathbf{j} + \mathbf{k}$

$\mathbf{u} \times \mathbf{v} = \begin{vmatrix} \mathbf{i} & \mathbf{j} & \mathbf{k} \\ 0 & 1 & 0 \\ 0 & 1 & 1 \end{vmatrix} = \mathbf{i}$

$A = \|\mathbf{u} \times \mathbf{v}\| = \|\mathbf{i}\| = 1$

29. $\mathbf{u} = \langle 3, 2, -1 \rangle$

$\mathbf{v} = \langle 1, 2, 3 \rangle$

$\mathbf{u} \times \mathbf{v} = \begin{vmatrix} \mathbf{i} & \mathbf{j} & \mathbf{k} \\ 3 & 2 & -1 \\ 1 & 2 & 3 \end{vmatrix} = \langle 8, -10, 4 \rangle$

$A = \|\mathbf{u} \times \mathbf{v}\| = \|\langle 8, -10, 4 \rangle\| = \sqrt{180} = 6\sqrt{5}$

31. $A(1, 1, 1,), B(2, 3, 4), C(6, 5, 2), D(7, 7, 5)$

$\overrightarrow{AB} = \langle 1, 2, 3 \rangle, \overrightarrow{AC} = \langle 5, 4, 1 \rangle, \ \overrightarrow{CD} = \langle 1, 2, 3 \rangle,$
$\overrightarrow{BD} = \langle 5, 4, 1 \rangle$

Since $\overrightarrow{AB} = \overrightarrow{CD}$ and $\overrightarrow{AC} = \overrightarrow{BD}$, the figure is a parallelogram. $\overrightarrow{AB}$ and $\overrightarrow{AC}$ are adjacent sides and

$\overrightarrow{AB} \times \overrightarrow{AC} = \begin{vmatrix} \mathbf{i} & \mathbf{j} & \mathbf{k} \\ 1 & 2 & 3 \\ 5 & 4 & 1 \end{vmatrix} = -10\mathbf{i} + 14\mathbf{j} - 6\mathbf{k}.$

$A = \|\overrightarrow{AB} \times \overrightarrow{AC}\| = \sqrt{332} = 2\sqrt{83}$

33. $A(0, 0, 0), B(1, 2, 3), C(-3, 0, 0)$

$\overrightarrow{AB} = \langle 1, 2, 3 \rangle, \overrightarrow{AC} = \langle -3, 0, 0 \rangle$

$\overrightarrow{AB} \times \overrightarrow{AC} = \begin{vmatrix} \mathbf{i} & \mathbf{j} & \mathbf{k} \\ 1 & 2 & 3 \\ -3 & 0 & 0 \end{vmatrix} = -9\mathbf{j} + 6\mathbf{k}$

$A = \dfrac{1}{2}\|\overrightarrow{AB} \times \overrightarrow{AC}\| = \dfrac{1}{2}\sqrt{117} = \dfrac{3}{2}\sqrt{13}$

35. $A(2, -7, 3), B(-1, 5, 8), C(4, 6, -1)$

$\overrightarrow{AB} = \langle -3, 12, 5 \rangle, \overrightarrow{AC} = \langle 2, 13, -4 \rangle$

$\overrightarrow{AB} \times \overrightarrow{AC} = \begin{vmatrix} \mathbf{i} & \mathbf{j} & \mathbf{k} \\ -3 & 12 & 5 \\ 2 & 13 & -4 \end{vmatrix} = \langle -113, -2, -63 \rangle$

Area $= \dfrac{1}{2}\|\overrightarrow{AB} \times \overrightarrow{AC}\| = \dfrac{1}{2}\sqrt{16{,}742}$

37. $\mathbf{F} = -20\mathbf{k}$

$\overrightarrow{PQ} = \dfrac{1}{2}(\cos 40°\,\mathbf{j} + \sin 40°\,\mathbf{k})$

$\overrightarrow{PQ} \times \mathbf{F} = \begin{vmatrix} \mathbf{i} & \mathbf{j} & \mathbf{k} \\ 0 & \cos 40°/2 & \sin 40°/2 \\ 0 & 0 & -20 \end{vmatrix} = -10\cos 40°\,\mathbf{i}$

$\|\overrightarrow{PQ} \times \mathbf{F}\| = 10\cos 40° \approx 7.66 \text{ ft} \cdot \text{lb}$

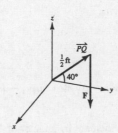

39. (a) Place the wrench in the *xy*-plane, as indicated in the figure.

The angle from $\overrightarrow{AB}$ to $\mathbf{F}$ is $30° + 180° + \theta = 210° + \theta$.

$\|\overrightarrow{OA}\| = 18$ inches $= 1.5$ feet

$\overrightarrow{AB} = 1.5[\cos 30°\mathbf{i} + \sin 30°\mathbf{j}] = \dfrac{3\sqrt{3}}{4}\mathbf{i} + \dfrac{3}{4}\mathbf{j}$

$\mathbf{F} = 60[\cos(210° + \theta)\mathbf{i} + \sin(210° + \theta)\mathbf{j}]$

$\overrightarrow{OA} \times \mathbf{F} = \begin{vmatrix} \mathbf{i} & \mathbf{j} & \mathbf{k} \\ 3\sqrt{3}/4 & 3/4 & 0 \\ 60\cos(210° + \theta) & 60\sin(210° + \theta) & 0 \end{vmatrix}$

$\qquad = [45\sqrt{3}\sin(210° + \theta) - 45\cos(210° + \theta)]\mathbf{k}$

$\qquad = [45\sqrt{3}(\sin 210° \cos \theta + \cos 210° \sin \theta) - 45(\cos 210° \cos \theta - \sin 210° \sin \theta)]\mathbf{k}$

$\qquad = \left[45\sqrt{3}\left(-\dfrac{1}{2}\cos \theta - \dfrac{\sqrt{3}}{2}\sin \theta\right) - 45\left(-\dfrac{\sqrt{3}}{2}\cos \theta + \dfrac{1}{2}\sin \theta\right)\right]\mathbf{k}$

$\qquad = (-90\sin \theta)\mathbf{k}$

Hence, $\|\overrightarrow{OA} \times \mathbf{F}\| = 90\sin \theta$.

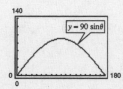

(b) When $\theta = 45°$: $\|\overrightarrow{OA} \times \mathbf{F}\| = 90\left(\dfrac{\sqrt{2}}{2}\right) = 45\sqrt{2} \approx 63.64$.

(c) Let $T = 90\sin \theta$.

$\dfrac{dT}{d\theta} = 90\cos \theta = 0$ when $\theta = 90°$.

This is what we expected. When $\theta = 90°$ the pipe wrench is horizontal.

41. $\mathbf{u} \cdot (\mathbf{v} \times \mathbf{w}) = \begin{vmatrix} 1 & 0 & 0 \\ 0 & 1 & 0 \\ 0 & 0 & 1 \end{vmatrix} = 1$

43. $\mathbf{u} \cdot (\mathbf{v} \times \mathbf{w}) = \begin{vmatrix} 2 & 0 & 1 \\ 0 & 3 & 0 \\ 0 & 0 & 1 \end{vmatrix} = 6$

45. $\mathbf{u} \cdot (\mathbf{v} \times \mathbf{w}) = \begin{vmatrix} 1 & 1 & 0 \\ 0 & 1 & 1 \\ 1 & 0 & 1 \end{vmatrix} = 2$

$V = |\mathbf{u} \cdot (\mathbf{v} \times \mathbf{w})| = 2$

47. $\mathbf{u} = \langle 3, 0, 0 \rangle$

$\mathbf{v} = \langle 0, 5, 1 \rangle$

$\mathbf{w} = \langle 2, 0, 5 \rangle$

$\mathbf{u} \cdot (\mathbf{v} \times \mathbf{w}) = \begin{vmatrix} 3 & 0 & 0 \\ 0 & 5 & 1 \\ 2 & 0 & 5 \end{vmatrix} = 75$

$V = |\mathbf{u} \cdot (\mathbf{v} \times \mathbf{w})| = 75$

49. $\mathbf{u} \times \mathbf{v} = \langle u_1, u_2, u_3 \rangle \cdot \langle v_1, v_2, v_3 \rangle = (u_2 v_3 - u_3 v_2)\mathbf{i} - (u_1 v_3 - u_3 v_1)\mathbf{j} + (u_1 v_2 - u_2 v_1)\mathbf{k}$

51. The magnitude of the cross product will increase by a factor of 4.

53. If the vectors are ordered pairs, then the cross product does not exist. False.

55. True

57. $\mathbf{u} = \langle u_1, u_2, u_3 \rangle$, $\mathbf{v} = \langle v_1, v_2, v_3 \rangle$, $\mathbf{w} = \langle w_1, w_2, w_3 \rangle$

$$\mathbf{u} \times (\mathbf{v} + \mathbf{w}) = \begin{vmatrix} \mathbf{i} & \mathbf{j} & \mathbf{k} \\ u_1 & u_2 & u_3 \\ v_1 + w_1 & v_2 + w_2 & v_3 + w_3 \end{vmatrix}$$

$$= [u_2(v_3 + w_3) - u_3(v_2 + w_2)]\mathbf{i} - [u_1(v_3 + w_3) - u_3(v_1 + w_1)]\mathbf{j} + [u_1(v_2 + w_2) - u_2(v_1 + w_1)]\mathbf{k}$$

$$= (u_2 v_3 - u_3 v_2)\mathbf{i} - (u_1 v_3 - u_3 v_1)\mathbf{j} + (u_1 v_2 - u_2 v_1)\mathbf{k} + (u_2 w_3 - u_3 w_2)\mathbf{i} -$$

$$(u_1 w_3 - u_3 w_1)\mathbf{j} + (u_1 w_2 - u_2 w_1)\mathbf{k}$$

$$= (\mathbf{u} \times \mathbf{v}) + (\mathbf{u} \times \mathbf{w})$$

59. $\mathbf{u} = \langle u_1, u_2, u_3 \rangle$

$$\mathbf{u} \times \mathbf{u} = \begin{vmatrix} \mathbf{i} & \mathbf{j} & \mathbf{k} \\ u_1 & u_2 & u_3 \\ u_1 & u_2 & u_3 \end{vmatrix} = (u_2 u_3 - u_3 u_2)\mathbf{i} - (u_1 u_3 - u_3 u_1)\mathbf{j} + (u_1 u_2 - u_2 u_1)\mathbf{k} = \mathbf{0}$$

61. $\mathbf{u} \times \mathbf{v} = (u_2 v_3 - u_3 v_2)\mathbf{i} - (u_1 v_3 - u_3 v_1)\mathbf{j} + (u_1 v_2 - u_2 v_1)\mathbf{k}$

$(\mathbf{u} \times \mathbf{v}) \cdot \mathbf{u} = (u_2 v_3 - u_3 v_2)u_1 + (u_3 v_1 - u_1 v_3)u_2 + (u_1 v_2 - u_2 v_1)u_3 = 0$

$(\mathbf{u} \times \mathbf{v}) \cdot \mathbf{v} = (u_2 v_3 - u_3 v_2)v_1 + (u_3 v_1 - u_1 v_3)v_2 + (u_1 v_2 - u_2 v_1)v_3 = 0$

Thus, $\mathbf{u} \times \mathbf{v} \perp \mathbf{u}$ and $\mathbf{u} \times \mathbf{v} \perp \mathbf{v}$.

63. $\|\mathbf{u} \times \mathbf{v}\| = \|\mathbf{u}\| \, \|\mathbf{v}\| \sin \theta$

If $\mathbf{u}$ and $\mathbf{v}$ are orthogonal, $\theta = \pi/2$ and $\sin \theta = 1$. Therefore, $\|\mathbf{u} \times \mathbf{v}\| = \|\mathbf{u}\| \, \|\mathbf{v}\|$.

Section 11.5   Lines and Planes in Space

1. $x = 1 + 3t, y = 2 - t, z = 2 + 5t$

(a)

(b) When $t = 0$ we have $P = (1, 2, 2)$. When $t = 3$ we have $Q = (10, -1, 17)$.

$$\overrightarrow{PQ} = \langle 9, -3, 15 \rangle$$

The components of the vector and the coefficients of t are proportional since the line is parallel to $\overrightarrow{PQ}$.

(c) $y = 0$ when $t = 2$. Thus, $x = 7$ and $z = 12$.
Point: $(7, 0, 12)$

$x = 0$ when $t = -\dfrac{1}{3}$. Point: $\left(0, \dfrac{7}{3}, \dfrac{1}{3} \right)$

$z = 0$ when $t = -\dfrac{2}{5}$. Point: $\left(-\dfrac{1}{5}, \dfrac{12}{5}, 0 \right)$

3. Point: $(0, 0, 0)$

Direction vector: $\mathbf{v} = \langle 1, 2, 3 \rangle$

Direction numbers: $1, 2, 3$

(a) Parametric: $x = t, y = 2t, z = 3t$

(b) Symmetric: $x = \dfrac{y}{2} = \dfrac{z}{3}$

5. Point: $(-2, 0, 3)$

Direction vector: $\mathbf{v} = \langle 2, 4, -2 \rangle$

Direction numbers: $2, 4, -2$

(a) Parametric: $x = -2 + 2t, y = 4t, z = 3 - 2t$

(b) Symmetric: $\dfrac{x + 2}{2} = \dfrac{y}{4} = \dfrac{z - 3}{-2}$

7. Point: $(1, 0, 1)$

Direction vector: $\mathbf{v} = 3\mathbf{i} - 2\mathbf{j} + \mathbf{k}$

Direction numbers: $3, -2, 1$

(a) Parametric: $x = 1 + 3t, y = -2t, z = 1 + t$

(b) Symmetric: $\dfrac{x-1}{3} = \dfrac{y}{-2} = \dfrac{z-1}{1}$

9. Points: $(5, -3, -2), \left(\dfrac{-2}{3}, \dfrac{2}{3}, 1\right)$

Direction vector: $\mathbf{v} = \dfrac{17}{3}\mathbf{i} - \dfrac{11}{3}\mathbf{j} - 3\mathbf{k}$

Direction numbers: $17, -11, -9$

(a) Parametric:
$x = 5 + 17t, y = -3 - 11t, z = -2 - 9t$

(b) Symmetric: $\dfrac{x-5}{17} = \dfrac{y+3}{-11} = \dfrac{z+2}{-9}$

11. Points: $(2, 3, 0), (10, 8, 12)$

Direction vector: $\langle 8, 5, 12 \rangle$

Direction numbers: $8, 5, 12$

(a) Parametric: $x = 2 + 8t, y = 3 + 5t, z = 12t$

(b) Symmetric: $\dfrac{x-2}{8} = \dfrac{y-3}{5} = \dfrac{z}{12}$

13. Point: $(2, 3, 4)$

Direction vector: $\mathbf{v} = \mathbf{k}$

Direction numbers: $0, 0, 1$

Parametric: $x = 2, y = 3, z = 4 + t$

15. Point: $(2, 3, 4)$

Direction vector: $\mathbf{v} = 3\mathbf{i} + 2\mathbf{j} - \mathbf{k}$

Direction numbers: $3, 2, -1$

Parametric: $x = 2 + 3t, y = 3 + 2t, z = 4 - t$

17. Point: $(5, -3, -4)$

Direction vector: $\mathbf{v} = \langle 2, -1, 3 \rangle$

Direction numbers: $2, -1, 3$

Parametric: $x = 5 + 2t, y = -3 - t, z = -4 + 3t$

19. Point: $(2, 1, 2)$

Direction vector: $\langle -1, 1, 1 \rangle$

Direction numbers: $-1, 1, 1$

Parametric: $x = 2 - t, y = 1 + t, z = 2 + t$

21. Let $t = 0$: $P = (3, -1, -2)$ (other answers possible)

$\mathbf{v} = \langle -1, 2, 0 \rangle$ (any nonzero multiple of $\mathbf{v}$ is correct)

23. Let each quantity equal 0: $P = (7, -6, -2)$ (other answers possible)

$\mathbf{v} = \langle 4, 2, 1 \rangle$ (any nonzero multiple of $\mathbf{v}$ is correct)

25. L_1: $\mathbf{v} = \langle -3, 2, 4 \rangle$ $\quad$ $(6, -2, 5)$ on line

L_2: $\mathbf{v} = \langle 6, -4, -8 \rangle$ $\quad$ $(6, -2, 5)$ on line

L_3: $\mathbf{v} = \langle -6, 4, 8 \rangle$ $\quad$ $(6, -2, 5)$ not on line

L_4: $\mathbf{v} = \langle 6, 4, -6 \rangle$ $\quad$ not parallel to L_1, L_2, nor L_3

Hence, L_1 and L_2 are identical.

$L_1 = L_2$ and L_3 are parallel.

27. At the point of intersection, the coordinates for one line equal the corresponding coordinates for the other line. Thus,

(i) $4t + 2 = 2s + 2$, (ii) $3 = 2s + 3$, and (iii) $-t + 1 = s + 1$.

From (ii), we find that $s = 0$ and consequently, from (iii), $t = 0$. Letting $s = t = 0$, we see that equation (i) is satisfied and therefore the two lines intersect. Substituting zero for s or for t, we obtain the point $(2, 3, 1)$.

$\mathbf{u} = 4\mathbf{i} - \mathbf{k}$ $\qquad$ (First line)

$\mathbf{v} = 2\mathbf{i} + 2\mathbf{j} + \mathbf{k}$ $\qquad$ (Second line)

$\cos \theta = \dfrac{|\mathbf{u} \cdot \mathbf{v}|}{\|\mathbf{u}\| \, \|\mathbf{v}\|} = \dfrac{8-1}{\sqrt{17}\sqrt{9}} = \dfrac{7}{3\sqrt{17}} = \dfrac{7\sqrt{17}}{51}$

29. Writing the equations of the lines in parametric form we have

$$x = 3t \qquad y = 2 - t \qquad z = -1 + t$$
$$x = 1 + 4s \qquad y = -2 + s \qquad z = -3 - 3s.$$

For the coordinates to be equal, $3t = 1 + 4s$ and $2 - t = -2 + s$. Solving this system yields $t = \frac{17}{7}$ and $s = \frac{11}{7}$. When using these values for s and t, the z coordinates are not equal. The lines do not intersect.

31. $x = 2t + 3 \qquad x = -2s + 7$

$\quad\;\; y = 5t - 2 \qquad y = s + 8$

$\quad\;\; z = -t + 1 \qquad z = 2s - 1$

Point of intersection: $(7, 8, -1)$

Note: $t = 2$ and $s = 0$

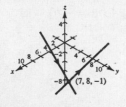

33. $4x - 3y - 6z = 6$

(a) $P = (0, 0, -1), Q = (0, -2, 0), R = (3, 4, -1)$

$\quad \overrightarrow{PQ} = \langle 0, -2, 1 \rangle, \overrightarrow{PR} = \langle 3, 4, 0 \rangle$

(b) $\overrightarrow{PQ} \times \overrightarrow{PR} = \begin{vmatrix} \mathbf{i} & \mathbf{j} & \mathbf{k} \\ 0 & -2 & 1 \\ 3 & 4 & 0 \end{vmatrix} = \langle -4, 3, 6 \rangle$

The components of the cross product are proportional to the coefficients of the variables in the equation. The cross product is parallel to the normal vector.

35. Point: $(2, 1, 2)$

$\mathbf{n} = \mathbf{i} = \langle 1, 0, 0 \rangle$

$1(x - 2) + 0(y - 1) + 0(z - 2) = 0$

$x - 2 = 0$

37. Point: $(3, 2, 2)$

Normal vector: $\mathbf{n} = 2\mathbf{i} + 3\mathbf{j} - \mathbf{k}$

$2(x - 3) + 3(y - 2) - 1(z - 2) = 0$

$2x + 3y - z = 10$

39. Point: $(0, 0, 6)$

Normal vector: $\mathbf{n} = -\mathbf{i} + \mathbf{j} - 2\mathbf{k}$

$-1(x - 0) + 1(y - 0) - 2(z - 6) = 0$

$-x + y - 2z + 12 = 0$

$x - y + 2z = 12$

41. Let $\mathbf{u}$ be the vector from $(0, 0, 0)$ to $(1, 2, 3)$:

$\quad \mathbf{u} = \mathbf{i} + 2\mathbf{j} + 3\mathbf{k}$

Let $\mathbf{v}$ be the vector from $(0, 0, 0)$ to $(-2, 3, 3)$:

$\quad \mathbf{v} = -2\mathbf{i} + 3\mathbf{j} + 3\mathbf{k}$

Normal vector: $\mathbf{u} \times \mathbf{v} = \begin{vmatrix} \mathbf{i} & \mathbf{j} & \mathbf{k} \\ 1 & 2 & 3 \\ -2 & 3 & 3 \end{vmatrix}$

$\qquad\qquad\qquad\quad = -3\mathbf{i} + (-9)\mathbf{j} + 7\mathbf{k}$

$-3(x - 0) - 9(y - 0) + 7(z - 0) = 0$

$3x + 9y - 7z = 0$

43. Let $\mathbf{u}$ be the vector from $(1, 2, 3)$ to $(3, 2, 1)$: $\mathbf{u} = 2\mathbf{i} - 2\mathbf{k}$

Let $\mathbf{v}$ be the vector from $(1, 2, 3)$ to $(-1, -2, 2)$: $\mathbf{v} = -2\mathbf{i} - 4\mathbf{j} - \mathbf{k}$

Normal vector: $\left(\frac{1}{2}\mathbf{u}\right) \times (-\mathbf{v}) = \begin{vmatrix} \mathbf{i} & \mathbf{j} & \mathbf{k} \\ 1 & 0 & -1 \\ 2 & 4 & 1 \end{vmatrix} = 4\mathbf{i} - 3\mathbf{j} + 4\mathbf{k}$

$4(x - 1) - 3(y - 2) + 4(z - 3) = 0$

$4x - 3y + 4z = 10$

45. $(1, 2, 3)$, Normal vector: $\mathbf{v} = \mathbf{k}, 1(z - 3) = 0, z = 3$

47. The direction vectors for the lines are $\mathbf{u} = -2\mathbf{i} + \mathbf{j} + \mathbf{k}$, $\mathbf{v} = -3\mathbf{i} + 4\mathbf{j} - \mathbf{k}$.

Normal vector; $\mathbf{u} \times \mathbf{v} = \begin{vmatrix} \mathbf{i} & \mathbf{j} & \mathbf{k} \\ -2 & 1 & 1 \\ -3 & 4 & -1 \end{vmatrix} = -5(\mathbf{i} + \mathbf{j} + \mathbf{k})$

Point of intersection of the lines: $(-1, 5, 1)$

$(x + 1) + (y - 5) + (z - 1) = 0$

$x + y + z = 5$

49. Let $\mathbf{v}$ be the vector from $(-1, 1, -1)$ to $(2, 2, 1)$:
$\mathbf{v} = 3\mathbf{i} + \mathbf{j} + 2\mathbf{k}$

Let $\mathbf{n}$ be a vector normal to the plane $2x - 3y + z = 3$:
$\mathbf{n} = 2\mathbf{i} - 3\mathbf{j} + \mathbf{k}$

Since $\mathbf{v}$ and $\mathbf{n}$ both lie in the plane P, the normal vector to P is

$\mathbf{v} \times \mathbf{n} = \begin{vmatrix} \mathbf{i} & \mathbf{j} & \mathbf{k} \\ 3 & 1 & 2 \\ 2 & -3 & 1 \end{vmatrix} = 7\mathbf{i} + \mathbf{j} - 11\mathbf{k}$

$7(x - 2) + 1(y - 2) - 11(z - 1) = 0$

$7x + y - 11z = 5$

51. Let $\mathbf{u} = \mathbf{i}$ and let $\mathbf{v}$ be the vector from $(1, -2, -1)$ to $(2, 5, 6)$: $\mathbf{v} = \mathbf{i} + 7\mathbf{j} + 7\mathbf{k}$

Since $\mathbf{u}$ and $\mathbf{v}$ both lie in the plane P, the normal vector to P is:

$\mathbf{u} \times \mathbf{v} = \begin{vmatrix} \mathbf{i} & \mathbf{j} & \mathbf{k} \\ 1 & 0 & 0 \\ 1 & 7 & 7 \end{vmatrix} = -7\mathbf{j} + 7\mathbf{k} = -7(\mathbf{j} - \mathbf{k})$

$[y - (-2)] - [z - (-1)] = 0$

$y - z = -1$

53. xy-plane: Let $z = 0$.

Then $0 = 4 - t \Rightarrow t = 4 \Rightarrow x = 1 - 2(4) = -7$ and

$\quad y = -2 + 3(4) = 10$. Intersection: $(-7, 10, 0)$

xz-plane: Let $y = 0$.

Then $0 = -2 + 3t \Rightarrow t = \frac{2}{3} \Rightarrow x = 1 - 2\left(\frac{2}{3}\right) = -\frac{1}{3}$ and

$\quad z = -4 + \frac{2}{3} = -\frac{10}{3}$. Intersection: $\left(-\frac{1}{3}, 0, -\frac{10}{3}\right)$

yz-plane: Let $x = 0$.

Then $0 = 1 - 2t \Rightarrow t = \frac{1}{2} \Rightarrow y = -2 + 3\left(\frac{1}{2}\right) = -\frac{1}{2}$ and

$\quad z = -4 + \frac{1}{2} = -\frac{7}{2}$. Intersection: $\left(0, -\frac{1}{2}, -\frac{7}{2}\right)$

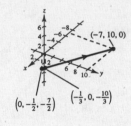

55. Let (x, y, z) be equidistant from $(2, 2, 0)$ and $(0, 2, 2)$.

$$\sqrt{(x - 2)^2 + (y - 2)^2 + (z - 0)^2} = \sqrt{(x - 0)^2 + (y - 2)^2 + (z - 2)^2}$$

$$x^2 - 4x + 4 + y^2 - 4y + 4 + z^2 = x^2 + y^2 - 4y + 4 + z^2 - 4z + 4$$

$$-4x + 8 = -4z + 8$$

$$x - z = 0 \qquad \text{Plane}$$

57. The normal vectors to the planes are

$\mathbf{n}_1 = \langle 5, -3, 1 \rangle$, $\mathbf{n}_2 = \langle 1, 4, 7 \rangle$, $\cos\theta = \dfrac{|\mathbf{n}_1 \cdot \mathbf{n}_2|}{\|\mathbf{n}_1\| \|\mathbf{n}_2\|} = 0$.

Thus, $\theta = \pi/2$ and the planes are orthogonal.

59. The normal vectors to the planes are

$\mathbf{n}_1 = \mathbf{i} - 3\mathbf{j} + 6\mathbf{k}$, $\mathbf{n}_2 = 5\mathbf{i} + \mathbf{j} - \mathbf{k}$,

$\cos\theta = \dfrac{|\mathbf{n}_1 \cdot \mathbf{n}_2|}{\|\mathbf{n}_1\| \|\mathbf{n}_2\|} = \dfrac{|5 - 3 - 6|}{\sqrt{46}\sqrt{27}} = \dfrac{4\sqrt{138}}{414} = \dfrac{2\sqrt{138}}{207}$.

Therefore, $\theta = \arccos\left(\dfrac{2\sqrt{138}}{207}\right) \approx 83.5°$.

61. The normal vectors to the planes are $\mathbf{n}_1 = \langle 1, -5, -1 \rangle$ and $\mathbf{n}_2 = \langle 5, -25, -5 \rangle$.
Since $\mathbf{n}_2 = 5\mathbf{n}_1$, the planes are parallel, but not equal.

63. $4x + 2y + 6z = 12$

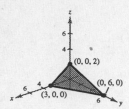

65. $2x - y + 3z = 4$

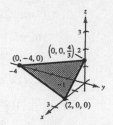

67. $y + z = 5$

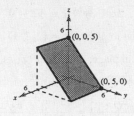

69. $x = 5$

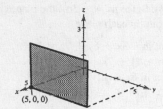

71. $2x + y - z = 6$

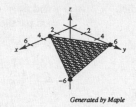

Generated by Maple

73. $-5x + 4y - 6z + 8 = 0$

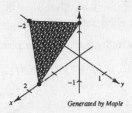

Generated by Maple

75. P_1: $\mathbf{n} = \langle 3, -2, 5 \rangle$ $(1, -1, 1)$ on plane

     P_2: $\mathbf{n} = \langle -6, 4, -10 \rangle$ $(1, -1, 1)$ not on plane

     P_3: $\mathbf{n} = \langle -3, 2, 5 \rangle$

     P_4: $\mathbf{n} = \langle 75, -50, 125 \rangle$ $(1, -1, 1)$ on plane

     P_1 and P_4 are identical.

     $P_1 = P_4$ is parallel to P_2.

77. Each plane passes through the points

     $(c, 0, 0)$, $(0, c, 0)$, and $(0, 0, c)$.

79. If $c = 0$, $z = 0$ is xy plane.

If $c \neq 0$, $cy + z = 0 \implies y = \dfrac{-1}{c}z$ is a plane parallel to x axis and passing through the points $(0, 0, 0)$ and $(0, 1, -c)$.

81. The normals to the planes are $\mathbf{n}_1 = 3\mathbf{i} + 2\mathbf{j} - \mathbf{k}$ and $\mathbf{n}_2 = \mathbf{i} - 4\mathbf{j} + 2\mathbf{k}$. The direction vector for the line is

$$\mathbf{n}_2 \times \mathbf{n}_1 = \begin{vmatrix} \mathbf{i} & \mathbf{j} & \mathbf{k} \\ 1 & -4 & 2 \\ 3 & 2 & -1 \end{vmatrix} = 7(\mathbf{j} + 2\mathbf{k}).$$

Now find a point of intersection of the planes.

$$6x + 4y - 2z = 14$$
$$x - 4y + 2z = 0$$
$$7x \qquad\qquad = 14$$
$$x = 2$$

Substituting 2 for x in the second equation, we have $-4y + 2z = -2$ or $z = 2y - 1$. Letting $y = 1$, a point of intersection is $(2, 1, 1)$.

$$x = 2, y = 1 + t, z = 1 + 2t$$

83. Writing the equation of the line in parametric form and substituting into the equation of the plane we have:

$$x = \frac{1}{2} + t, \ y = \frac{-3}{2} - t, \ z = -1 + 2t$$

$$2\left(\frac{1}{2} + t\right) - 2\left(\frac{-3}{2} - t\right) + (-1 + 2t) = 12, \ t = \frac{3}{2}$$

Substituting $t = 3/2$ into the parametric equations for the line we have the point of intersection $(2, -3, 2)$. The line does not lie in the plane.

85. Writing the equation of the line in parametric form and substituting into the equation of the plane we have:

$$x = 1 + 3t, \ y = -1 - 2t, \ z = 3 + t$$
$$2(1 + 3t) + 3(-1 - 2t) = 10, \ -1 = 10, \ \text{contradiction}$$

Therefore, the line does not intersect the plane.

87. Point: $Q(0, 0, 0)$

Plane: $2x + 3y + z - 12 = 0$

Normal to plane: $\mathbf{n} = \langle 2, 3, 1 \rangle$

Point in plane: $P(6, 0, 0)$

Vector $\overrightarrow{PQ} = \langle -6, 0\ 0 \rangle$

$$D = \frac{|\overrightarrow{PQ} \cdot \mathbf{n}|}{\|\mathbf{n}\|} = \frac{|-12|}{\sqrt{14}} = \frac{6\sqrt{14}}{7}$$

89. Point: $Q(2, 8, 4)$

Plane: $2x + y + z = 5$

Normal to plane: $\mathbf{n} = \langle 2, 1, 1 \rangle$

Point in plane: $P\langle 0, 0, 5 \rangle$

Vector: $\overrightarrow{PQ} = \langle 2, 8, -1 \rangle$

$$D = \frac{|\overrightarrow{PQ} \cdot \mathbf{n}|}{\|\mathbf{n}\|} = \frac{11}{\sqrt{6}} = \frac{11\sqrt{6}}{6}$$

91. The normal vectors to the planes are $\mathbf{n}_1 = \langle 1, -3, 4 \rangle$ and $\mathbf{n}_2 = \langle 1, -3, 4 \rangle$. Since $\mathbf{n}_1 = \mathbf{n}_2$, the planes are parallel. Choose a point in each plane.

$P = (10, 0, 0)$ is a point in $x - 3y + 4z = 10$.

$Q = (6, 0, 0)$ is a point in $x - 3y + 4z = 6$.

$$\overrightarrow{PQ} = \langle -4, 0, 0 \rangle, D = \frac{|\overrightarrow{PQ} \cdot \mathbf{n}_1|}{\|\mathbf{n}_1\|} = \frac{4}{\sqrt{26}} = \frac{2\sqrt{26}}{13}$$

93. The normal vectors to the planes are $\mathbf{n}_1 = \langle -3, 6, 7 \rangle$ and $\mathbf{n}_2 = \langle 6, -12, -14 \rangle$. Since $\mathbf{n}_2 = -2\mathbf{n}_1$, the planes are parallel. Choose a point in each plane.

$P = (0, -1, 1)$ is a point in $-3x + 6y + 7z = 1$.

$Q = \left(\frac{25}{6}, 0, 0 \right)$ is a point in $6x - 12y - 14z = 25$.

$$\overrightarrow{PQ} = \left\langle \frac{25}{6}, 1, -1 \right\rangle$$

$$D = \frac{|\overrightarrow{PQ} \cdot \mathbf{n}_1|}{\|\mathbf{n}_1\|} = \frac{|-27/2|}{\sqrt{94}} = \frac{27}{2\sqrt{94}} = \frac{27\sqrt{94}}{188}$$

95. $\mathbf{u} = \langle 4, 0, -1 \rangle$ is the direction vector for the line.

$Q(1, 5, -2)$ is the given point, and $P(-2, 3, 1)$ is on the line. Hence, $\overrightarrow{PQ} = \langle 3, 2, -3 \rangle$ and

$$\overrightarrow{PQ} \times \mathbf{u} = \begin{vmatrix} \mathbf{i} & \mathbf{j} & \mathbf{k} \\ 3 & 2 & -3 \\ 4 & 0 & -1 \end{vmatrix} = \langle -2, -9, -8 \rangle.$$

$$D = \frac{\|\overrightarrow{PQ} \times \mathbf{u}\|}{\|\mathbf{u}\|} = \frac{\sqrt{149}}{\sqrt{17}} = \frac{\sqrt{2533}}{17}$$

97. $\mathbf{u} = \langle -1, 1, -2 \rangle$ is the direction vector for the line.

$Q = (-2, 1, 3)$ is the given point, and $P = (1, 2, 0)$ is on the line (let $t = 0$ in the parametric equations for the line).

Hence, $\overrightarrow{PQ} = \langle -3, -1, 3 \rangle$ and

$$\overrightarrow{PQ} \times \mathbf{u} = \begin{vmatrix} \mathbf{i} & \mathbf{j} & \mathbf{k} \\ -3 & -1 & 3 \\ -1 & 1 & -2 \end{vmatrix} = \langle -1, -9, -4 \rangle.$$

$$D = \frac{\|\overrightarrow{PQ} \times \mathbf{u}\|}{\|\mathbf{u}\|} = \frac{\sqrt{1 + 81 + 16}}{\sqrt{1 + 1 + 4}} = \frac{\sqrt{98}}{6} = \frac{7}{\sqrt{3}} = \frac{7\sqrt{3}}{3}$$

99. The direction vector for L_1 is $\mathbf{v}_1 = \langle -1, 2, 1 \rangle$.

The direction vector for L_2 is $\mathbf{v}_2 = \langle 3, -6, -3 \rangle$.

Since $\mathbf{v}_2 = -3\mathbf{v}_1$, the lines are parallel.

Let $Q = (2, 3, 4)$ to be a point on L_1 and $P = (0, 1, 4)$ a point on L_2. $\overrightarrow{PQ} = \langle 2, 0, 0 \rangle$.

$\mathbf{u} = \mathbf{v}_1$ is the direction vector for L_2.

$$\overrightarrow{PQ} \times \mathbf{v}_2 = \begin{vmatrix} \mathbf{i} & \mathbf{j} & \mathbf{k} \\ 2 & 2 & 0 \\ 3 & -6 & -3 \end{vmatrix} = \langle -6, 6, -18 \rangle$$

$$D = \frac{\|\overrightarrow{PQ} \times \mathbf{v}_2\|}{\|\mathbf{v}_2\|} = \frac{\sqrt{36 + 36 + 324}}{\sqrt{9 + 36 + 9}} = \sqrt{\frac{396}{54}}$$

$$= \sqrt{\frac{22}{3}} = \frac{\sqrt{66}}{3}$$

101. The parametric equations of a line L parallel to $\mathbf{v} = \langle a, b, c, \rangle$ and passing through the point $P(x_1, y_1, z_1)$ are

$$x = x_1 + at, y = y_1 + bt, z = z_1 + ct.$$

The symmetric equations are

$$\frac{x - x_1}{a} = \frac{y - y_1}{b} = \frac{z - z_1}{c}.$$

103. Solve the two linear equations representing the planes to find two points of intersection. Then find the line determined by the two points.

105. (a) The planes are parallel if their normal vectors are parallel:

$$\langle a_1, b_1, c_1 \rangle = t \langle a_2, b_2, c_2 \rangle, \ t \neq 0$$

(b) The planes are perpendicular if their normal vectors are perpendicular:

$$\langle a_1, b_1, c_1 \rangle \cdot \langle a_2, b_2, c_2 \rangle = 0$$

107. An equation for the plane is

$$\frac{x}{a} + \frac{y}{b} + \frac{z}{c} = 1.$$

For example, letting $y = z = 0$, the x-intercept is $(a, 0, 0)$.

109. $0.04x - 0.64y + z = 3.4 \implies z = 3.4 - 0.04x + 0.64y$

(a)

Year	1994	1995	1996	1997	1998	1999	2000
x	5.8	6.2	6.4	6.6	6.5	6.3	6.1
y	8.7	8.2	8.0	7.7	7.4	7.3	7.1
z	8.8	8.4	8.4	8.2	7.8	7.9	7.8
Model, z'	8.74	8.40	8.26	8.06	7.88	7.82	7.70

The approximations are close to the actual values.

(b) According to the model, if x and z decrease, then so will y, the consumption of reduced-fat milk.

111. L_1: $x_1 = 6 + t, y_1 = 8 - t, z_1 = 3 + t$

L_2: $x_2 = 1 + t, y_2 = 2 + t, z_2 = 2t$

(a) At $t = 0$, the first insect is at $P_1 = (6, 8, 3)$ and the second insect is at $P_2 = (1, 2, 0)$.

Distance $= \sqrt{(6 - 1)^2 + (8 - 2)^2 + (3 - 0)^2} = \sqrt{70} \approx 8.37$ inches

(b) Distance $= \sqrt{(x_1 - x_2)^2 + (y_1 - y_2)^2 + (z_1 - z_2)^2}$

$$= \sqrt{5^2 + (6 - 2t)^2 + (3 - t)^2}$$

$$= \sqrt{5t^2 - 30t + 70}, \ 0 \leq t \leq 10$$

(c) The distance is never zero.

(d) Using a graphing utility, the minimum distance is 5 inches when $t = 3$ minutes.

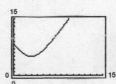

113. The direction vector $\mathbf{v}$ of the line is the normal to the plane, $\mathbf{v} = \langle 3, -1, 4 \rangle$. The parametric equations of the line are $x = 5 + 3t, y = 4 - t, z = -3 + 4t$.

To find the point of intersection, solve for t in the following equation:

$$3(5 + 3t) - (4 - t) + 4(-3 + 4t) = 7$$

$$26t = 8$$

$$t = \frac{4}{13}$$

Point of intersection is

$$\left(5 + 3 \left(\frac{4}{13} \right), 4 - \frac{4}{13}, -3 + 4 \left(\frac{4}{13} \right) \right) = \left(\frac{77}{13}, \frac{48}{13}, -\frac{23}{13} \right)$$

115. The direction vector of the line L through $(1, -3, 1)$ and $(3, -4, 2)$ is $\mathbf{v} = \langle 2, -1, 1 \rangle$. The parametric equations for L are $x = 1 + 2t, y = -3 - t, z = 1 + t$.

Substituting these equations into the equation of the plane gives

$$(1 + 2t) - (-3 - t) + (1 + t) = 2$$

$$4t = -3$$

$$t = -\frac{3}{4}.$$

Point of intersection is

$$\left(1 + 2 \left(-\frac{3}{4} \right), -3 + \frac{3}{4}, 1 - \frac{3}{4} \right) = \left(-\frac{1}{2}, -\frac{9}{4}, \frac{1}{4} \right)$$

117. True

119. True

Section 11.6   Surfaces in Space

1. Ellipsoid

Matches graph (c)

3. Hyperboloid of one sheet

Matches graph (f)

5. Elliptic paraboloid

Matches graph (d)

7. $z = 3$

Plane parallel to the xy-coordinate plane

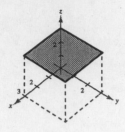

9. $y^2 + z^2 = 9$

The x-coordinate is missing so we have a cylindrical surface with rulings parallel to the x-axis. The generating curve is a circle.

11. $y = x^2$

The z-coordinate is missing so we have a cylindrical surface with rulings parallel to the z-axis. The generating curve is a parabola.

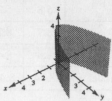

13. $4x^2 + y^2 = 4$

$$\frac{x^2}{1} + \frac{y^2}{4} = 1$$

The z-coordinate is missing so we have a cylindrical surface with rulings parallel to the z-axis. The generating curve is an ellipse.

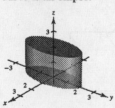

15. $z = \sin y$

The x-coordinate is missing so we have a cylindrical surface with rulings parallel to the x-axis. The generating curve is the sine curve.

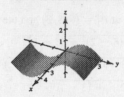

17. $z = x^2 + y^2$

(a) You are viewing the paraboloid from the x-axis: $(20, 0, 0)$

(b) You are viewing the paraboloid from above, but not on the z-axis: $(10, 10, 20)$

(c) You are viewing the paraboloid from the z-axis: $(0, 0, 20)$

(d) You are viewing the paraboloid from the y-axis: $(0, 20, 0)$

19. $\dfrac{x^2}{1} + \dfrac{y^2}{4} + \dfrac{z^2}{1} = 1$

Ellipsoid

xy-trace: $\dfrac{x^2}{1} + \dfrac{y^2}{4} = 1$ ellipse

xz-trace: $x^2 + z^2 = 1$ circle

yz-trace: $\dfrac{y^2}{4} + \dfrac{z^2}{1} = 1$ ellipse

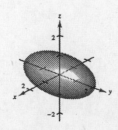

21. $16x^2 - y^2 + 16z^2 = 4$

$$4x^2 - \frac{y^2}{4} + 4z^2 = 1$$

Hyperboloid on one sheet

xy-trace: $4x^2 - \dfrac{y^2}{4} = 1$ hyperbola

xz-trace: $4(x^2 + z^2) = 1$ circle

yz-trace: $\dfrac{-y^2}{4} + 4z^2 = 1$ hyperbola

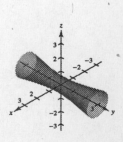

23. $x^2 - y + z^2 = 0$

Elliptic paraboloid

xy-trace: $y = x^2$

xz-trace: $x^2 + z^2 = 0$,
point $(0, 0, 0)$

yz-trace: $y = z^2$

$y = 1: x^2 + z^2 = 1$

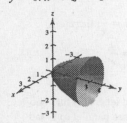

25. $x^2 - y^2 + z = 0$

Hyperbolic paraboloid

xy-trace: $y = \pm x$

xz-trace: $z = -x^2$

yz-trace: $z = y^2$

$y = \pm 1: z = 1 - x^2$

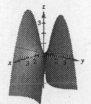

27. $z^2 = x^2 + \dfrac{y^2}{4}$

Elliptic Cone

xy-trace: point $(0, 0, 0)$

xz-trace: $z = \pm x$

yz-trace: $z = \dfrac{\pm 1}{2} y$

$z = \pm 1: x^2 + \dfrac{y^2}{4} = 1$

29. $\quad 16x^2 + 9y^2 + 16z^2 - 32x - 36y + 36 = 0$

$16(x^2 - 2x + 1) + 9(y^2 - 4y + 4) + 16z^2 = -36 + 16 + 36$

$16(x - 1)^2 + 9(y - 2)^2 + 16z^2 = 16$

$\dfrac{(x - 1)^2}{1} + \dfrac{(y - 2)^2}{16/9} + \dfrac{z^2}{1} = 1$

Ellipsoid with center $(1, 2, 0)$.

31. $z = 2 \sin x$

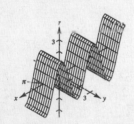

33. $z^2 = x^2 + 4y^2$

$z = \pm \sqrt{x^2 + 4y^2}$

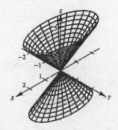

35. $x^2 + y^2 = \left(\dfrac{2}{z}\right)^2$

$y = \pm \sqrt{\dfrac{4}{z^2} - x^2}$

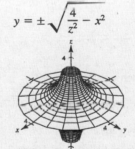

37. $z = 4 - \sqrt{|xy|}$

39. $4x^2 - y^2 + 4z^2 = -16$

$z = \pm \sqrt{\dfrac{y^2}{4} - x^2 - 4}$

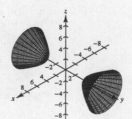

41. $z = 2\sqrt{x^2 + y^2}$

$z = 2$

$2\sqrt{x^2 + y^2} = 2$

$x^2 + y^2 = 1$

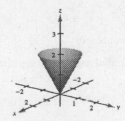

43. $x^2 + y^2 = 1$

$x + z = 2$

$z = 0$

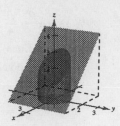

45. $x^2 + z^2 = [r(y)]^2$ and $z = r(y) = \pm 2\sqrt{y}$; therefore,

$x^2 + z^2 = 4y.$

47. $x^2 + y^2 = [r(z)]^2$ and $y = r(z) = \dfrac{z}{2}$; therefore,

$$x^2 + y^2 = \frac{z^2}{4}, 4x^2 + 4y^2 = z^2.$$

49. $y^2 + z^2 = [r(x)]^2$ and $y = r(x) = \dfrac{2}{x}$; therefore,

$$y^2 + z^2 = \left(\frac{2}{x}\right)^2, y^2 + z^2 = \frac{4}{x^2}.$$

51. $x^2 + y^2 - 2z = 0$

$x^2 + y^2 = \left(\sqrt{2z}\right)^2$

Equation of generating curve: $y = \sqrt{2z}$ or $x = \sqrt{2z}$

53. Let C be a curve in a plane and let L be a line not in a parallel plane. The set of all lines parallel to L and intersecting C is called a cylinder.

55. See pages 812 and 813.

57. $V = 2\pi \displaystyle\int_0^4 x(4x - x^2)\, dx$

$$= 2\pi \left[\frac{4x^3}{3} - \frac{x^4}{4}\right]_0^4 = \frac{128\pi}{3}$$

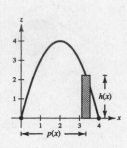

59. $z = \dfrac{x^2}{2} + \dfrac{y^2}{4}$

(a) When $z = 2$ we have $2 = \dfrac{x^2}{2} + \dfrac{y^2}{4}$, or $1 = \dfrac{x^2}{4} + \dfrac{y^2}{8}$

Major axis: $2\sqrt{8} = 4\sqrt{2}$

Minor axis: $2\sqrt{4} = 4$

$c^2 = a^2 - b^2, c^2 = 4, c = 2$

Foci: $(0, \pm 2, 2)$

(b) When $z = 8$ we have $8 = \dfrac{x^2}{2} + \dfrac{y^2}{4}$, or $1 = \dfrac{x^2}{16} + \dfrac{y^2}{32}$.

Major axis: $2\sqrt{32} = 8\sqrt{2}$

Minor axis: $2\sqrt{16} = 8$

$c^2 = 32 - 16 = 16, c = 4$

Foci: $(0, \pm 4, 8)$

61. If (x, y, z) is on the surface, then

$$(y + 2)^2 = x^2 + (y - 2)^2 + z^2$$

$$y^2 + 4y + 4 = x^2 + y^2 - 4y + 4 + z^2$$

$$x^2 + z^2 = 8y$$

Elliptic paraboloid

Traces parallel to xz-plane are circles.

63. $\dfrac{x^2}{3963^2} + \dfrac{y^2}{3963^2} + \dfrac{z^2}{3950^2} = 1$

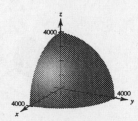

65. $z = \dfrac{y^2}{b^2} - \dfrac{x^2}{a^2}$, $z = bx + ay$

$$bx + ay = \frac{y^2}{b^2} - \frac{x^2}{a^2}$$

$$\frac{1}{a^2}\left(x^2 + a^2 bx + \frac{a^4 b^2}{4}\right) = \frac{1}{b^2}\left(y^2 - ab^2 y + \frac{a^2 b^4}{4}\right)$$

$$\frac{\left(x + \dfrac{a^2 b}{2}\right)^2}{a^2} = \frac{\left(y - \dfrac{ab^2}{2}\right)^2}{b^2}$$

$$y = \pm\frac{b}{a}\left(x + \frac{a^2 b}{2}\right) + \frac{ab^2}{2}$$

Letting $x = at$, you obtain the two intersecting lines
$x = at$, $y = -bt$, $z = 0$ and $x = at$, $y = bt + ab^2$
$z = 2abt + a^2 b^2$.

67. True. A sphere is a special case of an ellipsoid (centered at origin, for example)

$$\frac{x^2}{a^2} + \frac{y^2}{b^2} + \frac{z^2}{c^2} = 1$$

having $a = b = c$.

69. The Klein bottle *does not* have both an "inside" and an "outside." It is formed by inserting the small open end through the side of the bottle and making it contiguous with the top of the bottle.

Section 11.7   Cylindrical and Spherical Coordinates

1. $(5, 0, 2)$, cylindrical
$x = 5 \cos 0 = 5$
$y = 5 \sin 0 = 0$
$z = 2$
$(5, 0, 2)$, rectangular

3. $\left(2, \dfrac{\pi}{3}, 2\right)$, cylindrical
$x = 2 \cos \dfrac{\pi}{3} = 1$
$y = 2 \sin \dfrac{\pi}{3} = \sqrt{3}$
$z = 2$
$(1, \sqrt{3}, 2)$, rectangular

5. $\left(4, \dfrac{7\pi}{6}, 3\right)$, cylindrical
$x = 4 \cos \dfrac{7\pi}{6} = -2\sqrt{3}$
$y = 4 \sin \dfrac{7\pi}{6} = -2$
$z = 3$
$(-2\sqrt{3}, -2, 3)$, rectangular

7. $(0, 5, 1)$, rectangular
$r = \sqrt{(0)^2 + (5)^2} = 5$
$\theta = \arctan \dfrac{5}{0} = \dfrac{\pi}{2}$
$z = 1$
$\left(5, \dfrac{\pi}{2}, 1\right)$, cylindrical

9. $\left(1, \sqrt{3}, 4\right)$, rectangular
$r = \sqrt{1^2 + \left(\sqrt{3}\right)^2} = 2$
$\theta = \arctan\sqrt{3} = \dfrac{\pi}{3}$
$z = 4$
$\left(2, \dfrac{\pi}{3}, 4\right)$, cylindrical

11. $(2, -2, -4)$, rectangular
$r = \sqrt{2^2 + (-2)^2} = 2\sqrt{2}$
$\theta = \arctan(-1) = -\dfrac{\pi}{4}$
$z = -4$
$\left(2\sqrt{2}, \dfrac{-\pi}{4}, -4\right)$, cylindrical

13. $z = 5$ is the equation in cylindrical coordinates.

15. $x^2 + y^2 + z^2 = 10$,   rectangular equation
$r^2 + z^2 = 10$,   cylindrical equation

17. $y = x^2$,   rectangular equation
$r \sin \theta = (r \cos \theta)^2$
$\sin \theta = r \cos^2 \theta$
$r = \sec \theta \cdot \tan \theta$,   cylindrical equation

19. $y^2 = 10 - z^2$,   rectangular coordinates
$(r \sin \theta)^2 = 10 - z^2$
$r^2 \sin^2 \theta + z^2 = 10$,   cylindrical coordinates

21.

$$r = 2$$

$$\sqrt{x^2 + y^2} = 2$$

$$x^2 + y^2 = 4$$

23.

$$\theta = \frac{\pi}{6}$$

$$\tan \frac{\pi}{6} = \frac{y}{x}$$

$$\frac{1}{\sqrt{3}} = \frac{y}{x}$$

$$x = \sqrt{3}y$$

$$x - \sqrt{3}y = 0$$

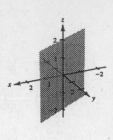

25.

$$r = 2 \sin \theta$$

$$r^2 = 2r \sin \theta$$

$$x^2 + y^2 = 2y$$

$$x^2 + y^2 - 2y = 0$$

$$x^2 + (y - 1)^2 = 1$$

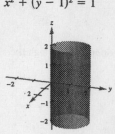

27.

$$r^2 + z^2 = 4$$

$$x^2 + y^2 + z^2 = 4$$

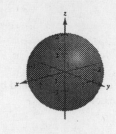

29. $(4, 0, 0)$, rectangular

$$\rho = \sqrt{4^2 + 0^2 + 0^2} = 4$$

$$\theta = \arctan 0 = 0$$

$$\phi = \arccos 0 = \frac{\pi}{2}$$

$$\left(4, 0, \frac{\pi}{2}\right), \text{ spherical}$$

31. $\left(-2, 2\sqrt{3}, 4\right)$, rectangular

$$\rho = \sqrt{(-2)^2 + \left(2\sqrt{3}\right)^2 + 4^2} = 4\sqrt{2}$$

$$\theta = \arctan\left(-\sqrt{3}\right) = \frac{2\pi}{3}$$

$$\phi = \arccos \frac{1}{\sqrt{2}} = \frac{\pi}{4}$$

$$\left(4\sqrt{2}, \frac{2\pi}{3}, \frac{\pi}{4}\right), \text{ spherical}$$

33. $\left(\sqrt{3}, 1, 2\sqrt{3}\right)$, rectangular

$$\rho = \sqrt{3 + 1 + 12} = 4$$

$$\theta = \arctan \frac{1}{\sqrt{3}} = \frac{\pi}{6}$$

$$\phi = \arccos \frac{\sqrt{3}}{2} = \frac{\pi}{6}$$

$$\left(4, \frac{\pi}{6}, \frac{\pi}{6}\right), \text{ spherical}$$

35. $\left(4, \frac{\pi}{6}, \frac{\pi}{4}\right)$, spherical

$$x = 4 \sin \frac{\pi}{4} \cos \frac{\pi}{6} = \sqrt{6}$$

$$y = 4 \sin \frac{\pi}{4} \sin \frac{\pi}{6} = \sqrt{2}$$

$$z = 4 \cos \frac{\pi}{4} = 2\sqrt{2}$$

$$\left(\sqrt{6}, \sqrt{2}, 2\sqrt{2}\right), \text{ rectangular}$$

37. $\left(12, \frac{-\pi}{4}, 0\right)$, spherical

$$x = 12 \sin 0 \cos\left(\frac{-\pi}{4}\right) = 0$$

$$y = 12 \sin 0 \sin\left(\frac{-\pi}{4}\right) = 0$$

$$z = 12 \cos 0 = 12$$

$$(0, 0, 12), \text{ rectangular}$$

39. $\left(5, \frac{\pi}{4}, \frac{3\pi}{4}\right)$, spherical

$$x = 5 \sin \frac{3\pi}{4} \cos \frac{\pi}{4} = \frac{5}{2}$$

$$y = 5 \sin \frac{3\pi}{4} \sin \frac{\pi}{4} = \frac{5}{2}$$

$$z = 5 \cos \frac{3\pi}{4} = -\frac{5\sqrt{2}}{2}$$

$$\left(\frac{5}{2}, \frac{5}{2}, -\frac{5\sqrt{2}}{2}\right), \text{ rectangular}$$

41.

$$y = 3, \quad \text{rectangular equation}$$

$$\rho \sin \phi \sin \theta = 3$$

$$\rho = 3 \csc \phi \csc \theta, \quad \text{spherical equation}$$

43. $x^2 + y^2 + z^2 = 36$, rectangular equation

$$\rho^2 = 36$$

$$\rho = 6, \quad \text{spherical equation}$$

45.

$$x^2 + y^2 = 9, \quad \text{rectangular equation}$$

$$\rho^2 \sin^2 \phi \cos^2 \theta + \rho^2 \sin^2 \phi \sin^2 \theta = 9$$

$$\rho^2 \sin^2 \phi = 9$$

$$\rho \sin \phi = 3$$

$$\rho = 3 \csc \phi, \quad \text{spherical equation}$$

47. $x^2 + y^2 = 2z^2$ rectangular coordinates

$\rho^2 \sin^2 \phi \cos^2 \theta + \rho^2 \sin^2 \phi \sin^2 \theta = 2\rho^2 \cos^2 \phi$

$\rho^2 \sin^2 \phi [\cos^2 \theta + \sin^2 \theta] = 2\rho^2 \cos^2 \phi$

$\rho^2 \sin^2 \phi = 2\rho^2 \cos^2 \phi$

$\dfrac{\sin^2 \phi}{\cos^2 \phi} = 2$

$\tan^2 \phi = 2$

$\tan \phi = \pm\sqrt{2}$ spherical equation

49. $\rho = 2$

$x^2 + y^2 + z^2 = 4$

51. $\phi = \dfrac{\pi}{6}$

$\cos \phi = \dfrac{z}{\sqrt{x^2 + y^2 + z^2}}$

$\dfrac{\sqrt{3}}{2} = \dfrac{z}{\sqrt{x^2 + y^2 + z^2}}$

$\dfrac{3}{4} = \dfrac{z^2}{x^2 + y^2 + z^2}$

$3x^2 + 3y^2 - z^2 = 0, z \geq 0$

53. $\rho = 4\cos \phi$

$\sqrt{x^2 + y^2 + z^2} = \dfrac{4z}{\sqrt{x^2 + y^2 + z^2}}$

$x^2 + y^2 + z^2 - 4z = 0$

$x^2 + y^2 + (z - 2)^2 = 4, z \geq 0$

55. $\rho = \csc \phi$

$\rho \sin \phi = 1$

$\sqrt{x^2 + y^2} = 1$

$x^2 + y^2 = 1$

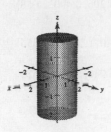

57. $\left(4, \dfrac{\pi}{4}, 0\right)$, cylindrical

$\rho = \sqrt{4^2 + 0^2} = 4$

$\theta = \dfrac{\pi}{4}$

$\phi = \arccos 0 = \dfrac{\pi}{2}$

$\left(4, \dfrac{\pi}{4}, \dfrac{\pi}{2}\right)$, spherical

59. $\left(4, \dfrac{\pi}{2}, 4\right)$, cylindrical

$\rho = \sqrt{4^2 + 4^2} = 4\sqrt{2}$

$\theta = \dfrac{\pi}{2}$

$\phi = \arccos\left(\dfrac{4}{4\sqrt{2}}\right) = \dfrac{\pi}{4}$

$\left(4\sqrt{2}, \dfrac{\pi}{2}, \dfrac{\pi}{4}\right)$, spherical

61. $\left(4, \dfrac{-\pi}{6}, 6\right)$, cylindrical

$\rho = \sqrt{4^2 + 6^2} = 2\sqrt{13}$

$\theta = \dfrac{-\pi}{6}$

$\phi = \arccos \dfrac{3}{\sqrt{13}}$

$\left(2\sqrt{13}, \dfrac{-\pi}{6}, \arccos \dfrac{3}{\sqrt{13}}\right),$

spherical

63. $(12, \pi, 5)$, cylindrical

$\rho = \sqrt{12^2 + 5^2} = 13$

$\theta = \pi$

$\phi = \arccos \dfrac{5}{13}$

$\left(13, \pi, \arccos \dfrac{5}{13}\right)$, spherical

65. $\left(10, \dfrac{\pi}{6}, \dfrac{\pi}{2}\right)$, spherical

$r = 10 \sin \dfrac{\pi}{2} = 10$

$\theta = \dfrac{\pi}{6}$

$z = 10 \cos \dfrac{\pi}{2} = 0$

$\left(10, \dfrac{\pi}{6}, 0\right)$, cylindrical

67. $\left(36, \pi, \dfrac{\pi}{2}\right)$, spherical

$r = \rho \sin \phi = 36 \sin \dfrac{\pi}{2} = 36$

$\theta = \pi$

$z = \rho \cos \phi = 36 \cos \dfrac{\pi}{2} = 0$

$(36, \pi, 0)$, cylindrical

69. $\left(6, -\dfrac{\pi}{6}, \dfrac{\pi}{3}\right)$, spherical

$r = 6 \sin \dfrac{\pi}{3} = 3\sqrt{3}$

$\theta = -\dfrac{\pi}{6}$

$z = 6 \cos \dfrac{\pi}{3} = 3$

$\left(3\sqrt{3}, -\dfrac{\pi}{6}, 3\right)$, cylindrical

71. $\left(8, \dfrac{7\pi}{6}, \dfrac{\pi}{6}\right)$, spherical

$r = 8 \sin \dfrac{\pi}{6} = 4$

$\theta = \dfrac{7\pi}{6}$

$z = 8 \cos \dfrac{\pi}{6} = \dfrac{8\sqrt{3}}{2}$

$\left(4, \dfrac{7\pi}{6}, 4\sqrt{3}\right)$, cylindrical

	Rectangular	*Cylindrical*	*Spherical*
73.	$(4, 6, 3)$	$(7.211, 0.983, 3)$	$(7.810, 0.983, 1.177)$
75.	$(4.698, 1.710, 8)$	$\left(5, \dfrac{\pi}{9}, 8\right)$	$(9.434, 0.349, 0.559)$
77.	$(-7.071, 12.247, 14.142)$	$(14.142, 2.094, 14.142)$	$\left(20, \dfrac{2\pi}{3}, \dfrac{\pi}{4}\right)$
79.	$(3, -2, 2)$	$(3.606, -0.588, 2)$	$(4.123, -0.588, 1.064)$
81.	$\left(\dfrac{5}{2}, \dfrac{4}{3}, \dfrac{-3}{2}\right)$	$(2.833, 0.490, -1.5)$	$(3.206, 0.490, 2.058)$
83.	$(-3.536, 3.536, -5)$	$\left(5, \dfrac{3\pi}{4}, -5\right)$	$(7.071, 2.356, 2.356)$
85.	$(2.804, -2.095, 6)$	$(-3.5, 2.5, 6)$	$(6.946, 5.642, 0.528)$

[**Note:** Use the cylindrical coordinates $(3.5, 5.642, 6)$]

87. $r = 5$

Cylinder

Matches graph (d)

89. $\rho = 5$

Sphere

Matches graph (c)

91. $r^2 = z, x^2 + y^2 = z$

Paraboloid

Matches graph (f)

93. Rectangular to cylindrical: $r^2 = x^2 + y^2$

$\tan \theta = \dfrac{y}{x}$

$z = z$

Cylindrical to rectangular: $x = r \cos \theta$

$y = r \sin \theta$

$z = z$

95. Rectangular to spherical: $\rho^2 = x^2 + y^2 + z^2$

$\tan \theta = \dfrac{y}{x}$

$\phi = \arccos\left(\dfrac{z}{\sqrt{x^2 + y^2 + z^2}}\right)$

Spherical to rectangular: $x = \rho \sin \phi \cos \theta$

$y = \rho \sin \phi \sin \theta$

$z = \rho \cos \phi$

97. $x^2 + y^2 + z^2 = 16$

(a) $r^2 + z^2 = 16$

(b) $\rho^2 = 16, \rho = 4$

99. $x^2 + y^2 + z^2 - 2z = 0$

(a) $r^2 + z^2 - 2z = 0, r^2 + (z - 1)^2 = 1$

(b) $\rho^2 - 2\rho \cos \phi = 0, \rho(\rho - 2 \cos \phi) = 0,$

$\rho = 2 \cos \phi$

101. $x^2 + y^2 = 4y$

(a) $r^2 = 4r \sin \theta$, $r = 4 \sin \theta$

(b) $\rho^2 \sin^2 \phi = 4\rho \sin \phi \sin \theta$,

$\rho \sin \phi (\rho \sin \phi - 4 \sin \theta) = 0$,

$\rho = \dfrac{4 \sin \theta}{\sin \phi}$, $\rho = 4 \sin \theta \csc \phi$

103. $x^2 - y^2 = 9$

(a) $r^2 \cos^2 \theta - r^2 \sin^2 \theta = 9$,

$r^2 = \dfrac{9}{\cos^2 \theta - \sin^2 \theta}$

(b) $\rho^2 \sin^2 \phi \cos^2 \theta - \rho^2 \sin^2 \phi \sin^2 \theta = 9$,

$\rho^2 \sin^2 \phi = \dfrac{9}{\cos^2 \theta - \sin^2 \theta}$,

$\rho^2 = \dfrac{9 \csc^2 \phi}{\cos^2 \theta - \sin^2 \theta}$

105. $0 \le \theta \le \dfrac{\pi}{2}$

$0 \le r \le 2$

$0 \le z \le 4$

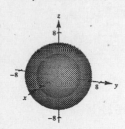

107. $0 \le \theta \le 2\pi$

$0 \le r \le a$

$r \le z \le a$

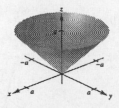

109. $0 \le \theta \le 2\pi$

$0 \le \phi \le \dfrac{\pi}{6}$

$0 \le \rho \le a \sec \phi$

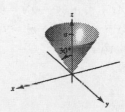

111. $0 \le \theta \le \dfrac{\pi}{2}$

$0 \le \phi \le \dfrac{\pi}{2}$

$0 \le \rho \le 2$

113. Rectangular

$0 \le x \le 10$

$0 \le y \le 10$

$0 \le z \le 10$

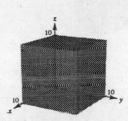

115. Spherical

$4 \le \rho \le 6$

117. Cylindrical coordinates:

$r^2 + z^2 \le 9$,

$r \le 3 \cos \theta$, $0 \le \theta \le 2\pi$

119. False. $\theta = c$ represents a half-plane.

121. False. $(r, \theta, z) = (0, 0, 1)$ and $(r, \theta, z) = (0, \pi, 1)$ represent the same point $(x, y, z) = (0, 0, 1)$.

123. $z = \sin \theta$, $r = 1$

$z = \sin \theta = \dfrac{y}{r} = \dfrac{y}{1} = y$

The curve of intersection is the ellipse formed by the intersection of the plane $z = y$ and the cylinder $r = 1$.

Review Exercises for Chapter 11

1. $P = (1, 2)$, $Q = (4, 1)$, $R = (5, 4)$ (a) $\mathbf{u} = \overrightarrow{PQ} = \langle 3, -1 \rangle = 3\mathbf{i} - \mathbf{j}$,   (b) $\|\mathbf{v}\| = \sqrt{4^2 + 2^2} = 2\sqrt{5}$

$\mathbf{v} = \overrightarrow{PR} = \langle 4, 2 \rangle = 4\mathbf{i} + 2\mathbf{j}$ (c) $2\mathbf{u} + \mathbf{v} = \langle 6, -2 \rangle + \langle 4, 2 \rangle = \langle 10, 0 \rangle = 10\mathbf{i}$

3. $\mathbf{v} = \|\mathbf{v}\| \cos\theta\, \mathbf{i} + \|\mathbf{v}\| \sin\theta\, \mathbf{j} = 8\cos 120°\, \mathbf{i} + 8\sin 120°\, \mathbf{j}$
$= -4\mathbf{i} + 4\sqrt{3}\mathbf{j}$

5. $z = 0$, $y = 4$, $x = -5$: $(-5, 4, 0)$

7. Looking down from the positive x-axis towards the yz-plane, the point is either in the first quadrant $(y > 0, z > 0)$ or in the third quadrant $(y < 0, z < 0)$. The x-coordinate can be any number.

9. $(x - 3)^2 + (y + 2)^2 + (z - 6)^2 = \left(\dfrac{15}{2}\right)^2$

11. $(x^2 - 4x + 4) + (y^2 - 6y + 9) + z^2 = -4 + 4 + 9$
$(x - 2)^2 + (y - 3)^2 + z^2 = 9$

Center: $(2, 3, 0)$

Radius: 3

13. $\mathbf{v} = \langle 4 - 2, 4 + 1, -7 - 3 \rangle = \langle 2, 5, -10 \rangle$

15. $\mathbf{v} = \langle -1 - 3, 6 - 4, 9 + 1 \rangle = \langle -4, 2, 10 \rangle$
$\mathbf{w} = \langle 5 - 3, 3 - 4, -6 + 1 \rangle = \langle 2, -1, -5 \rangle$
Since $-2\mathbf{w} = \mathbf{v}$, the points lie in a straight line.

17. Unit vector: $\dfrac{\mathbf{u}}{\|\mathbf{u}\|} = \dfrac{\langle 2, 3, 5 \rangle}{\sqrt{38}} = \left\langle \dfrac{2}{\sqrt{38}}, \dfrac{3}{\sqrt{38}}, \dfrac{5}{\sqrt{38}} \right\rangle$

19. $P = (5, 0, 0)$, $Q = (4, 4, 0)$, $R = (2, 0, 6)$
   (a) $\mathbf{u} = \overrightarrow{PQ} = \langle -1, 4, 0 \rangle = -\mathbf{i} + 4\mathbf{j}$,
   $\mathbf{v} = \overrightarrow{PR} = \langle -3, 0, 6 \rangle = -3\mathbf{i} + 6\mathbf{k}$
   (b) $\mathbf{u} \cdot \mathbf{v} = (-1)(-3) + 4(0) + 0(6) = 3$
   (c) $\mathbf{v} \cdot \mathbf{v} = 9 + 36 = 45$

21. $\mathbf{u} = \langle 7, -2, 3 \rangle$, $\mathbf{v} = \langle -1, 4, 5 \rangle$
Since $\mathbf{u} \cdot \mathbf{v} = 0$, the vectors are orthogonal.

23. $\mathbf{u} = 5\left(\cos\dfrac{3\pi}{4}\mathbf{i} + \sin\dfrac{3\pi}{4}\mathbf{j}\right) = \dfrac{5\sqrt{2}}{2}[-\mathbf{i} + \mathbf{j}]$

$\mathbf{v} = 2\left(\cos\dfrac{2\pi}{3}\mathbf{i} + \sin\dfrac{2\pi}{3}\mathbf{j}\right) = -\mathbf{i} + \sqrt{3}\mathbf{j}$

$\mathbf{u} \cdot \mathbf{v} = \dfrac{5\sqrt{2}}{2}\left(1 + \sqrt{3}\right)$

$\|\mathbf{u}\| = 5$

$\|\mathbf{v}\| = 2$

$\cos\theta = \dfrac{|\mathbf{u} \cdot \mathbf{v}|}{\|\mathbf{u}\|\, \|\mathbf{v}\|} = \dfrac{\left(5\sqrt{2}/2\right)\left(1 + \sqrt{3}\right)}{5(2)} = \dfrac{\sqrt{2} + \sqrt{6}}{4}$

$\theta = \arccos\dfrac{\sqrt{2} + \sqrt{6}}{4} = 15° \left[\text{or, } \dfrac{3\pi}{4} \cdot \dfrac{2\pi}{3} = \dfrac{\pi}{12} \text{ or } 15°\right]$

25. $\mathbf{u} = \langle 10, -5, 15 \rangle$, $\mathbf{v} = \langle -2, 1, -3 \rangle$

$\mathbf{u} = -5\mathbf{v} \implies \mathbf{u}$ is parallel to $\mathbf{v}$ and in the opposite direction.

$\theta = \pi$

27. There are many correct answers. For example: $\mathbf{v} = \pm\langle 6, -5, 0 \rangle$.

In Exercises 29–37, $\mathbf{u} = \langle 3, -2, 1 \rangle$, $\mathbf{v} = \langle 2, -4, -3 \rangle$, $\mathbf{w} = \langle -1, 2, 2 \rangle$.

29. $\mathbf{u} \cdot \mathbf{u} = 3(3) + (-2)(-2) + (1)(1)$

$= 14 = \left(\sqrt{14}\right)^2 = \|\mathbf{u}\|^2$

31. $\operatorname{proj}_{\mathbf{u}}\mathbf{w} = \left(\dfrac{\mathbf{u} \cdot \mathbf{w}}{\|\mathbf{u}\|^2}\right)\mathbf{u}$

$= -\dfrac{5}{14}\langle 3, -2, 1 \rangle$

$= \left\langle -\dfrac{15}{14}, \dfrac{10}{14}, -\dfrac{5}{14} \right\rangle$

$= \left\langle -\dfrac{15}{14}, \dfrac{5}{7}, -\dfrac{5}{14} \right\rangle$

33. $\mathbf{n} = \mathbf{v} \times \mathbf{w} = \begin{vmatrix} \mathbf{i} & \mathbf{j} & \mathbf{k} \\ 2 & -4 & -3 \\ -1 & 2 & 2 \end{vmatrix} = -2\mathbf{i} - \mathbf{j}$

$\|\mathbf{n}\| = \sqrt{5}$

$\dfrac{\mathbf{n}}{\|\mathbf{n}\|} = \dfrac{1}{\sqrt{5}}(-2\mathbf{i} - \mathbf{j})$, unit vector or $\dfrac{1}{\sqrt{5}}(2\mathbf{i} + \mathbf{j})$

35. $V = |\mathbf{u} \cdot (\mathbf{v} \times \mathbf{w})|$

$= |\langle 3, -2, 1 \rangle \cdot \langle -2, -1, 0 \rangle| = |-4| = 4$

37. Area parallelogram $= \|\mathbf{u} \times \mathbf{v}\| = \|\langle 10, 11, -8 \rangle\| = \sqrt{10^2 + 11^2 + (-8)^2}$ (See Exercises 34, 36)

$= \sqrt{285}$

39. $\mathbf{F} = c(\cos 20°\mathbf{j} + \sin 20°\mathbf{k})$

$\overrightarrow{PQ} = 2\mathbf{k}$

$\overrightarrow{PQ} \times \mathbf{F} = \begin{vmatrix} \mathbf{i} & \mathbf{j} & \mathbf{k} \\ 0 & 0 & 2 \\ 0 & c\cos 20° & c\sin 20° \end{vmatrix} = -2c\cos 20°\mathbf{i}$

$200 = \|\overrightarrow{PQ} \times \mathbf{F}\| = 2c\cos 20°$

$c = \dfrac{100}{\cos 20°}$

$\mathbf{F} = \dfrac{100}{\cos 20°}(\cos 20°\mathbf{j} + \sin 20°\mathbf{k}) = 100(\mathbf{j} + \tan 20°\mathbf{k})$

$\|\mathbf{F}\| = 100\sqrt{1 + \tan^2 20°} = 100\sec 20° \approx 106.4 \text{ lb}$

41. $\mathbf{v} = \langle 9 - 3, 11 - 0, 6 - 2 \rangle = \langle 6, 11, 4 \rangle$

(a) Parametric equations: $x = 3 + 6t, y = 11t, z = 2 + 4t$

(b) Symmetric equations: $\dfrac{x - 3}{6} = \dfrac{y}{11} = \dfrac{z - 2}{4}$

43. $\mathbf{v} = \mathbf{j}$

(a) $x = 1, y = 2 + t, z = 3$

(b) None

45. $3x - 3y - 7z = -4, \; x - y + 2z = 3$

Solving simultaneously, we have $z = 1$. Substituting $z = 1$ into the second equation we have $y = x - 1$. Substituting for x in this equation we obtain two points on the line of intersection, $(0, -1, 1)$, $(1, 0, 1)$. The direction vector of the line of intersection is $\mathbf{v} = \mathbf{i} + \mathbf{j}$.

(a) $x = t, y = -1 + t, z = 1$

(b) $x = y + 1, z = 1$

47. $P = (-3, -4, 2)$, $Q = (-3, 4, 1)$, $R = (1, 1, -2)$

$\overrightarrow{PQ} = \langle 0, 8, -1 \rangle$, $\overrightarrow{PR} = \langle 4, 5, -4 \rangle$

$$\mathbf{n} = \overrightarrow{PQ} \times \overrightarrow{PR} = \begin{vmatrix} \mathbf{i} & \mathbf{j} & \mathbf{k} \\ 0 & 8 & -1 \\ 4 & 5 & -4 \end{vmatrix} = -27\mathbf{i} - 4\mathbf{j} - 32\mathbf{k}$$

$$-27(x + 3) - 4(y + 4) - 32(z - 2) = 0$$
$$27x + 4y + 32z = -33$$

49. The two lines are parallel as they have the same direction numbers, $-2, 1, 1$. Therefore, a vector parallel to the plane is $\mathbf{v} = -2\mathbf{i} + \mathbf{j} + \mathbf{k}$. A point on the first line is $(1, 0, -1)$ and a point on the second line is $(-1, 1, 2)$. The vector $\mathbf{u} = 2\mathbf{i} - \mathbf{j} - 3\mathbf{k}$ connecting these two points is also parallel to the plane. Therefore, a normal to the plane is

$$\mathbf{v} \times \mathbf{u} = \begin{vmatrix} \mathbf{i} & \mathbf{j} & \mathbf{k} \\ -2 & 1 & 1 \\ 2 & -1 & -3 \end{vmatrix}$$
$$= -2\mathbf{i} - 4\mathbf{j} = -2(\mathbf{i} + 2\mathbf{j}).$$

Equation of the plane: $(x - 1) + 2y = 0$
$$x + 2y = 1$$

51. $Q(1, 0, 2)$ point

$2x - 3y + 6z = 6$

A point P on the plane is $(3, 0, 0)$.

$\overrightarrow{PQ} = \langle -2, 0, 2 \rangle$

$\mathbf{n} = \langle 2, -3, 6 \rangle$ normal to plane

$$D = \frac{|\overrightarrow{PQ} \cdot \mathbf{n}|}{\|\mathbf{n}\|} = \frac{8}{7}$$

53. The normal vectors to the planes are the same,

$$\mathbf{n} = \langle 5, -3, 1 \rangle.$$

Choose a point in the first plane, $P = (0, 0, 2)$. Choose a point in the second plane, $Q = (0, 0, -3)$.

$\overrightarrow{PQ} = \langle 0, 0, -5 \rangle$

$$D = \frac{|\overrightarrow{PQ} \cdot \mathbf{n}|}{\|\mathbf{n}\|} = \frac{|-5|}{\sqrt{35}} = \frac{5}{\sqrt{35}} = \frac{\sqrt{35}}{7}$$

55. $x + 2y + 3z = 6$

Plane

Intercepts: $(6, 0, 0)$, $(0, 3, 0)$, $(0, 0, 2)$

57. $y = \frac{1}{2}z$

Plane with rulings parallel to the x-axis

59. $\dfrac{x^2}{16} + \dfrac{y^2}{9} + z^2 = 1$

Ellipsoid

xy-trace: $\dfrac{x^2}{16} + \dfrac{y^2}{9} = 1$

xz-trace: $\dfrac{x^2}{16} + z^2 = 1$

yz-trace: $\dfrac{y^2}{9} + z^2 = 1$

61. $\dfrac{x^2}{16} - \dfrac{y^2}{9} + z^2 = -1$

$\dfrac{y^2}{9} - \dfrac{x^2}{16} - z^2 = 1$

Hyperboloid of two sheets

xy-trace: $\dfrac{y^2}{9} - \dfrac{x^2}{16} = 1$

xz-trace: None

yz-trace: $\dfrac{y^2}{9} - z^2 = 1$

63. $x^2 + z^2 = 4$. Cylinder of radius 2 about y-axis

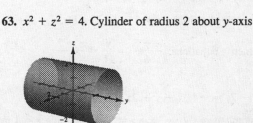

65. Let $y = r(x) = 2\sqrt{x}$ and revolve the curve about the x-axis.

67. $\left(-2\sqrt{2}, 2\sqrt{2}, 2\right)$, rectangular

 (a) $r = \sqrt{\left(-2\sqrt{2}\right)^2 + \left(2\sqrt{2}\right)^2} = 4$, $\theta = \arctan(-1) = \dfrac{3\pi}{4}$, $z = 2$, $\left(4, \dfrac{3\pi}{4}, 2\right)$, cylindrical

 (b) $\rho = \sqrt{\left(-2\sqrt{2}\right)^2 + \left(2\sqrt{2}\right)^2 + (2)^2} = 2\sqrt{5}$, $\theta = \dfrac{3\pi}{4}$, $\phi = \arccos\dfrac{2}{2\sqrt{5}} = \arccos\dfrac{1}{\sqrt{5}}$, $\left(2\sqrt{5}, \dfrac{3\pi}{4}, \arccos\dfrac{\sqrt{5}}{5}\right)$, spherical

69. $\left(100, -\dfrac{\pi}{6}, 50\right)$, cylindrical

$\rho = \sqrt{100^2 + 50^2} = 50\sqrt{5}$

$\theta = -\dfrac{\pi}{6}$

$\phi = \arccos\left(\dfrac{50}{50\sqrt{5}}\right) = \arccos\left(\dfrac{1}{\sqrt{5}}\right) \approx 63.4°$ or 1.107

$\left(50\sqrt{5}, -\dfrac{\pi}{6}, 63.4°\right)$, spherical or $\left(50\sqrt{5}, \dfrac{-\pi}{6}, 1.1071\right)$

71. $\left(25, -\dfrac{\pi}{4}, \dfrac{3\pi}{4}\right)$, spherical

$r^2 = \left(25 \sin\left(\dfrac{3\pi}{4}\right)\right)^2 \Rightarrow r = 25\dfrac{\sqrt{2}}{2}$

$\theta = -\dfrac{\pi}{4}$

$z = \rho \cos\phi = 25 \cos\dfrac{3\pi}{4} = -25\dfrac{\sqrt{2}}{2}$

$\left(25\dfrac{\sqrt{2}}{2}, -\dfrac{\pi}{4}, -\dfrac{25\sqrt{2}}{2}\right)$, cylindrical

73. $x^2 - y^2 = 2z$

 (a) Cylindrical: $r^2 \cos^2\theta - r^2 \sin^2\theta = 2z$, $r^2 \cos 2\theta = 2z$

 (b) Spherical: $\rho^2 \sin^2\phi \cos^2\theta - \rho^2 \sin^2\phi \sin^2\theta = 2\rho \cos\phi$, $\rho \sin^2\phi \cos 2\theta - 2\cos\phi = 0$, $\rho = 2 \sec 2\theta \cos\phi \csc^2\phi$

75. $r = 4 \sin\theta$ cylindrical coordinates

$r^2 = 4r \sin\theta$

$x^2 + y^2 = 4y$

$x^2 + y^2 - 4y + 4 = 4$

$x^2 + (y - 2)^2 = 4$ rectangular coordinates

77. $\theta = \dfrac{\pi}{4}$ spherical coordinates

$\tan\theta = \tan\dfrac{\pi}{4} = 1$

$\dfrac{y}{x} = 1$

$y = x, x \geq 0$ rectangular coordinates half-plane

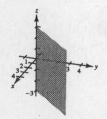

Problem Solving for Chapter 11

1. $\mathbf{a} + \mathbf{b} + \mathbf{c} = 0$

   $\mathbf{b} \times (\mathbf{a} + \mathbf{b} + \mathbf{c}) = 0$

   $(\mathbf{b} \times \mathbf{a}) + (\mathbf{b} \times \mathbf{c}) = 0$

   $\|\mathbf{a} \times \mathbf{b}\| = \|\mathbf{b} \times \mathbf{c}\|$

   $\|\mathbf{b} \times \mathbf{c}\| = \|\mathbf{b}\| \, \|\mathbf{c}\| \sin A$

   $\|\mathbf{a} \times \mathbf{b}\| = \|\mathbf{a}\| \, \|\mathbf{b}\| \sin C$

Then,

$$\frac{\sin A}{\|\mathbf{a}\|} = \frac{\|\mathbf{b} \times \mathbf{c}\|}{\|\mathbf{a}\| \, \|\mathbf{b}\| \, \|\mathbf{c}\|}$$

$$= \frac{\|\mathbf{a} \times \mathbf{b}\|}{\|\mathbf{a}\| \, \|\mathbf{b}\| \, \|\mathbf{c}\|}$$

$$= \frac{\sin C}{\|\mathbf{c}\|}.$$

The other case, $\dfrac{\sin A}{\|\mathbf{a}\|} = \dfrac{\sin B}{\|\mathbf{b}\|}$ is similar.

3. Label the figure as indicated.

From the figure, you see that

$$\overrightarrow{SP} = \frac{1}{2}\mathbf{a} - \frac{1}{2}\mathbf{b} = \overrightarrow{RQ} \text{ and}$$

$$\overrightarrow{SR} = \frac{1}{2}\mathbf{a} + \frac{1}{2}\mathbf{b} = \overrightarrow{PQ}.$$

Since $\overrightarrow{SP} = \overrightarrow{RQ}$ and $\overrightarrow{SR} = \overrightarrow{PQ}$, $PSRQ$ is a parallelogram.

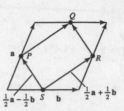

5. (a) $\mathbf{u} = \langle 0, 1, 1 \rangle$ direction vector of line determined by P_1 and P_2.

$$D = \frac{\|\overrightarrow{P_1Q} \times \mathbf{u}\|}{\|\mathbf{u}\|}$$

$$= \frac{\|\langle 2, 0, -1 \rangle \times \langle 0, 1, 1 \rangle\|}{\sqrt{2}}$$

$$= \frac{\|\langle 1, -2, 2 \rangle\|}{\sqrt{2}} = \frac{3}{\sqrt{2}} = \frac{3\sqrt{2}}{2}$$

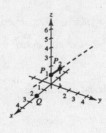

(b) The shortest distance to the line **segment** is $\|P_1Q\| = \|\langle 2, 0, -1 \rangle\| = \sqrt{5}$.

7. (a) $V = \pi \displaystyle\int_0^1 \left(\sqrt{z}\right)^2 dz = \left[\pi \frac{z^2}{2}\right]_0^1 = \frac{1}{2}\pi$

   Note: $\dfrac{1}{2}(\text{base})(\text{altitude}) = \dfrac{1}{2}\pi(1) = \dfrac{1}{2}\pi$

(b) $\dfrac{x^2}{a^2} + \dfrac{y^2}{b^2} = z$: (slice at $z = c$)

$$\frac{x^2}{\left(\sqrt{c}a\right)^2} + \frac{y^2}{\left(\sqrt{c}b\right)^2} = 1$$

At $z = c$, figure is ellipse of area

$$\pi\left(\sqrt{c}a\right)\left(\sqrt{c}b\right) = \pi abc.$$

$$V = \int_0^k \pi abc \cdot dc = \left[\frac{\pi abc^2}{2}\right]_0^k = \frac{\pi abk^2}{2}$$

(c) $V = \dfrac{1}{2}(\pi abk)k = \dfrac{1}{2}(\text{area of base})(\text{height})$

9. (a) $\rho = 2 \sin \phi$

   Torus

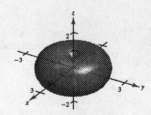

(b) $\rho = 2 \cos \phi$

   Sphere

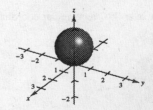

11. From Exercise 64, Section 11.4,

$$(\mathbf{u} \times \mathbf{v}) \times (\mathbf{w} \times \mathbf{z}) = [(\mathbf{u} \times \mathbf{v}) \cdot \mathbf{z}]\mathbf{w} - [(\mathbf{u} \times \mathbf{v}) \cdot \mathbf{w}]\mathbf{z}.$$

13. (a) $\mathbf{u} = \|\mathbf{u}\|(\cos 0\,\mathbf{i} + \sin 0\,\mathbf{j}) = \|\mathbf{u}\|\mathbf{i}$

Downward force $\mathbf{w} = -\mathbf{j}$

$\mathbf{T} = \|\mathbf{T}\|(\cos(90° + \theta)\mathbf{i} + \sin(90° + \theta)\mathbf{j})$

$\quad = \|\mathbf{T}\|(-\sin\theta\,\mathbf{i} + \cos\theta\,\mathbf{j})$

$\mathbf{0} = \mathbf{u} + \mathbf{w} + \mathbf{T} = \|\mathbf{u}\|\mathbf{i} - \mathbf{j} + \|\mathbf{T}\|(-\sin\theta\,\mathbf{i} + \cos\theta\,\mathbf{j})$

$\quad \|\mathbf{u}\| = \sin\theta\,\|\mathbf{T}\|$

$\quad\quad 1 = \cos\theta\,\|\mathbf{T}\|$

If $\theta = 30°$, $\|\mathbf{u}\| = (1/2)\|\mathbf{T}\|$ and $1 = (\sqrt{3}/2)\|\mathbf{T}\|$

$\quad \Rightarrow \|\mathbf{T}\| = \dfrac{2}{\sqrt{3}} \approx 1.1547\text{ lb}$

and

$$\|\mathbf{u}\| = \frac{1}{2}\left(\frac{2}{\sqrt{3}}\right) \approx 0.5774\text{ lb}$$

(b) From part (a), $\|\mathbf{u}\| = \tan\theta$ and $\|\mathbf{T}\| = \sec\theta$.

Domain: $0 \le \theta \le 90°$

(c)

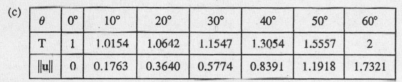

θ	0°	10°	20°	30°	40°	50°	60°
T	1	1.0154	1.0642	1.1547	1.3054	1.5557	2
$\|\mathbf{u}\|$	0	0.1763	0.3640	0.5774	0.8391	1.1918	1.7321

(d)

[graph: curves labeled T and $\|\mathbf{u}\|$, vertical axis 2.5, horizontal axis 0 to 60]

(e) Both are increasing functions.

(f) $\displaystyle\lim_{\theta \to \pi/2^-} T = \infty$ and $\displaystyle\lim_{\theta \to \pi/2^-} \|\mathbf{u}\| = \infty.$

15. Let $\theta = \alpha - \beta$, the angle between $\mathbf{u}$ and $\mathbf{v}$. Then

$$\sin(\alpha - \beta) = \frac{\|\mathbf{u} \times \mathbf{v}\|}{\|\mathbf{u}\|\,\|\mathbf{v}\|} - \frac{\|\mathbf{v} \times \mathbf{u}\|}{\|\mathbf{u}\|\,\|\mathbf{v}\|}.$$

For $\mathbf{u} = \langle \cos\alpha, \sin\alpha, 0 \rangle$ and $\mathbf{v} = \langle \cos\beta, \sin\beta, 0 \rangle$, $\|\mathbf{u}\| = \|\mathbf{v}\| = 1$ and

$$\mathbf{v} \times \mathbf{u} = \begin{vmatrix} \mathbf{i} & \mathbf{j} & \mathbf{k} \\ \cos\beta & \sin\beta & 0 \\ \cos\alpha & \sin\alpha & 0 \end{vmatrix} = (\sin\alpha\cos\beta \quad \cos\alpha\sin\beta)\mathbf{k}.$$

Thus, $\sin(\alpha - \beta) = \|\mathbf{v} \times \mathbf{u}\| = \sin\alpha\cos\beta - \cos\alpha\sin\beta.$

17. From Theorem 11.13 and Theorem 11.7 (6) we have

$$D = \frac{|\overrightarrow{PQ} \cdot \mathbf{n}|}{\|\mathbf{n}\|}$$

$$= \frac{|\mathbf{w} \cdot (\mathbf{u} \times \mathbf{v})|}{\|\mathbf{u} \times \mathbf{v}\|} = \frac{|(\mathbf{u} \times \mathbf{v}) \cdot \mathbf{w}|}{\|\mathbf{u} \times \mathbf{v}\|} = \frac{|\mathbf{u} \cdot (\mathbf{v} \times \mathbf{w})|}{\|\mathbf{u} \times \mathbf{v}\|}.$$

19. $x^2 + y^2 = 1$ cylinder

$z = 2y$ plane

Introduce a coordinate system in the plane $z = 2y$.

The new u-axis is the original x-axis.

The new v-axis is the line $z = 2y$, $x = 0$.

Then the intersection of the cylinder and plane satisfies the equation of an ellipse:

$$x^2 + y^2 = 1$$

$$x^2 + \left(\frac{z}{2}\right)^2 = 1$$

$$x^2 + \frac{z^2}{4} = 1 \quad \text{ellipse}$$

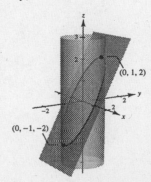